INTELLIGENT COMPONENTS AND INSTRUMENTS FOR CONTROL APPLICATIONS 1994

2nd IFAC Symposium, Budapest, Hungary, 8 - 10 June 1994

Edited by

Cs. BÁNYÁSZ
Computer and Automation Research Institute, Hungarian Academy of Sciences, Budapest, Hungary

Published for the

INTERNATIONAL FEDERATION OF AUTOMATIC CONTROL

by

PERGAMON
An Imprint of Elsevier Science Limited

UK Elsevier Science Ltd, The Boulevard, Langford Lane, Kidlington, Oxford OX5 1GB, UK

USA Elsevier Science, Inc., 660 White Plains Road, Tarrytown, New York 10591-5153, USA

JAPAN Elsevier Science Japan, Tsunashima Building Annex, 3-20-12 Yushima, Bunkyo-ku, Tokyo 113, Japan

First edition 1994

Library of Congress Cataloging in Publication Data

A catalogue record for this book is available from the Library of Congress

British Library Cataloguing in Publication Data

A catalogue record for this book is available from the British Library

ISBN 0-08-042234-9

This volume was reproduced by means of the photo-offset process using the manuscripts supplied by the authors of the different papers. The manuscripts have been typed using different typewriters and typefaces. The lay-out, figures and tables of some papers did not agree completely with the standard requirements: consequently the reproduction does not display complete uniformity. To ensure rapid publication this discrepancy could not be changed: nor could the English be checked completely. Therefore, the readers are asked to excuse any deficiencies of this publication which may be due to the above mentioned reasons.

The Editor

Printed in Great Britain

IFAC SYMPOSIUM ON INTELLIGENT COMPONENTS AND INSTRUMENTS FOR CONTROL APPLICATIONS

Sponsored by
International Federation of Automatic Control (IFAC)
Technical Committee on Components and Instruments

Co-sponsored by
IFAC Technical Committees on Applications, Computers, Education and System Engineering
IMEKO Technical Committees on Higher Education, Technical Diagnostics

Organized by
Computer and Automation Research Institute, HAS

Supported by
Hungarian Academy of Sciences
National Committee for Technological Development
Foundation for Industry

CONTENTS

PLENARY PAPERS

Intelligent Instrumentation

Fuzzy Logic Controllers

Sensors

Fault Diagnosis

Neural Networks

Low Cost Systems

Adaptive Control

Knowledge Based Systems

Identification

Drives

Control Strategies

Artificial Intelligence in Modeling

MODELS AND LANGUAGES FOR THE INTEROPERABILITY OF SMART INSTRUMENTS

M. STAROSWIECKI, M. BAYART

LAIL URA CNRS 1440D - UFR d'IEEA - Université des Sciences et Technologies de Lille, 59655 Villeneuve d'Ascq - France

ABSTRACT. The paper presents the functions of intelligent instruments in the frame of Real Time Process Operating Systems and shows that they have the same functional structure as the system itself. Thus, smart instruments based distributed architectures can be built provided they present interoperability properties. Some working groups and projects whose aim is to define interoperable field instruments are presented. In the last part, the paper proposes a smart instrument generic description.

Keywords. Smart sensors and actuators, Interoperability, Generic model, Operating modes

1. INTRODUCTION

During the last twenty years, the tremendous development of microelectronics has given rise to numerous applications which integrate data storage and processing capabilities. Their implementation in a microprocessor or a microcontroler-based system allows not only the miniaturization, the autonomy, the optimization of functions which were by the past completed by logic or analogic devices, but also the development of quite new ones. Smart sensors and actuators are an outcome of the microelectronics revolution.

In the MDA loop (Measure-Decision-Actuation), the Decision part has first taken advantage of the increase of computing capabilities : programmable controllers, computers, digital control systems have deeply modified the operators' tasks. The Measurement function has evolved by integrating signal processing in the sensors, as well as new functions in the transmitters. Actuators have recently began to benefit from local processing power which allows to increase their own performances and to provide them with new possibilities.

In fact, the impact of this revolution on sensors and actuators can be analyzed from three points of view :

- as dedicated tools, their performances can of course be improved due to their own processing capabilities.
- as automation systems components, they participate in the evolution towards increasing productivity and security. Automation systems become more and more complex, and need the processing of more and more information to meet the optimization constraints to which they are submitted. Classical architectures cannot easily take into account the resulting increase of the data and processing flows, and have to evolve towards distributed architectures. At the field level, smart sensors and actuators are the most natural devices in which some "intelligence" can be included, in order to locally perform some part of the overall functions of the automation system.
- as industrial products, sensors and actuators go through different phases throughout their life cycle. Their exploitation, which constitutes the phase for which they have been designed and produced, is not the only phase in which their intelligence can benefit to the users. In the configuration, implementation, testing or maintenance phases, smart sensors and actuators can provide different services to the different users.

In a first part of this paper, we will present the functions of intelligent instruments in the frame of Real Time Process Operating Systems, which are the systems that have to be designed in order to achieve the integrated automation of nowadays complex production processes. We will show that a smart instrument has the same functional structure as the overall system, and thus smart instruments based distributed architectures can be built provided they present interoperability properties. In a second part, we will briefly give an idea of the numerous working groups and projects whose aim is to define interoperable field instruments. The last part of the paper will propose a smart instrument generic description under the form of an external model whose structure will be developed.

2. INTELLIGENT INSTRUMENTS AND REAL TIME PROCESS OPERATING SYSTEMS

The automation of production processes calls for more and more complex realizations :

- an increased mastership of the physical transformations is needed for productivity and safety reasons. This leads to the implementation of increasing numbers of sensors and actuators within a given installation.
- the exploitation of such complex installations needs the development of integrated automation concepts : the design of the control and regulation functions of "classical" automation is now associated with the design of the supervision, the maintenance and the management functions which characterize the exploitation of the production system from a higher hierarchical point of view.

Increasing the number of sensors and actuators rises first a physical connexion problem. Different hardware architectures are possible for the information exchanges between the field instruments, the computing devices, the man/machine interfaces (Thomesse, 1991). Among them, local area networks and field buses are able to face the constraints which characterize the interconnexion of field instruments like sensors and actuators (Thomesse, 1993; Reeve, 1992). Connecting them on such a communication link, either directly or by the means of a concentrator gives them a first level of "intelligence".

The second problem is connected with the increased processing power which becomes necessary to control an increased number of sensors and actuators, and to perform the functions of integrated automation. Centralized architectures, in spite of their power, cannot take into account the whole of these functions nor can the centralized communication scheme transport such information flows unless their performances are drastically decreased. The necessity of minimizing the information flows leads to perform as much processing as possible at the level where the data are produced i.e. at the field instruments level. They are thus provided with a second level of "intelligence", which is the result of their local processing power. The so obtained parallelism participates to the necessary increase of the overall processing power.

Easy interconnexion and distributed processing characterize intelligent instruments based automation systems (Delcuvellerie, 1990). A Distributed Intelligence Automation System (DIAS) is built using a hardware architecture which allows the communication at the field level of sensors, actuators and data processing stations. The Fig.1 illustrates such an architecture.

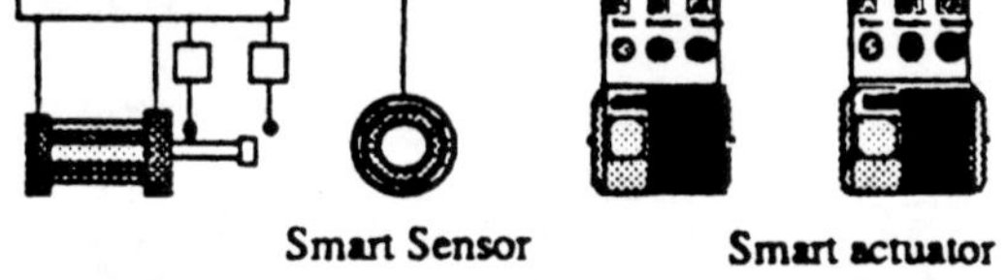

Fig. 1 : Illustration of the DIAS concept

2.1. The functions of a Real Time Process Operating System

The exploitation of complex installations needs the development of integrated automation concepts which consider the production system from a higher hierarchical point of view than the one only associated with the design of the control and regulation functions. Taking into account the supervision, the maintenance and the management functions which characterize the exploitation of the production system, as well as the installation and dismanteling services which appear when considering its whole life cycle, we see that what is needed is the design of a set of services for all the operators who will intervene during the system life, namely the design of a Process Operating System (POS). Restricting our attention to the exploitation phase in which the real time considerations are crucial, the set of the corresponding services will be called Real Time Process Operating System (RTPOS) (Bayart and Staroswiecki, 1993a).

The RTPOS Data Base. The RTPOS services are provided by a number of hardware or software processing. Each of them consumes input data, which represent the state of the physical process (open valve, closed valve), the state of the product it transforms (gas temperature, oil flow) or the value of more complex variables, produced by other services. Each of them also produces output data for the operators, the actuators, or for other services.
It can be shown (Bayart and Staroswiecki, 1992b) that logical as well as temporal coherence of these data are necessary conditions for the safe operation of the process. The RTPOS Data Base constitutes a common and homogeneous set of data to be used by all the RTPOS services (Bayart and Staroswiecki, 1992c).

Three subsets can be distinguished in the RTPOS Data Base (Bayart and *al.*, 1991) :

- Static data comprise the identification of the system elements, as well as their configuration and parametrization possibilities. These data characterize the DIAS and are never modified unless itself is changed.
- Semi-static data describe the configuration and the parametrization at a given time. They remain unchanged on a temporal window which corresponds to constant system operating conditions.
- Dynamical data represent the temporal evolution of the process and of the flow of the products it transforms. They constitute the "real time" part of the RTPOS data base.

Let us consider this last part of the Data Base. The only data which are available to describe the process in real time are provided by the inputs from sensors and human operators, and by the outputs of the control algorithms. We call "Raw Data Base" (RDB) this set of data. Since sensors, actuators, human operators, computers and the process itself may be faulty, the data from the RDB should not be directly used. The Fault Detection and Isolation (FDI) algorithms (Brunet and *al.*, 1990; Patton and *al.*, 1989) operate on the data of the RDB and have a double goal :

- the detection, isolation (and eventually the diagnosis) of the system faults creates new data on the system state and thus enriches the RTPOS Data Base.

- the data of the RDB are validated by the FDI algorithms either because their coherence has been "proved" (no fault has been detected) or because their coherence has been re-established (replacing an erroneous data by its most probable estimate, for example) (Ragot and *al.*, 1990).

The set of these enriched and validated data constitutes the Valid Data Base (VDB) of the RTPOS, which will be used by all the services it proposes (Cocquempot, 1993). The Fig. 2 shows the different layers of the RTPOS.

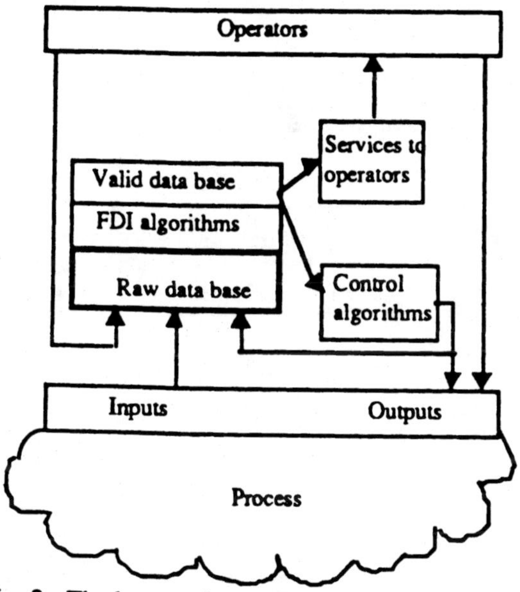

Fig. 2 : The layers of the RTPOS

Functional analysis of the RTPOS. The functional analysis of the RTPOS leads to distinguish four subsystems (Staroswiecki, 1992) :

- information subsystem
- decision subsystem
- application subsystem
- communication subsystem

The information subsystem. This subsystem is in charge of the collection, the elaboration and the validation of the data of the VDB. Its functions are the following (Gehin, 1994) :

- to INPUT : this function aims at the constitution of the RDB. It is supported from the hardware point of view by the sensors and the man/machine input interfaces.

- to VALIDATE : this function creates the VDB and is supported by the FDI algorithms

- to ELABORATE : this function generates data which cannot directly be measured (virtual sensors, state or output estimations) and creates specific data structures.

- to MANAGE THE DATA BASE : this function stores and makes available the data of the RDB and of the VDB. The management mechanisms may be quite complex in the case of distributed critical time data bases for example.

The decision subsystem. This system is in charge of two classes of decision procedures :

- on one hand, it decides about the operation of the production process. It proposes decision helps to the human operators and performs the automatic control of the process evolution. The corresponding function is named to DECIDE.

- on another hand, it has to decide about the operation of the RTPOS itself. The execution of the information, decision, application and communication subsystems have to be scheduled. This part of the decision subsystem is named to MANAGE.

The application subsystem. This subsystem applies to the process the decisions which have been taken (or suggested) by the decision subsystem (to DECIDE). According to the cases, it is supported by the actuators (automatic control) or by the chain : human operator-actuator (manual control). It performs the function to ACT.

The communication subsystem. This subsystem takes in charge all the information exchanges between the RTPOS and the external world. In the case of a distributed implementation of the RTPOS functions, the communication subsystem will also be in charge of the information flows between the components of the distributed architecture. The Fig. 3 shows the articulation of the four subsystems of the RTPOS while the Fig.4 represents its functional architecture.

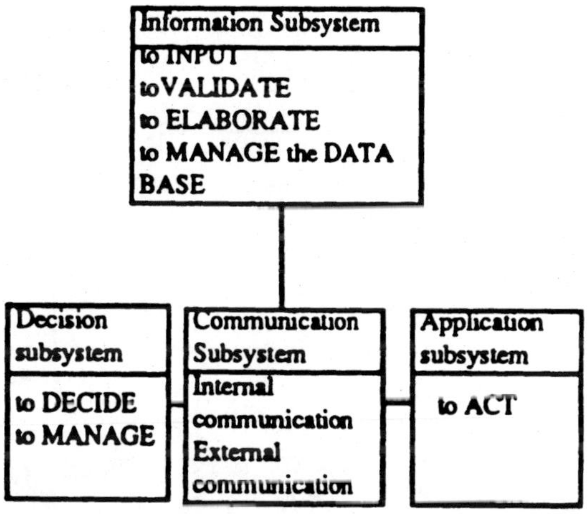

Fig. 3 : The subsystems of the RTPOS

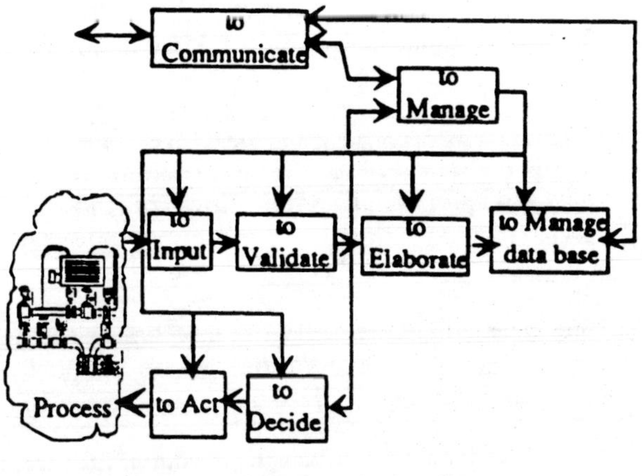

Fig. 4 : Functional structure of the RTPOS

2.2. RTPOS distribution and Intelligent Instruments.

The hardware part of the automation system includes sensors, actuators and data processing devices.
A first "intelligence" level provides the field instruments with numeric communication possibilities. A sensor will have the following functional structure :

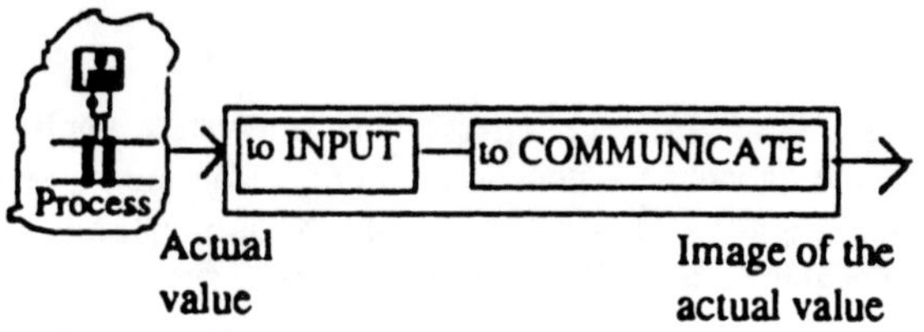

Fig. 5 : Communicating sensor

The functional structure of an open-loop actuator is given below :

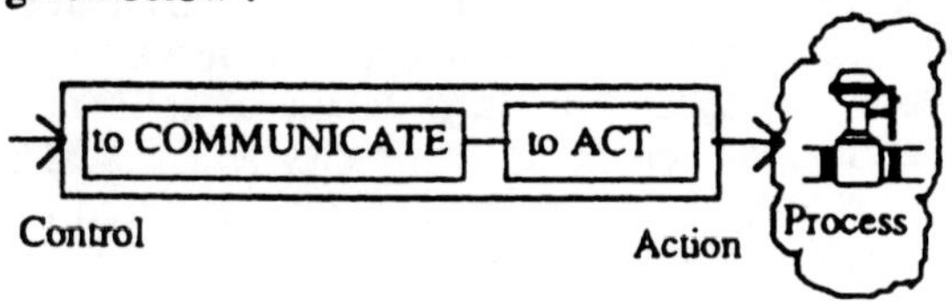

Fig. 6 : Communicating actuator (open loop)

More complex, a closed loop actuator will be described by :

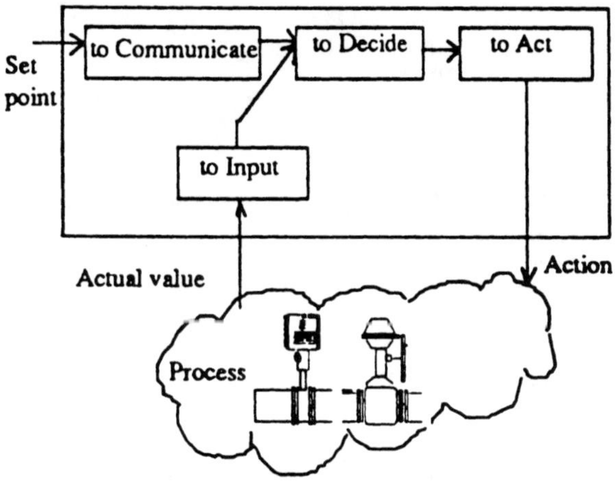

Fig. 7 : Communicating actuator (closed loop)

One sees from the preceding figures that the field instruments incorporate at the local level some of the functions of the RTPOS. Generalizing this remark, the local processing possibilities of the field instruments can be used to perform not only some but a great part of the RTPOS functions.

Thus an intelligent sensor will incorporate, besides the numeric communication possibility , the following functions (Gehin, 1994; Robert and *al.*, 1993) :

- to MANAGE, in order to obtain different services in different phases of the life cycle : sensor installation, configuration, parametrization, exploitation, maintenance...
- to VALIDATE, which distributes at the local level some procedures of the global FDI system (Staroswiecki and *al.*, 1992).
- to ELABORATE and to MANAGE the DATA BASE perform in the same way, at the local level, a part of these overall functions (virtual sensor, state estimation, maintenance history..).
- on another hand, the function to INPUT, which is always present, takes also some advantage of the presence of local processing possibilities (Najafi, 1991; Benoit, 1993).

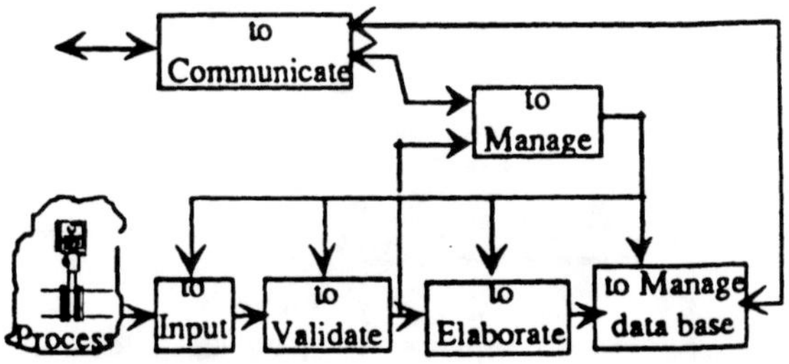

Fig. 8 : Functional structure of an intelligent sensor

The functional structure of an intelligent actuator is more complex, and covers entirely the structure of the RTPOS : to the functions to COMMUNICATE, to INPUT, to DECIDE, to ACT, which are already included in a closed loop communicating actuator are added, with the same justifications as for sensors, the functions to MANAGE, to ELABORATE and to MANAGE the DATA BASE (Staroswiecki and Bayart, 1994; Iserman and Raab, 1992; Bayart and Staroswiecki, 1992a)

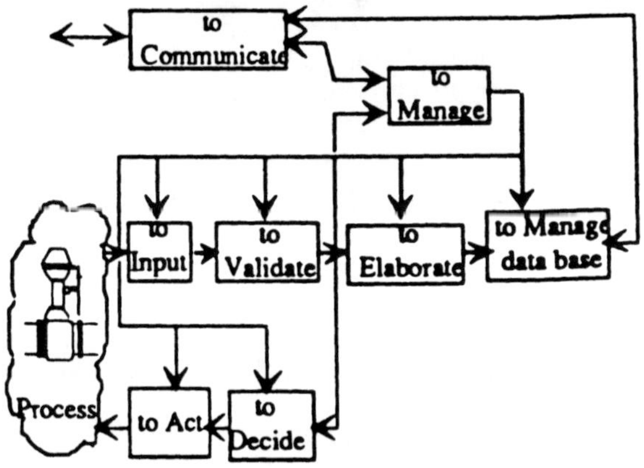

Fig. 9 : Functional structure of an intelligent actuator

2.3. Interoperability and the design of distributed architectures.

Since the functionnal structure of smart instruments and the functionnal structure of the overall RTPOS are the same, one sees that it should be possible to construct RTPOS architectures by means of the interconnexion of smart instruments and processing stations (PLCs, regulators, computers..). Each smart instrument or local station would perform a part of the overall RTPOS functions, and eventually operate on its own local data base which would be a part of the overall one. The concepts of distributed processing and distributed data bases result from the distributed architectures that local area networks allow (Thomesse, 1991; Staroswiecki, 1993). The interoperability problem arises as soon as :

- the data contained in the different local data bases have to be exchanged (what are the data contained in a smart sensor memory ? are they those I actually need ? how are they represented : which unit, which format ?...).
- local processing performs functions whose results will be remotely used (what is the precise semantics of the "data valid" bit issued from an intelligent sensor ? under which conditions does an intelligent actuator enter its "safe" mode ?).

One sees that the design of RTPOS using distributed intelligent sensors and actuators rests on the knowledge we have on their actual operation. It is often possible to design applications based on any kind of instruments, at the cost of implementing specific interfaces. The interoperability property is intended to avoid as far as possible these extra costs, through the definition of standard local data bases and standard local processing functions for intelligent sensors and actuators.

3. STATE OF THE ART

As soon as the beginning of the eighties, some companies have began to produce smart sensors, which mainly included parametrization and communication possibilities. This offer was a response to needs which were not very clearly expressed by the end users, and was mainly characterized by the fact it was "closed" : these instruments could only be implemented within the frame of the digital control systems which were proposed by their constructors (Honeywell, Fuji, Rosemount...). Smart actuators appeared on the market a few years later, at the time when the studies about the users needs and the field buses concepts began to be well structured, and when words like "interoperability", "interchangeability", "companion standards", "generic models" began to appear.

In fact, due to the infinite number of functions and implementations which can be performed, some design rules are necessary in order to propose "open" concepts which would be able to insure, in a distributed automation system, some operating compatibility between the field devices offered by different constructors.

A significative example is provided by the communication function. The "4-20 mA" standard, which is used to represent a [0% - 100%] interval is a typical means to unify the communication procedure between two field instruments, even issued from different constructors. Whatever the processing functions which are included in both instruments, the "4-20 mA" standard allows them to communicate in a univoque way. As far as the analogic communication is replaced by a digital one, which is able to transmit a large amount of multiplexed data on the same line, it seems clear that the transmitter and the receiver can exchange messages only if they submit themselves to the same rules for the representation and the manipulation of information. The general frame for communication rules is provided by the seven layers of the OSI (Open System Interconnexion (Zimmermann, 1980)) and some particular definitions have been proposed to normalization commitees (UTE, 1989).

Local area networks of the fieldbus type are well suited for the information exchange between sensors, actuators, field processors. A great number of proposals have been made these last years : HART (Highway Adressable Remote Transducer), FIP (Factory Instrumentation Protocol), Profibus (Process Field Bus), Modbus, Jbus (PLCs in process control), Batibus, Lonworks (domotic applications), VAN (Vehicule Area Network), CAN (Controller Area Network), ...

However, the possibility to design a distributed application with field instruments issued from different constructors is not only based on the use of a common communication standard. Another necessary condition is that the data which are transmitted by a given producer are well "understood" by their consumers, and more precisely that the control requests which are sent by a given instrument are executed by the receiver exactly in the way the sender expects him to perform them. Such a result needs the functions of field instruments, as well as the data they handle, to be precisely specified (Tinham, 1992; Groupo Italiano Fip, 1989). Moreover, the language which gives access to the field instruments services and data has also to be precisely defined. Such an approach constitutes in fact a modelization of field instruments and of their control language process. The proposed models have to be generic ones, which would apply to a broad class of instruments.

A formal and structured description of smart sensors and actuators, through the use of a generic model, constitutes a pertinent tool in order to evaluate and assess :

- the interoperability of a set of instruments in a distributed application, through the evaluation of the compatibility between the offered and required services, as well as between the produced and consumed data.
- the interchangeability between two instruments, through the verification that the second one provides the services and the data which were given and produced by the first one.

A great amount of work has been done in recent years about the interoperability concepts, which represent an enormous technical and economical challenge. The Interoperable System Project (ISP), and the WorldFIP one are the two leading working groups on the field, and mobilize major field instruments companies. In the frame of the EC programs, major research and developments have been performed under ESPRIT and EUREKA programs.

3.1. The Interoperable System Project

The Interoperable System Project (ISP) gathers approximately one hundred members among which ABB, AEG, Endress Hauser, Fisher, Foxboro, the Franhofer Institute, Hartman & Brown, Pepperl Fuchs, Rosemount, Siemens, Yokogawa.

The project aims are :

- to promote an interoperable fieldbus (the ISP Fieldbus), on the basis of the existing works (PROFIBUS, ISA-SP 50, FIP, HART)
- to develop specific tools and languages in order to achieve the interoperability of the ISP Fieldbus based equipments (the DDL - Device Description Language - has been proposed by Rosemount as a step for interoperability (Westbrock, 1992))
- to define and implement tools for the conformity and interoperability tests
- to propose a complete solution to the international normalization
- to support exchanges between the users and the suppliers of smart instruments.

Technological results are expected to be the following :

- a multi-vendors offer for communicating and interoperable instruments based on a normalized protocol
- hardware and software tools to help the suppliers integrate the protocol
- competence centers to support the users and suppliers
- certified organisms for the tests.

3.2. The WorldFip Project

The WorldFip Project is the other leading force in the interoperability challenge. Allen Bradley, Cegelec, Cise (ENEL), Elsag Bailey, Honeywell, ITT Canon, Square D, Télémécanique, Marconi Automazione, Merlin Gerin, Nuovo Pignone are some of the involved partners. The project aims are the same than the ISP ones, but they are based on the use of the FIP protocol, whose "Production, Distribution, Consumation" scheme and coherence ascertainment mechanisms seem to be well adapted to the specific needs of smart instruments communication and architecture distribution. The interoperability concepts are based on the enrichment of the ISA SP 50 proposals (ISA/SP50, 1993; Marti, 1993).

Rather than new normative research, the WorldFip project is based on existing norms or standards, and aims at providing for the short term a "WorldFip interconnexion standard" for the design of WorldFip interoperable intelligent instruments.

WorldFip Competence Centers already exist in Europe, North America and Japan.

3.3. ESPRIT and EUREKA European Programs

As soon as 1986, the DIAS program (Distributed Intelligent Sensors and Actuators) has been developed in the frame of the Esprit II program. It was essentially driven by end-users in the field of energy processes (EDF in France and ENEL in Italy), and brought experimental industrial offer by Bailey Esacontrol, Hartmann & Brown, Biffi,... The results mainly were the expression of the end-users needs in terms of data generation for control, management and maintenance purposes (the CMMS : Control, Management, Maintenance System concept) (DIAS; Poussin, 1991). The local data bases of smart sensors and actuators were defined as well as the local data processing, under the form of a list of CMMS dedicated functionalities. The DIAS concepts and specifications have been installed and experimented on three industrial sites, among which the Piombino electrical power plant in Italy (Fantoni and *al.*, 1992).

The PRIAM proposal (Prenormative Requirements for Intelligent Actuation and Measurement) constitutes the sequel of the DIAS project, and has been started within the Esprit III program. It deals whith functional standardization for intelligent sensors and actuators, and is based on a PRIAM Consortium (seven end users in the field of power plants, chemical and petrochemical processes, and eight suppliers of automation systems and field instruments) as well as a PRIAM User Group (PRIAM).

In the sequel of this paper, we report on some results we have developed within the frame of a cooperation in the European EUREKA project INCA (INCA, 1993) : Interface Normalisée Fonctionnelle pour Capteurs et Actionneurs (Normalized Functional Interface for Sensors and Actuators). The project included eight French and Italian partners (seven field instruments suppliers and one laboratory), and should continue with the development of an industrial offer for intelligent sensors and actuators. The results we report are based on the idea that interoperability languages can only be defined on the basis of a rigourous modelization of the intelligence of smart instruments. Such a modelization has to take into account the inner description of the instrument (its functional model) as well as an outer description (its external model).

4. GENERIC MODELS OF INTELLIGENT INSTRUMENTS

4.1. Functional model

The functional model describes the intelligent device from the point of view of the fuctions that it puts at work in order to contribute to the RTPOS distribution. The Fig. 8 and 9 of § 2.2 show the first decomposition level of this functional desciption. Using a hierarchical approach like SADT (IGL, 1989), detailed functional models of intelligent sensors and actuators have already been proposed

(Gehin, 1994; Bayart and Staroswiecki, 1993b). Such models make appear application specific functions (to INPUT, to VALIDATE,..) and generic ones (to COMMUNICATE, to MANAGE,..). Application specific functions may be considered as generic ones for certain application classes (to INPUT represents the same generic function for all the members of the specific class of pressure sensors which are based on the same physical principle for example). The standardization of the generic functions is one of the challenges of the interoperability working groups.

4.2. External model

The external model describes the device from the point of view of the services it is able to provide to an external entity (operator, other field instrument, computer..). Since a part of these services include data read/write operations in its local data base, the definition of the external model also includes the definition of the local data base.

The hardware architecture of an intelligent instrument includes :

- communication links with the operators and with the automation system
- processing unit(s) which allow the distribution of the RTPOS functions
- memories which allow the distribution of the RTPOS data base

A general representation of a smart instrument in control applications is given by Fig. 10.

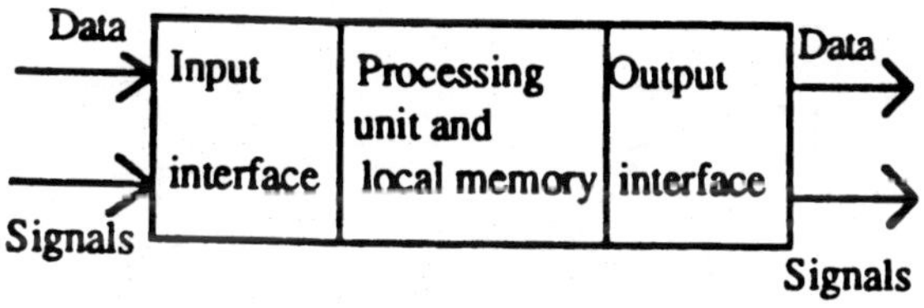

Fig. 10 : Smart instrument in a control application

The interfaces are the means by which the smart instrument interacts with its environment. They include interfaces with the operators, the automation system, the physical process. The information which cross these interfaces are <u>signals</u> or <u>data</u>.

The signals are pertinent through the evolution of their value, and are directely used by the system which consumes them (4-20 mA signal at the output of a sensor, 0-5 V at the input of a drive,.). Input and output signals cross the process interface of the intelligent instrument.

The data need some interpretation procedure before they can be used (RS 232, fieldbus frame,..). Input and output data cross the fieldbus interface and/or dedicated communication ones (pocket terminal, specific links,..).

<u>Services.</u> A service is defined as a procedure whose execution results in the modification of at least one data of the local data base, or/and at least one signal on the output interface.

A service is executed on the reception of its specific request by the intelligent instrument, according to its local scheduling policy. The association of a specific request to each possible service shows that the execution of a service cannot be split, nor can it be partly requested.

For example a "Write" service consumes input data which cross the input interface and memorizes them in the local data base. A "Regulation" service consumes signals which cross the process input interface and produces signals which cross the process output interface. It consumes also data (the set-point) which have been previously written in the local data base (using a "Write" request).

The description of a service consists first in the description of the result which is produced by its execution. The **produced variables** are those the value of which it modifies in the local data base or/and those it issues on the output process interface. In order to define the obtained values, one will have to describe the computations which are done (algorithmic or sequencial procedures, qualitative or fuzzy inferences,..) and the variables on which they are done (**consumed variables**). So, a service is first described by a triple :

< Service > ::= < Consumed variables, Produced variables, Procedure >

(more precisely, such a triple only defines one version of a service, as we will see later - § <u>Nominal or degraded service</u>).

Since critical time constraints may be considered, each consumed variable may be associated with a peremption delay, which will define the time limit after which the freshness of the variable is unsufficient for its consumation. So, the description of the variables which are consumed by a given service is the following :

< Consumed variables > ::= < (Variable name , [Peremption delay]) >

In the same spirit, the produced variables may be associated with their production time, which is the time at which their producing service has been executed :

< Produced variables > ::= < (Variable name, [Production time]) >

It can be noted that the peremption delay associated with a consumed variable is a constant, which characterizes the critical time constraints which apply to the corresponding service execution. The production time associated with a produced variable is of course up-to-dated at each execution of its producing service.

As an example, a service which would estimate the "true" value of a temperature measured by three redundant sensors would be defined by the triple :
- Consumed variables : (T1, D1); (T2, D2); (T3, D3)
- Produced variable : (T, D)
- Procedure :
a) IF peremption delays D1, D2, D3 not trespassed
THEN T = (T1 + T2 + T3) / 3
ELSE no estimation is possible (error message)
b) D = last firing time of the estimation service

where : - T1, T2, T3 are the three values of the same temperature transmitted under the form of signals (for sensors directly connected to the intelligent instrument trough its process input interface), or under the form of data (for sensors connected to the intelligent instrument through its communication interface - for example smart sensors).
- D1, D2, D3 are the three associated peremption delays.
- T is the estimated value of the true temperature.
- D is its production time.

Considering this example, it seems obvious that some degraded version of the service could nevertheless be provided, even in the case when one of the three temperature sensors (or the communication system), has not been able to transmit the temperature value in time. In order to increase the disponibility of intelligent instruments, we will introduce later (§ Nominal or degraded service) a more general definition of a service.

Control language and protocol. The execution of a service is obtained by means of a request. A request is defined by :
- its name, which allows to identify it.
- its execution parameters, which allow to modulate the results. The set of all the parametrized requests the intelligent instrument recognizes define its **control language**.
- its origin, which identifies the entity which produces it (control or maintenance operator, supervision device, control computer,..).
- the communication link through which it is transmitted. Due to the hardware wiring of the installation, the requests issued from the automation system are allways transmitted by the same medium (the fieldbus, for example, for the requests presented by a PLC to a smart valve). On the contrary, the requests issued from human operators may be transmitted by any of the means at their disposal (pocket terminal, fieldbus, control board, ...).

Associating to the set of the requests the set of their authorized origins and communication links defines the **control protocol**.

Services organization. The set of the available services characterizes, from an external point of view, an intelligent device. Keeping on the parallel with a digital processor, it can be compared with the instructions set, but the analogy stops here : while the instructions of a microprocessor have to be a priori organized within a program which will be executed later, the requests may be addressed to an intelligent device at any moment. However, one realizes that this liberty has to be limited : an intelligent device cannot simultaneously perform initialization and production services for example. For that reason, the set of the services has to be organized into coherent subsets, which we call User **Operating Modes (USOM).**

A USOM contains at least one service, and each service belongs at least to one USOM. So, the set of the USOMs constitutes a covering of the set of the services (Fig. 11).

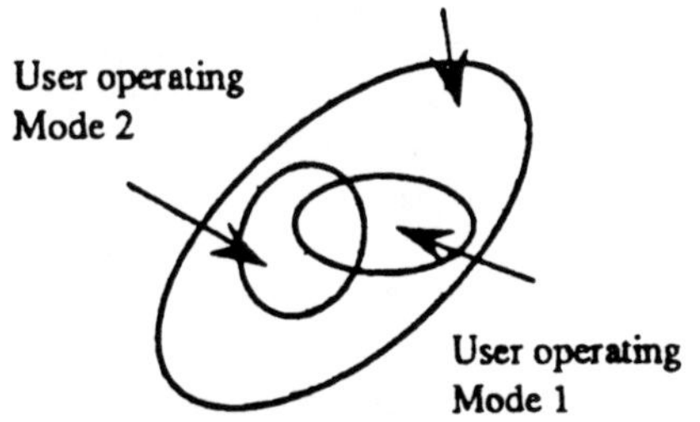

Fig. 11 : Services and USOMs

USOMs may also be represented by a table which indicates for each of them, if it includes (1) or not (0), a given service (see Fig. 12).

Modes / Services	USOM 1		USOM j	......	USOM n
S_1					
⋮			......		
Si			0 or 1		
⋮					

USOMs' description Table

Modes / Services	Automatic control	Manual control	Configuration
Settle setpoint	1	0	1
Configurate regulator	0	0	1
Open valve	0	1	0
Close valve	0	1	0

Example

Fig. 12 : USOM Definition and example

At each time, the intelligent device is in a current USOM : the only services it will accept to execute are those which belong to that USOM, in other words, any request for another service will be automatically rejected. Such an organization plays the role of a request filter and increases the operational safety of the intelligent device. However, in order to implement it, two questions still remain to be answered :

- how should we define the USOMs ?
- how to change the current USOM ?

Definition of the USOMs. A priori, the USOMs organization is arbitrary, since the definition gives no reason to prefer one covering of the services set to another one. However, the idea of the coherence of the services which constitute a USOM is justified by a rather general classification of the operating modes of industrial devices : off operation, configuration, manual or automatic mode, ...

Each kind of operation is associated with a given set of operators in a given situation of the life cycle. A USOM corresponds thus to a set of services which are at the disposal of these operators in that situation. The coherence of these services is a notion which is associated with the triple < Device, Operators, Situation >. It is a static notion, since the set of the USOMs of a given intelligent instrument is defined once for all.

One has to be aware that the existence of USOMs in different intelligent devices which cooperate in the same automated process introduces the need for another type of coherence : the (dynamic) system coherence.

In fact, the interoperability requirement supposes that when a device A presents a request to a device B, the latter is in a USOM such that this request is not rejected. This supposes a standardization of the intelligent devices USOMs, or at least, some parametrization which allows them to know the services organization of the devices with which they will have to interoperate. At a given time, the system coherence is relative to the coherence of the USOMs of the different devices which constitute the distributed application.

Changing the current USOM. Obviously, the list of the services of each USOM has to contain a "Set USOM" request, otherwise it would be impossible to leave the current USOM.

This request has to indicate the destination USOM (the origin being, of course the current one). However, it is clear that any couple < Origin, Destination > cannot be allowed, for security or operating consistency reasons. As an example, the request "Set Automatic USOM" presented to a regulation valve should be rejected as long as the regulator parameters have not been fixed (writing these parameters is one of the services of the "Configuration" USOM). For that reason, the intelligent instrument definition has to include the conditions under which the passage from one USOM to another one is possible.

The definition of the set of the USOMs and of the transition conditions completely specifies the intelligent instrument USOMs management system. This system can be described by a deterministic automata (see Fig. 13).

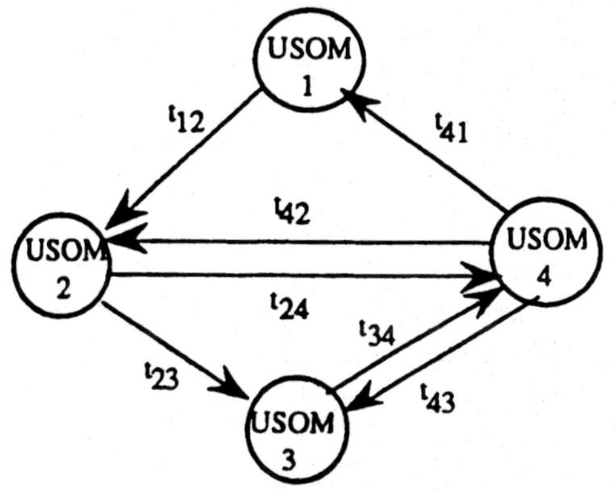

Fig. 13 : The USOMs deterministic automata

Services management : *Nominal or degraded service.* The execution of a service needs a set of ressources which includes hardware ressources (sensors, memory,..) and information ones (data with appropriate freshness status). These ressources may belong to the intelligent device or be external. At a given instant, the service will run in a **nominal** way if the set of the ressources it needs are able to perform normally. Unfortunately, this is not allways the case, and it may happen that some of the required ressources are faulty. Two situations are then possible :

- the intelligent instrument designer has implemented no replacement procedure. In that case, the service becomes unavailable, and it is taken away from the list of the services of the different USOMs in which it appeared. The question now arises to decide wether such an USOM keeps some sense in spite of this service absence. It will be answered later (cf. § Management of the actual operating modes.).
- the designer has implemented at least one replacement procedure. In the case when the ressources this procedure needs are effectively operational, it can be run to replace the nominal one at each request which concerns this latter. The list of the available services in the different USOMs remains unchanged, but the intelligent device operation is **degraded**.

So, introducing the notion of a degraded service increases the robustness and thus the disponibility of the intelligent instruments, since the faults which appear on some of the ressources they need do not necessarily interrupt the services they render. However, introducing this notion leads to modify the service definition which had previously been given.

Definition : a service is **an ordered list of versions**. Each version is characterized by the ressources it needs. In response to a request, the intelligent instrument runs the first version of the list whose ressources are all operational. The service becomes unavailable only when no version of the list can be run.

As an example, let us take the temperature estimation service already presented in § Services The nominal version needs, as we have seen, the three data T1,T2,T3, with convenient freshness status, but it is quite obvious that some degraded versions based on only two (even only one) data can be implemented. The estimation service might then be described by the following list of versions :

Versions	Procedures	Ressources
$V_{nominal}$	T = (T1 + T2 + T3) / 3	T1, T2, T3
V_1	T = (T2 + T3) / 2	T2, T3
V_2	T = (T1 + T3) / 2	T1, T3
V_3	T = (T1 + T2) / 2	T1, T2
V_{12}	T = T3	T3
V_{13}	T = T2	T2
V_{23}	T = T1	T1

Among the seven possible versions, the version which will be run at a given time will be the one whose ressources (the consumed data in our example) will all have the proper freshness status.

From a more general point of view, the management of the service versions supposes that the intelligent instrument is provided with **Fault Detection and Isolation capabilities**, which allow it to have a knowledge of the actual state of the ressources needed by its different services (Bayart and *al.* 1992; Cassar and *al.*, 1992; Bayart and *al.*, 1990).

Services execution. When a service of the current USOM is available, it may be in one among several states.

The **inactive** state is that of an available, but not requested service. The service becomes **active** as soon as its execution has been requested using a valid request with respect to the control protocol. However, since it is possible for several services to be simultaneously requested, the **waiting** and the **running** states have to be introduced to distinguish those requested services whose execution has not yet began from those which are currently running. The intelligent instrument **scheduler** is in charge of the management of the services execution. Different scheduling strategies may be employed (Lister, 1983). The figure 14 resumes the different services states.

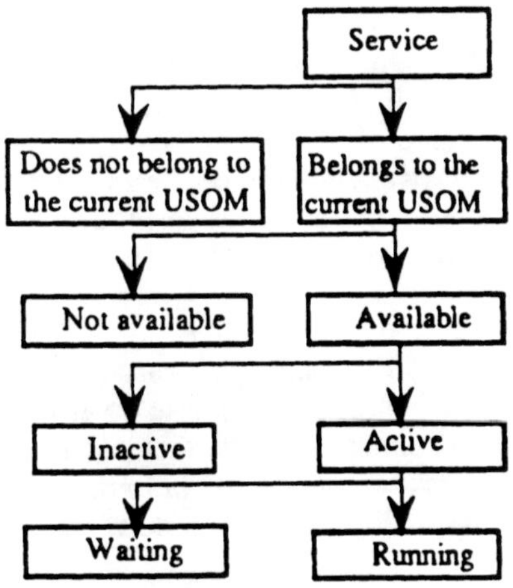

Fig. 14 : The different states of a service

The mechanisms which govern the transitions between the different states are represented on Fig. 15 and Fig. 16.

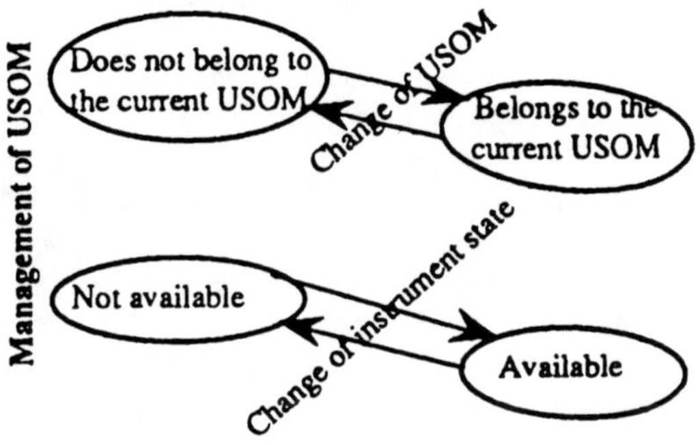

Fig. 15 : Services states transitions (management of the actual operating modes)

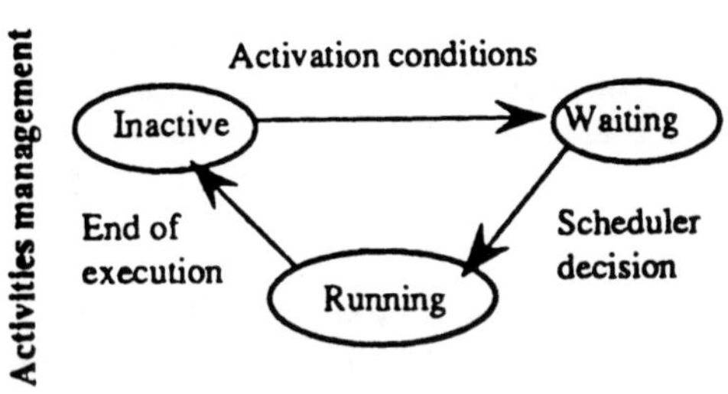

Fig. 16 : Services states transitions (management of the activities)

Management of the actual operating modes. For a service to be available, two conditions have to be fulfilled :

- on one hand, the service has to belong to the current USOM. This one results from the execution of a "Set USOM" request, according to the mechanism described by the USOMs state graph. The current USOM is the result of a **voluntary action** of the intelligent instrument users.
- on another hand, there must exist a version of the service which needs only non faulty ressources at the current time. The set of the faulty ressources is the result of events on which the users have no possible actions, and which they **not willingly endure.**

We define the **Actual Operating Mode** (AOM) of an intelligent device as the combination of the **desired** device USOM and of the **endured** device state.

This combination defines in a unique way the list of the available services, as well as, for each of them, the version under which it is available, at each instant of time.

The intelligent instrument is provided with an AOM Management System, which allows it to deliver the appropriate version of a requested service, and to give the proper information to the different operators in charge of the control, the maintenance, the management of the production process.

However, it might happen that, in the case of non-availability of some specific services, a given USOM has no longer any signification. The "Automatic" USOM of a control valve, for example, has no signification in the case when the "Regulation" service is not available because the closing loop sensor is faulty. This is the reason why each of the services which define a given USOM has to be associated with an identificator which identifies it as **compulsory or not** for that USOM.

The non availability of a non compulsory service leads to a degraded USOM : the whole of its constitutive services cannot be rendered, even under a degraded form. The non availability of a compulsory service renders the USOM meaningless. One has then to define the behaviour of the USOM and AOM Management System.

Two cases can be distinguished :

- starting from the current AOM, there is a "Set USOM" request whose destination is a meaningless one. Clearly such a request should be rejected. This means that the transitions of the USOMs state graph (Fig. 13), have to take into account the availabilty of the compulsory services of each destination USOM.
- some compulsory service of the current AOM suddenly disappear, making it meaningless. Clearly, some alarm should be dispatched to the instrument users. The AOM Management System could or not (depending on the specifications), automatically apply a "Set USOM" request, whose destination would be an appropriately defined safe USOM.

The AOM Management System, as we have seen, has to determine at each instant of time, the actual abilities of the intelligent instrument. This needs the determination of the set of the ressources which are effectively operational among those which are needed for the execution of the current USOM set of services. This possibility rests on the implementation of Fault Detection and Identification algorithms as part of the intelligence of the smart instrument. The USOM and AOM Management System covers the set of the mechanisms which appear on Fig. 15.

Activities Management. The Activities Management System covers the set of the mechanisms which appear on the Fig. 16.

Two levels are distinguished :

- the first one constructs the list of the services which are waiting for execution at a given instant. For that, it evaluates, for each of the available services of the current AOM, its activation condition, which is the product of three terms :
  + has the service been requested ?
  + was the entity who requested the service autorized to do it ?
  + did this entity use an autorized communication link ?
- the second level of the Activities Management System covers the services execution scheduling. It determines, among the services waiting for execution, which is (are) the running one(s).

5. CONCLUSION

The external model of an intelligent instrument defines its control language and its control protocol. The control language groups the set of the requests which are "understood" by the intelligent instrument while the control protocol defines the conditions under which they may be used.

The execution of the data processing associated with a request provides a service (under its nominal or one of its degraded versions). A service is perfectly determined as soon as its characteristics (consumed and produced variables, data processing) have been specified for each of its versions.

The special service which allows to choose a User Operating Mode gives a good means to structure the operation of a smart instrument. In fact, each USOM provides a partition of the set of the services into two classes :

- the services which do not belong to the USOM cannot be executed as long as this USOM is the current one : their requests are systematically rejected.
- the services which belong to the USOM can be executed at any time. Going further in the analysis of these services would lead to distinguish those whose requests are allways present (continuous execution of a service as long as the USOM is the current one) and those which have to be explicitely requested in order to be executed (Staroswiecki and Bayart, 1994).

The list of the services (the definition of their nominal and their degraded versions), as well as their logical organization (by means of the USOMs), constitute a generic model which allows to describe any smart instrument, using a formal description language. Furthermore, such a description could be used for the computer aided design of smart sensors or actuators software. However, the interoperability assessment has also to consider the temporal point of view, which is the consequence of the scheduling policy applied by the Activities management system (§ Activities Management). The models of the Actual Operating Modes and of the Activities management systems summarize the knowledge which is needed for the simulation and validation of smart instruments based distributed architectures.

6. REFERENCES

Bayart, M., P.H. Delmaire, M. Staroswiecki, (1990), Fault Detection and Isolation in Smart Actuators. Actuator 90, Brême, RFA, June.

Bayart, M., G. Blain, F. Beaudouin, D. De Buttet, J.M. Favennec, C. Georges, M. Lecoanet, J.L. Noizette, M. Staroswiecki, (1991), Fonctionnalités système des constituants intelligents. MESUCORA, Paris, November 1991.

Bayart, M. , A.L. Gehin, M. Staroswiecki, (1992), Fault Detection and Isolation and Mode Management in Smart Actuators. IFAC SICICA 92, Malaga, May.

Bayart, M., M. Staroswiecki, (1992a), Smart Actuators : Generic Functional Architecture, Service and Cost Analysis. IEEE SICICI 92, Singapore, February

Bayart, M. and M. Staroswiecki, (1992b), Coherence of a Distributed Application under Critical Time Constraints. IFAC/IFIP Workshop on Real Time Programming, Bruges, June.

Bayart, M. and M. Staroswiecki, (1992c), Information System in Distributed Intelligent Automation Processes. IMACS/SICE Symposium on Robotics, Mechatronics and Manufacturing Systems, Kobe, September.

Bayart, M. and M. Staroswiecki, (1993a), Hierarchical Data and Processing Structures for the Integration of Production Processes. IFAC Workshop on Production Control in Process Industry, Düsseldorf, March.

Bayart, M., and M. Staroswiecki, (1993b), A Generic Functional Model of Smart Instruments for Distributed Architectures. IMEKO Int. Symp. on Intelligent Instruments for Remote and On-site Measurements, Bruxelles, May.

Benoit, E. (1993), Capteurs symboliques et capteurs flous : un nouveau pas vers l'intelligence, Thèse de Doctorat, Université de Grenoble I, January.

Brunet, J., D. Jaume, M. Verge, M. Labarrere, A. Rault, (1990), Détection et diagnostic de pannes, Hermès, Paris.

Cassar, J.P., M. Bayart, M. Staroswiecki, (1992), Hierarchical Data Validation in Control Systems using Smart Actuators and Sensors. IFAC SICICA 92, Malaga, May.

Cocquempot V., (1993), "Surveillance de processus industriels complexes, Génération et optimisation des relations de redondance analytiques", Thèse de Doctorat, Université de Lille I, France, December.

Delculvellerie, J. L. (1990), L'impact de FIP sur la conception des architectures, Revue générale de l'électricité, pp.57-64.

DIAS (Distributed Intelligent Sensors and Actuators), Project Information, ESPRIT Project 2172.

Fantoni, D., G. Tedesco, S. Cancian, R. Fertino, (1992), L'application de FIP sur le site pilote industriel de la centrale ENEL de Piombino. Conférence Automation, Paris.

Gehin, A.L. (1994), "Analyse fonctionnelle et modèle générique des capteurs intelligents - Application à la surveillance de l'anesthésie", Thèse de Doctorat, Université de Lille I, January.

Groupo Italiano FIP, (1989) Gif Interoperability Guide Lines.

IGL, (1989), SADT : un langage pour communiquer, Paris, Eyrolles.

INCA, (1993) - Projet EUREKA EU 666 INCA (Interface Normalisée pour Capteurs et Actionneurs), Tâche 0B, Rapport final des travaux, Juillet 1993, 600 pages.

ISA/SP50 (1993), User Layer Technical Report for The Fieldbus Standard.

Iserman, R. and U. Raab (1992), Intelligent Actuators-Ways to Autonomous Actuating Systems, IFAC SICICA 92, Malaga, May.

Lister A.M. (1983), *Les principes fondamentaux des systèmes d'exploitation*, Eyrolles.

Marti, H., (1993) Interoperabilité WorldFIP SP50, AREMI, Lille, France.

Najafi, K. (1991), Smart Sensors, Journal of Micromechanics and Microengineering, Vol 1 N°2, June.

Patton R. J., P. M. Frank, R. N. Clark, (1989), Fault Diagnosis in Dynamic Systems, Theory and Applications, Prentice Hall.

Poussin, D.L.V. , (1991), DIAS (Distributed Intelligent Actuators and Sensors), ESPRIT PROJECT 2172. *Journée "L'instrumentation Intelligente"*, Paris.

PRIAM (Prenormative Requirements for Intelligent Actuation and Measurement), Project Information, ESPRIT Project 6188.

Ragot, J., M. Darouach, M. Maquin, G. Bloch, (1990), Validation de données et diagnostic, Hermès, Paris.

Reeve, A. (1992), Fieldbus-latest developments, Control and Instrumentation, Vol.24, n°2, April, pp.103-107

Robert, M. , M. Marchandiaux, M. Porte, (1993), Capteurs intelligents et méthodes d'évaluation, Hermès, Paris.

Staroswiecki M. (1992) Les systèmes et le concept d'instrumentation intelligente. Ingénieurs et Scientifiques de France, 2-92,pp. 35-42.

Staroswiecki, M. M. Bayart, J.P. Cassar, (1992), Repartition of Fault Detection Procedures in Distributed Intelligent Automation Systems. 8th IMEKO Symposium on Technical Diagnosis, Dresden, September.

Staroswiecki, M. (1993), Distributed System Supervision based on Intelligent Instruments. Int. Conf. on Fault Diagnosis, TOOL'DIAG 93, Toulouse, April.

Staroswiecki, M. and M. Bayart, (1994), Les actionneurs intelligents. Hermès, Paris, 130 pages,(to appear).

Thomesse, J.P. (1991),"Architectures des systèmes automatisés, la place des bus de terrain", Mesucora 91, November, Paris, France.

Thomesse, J. P. (1993), Le réseau de terrain FIP, Réseaux et informatique répartie, Vol. 3, n° 3, pp. 287-321

Tinham, B. (1992), What about HART and Open Systems ?, Control and Instrumentation, Vol.24, n°2, April, pp.76-77.

UTE (1989), NF C46-60X, Normative definitions of the different layers of the FIP fielbus, for information exchange between transmitters, actuators and PLCs.

Westbrock, J. (1992), Device Description Language Specification - Document COM-20, revision B, April.

Zimmermann, H. (1980), OSI Reference Model. The ISO Model of Architecture for Open System Interconnection, IEEE Trans. COM-28, n°4, April, pp. 425-432.

MECHATRONIC COMPONENTS
- Design and Control Aspects -

ROLF ISERMANN

Technical University of Darmstadt, Institute of Automatic Control
Laboratory for Control Engineering and Process Automation
Landgraf-Georg-Straße 4, 64238 Darmstadt, Germany
Tel.: (FRG) 6151/162114, Fax: (FRG) 6151/293445

Abstract. The integration of mechanical systems and microelectronics opens new possibilities for mechanical design and automatic functions. After a discussion of the mechanical and electronic design the organization of information processing in different levels is described. Within this frame "low-degree intelligent" mechatronic systems can be developed which comprise adaptive control, supervision with fault diagnosis and decisions with regard to further actions. This requires the realization of knowledge based systems with learning abilities. Some aspects of the design of information processing including modeling and estimation, control and supervision methods are considered. Finally as examples the adaptive nonlinear control of a pneumatic actuator and an adaptive semiactive shock absorber for vehicle suspension systems is shown.

Keywords. Mechatronics, adaptive control, intelligent control, fault diagnosis, pneumatic actuator, adaptive suspension.

1. INTRODUCTION

Mechanical systems produce motions or transfer forces or torques. For the control of positions, speeds or forces analog realized controllers are used since many years. Digital electronics now allows to perform much more functions. Fig. 1 shows a scheme of the mechanical energy flow influenced by actuators and the sensor based information flow for a mechanical-electronic system.

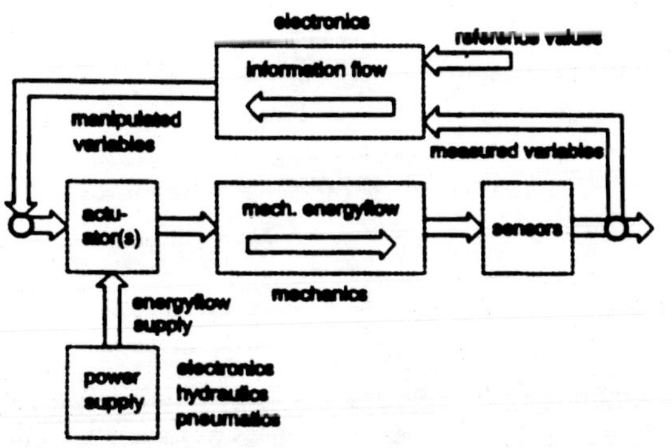

Fig. 1 Energy and information flow within a mechatronic system

Now a development can be observed that the mechanical systems, the actuators, sensors and the microelectronics are increasingly integrated, such forming a unit. This leads to *mechatronic systems* consisting of

- mechanics (mechanical engineering, engineering mechanics)
- electronics (microelectronics, power electronics, measurement and actuator technology)
- information technology (systems theory, automation, software-design, artificial intelligence)

see e.g. N.N.(1990); Bradley, et al (1991); McConaill et al (1991); Schweitzer (1992); Isermann (1993a).

1.1 Mechanical and electronic design

The solution of tasks within mechatronic systems is performed as well on the mechanical as on the digital-electronic side. Hereby interrelations during the design play an important role. Whereas the design and the local arrangement for a conventional system is done separately, the mechanical and electronic components of a mechatronic system must be considered as an integrated overall system from the beginning, Fig. 2. That means *simultaneous engineering* has to take place.

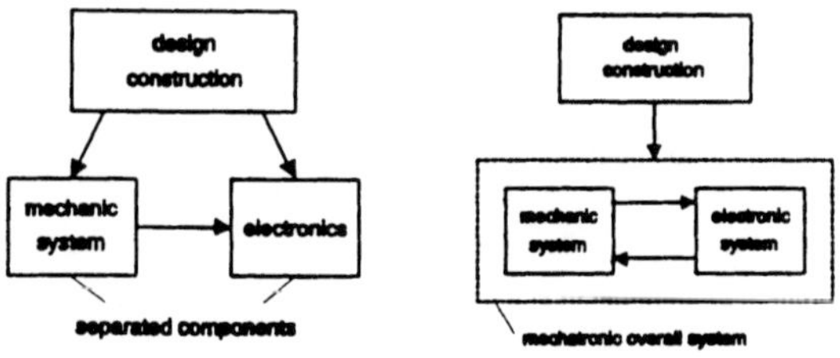

a) b)

Fig. 2 Relations during design of a mechatronic system
a) conventional b) mechatronic procedure

Mechatronic systems can be developed as well for mechanical elements as for machines and vehicles. Examples are:

a)<u>Mechanical elements with integrated electronics</u>

- suspension systems
- vibration dampers
- clutches, elastic or with friction
- bearings, mechanical or magnetic
- gears, mechanical (tooth-, chain-, belt-gears)

The properties which can be improved are:

- adaptive stiffness or damping
- adaptive backlash or pretension
- automatic kinematic functions
- feedback control
- supervision

b) <u>Machines with integrated electronics</u>

- *power producing machines* like electrical drives, pneumatic and hydraulic drives, water-, steam-, gasturbines, combustions engines, etc.
- *power consuming machines* like generators, pumps, compressors, machine tools, robots, printing machines, etc.
- *vehicles*, like automobiles, ships, aircraft

The integration of electronics into these machines allow many improved and new functions. Some examples are:

- precise speed control for all operating conditions
- control of nonmeasurable but reconstructed or estimated variables
- operation in unstable or dangerous boundary areas
- adaptive damping of oscillations or unbalance
- optimization of efficiencies or pollutions
- supervision with fault diagnosis
- anti blockage or slip control
- overall process management.

Fig. 3 shows a general scheme of a classical mechanical-electrical system. Such systems resulted from adding available sensors and actuators and analog or digital controllers to the mechanical components. The limits of this approach were given through the lack of suitable sensors and actuators, the unsatisfactory life time under the rough operating conditions (acceleration, temperature, contamination), the large space requirements, the required cables and slow data processing. Through proper design one has to avoid these disadvantages. This may be possible by creating more autonomous systems, e.g. in form of capsuled systems with bus connections or even wireless communication and robust microelectronics.

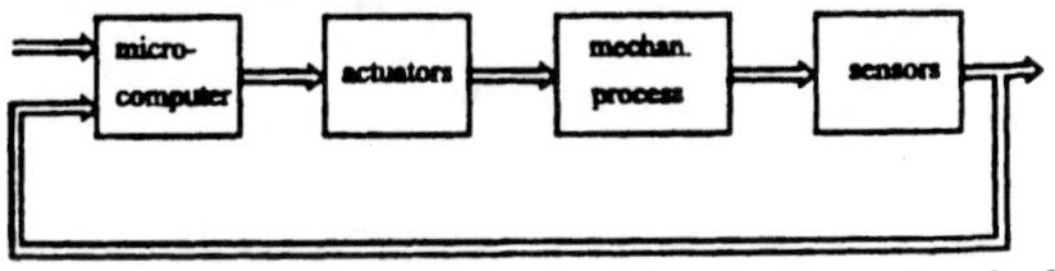

Fig. 3 General scheme of a (classical) mechanical-electrical system

1.2 <u>Ways of integration</u>

Typical for developing a mechatronic system is the idea of an *integrated overall system.* Two kinds of integration, the integration of components and the integration by information processing can be distinguished.

The *integration of components* (hardeware integration) results from designing the mechatronic system as an overall system and imbedding the sensors, actuators and microcomputers into the mechanical process, see Fig. 4. This spatial integration may be limited to process and sensor or process and actuator. The microcomputer can be integrated with the actuator, the process or sensor or be arranged at several places.

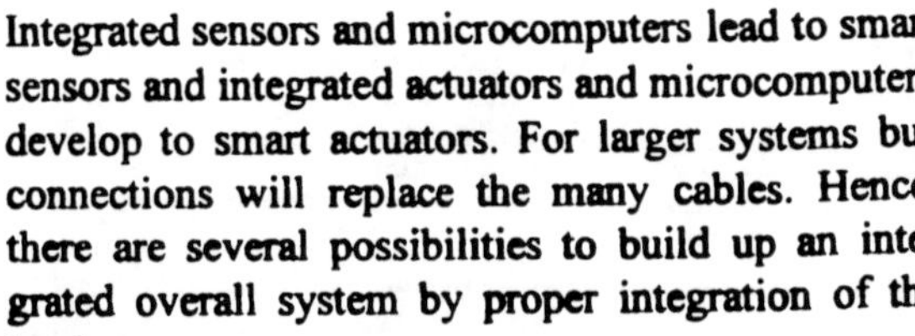

Fig. 4 Integration of components within a mechanical-electronic system

Integrated sensors and microcomputers lead to smart sensors and integrated actuators and microcomputers develop to smart actuators. For larger systems bus connections will replace the many cables. Hence, there are several possibilities to build up an integrated overall system by proper integration of the hardware.

The *integration by information processing* (software integration) is mostly based on advanced control functions. Fig. 5 shows that besides a basic feedback an additional influence may take place through the process knowledge and corresponding information processing. This means a processing of available signals in higher levels, as discussed in the next section, and includes the solution of tasks like supervision with fault diagnosis, optimization and general process management. The respective problem solutions result in realtime algorithms which must be adapted to the mechanical process properties, for example expressed by mathematical models in the form of static characteristics, differential equations, etc. By this way the mechanical parts are governed in various ways by higher level information processing with intelligent properties, possibly including learning, thus forming an integration with process adapted software.

In the following the integration through information processing will be further discussed.

Table 1 compares some properties of conventional and mechatronic designed systems.

Table 1 Properties of conventional and mechatronic designed systems.

CONVENTIONAL DESIGN	MECHATRONIC DESIGN
• **Added components**	• **Integration of components (hardware)**
bulky complex mechanisms cable problems connected components	compact simple mechanisms bus or wireless communication autonomous units
• **Simple control**	• **Integration by information processing (software)**
stiff construction feedforward control linear (analog) control precision through narrow tolerances nonmeasurable quantities change arbitrarily simple monitoring fixed abilities	elastic construction with electronic damping programmable feedback control (nonlinear) digital control precision through measurement and feedback-control control of nonmeasurable estimated quantities supervision with fault diagnosis learning abilitities

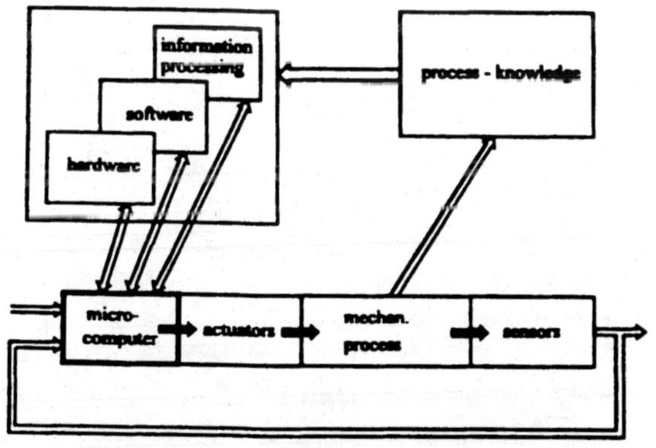

Fig. 5 Integration by information processing within a mechanical-electronic system

2. INFORMATION PROCESSING STRUCTURES FOR MECHATRONIC SYSTEMS

Governing mechanical systems by actuators results generally via changes of input variables like e.g. positions, speeds, flows, forces or torques. The directly measurable output quantities are usually positions, speeds, accelerations or forces.

2.1 Multi-level systems

The information processing of *direct measurable input and output signals* can be organized in several levels, compare Fig. 6.

Level 1:
- feedforward control
- feedback control
- feedback for stabilization or damping

Level 2:
- supervision with alarming
- supervision with automatic protection
- supervision with fault diagnosis
- supervision with redundancy actions and reconfiguration

Level 3:
- optimization (efficiency, wear, noise, emissions)
- coordination of subsystems
- general process menagment (adaptation to environment, commissions, schedules)

In general it holds:
- lower levels: react fast, act locally
- higher levels: react slow, act globally.

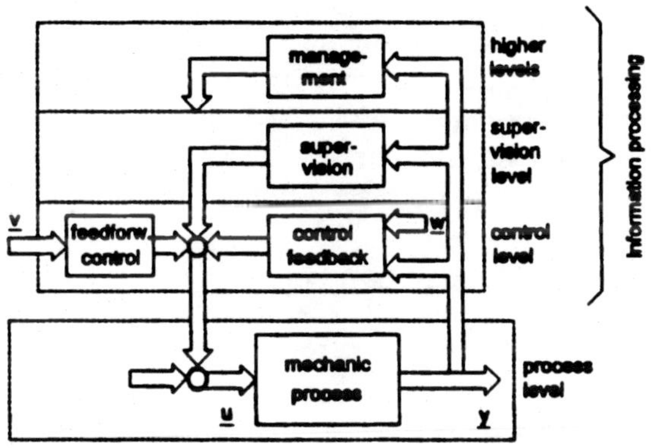

Fig. 6 Multi level information processing for mechanical systems

u: manipulated variables; v: input signals
y: measured variables; w:reference variables

Recent approaches for mechatronic systems mostly use signal processing in the lower levels, for example damping or control of motions or simple superversion. The digital information processing, however, allows the solutions of much more tasks, like adaptive control, learning control, supervision with fault diagnosis, decisions for maintenance or even redundancy actions, economic optimization and coordination. The tasks of the higher levels are sometimes summarized as "process management". The information processing in several levels under realtime conditions can also be described as "mechatronic process automation".

The described methods are partially also be applicable for *nonmeasurable quantities* which are reconstructed by using mathematical process models and e.g. state observers. By this way it is possible to control e.g. damping ratios, material and heat stress and slip or to supervise quantities like resistancies, capacitances, temperatures within components, or parameters of wear and contamination.

2.2 Model based and adaptive systems

The information processing is, at least in the lower levels, performed by simple algorithms or software-modules under real-time conditions. These algorithms contain free adjustable parameters, which have to be adapted to the static and dynamic behaviour of the process. In contrast to manual tuning by trial and error the use of mathematical models allows precise and fast automatic adaption.

The mathematical models can be obtained by identification and parameter estimation which use the measured and sampled input and output signals. These methods are not restricted to linear models, but also allow to identify several classes of nonlinear systems.

If the parameter estimation methods are combined with appropriate control algorithm design methods adaptive control systems results, which can be used for precise controller tuning continuously or only for commissioning (Isermann, Lachmann, Matko, 1992). This is described in more detail in section 3.

2.3 Intelligent systems

Improved information processing may create mechatronic systems with intelligent properties. Here "low degree intelligence" is meant with the ability "to model, reason and learn the process and its automatic functions within a given frame and to govern it towards a certain goal" (Isermann 1993b). This comprises, compare Fig. 7,

- Automation in multi levels
- Knowledge base
    - analytical knowledge (parameter/state estimation, controller design methods)
    - heuristic knowledge (normal features, fault-trees)
- Inference mechanism
    - decisions for adaptive control
    - decisions for supervision and fault diagnosis
- Communication
    - internally: with internal modules
    - externally: with other components

Hence, the intelligent mechatronic system adapts the controller to the mostly nonlinear behavior (adapta-

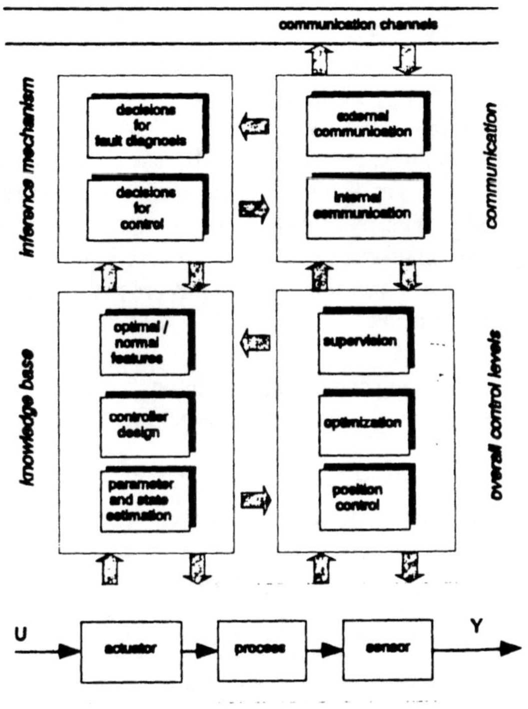

Fig. 7 General scheme of a (low-degree) intelligent mechanical system

tion) and stores its controller parameters in dependence on the position and load (learning), supervises all relevant elements and performs a fault diagnosis (supervision) to request for maintenance or if a failure occurs to fail safe (decisions on actions). In the case of multiple components supervision may help to switch off the faulty component and to perform a reconfiguration of the controlled process.

The intelligent system of Fig. 7 can also be considered as an on-line real time expert system or an autonomous system.

2.4 Electronic systems

The increasing integration density of microelectronics leads to improving solutions with regard to costs per calculations, larger mechanical robustness and less space requirements. Because of the fast mechanical process behaviour relatively high sampling rates in the range of 5 Hz < f_0 < 1000 Hz are necessary. It depends then on the amount of information processing, if the speed of calculations is sufficient with usual microprocessors or microcontrollers, or if signal processors, transputers or custom designed circuits like ASIC's must be used. If the microelectronics are directly assembled to the mechanical process the rough working conditions through vibrations, heat, fluids (oil, water) or dirt have to be taken into account. The mechanical integration is therefore very important.

A basic problem is the transfer of electrical signals to moving parts. Here wireless transmission (inductive, capacitive, optical) is of interest. Microelectronics can be attached decentralized at the sensors, actuators or process components. In many cases centralized solutions with the microelectronics in special casings as for machine tools and automobiles are standard. For larger systems hierarchical structured architectures with bus systems or local networks are in development.

3. ON THE DESIGN OF KNOWLEDGE BASED MECHATRONIC SYSTEMS AND COMPONENTS

Because of the integration of various functions the use of modern design tools plays an important role from the beginning. On the information side modelling, estimation and simulation methods are required for the development of mechatronic products but also sensors and actuators may contain smart functions.

3.1 Mathematical models

A precondition for precisely adapted algorithms for damping, control, fault detection and optimisation is the knowledge of mathematical process models for the static and dynamic behavior. For mechanical processes it can be recommended to obtain the *model structure* by theoretically modeling and the *parameters* by parameter estimation. For machines, consisting e.g. of a motor, drive chain and a working process the models show frequently following properties, (Isermann, 1993b):

- the drive chain elements are linearizable
- bearings can be linearized if the motion is unidirectional. For bidirectional motions the Coulomb friction generates a hysteresis, resulting in nonlinear models. However, it is possible to use direction dependent linear models,
- motors and working processes have often nonlinear behaviour
- Some parameters (e.g. mass, stiffness) are known, others (e.g. damping or friction coefficients, load parameters) are not known and change with time
- Model orders n may range within $3 \leq n \leq 20$. Hence, software tools and computer algebra are recommended, (Schumann, 1992) and model reduction is required.

3.2 Information on the process

The causal relations between measured input and output signals can be used to obtain information on the internal process behaviour. *Parameter estimation* with recursive estimation algorithms based on least squares methods allow to track time varying parameters of discrete time and continuous time models for linear and certain types of nonlinear processes. Suitable nonlinear models for mechatronic systems are for example nonlinear static polynomials or hysteresis curves with linear dynamics or nonlinear dynamics with linearity in the parameters. With known parameters *state estimation* provides information on internal variables which cannot be measured directly. Discrete-time models have the advantage to be better suited for the various algorithms. Continuous-time models should be performed if physical defined quantities (parameters, states) are essential. Parameter and state estimation methods are the basis for adaptive control or adaptive damping or for model based fault detection.

3.3 Model based nonlinear control

3.3.1 Correction of nonlinear static characteristics

Nonlinear static characteristics are present in many processes. The objective is to compensate the main static nonlinearity f by an approximate inverse function f^*, which can be implemented in the microprocessor. According to Fig. 8, the regular actuator input U is then substituted by the "corrected" value

$$U^* = f^* (U, x) \quad (1)$$

such that the I/O-behavior U-Y becomes (approximately) linear. G_{P1} and G_{P2} represent linear process parts.

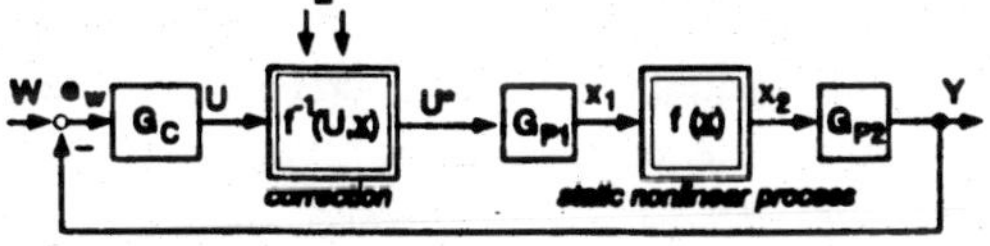

Fig. 8 General structure of a series correction (compensation) for static process nonlinearity

3.3.2 Friction compensation

Control problems with friction arise if high positioning accuracy is required. If the process stops within the hysteresis width before the set point is reached, only the integral part of the control algorithm can compensate for the offset. This leads to a significant loss of control performance and accuracy, especially during small position changes.

The basic idea of friction compensation is to compensate the relay function of the Coulomb friction by adding an adequate compensation voltage U_{comp} to the normal control action u, see Fig. 9. Different methods such as dithering, feedforward compensation and adaptive friction compensation and adaptive friction compensation will be described, compare Isermann (1993b).

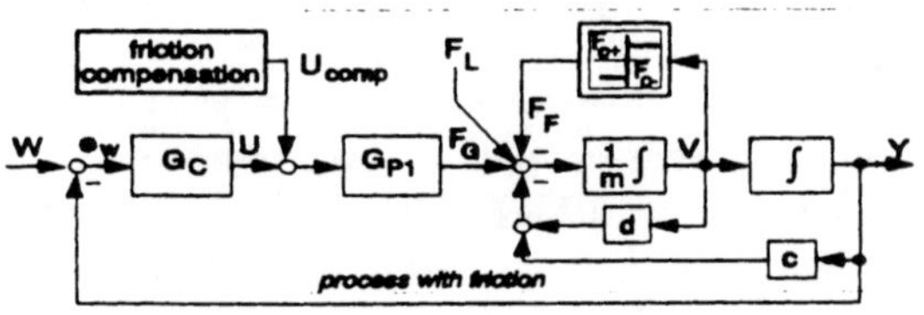

Fig. 9 General structure for friction compensation

Dithering. Dynamic linearization or so-called "dithering" is the classical way of analog and even digital friction compensation. By adding a high-frequency, periodic signal to the control action U, the friction is compensated during half the period, whereas during the second half friction is undercompensated. The method is quite robust with regard to the amplitude and frequency of the dither signal. However, if the amplitude is too large, the control performance deteriorates.

Feedforward compensation. This approach is from the theoretical point of view the ideal control strategy for friction compensation. By adding the compensation value

$$U_{comp}(k) = -\frac{f_{0\pm}}{K_{P1}} \cdot sign[e_w(k)] \qquad (2)$$

($f_{0\pm}$; *Coulomb friction*, K_{P1}; *gain of* G_{P1})

to the controller action U, an optimal inverse function of the Coulomb relay characteristic is obtained. Note that instead of the unknown velocity dY/dt the control error e_w is used for the sign of U_{comp}.

Adaptive Friction Compensation. In the preceding methods, the friction compensation was realized by a feedforward control strategy. Better results may be expected, if the actual friction value can be adapted in an additional feedback friction control loop. Therefore an adaptive friction compensation was developed, which interprets the deviation between the measured output Y(k) and a linear reference model $Y_M(k)$ as friction effect, (Isermann, Raab, 1993).

3.4 Adaptive control

During normal operation, most of the processes change their parameters in a determined way. An improved control performance over the whole range of operation as well as the lifetime may be obtained by adaptive control for the whole process.

Parameter scheduling. Parameter scheduling based on the measurement of varying operation conditions is an effective method to deal with known and approximately time-invariant process nonlinearities. Supposing measurable auxiliary variables $\underline{V}$, that correlate well with the process changes, the adaptation of the controller parameters $\underline{\Gamma}$ is performed as functions of $\underline{V}$ (parameter schedule).

Parameteradaptive control systems. Parameteradaptive control systems are characterized by using identification methods for parametric process models. The overall structure, performing online parameter estimation, controller design, supervision and coordination is shown in Fig. 10.

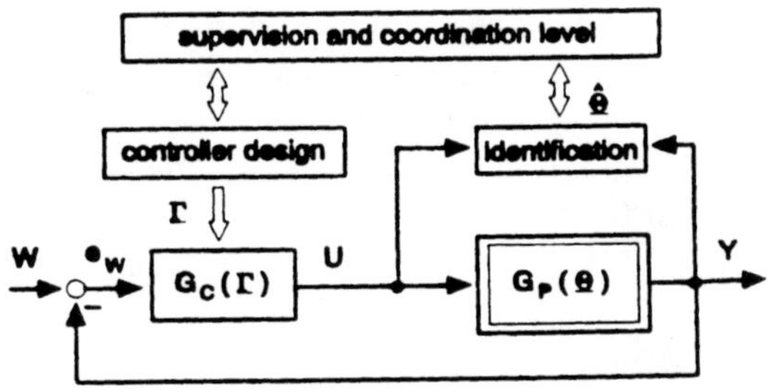

Fig. 10 Parameteradaptive control structure

Digital adaptive control works well if the assumptions for their design and convergence are satisfied. This includes proper excitation of the process dynamics as well as the appropriate model structure. For the cases where the assumptions are violated a supervision level is required which takes appropriate actions, (Isermann, Lachmann, Matko 1992).

3.5 Fault detection and supervision

An important feature of an intelligent system is the automatic supervision and fault diagnosis of its components. Figure 9 shows a component influenced by faults. External faults are for example caused by the power supply, contamination or collision, internal faults by wear, missing lubrication, actuator or sensor faults.

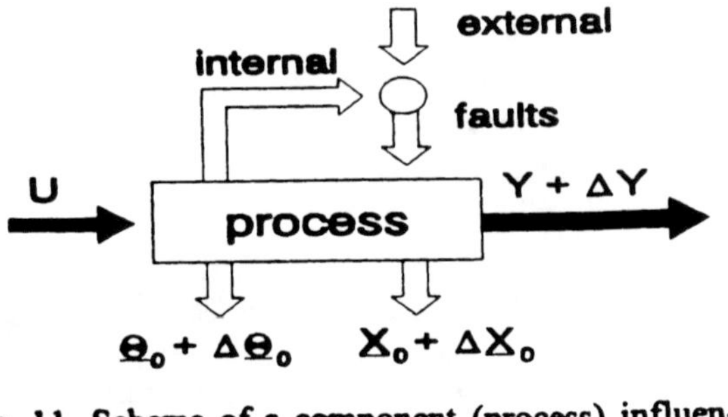

Fig. 11 Scheme of a component (process) influenced by faults

If the faults influence directly measurable output variables they may be detected by an appropriate signal evaluation. The corresponding functions are called monitoring, if the measured variables are checked with regard to a certain tolerance of the normal values and alarms are triggered if the tolerances are exceeded. In the cases where the limit value violation signifies a dangerous state an appropriate action can be indicated automatically. This is called *automatic protection.*

The classical ways of limit value checking of some few important measurable variables are appropriate for the overall supervision. However, developing component faults are only detected at a rather late state and the available information does not allow an in-depth fault diagnosis. Research efforts have shown that the use of process models allows an early fault detection in connection with normal measured variables (Isermann, 1993c). Then nonmeasurable quantities like state variables and parameters may be estimated. With this improved knowledge a *supervision with fault diagnosis* becomes possible.

Figure 12 shows the scheme of a knowledge-based fault detection and diagnosis with process models. The knowledge base consists of an automatic and a heuristic part. The analytic problem solution includes parameter or state estimation and generates symptoms as changes from the normal behaviour. An inference mechanism then performs the fault diagnosis. For more details see Isermann (1994).

A considerable advantage is, if the *same process model* can be used *for* both, the (adaptive) *controller design and the fault detection.* In general continuous time models are preferred, if fault detection is based on parameter estimation. For fault detection with state estimation or parity equations discrete-time models can be used.

4 SENSORS AND ACTUATORS

4.1 Sensors

The required measurement variables for mechatronic systems are electrical quantities like position, speed, acceleration, force, torque, pressure or thermal quantities like temperature or heat flow. A survey of known measurement principles is e.g. given in Profos, Pfeifer (1992), Tränkler (1990). Of special importance for mechatronic systems are the integration with the process, the dynamics, the resolution, mechanical and thermal robustness, wear resistancy, touchless transfer, miniaturisation and easy transfer to digital signal processing. Therefore incremental sensors for positions or speeds show an increasing trend. The integration of the sensor and the signal processing on a common carrier becomes more and more important. This integration, possibly on one chip, has following advantages: less cost and space, higher precision, less disturbances through digital transmission. The digital processing of sensor signal allows the programming of noise filtering, linearisation of nonlinear characteristics, correction of cross sensitivities or hysteresis, self calibration, compensation of dynamic lags and of drift, Tränkler (1992). Hence, a development can be observed from a classical measurement chain to a smart sensor component.

A drawback of sensors and microelectronics based on silicium technology is the limited temperature range until about 80 - 100° C. Gallium arsenide (GaAS) technology may allow to lift the range until 300° - 400° C.

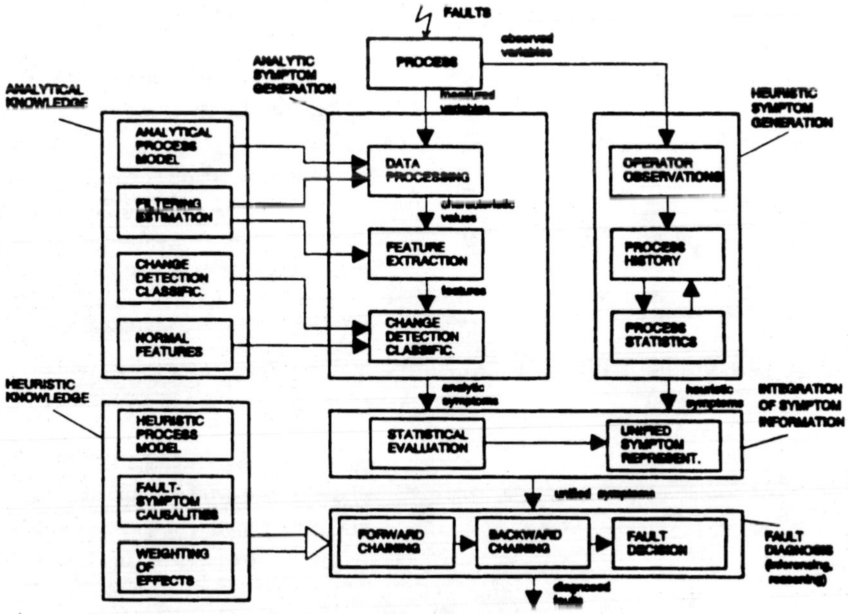

Fig. 12 Scheme of a knowledge-based fault detection and diagnosis

4.2 Actuators

Actuators in mechatronic systems transform electrical inputs into mechanical outputs such as position, angle, force or torque. The supporting energy is usually electrical, pneumatic or hydraulic. The actuator principles can be classified into three major groups:

- Electromagnetic actuators (e.g. electrical motors (d.c., a.c.)), step motors, electro-magnet)
- Fluidic actuators (hydraulic, pneumatic)
- Unconventional actuators (e.g. piecoelectric, magnetostrictive, memory metal).

An evaluation of typical application areas, control properties and robustness is given in Raab, Isermann (1990).

If a high precision of positioning and fast dynamics are required, generally a position control loop is necessary. However, following properties hinder a precise control: friction and backlash (hysteresis effects), nonlinear static characteristics, operating point and wear dependent parameters. It was shown in Isermann, Raab (1993) how these disadvantages can be compensated by a model based digital position control. Compensation of nonlinear characteristics and friction compensation with parameter adaptive control showed a considerable improvement of control performance for an electromagnetic and pneumatic actuator, see also section 6.

5. DESIGN TOOLS

The integration by information processing for high performance mechatronic systems requires that the static and dynamic behaviour of all components is well designed. This means a systematic development, starting with modeling and simulation and using methods for identification and computer aided control system design.

5.1 Software tools for the design

The computer aided development of mechatronic systems comprises:

a) constructive specification in the engineering development stage using CAD- and CAE-tools
b) model building for obtaining static and dynamic process models
c) transformation into computer codes for system simulation
d) programming and implementation of the final mechatronic software.

Some software tools are described for example in Otter, Gruebel (1993). A broad range of CAD/CAE tools is available for 2D- and 3D-mechanicel design, such as Auto CAD (Autodesk Inc) with a direct link to CAM (computer aided manufacturing); PADS (PADS Software Inc.) for multi-layer printed-circuit board layout, etc.

However, the efforts for computer aided modeling are comparable low (Fasol, Diekmann, 1990). A program system for computer aided modeling by using computer-algebra was developed by Schumann (1992). An object-oriented language Dymola for modeling of large combined systems is described in Otter, Gruebel (1993) and Elmquist (1993). These packages are based on specified ordinary differential equations, algebraic equations and discountinuities.

For system simulation (and controller design) a variety of program systems exist, like ACSL, SIMPAC-K, MATLAB/SIMULINK, MATRIX-X. These simulation techniques are valuable tools for the design, as they allow to study the interaction of components, and the variations of design parameters before manufacturing. However, they are in general not suitable for real-time simulation.

5.2 Hardware-in-the-loop simulation

For the further development of the final mechatronic system the hardware-in-the-loop simulation plays a special role. Various kinds can be realized, compare Fig. 13.

a) Simulation of electronics
Actuators, mechanics and sensors are real hardware. Information processing is realized in e.g. a process computer for development and also for testing, in case of malfunctions at the side of microcomputer, power supply, cable connections, etc.

b)Simulation of mechanics, actuator and sensor
Microcomputer hardware and software is real. Real-time simulation of the remaining system requires appropriate mathematical models, such that variations on the mechanical side can be tested.

c)Simulation of mechanics
Microcomputer hardware and software, actuators and sensors are real. Real-time simulation of the mechanics allows testing of the electronic components and effects in case of their malfunctions.
For the real-time simulation of mechanics, sensors and actuators transputers and signal processors are suitable, see e.g.Hanselmann (1993).

In the following two examples are described which show the development of mechatronic systems in some respects. The first example emphasizes model based nonlinear control, the second one the control of an estimated parameter.

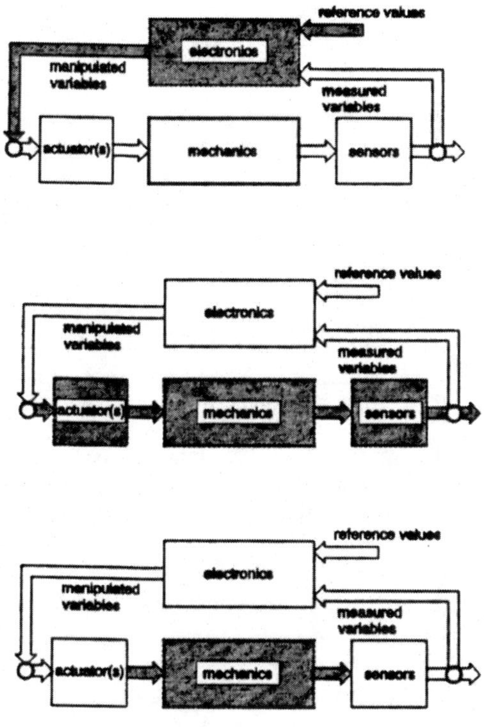

a)

b)

c)

Fig. 13 Different ways of hardware-in-the loop simulation
a) simulation of electronics
b) simulation of mechanics, actuator and sensor
c) simulation of electronics

6. ADAPTIVE NONLINEAR CONTROL OF A PNEUMATIC ACTUATOR

Pneumatic cylinder-piston actuators are mainly used for transportation and assembling tasks. Due to their nonlinear beahaviour they operate predominantly with limit stops or brakes. The nonlinear static and dynamic behaviour is due to the compressibility of the air, the friction of the piston and the characteristics of proportional control valves. Therefore precise motion control is rather involved and does not result in a sufficient control performance with linear controllers only.

In the following it will be shown how through a nonlinear model based adaptive control system it becomes possible to obtain a relatively high position accuracy. The integration of the new mechatronic system is, until now, performed by information processing.

The investigated pneumatic actuator is a standard pneumatic cylinder Fig. 14. The motion transmission of the piston to the outer roller is obtained by magnetic coupling. The positioning range is 200 mm and the piston diameter is 25 mm, which leads to a force of 213 N by applying a pressure of 6 bar.
Process input is a voltage U which manipulates the mass-flows $\dot{m}_1$ and $\dot{m}_2$ into the cylinder chambers through a control valve. The position Y can be measured by a linear potentiometer. In addition the pressures P_1' and P_2' are available.
Theoretical modelling, e.g. Nguyen (1987), of the energy transducer - the pneumatic part - leads to a model with several dynamic nonlinearities as it is shown in Fig. 15 for pressure P_1 of chamber 1.

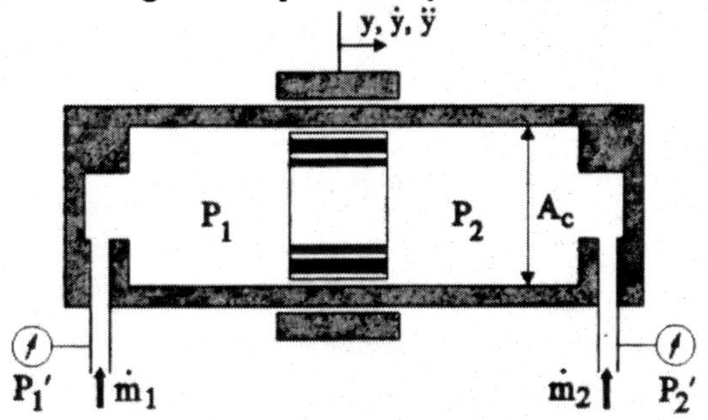

Fig. 14 Scheme of the investigated pneuamtic cylinder

Therefore a static nonlinear correction only will not work, but the implementation of an underlying differential pressure control loop with a simple P-controller has shown to work very well in order to obtain an adjustable input force F_G to the mechanical system. Actually this action requires the measurement of the differential pressure at the cylinder piston, which is not available, but can be computed approximately by using the pressure P_i' measured at the cylinder chamber connections.

A further significant nonlinearity of this actuator is friction. Typical stiction forces are roughly 45 N (21% of the nominal input force) and values of about 30 N have been estimated for Coulomb friction via mechanical system identification. Therefore friction compensation is required.

Fig. 16 depicts the implemented overall control strategy with state controller, underlying differential pressure control and adaptive feedforward friction compensation.

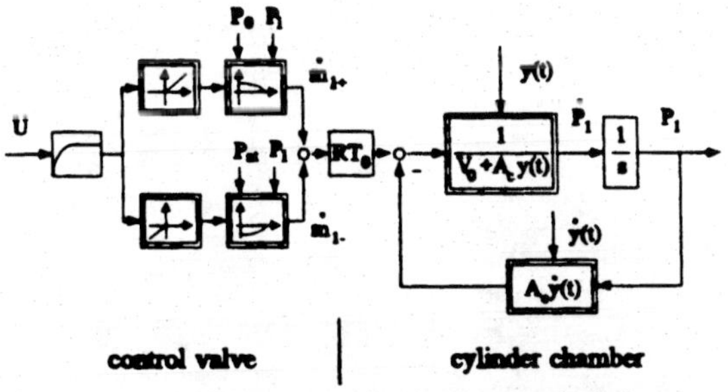

Fig. 15 Simplified model for the pneumatic part

Feedforward friction compensation as it is shown in Fig. 9 yields continuous oscillations with an amplitude of at least 0.5 mm. This occurs due to the non-negligible dynamics of the underlying differential pressure loop. Therefore the compensation has to be switched off if the control variable Y lies within a tolerance band (± 0.05 mm) of the set point W.

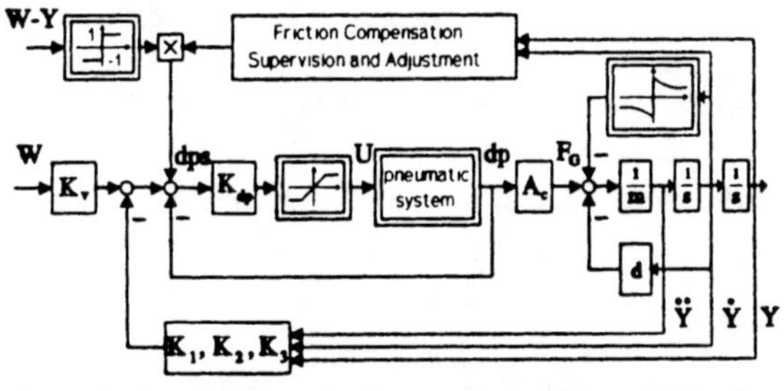

Fig. 16 Overall control structure for positioning control of a pneumatic actuator

As several investigations have shown, e.g. Rusterholz, Widmer (1985), Chen, Leufgen (1987), friction forces of pneumatic cylinders are highly position dependent and varying with the applied pressure as well as with the length of standstill.

Owing to this, under- and overcompensation is very likely and supervision of the steady state control behaviour must take place in order to adjust the amplitude of the compensation values. The amplitude will be reduced if oscillation is detected and increased if the control value doesn't lie in the set point tolerance band. This is done by "friction compensation supervision and adjustment" (FCSA).

Fig. 17 shows the influence of the friction compensation.

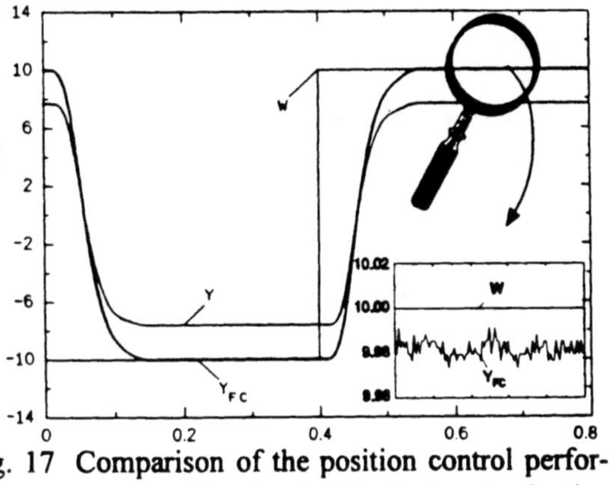

Fig. 17 Comparison of the position control performance without and with (index $_{FC}$) adaptive friction compensation, T_0=1 msec

The state controller values (K_1,K_2,K_3) are obtained by numerical optimization of a quadratic loss function using a nonlinear system model - with (Y_{FC}, dps_{FC}) and without (Y, dps) friction compensation. The computation is done with exactly the same model parameters and equal control value weighting.

Without friction compensation an offset of about 3 mm has been obtained. The dynamic features with friction compensation are satisfying and the positioning accuracy has been significantly improved. The comparison of the set points (dps) of the underlying diff. pressure loop clearly shows the influence of the d.c. compensation with compensation switch-off.

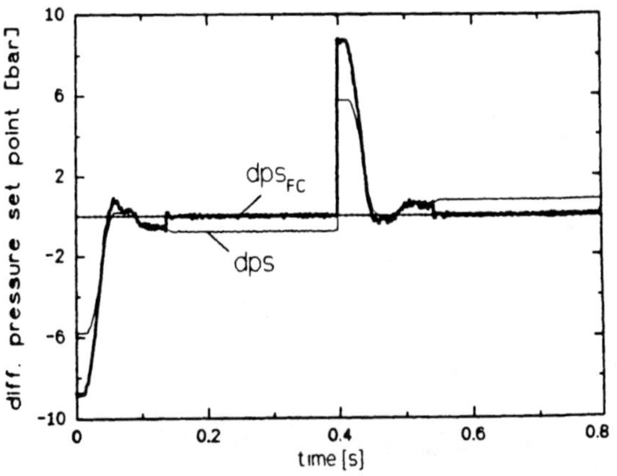

Fig. 18 Comparison of the position control performance without and with adaptive friction compensation

Using the above mentioned enlargement to the feed-forward friction compensation a positioning accuracy of at least 0.05 mm can be achieved.

The signals with FCSA are shown in Fig. 19. The d.c.-compensation value is shown for W = -10 mm.

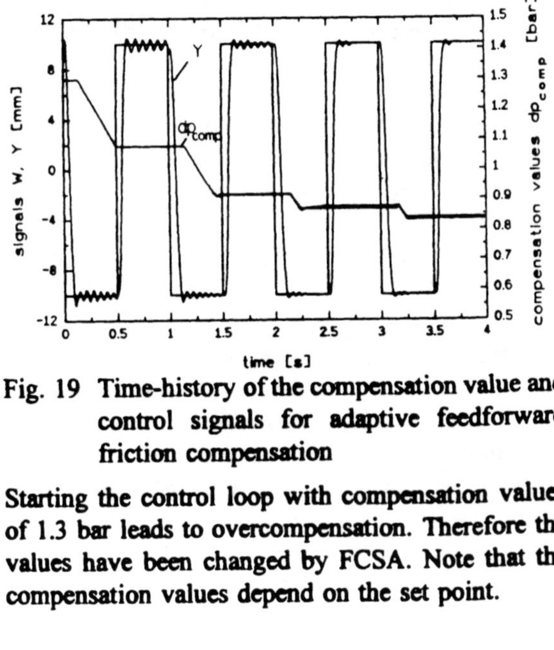

Fig. 19 Time-history of the compensation value and control signals for adaptive feedforward friction compensation

Starting the control loop with compensation values of 1.3 bar leads to overcompensation. Therefore the values have been changed by FCSA. Note that the compensation values depend on the set point.

7. ADAPTIVE SEMI ACTIVE SHOCK ABSORBERS FOR VEHICLE SUSPENSION SYSTEMS

Known adaptive and semi-active suspension control systems use fixed or scheduled controller parameters. Therefore some changes in vehicle conditions are not directly taken into consideration. Parameter adaptive suspension control systems were developed based on real-time parameter estimation which take into account changing vehicle conditions. The methods are based on a simplified model of the vertical dynamic behavior, Fig. 20. By measurement of e.g. the spring deflection (z_A-z_R) and the body acceleration $\ddot{z}_A$ the damping coefficient d_A, the body mass

m_A and the Coulomb friction F_R can be estimated for given spring stiffness c_A.

Nonlinear, direction dependent continuous-time models are used for parameter-estimation with discrete square root filtering (DSFI), Fig. 21. Fig. 22 shows a semi-active parameter-adaptive system with a feedback of the signals $(z_R - z_A)$ and $(\dot{z}_R - \dot{z}_A)$ and adaptation of the continuously adjustable damping coefficient d_A of the damper. By this adaptive shock absorber system an improvement of both, driving comfort and safety can be achieved, Fig. 23, (Bußhardt, Isermann 1992).

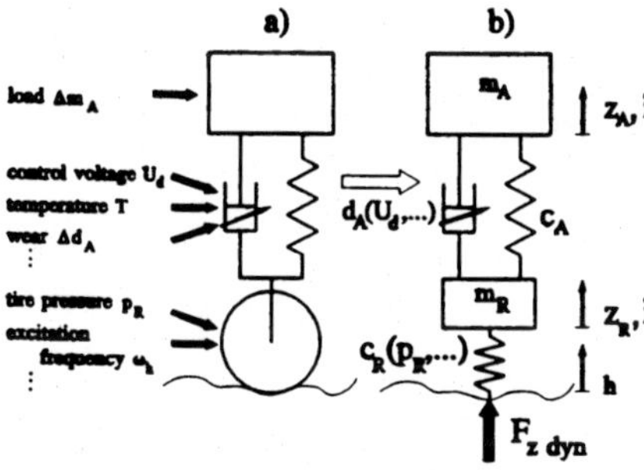

Fig. 20 Simplified model of the vertical dynamics (linear part)

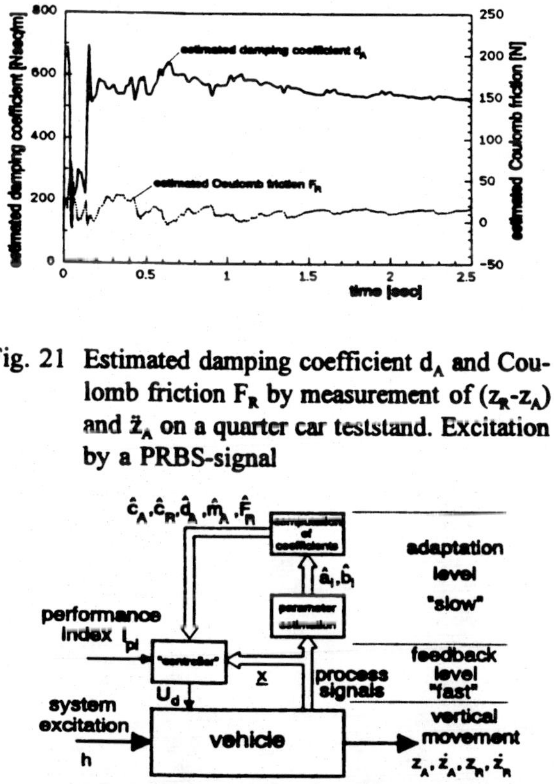

Fig. 21 Estimated damping coefficient d_A and Coulomb friction F_R by measurement of $(z_R - z_A)$ and $\ddot{z}_A$ on a quarter car teststand. Excitation by a PRBS-signal

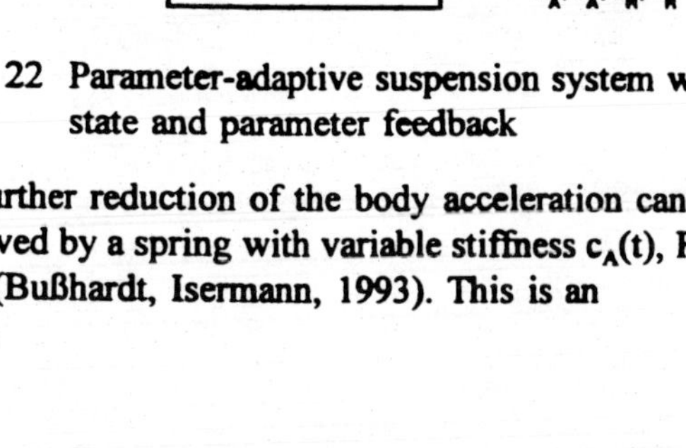

Fig. 22 Parameter-adaptive suspension system with state and parameter feedback

A further reduction of the body acceleration can be achived by a spring with variable stiffness $c_A(t)$, Fig. 23, (Bußhardt, Isermann, 1993). This is an

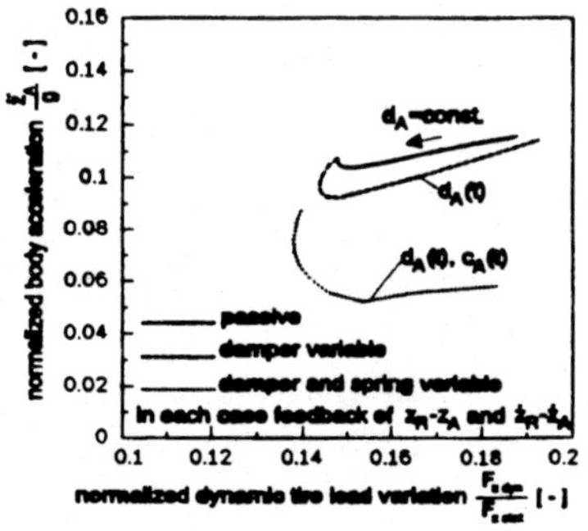

Fig. 23 Body acceleration $\ddot{z}_A$ versus dynamic tire load variation F_{zdyn} for passive (constant) and parameter-adaptive damping and parameter-adaptive damping and spring rate (simulations).

example for a mechatronic system which becomes adaptive by estimation of nonmeasurable quantities.

Remark: A part of the research results on mechatronic systems was obtained in a special research program (Sonderforschungsbereich 241 IMES) at the Tech. Univ. Darmstadt (Isermann, Ed.(1993)).

8. REFERENCES

Bradley, D.A., D. Dawson; D. Burd and A.J. Loader(1991). Mechatronics-Electronics in products and processes. Chapman and Hall, London.

Bußhardt, J. and R. Isermann, R. (1992). Realization of adaptive shock absorbers by estimation physical processes coefficients of a vehicle suspension system. ACC, Chicago.

Bußhardt, J.and R. Isermann (1993). Parameter adaptive semi-active shock absorber based on nonlinear models. 2nd European Control Conference. Groningen.

Chen, M.Sc.X., M. Leufgen (1987). Erfassung des Reibeverhaltens von Kolbendichtungen und deren Einfluß auf die Positionierung von pneumatischen Systemen. *o + p (ölhydraulik und pneumatik)* 31, Nr. 12.

Elmquist, H. (1993). Object-oriented modeling and automatic formula manipulation in Dymola. Scandin. Simul. Society SIMS 93, Kongsberg, June 1993.

Fasol, K.H., Diekmann (Ed.) (1990). Simulation in der Regelungstechnik. Springer, Berlin.

Hanselmann, H. (1993). Hardware in-the-loop simulation as a standard approach for development, customization, and production test. SAE 930207, March 1-5.

Isermann, R., (Ed.)(1993a). Integrierte mechanisch elektronische Systeme. Fachtagung Tech. Univ. Darmstadt, VDI-Fortschr. Ber., Reihe 12, Nr. 179. VDI Verlag Düsseldorf.

Isermann, R.(1993b). Towards intelligent control of mechanical processes. *Control Engineering*

Practice, Vol.1, No. 2, 232-252.
Isermann, R.(1993c). Fault diagnosis of machines via parameter estimation and knowledge processing. *Automatica* 29, No.4. 815-835.
Isermann, R. (1994). Integration of fault detection and diagnosis methods. IFAC-Symposium SAFE-PROCESS, Espoo, Finland.
Isermann, R., Lachmann, K.H. and D. Matko, (1992). Adaptive control systems. Prentice Hall, London.
Isermann, R. and U. Raab (1993). Intelligent actuators-ways to autonomous actuating systems. *Automatica* 29 (1993), No.5, 1315-1131.
MacConaill, P.A.; P. Drews and K.H. Robrock (1991). Mechatronics and Robotics I. IOS-Press, Amsterdam.
Nguyen, T.(1987). Verhalten servopneumatischer Zylinderantriebe im Lageregelkreis. Dissertation RWTH Aachen.
N.N.(1990). Mechatronics: Designing intelligent machines. Proc. IEE-Int.Conf. 12-13 Sept, Univ. of Cambridge, (1990).
Otter, M., G. Gruebel (1993). Direct physical mode ling and automatic code generation for mechatronics simulation. 2nd Conf. on Mechatronics and Robotics, Duisburg, Sep. 27-29, 1993 IMECH, Moers.
Profos, P. and T. Pfeifer (Ed.)(1992). Handbuch der indutriellen Meßtechnik. 3. Aufl., Vulkan-Verlag, Düsseldorf.
Raab, U. and R. Isermann (1990). Low power actuator principles. VDI/VDE Conf.-Actuator 90, Bremen.
Rusterholtz R, U. Widmer (1985).Grundlagenbetrachtung zur Auslegung pneumatischer Servoantriebe. *o + p (ölhydraulik und pneumatik)* 29, Nr. 10.
Schumann, A.(1992). Beiträge zur rechnergestützten Modellbildung dynamischer Prozesse. Dissertation Technische Hochschule Darmstadt. VDI-Verlag Düsseldorf, Reihe 8, Nr. 304.
Schweitzer, G.(1992). Mechatronics - a concept with examples in active magnetic bearings. *Mechatronics* 2, Nr. 1, p. 65-74.
Tränkler, H.R.(1990). Taschenbuch der Meßtechnik. Oldenbourg Verlag, München.
Tränkler, H.R.(1992). Information processing in sensing devices and microsystems. IFAC-Symposium on Intelligent Components and Instruments for Control Applications (SICICA), May 20-22, 1992, Malaga (Spain). Proc, Pergamon Press, London.

FUZZY CONTROL METHODOLOGY AND APPLICATIONS TO AUTOMOTIVE SYSTEMS AND MOBILE ROBOTS[1].

A. OLLERO[†], S. BOVERIE[††] and A. TITLI[†††].

[†]Departamento de Ingeniería de Sistemas y Automática.
Escuela S. Ingenieros, Universidad de Sevilla.Av. Reina Mercedes, 41012 Sevilla (Spain).
Fax:+34-5-4556849. E-mail:aollero@esi.us.es
[††]Systemes Moteurs Avancés, Siemens Automotive
B.P. 1149 Avenue du Mirail, 31036 Toulouse Cedex (France).
Fax: +33-62-141052
[†††]LAAS/CNRS
7 Av. du Colonel Roche, 31400 Toulouse Cedex (France).
Fax: + 33-61-553577

Abstract. This paper studies, from a control engineering perspective, some problems in fuzzy logic control: the equivalence between fuzzy controllers and conventional ones, fuzzy controllers for high order processes, fuzzy supervisory control, and the stability of fuzzy logic control loops. The paper also summarizes applications in two closely related fields: automotive systems and mobile robots. In the first case applications to chassis control (active suspension) and engine control (idle speed regulation, anti-knock) are reported.The second deals with automatic steering of wheeled mobile robots, and two different fuzzy logic strategies to solve the problem (direct fuzzy control and fuzzy auto-tuning of conventional path trackers) are presented.

Key Words: Fuzzy control, Control system analysis and design. Automotive Systems, Mobile robots.

1. INTRODUCTION

Following Zadeh and Mamdani, a fuzzy controller is a special knowledge-based system manipulating a set of IF-THEN rules in three steps: fuzzification, inference, and defuzzification. Figure 1 shows a basic fuzzy logic control loop.

The premisses of the rules contain statements on the inputs variables (for example the error and the change of the error in a regulation problem) and the conclusions express statements on the output of the controller (for example the control variable directly or the change in the control variable). A "projection" of the numerical value of the inputs on the linguistic variables of the premises gives a linguistic evaluation of the real situation of the process (it is the fuzzification phase). The linguistic evaluation is used in an inference engine to develop approximate reasoning on the basis of this evaluation. The conclusion of this reasoning is defuzzified in order to get a numerical value for the control to be applied on the process. This approach has several advantages when comparing with other conventional control strategies, a big one being to integrate, in an easy way, human expertise in the control law.

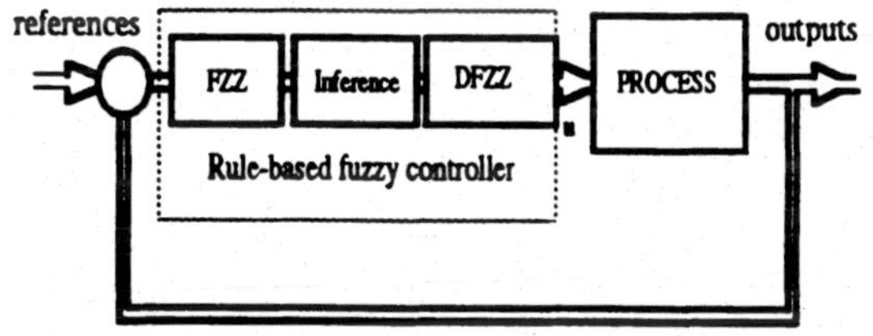

Fig. 1: Basic fuzzy logic control loop.FZZ: Fuzzification; DFZZ Defuzzification.

The section 2 of this paper is devoted to some topics in fuzzy control methodology. Particularly, the paper considers the equivalence between fuzzy controllers and conventional ones, fuzzy control for high order processes, fuzzy supervisory control, and the stability analysis. The third section deals with some specific applications developed by the authors in automotive systems and mobile robots. The conclusions and references are in sections 4 and 5.

2. FUZZY CONTROL METHODOLOGY.

2.1 Equivalence between fuzzy controllers and conventional ones (FC/CC).

Fuzzy control provides general controller structures

[1]This research has been partially supported by the EC ESPRIT Program (WG-6017), FALCON.

which can be particularized to meet the characteristics of conventional controllers. Thus, Foulloy (1993) demonstrated how automatically construct the rule base of a fuzzy Mamdani's type controller to get a perfect equivalence between PID type controllers and fuzzy controllers, extensions to some linear analytical controllers being easily possible.

Furthermore, various studies have shown that fuzzy controllers can generate a non-linear control surface (Fig 2).This non-linearity can be considered at two levels (Raymond *et al* 1993):

- a global non linearity due to the partitioning of the fuzzy subsets and to the rules on the overall domain,

- a local non-linearity due to the choice of the fuzzy operators (min, max, etc. . . .) on each elementary cell of the global non-linear surface (Fig 3&4). Local non-linearities have quite no influence in closed loop control and it would be possible to approximate the non-linear control surface on each cell by a linear one. So, a linear piecewise controller can be obtained.

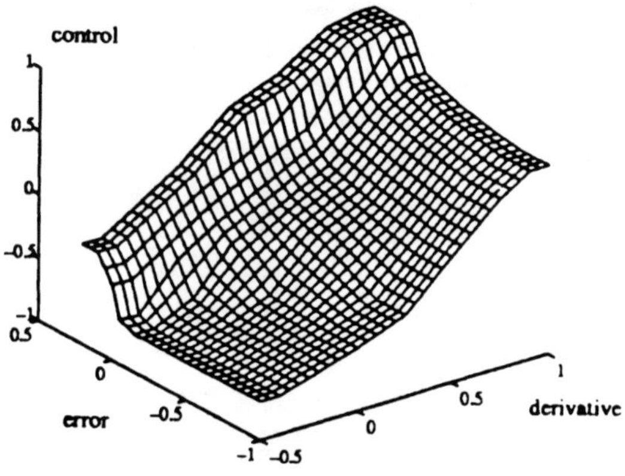

Fig.2 Control surface for a fuzzy controller

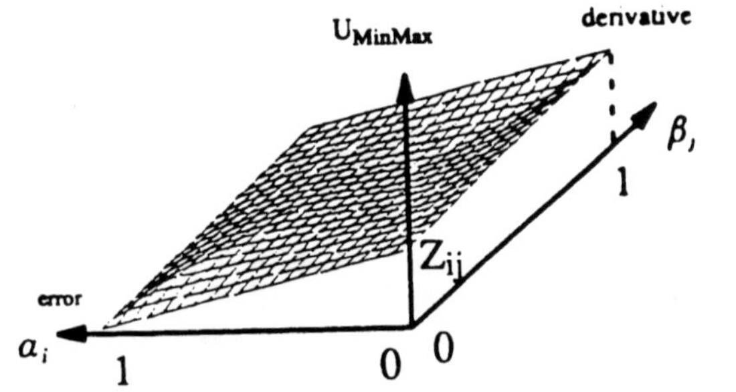

Fig.3 Control surface on one cell generated with min, max operators.

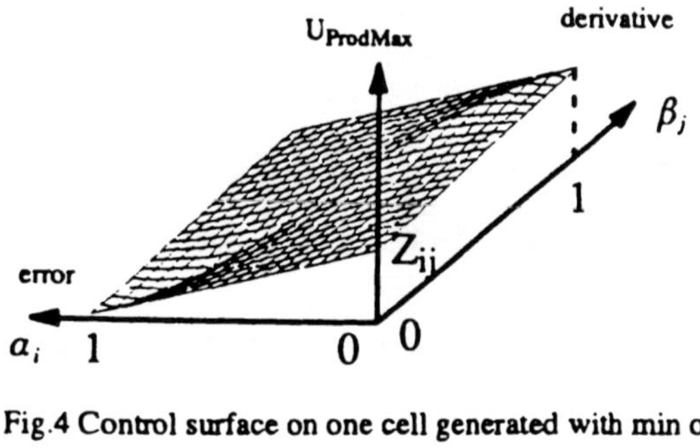

Fig.4 Control surface on one cell generated with min operators

Moreover, the equivalence between fuzzy controllers and sliding mode control has been studied (Palm, 1992; Cerf *et al*,1994). Many of the conventional fuzzy controllers are based on the Mac Vicar Weelan structure (Fig 5) which can be seen as an extension of the sliding mode control. A sliding curve where the applied control action is zero can be defined in the phase plane. :

ε	NB	NM	NS	ZE	PS	PM	PB
PB	ZE	PS	PM	PB	PB	PB	PB
PM	NS	ZE	PS	PM	PB	PB	PB
PS	NM	NS	ZE	PS	PM	PB	PB
ZE	NB	NM	NS	ZE	PS	PM	PB
NS	NB	NB	NM	NS	ZE	PS	PM
NM	NB	NB	NB	NM	NS	ZE	PS
NB	NB	NB	NB	NB	NM	NS	ZE

Fig.5 Structure of fuzzy rule based proposed by Mac Vicar-Whelan

When the state vector does not belong to this sliding curve, a control action which obliges the state vector to reach it is applied. Fuzzy control introduces the notion of non-linear sliding surface and also more smoothness in the control action around this sliding surface (Fig6&7).

An extension of this study has been done for large time varying processes. If we consider a process with time varying response this involves some constraints in the choice of the sliding curves. In order to take into account the different dynamics, it is necessary to rotate the sliding curve around the point (Fig 7). To do that a new control structure has been proposed (Fig 8).

2.2. Fuzzy controllers for high order processes.

Conventional control systems use generally as inputs the error ε and the variation of the error Δε. This structure is well suited to control processes when the order is less or equal to 2, but if the order increases, more information is needed as for conventional algorithmic control. For example, to control a third order system, informations about the state and its first and second derivatives are needed. Then the first idea is to design a nth input fuzzy controller in order to control a nth order process. Unfortunately, this will lead to an exponential increase of the rule number. In order to solve this problem original structures have been developed (Boverie *et al* 1992a,1993c; Demaya 1992):

- series high-order controller, which is based on a series association between conventional 2 inputs fuzzy controllers and an incremental controller (Boverie *et al*, 1993a),

-parallel high-order controller, which is based on a parallel association of conventional two inputs fuzzy controllers (Boverie *et al*, 1993b).

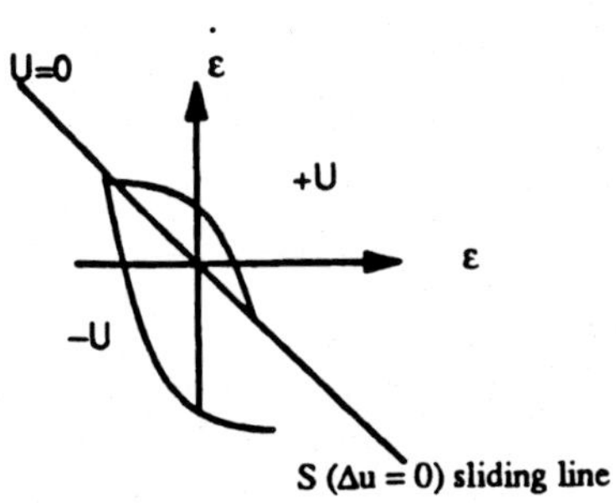

Sliding mode control with on/of technique

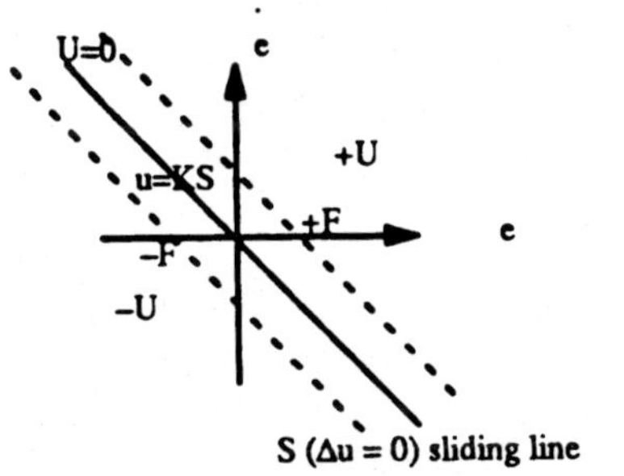

Extension of the sliding mode control technique
with proportional control around the sliding surface

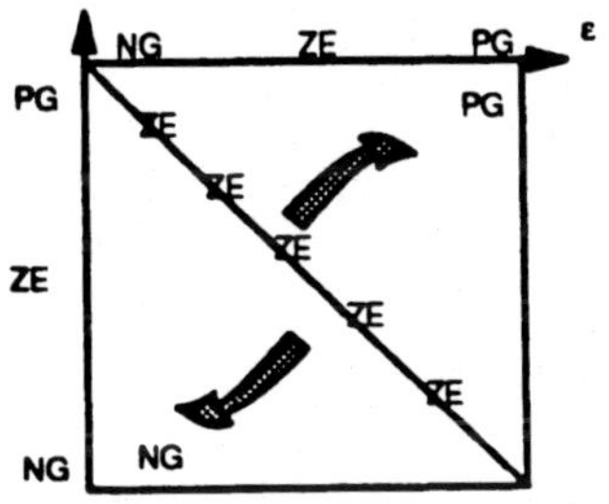

Fig.6 Extension of sliding mode control to fuzzy control techniques

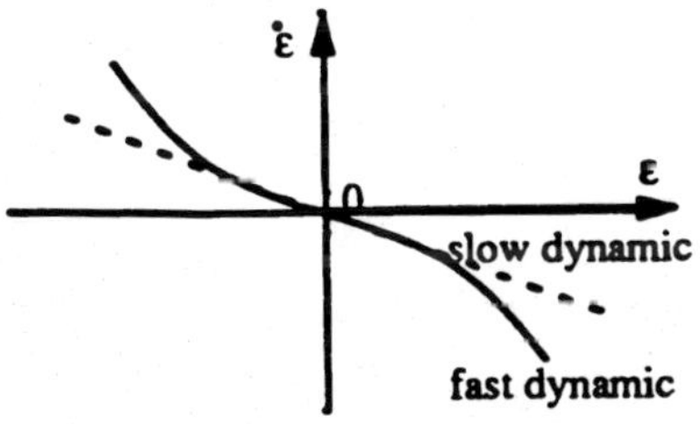

Fig.7 Representation in the phase plane of the sliding curve generated by the fuzzy controller

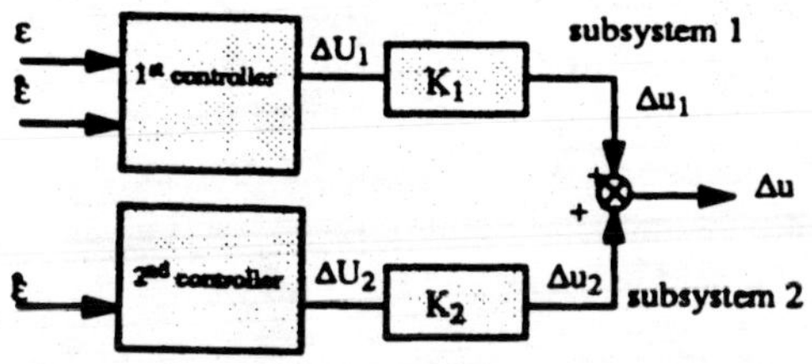

Fig.8 Fuzzy sliding mode controller structure

Multi-input fuzzy controller: Qualitatively, the structure of the fuzzy controller is equivalent to that of a $PID^{(n-1)}$ controller. If u is the value of the control action provided by the fuzzy controller, we can show that this value is a non-linear function of the error ε, of the integral of error, and its (n-1) successive derivatives.

$$u = F\left(\int \varepsilon, \dot{\varepsilon}, \ldots, \varepsilon^{n-1}\right) \qquad (1)$$

We find here an expression which is equivalent to those of a conventional controller. The control rules can be generated in a qualitative way, following the same approach as for the 1st order process. However, this method is complex to implement, because the number of rules increases in a near-exponential way with the order of the system. For example, for a first order system, the number of rules is equal to m^2 if m is the number of membership functions of the input variables. and for a nth order system, the number of rules is equal to m^{n+1}. We can easily understand how difficult it is to implement such controllers.

Synthesis of the fuzzy incremental controller (F.I.C):

This structure is based on the association between a conventional fuzzy controller and an incremental block (Fig 9), and allows the adaptation of the overall controller to the order of the process to be controlled.

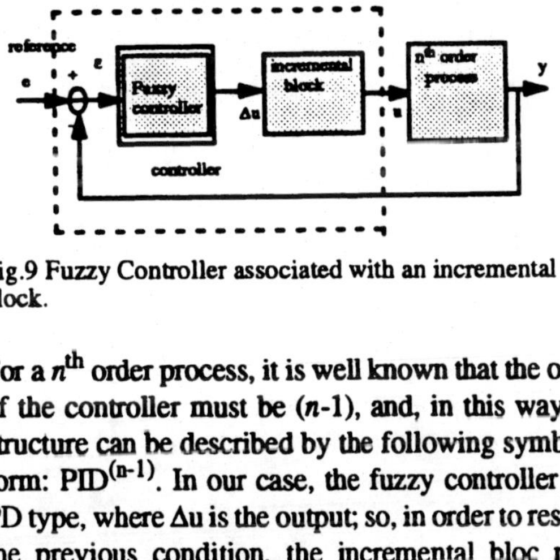

Fig.9 Fuzzy Controller associated with an incremental block.

For a nth order process, it is well known that the order of the controller must be (n-1), and, in this way, its structure can be described by the following symbolic form: $PID^{(n-1)}$. In our case, the fuzzy controller is a PD type, where Δu is the output; so, in order to respect the previous condition, the incremental bloc must have a $PID^{(n-2)}$ form. From this, we can deduce the main shape of this control block.

$$U = k_0\int_0^t \Delta u(\tau)d\tau + k_1\Delta\tau + k_2\Delta\dot{u} + \ldots + k_{n-1}\Delta u^{n-2} \qquad (2)$$

To ease the use of this formula and also its implementation in a digital unit, it is necessary to formulate a simplifying hypothesis and then, to get the discrete expression of the control:

$$U_j = k_0\sum_{i=0}^{j}\Delta u_i + k_1\Delta u_j + k_2\Delta\dot{u}_j + \ldots + k_{n-1}\Delta u_j^{n-2} \qquad (3)$$

with:

$$\Delta \dot{u}_j = \frac{\Delta u_j - \Delta u_{j-1}}{\Delta t} \quad (4)$$

and

$$\Delta u_j^{(2)} = \frac{\Delta u_j - 2\Delta u_{j-1} + \Delta u_{j-2}}{\Delta t^2} \quad (5)$$

$$\Delta u_j^{(n-2)} = \frac{f(\Delta u_j, \ldots, \Delta u_{j-n+2})}{\Delta t^{n-2}} \quad (6)$$

Finally, we obtain the control structure:

$$U_j = k_0 \sum_{i=0}^{j} \Delta u_i + k'_1 \Delta u_j - k'_2 \Delta u_{j-1} + \ldots + \quad (7)$$

$$+ (-1)^n k'_{n-1} \Delta u_{j-n+2};\ k'_i > 0$$

We notice that the incremental block has n degrees of freedom for a nth order process and therefore, we can suppose that this structure makes the tuning of the fuzzy controller more complex. But we have shown experimentally that the sensitivity of tuning parameters is very small thanks to the great robustness of the fuzzy controller and justify the association between a fuzzy controller and an incremental block.

Fuzzy controller with parallel structure of second order blocks (F.C.P.S.): We can easily show that a PD^{n-1} controller can be decomposed in a parallel association of (n-1) PD structures. Each elementary block can be described by the following equation:

$$\Delta u_i = \left(k'_{2i} + k'_{2i+1}\frac{d}{dt}\right)\varepsilon^i$$

The global control action can be given by:

$$\Delta u = \sum_i \Delta u_i = k_0\varepsilon + k_1\dot{\varepsilon} + \ldots + k_{n-1}\varepsilon^{(n-1)} \quad (8)$$

The global structure of such a controller is given on Figure 10.

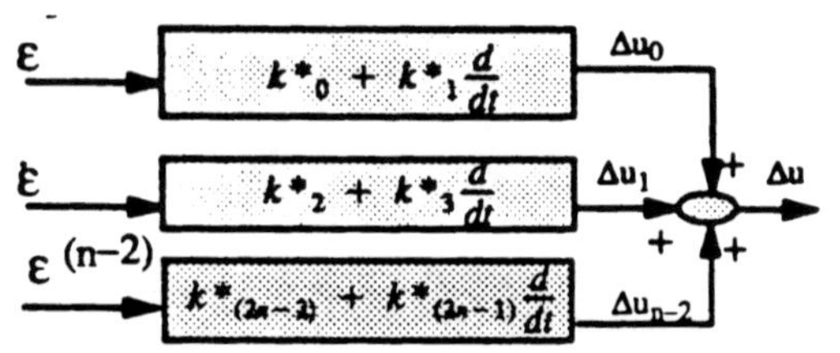

Fig.10 Parallel block structure for a linear controller

As for a PD^{n-1} linear controller, the use of elementary PD blocks can be extended to the fuzzy control of high order systems. This structure is described by the Figure 11. The equation of an elementary block will be:

$$\Delta u_i = Fuzz_i(\varepsilon_i, \frac{d\varepsilon_i}{dt}) \quad (9)$$

The general expression of the fuzzy control action is given by:

$$\Delta u = \sum_{i=0}^{(n-2)} \Delta u_i \quad (10)$$

The control rules can be expressed in a qualitative way in (n-1) two input tables.

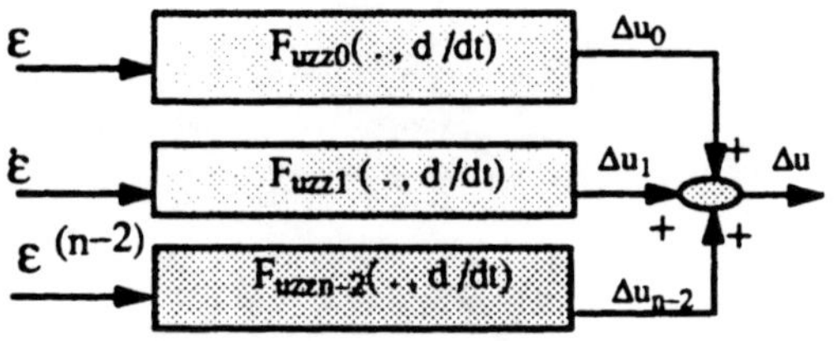

Fig.11 F.C.P.S. description

Note: This parallel structure of (n-1) elementary blocks has no integral term. This control is not able to suppress the static error. In order to take this into account we can use the following modified structure (Fig 12).

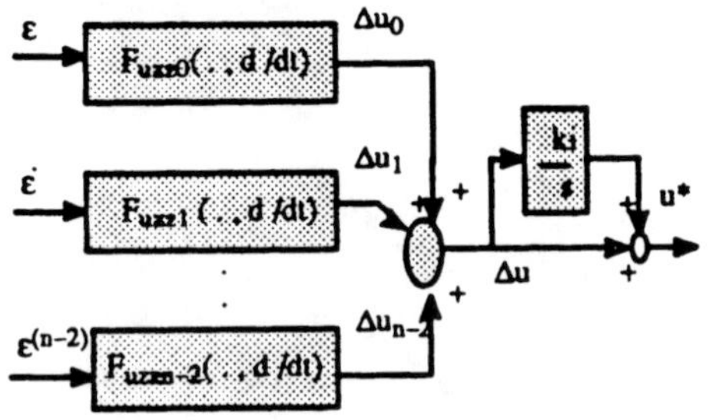

Fig.12 F.C.P.S. with integrator

2.4 Fuzzy supervisory control.

Fuzzy logic has been used in a supervisory level of conventional controllers to perform autonomously functions such as:

- tuning of control parameters: gains, weights, sample time, and others.

- switching between several control algorithms.

- on line learning and redesign based on the results of previous operations.

The inputs to the fuzzy system are the process inputs, process outputs, and some performance indices. The outputs of the level are the appropriate supervisory actions in the lower direct controller. Figure 13 illustrates the supervisory level.

The most common function implemented in the fuzzy supervisory level is the automatic tuning of conventional controller parameters to compensate external perturbations or variations of the parameters of the process to be controlled.Thus, consider a conventional non-fuzzy controller defined by:

$$u = \gamma(x, p) \quad (11)$$

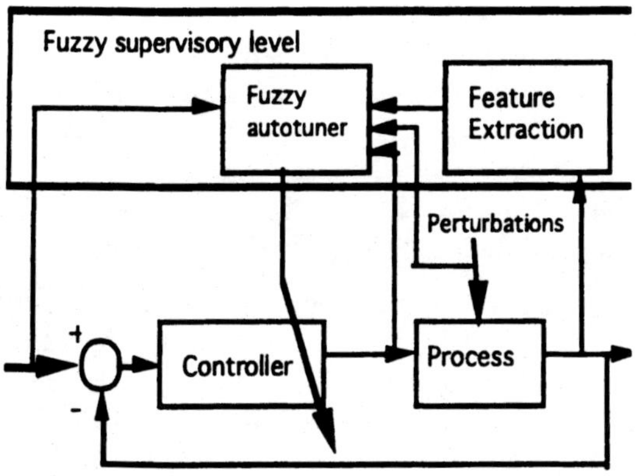

Figure 13: Fuzzy supervisory level

where x represents the process variables and p is a set of controller parameters. In this case the fuzzy component consist of a set of fuzzy rules to infer the appropriated value of each control parameter:

$$\text{IF } x_1 \, is \; lx_1, x_2 \, is \; lx_2, \ldots, x_n \, is \; lx_n \qquad (12)$$
$$\text{THEN } p \; is \; lp$$

where $lp \in \Lambda P$ is a linguistic label defining a region in the interval of variation of the controller parameter p (i.e. *positive_small* increment of the gain, or *positive_big* gain). For example, if we consider a closed loop control system and a linear state feedback controller, the control variable is given by

$$u = \boldsymbol{k}^T \boldsymbol{x} = k_1 x_1 + k_2 x_2 + \ldots + k_n x_n \qquad (13)$$

In this case, it is intended to adapt the parameters k_i, i=1..n, by using a set of heuristic rules. Two different structures have been considered. The additive structure and the multiplicative one (Ollero et al, 1991). In the former, the control law is defined by:

$$u = -(\boldsymbol{k}+\boldsymbol{s})^T \boldsymbol{x} = (k_1 + s_1) x_1 + \ldots + (k_n + s_n) x_n \qquad (14)$$

and the increments s_i are inferred from a set of linguistic rules defining non-linear functions

$$s = \Phi(x) \; ; \text{where } s_i = \Phi_i(x) \qquad (15)$$

In the multiplicative structure the control law is:

$$u = (k_1 c_1) x_1 + (k_2 c_2) x_2 + \ldots + (k_n c_n) x_n \qquad (16)$$

where c_i (i=1, ...n) are obtained from a set of rules. The same can be done with conventional PID regulators. In this case the control action can be written as:

$$u = K\left(e(t) + (1/T_i)\left(\int e(t)dt\right) + T_d \frac{d}{dt} e(t)\right)$$

where e(t) is the error between a reference signal and the output of the process. The parameters K, T_i and T_d can be automatically tuned taking into account rules relating its values with some characteristic of the process output. Fuzzy implementation of classical ***Ziegler-Nichols rules*** and other fuzzy rules to cope with the heuristic knowledge about the process being controlled have been applied. Fuzzy supervisory functions of PID controllers also include ***antireset windup*** (Garcia Cerezo et al, 1988).

2.5 Stability Analysis

The complexity of stability analysis of fuzzy control systems arises from two facts:

(a) The knowledge of the dynamic behavior of the process to be controlled is normally poor.

(b) Fuzzy control systems are essentially non-linear dynamic systems.

The fact (a) is in the origin of stability analysis methods based on the concept of dynamic fuzzy systems. It is associated with Zadeh's Extension Principle (1973). Work includes the stability criteria based on the concept of energy or the controllability of fuzzy systems (De Glass, Gupta and others,1987). These approaches normally lead to very conservative design solutions to guarantee stability.

Another different perspective is non linear system theory. This perspective arises from the fact (b) above. In this case, the system under control is a "non-fuzzy" system, and the "controller" a particular class of non-linear control system. In these methods, the basic assumption is that an approximate model of the process to be controlled is available. In fact, knowledge about the dynamic relations involved is normally available from process analysis and experimentation even in complex processes. Fuzzy (Sugeno, 1988) modelling techniques can also be used to obtain, through experimentation, a non-linear relation representing the process to be controlled. These non-linear stability analysis methods can be subdivided in state space approaches and input-output stability methods. In the first class the approximate model of the closed loop system is represented by:

$$\frac{dx}{dt} = f(x) + Bu \qquad (17)$$

where $x \in X \subset \Re^n$ is a n-vector of process variables, $f(x)$ is a non-linear function describing the open loop dynamic, and $u \in U \subset \Re^m$ represents the control variables. The closed loop fuzzy control system is represented by:

$$\frac{dx}{dt} = f(x) + B(\Phi(x)) \qquad (18)$$

with $x(0)=0$ being the equilibrium of the system.

The interpretation of trajectories in the phase plane provide the base for several analysis techniques. The concept of linguistic trajectories (Braae and Rutherford, 1979) is useful to determine the relation between the state space and control rules. This relation is also shown through the concept of the partition of the control protocol (Garcia-Cerezo, 1987). Analysis on the state space has also been shown as a simple but effective tool in simple systems. The "geometric method of stability analysis" (Aracil et al, 1988) is based on a study of the contributions of vectorial fields of the system and controller.

The stability and robustness indices (Aracil et al, 1989), based on concepts from the "qualitative theory of dynamic systems", formalize and extend the geometric approach mentioned above. The indices are defined using the model (18)- (19). This model should be considered as an approximate one. Then, if we apply any analysis and design technique, the robustness of the resulting control strategy must be studied. Two stability indices that provide a measure of the relative stability of the equilibrium point at the origin have been defined. These indices are related to the ways a system can loss its stability. Furthermore, a third stability index has been applied to study global stability. This index was formulated involving the conditions to preclude the appearance of other equilibria. The indices are not only useful to determine the stability and robustness of the system, but are also an indication of its dynamic behavior.

A.Ollero et al (1993) also presented the sensitivity indices, derived from stability and robustness indices, and their application to fuzzy control systems design. Furthermore, expressions to compute the stability at the origin indices for the fuzzy auto-tuning structures described above have also been obtained (Garcia-Cerezo et al, 1994). The results have been also extended to the analysis of fuzzy controllers with weighted outputs given by:

$$u = \frac{\sum_{i=1}^{nr} w^i (c_0 + c_1 x_1 + \ldots + c_n x_n)}{\sum_{i=1}^{nr} w^i} = \Phi(x_1, \ldots, x_n) \quad (19)$$

where c_j j=0, .. n are parameters for each rule, which can be obtained by means learning methods from the human operator control actions, and w_i i=1,..nr represent the truth value of each rule. In this casé, for membership functions of triangular type and symmetric around the origin, expressions for the indices at the origin have also been obtained (Garcia-Cerezo and others, 1994).

Input-Output stability approaches have been also used. Thus, Gosh and Majumder (1984) applied the circle criteria to stability analysis of fuzzy control systems. In Aracil et al (1990, 1991) reformulations of the conicity criteria for the analysis of fuzzy control systems are presented. Concepts such as rule conicity are also derived. This concept is used to define those rules considered most critical from a stability point of view.

3. APPLICATIONS

In this section vehicle´s control applications are considered. These applications are common in automotive systems and mobile robots. In fact the control of autonomous and semiautonomous vehicles is an emergent field in which fuzzy and other intelligent control technologies seems to be valuable tools to implement autonomous functions.

Automotive system applications include engine control, chassis control and vehicle management. The relations between vehicle management and the motion planning and control problems involved in mobile robots are obvious. Furthermore, engine control and chassis control (suspensions) are also related with velocity planning and control. In the following, fuzzy control applications both in automotive systems and mobile robots are described.

3.1 Automotive systems.

Automobile sub-system control is always considered as a difficult problem. The processes (engine, suspension, . . .) are often difficult to model due to numerous variables, high non-linearities, unobservability of the variables, . . . Moreover, they are time varying with respect to environmental conditions like temperature, water cooling temperature and ageing. And finally they show a large scattering between vehicles of the same class.

Therefore, today many vehicle functions are currently performed through strategies implemented in various types of electronic control units (ABS ECU, Injection ECU, . . .). The current control laws often use static or semi-static open loop approaches (for example: spark ignition control, . . .)or supervised PID controllers (Idle speed regulation). The development of these control laws is mainly empirical and needs a lot of tuning time where the role and the knowledge of "tuning engineers" is very important.

More recent studies have proposed very suitable solutions to improve the performances of vehicle sub-system control. These control strategies are based on more sophisticated control techniques like optimal control (application to idle speed regulation, to active suspension, ..), state feedback linearizing control and model based control techniques (application to engine control, . . .) , robust control techniques (vehicle dynamics control, ...). The results obtained are generally good. Unfortunately, these methods need generally a good mathematical model which could be difficult to perform. On the other hand, they are difficult to implement on real time ECUs, because they are time consuming and finally they lead to a cost increase.The techniques of fuzzy control, based on expert knowledge, and possessing intrinsic qualities of robustness and ease of implementation, seem to be an interesting alternative; these techniques can be considered as a good compromise between cost and performances.Thus, since a few years, a number of studies have already been performed and tested, related to the problem of engine control as well as chassis control and vehicle dynamic control (Boverie *et al,* 1992b, 1993d-e-f):

- **Engine**: idle speed regulation (Siemens Automotive, Fiat, . . .), anti-knock control (Siemens Automotive), injection and spark ignition control (Nissan).
- **Chassis**: active and semi-active suspension (Siemens Automotive, . . .).
- **Vehicle dynamics**: cruise control (Siemens Automotive, Nissan,..), intelligent cruise control (PSA, Renault,..), automatic gear box (Siemens Automotive, Nissan, ..).

All these application have been implemented on real time ECUs and tested on vehicles or on test benches. In the following section we give some applications examples studied by Siemens Automotive.

Fuzzy active suspension. The suspension systems have three fundamental functions a) to support vehicle at desired height above the ground b) to isolate the passengers from disturbances c) to guide the vehicle along the desired road with acceptable handling and stability.The conventional passive suspension is formed of two passive elements: spring and damper; in order to improve the comfort, the passive elements must be soft, but spring and damper must be hard in order to obtain good properties of road handling and holding. So, to obtain simultaneously excellent ride comfort and handling the suspension characteristics must be changed dynamically, according to the demands of the situation.

Here, we want to develop a robust non-linear controller based on the expertise of automotive engineers and not on mathematical models sometimes difficult to obtain or with limited validity. Moreover, it is easy to see by the open loop dynamical analysis that the physical system has two different dynamics, one corresponding to the wheel hop (the fast part of the system) the other one corresponding to the car body (the slow part of the system) and we will try to take profit of this dynamical property to design two parallel fuzzy controllers coordinated by a fuzzy supervisor (Roukieh *et al,* 1993a-b) (Fig 14)

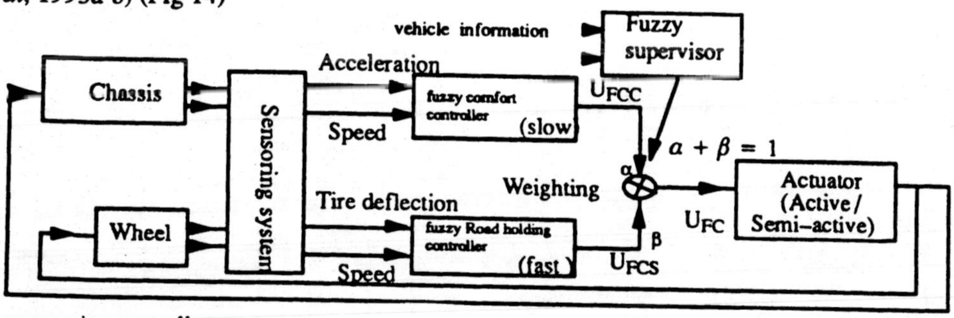

Fig.14 Fuzzy suspension controller.

A comparative study between passive suspension and active fuzzy suspension have been performed.Fig. 15 & 16 show the performances of the two systems. We can see the good behavior of the fuzzy controller (the maximum overshoot of the fuzzy controller is about 50% smaller than passive one, and the time response is lower than 0.8 sec). Moreover, a robustness analysis has been done with the fuzzy controller, changing the body mass (see Fig. 17) and reducing the damping coefficient by 50% with respect to nominal values. In the two cases, satisfactory results are obtained.

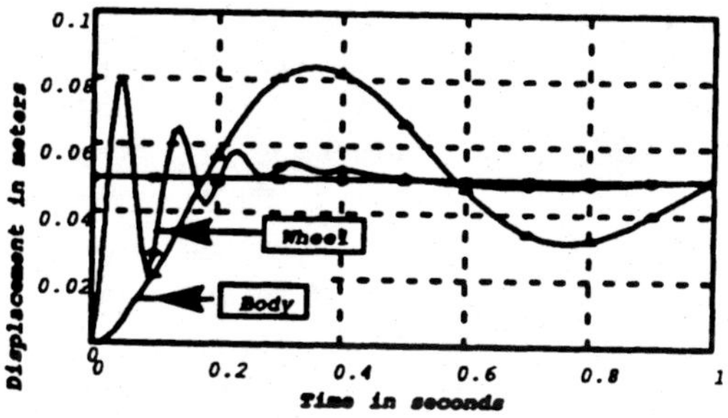

Fig.15 Passive controller: displacement of wheel/body as a response to a step of 5 cm

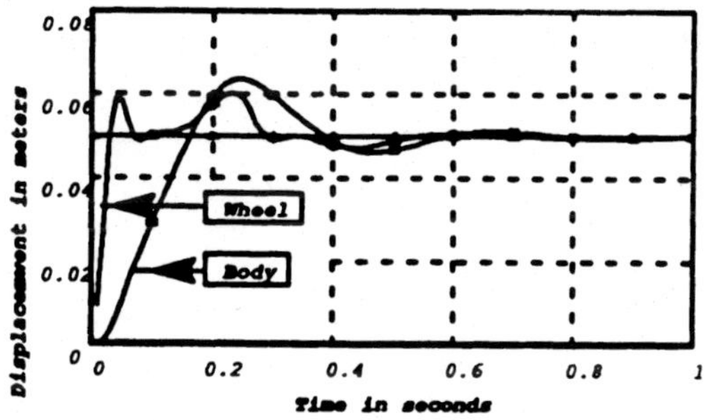

Fig.16 Fuzzy active controller displacement of wheel/body as a response to a step of 5 cm

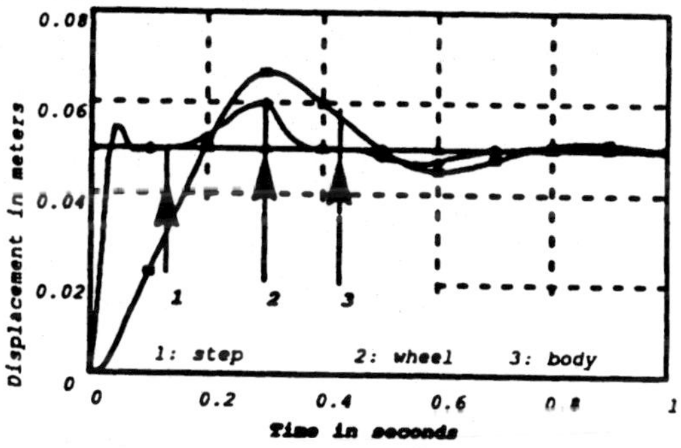

Fig.17 Robustness of Fuzzy controller with semi-active actuator for a body mass change of 50%

Fuzzy anti-knock control. Knock phenomenon is caused in internal combustion engines by a bad combustion of air/fuel mixture.

This causes a rough rise of the cylinder pressure and the vibration level of the engine. In addition, the knock phenomenon, once started as a tendency for being self-generating. Then, the consequences can be destructive for the engine. The occurrence of this phenomenon depends mainly on full loaded running conditions, air and engine temperatures, fuel quality, air humidity, design of the engine . . .

Conventional strategies to avoid knock are often based on real time ON/OFF techniques, using the energetic information of the vibration signal provided by an accelerometer sensor located on the engine. These strategies generally give satisfactory results. Unfortunately, due to their principle, they are not optimum and do not allow engine to run close to the knock limits. They also present some lack of robustness. The consequences are often a reduction of engine performances.

Fuzzy control seems to be a good solution to improve these performances (Boverie *et al,* 1992c). The proposed fuzzy controller takes into account the various energy levels measured by the accelerometer sensor and representative of the knock level. Related to the knock level, the controller calculates an adapted ignition correction offset. The Fuzzy controller (Fig 18)uses as inputs the 2 variables R_i and R_m which are representative of the knock phenomenon. R_i is the normalized instantaneous energetic level measured by an accelerometric sensor,R_m is the mean of R_i with a forgetting factor. The calculated variables R_i and R_m as well as a control variable R_a (here the ignition advance correction applied to each cylinder) are defined in qualitative way by using different fuzzy subsets to represent different linguistic values of these variables (Very Weak, Weak, Zero . . .).

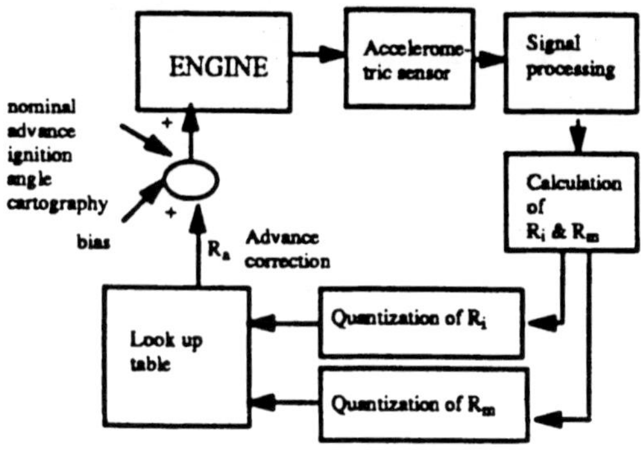

Fig.18 Anti-knock controller structure

Once the different variables of the fuzzy anti-knock controller have been defined, it is necessary to write a set of rules, expressed in a linguistic form, that allows the compensation of the knock phenomenon. These rules translate the experience and the expertise of engine engineers. The overview of one rule is given below:

If <the instantaneous value of the knock Ri is Very Weak> **And** <the mean value of the observation variable Rm is Zero> **Then** <the spark advance correction R_a is Zero>

All the expertise rules can be put together in the rule base table (Fig 19). The computational cost of the online inference, for high engine speeds, becomes a discriminatory parameter in real-time applications of anti-knock control. Therefore, it is interesting after quantization of the input and output variables to consider an off-line resolution and defuzzification, and to implement on the engine electronic unit control a look-up table containing the results of this off-line treatement. This look-up table has two inputs R_i and R_m and one output R_a.

R_m \ R_i	VW	W	O	S	M	B
ZE	ZE	ZE	ZE			
S	ZE	S	M	M		
M	S	M	M	M	B	
B	M	B	B	B	B	VB
VB	B	VB	VB	VB	VB	VB

unused zone (shaded cells)

Zone without knock (VW–O) — Zone with knock (S–B)

Fig.19 Fuzzy anti-knock set of rules

Fuzzy idle speed controller. Idle speed control is always considered as a difficult problem. It must be efficient every time and assuming a compromise between conflictual goals:

- for comfort, it must insure a good stability of the engine speed and a good robustness with respect to load disturbances (power steering, lights, air conditioner, . . .),
- for the minimization of the consumption and pollutant emissions, the regulation set point must be as low as possible.

On the other hand, the features of an automotive engine are very different according to its use (stationary and running vehicle). In the first case, the process to be controlled is only the engine (engine inertia); in the second case, the process to be controlled is the overall vehicle through the mechanical coupling (vehicle inertia), which corresponds to a tremendous change in the parameter values. The sensors located on the engine allow the detection of these different phases, which are taken into account in the control strategy. In order to control this large time varying system an original fuzzy controller structure has been designed (Fig. 20 & 21) (Cerf *et al,* 1993, 1994). This structure is based on two parallel fuzzy sliding mode controllers whose outputs are weighted by the gains K_1 and K_2. A supervision level modifies the values of the gains K_1 and K_2 in order to rotate the sliding curve and to adapt this sliding curve to the process dynamic.

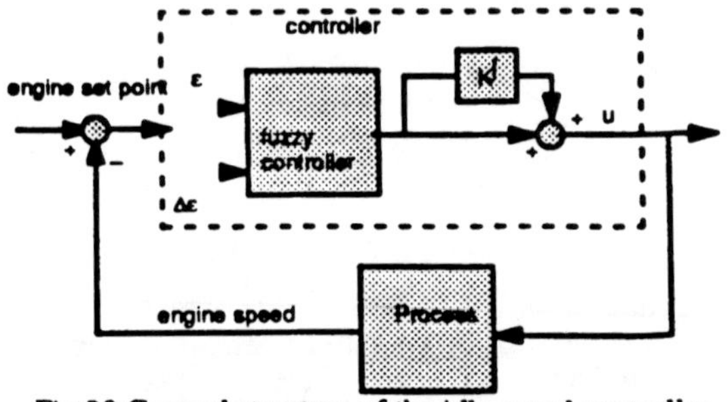

Fig.20 General structure of the idle speed controller

The results obtained show the good quality of the control: the decrease of the engine speed is smooth, without irregularities or undershoot (Fig 22 to 26).

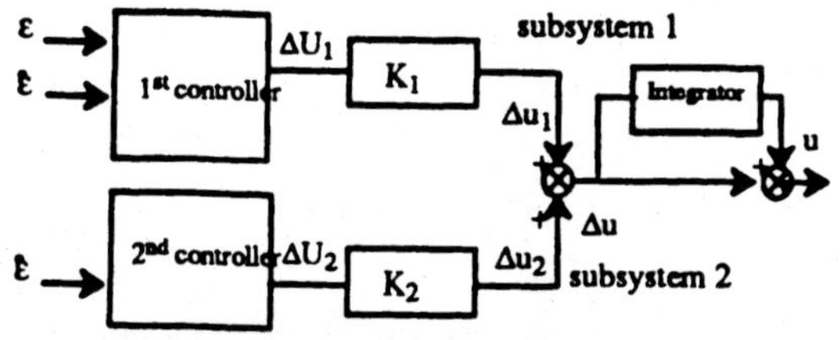

Fig.21 Adaptive fuzzy sliding mode controller structure

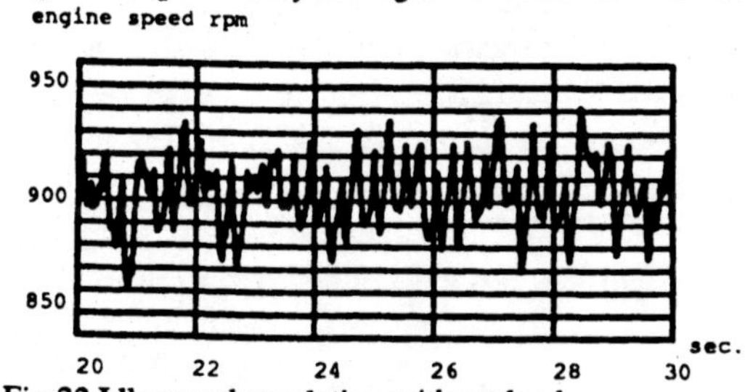

Fig.22 Idle speed regulation without load

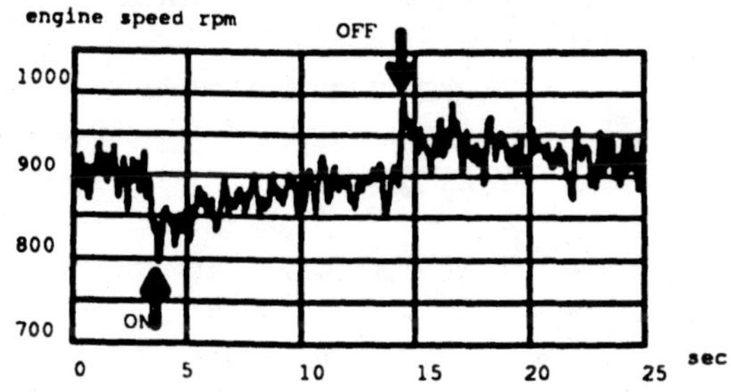

Fig.23 Idle speed regulation with load (power steering)

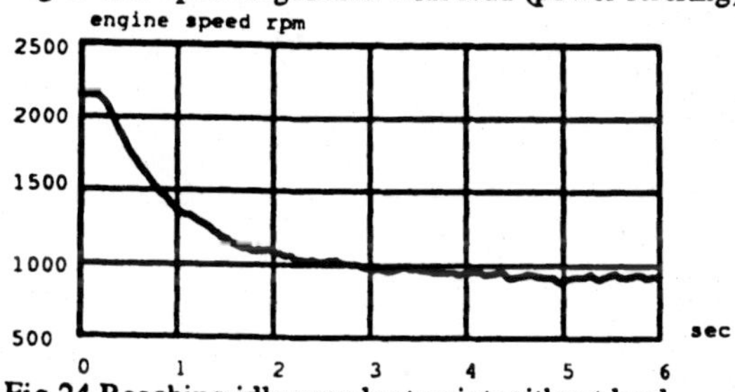

Fig.24 Reaching idle speed set point without load

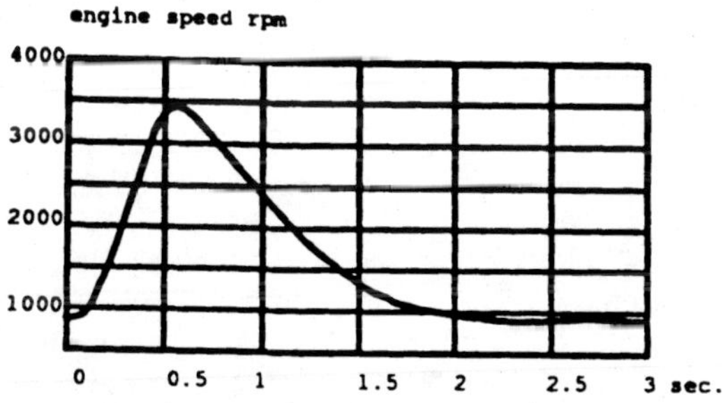

Fig.25 Reaching idle speed set point with load (power steering+ventilator)

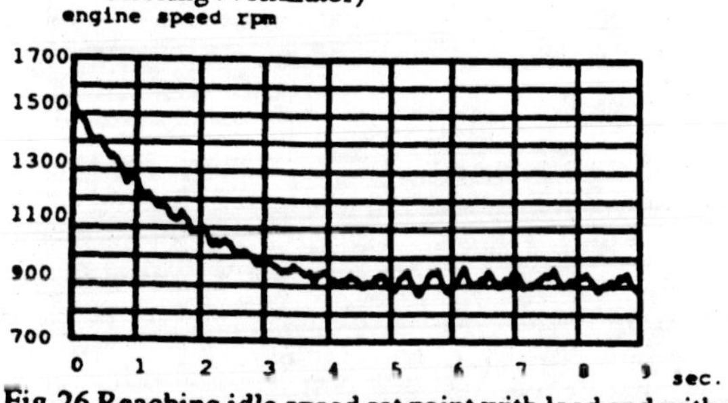

Fig.26 Reaching idle speed set point with load and with running vehicle

3.2 Mobile robot control

The control of a robotic system has been traditionally decomposed in several levels. Higher levels are related with perception and planning while lower levels are related with motion control to execute the plans.

Application of fuzzy control methods to solve these problems are motivated by:

- imprecise information from sensors,
- uncertainties in the environment knowledge,
- difficulties to apply in real time realistic control laws derived from complex kinematics and dynamic models.

At higher levels fuzzy logic can be applied to interpret plans expressed in qualitative linguistic terms. Moreover, fuzzy logic is also useful to generate direct actions from imprecise sensorial information. Thus, fuzzy logic seems to be a valuable tool in reactive control architectures in which the integration of the above mentioned functions for the generation of fast responses to external stimulous plays a central role. Fuzzy logic has been used to model the actions of an human driver relating environment sensor data with appropiated driving actions. Thus, fuzzy control systems to autonomous steering of a model car have been presented since 9 years ago (Sugeno and Nishida, 1985). In this case the controller inputs are distances to walls and corners and the current direction of the car. The output is the steering angle. In this application the vehicle´s velocity is assumed to be constant. There are also hierarchical fuzzy control system architectures with visual sensor subsystems, motor drive unit and fuzzy driving expert systems (Maeda *et al.*, 1991). In Tsukagoshi and Wakaumi (1990) fuzzy rules are applied to infer the velocities of the wheels of a differential steered vehicle from the angle desviation and its differences.

In all these methods the control signals are generated directly by the fuzzy controller from sensor signals without considering explicitly the kinematic and dynamic constraints of the problem. This strategy requires significant efforts in the controller design and callibration of sensors. Otherwise may give problems when implementing in a real vehicle, specially when looking for high performance navigation.

In the following, we only consider the problem of steering the vehicle to track a previously defined path in spite of external perturbations. This path can be defined off-line, prior navigation, or computed on-line by the autonomous navigation system. Today several path generation methods can be applied in real time providing a path with good kinematic properties to be followed by the real vehicle. Fuzzy logic can also be applied in these path planning and generation procedures as mentioned below (Martinez and Ollero,

1994).

High performance path tracking of mobile robots and autonomous vehicles is a difficult problem when considering vehicle's complex models and vehicle-terrain interaction. Simple geometrical techniques or methods based in simplified dynamic models can be used if the models are adapted to the current navigation condition (velocity), path characteristics, and actual state of the vehicle.

Fuzzy logic can be used to track a given path in the so called "fuzzy direct path tracking" scheme (Martinez and others, 1993). The inputs to the fuzzy controller are variables defining the state of the vehicle with respect to the path, and the output is the steering command to be executed by the low-level motion controllers (see Figure 28).

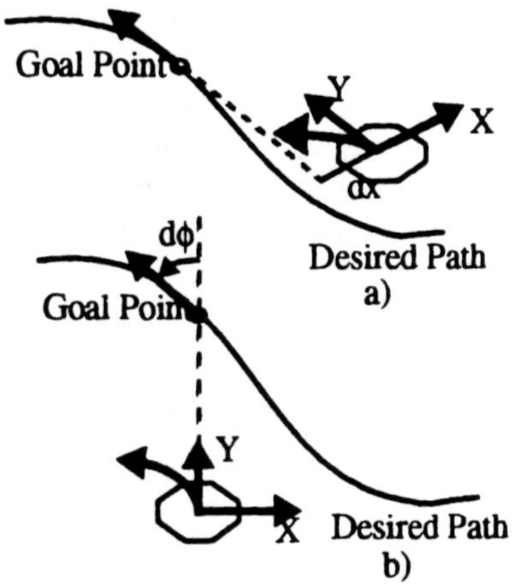

Figure 28: Fuzzy direct path tracking.

In this case we use a set rules such as

If Variable dx is Negative_Large **And** $d\phi$ is Near_Zero **And** L is Large, **Then**, γ_r is Positive_Large.

where L is the distance from the vehicle to the goal point ahead in the path (lookahead distance) and Negative_Large, Positive_Large, Near_Zero, are labels associated with the corresponding variables.

From the control law point of view, the controller is defined by the non-linear function: $\gamma = \Phi(l, dx, d\phi)$. The results of this control law depend on how the fuzzy controller is tuned. Furthermore, the selection of the goal point to follow has a significant effect on the tracking performance.

Furthermore, fuzzy supervisory levels have also been implemented for path tracking. The automatic tuning of control parameters of conventional path tracking methods is the most evident application. In Ollero et al (1993) the pure pursuit method (Amidi, 1991) and the Generalized Predictive Path Tracking method (Ollero and Amidi, 1991) are supervised autonomously by means of fuzzy logic. In the following we summarize the results when applying the pure pursuit technique. This technique is a very simple strategy based on geometric considerations. The path is tracked by repeatedly fitting circlers arcs to different goal points on the path as the vehicle moves forward. Thus, a new required vehicle's curvature (γ_r) is computed by:

$$\gamma_r = 2 \cdot \frac{dx}{L^2} \tag{20}$$

where dx is the x displacement of the goal point in vehicle coordinates and L is the distance between the current position of the robot and the desired point on the path to be tracked. This distance is called the lookahead distance. The pure pursuit is a proportional controller of the steering angle using dx as the error and the inverse of the lookahead square for the gain. The technique has proved efficiency but the results are critically dependent on the tuning of the lookahead for the current navigation conditions. For example, if the lookahead is too long the vehicle may cut corners and if too short oscillations may result. Furthermore, the best lookahead in the current vehicle's position and orientation depends on the vehicle's velocity, vehicle's heading $\Delta\phi$ with respect to the path, and its distance d from the path. Thus, the fuzzy supervisory level above described contains a set of fuzzy rules such as:

If d is Large, **And** v is Medium, **And** $\Delta\phi$is Very_-Large, **Then** L is Large.

For the generalized predictive path tracking method, the supervisory function is more complex. In this case, in addition of the inputs and output of the pure pursuit method it is necessary to consider the curvature of the goal point as input, and the number of points ahead, the number of control intervals and the weight of the control variable in the predictive cost function as outputs (see Ollero et al, 1993).

Both the fuzzy direct path tracking algorithm and the supervisory path tracking has been implemented for the steering of RAM-1. This is an autonomous mobile robot designed as a testbed for the automation of surveillance, manipulation and small part transportation (Ollero *et al.*, 1992). The top speed, that is 1.6 m/sec, only can be reached when the vehicle moves thought a straight line, and automatically decreases as the curvature of the vehicle increases. The steering time constant for the RAM-1 was measured to be 0.1 seconds. The path tracker (Ollero and others, 1993) generates the vehicle's steering command to track the path by taking into account the information provided by the state maintenance system (actual vehicle's posture and other state variables). These steering commands, as well as the velocity command, are sent to the low level motion controller which control the vehicle's motion through a firmware interface that perform the control of the driving and steering motors. Figure 29 shows the results of the implementation in the real vehicle, at v=1.2 m/sec., of the Pure Pursuit and Generalized Path Tracking (GPPT) with fuzzy supervision. The mobile robot starts far from the path to be tracked. However, the vehicle takes the path and follow it with good performance.

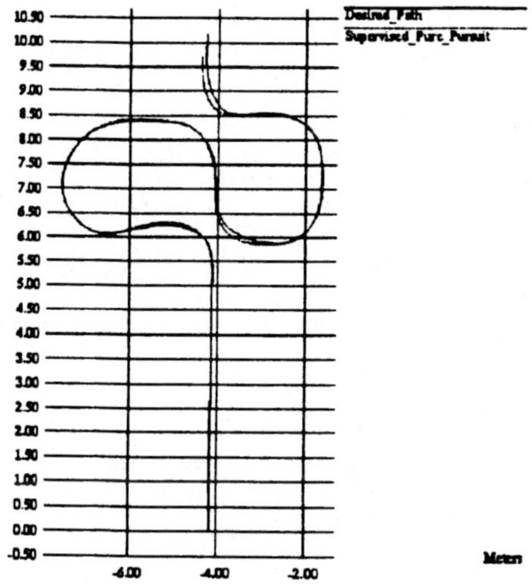

Fig.30: Path Tracking 1 with fuzzy supervision.

In Tables 1-2 we compare the results of the application of the proposed methods to path tracking of the mobile robot RAM-1. The performance measures consist of the computing of the following errors: Δx which is the x displacement of the nearest point in vehicle coordinates (absolute value); Δs which is the distance between the current position of the robot and the nearest point on the path to be tracked; $\Delta \phi$ which is the difference between the heading of the vehicle and the heading of the nearest point (absolute value) and $\Delta \gamma$ which is the steering increment (absolute value). The best results are obtained with a fuzzy direct controller with goal point at r=0.5 (D), and with the fuzzy supervision of the generalized pure pursuit method (C), which has less control activity. In these cases the mobile robot had been driven very close to the map. The fuzzy supervisory pure pursuit method (B), which is shown in Fig. 30, also provides a precise and stable tracking.

Table 1: Maximum values of errors

	Δx	$\Delta \phi$	Δs	$\Delta \gamma$
A	.228444	.416020	.231766	.348543
B	.133764	.331657	.137766	.370604
C	.087307	.286503	.091494	.280606
D	.029781	.151786	.055275	,456622

Table 2: Medium values of errors

	Δx	$\Delta \phi$	Δs	$\Delta \gamma$
A	.118806	.098616	.127429	.090586
B	.024110	.054256	.039252	.033100
C	.022713	.050890	.037327	.031770
D	.013748	.031591	.030457	.114997

A: Direct Fuzzy Control, r= 0.8 m. B: Fuzzy Supervisory Pure Pursuit. C: Fuzzy Supervisory GPPT. D: Direct Fuzzy control, r= 0.5 m.

3. CONCLUSIONS

Fuzzy controllers has more degrees of freedom than conventional controllers. That is one of the reasons that explain why fuzzy controllers may perform better that conventional controllers, but also complicates the design of an efficient fuzzy control system. Fuzzy logic controllers are usually designed to solve particular problems without using general analysis and design principles. Thus, it seems clear that there is a need of research efforts to conceive general design methodologies that simplify the task of control engineers. The first part of this paper studies, from a control engineering perspective, some topics related with fuzzy control methodologies pointing out the relation and equivalencies with non-linear controllers. The design of fuzzy controllers for high order processes is also considered.

In several applications conventional controllers, such as PID controllers or state controllers, are efficient enough to cope with normal working conditions but the dynamic behavior deteriorates when disturbances or unmodelled dynamics are present. These situations can be managed by tuning the parameters of the conventional controller to cope with the process variations. In these cases fuzzy logic can be used in a supervisory level for automatic tuning of conventional controller parameters. These supervisory functions are also considered in the paper. The last subsection of the first part of the paper is devoted to stability analysis. This important topic from a control theory/engineering point of view is revised mainly from the point of view of non-linear system theory.

The second part of the paper considers automotive and mobile robot applications. In these applications models are frequently very complex involving non-linearities and some parameters are unknown or very difficult to obtain. The paper shows how fuzzy control can be used for engine control, chassis control and automatic steering of semiautonomous or autonomous (robotized) vehicles. The results of the real experiments are quite satisfactory and suggest that more research could be very useful to improve the performance and safety of vehicles.

4. REFERENCES

Amidi O. (1990). Integrated Mobile Robot Control. Carnegie Mellon Univ. Robotics Institute. Technical Report CMU-RI-TR-90-17.

Aracil, J., A. Garcia-Cerezo and A. Ollero (1988). Stability Analysis of Fuzzy control Systems. A Geometric Approach, in Kulikowski et al (Ed.): Artificial Intelligence, Expert Systems and Languages in Modelling and Simulation. North-Holland, pp. 323-330.

Aracil, J., A. Ollero and A.Garcia-Cerezo (1989). Stability Indices for the Global Analysis of Expert Control Systems", IEEE Transactions on System, Man and Cybernetics", vol SMC 19, no. 5, pp. 988-1007.

Boverie S., Demaya B., Titli A., (1991). "Fuzzy logic control compared with other automatic control approach". 30th IEEE-CDC conference on decision control, Brighthon (U.K.), Dec. 11-13 1991.

Boverie S., Demaya B., Ketata R., Titli A. (1992a). "Performance evaluation of a fuzzy controller". IFAC Symposium on Intelligent Components and Instruments for Control Applications,MALAGA (S.P.), 20-22 1992.

Boverie S., Demaya B., Le Quellec JM., Titli A. (1992b)."Contribution of fuzzy logic to the improvement of modern car performances". IFAC symposium on Artificial Intelligence in real time control, Delft, The Netherlands, June 15-19 1992.

Boverie S., Demaya B., Titli A., (1992c). " Procédé de suppression d'un phénomène de cliquetis affectant le fonctionnement d'un moteur à combustion interne", SIEMENS Automotive SA, Toulouse, France. Patent application, Fr.9203645.

Boverie S., Demaya B., Le Quellec JM., Titli A. (1993a)."Fuzzy control for high order systems". FUZZ IEEE 1993 second international conference on fuzzy systems, San Francisco, California, march 28-april 1 1993.

Boverie S., Demaya B., Le Quellec JM., Titli A. (1993b)."Fuzzy control of high order systems using a parallel structure of second order blocks". IFAC , 12th World congress, Sydney, Australia 19th-23rd July 1993.

Boverie S., Demaya B., Le Quellec JM., Titli A. (1993c)."Performance evaluation of fuzzy control through an international benchmark". Fifth IFSA world congress, SEOUL, Korea, July 4-9, 1993.

Boverie S., Le Quellec JM., Titli A. (1993d)."Different applications of fuzzy logic control in the context of automotive systems". ISATA congress, Aachen sept. 93.

Boverie S., Demaya B., Le Quellec JM., Titli A. (1993e)."Contribution of fuzzy logic to the improvement of modern car performances". Control Engineering Practice,, April 1993

Boverie S., (1993f)":a comande floue dans les systèmes automobiles". Troisièmes journées nationales commande floue, Nimes, October 1993.

Braae M. and D.A. Rutherford(1979). Theoretical and Linguistic Aspects of the Fuzzy Logic Controller, Automatica, 15, pp. 553-577.

Cerf P., Le Quellec JM. Demaya B. (1993). Procédé et dispositif de commande du régime dún moteur à combustion interne en phase de ralenti. Application patent Fr 93/08275.

Cerf P., Boverie S., Vernières F. (1994), Fuzzy sliding mode control - application to large time varying system", SICICA, Budapest June 1994

Demaya B. (1992). Contribution des procédés de commande floue au contrôle. Deuxièmes journées nationales sur la commande floue, Nimes 2-3 novembre 1992.

García-Cerezo A., A. Ollero and J. Aracil. Dynamic Analysis of Fuzzy Logic Control Structures. Chapter 6 in "Fuzzy Control Systems". A. Kandel and G. Langholz editors. CRC press, 1994.

Kania A.A., J.B. Kiszka, M.B. Gorzalczany, J.R. Maj, and M.S. Stachowicz (1980).On stability of Formal Fuzzyness Systems, Inform. Sci. 22, pp.51-68.

Kickert W. and E.H. Mamdani (1978).Analysis of a Fuzzy Logic Controller, Fuzzy Sets and Systems, Vol.1, pp.29-44, 1978.

Kiszka J.B., M.M. Gupta, and P.N. Nikiforuk (1985). Energetic Stability of Fuzzy Dynamic Systems, IEEE Trans. SMC, Vol.15, no.6, pp.783-792.

Kumar S.R. and D.D. Majunder (1984). Application of Circle Criteria for Stability Analysis of Linear SISO and MIMO System Associated with Fuzzy Logic Controllers, IEEE Trans. Sist. Man. Cyber., Vol.SMC-14, 2, pp.345-349, 1984.

Langari G., and M. Tomizuka (1990). Stability of Fuzzy Linguistic Control Systems, Proc. 29th IEEE Conf. Decision and Control, pp. 2185-2190.

Maeda M, Y. Maeda and S. Murakami (1991). Fuzzy Drive Control of an Autonomous Mobile Robot. Fuzzy Sets and Systems 39, pp195-204.

Martínez J. and A. Ollero (1994). "Combined path generation and tracking method for Mobile Robots". Preprints of the SICICA'94.

Ollero A. and A.García-Cerezo (1989). Direct Digital control Auto-tunning and Supervision Using Fuzzy Logic", Fuzzy Sets And Systems, Intl. Journal, Vol.30, pp.135-153, 1989.

Ollero A., J. Aracil, A. García-Cerezo, and A. Barreiro (1993). "Stability of fuzzy control systems". An Introduction to Fuzzy Control. Chapter 7. D.Driankov et al. Springer Verlag.

Ollero A., García-Cerezo A., and Martinez J. (1994). "Fuzzy Supervisory Path Traking of Mobile Robots". To appear in Control Engineering Practice, Vol. 2, No.2.

Palm R. (1992) " Sliding mode fuzzy control" FUZZ-IEEE 92', San Diego, March 92

Raymond C., Boverie S., Le Quellec JM., "Practical realization of fuzzy controller- comparison with conventionnal methods". EUFIT Symposium Aachen, Sept 1993.

Roukieh S., Titli A.(1993a). " Design of active and semi-active automotive suspension using fuzzy logic". IFAC World congress, Sydney, 1993.

Roukieh S., Titli A.(1993b). " Dispositif de commande d'un ensemble de suspension interposé entre une roue et la caisse d'un véhicule automobile". Patent application 9305007.

Sugeno M. and Kang. G.T. (1988). "Structure identification of fuzzy models". Fuzzy Sets and Sistems. Vol. 28. pp 15-23.

Sugeno M., M.Nishida (1985).Fuzzy Control Model Car. *Fuzzy Sets and Systems 16*, pp. 103-113.

Tsukagoshi T., and H. Wakaumi (1990). Highly-reliable Semi-autonomous Vehicle Control on Lattice Lane. Proc. IEEE Robotics and Automation. pp. 731-738.

AKNOWLEDGMENT

The authors would like to thank the contribution of L. Foulloy, A.Garcia-Cerezo, J. Martinez and other coauthors of the referred papers and participants of the Cluster 3 of the Fuzzy Algorithm for Control (FALCON) ESPRIT Working Group.

FUZZY CONTROL METHODOLOGY IN THE PROCESS INDUSTRY

M. Martínez, P. Albertos, J.L. Navarro
Grupo de Automática e Informática Industrial (DISCA)
Universidad Politécnica de Valencia (UPV)
e-mail: mami@aii.upv.es
Apdo.22012, E-46071 Valencia, Spain

Abstract. Based on the authors experience in the development of intelligent control (IC) applications, mainly in the cement production area, a methology to design Fuzzy Logic Controllers is presented in this paper. Two alternative approaches are suggested: the fuzzyfication of a previously designed linear controller and the fuzzy implementation of the operator knowledge. Structural decisions and basic parameters are first assumed and final parameters are on-line tuned. The global procedure is iterative and some refinements based on the obtained practical results are required. Both approaches are applied in an industrial plant and some results are presented and discussed.

Keywords. Fuzzy Logic Control, Intelligent Control, Controller Tuning, Control Design Methodology, Cement Kiln Control.

INTRODUCTION

Control problems in the process industry are quite specific and different in both, models and criteria,[4] from other control application fields, such as manufacturing or aeronautics. These criteria are normally related with the more or less precise knowledge of the process and the control goals. If this knowledge is poor, approximated or heuristic, the Fuzzy Logic theory provides an option to design the control system [1], [11].

Moreover, the plant to be controlled is usually composed by a number of subprocesses and different operating modes or conditions are foreseen. That means the convenience of some kind of supervision or adaptation. For this purpose, Fuzzy Logic Controllers (**FLC**) are very flexible and easy to implement in Expert System environment, [7].

On the other hand, classical controllers, like PID or state feedback controllers, are well established and attractive to the end-users, but some problems of adaptation arise in the control of such processes.

In this paper some contributions to the design of FLC for the process industry are presented and discussed. One design approach is based on the experienced knowledge the operators have about the process behaviour, and another one deals with suggested refinements to improve the results obtained by a classical controller.

In order to illustrate the main issues of these designs, an industrial application, the control of a cement kiln, is presented. This development is based on the authors experience in the area of cement production. The cement kiln, is a very complex process due to its physical and chemical characteristics. The model structure of this process involves multivariable control, nonlinearities, time varying and distributed parameters, and there are a lot of disturbing variables. The classical control techniques have been shown not very satisfactory, to achieve a performant controlled process and even now, most kiln control strategies require the presence of skilled operators. These operators have a heuristic mental model of the process based on their experience and knowledge.

In this paper some technical characteristics of this process are summarized, allowing the description of an intelligent control structure based on a hierarchical state machine. The development of two FLC to control the kiln under normal operating conditions is reported and discussed. Finally, a comparative analysis between FLC and multistage PID (gain scheduling) is presented, the FLC showing smoother transition between states. This design may be applied to other industrial processes.

THE CONTROL PROBLEM

Control requirements can be stated at different levels, taking into account the process operating conditions. In the process industry, as previously mentioned, the operating environment is not well defined and many times the control objectives can be more easely expressed by heuristics than by analytical formulas. These criteria may involve, among others [9]: Disturbance rejection, Steady-state errors under changes in the set points, transient response characteristics, Sensitivity to parameter changes in the plant, and Optimization indices.

Solving the control problem generally involves a number of steps:

- Choosing sensors to measure the plant output. Some sensors may be unavailable, very expensive, or difficult to install.
- Choosing actuators to drive the plant, taking into account their adequacy to the variety of operating conditions.
- Developing the plant, actuator, and sensor equations (models). Owing to the complexity of the involved processes, simplified and/or qualitative models should be handled.
- Designing the controller based on the developed models and the control criteria, although these may be not precisely defined.
- Evaluating analytically the design, by simulation, and, finally, by testing the physical system. The unavailability of a good process model always requires a final test on the actual plant.
- If the control requirements are not achieved, iteration of these steps. The on-line tuning and refinement of the control is always required.

Owing to unaccuracies in the models, the control engineers must be able to iterate this design procedure, using all available tools, to improve the controlled system behaviour. The experimental knowledge about the process, based on practical experience on the physical plant, plays an important role in the design process. In practical applications, the major difficulties arise in the mathematical formulation of the problem and in translating the mathematical solution back to the physical world.

According to the experience of the designer, some of the steps above may be omitted in some control applications. In particular, many control systems are implemented by choosing standard forms of controllers and experimentally determining the parameters of the controller by following a step-by-step procedure without any analytical model. Sometimes, this type of procedure works very well.

One of the problems when using linear controllers is the plant model parameters variation. Parameters may vary due to nonlinearities in either the plant or the actuators, changes in the operating condition of the process, or nonstationary disturbances acting on the process. For instance, drying of material is a common basic process in many industries: paper, cement, grain, food, etc. To derive a good process model can be a very difficult task, because the process dynamics depend on many variables, and can change with the speed of the material through the dryer as well as the material flow or the content of the incoming material. Also, the time delay varies with the production rate, etc. [3]. As in the case of the rotary dryer, there are always many different sources of parameter variations, and usually there is a mixture of different phenomena, in most cases not fully understood.

The model of the process and the control goals determine the control design methodology. Classical control design methodologies are mainly based on a good process model, (linear/nonlinear, deterministic/stochastic, with bounded uncertainties), and functionally defined control objectives. Logical controllers, Continuous/discrete time controllers, and Adaptive controllers are examples of controllers obtained by these design methologies [1].

The design of a controller when the control goals are stated unprecisely, or the process model is basically a qualitative one, not well defined, the above techniques use to fail or to become extremely complex, inflexible and difficult to update.

FLCs, that is controllers where the control action is computed by means of fuzzy logic operations are an alternative to the classical controllers. Of course, fuzzy logic concepts are suitable at decision levels, as an improvement of the classical binary logic reasoning, [11]. But our main interest is focussed in control subsystems directly connected to the process, both, receiving crisp data from the plant data acquisition system and sending control actions or references to the manipulated variables. The real time operation requirements initially led to very simple reasoning structures involving only one level of implication [7]. This poor reasoning structure does not allow to deal with complex systems. If the number of fuzzy implication rules increases, the knowledge base appears quite difficult to understand or modify. Thus, selective reasoning as well as deep reasoning are necessary to handle the required control under a variety of uncertainly defined operating conditions.

IC STRUCTURE

The basic goals in any the industrial process control application can be summarized as:

- Quantity goal: associated to product number or volume, or production rate.
- Quality goal: associated to final product characteristics.
- Economic goal: dealing with production cost.

Moreover, these goals must be reached matching certain security and environment constraints.

The goals and constraints are the base for a more formal specification of the required control structure, in particular if an intelligent control strategy is implemented. The problem must be solved splitting the global problem into a combination of more simple problems and then apply some analytical technique to solve them. A possible functional decomposition tries to identify each process situation, the process keeping the same structure and assigning better defined control objectives, allowing partial process modeling.

We decide a top-down functional decomposition, beginning with the global process, and looking for situations or operating modes where the process presents a concrete behaviour. Each operating mode can be decomposed in other operating submodes. We stop this division procedure when a valid model for the described situation in the operating submode is obtained. It is important to emphasize that in any process there are always two distinguished situations: process stopped and process going on. Moreover there are another typical operating modes, such as incidence, as result of a component failure or the non-fulfillment of the environment constraints.

In the figure 1 the operating modes and transitions are presented.

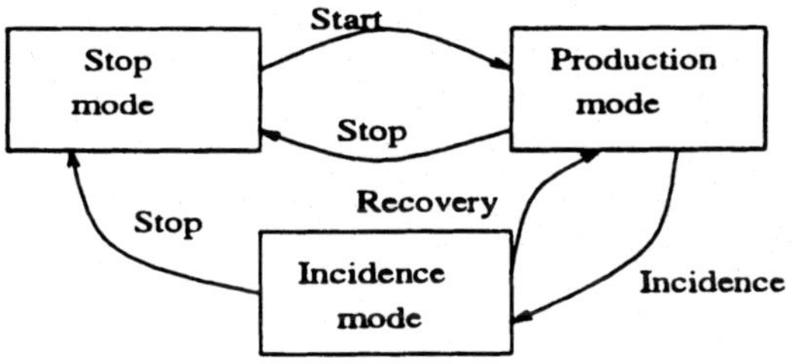

Figure 1: Basic operating modes and transitions.

In the following, we differentiate between stationary and transitory modes. If the process is in a stationary mode it remains in this mode in absence of external events. In transitory modes the process change with time and leaves the mode without external events. Transitory modes are commanded by the control system to carry the process from a mode to another one.

Following the top-down methodology, we can define more specifics modes inside the aforementioned ones:

- Stop mode: Compose by cold stop mode (all elements off) and ready mode (all subsystems ready to immediately pass to production).
- Production mode: Compose by nominal production mode (all control goals are achieved), and partial production modes. The production mode is reached from the ready mode. Both are stationary modes and are connected through a transitory mode. These modes can be divided in other submodes dealing with partial goals and constraints, leading to different control strategies. A transference or transitory mode should be considered between any pair of connected stationary modes.
- Incidence mode. This mode or set of modes, describe situations when the produc-

tion goals are less relevant because it is dangerous to maintain the nominal mode or the environment constraints are exceeded. Depending of the incidence relevance, this mode can be decomposed in: Recoverable incidence mode (it can maintain the control goals, althouh partially, while the incidence is counteracted), No recoverable incidence (the control goals can not be maintained, the process must be turned off to avoid any damage to the process or the environment). The time interval the process can stay in recoverable incidence mode must be limited (timeout).

By this analysis, it is possible to implement a hierarchical machine of states, identifying all the operating modes of the process, their control, and the conditions to transfer between modes and control strategies.

KB FLC DESIGN

Let us first consider a FLC design based on the operator knowledge. as this knowledge is always an approximated one, the fuzzy logic theory provides the tools to deal with the problem. The fuzzy concepts attached to any physical variable, such as fuzzy subsets or linguistic variables (LV), their membership functions (MF), and so on, must be defined. The physical variables range can be tuned according to the experience. Also, MF parameters like shape, support, or gain will allow to establish smoothed or sharper control actions. The rules to carry out this tuning are approximated, as the total approach is.

One of the FC options is to change the control based on the results obtained. In some sense it could be compared with the multimode classical control. Some changes in the controller structure can be easily implemented: Removing LV not frequently used, addition/deletion of rules or facts, activation of guard antecedents in the event of new operating conditions, and so on.

The design of a FC always follows a trial and error procedure. If the plant to be controlled has been running and some data records are available, the FC design and parameters tuning may be done off-line. If not, an iterative approach will lead to the most convenient controller.

Let us summarize the design methodology:

a) Structural parameters:

To be decided based on the process/controller *a priori* knowledge, or as a result of on plant preliminary experiments:

- Number of output variables.
- Number of input variables.
- Number of input trends, accumulations, ...
- Linguistic Variables.
- Rule base structure.
- Basic rule set.

Also, from a technical point of view, we must decide how to implement the fuzzy logic operations, such as: fuzzy implication rule, fuzzy subsets matching in approximated reasoning, the also connective operation, and the defuzzification method.

One of the main problems in the FC Design is to capture the expert Knowledge and to translate it into LV, MF, rules, and so on. Rule base properties must be tested: completeness, consistency, interaction, and robustness. The last one is related to the sensitivity of the control to some unmodeled behaviour or disturbance.

b) Tuning parameters:

Among the parameters to be on-line updated are:

- Universe of discourse of the dealt with variables.
- MF parameters, such as peakvalue, bandwidth, or support.
- Input and/or output static gains.

If it is not possible to get the control goals, we must go back, to the previous step, trying to:
- Add new LV, with partial validity.
- Perform an extra input data preprocessing, creating new physical variables, such as new trend, a non linear function of some measurements, and so on.
- Introduce new measurements.
- Refine the rule base structure.

All these changes should be reflected in changes in the rule and data base.

If some past experience over the controlled process is available, in the form of variables and events records, and a controller structure has been decided as convenient, the controller parameters better fitting these records can be estimated. In [1] an experimental approach to FLC design has been outlined.

Anyway, in control of complex processes, the controlled process operating conditions are frequently changing and control updating and tuning is a must.

FLC FROM LINEAR ONES

As previously mentioned, one of the most important challenges in FLC design is how to adjust the rule base and the fuzzy set MF to reach some predefined control goals. This has led a lot of research in this area. One approach is to adjust the parameters by trial and error and compare their performance with the classical PID controllers [10]. Our proposal is to construct a FLC behaving similarly to an already existing linear controller, or a controller designed on the basis, of an approximate model of the plant. Later on, some methods can be used to refine the FLC and to improve the controlled system performance.

First let us look for a FLC equivalent to a PID one. For this application, some preprocessing to the error signal should be added to the basic structure of a FLC, [1]. Normalized membership functions with a scale factor are used.

The model used for PID fuzzy controller is as depicted in figure 2. The parameters and design factor used in this model are:

- e_k is the input variable to the controller.
- K_i are scaling factors.
- $e_k^{'}$, $e_\Delta^{'}$, $e_\Sigma^{'}$ are the normalized inputs.
- fe is the module transformating the inputs variable to its linguistic variables V_k, V_Δ, V_Σ. Each linguistic variables has its own linguistic terms.
- *RuleBase* Maintains the knowledge of the regulator. It is a set of IF-THEN rules.
- V_u is the linguistic variable of the output variable.
- NM is the defuzzification method used to obtain a number from the output linguistic variable.
- fu is the module transformating the linguistic outputs values to its normalized numeric value $u^{'}$.
- K_u is the output scale factor.
- u is the output variable of the controller.

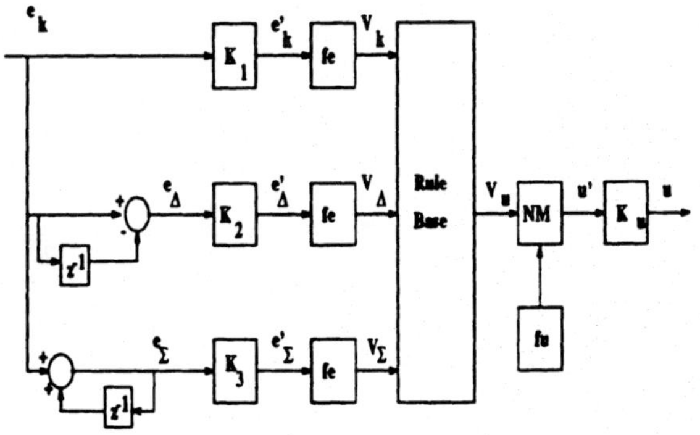

Figure 2: PID FLC structure.

The output of a discrete PID controller can be expressed by:

$$u_k = K_p[e_k + T_d\frac{e_k - ek - 1}{T_s} + \frac{1}{T_I}\sum_{i=0}^{k-1} T_s e_i] \quad (1)$$

Rewriting this expression to fit it into the proposed fuzzy structure:

$$u_k = K_e e_k + K_\Delta(\Delta e)_k + K_\Sigma(\sum e)_k \quad (2)$$

This equation is a linear manifold of the input variables to the FLC, thus, a linear equivalence can be found. The advantage of using normalized functions is that only the scaling factors K_1, K_2, K_3, K_u need to be adjusted. The normalized functions are uniformly distributed in the range [-1,1] for the input variables, and in [-1.2,1.2] for the output one, as seen in figure 3. In this case, five triangular functions have been selected. It is also necessary to define the inference matrix. In the case of a PD controller the matrix is as in table 1. Every linguistic input variable V_k, V_Δ, and output variable V_u has the linguistic terms A_1^i, A_2^j, A_u^k, respectively.

	A_2^1	A_2^2	A_2^3	A_2^4	A_2^5
A_1^1	A_u^1	A_u^1	A_u^1	A_u^2	A_u^3
A_1^2	A_u^1	A_u^1	A_u^2	A_u^3	A_u^4
A_1^3	A_u^1	A_u^2	A_u^3	A_u^4	A_u^5
A_1^4	A_u^2	A_u^3	A_u^4	A_u^5	A_u^5
A_1^5	A_u^3	A_u^4	A_u^5	A_u^5	A_u^5

Table 1: FC same PD controller.

Similar matrices can be defined for other controllers, but in the case of a PID, a 3D matrix results [8].

The rules for selecting the scale factors are:

- The output scale factor K_u, must be fixed depending on the saturation limits of the actuator, whenever this range is symmetric.

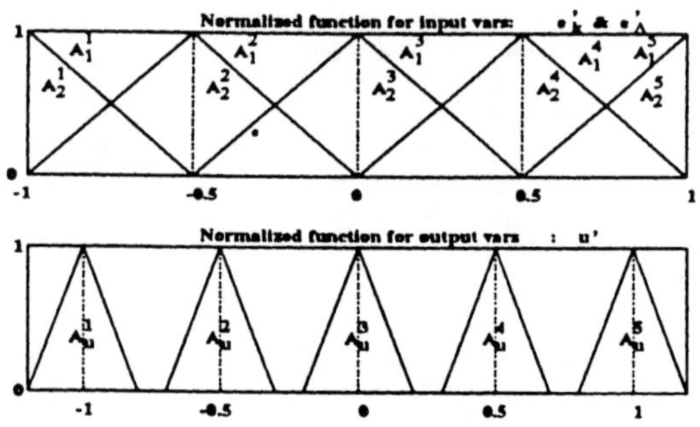

Figure 3: Normalized function for PID FLCs.

- If the regulator is just a proportional one: $K_1 = \frac{K_p}{R_e K_u}$, where R_e is the range of relevant values of the error signal.

- For a PD regulator: $K_1 = \frac{K_p}{R_e K_u}$, $K_2 = \frac{K_p T_d}{T_s R_\Delta K_u}$.

- For a PID regulator: $K_1 = \frac{K_p}{R_e K_u}$, $K_2 = \frac{K_p T_d}{T_s R_\Delta K_u}$, $K_3 = \frac{K_p T_s}{T_I R_\Sigma K_u}$

In order to verify this approach, a comparison between a linear PD and the equivalent FLC has been implemented. The results, shown in figure 4, are similar but there are some differences. Our interpretation is that the number of fuzzy sets of the output variable is very low and this leads to a large difference between both controllers.

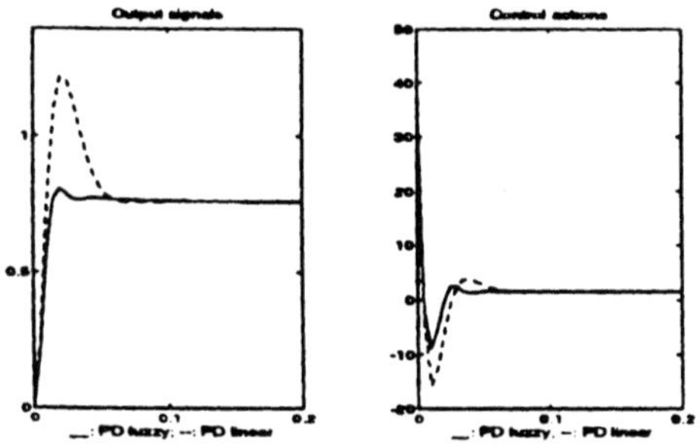

Figure 4: Comparing the response of a linear PD with a linear PD FLC.

One of the main advantages of the FLCs versus the linear ones is that they can implement non-linear control functions, and at least in theory, a better response can be achieved. But, the most important difficulty is how it can be done, because it is very difficult to establish an analytical model of this controller. However, there are many applications where this controller has been successfully used [6], improving the results got by a linear controller.

THE CEMENT KILN PROCESS

The kiln process heats the inlet flow of raw material (*lime*, *sand*, *pyrite*, *clime*) until the clinker is obtained. The process is formed by three step: the **preheater**, where the material is heated using the residual heat of the combustion fumes; the **kiln**, a rotary large cylinder where the clinker is formed; and the **cooler**, where the clinker interchanges its heat with the combustion air.

Some of the characteristics of this process are: i) It is a distributed process, ii) the system can operate in several different situations, iii) high number of variables (between 40 and 50): pressures(9), temperatures(14), gas concentration(6), laboratory analysis, etc, iv) large delay in some variables due to the kiln condition, v) due to the process complexity, there is not a good process model, and vi) only few people know how the system works (expert operators).

The proposed control goals, and in this order, are: Maximum quality, Maximum production, and Minimum production costs.

The first one is the most important and difficult to achieve owing to the delay between the clinker formation and its quality measurement. There is a delay in the sample preparation, transportation, and X-Ray analysis (about 40 minutes). Another difficulty is due to the impossibility of the clinker formation temperature measurement, which is directly related to the final quality. This question forces to estimate the temperature using other variables (kiln torque, kiln temperatures, preheater temperatures, etc.).

Maximum production is achieved maintaining the control variables at their nominal values whenever it is possible.

Minimum production costs are obtained minimizing O2 concentration in the fumes, CO concentration between two limits, and avoiding abnormal situations.

The variables used to control the kiln are: *raw material*, *coal*, and *fumes* flows, and kiln rotation speed. Thus the controller has to conclude a value for each variable to achieve the control goals.

The kiln process can operate in different situations: *normal* situation, the desirable one, *abnormal* situation, if the basic control goals are not fully reached, and *exceptional* situations, in which the control goal is to return under normal operating conditions. Some of these situations are:

repairing ring, *inlet coolers obstruction*, *cyclone obstruction*, or *raw cut off*. For each situation a different control policy must be implemented.

In the cement kiln, the hierarchical machine of states comes defined with the following operating mode:

- Kiln stopped.
- Kiln starting. The kiln is heated progressively following a curve of heating until to reach the conditions of nominal production.
- Kiln ready (prepared mode).
- Increase of production: Transitory mode than increase the production goal.
- Decrease of production: Transitory mode than decrease the production goal.
- Kiln stopping. Transitory mode than realize the necessary actions to reach the kiln stopped mode.
- Nominal production: In this mode the process reaches the proposed goals. It is the most important mode. In this one, a double FLC has been designed, and the coal and exhaustor opening are determined to respectively obtain the optimal production and quality.
- No recoverable incidence modes: repairing ring, inlet coolers obstruction, cyclone obstruction, etc. For different reasons, transitions corresponding to some process variables (temperature, pressure, fumes flow analysis) are overpassed and it is necessary to move to manual control.
- Recoverable incidence modes: They are related with the trespassing of some limits associated to production variables (free lime, kiln torque), temperature, pressure and fumes flow analysis (CO). These incidents have influence in the quality production but they can not danger the process or the environment. The control actions to counteract these incidences are based on raw material and kiln rotation speed modifications.

With the state diagram of the global system discussed in the past section, control structure design of each operating mode must be calculated. It is preferable to dispose of mathematical models (logical, continuous, time discrete, stochastic, etc.) for each mode.This modeled, in the industrial problem of the cement kiln process is very difficult, how we saw in the previous section, and even this moment the kiln control process in the cement factory is bases in experience and knowledge of technical operator.

In this difficult case, the elaboration of inquiries to the technical operator allows qualitative or heuristic process modeled , with rules of thumb if-then. This rules model is very suitable for the application of fuzzy set theory and FLC definition.

The heuristic modeled of nominal mode in the kiln control process is presented. In this description the physical input, output variables, internal variables, structural parameters and refinamets. Previously a methodology summary is presented.

CEMENT KILN CONTROL

We first define the control domains or modes. A fault detection system, based on a large amount of data, will determine if an emergency situation happens. Under each predefined situation, a FLC will try to recover the kiln situation or will fire the actions to safely stop the kiln. The fault detection system is very well suited to be implemented by a neural network, dealing with measurements coming from process internal temperature sensors, external temperature scanner, motor torque and consumption, and so on.

Under normal and abnormal conditions, two FLC have been designed. There are two main control loops: O_2 concentration controlled by fumes flow, and clinker quality by coal flow. The first one was initially operating under PID control. The second one was mainly under the operator responsability. Raw materials flow and motor speed have been devoted to optimize the production.

Heuristic FLC

Let us review the clinker quality control. The control structure, as finally shown, includes a first level of measurement assessment, a second level of kiln state estimation, and a final level of coal flow computation. That means a splitting of the KB, per level, easier to understand, to access, and to update.

Measurement assessment Free lime clinker content is very difficult to on-line measurement. Laboratory analysis provides a good measurement, but a long delay is required. From a kiln process knowledge we can derive that free lime content in the clinkerization zone is related with the motor torque. Direct torque measurement is not very reliable owing to noise measurement disturbances. Moreover, the absolute torque value is strongly dependent on the kiln operating conditions. For these reasons, a qualitative measurement of the averaged torque trend is considered as the most relevant input to estimate the variations in the free lime.

The first attempt was to estimate free lime content, (Increment in estimated free lime, IEFL) from the torque Trend, (TT). The LV were: IEFL (Put, Let, Take-off), and TT (Down, Steady, Up). The rule base was chosen "linear" and symmetric:

```
If TT Up then IEFL T-off
If TT Steady then IEFC Let
If TT Down then IEFL Put
```

Experiments carried out in the plant pointed out a mismatch in the edges of the considered TT behaviour, the relevance of the information provided by the laboratory, referred to the past value of the free lime, as well as the convenience to modify the initial "linear" relationship.

Under trial and error approach, a new physical variable was introduced, the laboratory free lime content (LFL), with three LV (Low, Normal, High) and the number of LV attached to the TT was increased to five. The rule base structure appears in Table 2.

IEFL		LFL		
		Low	Normal	High
TT	V-Down	Put	Put-M	Put-M
	Down	Let	Put	Put
	Steady	Let	Let	Let
	Up	Let	Let	T-off
	V-Up	T-off	T-off	T-off-M

Table 2: Rule base: Free lime Estimation

Kiln State Estimation This FC level has the main goal of determining the state of the kiln, in the same way the expert kilner does. Thus, for the output variable, a set of nine LV has been decided. Nevertheless, in order to make a difference based on the user confidence in his/her conclusion, certainty coefficients are attached to each conclusion.

The first attempt to define the KB was to use the Error in Free lime, EFL, (Negative, Zero, Positive),and the Torque Trend, TT (VeryDown, Down, Steady, Up, VeryUp), as input variables, and the Kiln State, KS, (Hh, H, Nh, Hc, N, Ch, Nc, C, Cc), where H/h stands for Heat/heating, N for Normal, and C/c for Cool/cooling, as the FLC output.

The rule base is as shown in the Table 3:

	EFL	Neg	Zero	Pos
TT	V-Down	Hc,1	Nc,1	Cc,1
	Down	Hc,0.7	Nc,0.7	Cc,0.7
	Steady	H,1	N,1	C,1
	Up	Hh,0.7	Nh,0.7	Ch,0.7
	V-Up	Hh,1	Nh,1	Ch,1

Table 3: Rule Base: Kiln State Estimation.

As KS, the output variable, is a pure linguistic one, non discriminant MF have been chosen for the antecedents in order to establish dead zones and to avoid KS conclusion instability (oscillations).

This KB initial proposal was shown to be unappropriated in the table edges. This was motivated by the non-linear effect of the heat capacity and the materials melting heat in the current kiln state.

For that purpose, a new input variable, the Trend in Coal Flow, TCF, averaged over the past three hours, was considered. This information, in some sense, reflects a pseudo derivative effect on the kiln state, leading to an overheating or undercooling. This new variable only was relevant if the error in the free lime was negative. Thus, the new rule base appears in the Table 4.

If the coal trend was considered under any conditions an extended table with 9 columns instead of only 5 would result.

Control Action Computation The goal of this control level is to determine the action to be taken in the coal flow, that is, to increment, reduce or maintain the coal flow, based on the kiln state obtained as a conclusion, with an attached certainty coefficient, in the previous level. Although the reasoning is very simple, to get the final control action as a conclusion of this levels allows an easy understanding as well as the possible improvement if the expert would decide that some data, other than the kiln state, should be taken into account to determine any change in

	EFC	Neg	Neg	Neg	Zero	Pos
	TCF	Steady	Down	V-Down		
TT	V-Down	Hc,1	Nc,1	C,1	Nc,1	Cc,1
	Down	Hc,0.7	Nc,0.7	C,0.7	Nc,0.7	Cc,0.7
	Steady	H,1	Hc,1	Nc,1	N,1	C,1
	Up	Hh,0.7	H,0.7	Hc,0.7	Nh,0.7	Ch,0.7
	V-Up	Hh,1	H,1	Hc,1	Nh,1	Ch,1

Table 4: Rule base: Kiln State Estimation Refinement.

the coal flow.

The Increment in Coal Flow, ICF, with LV: VP, P, L, T, VT. (where P stands for Put, or increment, L for Let, or maintain, T for Take-off, or reduce, and V for very, showing a strong action) is determined from the kiln state. The Rule base is as in Table 5.

It is realized that the MF Static gain is very critical. Its tuning was done experimentally, starting with a low value.

According with real tests in the factory, two kind of refinements have been carried out: 1) A new variable (fumes temperature) was considered because this variable shows the future evolution of the kiln state without delay, and so, it provides a good information in order to take the appropriate action. 2) It is known that the kiln has a big thermal inertia, so, the consideration of the coal trend seems a solution in order to reduce this physical property. In this way, a new variable was considered, the coal trend, as an antecedent in the rules.

A graphical interpretation of these relationships is reported in [2].

Fuzzy PID controller

The goal of this controller is to determine the optimal combustion conditions by controlling the O_2 and CO concentrations. At a first stage a linear PID controller was designed, but when implemented its performance was not the desired one owing to the non-linear behaviour of the actuator, like dead-zone, saturation and others. Then, a fuzzy PD was designed to improve the controller response. The initial regulator was equivalent to the linear one, using the relations already mentioned. In this case, three normalized membership functions were considered for the input variables, the CO concentration and its trend. The output, the increment of the exhaustor aperture, has five normalized functions and the coefficients K_u, K_1 y K_2 were adjusted based in the parameters of the linear controller. In a third stage, this membership functions were modified to cope with the non-linear behaviour of the process: the overlap between the input functions was reduced to create a dead zone, and the area and the position of the output membership functions were changed to incorporate non-linear input-output relationship. The resulting membership functions distribution is shown in figure 5. The controller is defined by:

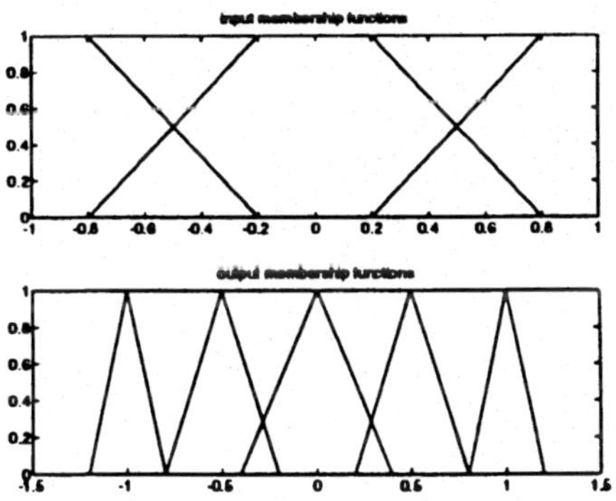

Figure 5: Membership functions for O_2 fuzzy PID controller.

- *Output variable: Increment in exhaustor aperture: IEXH, (Take-off -Much, Take-off, Let, Put, Put-Much).*

- *Input variables: 2, Error in CO, ECO, (Positive, Zero, Negative), and CO Trend, TCO, (Down, Steady, Up).*

- *Rule base: as shown in Table 6.*

- *MF Parametrization: Non discriminat MF have been chosen for the antecedents in order to establish dead zones to avoid output variable oscillations.*

	KS								
	Cc	C	Nc	Ch	N	Hc	Nh	H	Hh
ICF	VP,1	P,1	P,0.5	L,0.7	L,1	L,0.7	T,0.5	T,1	VT,1

Table 5: Rule base: Coal Flow Computation.

IEXH		ECO		
		Pos	Zero	Neg
TCO	Down	T-off-M	T-off	Let
	Steady	T-off	Let	Put
	Up	Let	Put	Put-M

Table 6: Rule base: Fuzzy PID controller

FLC Implementation

Owing to the high cost of a FLC development, the first question is when it is justified to use such a controller. A possible answer is that the FLC is a controller candidate whenever the process model is partially or totally unknown and there is some knowledge about how to control the process. This knowledge can be got from:

- A linear regulator that controls the process, providing unsatisfactory results.
- An expert in the process, able to express his/her knowledge by rules.
- Modeling the dynamics of the human operators of the plant, using some fuzzy identification technique.

Usually, the result of this stage is an initial FLC whose response must be improved, due to errors or unmodeled dynamics. So, a systematic approach to refine this controller to reach a controller fitting the control goals is required. The proposed methodology is the following:

1. First of all, the number of functions and distribution are fixed. In the zones where there is no overlap between the membership functions, only one rule is fired and the output of the regulator will be the centroid of the concluded fuzzy membership function. If this value is nor correct, then three possible actions can be done:
   - Movement: you can move the function to fit the desired value, but it can affect to other rules.
   - Growth: If the movement of a function is not admissible because of this interaction, the solution is to increase the number of fuzzy sets of the output variable. A new rule must be add to the rule base.
   - Modification: When you can find another function, already defined, that fits the desired value. A rule is changed.
2. If the transition between the function centroid of the output variable is not the one it is required, the overlap in the input variables must be changed.
3. If the response is not the correct, the number of fuzzy sets of the input variables can be increased, starting in point 1 with the new distribution.
4. Finally, perhaps it is necessary to use another variables to adjust the system response. For instance, in nonlinear system it is important to know the absolute value of the reference besides the error signal. This procedure has been used to design the previous FLC and the results obtained have been very satisfactory.

CONCLUSIONS

FLC have shown very appropriated features to be applied in process control. In this control application field, both models and control requirements are handled in an approximated way, and the possibility to understand what is happening in the control loop and the rules of thumb to tune the control are more relevant than the accuracy of the results.

The traditional approach to control this kind of processes is by either the use of simple controllers, like the PID controllers, or any control under the direct supervision of the operator.

In the paper, this is the starting point to design a FLC: the previous linear unsatisfactory controller, or the skilled operator. Two methodologies have been summarized and the approaches have been applied to the control of a complex industrial process, the cement kiln. The results obtained in a life application are really good.

References

[1] Albertos, P. (1992). *Fuzzy Controllers. Applications of Artificial Intelligence in Process Control.* Ed. Pergamon Press. ISBN 0-08-042016-8.

[2] Albertos, P., Martínez, M., Navarro, J.L., Morant F. (1993). *Fuzzy Controllers Design: a Methodology.* 2nd IEEE Conference on Control Applications. pp. 69-76. Vancouver, Canadá, 1993.

[3] Aström, K.J., Wittenmark, B. (1989). *Adaptive Control.* Addison Wesley.

[4] Coughanowr, D. R. and Koppel, L.B. *Process System Analysis and Control* Mac Graw-Hill. N.Y. 1965

[5] Laffey, T.J; Cox, P.A. (1988) *Real Time Knowledge Based Systems,* AI Magazine, pp 27-45, spring 1988.

[6] Maiers, J., Sherif Y.S. (1985), *Applications of Fuzzy Set Theory.* IEEE Transactions on System, Man ad Cybernetics, 23(3): 665-685.

[7] Mamdami, E. (1977),*Application of Fuzzy Set Theory to Control Systems. A Survey,* Fuzzy Automata and Decision Processes, M. Gupta, G. Saridis and B. Gainer Eds., North Holland, 1977.

[8] Navarro, J.L., Albertos, P., Martínez, M. (1994). *Some Experiences in the Design of Industrial Fuzzy Controllers.* International Workshop on Fuzzy Technologies in Automation and Intelligent Systems. Duisburg, spring 1994.

[9] Phillips, C.L., Harbor, R. (1985). *Feedback Control Systems.* Prentice Hall.

[10] Tang K., Mulholland R.(1987),*Comparing Fuzzy Logic with Classical Controller Designs.* IEEE Transactions on System, Man and Cybernetics, 17(6): 1085-1087.

[11] Zadeh, L.A. *Fuzzy sets,* Informat. Control. vol 8, pp. 338-353. 1965

AN INTELLIGENT SENSOR FOR ASSESSMENT OF THE QUALITY OF FISH PROCESSING

N. WICKRAMARACHCHI* and C. W. DE SILVA†

Industrial Automation Laboratory, Department of Mechanical Engineering, The University of British Columbia, Vancouver, Canada.

Abstract. An intelligent sensor, based primarily on computer vision, for the assessment of quality of processed fish, as applied to an automated fish processing plant, is discussed. The role of information resolution on the degree of intelligence required for data processing is given major consideration. A hierarchy of data processing is introduced where the crisp, high resolution data of an image is transcribed into low resolution context that is more amenable to further intelligent processing at a higher level of the hierarchy. Experiments carried out in the laboratory with a prototype fish processing workcell are described and the results are summarised.

Key Words. Intelligent machines; hierarchical decision making; quality control; image processing; fuzzy systems.

1. INTRODUCTION

The assessment of quality of a food product such as processed fish, and also the performance of the associated processing machinery may be carried out by monitoring and evaluating a multitude of parameters such as geometrical accuracy, surface texture, firmness, smoothness of edges, coloration, and presence or absence of stray objects. While certain parameters like geometrical accuracy could be easily measured and evaluated, the overall quality of the fish product is mostly determined through subjective assessment of the end product by industry experts who have to visually examine each and every item that is produced. Automation of such a processing plant must necessitate the use of an *intelligent sensor* approach in order to incorporate the assessment of quality as part of the control activity.

A hierarchical architecture with embedded intelligence in appropriate levels has been developed for workcell monitoring and assessment of quality. A series of sensors, ranging from optical encoders to CCD cameras, provides low level information for intelligent preprocessors through which a more meaningful high-level context is constructed. A top-level knowledge base operates on this context and derives the final assessment of quality.

Conventional image processing techniques are extremely time consuming due to the large amount of information available in a single digital image. Process automation, on the other hand, severely restricts the time available for on-line processing, in view of the requirement of high throughput rates. The present paper emphasizes that vast quantities of high resolution information, obtained from the low-level process sensors, are not needed in high level control tasks such as system tuning, reconfiguration, and self-organisation. Instead, an *Intelligent Sensor* which incorporates the low-level sensors and a set of "intelligent" preprocessors may be employed. In this manner, unwanted information could be discarded and only a set of high-level context information that has a significantly lower resolution would be presented to the reasoning mechanism of the high-level knowledge base. This is made possible by the use of *fuzzy linguistic variables* to quantify *quality* at the preprocessing stage. Then the decisions can be made quite efficiently at a high level of intelligence using the low-resolution context information. This approach is demonstrated through the use of a CCD camera, image processor, and intelligent preprocessor which comprises an intelligent sensor, for obtaining quality information on processed fish in a laboratory workcell.

2. THE HIERARCHY

The research described in this paper is based on experiments carried out on a prototype workcell that has been developed in the Industrial Automation Laboratory for automated processing of

*Graduate Research Associate.

†NSERC Professor of Industrial Automation.

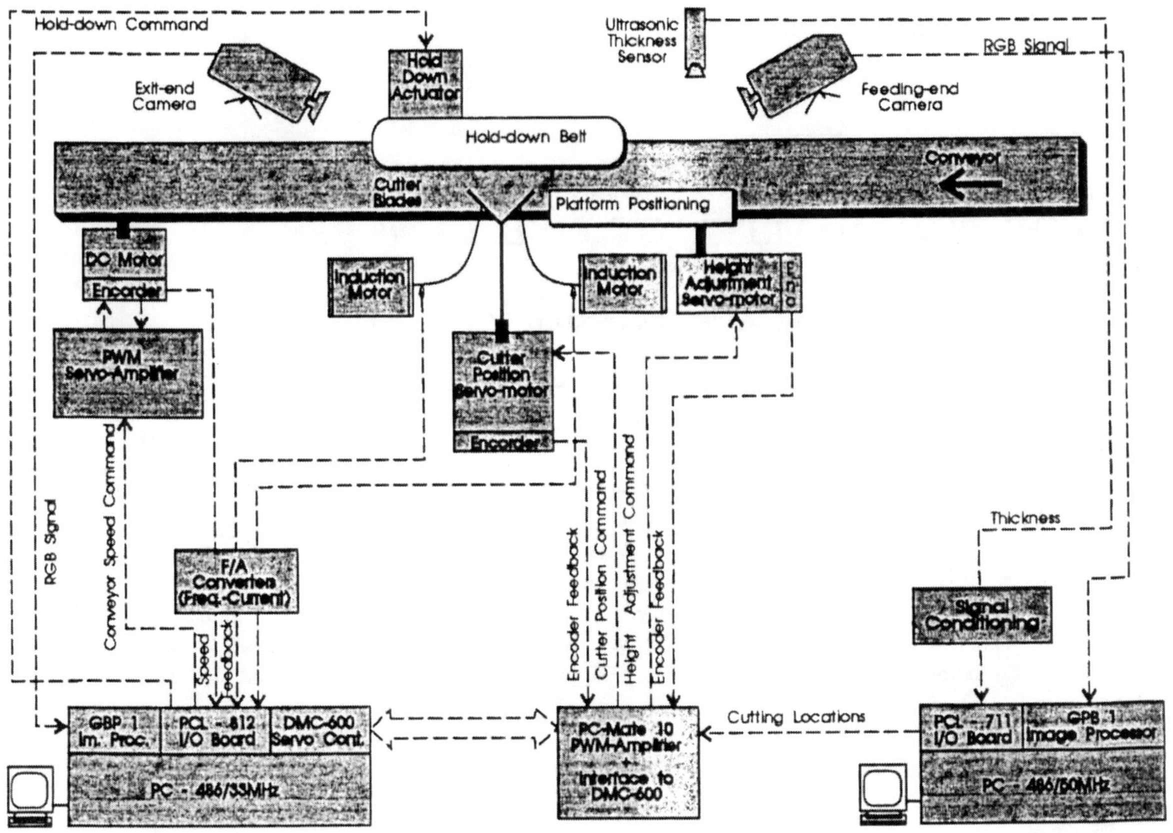

Figure 1: Prototype Workcell for Automated Processing of Fish.

fish (de Silva, 1990). The laboratory implementation of the prototype workcell is shown schematically in Figure 1. There are two imaging stations in the system, one before the cutter and the other at the exit end of the processed fish. The exit-end camera which captures images of fish following the cutting process, is the primary source of visual information that is received by the monitoring and quality control system that runs on the second 486 based PC. The block diagram in Figure 2 shows the hardware and communication interfaces of the PC that enable on-line determination of the system status by the monitoring and quality control system.

In order to make the decision making process simpler and tractable, the monitoring and quality control system is organized in a functional hierarchy as illustrated in Figure 3. The hard-wired controllers and the direct sensors that provide crisp information are delegated to the lowest level of the hierarchy. This level is characterised by the crisp nature of the information handled and the correspondingly large control bandwidth. The CCD camera that provides sharp, quantitative information in rapid succession is a typical example for the inherent crisp nature of information generated at this level of hierarchy. Similarly, the PID controllers for the servo-motors as well occupy the same level for reasons of precision and higher bandwidth of information required by

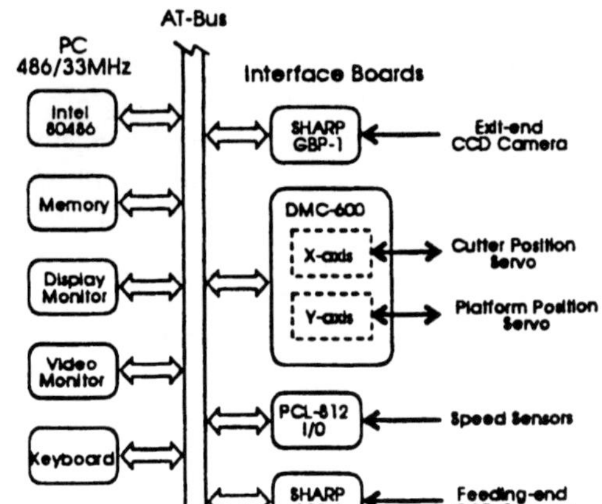

Figure 2: Hardware Configuration for Component Interfacing.

the controllers. As a consequence of this precision of the information that is processed at the lowest level, the degree of intelligence required to interpret data will be either minimal or none at all.

In keeping with the principle of "increasing precision with decreasing intelligence" (Saridis, 1989), and also to make the low-level sensory information compatible and consistent with the top-level knowledge base, a preprocessor stage is in-

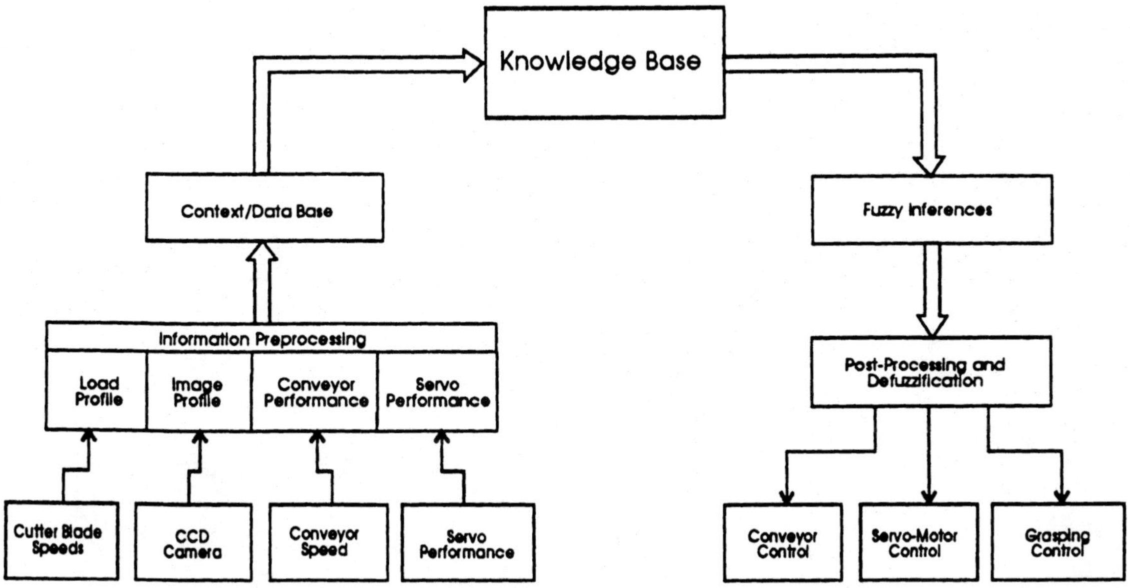

Figure 3: Hierarchy for Quality Control.

troduced as the intermediate level of the hierarchy. The preprocessor level is primarily concerned with the translation of crisp, high resolution information from the sensors into a more intelligible, low resolution format at the top level of the hierarchy. This translation could be considered basically as an abstraction of data from high resolution to low resolution where the low resolution information constitutes the context for a knowledge-base at the top level of the hierarchy. In this manner, the low-level sensors together with their preprocessors may be interpreted as an intelligent sensor.

The knowledge-base at the top level of the hierarchy has to operate in the context of non-crisp, vague, incomplete, imprecise or approximate information, necessitated by the nature of the data that are processed by the workcell. For example, the visual quality of processed fish could not generally be quantified in a precise manner by observing a few parameters such as dimensions, shapes, and surface defects. However, at least in theory, the visual quality could be considered as a complex combination or interpretation of many such features. In recognition of this pervasive fuzziness of information at the higher levels of the hierarchy, the knowledge base has been implemented as a *fuzzy linguistic rule-base* and the context is constructed in terms of *fuzzy variables*. Inferences are made by applying the compositional rule of inference (Zadeh, 1979) of fuzzy logic to the knowledge base, on the basis of context information provided by the intelligent sensor.

3. ASSESSMENT OF QUALITY

Experts in fish processing industry usually have to examine each and every product laboriously in order to assess its quality. The subjectivity of such assessments as well as the labor intensive nature of the process have necessitated the incorporation of quality assessment into the automated workcell itself. The main features in a processed fish that are considered for the current purpose of quality assessment are:

(i) Depth of solid cut expressed as a ratio to the width of body (D_c).
(ii) Smoothness of the cut contour (S_c).
(iii) Smoothness of the surface of cut (S_s).

These features require only a minimal amount of preliminary image processing to be carried out on the source images as described in the next section, and the parameters derived are used as the context of a quality assessment rule-base. This forms the intelligent sensing process. The quality that is "measured" in this manner may form an intelligent feedback to the system adjustment (tuning, reorganizing, etc.) knowledge base.

3.1 *Preliminary Image Processing*

Figures 4 and 5 show two typical images of fish as captured by the exit-end camera; the former illustrating a good quality cut and the latter somewhat unacceptable cut. In order to reduce the effects of noise and irregularities the image is first smoothed by a 3×3 binomial convolution kernal. Since smoothing can wipe out some cutting edge irregularities that are also of interest, convolution with the smoothing kernal is applied only twice over the entire image. The image is then thresholded at a fixed gray level of mid-range (128) to accentuate the body of the fish and the cutting region. Sobel filtering is then carried out on the thresholded image in order to highlight edge features in the directions of both cutter movement and conveyor movement. The regions of interest of each image after these operations are shown in figures 6(a) and 7(a) respectively. Segmentation and labeling are then performed over the area of interest and the cutting contours shown in figures 6(b) and 7(b) respectively, can be easily extracted as the labeled segment with the largest area. All these operations are carried out using the embedded functions of the SHARP GPB-1 image processor and as a result the processing time required would be adequately low.

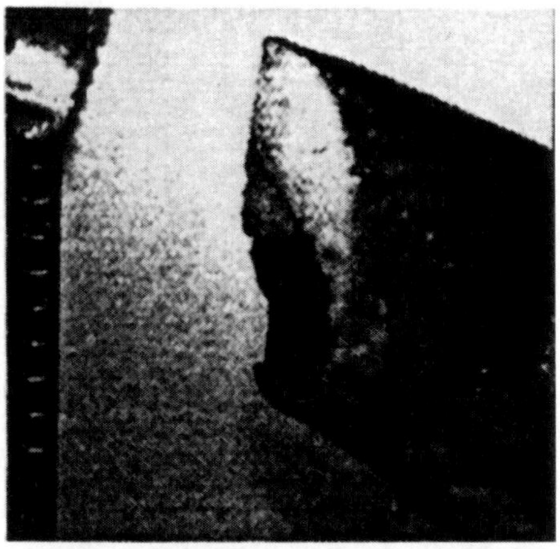

Figure 4: Raw image of Fish 1.

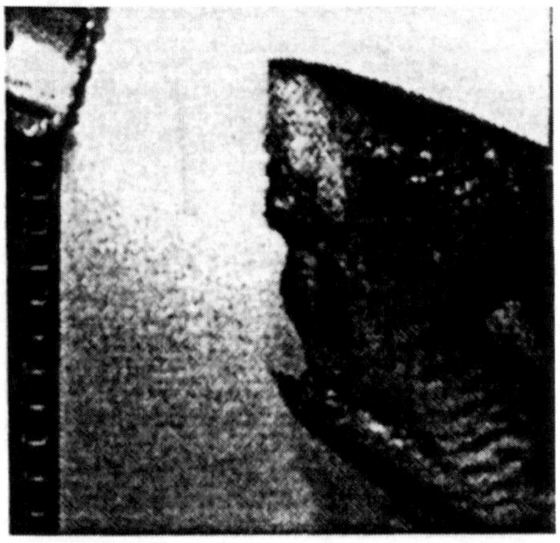

Figure 5: Raw image of Fish 2.

The final step in image analysis is to recognize and compute the features D_c, S_c, and S_s which were defined in the previous section, from the processed images of Figure 6 or 7. Instead of employing a pattern recognition algorithm on the image directly, first the cutting contours of figures 6 and 7 are converted into contour-length profiles as shown in figures 8 and 9 respectively. The contour-lengths are computed on a Cartesian basis along the contour with the length between adjacent pixels in both horizontal and vertical directions being taken as the unit length and hence in diagonal direction as $\sqrt{2}$ of unit length. The resulting contour-length profiles retain some of the angular properties of the original cutting contours and can be employed to determine the features D_c, and S_c. The contour-length profiles can be readily derived from the chain codes for the corresponding cut contours of figures 6(b) and 7(b). Figures 8 and 9 illustrate contour-length profiles against the y-axis pixel coordinates since we are primarily interested in the lengths measured across the body of the fish. The characteristic contour-length profile of a good symmetric cut is shown in Figure 8 while Figure 9 characterises an asymmetric and jagged cut contour.

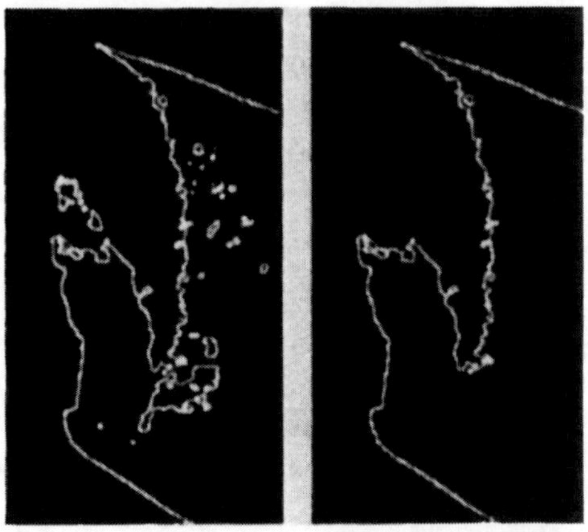

Figure 6: (a) Region of interest after filtering image of Fish 1. (b) Cut contour for Fish 1.

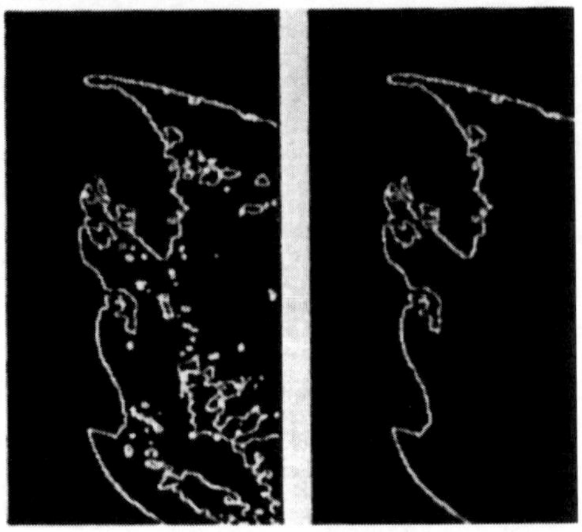

Figure 7: (a) Region of interest after filtering image of Fish 2. (b) Cut contour for Fish 2.

3.2 *Intelligent Preprocessing*

The preprocessor stage plays the most important role of the intelligent vision sensor in that it translates the raw image data into a more meaningful context for the quality control rule-base. First the points of interests for the purpose of assessing quality are recognized from the figures 8 and 9 by using an an algorithm which searches for gross maxima and minima. These points are namely, Point *a* which signifies the beginning of the cut contour, Point *b* - the beginning of the solid cut and Point c - the end of the solid cut. The following parameters can now be computed:

$$\text{Index of depth of solid cut } D_c = \frac{y_{bc}}{y_{af}}$$

$$\text{Index of smoothness of cut contour } S_c = \frac{y_{bc}}{l_{bc}}$$

where, y_{bc} = Depth of solid cut from b to c in pixel coordinates, y_{af} = Height of fish from a to f in pixel coordinates, and l_{bc} = Contour-length from b to c.

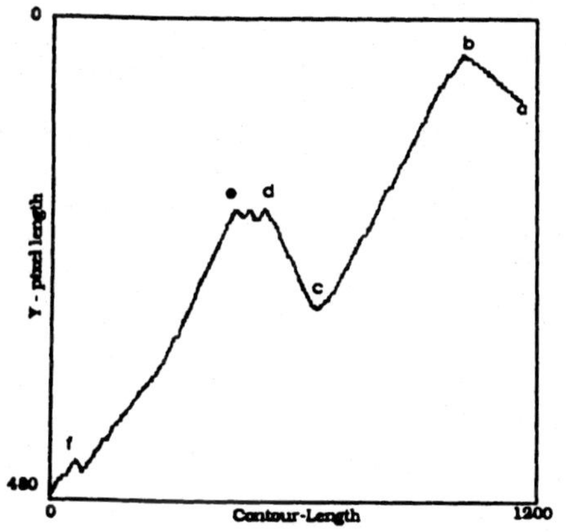

Figure 8: Contour-Length profile for Fish 2.

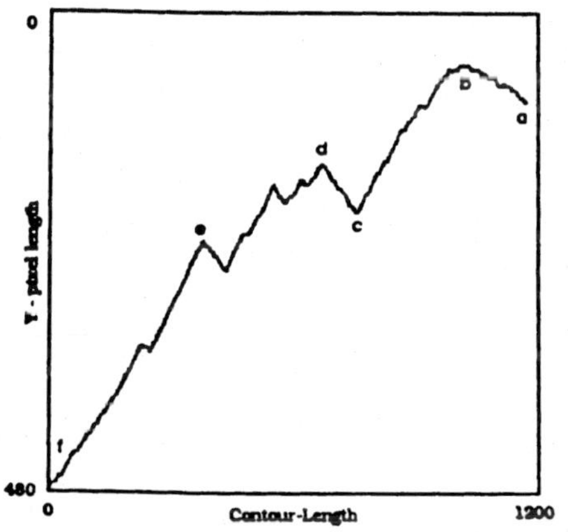

Figure 9: Contour-Length profile for Fish 2.

A larger value for D_c signifies a better cut while a larger value for S_c signifies a smoother cut. The smoothness of the surface of cut is strongly indicated by the number of labeled image object in the region of interest as illustrated by the figures 6(a) and 7(a). Therefore, the index for the smoothness of surface (S_s) is expressed as a percentage of a nominal value (100) determined experimentally.

$$S_s = \frac{100 - N_l}{100}$$

where, N_l = Number of labeled image objects within the cut contour. For our example raw images given in figures 4 and 5, the computed parameters are shown in the Table 1.

Table 1: Parameters for the Assessment of Quality.

Parameter	Fish 1	Fish 2
D_c	0.63	0.34
S_c	0.70	0.56
S_s	0.73	0.45

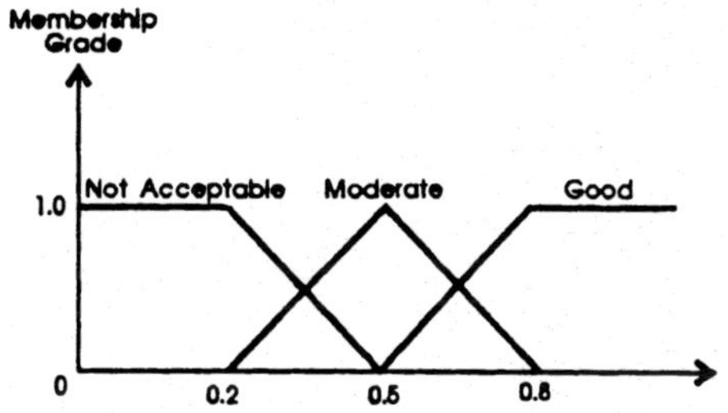

Figure 11: Membership Function for the Inference of Quality.

Computation of indexing parameters is only the first step of evaluating quality. The next step would be the assignment of proper membership functions for the fuzzy sets characterizing the relative importance of each parameter and building a fuzzy linguistic rule base with quality-indicative parameters as its antecedent variables and quality as the consequent variable. Each of the above parameters is fuzzified on a universe of discourse defined over [0.0, 1.0] and partitioned into three primary fuzzy states namely, poor, moderate, and good; i.e., a fuzzy variable with resolution 3 (de Silva, 1991). The resolution of the fuzzy partitioning is usually determined by the information resolution possessed at that particular level of hierarchy. Figure 10 shows the shape and the resolution adopted for the individual fuzzy sets in the universe of discourse for the attributes of quality.

Table 2 shows the rule-base developed for the assessment of quality from the three(3) fuzzy context variables, namely: Depth - Depth of solid cut, Contour - Smoothness of cutting contour, and Surface - Smoothness of the surface of cut. The consequent fuzzy variable is Quality and its scope and partitioning are shown in Figure 11, where a fuzzy resolution of three has been chosen as well.

Application of the rule-base in Table 2 to the images given in figures 4 and 5 results in the fuzzy inferences illustrated in the figures 12(a) and 12(b), respectively. The shaded region of each figure represents the membership of visual

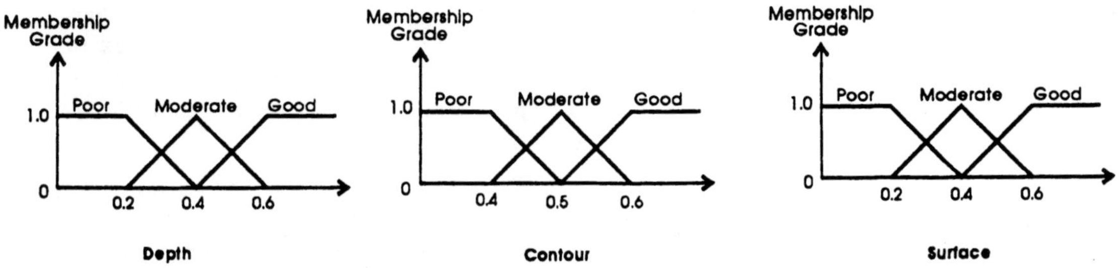

Figure 10: Membership Function for the Inference of Quality.

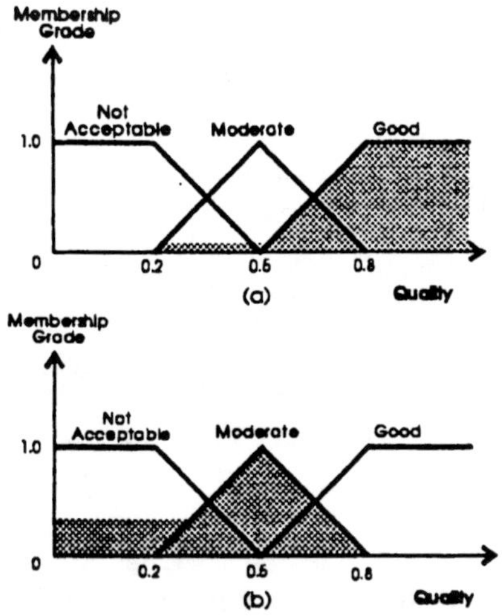

Figure 12: (a) Inference for Fish 1. (b) Inference for Fish 2.

Table 2: Rule-base to derive quality.

i). If (Depth=poor) then (Quality=poor).

ii). If (Contour=poor) then (Quality=poor).

iii). If (Surface=poor) then (Quality=poor).

iv). If (Depth=good) and (Contour=good) and (Surface=good) then (Quality=good).

v). If (Depth=mod.) and (Contour≠poor) and (Surface≠poor) then (Quality=mod.).

vi). If (Depth≠poor) and (Contour=mod.) and (Surface≠poor) then (Quality=mod.).

vii). If (Depth≠poor) and (Contour≠poor) and (Surface=mod.) then (Quality=mod.).

quality in each of the primary fuzzy sets of Unacceptable Quality, Moderate Quality, or Good Quality. The image of Fish 1 returns mostly Good Quality while Fish 2 returns Moderate to Unacceptable quality. These inferences compare quite preferably with the expected assessment of quality by human experts. The fuzzy inference of quality has low resolution but amenable to further intelligent processing and therefore, particularly suitable for use in a knowledge base for system tuning or adjustment.

ACKNOWLEDGMENTS

This work was supported by grants to C. W. de Silva from the *Advanced Systems Institute of B. C.* and from the *Natural Sciences and Engineering Research Council of Canada* for a fellowship and an NSERC Research Chair in Industrial Automation.

REFERENCES

De Silva, C. W. (1990). Research Laboratory for Fish Processing Automation. *Robotics and Manufacturing*, Vol. 3, ASME Press, New York, pp. 935-940.

De Silva, C. W. and Nalin Wickramarachchi (1992). Knowledge Base Development for an Intelligent Control System of a Process Workcell. *Intelligent Control systems*, DSC-Vol. 45, pp. 113-119, ASME, New York.

De Silva, C. W. (1991). An Analytical Framework for Knowledge-Based Tuning of Servo Controllers. *Engineering Applications of Artificial Intelligence*, Vol. 4, No. 3, pp. 177-189.

Saridis, G. N. (1989). Analytical Formulation of the Principle of Increasing Precision with Decreasing Intelligence for Intelligent Machines. *Automatica*, Vol 25, No. 3, pp. 461-467.

Wickramarachchi, N. and C. W. de Silva (1993). Vision-Based Product Quality determination in a hierarchical system for fish processing. *Proc. IEEE Pacific rim Conference*, Vol. 2, pp. 710-713, Victoria, Canada.

Zadeh, L. A. (1979). A Theory of Approximate Reasoning. *Machine Intelligence*, (Hayes J. et al., Eds), Vol. 9, pp. 149-194.

ROBUST TRAJECTORY TRACKING BY USING FUZZY SERVO CONTROL OF AN ARTICULATED ROBOT ARM

Z. KOVAČIĆ*, S. BOGDAN*, V. LACI*, R. KRISHNAN**

*University of Zagreb, Faculty of Electrical Engineering, Unska 3, Zagreb, CROATIA
**Virginia Polytechnic Institute and State University, The Bradley Department of Electrical Engineering, Blacksburg, VA 24061, USA

Abstract. The paper deals with a robust trajectory tracking by using fuzzy joint servo control of a five axis electric-driven articulated robot arm with tool pitch and roll motions. Trajectory planning ensures accurate approximation of the desired path and includes limits of the velocities and accelerations of each robot joint. The same fuzzy control algorithm is used in all joint servo loops, requiring only multiplexing and proper scaling of the fuzzy controller inputs and outputs. The comparative simulation experiments were also performed with a PD and a robust controller in the presence of white measurement noise, load change and nonlinear effects of robot dynamics. The results obtained have confirmed a very good performance and fairly robust trajectory tracking in case of using a fuzzy control method.

Key Words. Fuzzy control, industrial robots, robust control.

1. INTRODUCTION

Very often, a target control system cannot be mathematically modelled because of its complexity, irregular form or just because such a modelling, if used in control, cannot be done within the given control interval. In such case, traditional control methods usually fail because of their dependence on the exact mathematical description. Robots and robotic mechanisms are objects which belong to the group of very demanding control objects. They are rather difficult to be modelled, very often they have nonstandard extraordinary forms and some calculations, such as continuous solving of the inverse kinematics problem and estimation of the moments of inertia, require excessive time consumption which cannot be afforded within the given control intervals.

Intelligent control has appeared as a very challenging and innovative approach to solving the wide spectrum of control problems. Knowledge-based control techniques using, for example, fuzzy logic, neural network and genetic algorithms, have enabled a design of application-specific controllers. These methods tend to emulate the desired controller dynamics rather than to follow the exact mathematical description, thus changing completely the methodology of controller design. Namely, intelligent control techniques are focused more on behavioural aspect of the target system, while traditional techniques prefer mathematical description (linearization as the best example). Fuzzy control algorithms contain elements of human expert behaviour which are expected to bring flexibility and improved characteristics of the target system. Robotics is the field where success of the fuzzy logic algorithms was reported very frequently (Arciniegas, 1993; Banerjee and Woo, 1993; Cela and Hamam, 1992; Kelemen et al., 1993; Kumbla et al., 1992; Lim and Iliyama, 1991; Nedungadi, 1992; Scharf and Mandić, 1985; Tascilo et al., 1993; Tzafestas, 1991).

The paper addresses some issues of introducing fuzzy control algorithms in joint servo control of a five-axis electric-driven articulated robot arm. The comparative simulation experiments were also done with a PD and a robust controller developed by Hsia (1989) in the presence of white measurement noise, load change and nonlinear effects of robot dynamics.

2. ROBOT JOINT SERVO CONTROL

The Lagrange-Euler method was used for dynamic modelling of the manipulator (Fig. 1). The closed-form dynamic equation of the manipulator which can move through its workspace without restrictions attains the following form

$$\sum_{j=1}^{n} [D_{ij}(q)\ddot{q}_j] + \sum_{k=1}^{n}\sum_{j=1}^{n} [C^i_{kj}(q)\dot{q}_k\dot{q}_j] + h_i(q) + b_i(\dot{q}) = \tau_i, \quad 1 \le i \le n, \tag{1}$$

where q is the vector of joint variables, τ is the vector of actuator torques of an n-axis robotic arm, D(q) denotes the manipulator inertia tensor connected to the forces and torques generated by the motion of the links of the arm, C(q) is the velocity coupling matrix associated with Coriolis and centrifugal forces, h(q) is the gravity loading vector and b(dq/dt) denotes friction opposing the motion of the arm.

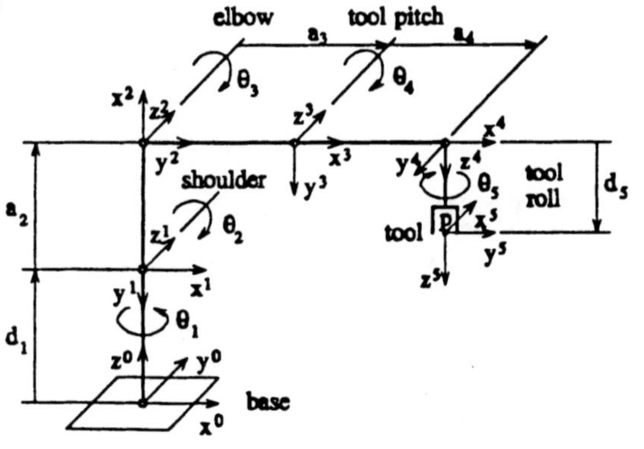

Fig. 1. Link coordinates of a five-axis robot.

Frictional force model for joint k is described with the following equation

$$b_k(\dot{q}) = b_k^v \dot{q}_k + sgn(\dot{q}_k)\left[b_k^d + (b_k^s - b_k^d)e^{-\frac{|\dot{q}_k|}{\epsilon}}\right], \quad 1 \le k \le n \tag{2}$$

where dq_k/dt is the velocity; b_k^v, b_k^d, b_k^s are the coefficients of viscous, dynamic and static friction, respectively; ε is a small positive parameter.

The parameter values of the studied robot (Fig. 1) were as follows: d_1=0.28575 m, a_2=a_3=0.2286 m, a_4=0.0095 m, d_5=0.15875 m, m_1=1.2 kg, m_2=0.7 kg, m_3=0.7 kg, m_4=0.15 kg, m_5=0.25 kg, $J_{n1}=J_{n2}=J_{n3}=$ $=J_{n4}=J_{n5}=2.63E^{-6}$ kgm^2, $T_{em1}=T_{em2}=T_{em3}=T_{em4}=T_{em5}=$ 0.0305 Nm, gear ratio r_1=264.4, $r_2=r_3=r_4$=528.8, r_5=384, b_1^v=0.5, b_2^v=0.25, b_3^v=0.2, b_4^v=0.5, b_5^v=0.25, b_1^d=0.2, b_2^d=0.1, b_3^d=0.05, b_4^d=0.1, b_5^d=0.2, b_1^s=2, b_2^s=1, b_3^s=0.5, b_4^s=1, b_5^s=0.5, $\varepsilon_1=\varepsilon_2=\varepsilon_3=\varepsilon_4=\varepsilon_5$=0.1, g_0=9.80665 ms^{-2}.

A robotic manipulator is able to perform the given manipulation tasks only if trajectory planning ensures accurate approximation of the desired path and includes limits of the velocities and accelerations of each robot joint. Manipulation tasks are normally given in terms of the desired tool position and orientation. If knot points along the path such as the end points and intermediate via-points are specified, segments of the path between them have to be interpolated with spline functions to produce a smooth trajectory, i.e. to maintain the continuity of at least first two derivatives (Bogdan et al., 1993; Laci and Kovačić, 1994). Once the nominal trajectory is determined, there remains a job to design a servo controller which would provide fast and accurate joint variable responses, and accordingly, robust and effective trajectory tracking. A joint servo control loop containing a servo controller, a servo amplifier, a dc motor and a gear is shown in Fig. 2.

In order to compare the performance of the proposed fuzzy controller with performance of other types of controllers, a PD and a robust controller developed by Hsia (1989) were chosen for this purpose. A servo controller depicted in Fig. 2 can be either a PD or a fuzzy controller, while the robust controller, as shown in Fig. 3, contains a PD controller for asymptotic tracking of the path and an auxiliary controller for system parameters uncertainty and disturbance effects cancellation. In both cases, the parameters of the PD controller were as follows: K_p = 277.7, K_d = 33.3.

3. DESIGN OF A FUZZY SERVO CONTROLLER

The fuzzy controller used in each joint servo control loop has two inputs, joint position error signal $e_i(k)$ and change in joint position error signal $de_i(k)$. Error $e_i(k)$ is defined by

$$e_i(k) = q_{ri}(k) - q_i(k) \tag{3}$$

where: $q_{ri}(k)$ - position reference of the ith joint,
$q_i(k)$ - measured position of the ith joint.

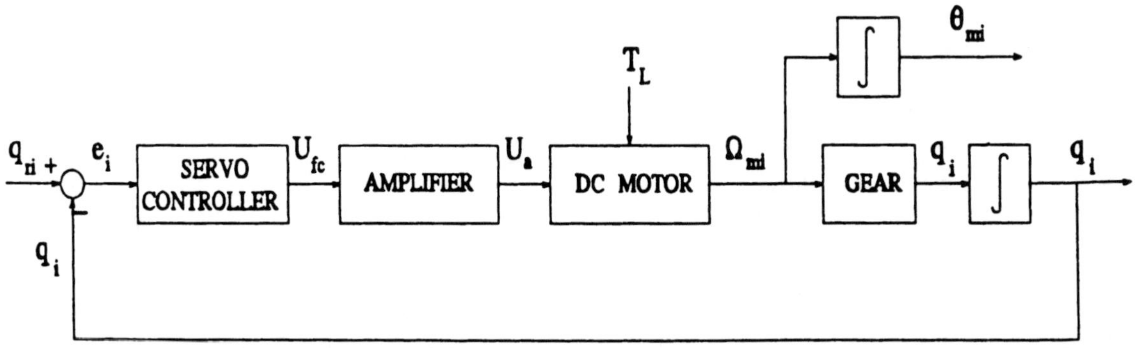

Fig. 2. A structure of a robot joint servo control scheme.

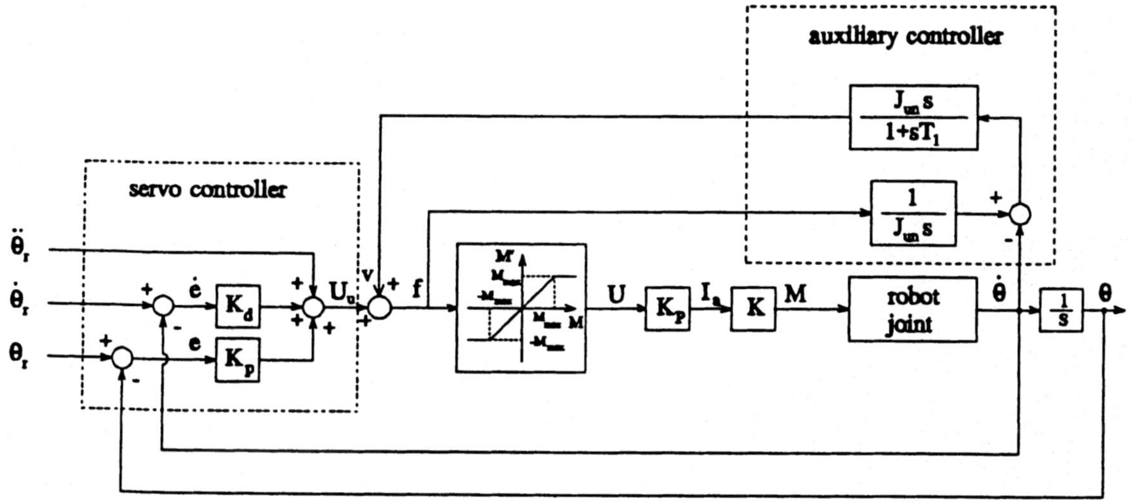

Fig. 3. A structure of a robust robot joint servo control scheme (Hsia).

Change of error signal $de_i(k)$ is defined by

$$de_i(k) = e_i(k) - e_i(k-1) \qquad (4)$$

Because of astatism in the servo loop, the fuzzy controller operates in the nonincremental mode, i.e. its current output value does not depend on the previous output values. Seven linguistic subsets have been defined for both inputs (universe of discourse E and DE): large (LN, LP), medium (MN, MP), small (SN, SP) and zero (Z). The maximum values of fuzzy controller inputs and outputs were estimated separately for each joint of the considered manipulator. The distribution of membership functions for both inputs is shown in Fig. 4.

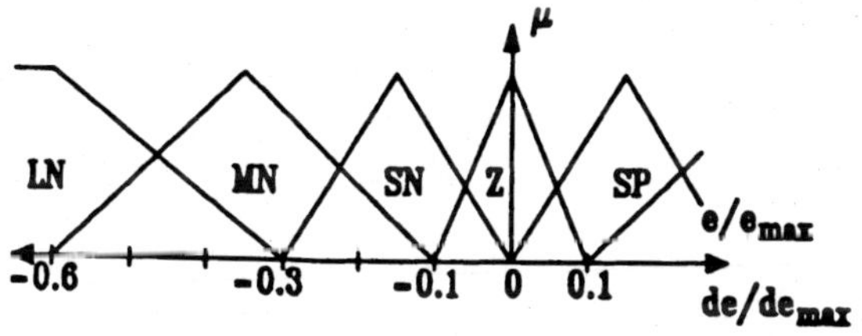

Fig. 4. The distributions of the fuzzy controller input membership functions.

The universe of discourse of the fuzzy controller output is discrete and divided into thirteen regular and unimodal fuzzy subsets, Kosko (1992). The distribution of corresponding fuzzy subsets is nonlinear due to higher subset density towards the zero controller output value. The controller output is calculated according to the centre of gravity principle. In case of using regular and unimodal controller output fuzzy subsets, they can be substituted in the fuzzy rule-table with their centroid, as shown in Tab. 1. During the creation of the fuzzy rule-table, we had to take the torque limits into account in order to keep the tracking error below the maximally allowed tracking error value.

Tab.1. Seven by seven fuzzy rule-table.

	LNE	MNE	SNE	ZE	SPE	MPE	LPE
LNDE	-8.	-4.	-2.	-1.	-0.5	-0.25	0.0
MND	-4.	-2.	-1.	-0.5	-0.25	0.0	0.25
SNDE	-2.	-1.	-0.5	-0.25	0.0	0.25	0.5
ZDE	-1.	-0.5	-0.25	0.0	0.25	0.5	1.0
SPDE	-0.5	-0.25	0.0	0.25	0.5	1.0	2.0
MPD	-0.25	0.0	0.25	0.5	1.0	2.0	4.0
LPDE	0.0	0.25	0.5	1.0	2.0	4.0	8.0

In case of the five-axis manipulator, only one fuzzy rule-table containing forty nine fuzzy control rules has been proposed for all five servo loops, but in this way, multiplexing and proper scaling of fuzzy controller inputs and outputs for each joint is required. The principle scheme of the proposed fuzzy servo control is shown in Fig. 5.

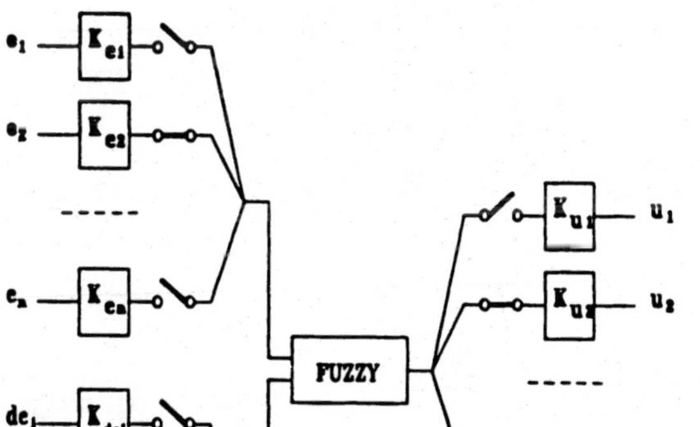

Fig. 5. A principle scheme of a multiple joint fuzzy servo control.

4. SIMULATION EXPERIMENTS

The proposed fuzzy control has been tested in case of a straight move of the robot tool between two points at the given path (in meters): A (0.467, 0, 0.127); B (0.238, 0.114, 0.356). The given path can be transformed into the joint coordinates (Fig. 6). Comparative experiments were also performed with the PD and the robust controller to enable better evaluation of the proposed fuzzy control scheme.

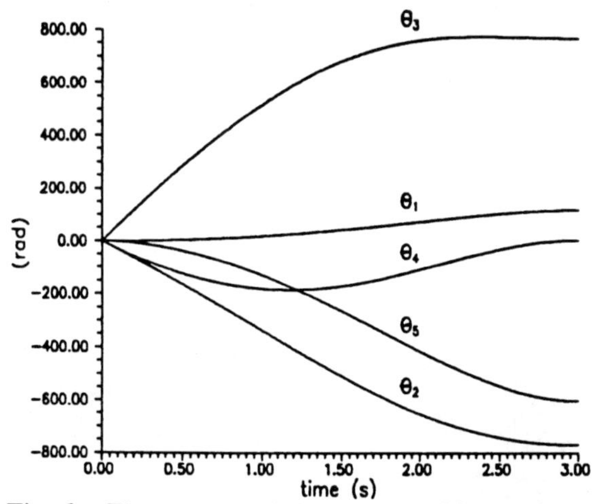

Fig. 6. The commanded motor positions of the joints for the given path.

Figs. 7, 8, 9, 10 and 11 show the responses of the motor position trajectory tracking errors of all five joints, respectively, in case of using three different types of controllers. As may be seen, the best results were obtained with the fuzzy controller which kept the trajectory tracking errors within a 5% of the imposed change in reference. Figs. 12, 13 and 14 show the controller output responses for the first three robot joints. The robust controller has the fastest reaction in the initial part of the trajectory due to the derivative type of the auxiliary controller, while in the final part, all three controller output signals have the similar form. Figs. 15, 16 and 17 show the respective error responses of the Cartesian x, y, z coordinates. It may be noticed that the fuzzy controller performed much better than the PD controller, but it was not better than the robust controller.

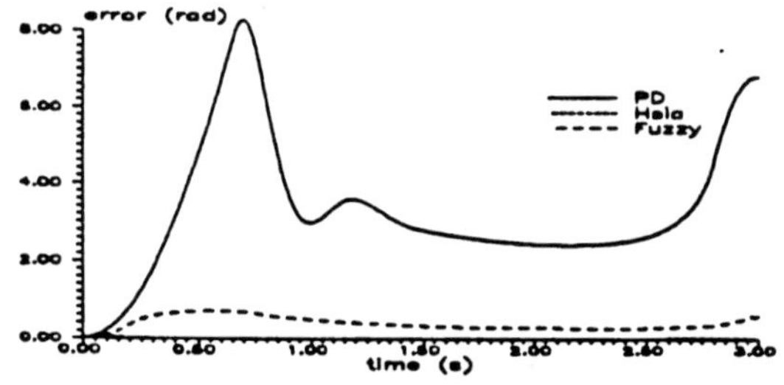

Fig. 7. The trajectory tracking error responses of the motor position of the joint 1 (base).

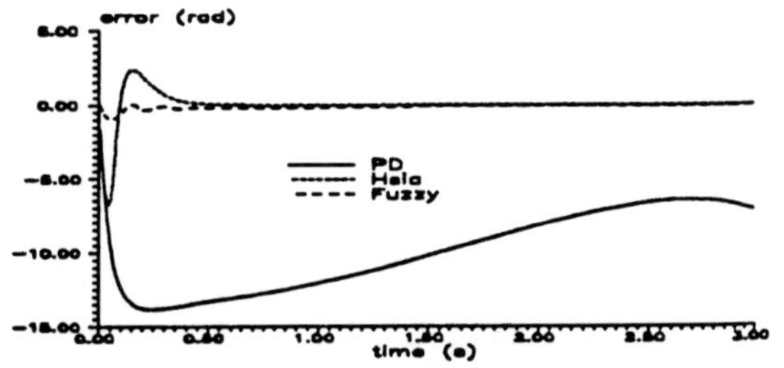

Fig. 8. The trajectory tracking error responses of the motor position of the joint 2 (shoulder).

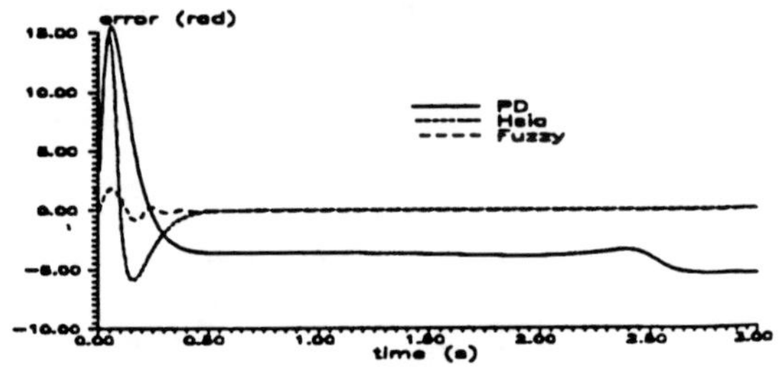

Fig. 9. The trajectory tracking error responses of the motor position of the joint 3 (elbow).

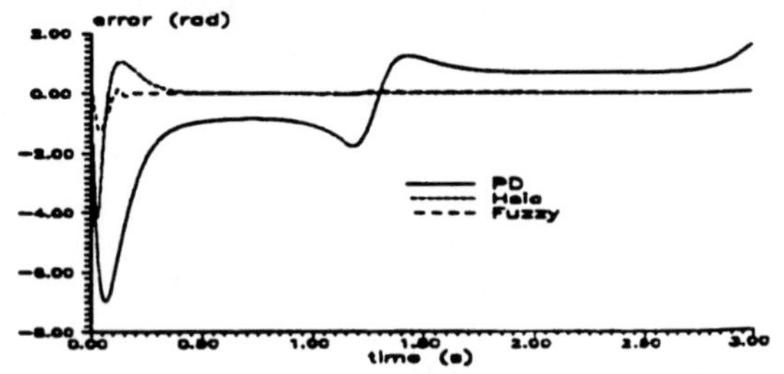

Fig. 10. The trajectory tracking error responses of the motor position of the joint 4 (tool pitch).

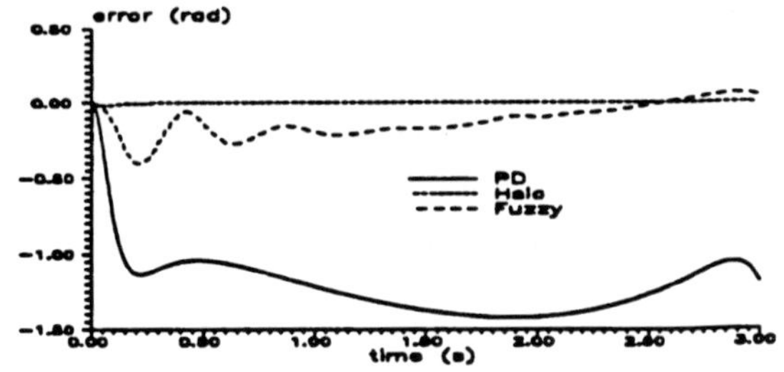

Fig. 11. The trajectory tracking error responses of the motor position of the joint 5 (tool roll).

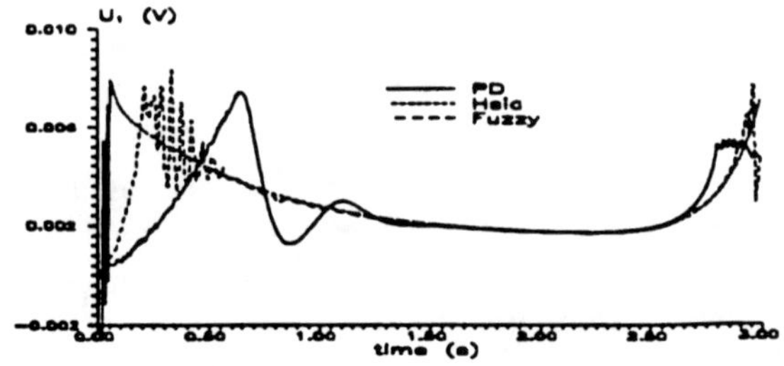

Fig. 12. The controller output responses of the joint 1 (base).

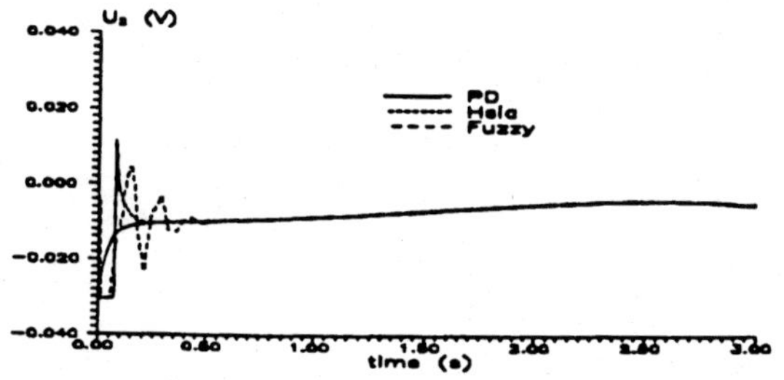

Fig. 13. The controller output responses of the joint 2 (shoulder).

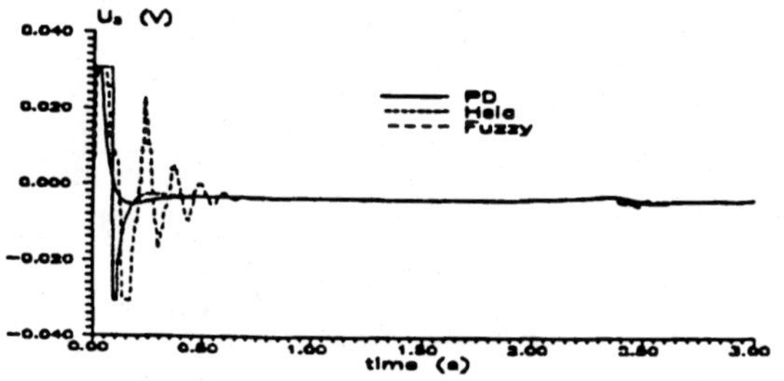

Fig. 14. The controller output responses of the joint 3 (elbow).

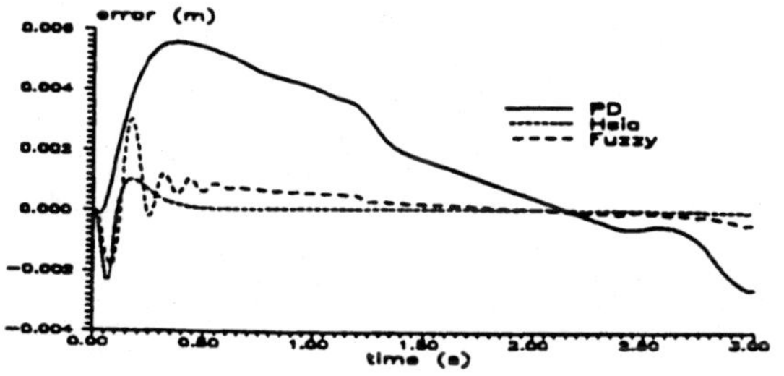

Fig. 15. The tracking error responses of the x coordinate of the tool position.

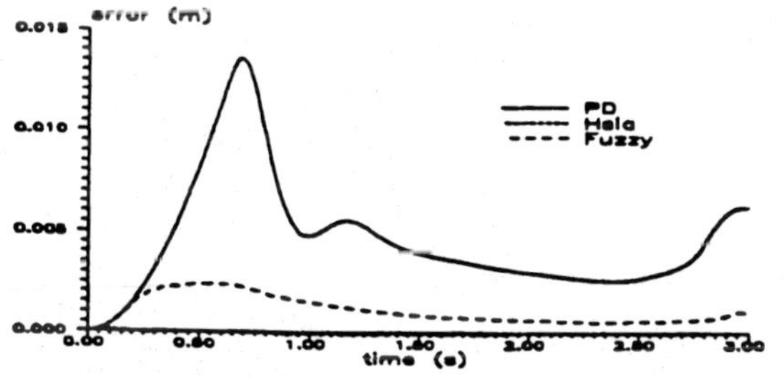

Fig. 16. The tracking error responses of the y coordinate of the tool position.

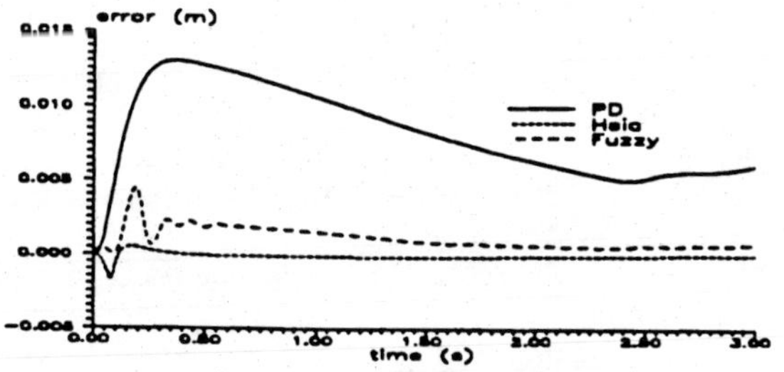

Fig. 17. The tracking error responses of the z coordinate of the tool position.

The performance of the robot was also tested in case of change of the manipulated load and in the presence of white measurement noise.

Change of the manipulated load. The load carried in the gripper was changed from no load to 2 kg. Fig. 18 shows the trajectory tracking motor position error responses of the most affected joint (elbow). It may be seen that the fuzzy controller still kept the tracking error within a 5% of the imposed change in reference, thus showing a very good robustness. Fig. 19 shows the controller output responses which indicate an abrupt form of the robust controller output which imposes higher enforcement to the joint actuators, and a smooth form of the fuzzy controller output, which is suitable for servo application. Fig. 20 shows the error responses of the Cartesian z coordinate.

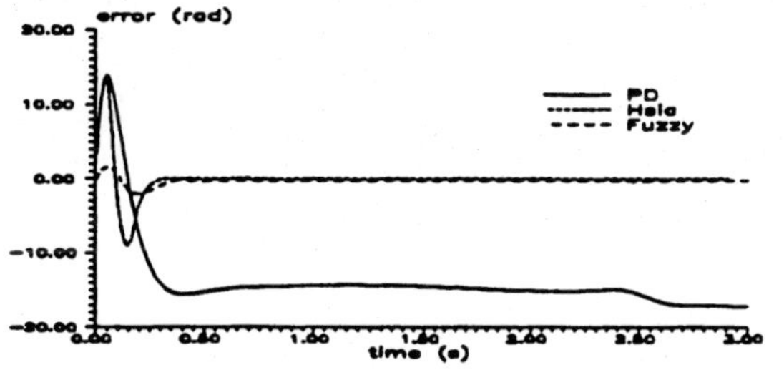

Fig. 18. The error responses of the motor position of the joint 3 in case of a load change.

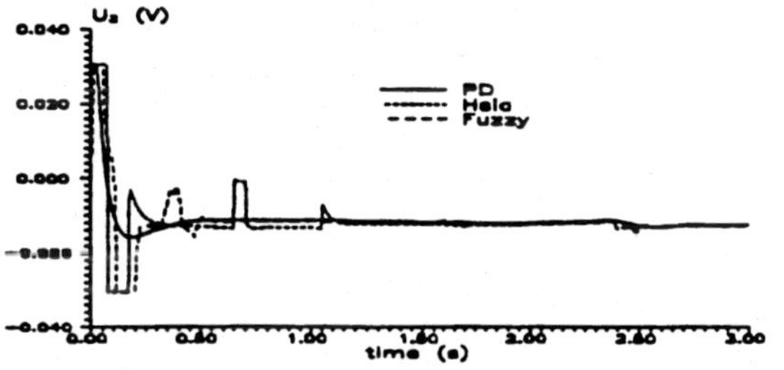

Fig. 19. The controller output responses of the joint 3 in case of a load change.

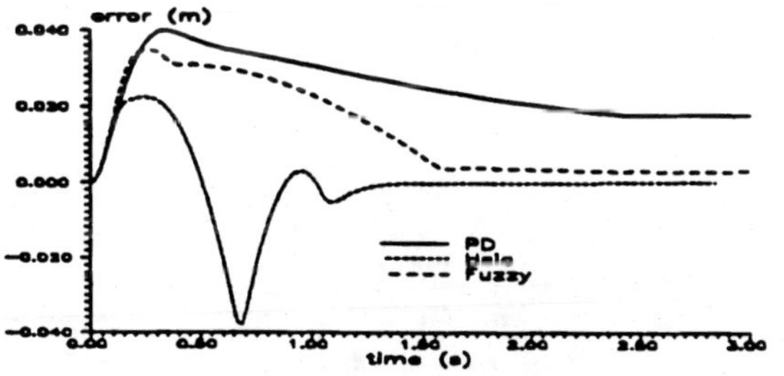

Fig. 20. The tracking error responses of the z coordinate in case of a load change.

Presence of white measurement noise. A white measurement noise with the maximum of 10 rad/s has been applied to each joint servo loop under no load condition. Fig. 21 shows the trajectory tracking motor position error responses of the third joint. The fuzzy controller was not affected by the noise because the tracking error is still kept within a 5% of the imposed change in reference. Fig. 22 shows the controller output responses which indicate high sensitivity of the robust controller to the measurement noise (peaks are noticed), while the fuzzy controller behaves more like a low-pass filter. Fig. 23 shows the error responses of the Cartesian z coordinate.

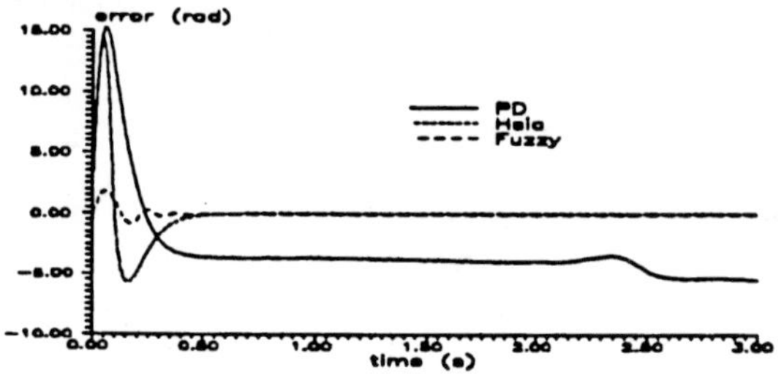

Fig. 21. The error responses of the motor position of the joint 3 in the presence of the measurement noise.

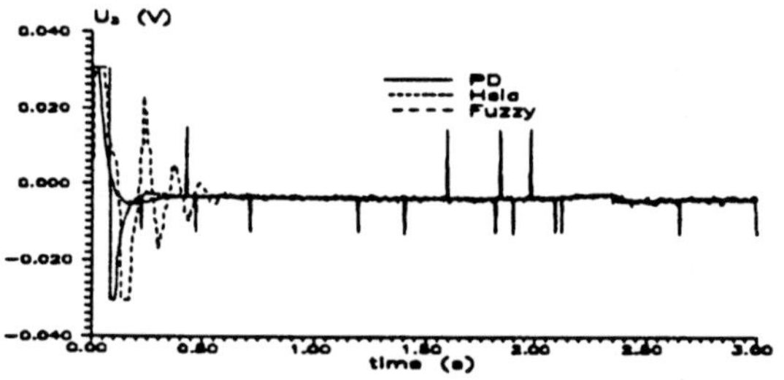

Fig. 22. The controller output responses of the joint 3 in the presence of the measurement noise.

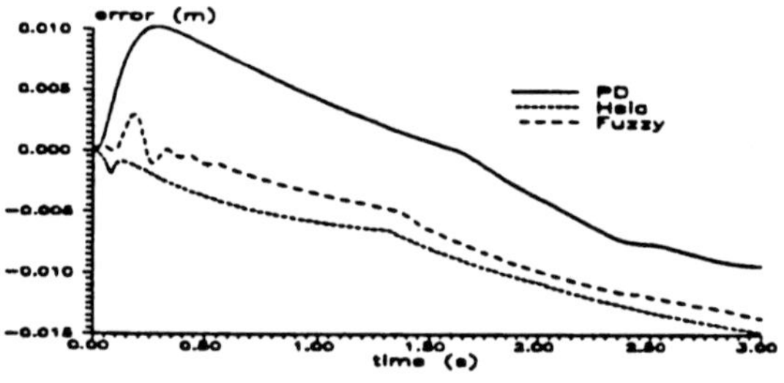

Fig. 23. The tracking error responses of the z coordinate in the presence of the measurement noise.

5. CONCLUSIONS

The paper has presented the fuzzy control scheme which has been applied to the dc motor driven joints of a five-axis robot arm. The fuzzy controller was designed to provide fast and accurate joint position response without overshoot. The proposed fuzzy control algorithm has a nonincremental character. It has been shown that emulated nonlinear behaviour of the fuzzy controller has contributed to very good trajectory tracking with form of the controller output which is very suitable for servo applications. Thanks to the moderate gear ratio values and to lower sensitivity of the fuzzy controller to parameter variations, almost decoupled control of each robot joint was obtained. As a result, creation of only one fuzzy rule-table was sufficient for all five servo loops, but multiplexing and appropriate scaling of the fuzzy controller inputs and outputs for each servo loop was required.

The simulation experiments have also confirmed the effectiveness of the proposed fuzzy control method in case of a load change and in case of a white measurement noise. Comparative experiments with the PD servo controller and the robust controller (Hsia, 1989) have shown the supremacy of the fuzzy controller over the PD controller, while the robust controller performed better in the tool configuration frame thanks to the higher enforcements of the actuator. On the other hand, fuzzy controller performed better in the presence of the measurement noise, thus showing a marked robustness which has in turn provided a very good trajectory tracking.

6. ACKNOWLEDGMENTS

This paper was performed within the framework of the 2-07-280 research project entitled "Adaptive and Optimal Control of Electric Drives" which was supported by a grant from the Ministry of Science of the Republic of Croatia.

7. REFERENCES

Arciniegas, J.I. (1993). Fuzzy Inference Systems and the Control of Flexible Robotic manipulators", In: *Proc. of the 8th Int. Symp. on Intelligent Control*, pp. 250-254, Chicago.

Banerjee, S., and P.Y. Woo (1993). Fuzzy Logic Control of Robot Manipulator. In: *Proc. of the 2nd IEEE Conf. on Control Applications*, Vol. 1, pp. 87-88, Vancouver.

Bogdan S., V. Laci, and Z. Kovačić (1993). Fuzzy Control of a Three Degrees of Freedom Manipulator. In: *Proc. of the 2nd Slovenian Conference ERK'93*, Vol. A, pp. 197-200, Portorož.

Cela, A., and Y.Hamam (1992). A Fuzzy Robust Controller for Trajectory Tracking of Robotic Systems. In: *Proc. of the IEEE/RSJ Int. Conf. on Intelligent Robots and Systems*, Vol. 2, pp. 767-773, Raleigh.

Kelemen, A., M. Imecs, and I. Birou (1993). Fuzzy controller for field-oriented robot drive with induction motor. In: *Proc. of the IEEE Int. Symp. on Industrial Electronics*, pp. 695-701, Budapest.

Kosko, B. (1992). Neural Networks and Fuzzy Systems. Englewood Cliffs, NJ: Prentice Hall, Inc.

Kumbla, K. et al. (1992). Fuzzy Control of Three Links of a Robotic Manipulator. In: *Proc. of the 4th Int. Symp. on Robotics and Manufacturing*, Vol.4, pp. 687-694, Santa Fe.

Laci, V., and Z. Kovačić (1994). Trajectory Planning, Dynamic Modelling and Robust Control of Articulated Robot Arm. In: *Proc. of the IMACS Symp. on Mathematical Modelling*, Vol. 4, pp. 616-617, Vienna.

Lim, C.M., and T.Iliyama (1991). Application of Fuzzy Control to a Manipulator. *IEEE Trans. on Robotics and Automation*, Vol. 7, 5, pp. 688-690.

Nedungadi, A. (1992). A Fuzzy Logic Based Robot Controller. In: *Proc. of the 4th Int. on Robotics and Manufacturing*, Vol. 4, pp. 631-636, Santa Fe.

Scharf, E.M., and N.J. Mandić (1985). The application of a fuzzy controller to the control of a multi-degree-freedom robot arm. In: *Industrial Applications of Fuzzy Control*, (M.Sugeno, Ed.), North-Holland, pp. 41-62.

Tascilo, A. et al. (1993). Intelligent Control of a Robotic Hand with Neural Nets and Fuzzy Sets. In: Proc. of the 8th Int. Symp. on Intelligent Control, pp. 232-237, Chicago.

Tzafestas, S.G. (1991). Adaptive, Robust, and Fuzzy Rule-Based Control of Robotic Manipulators. In: *Intelligent Robotic Systems*, (S.G. Tzafestas, Ed.), Marcel Dekker, Inc., pp. 313-419, 1991.

AN ADVISOR-ENHANCED, SIGNATURE-TABLE CONTROLLER FOR REAL-TIME CONTROL OF ILL-DEFINED SYSTEMS

D. W. RUSSELL

Penn State Great Valley, School of Engineering, Malvern, PA 19355, USA

Abstract. This paper addresses a novel method for control of ill-defined systems that utilize bang-bang actuation. The system described is a hybrid signature-table controller that uses advisor logic in its decision making and statistical inference processes. The automaton reacts to changes in state space with rapid-fire control decisions and uses timeouts or failures to initiate learning interludes. This post-processing algorithm improves upon the values in the control matrix by comparing decision strengths computed from data collected for each state and its neighbors. The familiar inverted pendulum is used to highlight the effects of advisors on overall system merit.

Key Words. Adaptive control; Advisor logic; BOXES methodology; Inference processes; Intelligent machines; Learning systems.

1. INTRODUCTION

The taxonomy of intelligent process controllers takes on increased significance when the process to be controlled is ill-defined, mathematically speaking. It is extremely difficult to implement control using conventional methods because they need a mathematical profile of the process in order to determine adaptive control manoevres. Control systems must navigate the system towards an optimal hyper-surface of solutions. If a system is ill-behaved, the adaptive controller may also thrash between multiple trajectories in its attempt to cause a closure of the system into a solution space. Intelligent control schemas are natural candidates for a solution to this dilemma. Their basic *modus operandi* is to learn appropriate control actions in response to stimuli from the system--using some sort of cause and effect methodology. However, it is often the case that algorithms such as these require many iterations before a solution is reached, and their attachment to the relentless real-world may prove infeasible.

The paper discusses several key issues relevant to this type of control uses results obtained from simulations of an inverted pendulum, a well-known dynamical system, to show the profound impact that advisor logic has on the learning performance of a familiar learning algorithm. The article concludes with some thoughts on how the methodology might be applied to the control of ill-defined systems.

2. MACHINE LEARNING

The mechanics of learning has been the topic of much research and experimentation. The understanding and mimicking of the mind and its processes, such as skill, adaptability, gifting and genius have been considered relevant in the design of computer systems that, apparently, learn to perform tasks of varying difficulty. The learning experience is widely accepted as being highly influenced upon the environment in which it takes place. In the case of real-time, heuristic, systems, the learning processes can be erratic and non-monotonic.

Machine learning can be accomplished by using algorithms that are designed to search out solutions by dividing the overall problem into sets of local sub-problems. For example, the "learning with a critic" paradigm is concerned with tutoring, on a state-by-state basis, under the belief that the overall purpose of "education" is attained by solving and aggregating knowledge gained from individual "tests." Transposing the metaphor back into process control, a real-time direct digital control schema can be envisaged, in which control decisions are read from memory in response to a pointer value, that encodes the current state of the system. After some interrupt condition occurs, the matrix of decision values is updated by a learning algorithm. Canudas de Wit (1988) calls this memory-element a "signature-table," because the data within each of its elements bears the signature of the system under control. Once the learning process is over, the system is allowed to bumplessly restart, marking the beginning of a new set of "adventures" in control space, using its newly aquired skills.

2.1 Signature Table Methodolgy

Figure 1. shows a schematic of a typical signature-table architecture for in-line computer control of a real-time system.

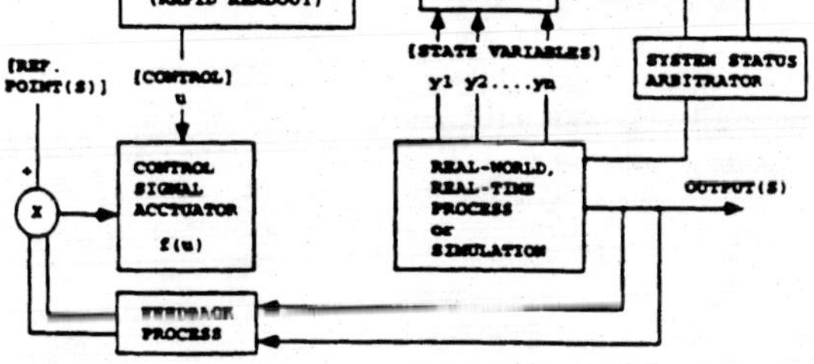

Fig. 1. A Signature-Table Adaptive Controller.

It is apparent that the success of the method depends upon the algorithm:-

- being able to identify the state of the system and encode it into a unique number.
- having a realistic attachment of the decision to the process being controlled.
- possessing a learning algorithm that improves the performance of the system over time.

It is implied that some metrication of the level of success should be incorporated in the learning system and that aged data should be given less credence than newer information.

2.2 The BOXES Learning Taxonomy

The BOXES method (Michie & Chambers, 1968; Russell, 1993) is an example of a signature-table approach to machine learning. In the original work the table contained LEFT/RIGHT decision values that were attached to the direction switch of the motor that drove the trolley that sought to balance a freely hinged pole. This is the familiar inverted pendulum problem. Russell *et al.* (1977) built a real-life trolley and pole system and successfully implemented balancing patterns using the method. The BOXES system outplayed a human operator in this "balancing act"", often controlling the rig for several minutes as compared to (a rather dismal) five to ten seconds by a trained operator. The following sections describe the data processes necessary to implement statistical inferencing in a BOXES control system for the trolley and pole--and may be ommitted by readers who are familiar with its algorithms. The equations are included for reference purposes and are necessary for understanding of the contribution of advisor logic later on in the paper.

System Identification. Each state variable is normalized and assigned an integer number, based the region in which it falls within arbitrary but fixed state boundaries. If any variable falls outside of its limits, the system is said to have "failed." The state numbers for each variable are then combined into a unique overall state number, **m**, using simple combinational arithmetic as follows: If each of **n** variables, $y_1..y_n$, is divided into $z_1..z_n$ legal zones, at any moment, **tm**, there must exist state numbers for every variable. If these integers are $i_1..i_n$, the corresponding value of **m** is computed using (1).

$$m = i_1 + z_1.(i_2-1) + z_1.z_2.(i_3-1)..etc. \quad (1)$$

Data Handling During the Run. At any instant during the run, the system is identified (1) by **m** and **tm**, and the appropriate control value **u** is returned from the m-th tuple of the signature table using (2).

u = sig_table [m] (2)

Statistical data is collected on every entry into any state, **m**; for example, the time of entry, **tm** (3), and it is customary to also increment a local counter (4).

timer_m = timer_m + tm (3)
count_m = count_m + 1 (4)

Performance and Global Data Algorithm. If the system fails at some (real) time, **TF**, or a timeout occurs, the post-processor computes a new system **MERIT** value (5)-(7) by combining the data from all prior runs with the one just ended.

RUN_COUNT = RUN_COUNT * dk +1 (5)
GLOBAL_TIMER = GLOBAL_TIMER * dk + TF (6)
MERIT = GLOBAL_TIMER /RUN_COUNT (7)

The **dk** term is a historical weighting factor that introduces an element of "forgetfulness" which, in agreement with Kubat & Krizakova (1992), is necessary in the learning process. Without this the algorithm over-compensates for early run data.

New Historical State Statistics. The post-processor now updates historical statistics for each state using the data collected during the run. Duplicate sets of values are kept based on the signature table value used during a run. This explains the L(EFT) and R(IGHT) dataset names used below.

For each state number, **m**, *all* current statistics are aged in (8)-(12), where **dk** is defined above.

L_count_m = L_count_m*dk (8)
L_timer_m = L_timer_m*dk (9)
R_count_m = R_count_m*dk (10)
R_timer_m = R_timer_m*dk (11)

The state statistics are now updated using (12)-(15) depending upon the present value in the signature table:

IF u = "LEFT" THEN
L_count_m = L_count_m + count_m (12)
L_timer_m = L_timer_m + count_m*TF-timer_m (13)
IF u = "RIGHT" THEN
R_count_m =R_count_m + count_m (14)
R_timer_m =R_timer_m + count_m*TF- timer_m (15)

In the above algorithm, the counters reflect historical experience and the timers the notion of "time to failure" which translates to an acquired skill level.

The Inference Algorithm. It is now possible to construct a measure of the strength of the "LEFT" and "RIGHT" decision for each state using (16)-(18). In (16) **dla** corresponds to a measure of "desired level of achievement". **C0, C1** and **K** are learning coefficients and **MERIT** is defined in (7) above.

dla = C0 + C1* MERIT (16)
Lsm = (L_timer_m + K.dla)/ (L_count_m + K) (17)
Rsm = (R_timer_m + K.dla)/ (R_count_m + K) (18)

The values in the control matrix are updated according to rules (19) & (20).

IF Lsm > Rsm THEN new_matrix_value_m = LEFT (19)
IF Rsm > Lsm THEN new_matrix_value-m = RIGHT (20)

If **Lsm** and **Rsm** are equal, or if a state has no experience of entry, then the matrix value for that state is to be left unaltered.

Fig. 2 shows a typical learning sequence for the BOXES algorithm working on a simulated trolley and pole system. The x-axis represents number of test runs and the y-axis the system merit.

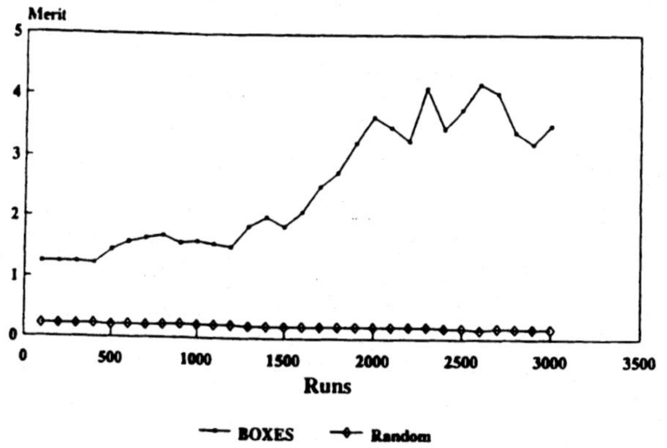

Fig.2. MERIT versus Experience for a Trolley & Pole Controlled by the BOXES Algorithm.

RANK-2 Advisor j+1,k-1	RANK-1 Advisor j+1,k	RANK-2 Advisor j+1,k+1
RANK-1 Advisor j,k-1	Selected State m F(j,k,)	RANK-1 Advisor j,k+1
RANK-2 Advisor j-1,k-1	RANK-1 Advisor j-1,k	RANK-2 Advisor j-1,k+1

Fig.3. Strong and Weak Advisors for a 2-Variable System.

3. ADVISOR STRATEGIES

The paper proposes that the BOXES algorithm can be modified to include elements of advisor logic. The paradigm is designated "SUPER-BOXES" in order to distinguish it from the original.

3.1 Introductory Remarks

If the notion of peer advising is to be considered, each state must maintain advising information in addition to its normal counter and timer data. Care must be taken to select, screen and validate advisor states before including them in the process and also ensure that they can be held accountable statistically for any "advice" given. On a cautionary note, advising can cause a group of strong decision cells to permeate the control matrix, and render the individual (natural) state impotent. In order for this majority rule effect to be nullified, the researcher must be prepared for the appearance of a (rather disturbing) group of untrainable, ambivalent states, lying along the edges of the solution hyper-surface.

3.2 Selection and Ranking of Advisors

As the state variables alter in real-time, so the state number, **m**, varies. In order for peer advising to be possible, the cell must first identify and select potential advisors, using the principle of nearest neighbor classifiers (Castillo & Alvarez, 1991), and then rank them according to positional phyla. It is obvious that there always exists a set of juxtaposed states that vary by +/- one region in any combination of the **n** state quanta. Because a selected advisor state may not be legal, i.e. outside of the allowable maxima or minima, the number of advisors will vary according to the operational domain of the system at any time.

During the algorithmic selection process, which essentially just ranges the component state quanta, it is a simple matter to rank each advisor cell. One scheme is to count how many state boundaries would need to be crossed for the advisor state to become the next operational state. If there are **n** variables, then this ranking would be from **1** to **n**. For each rank, the SUPER-BOXES algorithm assigns weighting factors that are inversely proportional to this positional ranking and computed to include the notion of "voting strength." In this way, the direct connectivity of the advisor to the natural cell affects its voting power. As an illustration, Fig.3 shows the eight possible advisors for a two-variable system, and indicates how strong and weak advisor states occur.

The selection and ranking process is designed to aggressively seek to discourage states from offering advice, by progressively decreasing their ranking weights based on positional rules.

3.3 Advising Schemas.

The SUPER-BOXES algorithm enforces advising at either the operational or statistical level, or both. In the operational schema, the new algorithm utilizes on-line, real-time voting. In the statistical advising mode, the advisor states are off-line contributors to the inference process.

Advising by Run Time Voting. Before a control decision is returned for any state, **m**, the system selects and examines a set of advisor states, and lets each cast a vote, using their matrix values. The vote value is modified by the ranking weights, discussed above. The natural state is included in the tally and can be given preferred voting power. Using a "winner takes all" philosophy, the votes for each control value ("LEFT" or "RIGHT") are counted and the successful control value returned accordingly. In the event of a tie, the value corresponding to the natural vote prevails.

Statistical Advising. The statistical knowledge of each advisor is included in the determination of updated values for the control variable in the post-processing phase of the SUPER-BOXES algorithm.

For any state, **m**, that selects **na** advisor states, **ai**, each possessing advisor weight, **w(ai)**, the post-processor computes modified left and right strength components. Advisor components of inference are computed (21)-(24) using accrued statistical data for the advisor states.

For each state (m)

```
Clear Lcm, Ltm, Rcm, Rtm
  For each legal advisor, ai {i=1..na}
      IF u(ai) = "LEFT" THEN
        Lcm = LCm + w(ai)*L_count_ai                    (21)
        Ltm = Ltm + w(ai)*L_timer_ai                    (22)
      IF u(ai) = "RIGHT" THEN
        Rcm = Rcm + w(ai)*R_count_ai                    (23)
        Rtm = Rtm + w(ai)*R_timer_ai                    (24)
  Next advisor.
```

(17) and (18) are modified to (25) and (26) to yield advisor strengthened values of **Lsm** and **Rsm**.

Lsm=(L_timer_m+Ltm+K.dla)/(L_count_m+Lcm+K) (25)
Rsm=(R_timer_m+Rtm+K.dla)/(R_count_m+Rcm+K) (26)

The new control value is established using (19) and (20) in the normal manner.

4. RESULTS

Fig. 4. shows the effects that simple voting and statistical augmentation have on performance merit compared to results obtained for identical runs using the original BOXES algorithm.

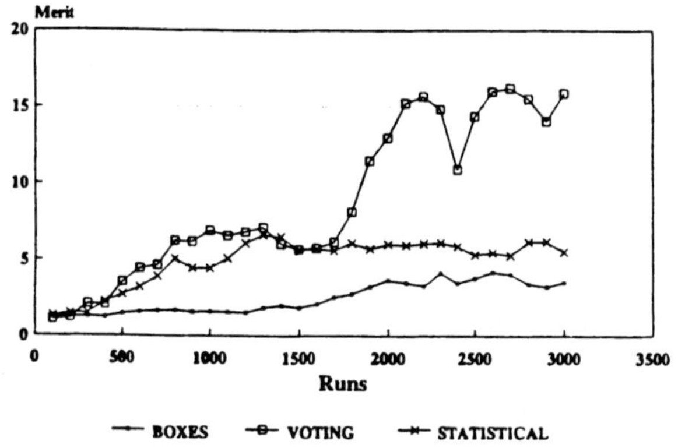

Fig. 4. Merit with Voting and Statistical Augmentation.

The statistical schema seems to promote slower learning, which is not totally unexpected seeing that the initial control matrix is randomly generated. The lack of cogent data will render any statistical advising useless. Fig. 5. shows the same test sequence when the statistical advising procedure is suspended until the system control matrix has some maturity. The performance improvements are startling.

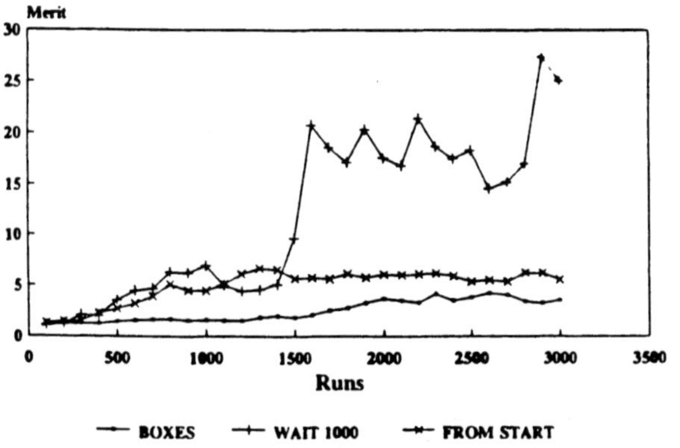

Fig.5. Effect of Mature Statistical Advising on Merit.

5. ADVISOR ACCOUNTABILITY

At the end of a run, the advisor states must be made accountable for their contribution to decisions made inside the run. Because matrix values and cell statistics are fixed during a run, it is only possible to use current run values, advisor overrides and the merit of the system off-line to achieve this result. In (27), Q reflects the system's performance gain at the end of a run and is a good starting place in evaluating advisors.

Q = (new_merit - old_merit) / new_merit (27)

This can now be used, (28) & (29) in conjunction with every state's advisors and their ranking, in modifying historical statistics.

For each state, m

For each advisor, ai:
IF u(ai) = "LEFT" THEN
L_timer_ai= L_timer_ai *F*(1+Q)*w(ai) (28)
IF u(ai) = "RIGHT" THEN
R_timer_ai= R_timer_ai *F*(1+Q)*w(ai) (29)
Next advisor

Next state.

Where the "F" term is an overall scale factor. The statistics now include a "penalty and reward" measure of accountability.

6. CONCLUSIONS

In the pole and cart work by Russell and Rees (1977), it was discovered that the map of LEFT/RIGHT values approximated to the optimal control surface obtained from a Pontriagin analysis (Rozonoer 1959). Over time, the learning and advising process will tend to spawn groups of decisions. This is good and to be expected in secure regions, where strong LEFT/RIGHT decisions should belong. It is highly unlikely that a control surface will exactly coincide with a state boundary, so values on one side of this (unknowable in some cases) line will show predilection to LEFT and values on the other side to RIGHT--and all within the same state. With this in mind, it is possible that the states that lie along these fronts are untrainable. This ambivalence is necessary, and unavoidable, and will lead to groups of states that will constantly be in an indecisive flux. This may be not such a bad thing when one considers how, in real systems, control surfaces drift and require rapid adjustments.

The object of this paper has been to postulate a taxonomy for peer advising that serves to accelerate the process of machine learning in a well-known paradigm. Advisor states are made to affect the decision and inferencing processes and are themselves adjudicated on the efficacy of their actions.

The advising methods, described above, apparently reduce the learning curve in the attainment of acceptable levels of performance. It is conceivable that control matrices for ill-defined processes can be computed rapidly in much the same manner. Russell (1991) proposes an attachment of a BOXES algorithm to a continuous system and some successful control of transients without use of the process' characteristic equations. For safety reasons, in continuous processes, the learning algorithm will achieve fine tuning rather than primary control.

7. REFERENCES

Canudas de Wit, C. A. (1988). *Adaptive Control for Partially Known Systems*. Elsevier Press. New York. Chap. 5.

Castillo, E. & Alvarez, E. (1991) *Expert Systems: Uncertainty and Learning*. Elsevier Applied Science. New York. p.177.

Kubat, M. & Krizakova, I. 1992. "Forgetting and Aging of Knowledge in Concept Formation." *Int. J- Applied Artificial Intelligence*, Vol 6. No. 2. Taylor & Francis, Washington DC. pp.195-206.

Michie, D, and R . A. Chambers (1968). In: *Machine Intelligence 2* -BOXES; an Experiment in Adaptive Control, Oliver & Boyd, London.

Rozonoer, L. (1959). Pontriagin's Maximum Principle in its Application to the Theory of Optimal Systems. *Automation Remote Control*. Vol.20.

Russell, D.W and S. J. Rees. (1975). System Control - a case study of a statistical learning automaton. *Progress in Cybernetics & Systems Research*. Vol 2, Vienna. Hemisphere Publishing Corporation.

Russell, D.W., Rees, S.J & Boyes, J. (1977). A Microsystem for Control by Automata in Real-Life Situations. *Proc. Conference on Information Sciences and Systems*, Johns Hopkins University, Baltimore, pp.226-230.

Russell, D. W. (1991). AI Augmented Process Control using the BOXES Methodology. *Proc. 6th Int Conf. on Applications of AI in Engineering, AIENG'91*, Oxford, July 2-4, 1991, Elsevier Applied Science, London. pp. 611-622.

Russell, D. W. (1993). A Critical Assessment of the BOXES Paradigm. *Int. J- Applied Artificial Intelligence*, Vol 6. No. 2. Taylor & Francis, Washington DC. pp.383-398.

DYNAMIC RESTRUCTURING OF FLEXIBLE PRODUCTION SYSTEMS

JIANHUA GU* and CLARENCE W. DE SILVA†

Industrial Automation Laboratory, Department of Mechanical Engineering, University of British Columbia, Vancouver, Canada

Abstract. The dynamic restructuring of a flexible production system (FPS) is a high-level control technique in computer-integrated systems. It enables the system to change its configuration automatically to adapt to new situation and new demands. In this paper, a knowledge based decision making system is developed, which reconfigures the flexible production system according to the information of its operation and production demand. Fuzzy rules and associated belief values are used in the decision making related to restructuring actions. The paper emphasises on the practical system development as well as methodology study. Finally, some restructuring simulations are given and studied for an automated fish processing plant.

Key Words: Integrated plant control; artificial intelligence; fuzzy set theory; blackboard model; mechanical engineering

1. INTRODUCTION

In dynamic restructuring of production systems, the components of the systems are reorganised depending on the demands for these components under various operating conditions, in order to achieve some form of uniform loading or optimality of operation of the system (de Silva, 1993a, 1993b).

The need for high efficiency in automated flexible production systems (FPS), may call for dynamic restructuring. A system cannot operate efficiently when the overall workloads are not uniformly distributed among the components, or the system resources are not used properly. This situation arises when the status of system operation changes, say, when some components are damaged or partially damaged, or when the production demand of the system changes. As a result, some components could be overloaded or capacity shortages could result in some workcells, while some other components of the same type could be below their full capacity. Under such conditions, the entire system could be restructured, in principle, to improve the system performance, increase efficiency and to decrease operating cost.

Particularly, we consider dedicated multiple-workcell flexible production systems(FPS). A dedicated FPS is different from a general-purpose production system that is usually designed to handle extra capacity. A dedicated FPS, such as a fish processing system, deals only with several special tasks using somewhat similar components, such as vision stations for image processing and, robots for material handling and processing, after augmented by humans for quality supervision and raw material feeding. Usually, these components are reassignable from one workcell to another, and even sharable between two or more workcells. It is these components that determine the processing capability of workcells. In dynamic restructuring, these components may be dynamically rearranged to make the system adapt to a new situation and operate in an optimal manner.

*Graduate Research Associate.

†NSERC Professor of Industrial Automation.

2. RESTRUCTURING SYSTEM

The restructuring system can be viewed as a closed-loop feedback system as shown in Figure 1. On the basis of production demand, the capacity requirements of the workcells are generated by the functional procedure of capacity planning. A knowledge-based decision making system can be used for this purpose. This has been discussed in detail in Gu and de Silva (1994). Capacity requirements are a necessary piece of information for restructuring decision making. Information

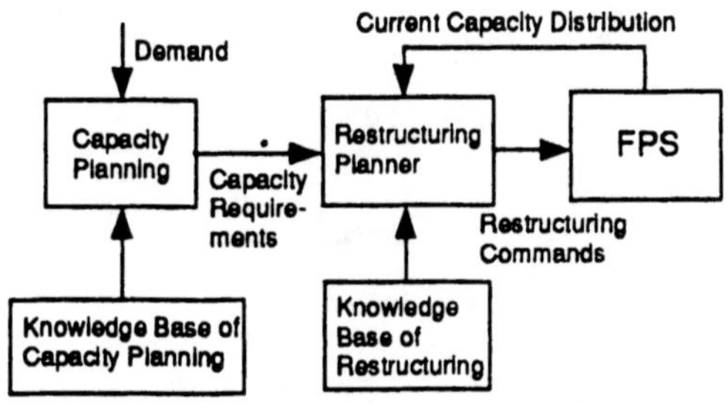

Figure 1: Framework of the Restructuring System

on the current status of the system is available as feedback from the plant. By comparing the capacity requirements and current capacity distribution, the restructuring planner determines the component status and capacity shortages in the workcells. Using these data, it makes decisions on how to restructure the FPS to achieve optimal or at least more uniform operations, and issues restructuring commands. This is a high level control task in a computer integrated system. It can be automated through intelligent sensing of the system operation and performing the restructuring actions automatically through the use of programmable components such as mobile robots and control devices. A knowledge-based decision making system is useful in this restructuring process as well.

2.1 Characteristics of High Level Control

In the design of the dynamic restructuring system, it is important to distinguish between high level control and low level control.

- **Control bandwidth and dynamic response:** (de Silva, 1991, 1993c) The control cycle of a high level restructuring is sufficiently long that the system may be considered to be in a steady state (normal operation) most of the time. The steady state, therefore, is the most important consideration. What we optimize is the FPS operation rather than the dynamic response during the transient stage of performing a restructuring operation.

- **Intelligence and accuracy:** Intelligence is high and precision is low in a high level system. The restructuring system should be designed to make intelligent decision and to deal with uncertainty of data and knowledge. These two aspects will be further discussed and developed in this paper.

2.2 Restructuring Methods

After capacity planning and the feedback of component status information, the central activity of restructuring is a planning procedure with the following conditions:

- An initial world (here the term "world" is used to devote a certain FPS situation) in which overloads may exist;

- A restructuring goal that typically states "There should be no component overloads or capacity shortages in the FPS"; and

- Two types of action which transfer workloads from some components to some other components (of the same type) and shift components from some workcells to other workcells.

The planning problem then is to find a sequence of actions which can transfer an initial world into one in which the desired goal is achieved (Poole, *et al.*, 1993). Every action changes the world to some extent, and eventually, by completing the entire set of actions, the system will reach the final world, in which no component overloads or capacity shortages exist. If such a final world does not exist after all the actions have been tried, then the restructuring planning fails.

2.3 World Representation

Frame is a general way to represent a situation of a flexible production system (FPS). A typical component of a workcell has the following properties: its type (group), its current position, the workcells it is working for, its capacity, its workload status, and its sharability. These properties are put as entities in a frame as shown below:

Object:
group;
position;
work_for;
capacity;
work_load;
sharability.

This frame is represented by using the general property predicate $holds(O, P, V, W)$ which is intended to mean that an object O holds property P at value V in a world W. For example, the situation of component *vision station* 1 in the initial world can be represented as:

$holds(vision1, group, vision, init).$
$holds(vision1, position, cutting, init).$
$holds(vision1, work\_for, [cutting, grading], init).$
$holds(vision1, capacity, 100\%, init).$
$holds(vision1, workload, [60\%, 20\%], init).$
$holds(vision1, sharability, yes, init).$

where capacity value is represented in percentage of that of a standard component of the same type, and the workload is expressed as a percentage of the full capacity of the component. The values of *position* and *work_for* are workcell names.

2.4 Action Representation

We have two kinds of actions *shift_a_component*, and *transfer_workload*:

shift_a_component$(C, from(OP), to(NP))$.
transfer_workload$(from(OC), to(NC), LD)$.

where the first action operation means to shift a component C from an old position OP to a new position NP; and the second one means to transfer a workload LD from an old component OC to a new component NC. The *transfer_workload* also means to share a component if OC is a virtual component whose real capacity is zero, when a capacity shortage exists in a workcell. The NC will be shared between its own workcell and the workcell to which OC belongs.

2.5 Planning Procedure

By considering the complexity of a real flexible production system, and all possibilities of actions (transfer 10% load from a component to another, or 15%, 30% *et al.*), It is unrealistic to apply general planning methods, say STRIPS, to solve the restructuring problem by searching and backtracking. A blackboard based planning method is developed for the case (Gu and de Silva, 1994). The basic idea of the blackboard-based planning is as follows: Put the initial world in a blackboard which is a common data area (Engelmore, *et al.*, 1988); then change the world by performing some actions until the final world is found, in which the desired goal is true. Here, a significant representation change is that object properties are represented without specifying the world: *holds*(*Object*, *Property*, *Value*). This change does not mean we treat an FPS as a static world. Instead, we implicitly assume that if an object property is not changed by an action, then the object will still keep the old property value in the new world.

Then the planning procedure is illustrated below. The planner tries to find an overloaded component: if found, the current world is not the goal world, and some action should be applied to change it; if not found, every component is in its proper status, and the final world is reached. This procedure is designed as a predicate *planning*.

```
planning ⟵
    maximum_overload(C),  %find overload.
    select_action(C, A), !,  %select an action.
    A,                       %execute the action.
    planning.                %continue.

planning ⟵
    maximum_overload(C), !, %still overload.
    report_restructuring_failure.

planning ⟵
    report_restructuring_plan.
```

Here, *maximum_overload*(C) means that the component C has the highest overload in its group. This corresponds to *select_action*(C, A) which is to make a decision to select a best action to resolve this overload. The decision maker considers the largest overload at its highest priority. The quality of a plan depends on the effectiveness of the decision maker. We shall address this problem using heuristics in the next section.

3. DECISION MAKING

Given a component with an overload, the decision maker has to choose a feasible (optimal) action towards resolving this overload. The feasibility should be considered by checking the preconditions of the action. The *select_action*(C, A) operation fails when the decision maker cannot find an action A to satisfy the above requirements. This will lead to a restructuring failure.

3.1 Heuristics

The priority of an action is determined by the restructuring heuristics. The following are three typical heuristics:

- **Sharing within a workcell:** This is an action to transfer workload between two components in the same workcell. This heuristic has the 2nd highest priority (the highest priority is assigned to "release-a-sharing") in the restructuring knowledge base, and is given by:

  ```
  select_action(C, A) ⟵
      select_sharable_component1(C, NC),
      load_transferred(C, NC, LD),
      A =transfer_workload(from(C),
          to(NC), LD).
  ```

  where, the operation *select_sharable_component1*(C, NC) selects a component NC to share the workload from component C (of the same type), and the operation *load_transferred(C,NC,LD)* decides how much load should be transferred between components C and NC.

- **Shifting a component:** This heuristic usually generates an auxiliary action first to make a component load-free by transferring its load to another component of the same type in the same workcell. Next, the freed component is assigned to another workcell to take up the load in an overloaded component of the same type. This heuristic is given by:

  ```
  select_action(C, A) ⟵
      select_shiftable_component1(C, NC),
      pre_action(NC),
  ```

$holds(C, position, NP)$,
$holds(NC, position, OP)$,
$A = shift\_a\_component(NC, from(OP), to(NP))$.

where the operation *select_shiftable_component*(C, NC) selects a component NC of same type as C to be shifted to the workcell which C belongs to, and the operation *pre_action*(NC) makes NC load-free by transferring its workload to other components within its workcell. Finally, this free component NC is released from its present workcell OP and assigned to the workcell NP to which the overloaded component C belongs, so as to share the workload of C.

- **Sharing between workcells:** This heuristic has the lowest priority in the knowledge base, which means it is the worst acceptable action for the restructuring. This heuristic is expressed as:

$select\_action(C, A) \longleftarrow$
$select\_sharable\_component2(C, NC)$,
$load\_transferred(C, NC, LD)$,
$A = transfer\_workload(from(C), to(NC), LD)$.

where, the operation *select_sharable_component2*(C, NC) is similar to *select_sharable_component1*(C, NC), only difference being the former operation looks for NC within the entire system whereas the latter one does it only within its own workcell.

The operations *select_sharable_component* (C, NC), and *select_shiftable_component*(C, NC) are decision making procedures which will be discussed in the following section. Generally speaking, if more restructuring factors are considered, such as the FPS layout, more heuristics should be augmented.

3.2 Selection of Components

When more than one component could be shared for transferring a load, a decision should be made in order to select the best one. The restructuring decision itself could be subjective and should be carried out by weighing the associated costs and benefits. For example, it might be more derivable to maintain a shift overload rather than to achieve a significant under-capacity among components. Fuzzy logic (Zadeh, 1965) may be used to represent component status and certainty factor to represent belief of our knowledge. Also, it may be used to represent the non-crisp and subjective rather of the restructuring decisions. Some of the reasons for using fuzzy logic are as follows.

- Component status (especially manpower) cannot be measured exactly. We do not have enough confidence to make a measurement of the workload as, for example, 78% of the full capacity of a component;
- It may not be feasible to share, for example, the remaining 22% of its full capacity exactly with another component;
- A workload is measured typically by some intelligent system, with associated fuzziness. It is better to represent the component status by a fuzzy descriptor under such circumstances.

As an initial step, provide an extra property for component ***status*** which is derived by considering the workload as a fuzzy descriptor with respect to the capacity of a standard component. Here ***status*** is represented by a fuzzy variable having five levels of resolution (de Silva, 1991b): {great-surplus (*gs*), surplus (*sp*), just-ok (*ok*), overload (*ol*), great-overload (*go*)}, with its membership function as shown in Figure 2.

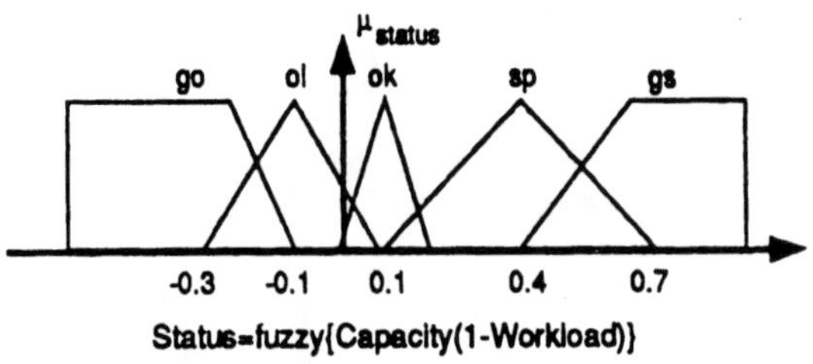

Figure 2: The Membership Function of Component Status

For a standard component, the ***status*** is ***ok*** when it works at around 90% of its full capacity; a capacity surplus of ***sp*** exists at about 60%; a great surplus of ***gs*** exists below about 30%; an overload of ***ol*** exists when operating at about 110% of its full capacity; and a great overload of ***go*** exists when operating above roughly 130% of the full capacity. Now the ***holds*** predicate is $holds(O, status, [go(GO), ol(OL), ok(OK), sp(SP), gs(GS)])$, where GO, OL, OK, SP, and GS are variables of belief corresponding to members *go*, *ol*, *ok*, *sp*, and *gs*, respectively. Then the actions, for example ***transfer_workload***, are designed in fuzzy rules. What should be emphasised is that we associate a belief factor to each rule such that for the given fuzzy context, the associated "crisp" action may be carried out with a belief level of success as given by the belief factor. For example, suppose that OC is a component with overload, and NC is a component of the same type. Then the following rules may be used for making a decision as to whether a load should be transferred between these two components.

IF OC is ol AND NC is gs
THEN do transfer WITH belief=1.0
IF OC is ol AND NC is sp
THEN do transfer WITH belief=0.8
IF OC is ol AND NC is ok
THEN do transfer WITH belief=0.3
IF OC is ol AND NC is ol
THEN do transfer WITH belief=0.0
IF OC is ol AND NC is go
THEN do transfer WITH belief=0.0
... ...
IF OC is go AND NC is gs
THEN do transfer WITH belief=1.0
IF OC is go AND NC is sp
THEN do transfer WITH belief=0.5
... ...

Note that in this ruleset, OC and NC assume fuzzy values of status, but action *transfer* is crisp, with an associated belief value. The priority degree to transfer a load from component OC to component NC is determined not only by the status of the two components, but also by the belief values of our restructuring rules (refer to Figure 3). We may adjust the rule beliefs for different circumstances (for example, for different types of components). Also, belief values of the rules may be trained from known data sample, say using a neural network, to make better decisions.

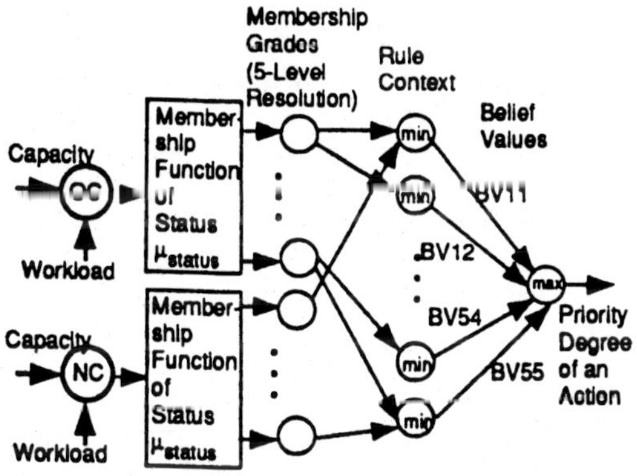

Figure 3: Computation of Priority Degree of an Action

In a decision making procedure, all the rules are fired, and the priority degree of an action is calculated by:

$$PD = \max_{\substack{OC \in [ol, \cdots] \\ NC \in [gs, \cdots]}} min[f(OC), f(NC)] b(R(OC, NC)) \tag{1}$$

where PD = priority degree of this action; $f(OC)$ = membership grade of OC; OC = fuzzy level of the status of a component that is overloaded; $f(NC)$ = membership grade of NC; NC = fuzzy level of the status of a component to be shared; $b(R(OC, NC))$ = belief of the rule of context OC and NC. Note that we use $min\{f(OC), f(NC)\}$ as the fuzzy logic operation for conjunction and max is used for union and combination of individual inference values in composing the priority degree (PD) for the particular action, as shown in Figure 3.

Finally, an action with the highest PD will be chosen. Of course, if the highest PD is less than a certain threshold, say 0.6, the decision making should fail. The same principle can be applied to *shift_a_component* actions, but with different rules.

4. AN EXAMPLE

An example of a simple fish processing system is used to illustrate how planning and decision making procedures work. The system has 2 workcells: a fish cutting workcell, and a fish filleting workcell. Manpower, vision-station and robot are the 3 groups of components which are sharable and shiftable (for simple, assume all standard components). The overall system is illustrated in Figure 4, in which phase-I shows the component workloads under a certain operating condition.

Now suppose that the cutting demand is increased by 50%, and the filleting demand is decreased by 40%. Then, it is clear that the cutting workcell is overloaded, and the filleting workcell operates below its full capacity. This situation is shown in the "transition" phase of Figure 4.

In *robot* group, *robot*1 is short of capacity by about 15%. Its status in fuzzy logic is $[ol(0.75), go(0.3)]$ according to the membership function in Figure 2. Our first heuristic can be used here since *robot*2 is in the same workcell of *robot*1 and has capacity redundancy. *robot*2 works at 60% of its full capacity. Its status is $[sp(1.0)]$. The PD of "transferring workload from *robot*1 to *robot*2" can be calculated by firing all *transfer_workload* rules,

$$\begin{aligned} PD &= max\{min\{.75, 1.0\} * 0.8, \\ &\quad min\{0.3, 1.0\} * 0.5\} = 0.6 \end{aligned}$$

It is not less than 0.6 which we set up as a lowest acceptable value. Thus this action is accepted in a plan, and corresponding blackboards (world) are updated.

In *man* group, *man*1 is heavily overloaded by about 20%. The first heuristic is not applicable in this case since nobody else is working in the same (cutting) workcell. But *man*2 and *man*3 have very light workloads, at 30% and 50% respectively. Here, *man*2 is selected to shift to the cutting workcell. Next time, when *man*1 is found overloaded again (The planning procedure did not transfer workload this time, as it continues, *man*1 is definitely found the second time.), its workload can be transferred to *man*2 by using the first heuristic (because *man*2 is now in the cutting workcell).

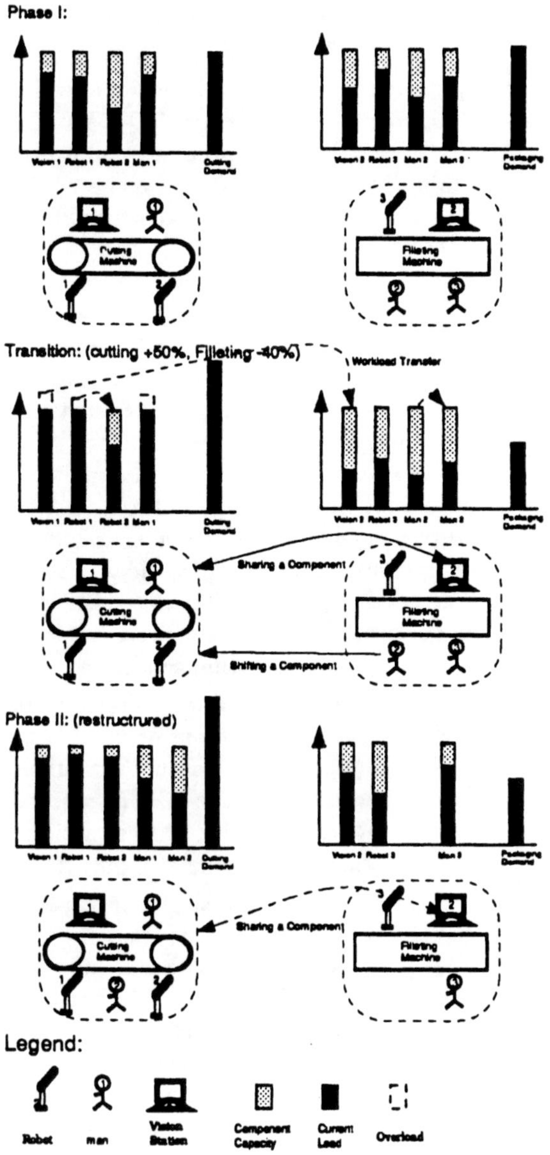

Figure 4: Restructuring of a Sample Fish Processing System

In *vision* group, the cutting workcell needs more image processing capacity to help *vision*1. This time, first and second heuristics both fail. The third one "sharing between workcells" is fired. Here *vision*1 is overloaded by about 20%, and its membership grades are $[ol(0.5), go(0.6)]$; and *vision*2 has a workload of about 36% of its capacity, and its membership grades are $[gs(0.9), sp(0.2)]$. We compute the PD to make sure that we can transfer some load from *vision*1 to *vision*2.

$$\begin{aligned} PD &= max\{min\{0.5, 0.9\} * 1, \\ &\quad min\{0.5, 0.2\} * 0.8, min\{0.6, 0.9\} * 1, \\ &\quad min\{0.6, 0.2\} * 0.5\} = 0.6 \end{aligned}$$

Again the value is not less than the threshold of 0.6.

In the final plan, *robot*2 shares some load of *robot*1; *man*2 is shifted to the cutting workcell with its load transferred to *man*3; and *vision*2 works for both cutting and filleting workcells. Phase-II in Figure 4 illustrates the system conditions after restructuring. The dotted line connecting *vision*2 to cutting workcell indicates that *vision*2 is shared for cutting as well.

ACKNOWLEDGMENT

This work was supported by grants to C.W. de Silva from the *Advanced Systems Institute of B.C.*, and from the *Natural Sciences and Engineering Research Council of Canada* for a fellowship and an NSERC Research Chair in industrial automation.

REFERENCES

de Silva, C.W. (1991a). Fuzzy information and degree of resolution within the context of a control hierarchy. *Proc. IECON'91*, Vol.2, pp. 1590-95.

de Silva, C.W. (1991b). An analytical framework for knowledge-based tuning of servo controllers. *Engg. Appl. AI*, Vol. 4, No. 3, pp. 177-89.

de Silva, C.W. (1993a). Soft automation of industrial processes. *Engg. Appl. AI*, Vol. 6(2), pp. 87-90.

de Silva, C.W. (1993b). Knowledge-based dynamic structuring of process control systems. *Proc. Fifth Intel. Fuzzy Sys. Assoc. World Congress*, Seoul, Korea, Vol.II, pp. 1137-1140.

de Silva, C.W. (1993c). Hierarchical processing of information in fuzzy logic control applications. *Proc. 2nd IEEE Conf. Ctrl Appl.*, Vancouver, B.C., Vol.1, pp. 457-461.

Engelmore, R.S., T. Morgan (eds.) (1988). *Blackboard Systems*. Addison-Wesley.

Gu, J., C.W., de Silva, (1994). A procedure for dynamic sharing of components in multiple-workcell process plants. *Proc. 1994 American Ctrl Conf.*, Baltimore, MD. (in press)

Poole, D., A. Mackworth, R. Goebel (1993). *Computational Intelligence : A Logical Approach*, (draft). Univ. BC.

Zadeh, L.A. (1965). Fuzzy sets. *Inf. Control*, Vol. 8, pp. 338.

STABILITY ANALYSIS OF FUZZY CONTROLLER AND IMPLICATIONS FOR DESIGN

L.JURIŠICA and M.SEDLÁČEK

Slovak Technical University, Faculty of Electrical Engineering, Department of Automation and Control, 3 Ilkovičova St., Bratislava 812 19, Slovak Republic

Abstract. Some techniques for the stability analysis of the control system with fuzzy logic controller are presented. The techniques are based on classical methods and they require the fuzzy controller, satisfying the sector condition. The structure of fuzzy controller, satisfying this requirement is proposed. For this controller, the analytical dependence between the stability margin and controller parameters is developed and the technique for tuning the controller parameters is also developed. The analogy with switching curve is used to tune the remaining degrees of freedom of the controller. The method is documented on a d.c. velocity servosystem.

Key Words. Fuzzy control; stability criterion; parameter tuning; d.c.motor; uncertain system.

1. INTRODUCTION

One of the major problems in the field of fuzzy logic controllers is the lack of general methodology for their design and stability analysis. This is due to their non-parametric nature as linguistic knowledge based controllers. Much effort has been spent on the problem of synthesis of the fuzzy controller. The basic methodologies for their design can be found in (Lee, 1990); in (Albertos, 1992).

Stability analysis of the fuzzy logic controller is a problem, not solved in general until now. Particular solutions include linguistic phase plane (Braae, 1979), energetic criterion (Kiszka 1985), Lyapunov method, circle criterion (Ray, 1984), Popov criterion (Yamashita, 1991) , and others. Most of these solutions suffer either from their enormous complexity or from their intuitive nature.

2. MAIN TOPICS

It is first, in Propositions 1 and 2, shown the possibility of the application of two classical stability analysis methods (the Lyapunov method and the Small Gain Theorem) in a system with fuzzy logic controller. Then, in Definitions 1, 2 and 3, the structure of the fuzzy logic controller is generated, satisfying stability requirements. Finally, in Theorems 1 and 2 it is shown that the system with this controller is stable and the bounds for parameters of the controller are derived in Theorem 4 and Fact 1. Finally, extension of method of Lim (1992) is proposed in Theorem 3, allowing the stability analysis of general nonlinear plant and non-memoryless fuzzy logic controller.

The parallel with a switching curve in a phase plane is outlined and used to synthesize the remaining degrees of freedom. The stability-based design is documented in an example of a velocity d.c. servosystem with changing moment of inertia.

3. FORMULATION OF THE PROBLEM

The paper is dealing with the feedback structure of a linear time invariant (LTI) plant (operator G) and a fuzzy logic feedback controller (Φ). For clarity, it will be further restricted G to a second order proper system, and fuzzy controller to be with input error and derivative of error. Then:

$$\mathbf{dx} / \mathrm{dt} = \mathbf{Ax} + \mathbf{bu} \quad (1)$$
$$\mathbf{y} = \mathbf{c}^T \mathbf{x}$$
$$\mathbf{u} = \Phi(\mathbf{x})$$
$$\mathbf{x} = (x_1, x_2)$$

3.1 Notation of the Fuzzy Controller

There is no widely used notation for the description of functioning of the fuzzy controller. In this paper it is introduced the method, which is a combination of the index method and the parametric function method. These methods have the advantage to be more rigorous, but less close to human reasoning. Let $\mathbf{x} \in X^n$ are the inputs of the fuzzy controller:

$$X^n = [x_{1min}, x_{1max}] \times \ldots \times [x_{nmin}, x_{nmax}] \quad (2)$$

Then the fuzzy controller is described by the operator, in general nonlinear:

$$u = \Phi(\mathbf{x}, t) \qquad (3)$$

where u is an output of the controller. For the internal structure of the fuzzy controller holds the following notation:

Let the j-th fuzzy set of the i-th input in X^n has assigned the index z_i^j. For the index holds:

$$z_i^j \in [p_i, q_i] \qquad (4)$$

where p_i and q_i are the limits. Then the j-th fuzzy set of the i-th input has associated the membership function $\mu_i^j(x_i)$.

Following the same reasoning, let the j-th fuzzy set of the output is denoted by the index z_{n+1}^j.For the index holds: :

$$z_{n+1}^j \in [p_{n+1}, q_{n+1}] \qquad (5)$$

where p_{n+1} and q_{n+1} are the limits.

Then the j-h fuzzy set of the output has associated the membership function $\mu_{n+1}^j(\mathbf{x})$.

The rule base of the fuzzy controller is given by the hypercube of rules Θ, which is assigning to every combination of the indexes of input fuzzy sets the corresponding index of the output fuzzy set. Because of poor accuracy in using Θ as super- and subscripts it will be used the symbol o instead of Θ.

Example of the used notation: Consider the fuzzy controller with two inputs x_1 and x_2 and the output x_3=u. Both inputs and output have the fuzzy sets: NB (-2) - negative big, NM (-1) - negative medium, Z (0) - zero, PM (+1) - positive medium, PB (+2) - positive big. The membership functions are in Fig.1. The rule base is given in Table 1.

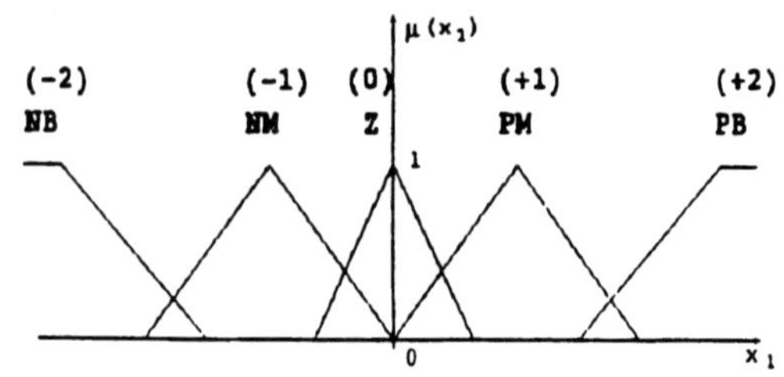

Fig.1. Index notation of membership functions

Table 1 Example of the rule base

→ x_1

	NB	NM	Z	PM	PB
NB	NB	NB	NB	NM	Z
NM	NB	NB	NM	Z	PM
Z	NB	NM	Z	PM	PB
PM	NM	Z	PM	PB	PB
PB	Z	PM	PB	PB	PB

↓ x_2

Using the proposed notation:

$$z_1 = z_2 = z_3 = (-2, -1, 0, +1, +2) \qquad (6)$$

and the limits are:

$$p_1 = p_2 = p_3 = -2 \qquad (7)$$

$$q_1 = q_2 = q_3 = +2 \qquad (8)$$

From the Table 1 follows, that the rule base can be expressed as:

$$z_3(z_1, z_2) = \mathrm{sat}(z_1 + z_2) \qquad (9)$$

where:

$$\mathrm{sat}(i) = \begin{cases} p_3 & \text{for } i < p_3 \\ i & \text{for } p_3 \leq i \leq q_3 \\ q_3 & \text{for } q_3 < i \end{cases} \qquad (10)$$

4. CONTRIBUTIONS

4.1 Stability Theorems for the Fuzzy Controller

Proposition 1 (Stability of 2nd order process):
If $x_2[\Phi(x_1, x_2) - \Phi(x_1, 0)] < 0$, then the second-order system (1) is stable.

Proof: The proof is based on the Lyapunov direct method with quadratic-integral energy function.

$$V(\mathbf{x}, t) = \alpha(x_1^2 + \beta x_2^2) + \gamma \int_0^{x_1} \Phi(\sigma, 0) d\sigma \qquad (11)$$

$\alpha, \beta, \gamma > 0$

From the conditions of Lyapunov stability follows:

$$0 \leq \Phi(x_1, x_2)/x_1 \leq k \qquad (12)$$

$$x_2(\Phi(x_1, x_2) - \Phi(x_1, 0)) \geq 0 \qquad (13)$$

Q.E.D.

Requirements of the Lyapunov method (C^1class) have been now omitted, but are of significance for the fuzzy controller and are satisfied using the further proposed specialised fuzzy controller.

Ray (1984) and others suggested to use the circle criterion to analyse the stability of the system with the fuzzy controller. This conjecture has only intuitive nature, since the original version of the circle criterion was derived for the static nonlinearity and the extension to the dynamic nonlinearity, which the fuzzy controller is, is not straightforward. The circle criterion can be proved using the small gain theorem (Zames, 1966), omitting the requirement to use static nonlinearity. Thus the circle criterion is proved to be suitable for the stability analysis of the system with a fuzzy controller. The circle criterion for the system with fuzzy controller is a subject of Proposition 2.

Proposition 2 (Circle criterion): If

(i) $k_1 \leq \Phi(x)/x_1 \leq k_2$ (14)

(ii) Nyquist curve of G avoids circle $(-1/k_1, -1/k_2)$

(iii) c-shifted loop of G (denoted G_c) is stable

then the system (1) is stable.

Proof: The proof is based on the Small Gain Theorem and the conicity concept (Zames, 1966). The feedback system is considered:

$$y=G(r-\Phi(y)) \quad (15)$$

where G is a linear plant operator, Φ is a nonlinear controller. For the norm limits holds:

$$\|G(e_1)\| \gamma_G \|e_1\| \quad (16)$$

$$\|\Phi(.)\| \leq \gamma_\Phi \|y\| \quad (17)$$

where $\|.\|$ is norm, γ_G and γ_Φ are induced norms. Then the transformation to the equivalent c-shifted loop is done.

Then $y=G(r-\Phi(y)-cy+cy)= G^*(r-\alpha)$ (18)

$G^*=G/(1+cG)$

Consideration that the gain is to be limited

$$\|\alpha(.)\| / \|y\| \leq \delta \quad (19)$$

is leading to the condition i) of Proposition 2.

From the Small gain theorem follows the requirement

$$\|G^*\| \leq 1/\delta \quad (20)$$

Considering $G(jw)=x+jy$ follows

$$[x+1/2(1/\alpha+1/\beta)]^2 + y^2 \geq [1/2\,(1/\alpha-1/\beta)]^2 \quad (21)$$

which is a circle criterion.

Q.E.D.

4.2 Specialised Fuzzy Controller

Let us consider the notation of membership functions as in Fig. 2. Following a similar reasoning to Tang (1989) we have defined the so called specialised fuzzy controller (SFLC).

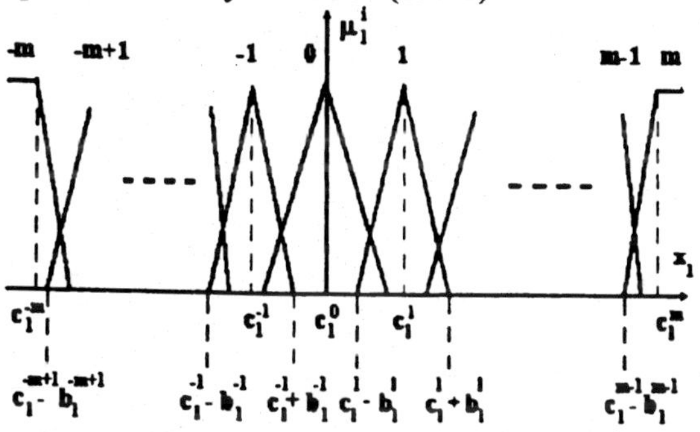

Fig. 2. Notation of the input membership functions

First, regular membership functions are introduced.

Definition 1: (Regular membership functions)
Regular membership functions are triangular membership functions and for the FLC holds:

i) $\mathbf{x} = (e, de / dt)$

ii) $q_1 = q_2 = -p_1 = -p_2 = m;\ p_3 = 2p_1 = -q_3;$

iii) $c_j^i \leq c_j^{i+1} - b_j^{i+1} < c_j^i + b_j^i \leq c_j^{i+1}$ (22)
for $j = 1,2;\ i \in [p_j, q_j]$

iv) $c_i^{r_j} = x_{jmin};\ c_i^{q_j} = x_{jmax};$ for $j = 1,2$

v) $c_j^i = -c_j^i \geq 0$ (23)
$b_j^i = -b_j^i \geq 0$ for $j = 1,2;\ i \in [0, q_j]$

Definition 2: (Specialised knowledge base)
Specialised knowledge base for the fuzzy controller with regular membership functions is defined as:

- $o(i, j) = 0$ for $i \geq 0$ $j \leq -i$
- $o(i, j) = 0$ for $i \leq 0$ $j \geq -i$
- $o(i, j) = o(i, j+1) - 1$ for $i > 0$ $j \geq -i$
- $o(i, j) = o(i, j-1) + 1$ for $i < 0$ $j \leq -i$

where o(i,j) is an index method for the notation of the rule base.

An example of the Specialised knowledge base is the structure depicted in Fig. 3.

x_2 \ x_1	-m	-m+1	:	-2	-1	0	+1	+2	:	m-1	m
-m	-2m	-2m+1	:	-m+2	-m+1	0	0	0	:	0	0
-m+1	-2m+1	-2m+2	:	-m+1	-m	0	0	0	:	0	+1
	..	..	:	..	..	..	..	..	:	..	..
-2	-m-2	-m-1	:	-4	-3	0	0	0	:	m-3	m-2
-1	-m-1	-m	:	-3	-2	0	0	+1	:	m-2	m-1
0	-m	-m+1	:	-2	-1	0	+1	+2	:	m-1	m
+1	-m+1	-m+2	:	-1	0	0	+2	+3	:	m	m+1
+2	-m+2	-m+3	:	0	0	0	+3	+4	:	m+1	m+2
	..	..	:	..	..	..	..	..	:	..	..
m-1	-1	0	:	0	0	0	m	m+1	:	2m-2	2m-1
+m	0	0	:	0	0	0	m+1	m+2	:	2m-1	2m

Fig. 3. Specialised knowledge base

Definition 3 (SFLC): The specialised fuzzy logic controller (SFLC) is with:
(i) Specialised knowledge base
(ii) t-conorm is a product operation

Lemma 1:(Sector condition)
Let the $u = \Phi(x)$ is a specialised FLC. Then:

i) $0 \leq \Phi(x_1, x_2)/x_1 \leq k$ (24)
for $\mathbf{x} \in X^2$, $k>0$

ii) $\Phi(0, x_2) = 0$ (25)

Proof: The proof is based on Definitions 1, 2, 3 and systems of inequalities. The proof is restricted to the interval: $0 \leq x_1 \leq x_{1max}$
because of $sgn(\Phi(x_1, x_2)) = sgn(x_1)$
First, it can be shown, that
$\Phi(0, x_2) = 0$
Further, it can be proved that
$\Phi(x_1, x_2) \geq 0$
And finally, it can be proved that
$\Phi(x_1, x_2^*) \geq \Phi(x_1, x_2)$ for $x_2^* = x_{2max}$
It all implies the condition:
$0 \leq \Phi(x_1, x_2)/x_1 \leq k$
Q.E.D.

Fact 1 (Upper stability margin):
The bounds for SFLC are $k_1 = 0$; $k_2 = c_3^{o(m,m)}/c_1^1$

Proof: The proof is based on Definitions 1, 2, 3 and systems of inequalities. From the Lemma 1 follows:
$\Phi(x_1, x_{2max}) \geq \Phi(x_1, x_2)$
It is searched for $k \geq 0$ such that:

$0 \le \Phi(x_1, x_2)/ x_1 \le \Phi(x_1, x_{2max})/ x_1 \le k$

Analysing two possible situations:

i) $0 \le x_1 \le c_1^1$

ii) $c_1^1 < x_1 \le c_1^m$

and considering regular membership functions, the worst case is approached. It is leading to the result:

$0 \le \Phi(x_1, x_2)/x_1 \le \Phi(x_1, x_{2max})/x_1 < c_3^{o(m,m)}/ c_1^1$ (26)

Q.E.D.

In Theorem 1 it will be shown the suitability of the SFLC for stability analysis using the Proposition 1.

Theorem 1: If G is a second order system and $u= \Phi(\mathbf{x})$ is the SFLC, then the system (1) is stable.

Proof: It is shown that G and Φ satisfy conditions of Proposition 1. The sector condition was proved in Lemma 1. Remains to prove the second condition of Proposition 1. From Definition 3 follows:

i) for $x_1 \ge 0$ and $x_2 \ge 0$ is $\Phi(x_1,x_2) \ge \Phi(x_1,0)$

ii) for $x_1 \ge 0$ and $x_2 \le 0$ is $\Phi(x_1,x_2) \le \Phi x_1,0)$

iii) for $x_1 \le 0$ and $x_2 \ge 0$ is $\Phi(x_1,x_2) \ge \Phi(x_1,0)$

iv) for $x_1 \le 0$ and $x_2 \le 0$ is $\Phi(x_1,x_2) \le \Phi(x_1,0)$

follows $x_2 (\Phi(x_1,x_2)-\Phi(x_1,0)) \ge 0$

Q.E.D.

In Theorem 2 it will be shown the suitability of the SFLC for stability analysis using the Proposition 2.

Theorem 2: If G satisfies conditions of Proposition 2 and $u= \Phi(\mathbf{x})$ is the SFLC, then system (1) is stable.

Proof: In Lemma 1 it is shown that the SFLC satisfies the condition (i) of Proposition 2 and thus the overall system is stable.

The stability analysis of the system with nonlinear process and the FLC is rare in literature. As an extension of method of Lim (1992) it is developed and presented in Theorem 3 the suitable criterion.

Theorem 3 (Nonlinear plant and FLC): Denote

Plant: $d\,\mathbf{x}/dt = F(\mathbf{x})\,\mathbf{x} + g(\mathbf{x})\,u$ (27)

$y = H(\mathbf{x})\,\mathbf{x}$

Controller: $u = -\Phi(y,\ dy/dt)$ (28)

$0 \le \Phi(y, dy/dt) / y \le k$

$-\alpha \le dy/dt \le \alpha$ (29)

$\Phi(y, dy/dt) \ge \Phi(y, -\alpha)$ for $y \ge 0$ (30)

$\Phi(y, dy/dt) \le \Phi(y, \alpha)$ for $y<0$

Let P and Q are positive definite matrices such that $R(x) = -(F^T(x)P+PF(x)+Q)$ is positive definite. If exists $\beta>0$ such that

$(H-g^TP^T)Q^{-1}(H-g^TP^T)^T < 1/k$ (31)

and

$-\beta^2/(4k)\ d/dx\ [Hx]\ FF^T d/dx\ [Hx]^T - 1/k\ HH^T + 2/k\ HPg - 1/k\ (Pg)^TPg > 0$ (32)

Then the system is globally asymptotically stable.

Proof: Proof is based on Lyapunov direct method with energy function:

$$V(\mathbf{x}) = \mathbf{x}^T P \mathbf{x} + \beta \int_0^y \Phi(\sigma,-\alpha\ \mathrm{sgn}(\sigma))\, d\sigma \quad (33)$$

Then: $dV/dt = - \mathbf{x}^T R \mathbf{x} - \mathbf{x}^T Q \mathbf{x} + 2u\mathbf{x}^T(Pg-H^T) - \beta\Psi\ \delta/\delta\mathbf{x}[H\mathbf{x}](F\mathbf{x}+gu) + u(2y+u/k) - u^2/k$ (34)

where

$\Psi = -\Phi(y,-\alpha\ \mathrm{sgn}(y))$ (35)

Defining

$\Omega = [\mathbf{x}^T, u, \Psi]\ \Theta\ [\mathbf{x}^T, u, \Psi]^T$ (36)

$dV/dt = - \mathbf{x}^T R \mathbf{x} - [\mathbf{x}^T, u, \Psi]\ \Theta\ [\mathbf{x}^T, u, \Psi]^T + u(2y+u/k) + \Psi^2/k$ (37)

which gives the conditions of the theorem.

Q.E.D.

5. DESIGN PROCEDURE AND APPLICATIONS

There is no general method for the design of FLC today. It is necessary to answer following questions:

1) What are the inputs of FLC
2) What is the output of FLC
3) What rules are in the knowledge base
4) How many input membership functions
5) What are the parameters of them
6) How many output membership functions and what parameters do they have
7) The shape of membership functions
8) The type of fuzzy operators
9) The choice of defuzzyfication strategy

In the proposed design technique the preliminary answers are following. The details are afterwards.

ad 1) It is considered the most widely used choice of e and de/dt as the inputs of the FLC. The reason comes from the intuitive design of the rule base in the phase plane.

ad 2) The choice of output influences the type of FLC-either PI or PD. With no loose of generality the case $\Phi(\mathbf{x})=u$ is considered..

ad 3) The choice of rule base is a crucial element of the design. In the specialised FLC the rule base is strictly limited to the structure of Definition 2.

ad 4) In the following it will be shown, that the input membership functions form the area u=0 in the phase plane. According to fuzzy VSS this area will be called switching curve. Then the number of input membership functions determines the accuracy of approximation of the switching curve.

ad 5) According to the previous point the parameters of input membership functions determine the shape and slope of the switching curve. The analytical dependence between parameters and the curve will be shown. Thus from the desired closed loop response the parameters of input membership functions can be directly found.

ad 6) The number of output membership functions is determined by the Definition 2 of the specialised fuzzy controller. In Lemma 1 it was shown that for the SFLC holds the sector condition

$0 \le \Phi(x_1, x_2)/x_1 \le k$

and further from the Fact 1 follows the slope

$0 \le \Phi(x_1, x_2)/ x_1 \le c_3^{o(m,m)}/c_1^{1}$

The biggest slope of the sector, satisfying the circle criterion, is a stability margin. For the SFLC the one parameter goes to minus infinity and the circle degenerates to the line, perpendicularly crossing the real axis. The biggest slope is:

$$k_{max} < -1/\inf[\mathrm{Re}\{G(jw)\}] \qquad (38)$$

This value can be directly used for tuning the output membership functions. From the known c_1^{1} and the Fact one has $c_3^{o(m,m)}$:

$$c_3^{o(m,m)} \le k_{max}\, c_1^{1} \qquad (39)$$

and the rest of parameters comes from Definition 1:

$$c_3^{i} = c_3^{-i} \qquad (40)$$

$$c_3^{i} < c_3^{o(m,m)}$$

ad 7) Regular membership function are considered.

ad 8) The operator of logical AND has significant influence on the control performance. From Definition 3 an algebraic product is used.

ad 9) The method COG (Centre of Gravity) is considered, because of simplicity in the case of fuzzy singletons as output membership functions.

What remains is the analytical dependence between the switching curve and the parameters of input membership functions. It is a subject of Theorem 4.

Theorem 4: (Switching curve for the SFLC)
The area $\xi = \{e, de/dt\ ;\ u=\Phi(\mathbf{x})=0\}$ at the SFLC is bounded by a staircase function:

For $i \ge 0$: $\quad 0 \le e \le c_1^{i+1} - b_1^{i+1} \qquad (41)$

and $\quad (de/dt)_{min} \le de/dt \le c_2^{i+1} - b_2^{i+1}$

Proof:
We are looking for $u=\Phi(\mathbf{x})=0$. Since $c_3^{0} = 0$, we search z_1^{i} and z_2^{j} that have corresponding $z_3^{k}=0$. From the Definition 2 follows that it holds only for $i=j$. For the rest of them is $z_3^{k} \ne 0$, i.e. $u=\Phi(\mathbf{x}) \ne 0$. This is leading to the inequalities:

for $i \ge 0$: $\quad c_1^{i-1} + b_1^{i-1} \le e \le c_1^{i+1} - b_1^{i+1}$

and $\quad c_2^{i-1} + b_2^{i-1} \le de/dt \le c_2^{i+1} - b_2^{i+1}$

Further holds

$-\ o(i, j) = 0 \qquad$ pro $i \ge 0;\quad j \le -i$

$-\ o(i, j) = 0 \qquad$ pro $i \le 0;\quad j \ge -i$

From this follows, that $z_3^{k}=0$ also for $i \ne j$, i.e. for $i>0$ in interval $i \in [0,j]$ or $j \in [-m, i]$. This is leading to the restrictions:

for $i \ge 0$: $\quad 0 \le e \le c_1^{i+1} - b_1^{i+1}$

and $\quad (de/dt)_{min} \le de/dt \le c_2^{i+1} - b_2^{i+1}$

Q.E.D.

The final design algorithm is as follows:

1. Finding the maximum slope for the sector condition, given by the circle criterion.
2. Choosing FLC, satisfying the sector condition. According to Lemma 1 the SFLC is suitable.
3. Selection of the switching curve in a phase plane.
4. Selection of number of membership functions, according to desired accuracy of approximation of the switching curve and real-time requirements.
5. Selection of the parameters of the input membership functions, which approximate the given switching curve. The selection is based on the dependence, given analytically in Theorem 4.
6. Choosing the parameters of output membership functions, satisfying stability requirements.

6. EXAMPLE

The technique is documented step by step in an example of the d.c. velocity servosystem. The d.c. motor Servalco 300 W is considered with the current loop as in Jurisica (1992).The overall process has additive signal disturbance-load torque M_Z and the parameter uncertainty-changing moment of inertia J in a range 1:10. Considering the multiplicative model of uncertainty, the control system is depicted in Fig.4.

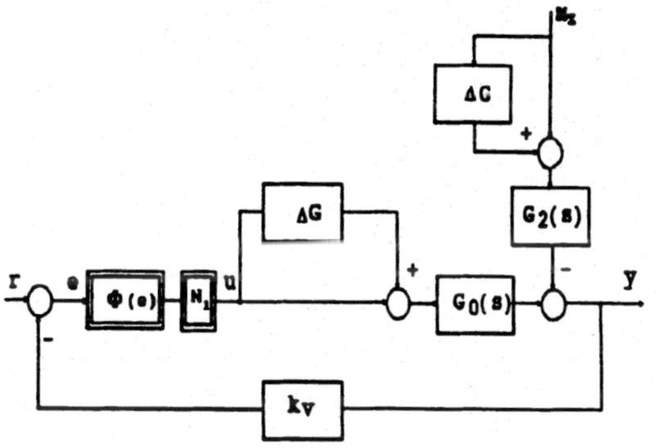

Fig.4. Control system

$$\text{where } G_0(s) = \frac{C_u}{10JK_T s(1+2T_e s)} \qquad (42)$$

is a nominal process and

$\Delta G=[0, 9]$

is the uncertainty. Further:

$$G_1(s)=G_0(s)\,(1+\Delta G) \qquad (43)$$

$$G_2(s)= \omega(s)/M_Z(s) = \frac{-1}{10Js}$$

The design is done using the proposed methodology.

1) Finding the stability limit: According to the circle criterion the value $\lambda = \inf\{\mathrm{Re}\ [G_1(s)]\}$ is searched. It gives $\lambda = -0.0367$; $k_{max} = 1/\lambda = 27.22$

$$0 \le \Phi(x_1, x_2)/x_1 \le 27.22 \qquad (44)$$

2) The SFLC is chosen.

3) Selection of the switching curve: The selection is done according to the general methodology for the time-optimal control from the literature.

4) Selection of number of membership functions: As a compromise between the demand for accurate approximation of the switching curve and the real-time requirements, the number m=5 was chosen.

5) Selection of input membership function parameters: The parameters of input membership functions are in Table 2, chosen according to Theorem 4 to approximate the chosen switching curve.

Table 2 Input membership function parameters

i	c_1^i	b_1^i	c_2^i	b_2^i
0	0	0.5	0	500
1	1.5	1.5	1 500	1 500
2	3.5	1.5	3 000	1 000
3	6	1.5	4 000	500
4	7.5	1	5 000	750
5	10	2	6 000	500

The area ξ:{e, de; $\Phi(\mathbf{x})=0$} for the given parameters is depicted in Fig.5.

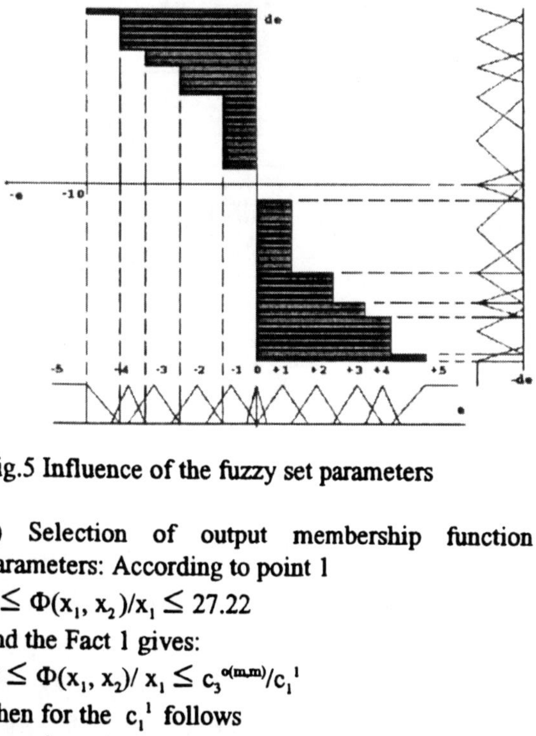

Fig.5 Influence of the fuzzy set parameters

6) Selection of output membership function parameters: According to point 1

$0 \leq \Phi(x_1, x_2)/x_1 \leq 27.22$

and the Fact 1 gives:

$0 \leq \Phi(x_1, x_2)/\, x_1 \leq c_3^{\alpha(m,m)}/c_1^1$

Then for the c_1^1 follows

$40.83 \geq c_3^{\alpha(m,m)}$

and the output membership function parameters are in Table 3.

Table 3: Output membership function parameters

i	0	1	2	3	4	5	6	7	8	9	10
c_3^i	0	15	17	19	20	22	24	28	30	35	40

Thus the design is completed. Results in simulations shown very good performance of the system.

7. SUMMARY

Considering two classical approaches - the Lyapunov method and the Small Gain Theorem - two propositions are derived, resulting in the structure of the so-called specialised fuzzy logic controller. This is a fuzzy controller which stabilises given LTI plant and generates bounds for tuning of the controller parameters. Using the proposed design procedure, one can guarantee stability of the system and enormously decrease tuning effort and time. Method for the nonlinear plant with nonmemoryless fuzzy controller is finally presented. Results are documented by a design of velocity d.c. servosystem

Acknowledgement: The research of M.Sedláček was supported by the Daimler-Benz Foundation, Germany. The support is greatly acknowledged.

8. REFERENCES

Albertos, P. (1992): Fuzzy neural control. In: IFAC Symp. *Low Cost Automation*, Wien, p.143-156.

Braae, M. and D.A.Rutherford (1979): Theoretical and linguistic aspects of the fuzzy controllers. *Automatica*, **15**, pp.553-577.

Jurisica L. and M.Sedlacek (1992): Implementation of fuzzy controlled d.c. servosystem. In: IFAC Workshop on *Intelligent Motion Control, Perugia*, pp. II-107-112.

Lee Ch.Ch. (1990): Fuzzy logic in control systems - I,II. *IEEE Transactions on Systems, Man and Cybernetics*, **20**, No.2, pp.404-435.

Lim J.T. (1992): Absolute stability of class of nonlinear plants with fuzzy logic controllers. *Electronic Letters*, **28**, No.21, pp.1968-1970.

Kiszka J.B, Gupta M.M. and P.N.Nikiforuk (1985): Energetistic stability of fuzzy dynamic systems. *IEEE Transactions on Systems, Man and Cybernetics*, **5**, No.6, pp.783-792.

Ray K.S. (1984): L_2- stability and the related design concept for SISO linear system with fuzzy controller. *IEEE Transactions on Systems, Man and Cybernetics*, **14**, No.6, pp.932-939.

Tang K.L. and R.J.Mulholland (1987): Comparing fuzzy logic with classical controller design. *IEEE Transactions on Systems, Man and Cybernetics*, **17**., NO.6, pp. 1085-1087.

Yamashita Y. and T.Hori (1991) : Stability analysis of fuzzy control system. Proceedings *IECON'91*, Kobe, October, Vol.2, pp.1579-1584.

Zames G. (1966): On the input-output stability of time-varying nonlinear feedback system. Part I and II. *IEEE Transactions on Automatic Control*, **11**, pp.228-238 and pp. 465-476.

AN ESTIMATOR OF THE NUMBER OF RULES IN FUZZY MODELS

L. SANCHEZ, J.TUYA, H. LOPEZ, J.A. CORRALES

E.T.S. Ingenieros Industriales de Gijón. Viesques, Gijón. Asturias. Spain

Abstract. The proposed method builds automatically a fuzzy rules based description of an arbitary process. All the parameters defining such description can be estimated, including the number of fuzzy rules needed to build a *natural* model of the aforementioned system. This method assumes that the input-output relation of the system being modelled is known in a number of situations. This knowledge can be imprecise up to a certain degree. It won't be needed a stochastical characterisation of that imprecision.

Key Words. Learning Systems, Fuzzy Systems, Fuzzy Control, Entropy, Minimum Description Length Theory, Systems Complexity

1. INTRODUCTION

The description of Complex Sistems based on fuzzy sets and rules was proposed by Zadeh (1973). The aim of this concept is to make use of the *tolerance to the imprecision* in the description of a process, with the purpose of reducing the complexity of its model.

Fuzzy models are usually based upon the definition of various different input-output relations. Each of this relations is valid in a limited number of situations. The primary role of fuzzy sets in these models consists in characterizing these situations, along with the admissible tolerance in the definition of each one.

Fuzzy models designing is an heuristic task, up to a certain degree. At the moment, there is not any *completely* automatic design procedure of fuzzy models (Lee, 1990). In this paper it will be proposed an information-based procedure able to infer all parameters that define a fuzzy model.

2. EXPOSITION

This method relies on a well known idea: a model of a system can be simplified allowing to grow the degree of imprecision in its definition (Klir, 1988). The procedure deduces a fuzzy model from a certain number of process input-output pairs –so called *examples*–. It is divided into three phases (see Fig. 1):

1. Elaboration of an initial control system. This system will contain enough rules to include the whole declarative knowledge described by the mentioned set of examples.
2. Iterative simplification procedure of the model. The aim of this simplification is to reduce the control system degrees of freedom –i.e., the number of rules– .
3. Selection of a *natural* number of rules. This number will be selected to provide an equilibrium point between performance quality and system complexity.

During the second step a certain number of different simplifications will be produced.

[1]Please forward correspondence to the first author. e-mail luciano@etsiig.uniovi.es

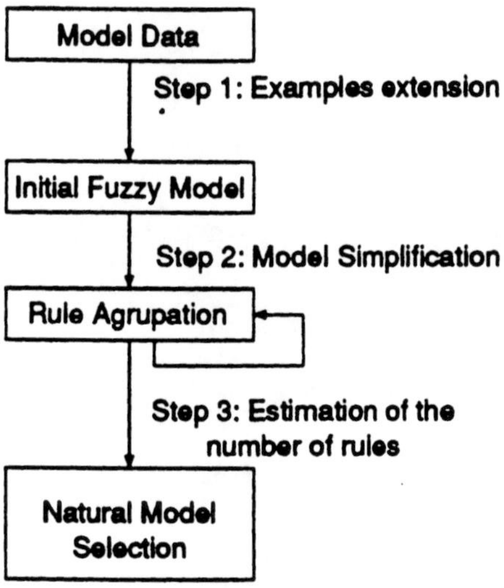

Fig. 1. Outline of the method

One of these simplifications will be selected as the natural model. The criterium devised to make this decision is exposed below.

2.1. Fuzzy Minimum Description Length

The selection of a natural number of rules has been solved by means of a fuzzy extension of the Minimum Description Length Principle. This theory has been applied to similar problems, using probabilistic methods (Wallace, 1989; Sanchez, 1993). In this work, it is proposed a fuzzy entropy based formulation of this principle.

This criterium is stated informally as follows: Given a model that comprises N rules, and a set S of examples, the amount of information of both system and examples is

$$I(S, N) = I(N) + I(S|N)$$

where $I(N)$ is named *sintactic information* and $I(S|N)$ *semantical information*. $I(N)$ measures the structural complexity of the model and $I(S|N)$ the matching between model and examples. The sum of both quantitities is named *description length* of the model. If there are different models of the same reality, it will be considered that the natural model is the one with the lowest description length.

The proposed expression for sintactic information is

$$I(n) = \log_2(n) \tag{1}$$

$I(n)$ does not measure the complexity but it can be used to *compare* complexities of models whose structure is the same.

On the other hand, an example contribution to total semantical information will be approximated with an amount of information depending on the fuzzy distribution of its most related rule. This quantity is (Sanchez, 1994)

$$I(r_i) = \int \log_2 |A_\rho| d\rho \tag{2}$$

where A is a fuzzy set that covers the rule –the meaning of this set will be stated in section 2.4.–. $|A_\rho|$ is the Lebesgue measure of the domain of the $\rho - cut$ of A. This entropy is based in Klir's measure of nonspecifity (Klir, 1988).

Then, the description length of a N rules model is defined as

$$\text{TDL} = I(N) + I(S|N) = log_2 N + \sum_{i=1}^{N} n_{pi} I(r_i) \tag{3}$$

where n_{pi} is the number of examples belonging to rule r_i.

In the figure 4 are plotted the sintactic complexity, the semantic complexity and the description length, vs. the number of rules, in a test problem. It can be seen that semantic complexity decreases with the number of rules: a model can be better fitted when its number of degrees of freedom is high. The sintactic complexity shows the opposite evolution. The description length has a minimum somewhere between 1 and N rules.

2.2. Types of Fuzzy Rules

Rule-based fuzzy systems include two main types of rules:

1. Those in which both antecedent and consequent are fuzzy propositions, i.e. rules of a Mamdani type system or *type I* rules.
2. Those in which the antecedent is a fuzzy proposition and the consecuent

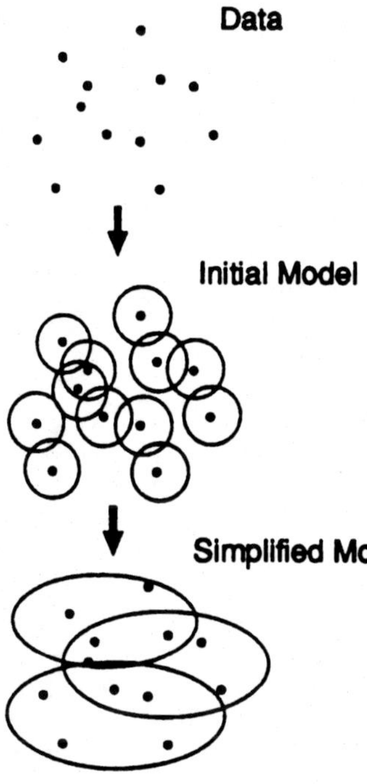

Fig. 2. Examples extension

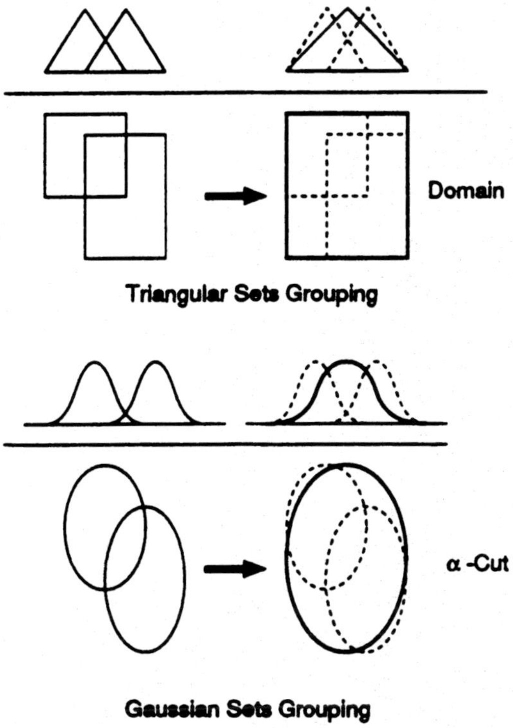

Fig. 3. Rule Grouping Operations

is a linear equation, i.e. Sugeno type rules or *type II* rules.

The proposed method is valid in both cases. The only differences exists in the last stage (see Subsection 2.6.).

2.3. Examples Extension

If input and output state dimensions are m and n, an example will be a vector of dimension $n = m + p$. Every example is extended with a small fuzzy set that enlarges it to its neighborhood. Then it is defined an initial fuzzy system, composed by a number of rules equal to the number of examples (see Fig. 2).

The area of an extended example is a design parameter. It will be related with the criterium of similarity of fuzzy sets used by SIMP, so this number determines the behaviour of this method when data is noisy.

2.4. Proposed Rules Grouping Operations

The initial model is simplified applying a *grouping* operation. A grouping operation is intended to extend the description of two rules to an area that includes the domain of both ones (see Fig. 3). Fuzzy rules will be derived from the fuzzy sets formed when the grouping operation is repeated over the initial model.

It will be considered two kinds of fuzzy membership function parameterising: triangular and gaussian.

Triangular sets Given two sets $A^1 \subset X_1^n$, $A^2 \subset X_2^n$ where

$$A = A_{X_1 \times \ldots \times X_n}(x_1, ..x_n) = A_{X_1}(x_1) \wedge \ldots \wedge A_{X_n}(x_n)$$

and

$$A_{X_i} = A_{X_i}(c_i^A, \alpha_i^A, \beta_i^A)$$

where α and β are the coordinates of the fuzzy number lower and upper bounds, the compound set $A^1 \bullet A^2$ is defined by

$$(A^1 \bullet A^2)_{X_1 \times \ldots \times X_n}(x_1, ..x_n) = (A^{1\bullet 2})_{X_1}(x_1) \wedge \ldots \wedge (A^{1\bullet 2})_{X_n}(x_n)$$

and

$$A_{X_i}^{1\bullet 2} = A_{X_i}^{1\bullet 2}(c_i^{1\bullet 2,A}, \min(\alpha_i^{1,A}, \alpha_i^{2,A}), \max(\beta_i^{1,A}, \beta_i^{2,A}))$$

The two centers $c_i^{1\bullet2,A}$, $c_i^{1\bullet2,C}$ are determined via a convex combination of initial centers. If set A^1 covers n_1 examples and set A^2 covers n_2:

$$\begin{aligned} c_i^{1\bullet2,A} &= \omega_1 c_i^{1,A} + \omega_2 c_i^{2,A} \\ c_i^{1\bullet2,C} &= \omega_1 c_i^{1,C} + \omega_2 c_i^{2,C} \end{aligned} \tag{4}$$

where ω_1, ω_2 are

$$\omega_1 = \frac{n_1}{n_1+n_2}; \quad \omega_2 = \frac{n_2}{n_1+n_2} \tag{5}$$

<u>Gaussian sets</u> Given two sets $A^1 \subset X_1^n$, $A^2 \subset X_2^n$, defined by the membership functions

$$\begin{aligned} \mu_{A_1}(\mathbf{x}) &= e^{-k(\mathbf{x}-\mathbf{x}_{c1})^T C_1^{-1}(\mathbf{x}-\mathbf{x}_{c1})} \\ \mu_{A_2}(\mathbf{x}) &= e^{-k(\mathbf{x}-\mathbf{x}_{c2})^T C_2^{-1}(\mathbf{x}-\mathbf{x}_{c2})} \end{aligned} \tag{6}$$

the compound set $A^1 \bullet A^2$ is defined by

$$\mu_{A_{1\bullet2}}(\mathbf{x}) = e^{-k(\mathbf{x}-\mathbf{x}_{c1\bullet2})^T C_{1\bullet2}^{-1}(\mathbf{x}-\mathbf{x}_{c1\bullet2})} \tag{7}$$

where

$$p = \frac{n_1}{n_1+n_2} \tag{8}$$

$$\mathbf{x}_{c\,1\bullet2} = p\,\mathbf{x}_{c1} + (1-p)\,\mathbf{x}_{c2} \tag{9}$$

$$C_{1\bullet2} = p\,C_1 + (1-p)\,C_2 + p(1-p)(\mathbf{x}_{c1}-\mathbf{x}_{c2})(\mathbf{x}_{c1}-\mathbf{x}_{c2})^T \tag{10}$$

2.5. Simplification Process

Each time the grouping operation is applied, the number of sets decreases. Before taking this step, two questions must be answered:

1. Which are the best two rules to be grouped?
2. Is desirable or not to make a new simplification?

The second point has been discussed before. In the first one, this method will try to group the nearest sets. The decision about the proximity of two fuzzy sets depends on a similarity function. This kind of functions are entropy measures (Xuecheng, 1992).

The proposed solution does not search directly the nearest sets, but builds a partially

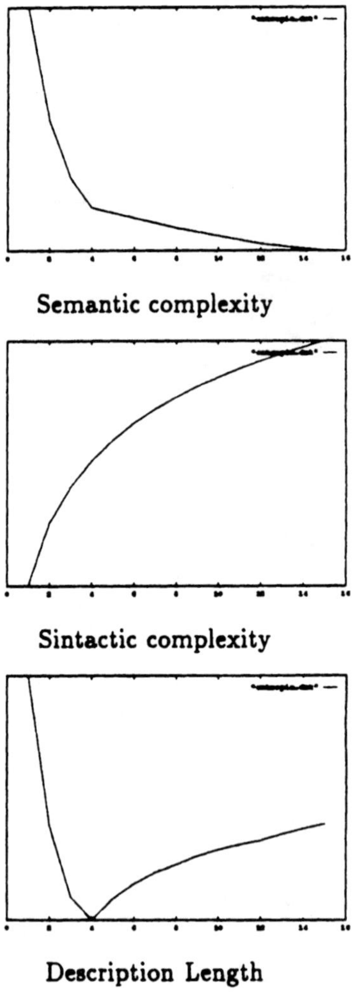

Fig. 4. Rule Number Estimation

optimal tree of examples, miminizing an energy function (see Fig. 5). The final nodes of this tree are the initial extended examples. Every internal node embodies the grouping of their final nodes. The mentioned energy is defined as the sum of values of the entropy (see Eq. 2) in all nodes of the tree. From this tree, it can be obtained a level clustering for any number of clusters between 1 and total number of examples. Each one of these clusters will produce a fuzzy rule.

The process is finished when the description length of every level clustering is calculated. The one with the lowest description length will determine the natural model of the system.

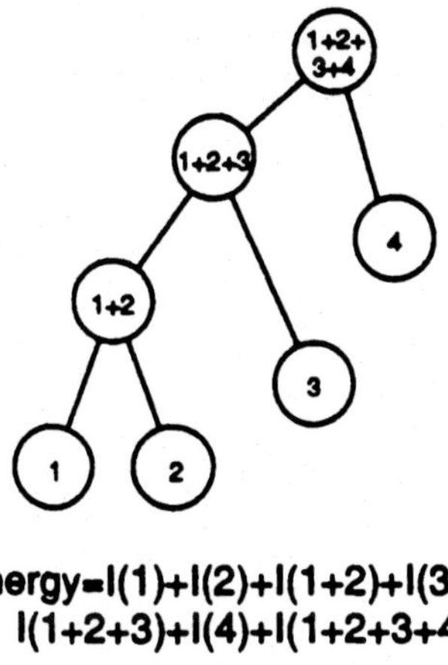

Fig. 5. Tree of examples and energy

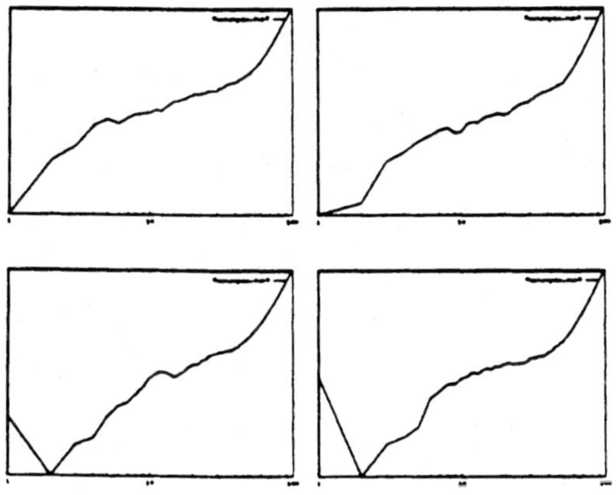

Fig. 6. M.D.L. and noisy data

2.6. Rule extraction

The extraction of a rule from one of the fuzzy sets obtained is direct. The antecedent is built applying the grouping to the first m coordinates of every example. Type I rule consecuents are defined by the same grouping, but applied to the last p coordinates. Type II rule consecuent are the minimum squares fitted planes to these last coordinates of examples.

3. RESULTS

3.1. Noise filtering

This method is quite effective when initial data contains noise, because there are not *a priori* assumptions about noise distribution. To show this effect, data from four theoretical plants is used. Each plant characteristic surface is defined by means of two horizontal displaced semiplanes –an step–. The distance between planes is lower than the noise margin in the two first plants, and greater in the last ones. The data was corrupted with an uniform noise in all cases.

The method correctly detects the presence of two regions in the latter plants, and also decides that the natural number of rules is 1 in the two first ones. The evolution of description length is shown in figure 6. It's noticeable the decreasing of the description length with the growing of the step amplitude.

3.2. Example of test plant

One of the theoretical plants used in tests was the cubic parabola $z = x^3$. The approximation is shown in figure 7. The initial data is plotted in the upper part of the figure. The central part of the same one shows α-cuts of the domain of the estimated sets. It can be seen the simplified model is composed by three rules. The response of the simplified model is plotted in the lower part. The rules were of type II, with gaussian parameterising.

4. CONCLUSIONS

This method uses an information-theory based process that determines automatically all the parameters that define a fuzzy model of a system. Its behaviour is good even when data is corrupted by noise.

Its implementation is easy and it is able to work in real time. The most intensive computing effort is made during the construction of the tree of examples, but the wasted time is minimal compared with other methods, like gradient-descent trained connectionist architectures.

REFERENCES

Klir, G., and Folger, T. (1988). *Fuzzy Sets, Uncertainty, and Information*. Prentice Hall.

Lee, Ch, (1990). Fuzzy Logic in Control Systems: Fuzzy Logic Controller. IEEE

Transactions on SMC, Vol 20, no. 2, pp 404-418 and 419-435

Sánchez, L. (1993). Extracción Automática de Reglas en un Sistema de Control Borroso. III Spanish FLAT Congress. Santiago de Compostela. pp 243-250.

Sánchez, L. (1994). *Control Difuso de Procesos Industriales mediante una Arquitectura Paralela y Distribuida, tipo Red Neuronal.* Ph. D. Thesis. Universidad de Oviedo.

Wallace, R.S. (1989). *Finding Natural Clusters through Entropy Minimization.* Ph. D. Thesis. School of Computer Science. Carnegie Mellon University.

Xuecheng, Liu. (1992) Entropy, distance measure and similarity measure of fuzzy sets and their relations. Fuzzy Sets and Systems.,**52**, 305-318. North-Holland.

Zadeh, L. (1973). Outline of a New Approach to the Analysis of Complex Systems and Decision Process. IEEE trans. on SMC, vol SMC-3, no. 1, pp 28-44

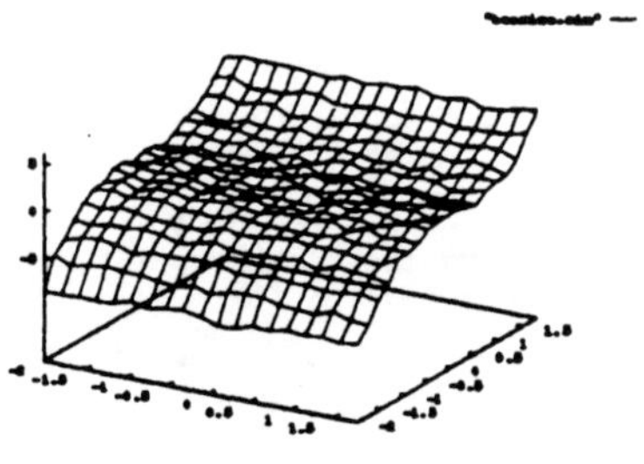

Model Data

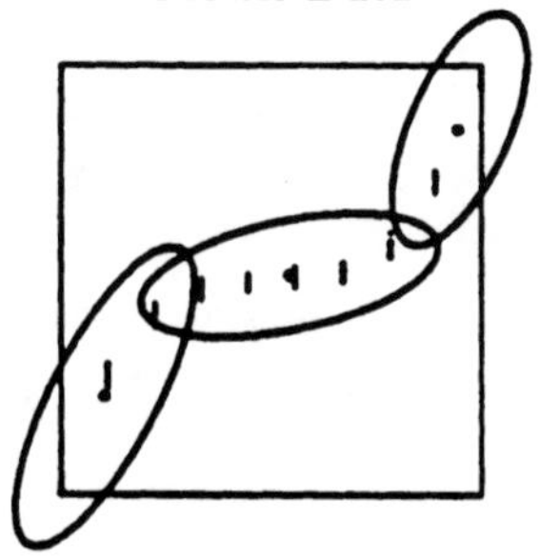

Rules Estimated

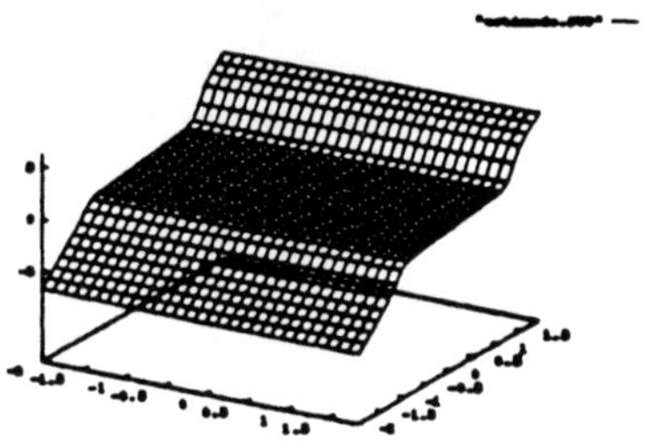

Estimated Model

Fig. 7. Test Plant

TEMPORAL MECHANISMS IN COMMUNICATION MODELS APPLIED TO COMPANION STANDARDS

P. LORENZ and Z. MAMMERI

CRIN (CNRS URA 262), 54516 Vandoeuvre Cedex, FRANCE, lorenz@loria.fr

Abstract. Models for communication between distributed application processes are presented. This paper analyses the way the timing constraints are handled by the communication models. It shows how some temporal mechanisms integrated into the producer/consumers communication model may be used at the companion standard level to check if the timing constraints of a real-time application are met or not.

Key Words. Distributed control, Intelligent machines, Real-time computer systems, Models, Communication models, Timing constraints, Companion standards, Temporal mechanisms.

1. INTRODUCTION

A real-time system is defined as one in which the correctness of its results depends not only on logical computation carried out, but also on the time at which the results are delivered. The built-in notion of time and the way it is used in the system is the basic difference between real-time and non-real-time systems. When a real-time application is composed of several application processes (or tasks) distributed among a set of processors, the communication system interconnecting the processors must be able to manage the temporal mechanisms (LeLann, 1986; Lorenz *et al.*, 1991; Lorenz *et al.*, 1993a). As the APs (AP = application Process) are implemented on intelligent machines, some temporal mechanisms must be integrated to these machines to manage the timing constraints.

Several models enabling communication between tasks in distributed applications have been studied. In this paper, we are especially interested in the way these models manage the timing constraints related the temporal validity of data exchanged between distributed tasks. In Lorenz *et al.* (1994), we have presented an approach to integrate temporal mechanisms into communication protocols. In this paper, we show that these mechanisms must be also integrated at the companion standard level.

This paper is structured as follows: first, the functional, the material and the operational architectures are introduced. In the second section, usual communication models (client/server, multiclients/multiservers, producer(s)/consumer(s)) are presented. In section 3, some temporal mechanisms for the companion standards suitable for the producer/consumer model are presented.

2.COMMUNICATION ARCHITECTURE

A great number of problems of manufacturing companies may be solved by integrating the computer in the process control, in the product design, in the economic decisions, ... This computer integration is called "Computer Integration Manufacturing" (or CIM) (Lavie, 1988). A CIM architecture allows to introduce the notion of functional, material and operational architecture.

A functional architecture is composed of the different functions and the relationships between these functions.
The material architecture is composed of the list of the different equipments (processors, networks, ...) used by the functions.

The operational architecture is obtained by the mapping of the functional architecture on the material architecture.

Some constraints can be associated to the functional architecture (as timing constraints) and to the material architecture (as constructor constraints). The timing constraints of the functional and material architectures must be met in the operational architecture. In the functional architecture, the temporal requirements can be translated into timing constraints. In the operational architecture, these timing requirements can be expressed in terms of communication model choice.

In the operational architecture, a communication model can be translated into temporal mechanisms to verify the timing constraints.

The following scheme summarizes the different timing concepts introduced in the functional and in the operational architectures:

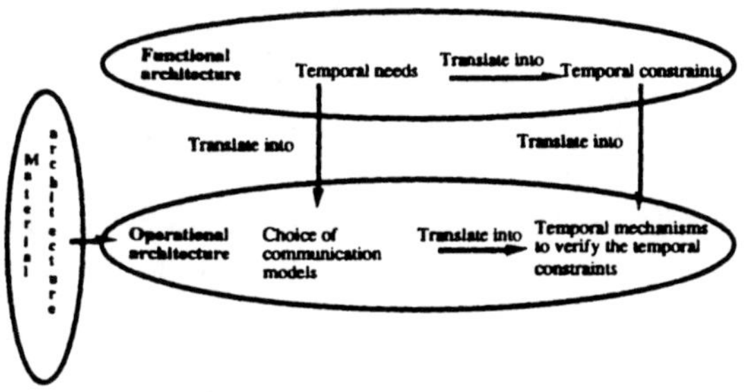

Fig. 1. Temporal aspects in the functional and operational architectures

According to the selected communication models, temporal mechanisms will verify or not the timing constraints.

3. COMMUNICATION MODELS

In this section, communication models of the operational architecture will be studied: the temporal aspects of the client/server, the client/multiservers, the multiclients/multiservers, the producer/consumer and the producers/consumers model (Thomesse *et al.*, 1991). The management of the time is different according to the selected communication model.

Client/Server model

The client/server cooperation model is a point-to-point model. Two APs are concerned: the client and the server (see fig. 2). The client sends a service request to the server. According to its resources, the server performs the request and sends back the response to the client. This is a very general model. A great number of services can operate according to this model. Most of existing services and protocols at the application layer are based on this model. For example, MMS which is the most known service in manufacturing systems is a client/server oriented service (MMS, 1990). The full duration of the transaction is unforeseeable except by making hypothesis on the server and on the network availabilities

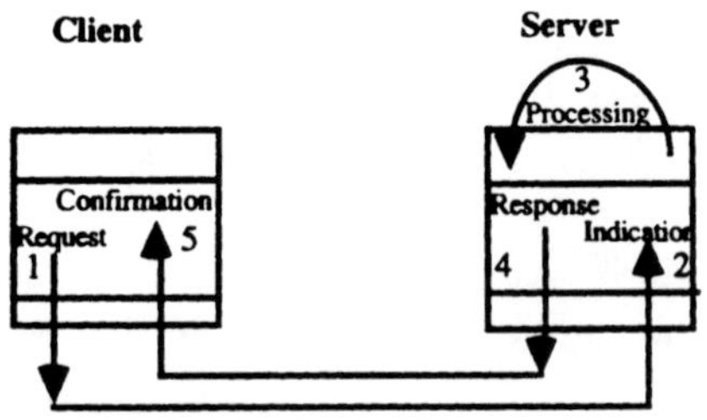

Fig. 2. Client/server model

If at a given time, several clients want to read the same value of a server (for example, several regulation tasks want to read the temperature of a given device), then each client must establish an association with the server. The sequential management of the requests implies that the server sends different values to each client. So, it is not possible to know if the responses received by the clients are identical or not, because each request is performed by the server at a different time.

Afterwards the spatial consistency statuses will be used to characterize the fact that several clients receive exactly the same value.

Client/Multiservers model

In the Client/Multiservers communication model, different communication standards are possible:

1°) Several requests from a client are sent

to several servers unknown to the client,
2°) A request from a client is sent to several servers unknown to the client,
3°) Several requests from a client are sent to several servers known to the client,
4°) A request from a client is sent to several servers known to the client.

Only the first case will be studied here because the other ones can be deduced from this first model.
The client sends a request to a server that becomes a client regarding the other servers which are known from this later client.
Then, when a server is unknown to the client, it is necessary to add a step for the exchange of information between a client and a server.

This case can be represented as follows:

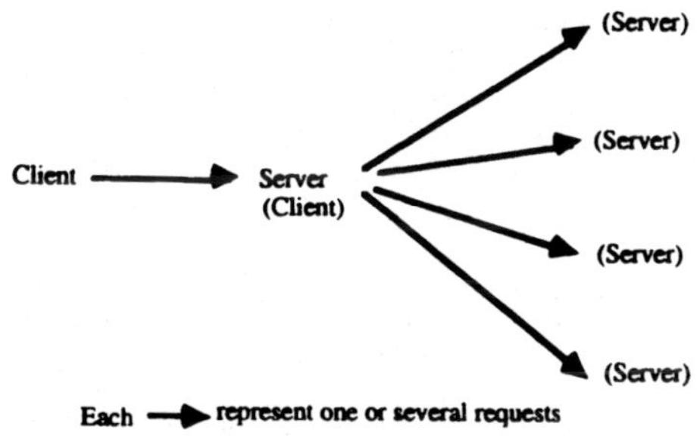

Fig. 3. Client/multiservers model

If the communication must be performed with respect to timing constraints, it is necessary to verify that each server has responded in the delays.

An example of the different communication steps between a client and two servers in the client/multiservers model can be represented as follows:

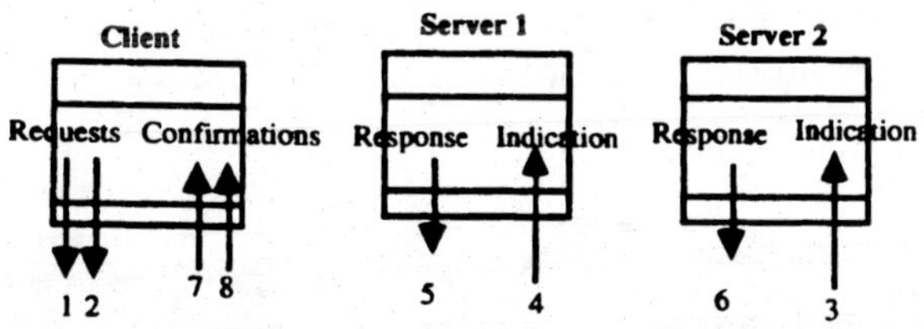

Fig. 4. Exchange of information between a client and two servers

Multiclients/Multiservers model

In the Multiclients/Multiservers communication model, a transaction is performed by different clients. As in the client/multiservers model, two cases can be studied:

1°) Several requests are issued from several clients and sent to several servers unknown to the clients,
2°) Several requests are issued from several clients and sent to several servers known to the clients,

Only the first case will be studied here because the other one can be deduced from the first model.
The clients send their requests to a known server and this server becomes a client regarding to other knowing servers.

This case can be represented as follows:

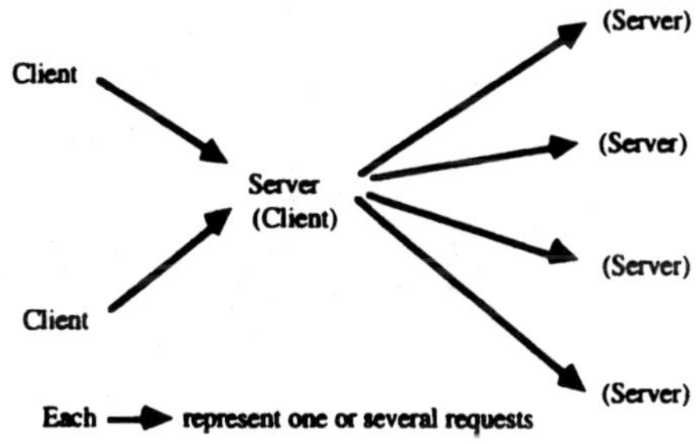

Fig. 5. Multiclients/multiservers model

Two clients sending the same requests at the same time to two servers may be represented as follows:

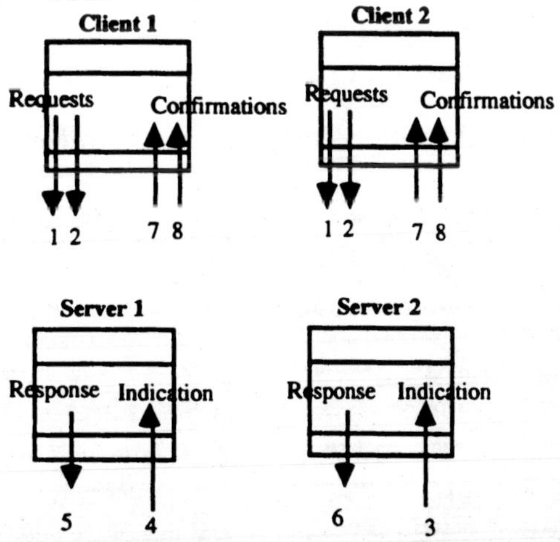

Fig. 6. Exchange of information between two clients and two servers

The previous example shows that the multiclients/multiservers model allows different clients to receive an identical response from different servers.

Producer/consumers model

The producer/consumers model is a multipoint model. The producer sends each produced data to one (or several) consumer(s). Once a data is produced, it is automatically sent to the consumers without request.

Each communicating entity (or station) may be at a given time producer or consumer of a data. Each connected station can consume a data or a list of data at the same time. When a list of variables has to be consumed by an entity, this entity may be interested by the temporal consistency of the variables of the list. For example, when a regulator algorithm using speed and pressure measurements requires that the measurements must be captured at the same time window. On the other hand, when a list of variables has to be used by several consumers, these consumers may be interested by the spatial consistency of the list of variables (i.e., the values of the variables composing the list must be the same at the different consumer stations).

Producer/distributor/consumers model

The producer/distributor/consumers model is an extension of the previous one.
The distributor is an AP in charge of the data transfer from the producer to the consumers. It requests, at the adequate moments, the producer to transmit the produced data.

This model can be represented as follows:

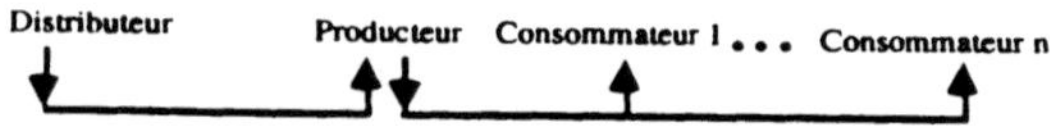

Fig. 7. Producer/Distributor/Consumers model

The main difference between the producer/consumer model and the client/server model is due to the fact that in the producer/consumer model, the confirmation mechanism does not exist. Then if a consumer has not consumed a value, this value is retransmitted only when there is a retransmission request.

To realize a client/server behaviour, one has to define the server as a producer and a consumer. In such a case, the server consumes the messages containing the requests and produces the messages containing the response to the previous requests. In other terms, the server behaves as a producer and as a consumer.

The producer/distributor/consumers model can be used instead of the client/server model if the distributor knows a priori the moments at which the clients will issue their requests. With regard to the client/server model, the producer (which can be considered as a server) has several simultaneous clients (i.e., the consumers). If several entities which are clients of a same server use a service at the same time for coherency reasons, then the producer/distributor/consumers model allows to fulfill the request of all entities within one single transaction. In consequence, the producer/distributor/consumers model is suitable for the elaboration of temporal consistency for variable list handling.

Distributor/producers/concentrator/consumers model

This model enables several consumers to receive at the same time information from several producers.

The distributor sends production orders to different producers. Then at the same time, all the producers send the information to a concentrator entity. As soon as the concentrator entity receives all the information from the different producers, it sends at the same time the received information to the consumers.

The distributor/producers/concentrator/consumers model can be represented as follows:

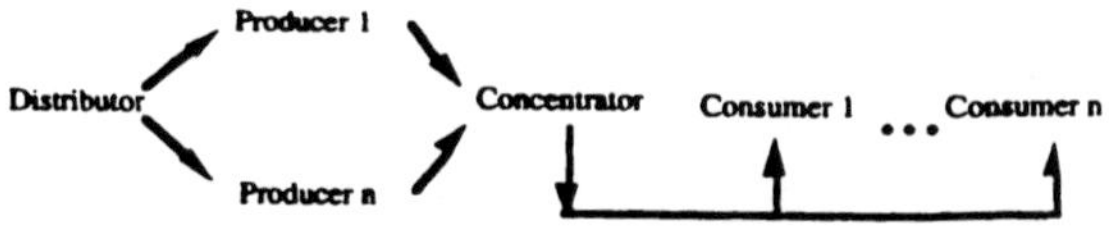

Fig.8. Distributor/Producers/Concentrator/Consumers model

4. TEMPORAL MECHANISMS OF THE PRODUCER/CONSUMER MODEL APPLIED TO THE COMPANION STANDARDS

The companion standards are located between a physical process and the application layer of the communication system (ISA, 1992).
In the production entity, the companion standards give an environment enabling to transform the physical data from the physical process in quantitative data, to perform and to encode the data in frame which can be interpreted by the application layer.
In the consumption entity, the companion standards decode the frame from the application layer, perform and transform it in physical data which can be used by the physical process.

Depending on the intelligence introduced in the companion standards, it is possible to process more or less complex operations. At this level the terms of smart (or intelligent) machines are used (Favennec, 1987; Lorenz *et al.*, 1993b).

As the data received from (or sent to) the physical process is subject to timing constraints, it is necessary to know if the communication delays have been respected, or not, by the operations related to:

- the conversion of physical data from the physical process to numerical data,
- the processing of the data,
- the coding and decoding of data by the application layer of the communication network (UTE, 1993).

The three steps used by the companion standards can be represented as follows:

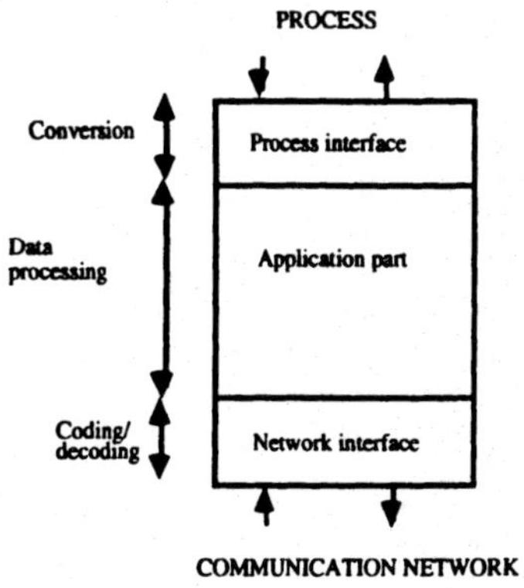

Fig. 9. Description of the different parts composing the companion standards

To know if the operations have respected the timing constraints, temporal statuses in the companion standards can be introduced.

If the production of an information has not respected the timing constraints, then it is necessary to study the statuses of the companion standards to know the reasons of the non respect of the timing constraints.
Three different temporal statuses may be used to know if the conversion, the process and/or the coding/decoding of the data have respected, or not, the fixed delays.

The more an equipment is "intelligent", the more data processing is complex and the more is long the time necessary for the execution of the different algorithms. If an AP uses heuristics, then it is not possible to know the exact duration of the algorithm execution of the algorithm associated to this AP.
The conversion and the coding/decoding parts can be considered as deterministic and require a short execution time.

Producing in the delay implies that the AP in the production entity has respected the timing constraints.
Consuming in the delays implies that the information has been received in the fixed delays by the AP interface and that the AP in the consumption entity has respected the timing constraints.

The timing constraints can be represented in terms of temporal intervals (Allen, 1984; ISO, 1993). The temporal intervals between two communication entities can be represented as follows:

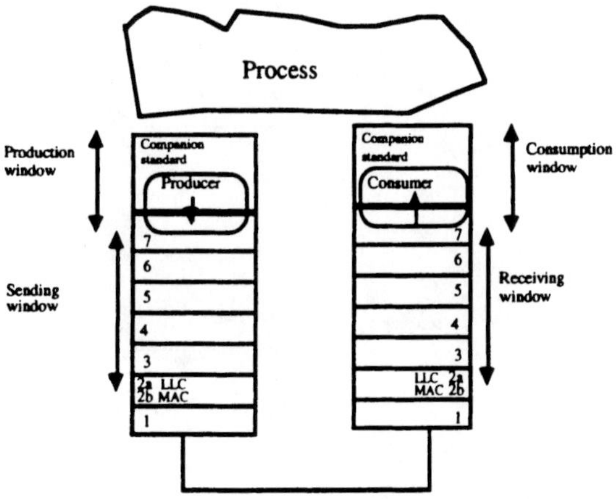

Fig. 10. Representation of the time-windows common to the companion standards and to the producer and consumer communication entities

To elaborate the temporal statuses in the companion standards, it is possible to use a temporal state machine as the one introduced in Lorenz *et al.* (1994). This state machine allows to generate temporal statuses which indicate if the conversion, process and coding/decoding steps have been achieved in the adequate temporal intervals.

5. CONCLUSION

Different communication models have been presented, showing their capabilities with regard to the timing constraints handling.

The producer/consumer model has been used to integrate the temporal mechanisms in the companion standards. Temporal statuses are used to know if the timing constraints are met or not. When the timing constraints are not met, the reason of the non-respect is specified.

We have shown that the client(s)/server(s) models can be deduced from the producer(s)/consumer(s) model. In consequence, the application of temporal mechanisms based on the produce/consumers model may be easily extended to deal with operational architectures based on other communication models.

6. REFERENCES

Allen, J.F. (1984). Toward a general theory of action and time. *Artificial Intelligence*, **23**, 123-154.

Favennec (1987). Smart sensors in industry, *Eurosensor Communication.*

UTE (1993). FIP bus for exchange of informations between transmitters, actuators and programmable controllers - Guide for the redaction of companion standard. *Union Technique de l'Electricité*, **UTE C46-645**, Courbevoie.

ISA (1992). User layer. *Instrument Society of America*, **ISA 389E**.

MMS (1990). Manufacturing Message Specification, Part 1: Service Definition, Part 2: Protocol Specification. *ISO*, **ISO 9506**.

ISO (1993). User requirements for systems supporting time-critical communications. *ISO*, **ISO TR 12178**.

Lavie, R. (1988). The CIM Integrated Data Processing Environment in the European open systems architecture CIM-OSA. *MAP/TOP COS Enterprise Networking Event '88 International, Baltimore*, USA.

LeLann, G. (1986). Distributed real-time processing, *Computer systems for process control*, Plenum Press.

Lorenz, P. , Ż. Mammeri and J.-P. Thomesse (1991). Time-critical network management for FIP fieldbus in industrial applications. *Workshop "CIM in the Process Industry"*, Commission of the European Community, Athens, Greece.

Lorenz, P., Z. Mammeri and J.-P. Thomesse (1993a). Time management in communication for real time applications. *Proceedings of the Ninth CIM-Europe Annual Conference*, Amsterdam, 161-169.

Lorenz, P., F. Beaudouin and M. Castello (1993b). Intelligent Transmitters: Needs for Interoperability and Interchangeability. *6 th International Exhibition and Congress for Sensor and System Technology*, Germany.

Lorenz, P., Z. Mammeri and J.-P. Thomesse (1994). A State-Machine for Temporal Qualification of Time-Critical Communication. *26th IEEE Southeastern Symposium on System Theory*, USA.

Thomesse, J.-P. and P. Lorenz (1991). Fieldbus - Communications models - Real Time Data Base. *ANIPLA*, Milano.

INTERWORKING OF FIELDDEVICES

M. SIEBERT* and J.P. THOMESSE**

* Association pour la Formation Professionnelle des Adultes, rue de l'Embanie,
54390 FROUARD, FRANCE
** Centre de Recherche en Informatique de Nancy, ENSEM 2 avenue de la Forêt de Haye,
54516 VANDOEUVRE-lès-NANCY, FRANCE

Abstract. This paper allows to distinguish conformance, interoperability, interworking and interchangeability. The use of functional, hardware and operational architectures helps to clarify a lot of interworking and/or interoperability problems. Two usual approaches (common language to describe devices and companion standards) to obtained interworking are described. The interworking problems, specially for fielddevices and a new approach to verify the interworking capability of fielddevices are presented.

Key Words. Fieldbusses; interoperability; interworking; companion standards; function blocks

1. INTRODUCTION

The industrial applications complexity implies the use of a growing number of fielddevices such sensors or actuators. The fielddevices are able to assure elaborated functions but they must also be able to communicate and work with other equipments, according to the end users requirements. The equipments must at first be conformed to communication standards. They must have satisfied firstly conformance testing, but it is not sufficient because of conformed implementations may be unable to cooperate; they must then secondly offer properties as so-called interoperability or interworking.

What is interoperability ? What is interworking? What is interchangeability ? How to obtain these properties or to construct applications with such properties? And how to test or to verify them?

This paper has a main goal which is to clarify these concepts and to present different approaches to answer the previous questions. In the following section (§2), we will define informally the different concepts. The section 3 will focus on the interworking problems of fielddevices and their origins The last two sections are related to the ways. firstly to obtain interoperability and interworking properties by construction. thanks to the definition of companion standards or to the use of special languages or mechanisms and secondly to test or verify them. An approach for detecting non-interworking is finally presented ; it is based on the use of Prolog clauses which describe the concerned implementations.

2. CONFORMANCE, INTEROPERABILITY, INTERWORKING OR INTERCHANGEABILITY

A functional architecture of a fielddevice consists of four functional parts which allow to situate all these aspects (see Fig. 1).

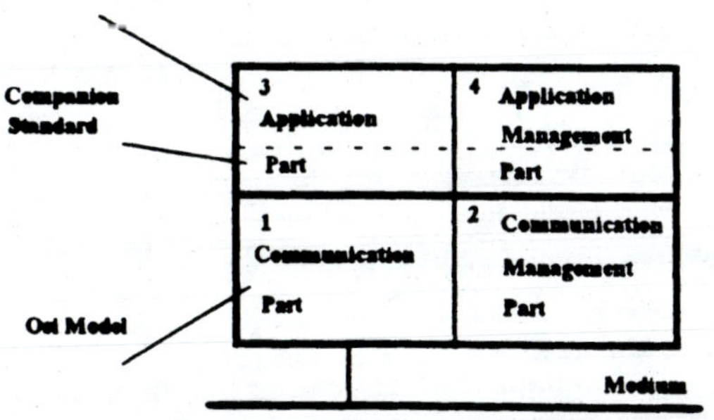

Fig. 1. The four functional parts of fielddevice

The parts 1 and 2 represent all the networks (or fieldbusses) communication aspects. The part 1 represents more particularly a communication stack in the meaning of the OSI model, while the part 2 is related to the communication management. The parts 3 and 4 cover the whole application through application processes (AP). They access to the last OSI layer services. In case of a classical computer, an application process is for instance a software, which uses directly the application layer services (MMS, 1990).

Above the application layer and management, indicated by the continuous line figure 1, are applications processes and their management (noted part 3 and 4). They are normally specific and not standardized. But for fielddevices interworking improvement, it is possible to define general functions which may be standardized (so called companion standards or function blocks). They are represented between the continuous and dotted lines of figure 1. Their rules cover not only the syntax but also the semantics of exchanged application PDU (Protocol Data Unit). This is the aim of the MMS companion standards or the ISA/IEC functional blocks, which enrich the application layer services.

The distinction between the third and fourth parts may be for a fielddevice :
- in 3, all functions which need real time communication
- in 4, all parametrization function, management function, and so on.

2.1. Conformance

The conformance is relative to the standardized parts (1 and 2 of the first figure). At the time that networks was private, a manufacturer defined its communication protocols for all equipments and eventually for partners. Today with the emergence of open system concepts and specially for fielddevices, considering the great number of manufacturers, the services and protocols must be standardized. ISO defines the conformance test concept (ISO, 1987) and also the manner to lead the test. The conformance test is an operation which allows to say if a particular implementation of a protocol or a protocol stack, conforms to the claimed standards. Even if N equipments are declared conformed, it is possible that they will not cooperate in the same application.

The reasons are simple. Firstly the standards present options, are sometime ambiguous and different choices made that products may be conformed without to be compatible between them. Secondly, the test is never exhaustive. Thirdly, time considerations and resources are not taken into account in conformance testing. This are the reasons why the interoperability notion has been introduced.

2.2. Interoperability

The interoperability of devices conformed to a same standard is a property that expresses the capacity of these devices to communicate for cooperation and to participate to a common objective. The reasons of non-interoperability have been identified (Benkhellat *et al.*, 1992). They are generally bound to imprecise, ambiguous or wrong standards specifications, to implementations with different temporal performances or to non-exhaustive conformance testing.

The interoperability test is viewed as a complement to the conformance test and the first approach was to gather real and conformed equipments to the same platform and test if all the equipments were able to communicate correctly and if they could then interoperate (Rafiq, 1991). This method allows only to verify that given products can work together under chosen hypothesis. If the end user take other products with others configurations, it is clear that the previous test cannot guarantee the interoperability in all cases. The communication interoperability is not yet defined in a standard manner.

Some confusions about interoperability are coming from the fact that this property is partially obtained from a global test involving the four parts including necessarily some APs. We will then introduce interworking to be distinguished from interoperability. The interoperability is reserved to the communication layers and to the profiles of all standardized parts (independently of the APs). The interoperability is so related to the parts 1 and 2 while companion standards don't exist. It's why the interoperability can be distinguished from interworking which is related to the parts 3 and 4 (specific parts).

2.3. Interworking

The interworking is the property that APs can cooperate to satisfy a mission (the system one). It supposes the interoperability and then the conformance to standards or to common specifications.

The interworking property is obtained when
- servers APs interpret correctly the services requests issued from client APs,

- clients APs interpret correctly the services responses issued from servers,
- exchanged data issued from producers are correctly interpreted by consumers.

An interpretation of information is said correct, if the syntax is well recognized by all the parties, if the semantics is well understood and if the common time characteristics are respected.

Today, it cannot exist conformance test for the parts 3 and 4, so that this situation leads to no interworking cases, similar to the interoperability problems (Benkhellat, 1991). This cases are presented in the next section. The interworking test is achieved today in the application integration stage.

2.4. Interchangeability

The interchangeability is the property of two functionally similar equipments (for example two pressure sensors from two suppliers), that an equipment can be replaced by the other without to modify the system. Each equipment must obviously be interoperable with the other equipments of the global system. The APs of each one must present the interworking property with the other system APs. If all the characteristics of the equipments A and B are similar, then A and B are interchangeable. Now, if the characteristics of A are included in B characteristics, then A may be substituted by B, but B is interchangeable with A if the only characteristics of A are used in the system interoperability and interworking relations.

Another confusion is sometimes made between interoperability and interchangeability. For example, in Fieldtalk (1994) it is written "Interoperability is the ability to substitute a device from one supplier for a similar device from another supplier without limiting functionality or sacrificing the degree of integration with the control or host system". This definition is more the one of interchangeability rather than interoperability.

3. FIELDDEVICES INTERWORKING PROBLEMS

The origins of possible problems are presented. In this section, it will be supposed that the equipments use the same communication standards, the same profiles and are conformed and interoperable.

3.1. Some examples

Unit measure. If a sensor provides a pressure in bar and an other equipment consume a pressure only in Pascal, then the equipments may not interwork together. In this case, the interworking is possible only if one (or several equipments) is (are) able to adapt their characteristics to the others with a parametrization operation. The end user can have interworking problems such a predefined value parameter, or a required parameter value is out of the equipment capacity, or the required parameter does not exist, or yet the parameter has not the same syntax or semantic.

Tag-name. The tag-name is often a character string and in the FIP network management standard C46-605, the tag-name must have a length between 1 and 32 characters. Nothing prevents a manufacturer to fix the maximum length to 8 characters conformly to the standard and the end user to affect a tag-name with 32 characters informally to the same standard. It is clear that this device is not adapted to the user application constraints and cannot work correctly with other devices. Anyway, a parameter whose tuning band is restricted, can lead to interworking problems, if the choices are effectively out of range.

Temporal constraints. A functional architecture specifies time constraints on services or on application processes. A given hardware may meet or not such constraints. Even if functionally speaking a distribution of functions may be done, it is possible that timing or lack of performances lead to not interworking architectures.

Material components. The market communication components can present different characteristics such services (conformance class) and available resources. The FIPIU (made by TELEMECANIQUE) can process for example 2000 variables with a maximum length of 128 bytes, or 4000 variables and a maximum length of 64 bytes or yet 8000 variables and a maximum length of 32 bytes. On the other hand, the FULLFIP (made by CEGELEC) has the capacity to process 4096 variables with a maximum length of 256 bytes. In this example, it is possible that two devices don't interwork if the maximum length are different.

Others problems can also be bound to parameters for PID regulators, for filtering algorithms, for ADC resolution and gain, and so on.

All exchanged information must be understood in the same way by all the APs that participate in the exchange.

Setting up. Before running, it is necessary to set up the devices. Hardware architecture of each of them induces different capabilities :
- no parameters may be chosen by the end user. So, the communicating APs have to be adapted if possible to the device characteristics.
- parameters may be chosen in same ranges and fixed thanks to a local operation or a distant one.
- the device may be downloaded by programs through the network or by a local operation (change of a ROM for example).

In all these cases, the fielddevice and the cooperating APs have to be adapted in order to work together. This adaptation, usually realized at the integration stage if possible, is one of the main reasons for non-interworking. It is then necessary to detect it as soon as possible.

Others problems can be bound to programs or parameters that are downloadable or not. In all cases, the characteristics (fixed or not) of a device must be taken into account by others devices for interworking. Hence, it's very important to detect if a fielddevice can be adapted or not with remote or local operations, to the end user application.

3.2. *Architectures for interworking problems representation*

To clarify these problems, it is possible to take three architectures into consideration.

Functional architecture. The functional architecture of a distributed application is composed :
- of functions specification, which will be implemented as tasks or application processes :
- of relations between these functions services requests, exchanges of data.
- of time constraints on functions and exchanges.

In these functions we can identify the real time ones in different running modes and all the ones which are related to the configuration and setting up of the devices and of the global application. These functions will be realized according to the different capabilities of each device.

Hardware architecture. The hardware architecture of a a device is the description of all components with their characteristics (CPU, memory and size, communication processors with their protocols, switches for configuration and so on). It must be interoperable. If devices are not interoperable by construction, they must be adapted to provide this property before thinking to the interworking problem.

Example. Unlike OSI products, that know to identify exactly every exchanged PDU (Protocol Data Units) using all possibilities of the Abstract Syntax Notation number 1 (ASN.1), the fieldbusses propose a restricted PDU description. The first reason is a technological order (power necessary to encode under hard time constraints). An other reason is the degradation of the network communication capacity in the case of a cyclical traffic, if the useful information are compared with the description information included in the PDUs.

Operational architecture. The operational architecture of a distributed application is the result of the mapping of the functional architecture on a set of hardware architectures which represent the devices, the PLCs, the networks and so on.

3.3. *Conclusion*

To obtain interworking property, it is usually necessary to adapt some components of the functional architecture according to the constraints of the material architecture. The objective of a lot of current works is to simplify this problem of building an operational architecture and/or to verify, to test the validity of such architectures.

4. DIFFERENT APPROACHES FOR INTERWORKING CONSTRUCTION IMPROVEMENT

If the communication standards from different fieldbusses are well advanced, the works to standardize fielddevices are still in progress. They can take several forms (Laine, 1989). The companion standard to MMS, the intelligent FIP model station (UTE, 1992), the functional blocks by ISA/IEC, the language to describe fielddevices like DDL (Device Description Language) (Westbrock, 1992) or yet the results of European projects like DIAS (DIAS, 1992).

4.1. *Development tools*

The quality of services in relation to the service primitives defined in a standard can lead to interworking problems. It's not impossible that some parameters are ignored at the implementation stage. If all devices are developed with the same tool, the standardization by the furnished services, can help to eliminate some interworking problems.

4.2. *Companion standards*

The companion standards provide a device standardized model (for sensors, actuators, PLCs, robot and numerical commands) and specialized the services and the application layer objects. The application layer semantics will make richer and uniform the presentation of its services and objects, despite the various device structures. Each companion standard addresses a particular device or application class. It exists two principal models for companion standard of fieldbusses. The first model defined by ISA is service oriented and use function blocks. The second mode is data flow oriented. It is specified in the companion standard implementation guide for FIP. A companion standard must contain a device description. (a device model with the visible objects and functions and the action on the functions and objects), a definition of the model entities (definition of the objects, functions and actions) and the projection of every object on the application layer with the definition of the associated PDUs.

A companion standard allows to simplify the interworking problem, by taking it back to the interoperability problem. The specific part is then reduced. In any case, the specifics APs must be written taking account of their specification. Interworking will be really improved by companion standard definition if all the functions of parts 3 and 4 could be integrated in them.

4.3. *Device Description Language*

The Device Description Language (DDL) is a language designed to describe the external fielddevice behaviour, as it is seen by the cooperating APs on the network. A fielddevice is described by a variable set with attributes accessible by the hosts (PLC, regulators, ...), and by services requests. The device specification is supplied by a host system. This solution is really adapted if only point to point relations between a host and the concerned device are used. If more than a host has to access a given device, (for example other devices or hosts), the description of this device must be integrated into all of them. The DDL original feature is that the host adapt its behaviour, when it sends a command to the fielddevice (read or write typically). The host can operate without a particular device knowledge.

This technique is a solution to improve interworking in the sense that without standardization, a device D from a supplier X may be integrated in a system S of a supplier Y giving the description of D in DDL for a DDL compiler for S. When it is said that DDL improves interoperability in the sense of interchangeability, that's only if two devices D1 and D2 from two suppliers X1 and X2, have the same description and present the same interface in the DDL.

4.4. *Companion standard and DDL.*

DDL is a solution to take into account specific APs (parts 3 and 4) of a device in hosts. Companion standard is a solution to define in a standard way some parts of specific APs. Companion standard may be define in DDL or in other languages.

5. INTERWORKING TEST OR VERIFICATION

5.1. *Principle*

The usual approach to test interoperability and interworking is issued from the conformance testing methods. It consists of gathering devices on a test platform. It is an expensive and difficult operation, in view of the heterogeneous devices, the number of equipments and the lack of conformance tests for the part 3 and 4 (see Fig. 1).

We developed another approach based on a verification a priori of some devices properties or characteristics to avoid to detect non-interoperability and non-interworking cases before integration stage. The verification method is static and is built on a fielddevice description model (Siebert, 1993). To carry out different characteristics and properties comparisons, it is possible to conclude on partial non-interworking and on their reasons.

5.2. *Results*

The description model allows to describe the functional and operational features of a fielddevice. It is compound by three parts (equipment identification, functional and operational architecture). The verification prototype use Prolog predicates for all devices based on FIP communication standards and DIAS projects specification. The different parts of a device are

described in Prolog clauses according to the following points :
- profile, each service and protocol layer is defined according to the results of conformance testing and of complementary measures.
- specific APs are described in terms of input/output data. Their types and all their characteristics are defined.

The a priori verification consists of applying rules on the compatibility profiles and DIAS specifications, configurations and setting-up capabilities. It's possible to detect easily some non-interworking cases specially related to the syntax, semantics and time validation of data.

Prolog rules allows to lead different static verification operation by simple comparisons. The acquisition of the information necessary for a verification are carried out by PICS (Protocol Implementation Conformance Statement) and specific questionnaire elaborated from Prolog predicates. The knowledge base is represented by Prolog clauses and is supplied by the different questionnaire data, normally fulfilled by manufacturers, but also by the results of conformance tests (part 1 and 2) or unit tests (performance measure, time limit measure, ...)

6. CONCLUSION

This paper has presented several properties participating to interworking of fielddevices in distributed applications. The main important point is to distinguish conformance, interoperability, interworking and interchangeability. Two complementary approaches must be set up to obtain interworking applications.
- building interworking by definition of companion standards or function blocks.
- definition in a standardized language (to be defined) of specific parts of applications in order to be able to verify statically (as much as possible) the interworking property before integration.

7. REFERENCE

Benkhellat, Y. (1991). Etude du test d'interopérabilité. Application à MAP (MMS) et à FIP, Rapport de DEA . *CRIN/ENSEM.*

Benkhellat, Y., Siebert, M., and Thomesse, J.P. (1992). Interoperability of sensors and distributed systems, *Eurosensor 92* - San Sebastian.

DIAS (1992). Distributed Intelligent Actuators and Sensors, . GIF Interoperability Guide-Line. *GIF Gruppo Italiano FIP.*

FieldTalk (1994) December/January 1994, number 9, *Interoperable Systems Project*

FIP (1992). FIP Toolbox. Manuel de référence, version 3,0. *Centre de compétence FIP*, Nancy.

Laine, T., Elloy, and J.P. Thomesse, J.P. (1989). Le bus de terrain FIP. La couche application et les standards d'accompagnement *Conférences CAPTEURS 89.*

ISO (1987). OSI Conformance Testing Methodology and Framework: Part 1. General Concepts. ISO, **DP 9646-1**

MMS (1990). Manufacturing Message Specification, Part 1: Service Definition, Part 2: Protocol Specification. *ISO*, **ISO 9506**.

Rafiq, O. (1991). Le test d'interopérabilité des protocoles, CFIP'91 *Ingénierie des Protocoles,* Edition HERMES.

Siebert, M. (1993). L'interopérabilité des équipements de terrain. Vers une méthode de vérification. Application au réseau de terrain FIP, Mémoire d'ingénieur *CNAM*, NANCY.

UTE (1992). Bus FIP pour échange d'informations entre transmetteurs, actionneurs et automates, Guide de rédaction des normes d'accompagnement, *Union Technique de l'Electricité,* **UTE C 46-645** - .

Westbrock, J. (1992) Device Description Language Specification, *Rosemount*, Document COM-20 Rev.B.

INTELLIGENT SENSOR : OBJECT APPROACH

A. FILALI, S. ROTH, M. ROBERT and C. HUMBERT

Centre de Recherche en Automatique de Nancy, CNRS URA D821
Université de Nancy I, ESSTIN, Rue Jean Lamour, 54500 Vandoeuvre ,France

Abstract. By way of introduction, the first section of this paper is about the need of models to represent Intelligent Sensor (I.S.) concepts. Through the second section, the involved benefits brought by the use of Object Approach are discussed. The third section gives a brief description of the chosen method, the O.M.T. methodology. In section four, an application example is proposed to illustrate our approach for modelling Intelligent Sensor Measurement activity. Finally as a conclusion, the required needs are shown to reach our final objective, the design of a CAD tool to provide more assistance for I.S. design.

Key Words. Sensors, instrumentation, intelligent machines, computer-aided design, modelling, models, object approach

1. INTRODUCTION

Smart or Intelligent Sensors are becoming reality in process control and industrial production systems. In the eighties, the demands for reliable measurements expressed by engineers, have led them to seek solutions that would no longer allow to systematically put the fault on the sensor whenever a problem arises (Ciame 1987). It is obviously the progress made in related fields of microelectronics and microcomputer engineering, in association with the developments of microprocessors, microcontrollers or digital signal processors, that subtend the recent developments in the field of Intelligent Sensor (I.S.).

To integrate them in automation systems, it is not so easy then to connect a classical sensor because Intelligent Instruments must be able to collaborate and to cooperate in order to make up a real Distributed Database, which must contain consistent information. So, it is necessary to build models for I.S. to define as precisely as possible the services offered by Intelligent Instrument (I.I.), the kind of information they are able to product or consume, and their behaviour, in order to reach interoperability (Robert *et al.* 1993b).

From a constructor or vendor point of view, an approach to the construction of an equipment where only the designer is involved, expressing HOW to "technologically" realise the equipment, cannot be satisfactory applied to the realisation of an I.S. It will end in realising a device offering a certain number of services to the user but not necessarily those really desired. The construction should be based on standards, in fact, they don't exist. So, the design of I.S. could be done according to Reference Models. In this paper the proposed Object Approach can be suitable to formalise the I.S. concepts and to build evolutive models.

2. OBJECT APPROACH : WHY ?

The main reasons to choose the object paradigm for modelling the I.S. concepts are presented in this section. Fig. 1 describes the thought processes for modelling. This methodology is based on a classical design life cycle, including six steps :

- building of the Practical Model to define specifications in an informal way,
- building of the Cognitive Model to formalise and validate the knowledge,
- building of the Computerisable Model, supported by a computer tool, to simulate behaviours,
- instanciation of the model,
- building of a laboratory mock-up, which is a first technological validation,
- including the prototype on site, to validate relationships between the system and its environment.

As it is defined in CIM-OSA architecture by Panetto *et al.* (1993), an other point of view could be added to this design life cycle, it allows to build a nest of models, from generic to particular.

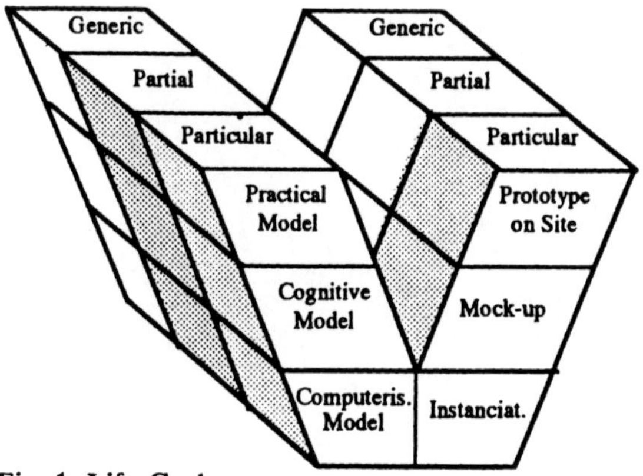

Fig. 1. Life Cycle

The Object Approach (O.A.) is interesting in modelling processes because it allows a link between the four first steps previously described.
In addition, the O.A. concerns the global process of development for building systems by the way of three phases :

Object-Oriented Analysis, (O.O.A.),
Object-Oriented Design, (O.O.D.),
Object-Oriented Programming, (O.O.P.).

But, it is obviously necessary to distinguish these steps. This paper deals only with O.O.A., used to obtain a Cognitive Model of a partial class of I.I., the I.S. family (I.S. represents a class of I.I., because Intelligent Actuators for instance belong also to the I.I. meta-family). This orientation emphasises that Object-Oriented (O.O.) methods are useful to extract the "invariants" as "Object-Classes". These "invariants" are essential for the constructor. Because I.S. is the association of hardware and software, the constructor is able to conceive his application as a set of re-usable and evolutive modules, so the time to develop a new product could be minimised and the vendor will adapt more quickly to users' needs.

O.O.D. could be used to design the application and O.O.P. to implement the design choices, in order to emphasise software evolution and maintenance.

3. WHAT IS OBJECT APPROACH ?

3.1. Object approach

In this part, the fundamental basis of the object technology (Afcet, 1991; Booch, 1986; Masini *et al.*, 1990; Pernot and Wolinski, 1992; Wasserman *et al.*, 1989) is presented. An object is an entity containing a set of data attributes which define its characteristics and its relationships with other objects and a set of methods (operations) which describe its behaviour. Class is a technical word to describe a set of objects which have similar properties. In this context, class corresponds to generic model which defines the family of analogous objects.

The basic idea of object approach is to organise a system from entities in which data and their associated methods are encapsulated. Abstraction and encapsulation are the fundamental mechanisms of the object concept :

- abstraction consists in focusing on the essential or main characteristics of an entity and disregarding particular properties,
- encapsulation consists in distinguishing the data and their associated methods of an object which are accessible to other objects of the system. Abstraction allows to build an entire object and to reduce it from a set of values and operations, which are used and understood by external real world.

Abstraction and encapsulation are both complementary concepts. Abstraction focuses on the external view of an object and encapsulation prevents users to change or to modify its internal orgahisation. There are many tools to structure a system by an O.O.A. point of view : inheritance, association and aggregation.

Inheritance allows to facilitate the generalisation/specialisation mechanism. A subclass is able to inherit data attributes and associated methods of its "parent-class". Association is another mechanism which defines a relationship among instances of two or more classes describing a group of links with common structure and common semantics. Aggregation is a special link between a whole set and its parts.

3.2 O.M.T. methodology

Building models with traditional methods such as SADT, is not sufficient because they don't support in particular the encapsulation mechanism. The object - oriented methods allow to define a system as a collection of reusable objects. Nevertheless, different methods exist with different graphical and syntactical notations and associated meanings like Shlaer and Mellor, Hood, Coad and Yourdon methods (Graham, 1991). One of the main difficulties to select an Object-Oriented Method comes from the choice of an appropriate method for a particular application.

The Object Modelling Techniques (O.M.T.) (Rambaugh *et al.*, 1991) have been chosen to describe the I.S. concept. One of the benefits of O.M.T. method is assuredly its ability to consider three aspects of a system : the information aspect which introduces the data structure, the functional aspect which concerns the semantic of object activities, and the behaviour aspect is interested with temporal model of system's evolution.

The O.M.T. methodology uses three kind of models which are the following :

- the object model which allows to describe the static structure of the system by defining potential objects of the systems, the relationships between the objects, the data attributes and their associated operations which characterise each class.

- the dynamic model which describes the aspects of a system that change over time. It is used to specify and to implement temporal behaviour by the description of state diagram that is a graph whose nodes are states and arcs are transitions between states caused by events.
- the functional model which allows to describe transformational processes of data.

4. I.S. OBJECT APPROACH

Previous works (Robert *et al.* 1993a) have proposed the Virtual Instrument concept, which is an instanciation of an O.O. model, including six "domains" initially proposed by Albus (1991) to define an intelligent system in a robotic context, but also available in more general industrial production systems. Each of them gathers "classes" as proposed Fig. 2.

If the "Data acquisition domain" is only considered, the Measurement activity could be represented by an unformal model described Fig. 3. The aim of the next section is to show the way to define more precisely classes into a domain.

The hierarchy which allows to elaborate the *Operational* Measurement (O.M.) which is a *valid* data, issued from ***Primary*** Measurements (P.M.), eventually completed by ***Auxiliary*** Measurements (A.M.) and under the control of ***Technological*** Measurements (T.M.).

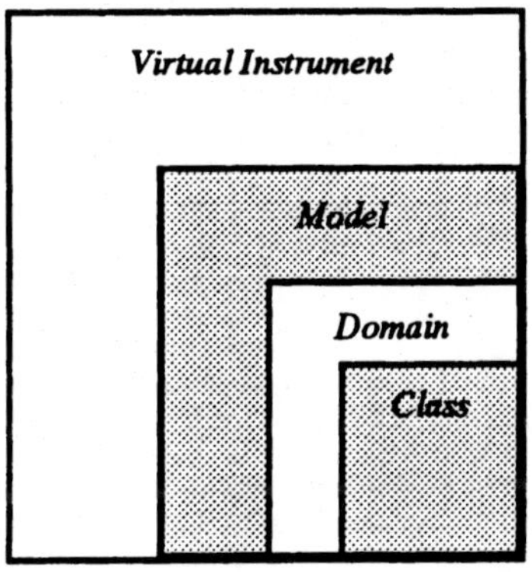

Fig. 2. From Classes to the Virtual Instrument

Delivered through a transmission medium, the Operational Measurement contains data that can be immediately exploited by the user. This measurement is obtained from a Functional Measurement involving the influence quantities through more or less sophisticated mathematical models or by taking into account the different non-linear features characterising the interaction between the measurand and the sensitive cell.

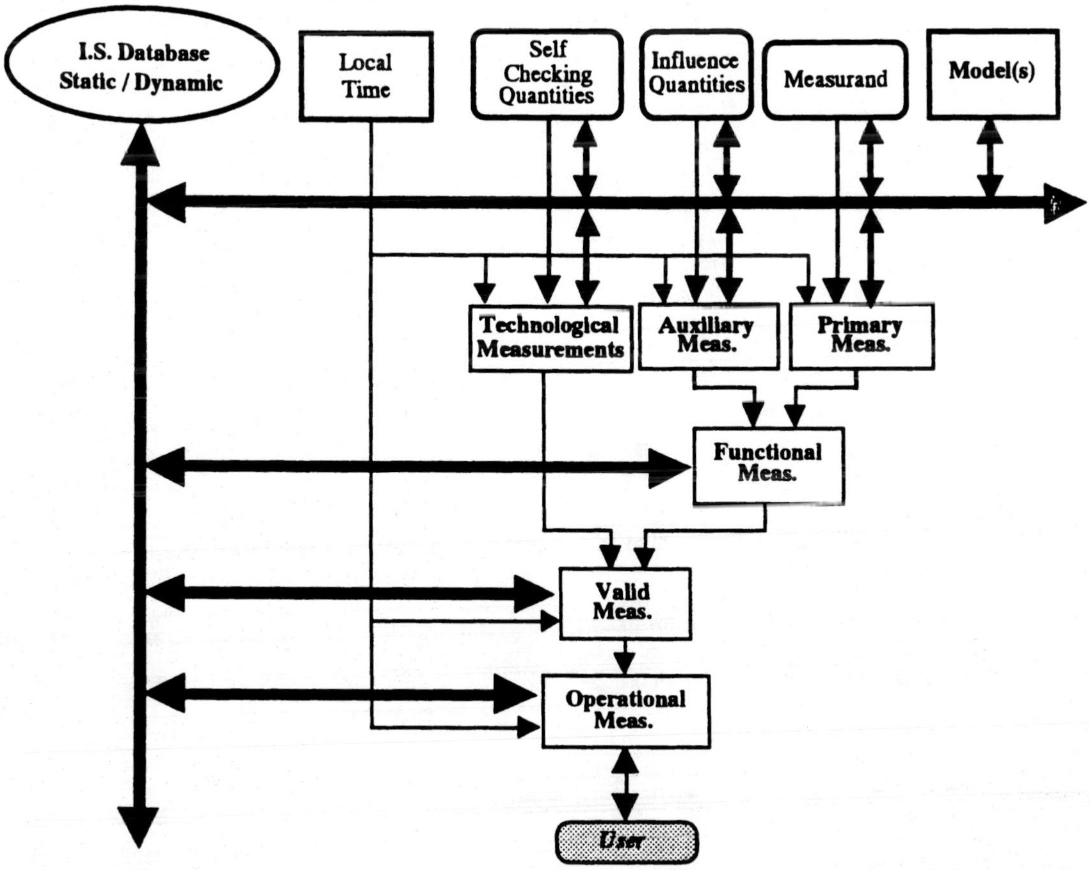

Fig. 3. Measurement activity

These measurements must be associated with :

- Information relating to local time, in order to ensure consistent datation,
- Information stored in an internal database which can be *Static* for instance : constructor's and user's identifiers or *Dynamic* , i.e. liable to change along the life cycle. Note that the speed of variations may be extremely variable.

5 I.S. OBJECT MODEL

To illustrate our approach, a particular application example is provided here, involving some concepts of I.S. in order to model the operation "to elaborate an Operational Measurement" (O.M.).

In the first step, classes could be defined. Thus, links between some identified classes are extracted. Finally, by the use of inheritance property, "invariants" are proposed in this object diagram.

5.1. Identification of classes

At first, the analysis problems are focused on what sort of classes are able to separate hardware and software system and how do classes collaborate to perform an O.M.

On the one hand, the physical quantities as the Self-Checking Quantities, the Influence Quantities and the Mesurand can be listed (see Fig. 4). Those quantities introduce the physical data which come from hardware system. However, it is necessary to transform those data into usable information. Those quantities must be translated in information to be used by some treatments in order to product the O.M.

On the other hand, the data attributes and their associated methods for each class are described. According to the class "Functional Measurement", the method "to use model" appeals to the class "Models" to obtain the required structure and coefficients of the desired model.

5.2. Identification of the links between classes

This part deals with the relationships between the previous classes.

At the first time, association and aggregation links to sum up the classes are only used (see Fig. 5). The association link described by an half circle in the object diagram allows to combine the Class "Functional Measurement " and the class "Valid Measurement" throughout the class "Technological Measurement".

The aggregation link mentioned by a diamond shaped is use to describe the building of the class "Functional Measurement" through the classes "Auxiliary Measurement" and "Primary Measurement".

The link between the classes "Valid Measurement" and "Operational Measurement" is a single association. Actually, a measurement which is validated becomes an operational one.

As far as the classes "Models" and "Local Time", they are also linked with all the other classes with a single association.

5.3. Inheritance link

According to similar data attributes and methods in some classes of the model, the inheritance mechanism can be shown.

The "invariants" (see Fig. 6) can be identified and the inheritance link described by a triangle in the object diagram, is defined to formalise an generalisation. Classes are organised in a generalisation hierarchy in which subclasses (A.M., T.M., O.M.) inherit structural and operational properties of their superclass "Measurement".

6. CONCLUSION

The object techniques has been investigated in this article for modelling the I.S. concept. The problems related to knowledge representation are illustrated by an particular example to show the way to model the whole concept I.S. According to section 3.2, this Object Model must be completed with a functional one and a dynamical one. Functional models have already been produced by Rivière *et al.* (1993) or Bayart and Staroswiecki (1993). Dynamical model is going to be realised in order to verify the feasibility of the global model, in a Real Time context. The next step in this research work will be toward a CAD tools. This I.S. CAD toolbox could be obviously helpful for the designer to gather modules in order to reply user's needs. But, this tool could also be used by the user to define more precisely the specifications of I.S.

7. ACKNOWLEDGEMENT

The authors gratefully acknowledge CRAN AMI Working Group, VERILOG company who puts the LOV Object tool to our disposal. They also thank CIAME Working Groups members for their enriching discussions around Intelligent Sensors models.

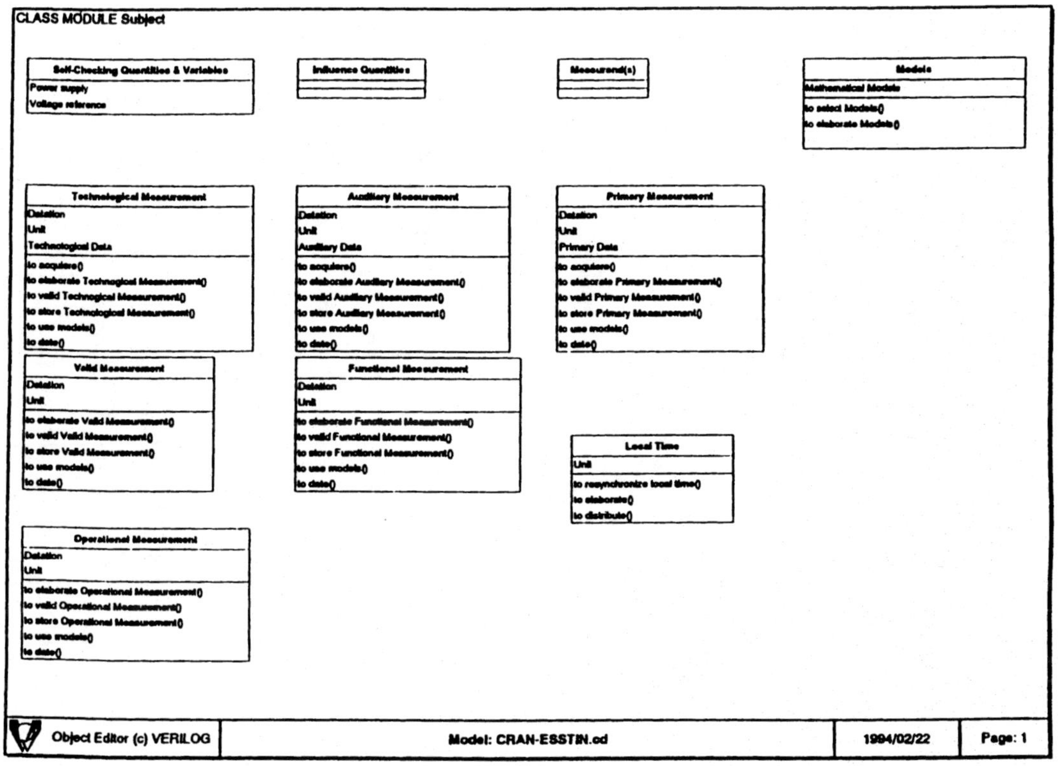

Fig. 4. Classes model

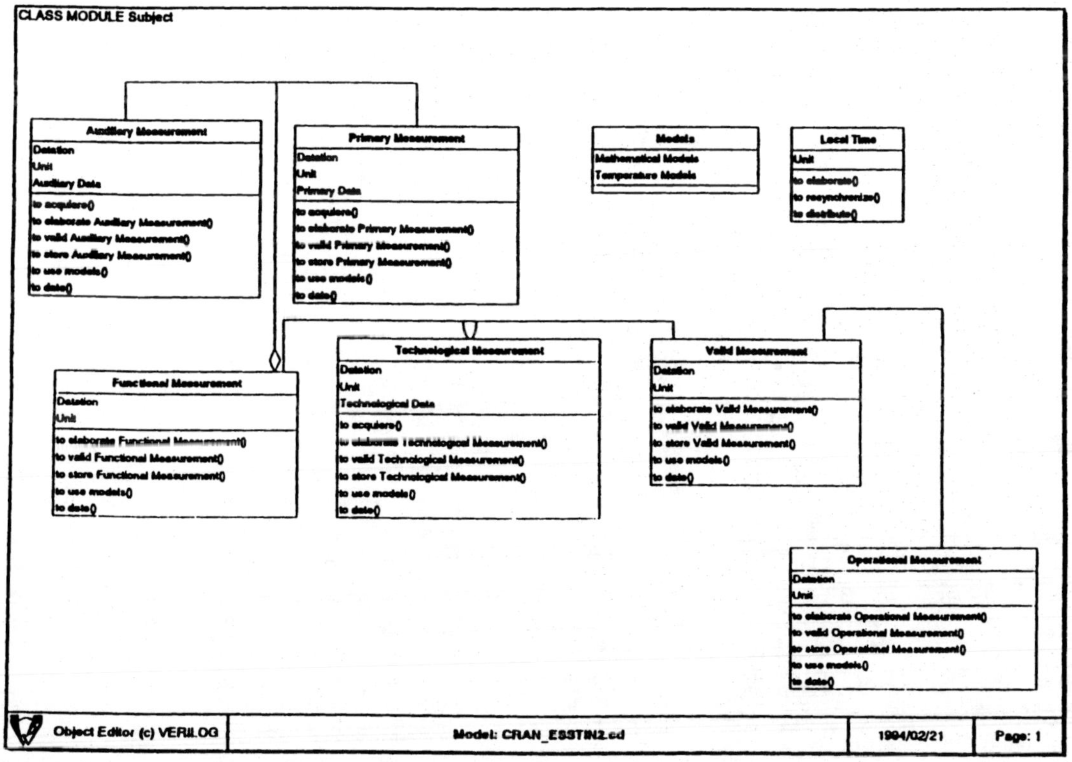

Fig. 5. Links between classes

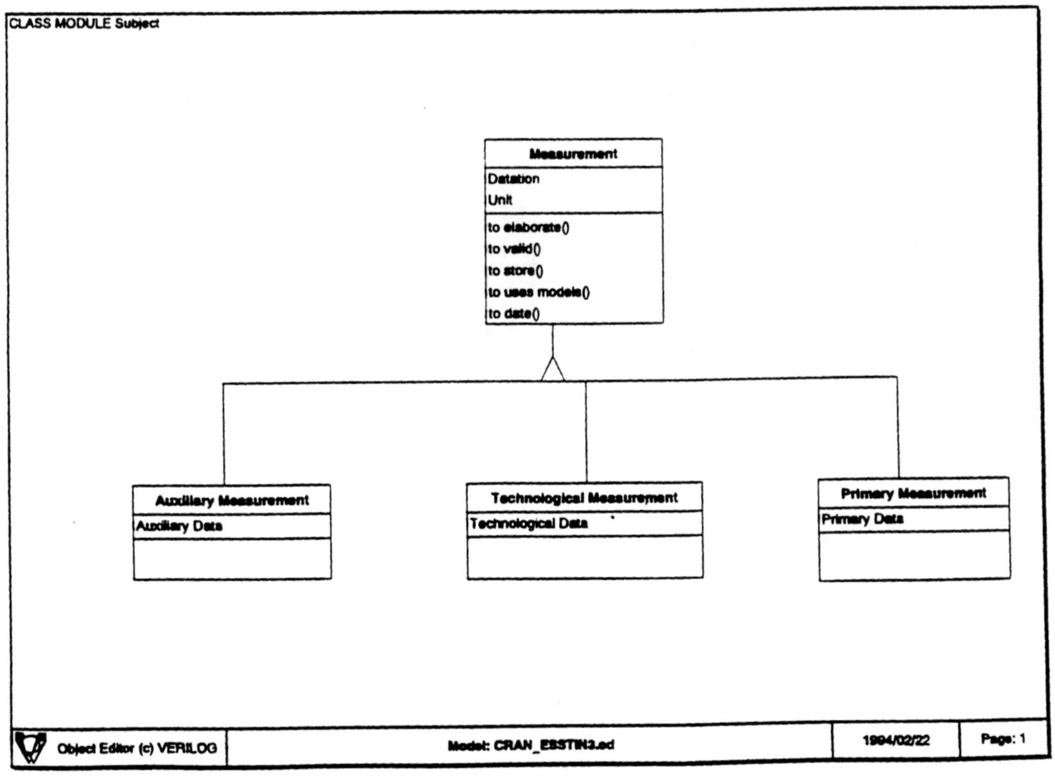

Fig. 6. Inheritance link

8. REFERENCES

Afcet (1991). Objectifs objets!. *Afcet interfaces*, May-June, 103-104.

Albus, J.S. (1991). Outline for a theory of intelligence. *IEEE transactions on Systems, Man and Cybernetics*, Vol. 221, **3**, 473-509.

Bayart, M., and M. Staroswiecki (1993). A generic functional model of smart instrument for distrubuted architecture. Imeko TC4 Symposium Intelligent instrumentation for remote and on site measurement, 231-238, May 12-13, Brussels, Belgium.

Booch, G. (1986). Object-Oriented development. *IEEE Trans. on software engineering*, vol. SE-12, **2**, 28-38.

Ciame, (1987). Les capteurs intelligents - Réflexions des utilisateurs. Ciame Afcet, Livre Blanc.

Graham, I. (1991). Object Oriented Methods. Addison-Wesley Publishing Company.

Masini, G., Napoli, D. Colnet, D. Léonard and K. Tombres (1990). Les langages à objets. Interéditions.

Panetto, H., J.M. Rivière and J.F. Pétin, (1993). Towards a unified approach for intelligent actuators and sensors. IEEE International Conf. SMC, vol 5, 177-182, October 17-20, Le Touquet, France.

Pernot, J.F. and F. Wolinski (1992). Modélisation par objets en robotique. *Technique et Science Informatique*, Vol 11, **1**, 97-115.

Rambaugh, J., M. Blaha, W. Premerlani, F. Eddy and W. Lorenson, Object oriented modeling and design. Prentice Hall International Editions.

Rivière, J.M., M. Robert, F. Hermann and C. Aubrun (1993). Modelling of smart sensors : application to a smart temperature transmitter. Imeko TC4 Symposium Intelligent instrumentation for remote and on site measurement, 173-180, May 12-13, Brussels, Belgium.

Robert, M., J.M. Rivière, J.L. Noizette and F.Herman (1992). Smart sensors in Flexible Manufacturing Systems. *Sensors & Actuators Physics*, **39-38**, 239-246.

Robert, M., J.M. Rivière, D. Theillol and P. Geoffroy (1993a). Smart or intelligent instruments in industry, past, present and future. IAR Kolloquim, 29-32, November 16, Duisburg Germany.

Robert, M., M. Marchandiaux and M. Porte (1993b). Capteurs Intelligents et Méthodologie d'Evaluation. Edition Hermès, Paris.

Wasserman, A., P. Pircher and R.J. Muller (1989). Concepts of Object -Oriented Design. Tool'89, 269-280.

ROBLIN LINEAR TRANSPORT ROBOTS COMMAND USING FUZZY LOGIC CONTROLLER

L. MASTACAN, I. OLAH and C. CALISTRU

Technical University of Iasi, Department of Automatic Control and Industrial Informatics, str. Horia 7-9, Iasi 6600, ROMANIA

Abstract. A speed fuzzy logic controller is proposed for transport robots with ac linear induction motors drives. To control the robot displacements the controller reference signal is changing continuously and robot speed output signal must track it. The performance of the fuzzy logic controller is improved compared to that of the conventional PI-action controller in spite of the robot command saturation. To show that, some simulation results are illustrated.

Key Words. Fuzzy control; conventional control; speed control; tracking systems; modeling; performance bounds

1. INTRODUCTION

To obtain long linear displacements for transport robots, a possibility is to utilize for driving ac linear induction motors. A group from Technical University of Iasi (Cantemir *et al.*, 1986; Livint *et al.*, 1986) applied this idea for the ROBLIN linear transport robots drive. To command an ac linear induction motor, usually its speed is controlled by the modification of the electric tension amplitude and frequency (Fransua and Magureanu, 1984). A transport robot has a good performance drive if the response time between two operation is the smallest. A transport robot conventional control is a cascade regulated system of the position and the speed (Mastacan, 1991). To control the transport robot displacements the speed-loop output signal continuously tracks the variable speed-loop reference signal. The command of an electric robot is constructive limited between ($-U_{max} \cdots +U_{max}$) Volts. Unfortunately, in PI conventional control the command saturation will cause the time response increase and large overshoots.

In this paper, to overcome the difficulties due to the robot command saturation, a speed fuzzy logic controller is proposed. The main goal is to get good performance bounds with respect to the limited robot command. Recent works (Li and Lau, 1989; Titli and Ketata, 1991) had as objective to implement fuzzy logic controllers for fast dynamically systems. These fuzzy controllers have good performances if their reference signals are constant (Mastacan *et al.*, 1994). The object of this paper is to solve a servosystem problem using fuzzy control. That means the transport robot output speed must track continuously the input reference signal variations of the fuzzy controller.

In the sequel, the following notational conventions will be used:

$\underline{u}_1$	Stator voltage space phasor
$\underline{i}_1$	Stator current space phasor
$\underline{\Psi}_1$	Stator magnetic flux space phasor
$\underline{i}_2$	Armature current space phasor
$\underline{\Psi}_2$	Armature magnetic flux space phasor
$\underline{i}_2^*$	i_2 complex conjugated value
v_1	Stator magnetic field synchronous speed
v	Armature speed or stator speed
τ	Stator pole-pitch
f	Stator tension frequency
f_m	Electromagnetic force
f_r	Load force
Im	Imaginary component
m	Robot mass
R_1	Stator-phase resistance
R'_2	Armature-phase resistance referred to stator
$L_{\sigma 1}$	Stator-phase leakage inductance
$L_{\sigma 2}$	Armature phase leakage inductance referred to stator
L_m	Motor magnetization inductance
s'	Linear slip

U_{1max} Stator phase voltage maximum instantaneous value
K_R Robot gain
T_m Mechanics time constant
T_e Electric time constant
T_s Sampling time
Δx Variable variation
$p=d/dt$ Differential operator
s Laplace operator
ΔV_1 Laplace transform of Δv_1
ΔV Laplace transform of Δv
ΔU_1 Laplace transform of Δu_1
ΔF_m Laplace transform of Δf_m
ΔF_r Laplace transform of Δf_r
G_X Transfer function with ΔX input signal
v_r Speed reference input signal
$e=v_r-v$ Speed error
Δe Speed error change
c Command signal
k Gain for command c
Δc Command change signal
$\tilde{x}$ Fuzzy variable
x_c Coarse variable
x_f Fine variable

2. ROBLIN TRANSPORT ROBOT MODEL

The ROBLIN transport robot drive is a bilateral type three-phase induction linear motor. Its double stator is mounted on the mobile side of the robot. The non-magnetic linear motor armature is an aluminum or copper plate placed along the rolling-track (Cantemir *et al.*, 1986). The ROBLIN transport robot model is obtained starting from the biphase induction linear motor representation in an orthogonal axis system (Livint *et al.*, 1987). The coordinates system is linear synchronous moving together with stator magnetic field of the motor having the speed given by :

$$v_1 = 2\tau f \tag{1}$$

The biphase induction linear motor representation is given by:

$$\underline{u}_1 = R_1 \underline{i}_1 + p\underline{\Psi}_1 + j(\pi/\tau)v_1\underline{\Psi}_1$$

$$0 = R_2' \underline{i}_2 + p\underline{\Psi}_2 + j(\pi/\tau)(v_1 - v)\underline{\Psi}_2$$

$$f_m = (3\pi/2\tau)\,\mathrm{Im}(\underline{\Psi}_2 \underline{i}_2^*) = f_r + mpv \tag{2}$$

in which:

$$\underline{\Psi}_1 = L_1 \underline{i}_1 + L_m \underline{i}_2;\; \underline{\Psi}_2 = L_m \underline{i}_1 + L_2 \underline{i}_2;$$

$$L_1 = L_m + L_{\sigma 1};\; L_2 = L_m + L_{\sigma 2} \approx L_m \tag{3}$$

The leakage inductance of the non-magnetic linear induction motor armature is sufficiently small and is neglected ($L_{\sigma 2} \approx 0$). In the relations (2), $\underline{i}_1$, $\underline{\Psi}_1$, $\underline{i}_2$ and $\underline{\Psi}_2$ are eliminated and the small-variation approach for all the variables is applied. Using the Laplace Transform one can get for the electromagnetic force variation of the robot linear motor the form:

$$\Delta F_m = G_{U1}\Delta U_1 + G_{V1}\Delta V_1 - G_V \Delta V \tag{4}$$

Let G_F be the transfer function connected with the variation of the tractive force ΔF_t:

$$\Delta F_t = \Delta F_m - \Delta F_r \tag{5}$$

The ROBLIN transport block diagram is illustrated in fig. 1:

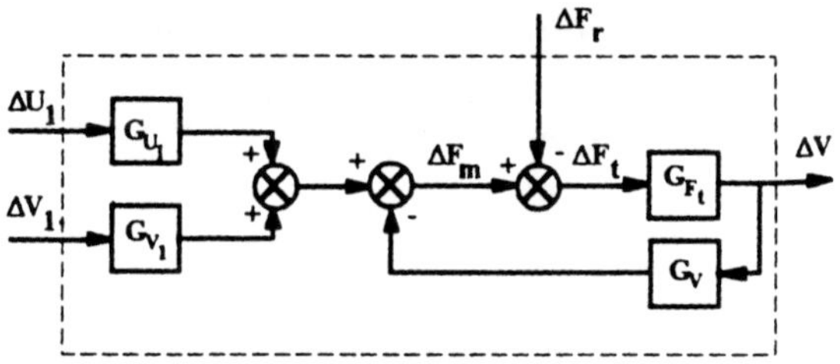

Fig. 1 Basic ROBLIN block diagram

ΔV is obtained with relation:

$$\Delta V = G_1 \Delta U_1 + G_2 \Delta V_1 - G_3 \Delta F_r \tag{6}$$

in which:

$$G_1 = \frac{G_{U1}}{1/G_{Ft} + G_V};$$

$$G_2 = \frac{G_{V1}}{1/G_{Ft} + G_V};$$

$$G_3 = \frac{1}{1/G_{Ft} + G_V} \tag{7}$$

The robot can be controlled with one of the following procedures:
- the command voltage amplitude is modified at constant frequency ($\Delta U_1 \neq 0$; $\Delta V_1 = 0$);
- the command voltage frequency is modified at constant amplitude ($\Delta V_1 \neq 0$; $\Delta U_1 = 0$);
- the command voltage frequency and amplitude are simultaneously and independently modified ($\Delta U_1 \neq 0$; $\Delta V_1 \neq 0$);
- the command voltage frequency and amplitude are simultaneously and independently modified at constant motor flux ($\Delta U_1 \neq 0$; $\Delta V_1 \neq 0$; $\Delta U_1/\Delta V_1 = U_1/V_1 = \text{const.}$).

The transfer function G_1, G_2 and G_3 in (7) are polynomial fractions with maximum fourth polynomial numerators and fifth polynomial denominators. For practical design methods the reduced form can be used for these transform functions. In this respect the ROBLIN robot transfer function controlled by changing the command voltage amplitude is given by (Mastacan, 1991):

$$G_1 = \frac{K_R(T_e s+1)}{2T_m T_e s^2 + T_m s + 1} \quad (8)$$

in which:

$$K_R = 2v_1 s' / U_{1\,max} \quad (9)$$

$$T_m = 2mR_2' v_1^2 L_1^2 / 3U_{1\,max}^2 L_m^2 \quad (10)$$

$$T_e = L_m L_{\sigma 1} / L_1 R_2' \quad (11)$$

3. ROBOT SPEED FUZZY CONTROL

3.1. Basic Problems

The knowledge and expertise of process operators are utilized in fuzzy control to obtain the control rules of the if ... then form. To control a low dimension process an operator takes a decision on the command signal c after observing the error signal e (the difference between reference signal and the process output signal) and the rate of change of the error Δe (King and Mamdani, 1977; Sugeno, 1985; Lee, 1990a). In control rules terms these actions can be formulated as:

if e is A and Δe is B then c is C (12)

A robot speed fuzzy controller can be realized following the approach described above.

To control the transport robot speed the command law is given by:

$$c = f(e, \Delta e, \Delta c) \quad (13)$$

The control algorithm is:

$$c_i = c_{i-1} + K_i \Delta c_i \quad (14)$$

The problem is to minimize the error signal e when the controller input reference signal v_r is variable. To solve this problem in the control algorithm (14) two variables are introduced; Δc and K:

$$\Delta c = g(e, \Delta e) \text{ and } K = h(e, \Delta c) \quad (15)$$

To implement the control algorithm (14), four fuzzy variables are defined: $\tilde{e}$, $\Delta\tilde{e}$, $\Delta\tilde{c}$ and $\tilde{K}$. All the variables e, Δe, Δc and K are quantized into 11 points ranging from a change of -5 steps through 0 to +5 steps. The quantized values are associated to a fuzzy partition with six terms: LARGE NEGATIVE, SMALL NEGATIVE, ZERO, SMALL POSITIVE, LARGE POSITIVE and ANY. A trapezoidal form of the membership functions is chosen.

Table 1 Fuzzy sets and their membership functions

	-5	-4	-3	-2	-1	0	1	2	3	4	5
LP	0	0	0	0	0	0	0	0	.4	.8	1
SP	0	0	0	0	0	.4	.8	1	.8	.4	0
ZE	0	0	0	.4	.8	1	.8	.4	0	0	0
SN	0	.4	.8	1	.8	.4	0	0	0	0	0
LN	1	.8	.4	0	0	0	0	0	0	0	0

The membership function of fuzzy set ANY equals to 1 for the whole universe of discourse. Two sets of decision rules are defined to obtain the fuzzy variables $\Delta\tilde{c}$ and $\tilde{K}$. The inference algorithm is based on fuzzy inference of Mamdani reasoning (Mamdani, 1977) and the output defuzzification strategy on the center of area method (Lee, 1990b).

The values of the command change and the command gain are computed off-line to form two look-up tables. These look-up tables are stocked in memory. In real time control, only quantization is performed and the corresponding look-up table value is obtained. Output quantization is linear, which means that values contained in the look-up tables are multiplied by different constants and then applied to the process as control variables. For fast dynamically plants, the fuzzy controller must to manipulate two look-up tables for each command variable, one for a coarse control and the other for a fine control (Li and Lau, 1989). These look-up tables differ each other for a command variable by the step of quantization.

3.2 Fuzzy Logic Controller

The basic consideration of the fuzzy logic controller for robot linear speed command is:

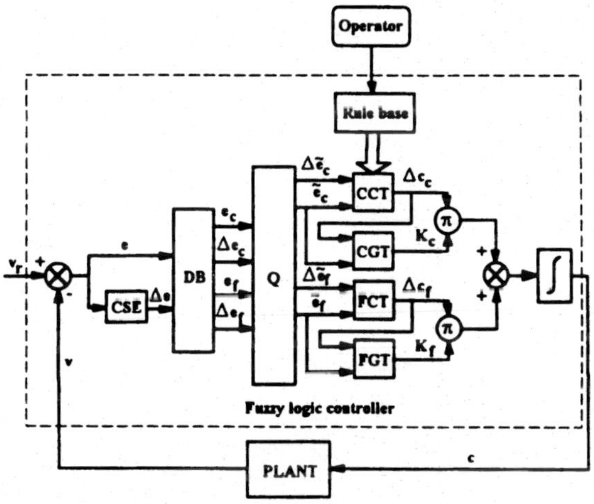

Fig.2 Fuzzy controller block diagram

The above fuzzy logic controller comprises the following blocks: CSE (change in speed error); DB (decision block); Q (quantization); CCT (coarse command change look-up table); CGT (coarse command gain look-up table); FCT (fine command change look-up table); FGT (fine command gain look-up table); ∫ (integral block). The plant comprises a digital-analog converter, an electric inverter, the ROBLIN transport robot, a linear speed transducer, and an analog-digital converter. The fuzzy logic controller is implemented on an IBM-PC.

4. APPLICATION

4.1 ROBLIN Linear Transport Robot Dynamics

To evaluate the performances of the speed fuzzy logic controller some simulation experiments are conducted for a ROBLIN linear transport robot with parameters in table 2.

Table 2 ROBLIN linear robot specification

Item	Value	Unit
Rated Power	4500	Watt
U_{1max}	$220\sqrt{2}$	Volt
R_1	1.909	Ω
$L_{\sigma 1}$	32.2	mH
L_m	13.5	mH
R'_2	2.615	Ω
m	475	Kg
τ	0.072	m
T_e	0.0036	sec
T_m	5.082	sec
K_R	0.576	m/sec/Volt

Using the command by the linear motor voltage amplitude changing the ROBLIN robot transfer function is given by:

$$G_R = \frac{0.002\,s + 0.576}{0.0369 s^2 + 5.082\,s + 1} \qquad (16)$$

Because $T_m >> T_e$ the dynamics of the transport robot at unit step input signal is similarly to a first order system with T_m and K_R parameters.

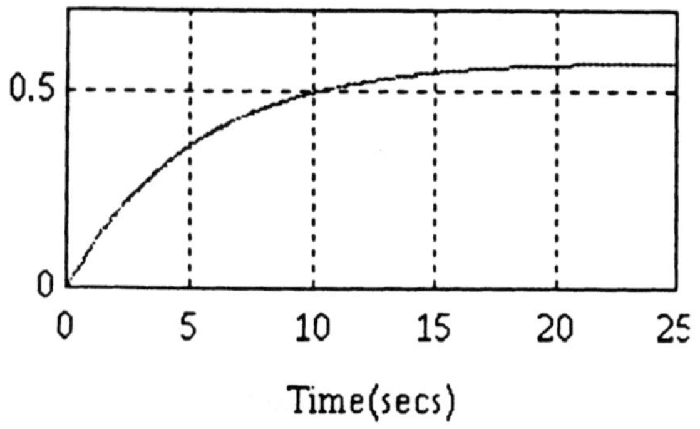

Fig.3 ROBLIN robot step response

The ROBLIN transport robot discrete time model is:

$$v_{i+1} = a\,v_i + b\,c_i \qquad (17)$$

where:

$$a = e^{-T_s/T_m};$$

$$b = K_R(1 - a);$$

$$T_s = 24 \text{ msec.} \qquad (18)$$

4.2 Fuzzy Control Algorithm and Look-up Tables

The following algorithm is utilized for the adjustment of the robot liner speed:

- measure of the variable reference input v_{ri};
- computation of the speed error: $e_i = v_{ri} - v_i$;
- computation of the speed error change: $\Delta e_i = e_i - e_{i-1}$;
- decision for the coarse control or the fine control;
- quantization of the e_i and Δe_i values;
- reading in the look-up tables of the associated quantized command change Δc_i and command gain K_i;
- computation of the control variable: $c_i = c_{i-1} + K_i \Delta c_i$;
- computation of the robot speed: $v_i = a v_{i-1} + b c_i$.

This algorithm is written in Turbo C language and is implemented on an IBM-PC.

The look-up tables for coarse and fine command variables are developed in the following tables.

Table 2 Coarse command change look-up table

e_c	Δe_c										
	-5	-4	-3	-2	-1	0	1	2	3	4	5
-5	-5	-5	-5	-5	-4	-2	0	1	1	2	1
-4	-5	-5	-5	-5	-3	-2	0	1	2	2	2
-3	-5	-5	-5	-4	-3	-1	0	1	2	2	2
-2	-5	-5	-4	-4	-3	-1	1	2	2	3	3
-1	-5	-4	-4	-3	-2	0	1	2	3	3	4
0	-4	-4	-4	-2	-2	0	1	2	3	4	4
1	-4	-4	-3	-2	-1	0	2	3	4	4	4
2	-4	-4	-3	-2	-1	1	2	3	4	5	5
3	-4	-3	-2	-1	0	1	2	3	4	5	5
4	-3	-2	-2	-1	0	1	3	3	5	5	5
5	-3	-2	-1	-1	0	1	3	3	5	5	5

Table 3 Fine command change look-up table

e_f	Δe_f										
	-5	-4	-3	-2	-1	0	1	2	3	4	5
-5	-5	-5	-5	-4	-4	-3	-2	-1	0	1	2
-4	-5	-4	-4	-3	-3	-2	-1	-1	0	1	2
-3	-5	-4	-3	-3	-2	-2	-1	0	1	1	2
-2	-4	-3	-2	-2	-1	-1	-1	0	1	2	2
-1	-3	-2	-2	-1	0	1	0	1	2	2	3
0	-3	-2	-2	-1	0	0	0	1	2	3	3
1	-3	-2	-2	0	1	1	1	2	3	3	4
2	-2	-1	-1	0	1	1	1	2	3	4	4
3	-1	0	0	1	2	2	2	3	3	4	5
4	-1	0	0	1	2	3	3	3	4	4	5
5	0	0	1	2	3	3	3	4	5	5	5

Table 4 Coarse command gain look-up table

e_c	Δc_c										
	-5	-4	-3	-2	-1	0	1	2	3	4	5
-5	5	4	3	2	1	0	-1	-2	-3	-4	-5
-4	4	4	3	2	1	0	-1	-2	-3	-4	-4
-3	3	3	3	2	1	0	-1	-2	-3	-3	-3
-2	2	2	2	2	1	0	-1	-2	-2	-2	-2
-1	1	1	1	1	1	0	-1	-1	-1	-1	-1
0	0	0	0	0	0	0	0	0	0	0	0
1	-1	-1	-1	-1	-1	0	1	1	1	1	1
2	-2	-2	-2	-2	-1	0	1	2	2	2	2
3	-3	-3	-3	-2	-1	0	1	2	3	3	3
4	-4	-4	-3	-2	-1	0	1	2	3	4	4
5	-5	-4	-3	-2	-1	0	1	2	3	4	5

Table 5 Fine command gain look-up table

e_f	Δc_f										
	-5	-4	-3	-2	-1	0	1	2	3	4	5
-5	5	5	4	3	1	0	-1	-3	-4	-4	-5
-4	5	4	3	2	1	0	-1	-2	-3	-4	-5
-3	4	3	3	2	1	0	-1	-2	-3	-3	-4
-2	3	2	2	2	1	0	-1	-2	-2	-2	-3
-1	2	1	1	1	1	0	-1	-1	-1	-1	-2
0	0	0	0	0	0	0	0	0	0	0	0
1	-2	-1	-1	-1	-1	0	1	1	1	1	2
2	-3	-2	-2	-2	-1	0	1	2	2	2	3
3	-4	-3	-3	-2	-1	0	1	2	3	3	4
4	-5	-4	-3	-2	-1	0	1	2	3	4	5
5	-5	-4	-4	-3	-1	0	1	3	4	5	5

4.3 Simulation Results

A conventional PI control of the ROBLIN robot linear speed with unlimited command value is illustrated in fig.4:

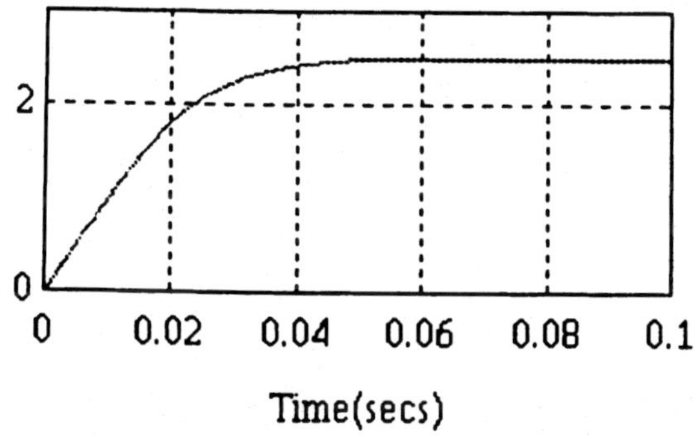

Fig.4 Robot unlimited PI command

The best control performances are obtained. The time response is very short, about 50 msec, and no overshoot is obtained. The ROBLIN robot command is constructive limited in the interval [-10 ... +10] Volt. If the conventional PI control works with this limited command the real time response increases up to 15 sec and a large overshoot is obtained.

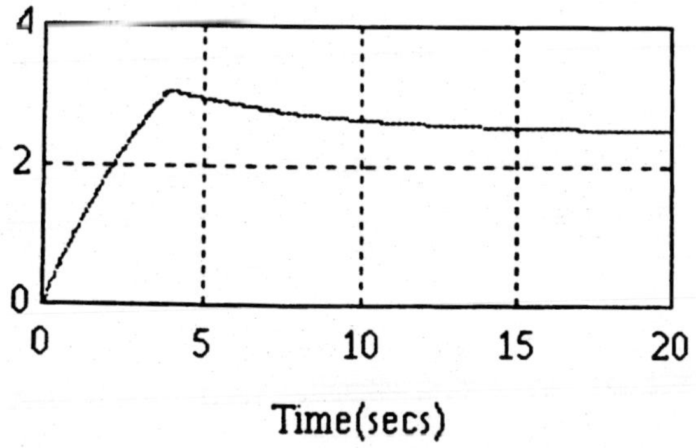

Fig.5 Robot limited PI command

Also, bad performances are obtained for tracking systems with PI control and limited command. The robot speed fuzzy control with variable speed reference signal and limited command is illustrated in fig.6.

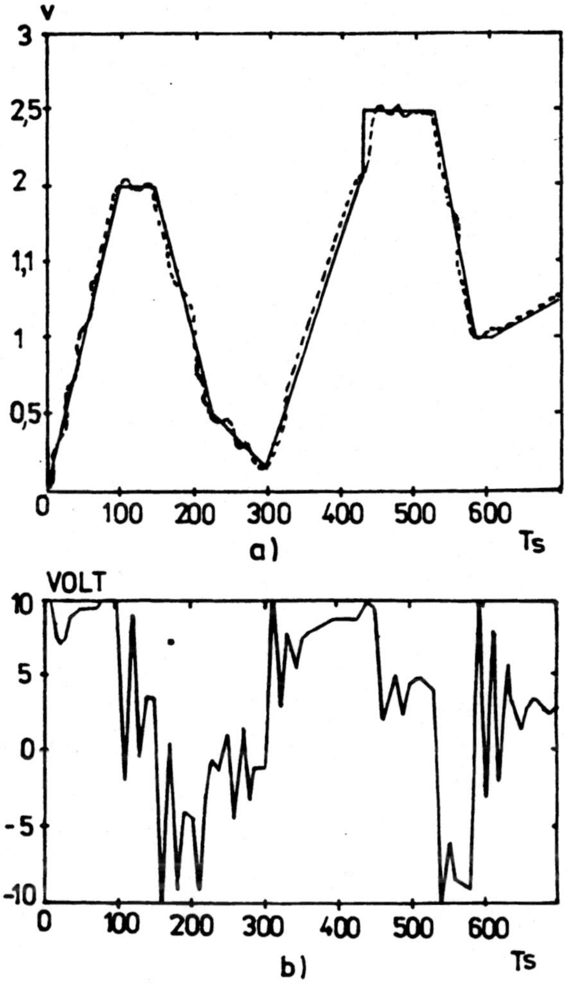

Fig.6 Fuzzy robot speed command

The variable speed reference of fuzzy controller and the output robot linear speed are indicated in fig.6a by a full line and doted line, respectively. The fuzzy controller limited command is represented in fig.6b. Introducing two control modalities (coarse and fine) and in the control law two variables (command change Δc and command gain K) good performances in robot speed control are obtained. The reference speed tracking error of the fuzzy controller is limited at maximum $\pm 6\%$.

5. CONCLUSIONS

The fuzzy control can be implemented not only for slow dynamics plant but also for fast dynamics plant. It was shown via simulated example that a tracking problem can be solved by fuzzy control. The output speed of a linear transport robot tracks the fuzzy controller variable reference input signal. In this respect good performances are obtained.

6. REFERENCES

Cantemir, L., G. Livint, L. Ciobanu, L. Mastacan, M. Boita (1986). The linear induction motors drive of the ROBLIN type manipulators and transport robots. In: *Symp. on Electrical Energy Utilization*, preprints (in Romanian), pp. 317-322, Technical University, Iasi.

Fransua, A., and R. Magureanu (1984). *Electrical machines and drive systems*. Technical Press Oxford, Bucharest.

King, P.J., and E.H. Mamdani (1977). The applications of fuzzy control systems to industrial processes. *Automatica*, **13**, 235-242.

Lee, C.C. (1990a). Fuzzy logic in control systems: fuzzy logic controller - part I. *IEEE Trans. on Systems Man and Cybernetics*, **20**(2), 404-418.

Lee, C.C. (1990b). Fuzzy logic in control systems: fuzzy logic controller - part II. *IEEE Trans. on Systems Man and Cybernetics*, **20**(2), 419-435.

Li, Y.F., and C.C. Lau (1989). Development of fuzzy algorithms for servo systems. *IEEE Control Magazine*, **9**(3), 65-72.

Livint, G., L. Ciobanu, L. Cantemir, and M. Cleju (1986). Driving of transport and positioning robots by linear induction motors. In: *The Fifth National Conference on Electrical Drives*, preprints, Vol.2, pp. 103-108, Technical University, Iasi.

Livint, G., M. Cleju, and D. Zamfir (1987). Considerations on the linear induction motors transfer functions (in Romanian). *EEA Electrotehnica*, **35**(2), 84-90.

Mamdani, E.H. (1977). Applications of fuzzy logic to approximate reasoning using linguistic synthesis. *IEEE Trans. on Computers*, **C-26**, 1181-1191.

Mastacan, L. (1991). Considerations on the position control of transport robots with induction linear motors. In: *SACCS '91 Symp. on Automatic Control and Computer Science*, preprints (in Romanian), pp. 273-278, Technical University, Iasi.

Mastacan, L., I. Olah, and C. Calistru (1994). Robot speed fuzzy control with minimum time response. In: *International Workshop on Fuzzy Technologies in Automation and Intelligent Systems*, preprints (submitted), University of Duisburg.

Sugeno, M. (1985). An introductory survey of fuzzy control. *Information Sciences*, **36**, 57-85.

Titli, A., and R. Ketata (1991). Design and applications of fuzzy algorithm for the speed control of a motor-alternator group. In: *The 13th IMACS World Congress in Computation and Applied Mathematics*, preprints, Trinity College, Dublin.

Fuzzy Components for Fuzzy Control

L. Foulloy, S. Galichet, J. F. Josserand

Laboratoire d'Automatique et de MicroInformatique Industrielle
LAMII / CESALP, Université de Savoie,
41 Avenue de la Plaine, BP 806,
74016 ANNECY CEDEX,
Tel : 50.66.60.00; Fax 50.66.60.20

Abstract:**Fuzzy components are characterized by their ability to handle fuzzy symbolic information. This paper presents their use in fuzzy control. First, we present a general typology of fuzzy controllers. Fuzzy components like fuzzy sensors, fuzzy actuators and fuzzy inference components are then defined. The use of fuzzy components in fuzzy control is finally explained with respect to the typology.**

I. INTRODUCTION

Several methods can be used to implement fuzzy controllers. The most classical one is to use a hight-level programming language to encode the fuzzy controller in a processor or a microcontroller. In order to implement fuzzy controllers a software developing environement like TillShell [1] or Fuzzy Tech [2] could be used. The second approach is to use fuzzy processor such as FP-3000 of OMRON [3], FC110 of Togai Infra Logic [4], Fuzzy 166 of Siemens Inform [5], NeuraLogix NLX230 [6]. Fuzzy co-processor have also been developped (see for example [7]).

In this paper, we develop a new concept called *"fuzzy components"*. Fuzzy components can be used to implement one or several elements of a fuzzy controller among the fuzzification, the inference and the defuzzification.

The paper is organized as follows. In section II., a typology of fuzzy controller is presented. Fuzzy components such as fuzzy sensor, fuzzy inference component and fuzzy actuator are detailled in section III. Finally, an example is presented in section IV.

II. TYPOLOGIE

Let X be a finite set of the numerical values, $\mathcal{L}(X)$ denotes the set of linguistic values associated with the set X. Let Y be any set, $\mathcal{F}(Y)$ denotes the set of the fuzzy subsets of Y. A fuzzy controller is generally separated into three processes known as fuzzification, inference and defuzzification defined as follows:

- Fuzzification is an interface that produces a fuzzy subset from the measurement, i.e. it is a mapping from the set X to a set of fuzzy subsets associated with the controller input, denoted $\mathcal{F}(Y)$.
- The inference is an interface that produces a new fuzzy subset from the result of the fuzzification using the set of rules. The result of the inference is a fuzzy subset associated with the controller output. It can be represented by a mapping from the set $\mathcal{F}(Y)$, associated with the input, to a set of fuzzy subsets associated with the output and denoted $\mathcal{F}(Z)$.
- The defuzzification is an interface that produces a crisp output from the result of the inference, i.e. it is a mapping from the set $\mathcal{F}(Z)$, associated with the output, to the set U.

The set Y related to the controller input can either be the set of numerical values, i.e. X, or the set of linguistic values, i.e. $\mathcal{L}(X)$. In the same way, the set Z related to the controller output can either be the set of numerical values, i.e. U, or the set of linguistic values, i.e. $\mathcal{L}(U)$. Therefore, the previous definitions can be used to provide a nice classification, i.e. a typology, of fuzzy controllers in four classes as shown in Fig. 1 We propose to call a fuzzy controller using the inference g_i, a type i fuzzy controller or simply a FC*i*.

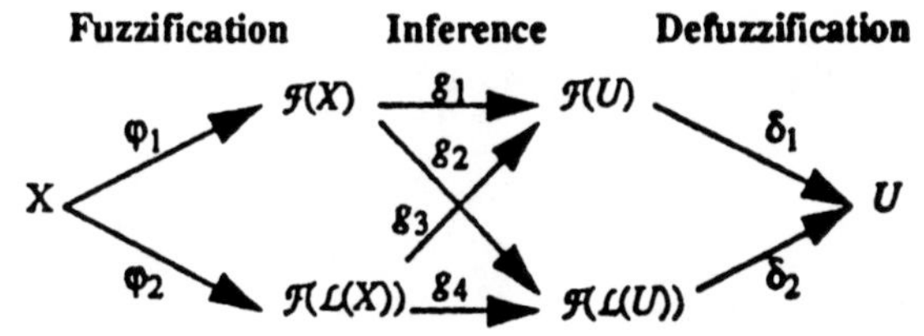

Fig. 1 Typology of fuzzy controllers.

From the general point of view, *fuzzy components* implement one of the previous functions. Therefore, we have to distinguish between two types of components: components dealing with fuzzy sets of numbers and components dealing with fuzzy sets of symbols. We propose to call the former *numerical fuzzy components* and the latter *fuzzy symbolic components*. Because of the small number of symbols used in many applications, fuzzy symbolic components seem more promising to be developed. In the following, we consider only fuzzy symbolic components and we call them simply *fuzzy components*.

III. FUZZY COMPONENTS

III.1. Fuzzy sensors

We call a *fuzzy symbolic sensor* or simply *a fuzzy sensor*, a sensor which is able to perform a *symbolic fuzzification* [9], [10]. From a formal point of view, the symbolic fuzzification can be represented by a mapping from the set X to the set of fuzzy subsets $\mathcal{F}(\mathcal{L}(X))$. In the typology represented in Fig. 1, the symbolic fuzzification is the mapping denoted φ_2.

Using the concept of *fuzzy meaning* [12], we can define a symbolic fuzzification. The fuzzy meaning of a symbol is defined by a mapping $\tau: \mathcal{L}(X) \rightarrow \mathcal{F}(X)$. This mapping links the symbolic and the numerical representation. It is now quite easy to defined a new mapping $\iota: X \rightarrow \mathcal{F}(\mathcal{L}(X))$, called *a fuzzy description*, whose membership function is:

$$\mu_{\iota(x)}(L) = \mu_{\tau(L)}(x) \quad (1)$$

The crisp interpretation of Eq. (1) is that a numerical value is described by a symbol if it belongs to the meaning of this symbol, i.e. $\iota(x) = \{L \mid x \in \tau(L)\}$. The fuzzy description performs a symbolic fuzzification. An example of such a process is given in Fig. 2 where the fuzzy description of the measurement $x = 12$ cm is obtained from the fuzzy meanings.

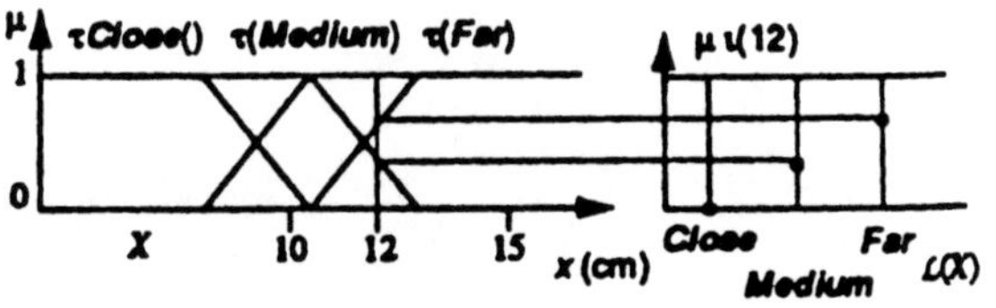

Fig. 2 : Fuzzy meanings and fuzzy description of $x = 12$ cm.

III.2. Fuzzy actuators

We call *a fuzzy symbolic actuator*, or simply *a fuzzy actuator*, an actuator able to perform *a symbolic defuzzification*. From the formal point of view, symbolic defuzzification is a mapping from the set of fuzzy subsets $\mathcal{F}(\mathcal{L}(U))$ to the set U. In the typology given in Fig. 1, the symbolic defuzzification is the mapping δ_2.

Several methods of symbolic defuzzification can be envisaged. We describe here an indirect method where we consider the mapping δ_2 as the composition of two mappings. The first one, denoted δ_3: $\mathcal{F}(\mathcal{L}(U)) \rightarrow \mathcal{F}(U)$, converts the result of the inference (which is a fuzzy subset of $\mathcal{L}(U)$) into a fuzzy subset of U. The second one is just an usual numerical defuzzification δ_1. By numerical defuzzification, we mean classical methods like the centre of gravity (COG) or the mean of maxima (MOM).

Knowing the meanings of the symbols of $\mathcal{L}(U)$, the mapping δ_3 can be computed as the image of fuzzy set of symbols by the meanings of these symbols. If the membership function of input is denoted μ_F, the combination-projection principle gives:

$$\forall u \in U, \mu_G(u) = \mu_{\delta_3(F)}(u) =$$
$$\sup_{L \in \mathcal{L}(U)} T(\mu_F(L), \mu_{\tau(L)}(u)). \quad (2)$$

Using the COG method and the triangular norm T=*min*, the numerical value computed by the defuzzification interface of the fuzzy actuator is given by Eq. (3).

$$u = \delta_2(F) = \frac{\int_{u \in U} g(u) \times u \times du}{\int_{u \in U} g(u) \times du} \quad \text{with} \quad (3)$$

$$g(u) = \sup_{L \in \mathcal{L}(U)} (\min(\mu_F(L), \mu_{\tau(L)}(u)))$$

Two examples are shown in Fig. 3 The fuzzy meanings of the symbol N, Z, P are respectively labelled ❶, ❷ and ❸. The membership functions μ_G are represented in bold. The first one is obtained using the *min* as the T-norm. It can be used to derive several equivalent representation of Mamdani's controllers [8]. The second is obtained using the smallest T-norm (i.e T (x, 1) = x, T (1, y) = y otherwise T (x, y) = 0). One can recognize the defuzzification method called the method of the heights which is, from our point of view, difficult to justify when it is presented as a numerical defuzzification.

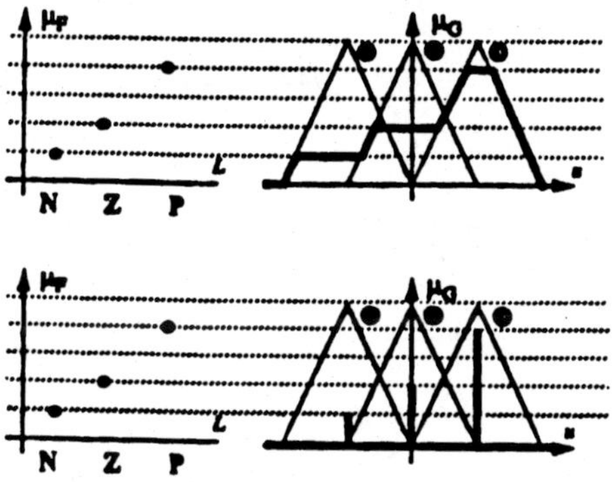

Fig. 3 : Examples of mapping δ3.

III.3. Inference components

A fuzzy symbolic inference component, or simply *a fuzzy inference component*, is able to perform fuzzy symbolic inference. It is repsented in the typology by the mapping g_2. A fuzzy inference component receives fuzzy subsets of symbols. Given a set of rules, these fuzzy subsets are processed by means of the classical methods of inference, called the generalized modus ponens. This component produces a fuzzy subset of symbols as the result of the inference. Let $\mathcal{L}(Y)$ and $\mathcal{L}(Z)$ be two sets of symbols. From the formal point of view, an inference component is defined by a mapping from a set of symbolic fuzzy subsets, e.g. $\mathcal{F}(\mathcal{L}(Y))$, to another set of symbolic fuzzy subsets, e.g. $\mathcal{F}(\mathcal{L}(Z))$. Let E be a fuzzy subset of $\mathcal{F}(\mathcal{L}(Y))$. Let Γ be a fuzzy graph of the cartesian product $\mathcal{L}(Y) \times \mathcal{L}(Z)$ representing the set of rules. The result of the inference is a fuzzy subset F of $\mathcal{F}(\mathcal{L}(Z))$ defined by Eq. (4) where T denotes a triangular norm.

$$\forall L' \in \mathcal{L}(Z), \mu_F(L') = \sup_{L \in \mathcal{L}(Y)} T(\mu_E(L), \mu_\Gamma(L, L')) \quad (4)$$

IV. APPLICATION TO FUZZY CONTROL

Applications of fuzzy components to fuzzy control are straightforward. For example, connecting a fuzzy sensor to a fuzzy inference component which is itself connected to a fuzzy actuatur leads to a fuzzy controller classified as FC4 in the typology. Fig. 4 shows this representation and the type of components to be used. However, this is only one possibility. Depending on the configuration, a fuzzy component is able to perform one or several functions for example, a fuzzy sensor and a fuzzy inference can be implemented in the same component. Fuzzy sensors and fuzzy actuators can also be used with classical controllers such as Mamdani's fuzzy controllers or Sugeno's fuzzy controllers as shown in [11], [8].

Fuzzy components are implemented as generic fuzzy cells. In our application, fuzzy cells are realized by means of a single 80C196 microcontroller. Communications are performed using an I2C multimaster bus. The format of the data exchange is given Fig. 5.

Receiver Address	Sender Address	Data Type	Identifier	Data . . .

Fig. 5 .: The format of data exchange.

Each cell has an identification address which is used in the communication protocols. For example, in Fig. 4, the fuzzy sensor is implemented by means of a fuzzy cell whose address is 20. The others are implemented by means of fuzzy cells whose address are respectively 40 for the fuzzy inference, 30 for the fuzzy actuator and 60 for the supervisor.

Fuzzy cells are characterized by their ability to handle and exchange fuzzy symbolic information. The

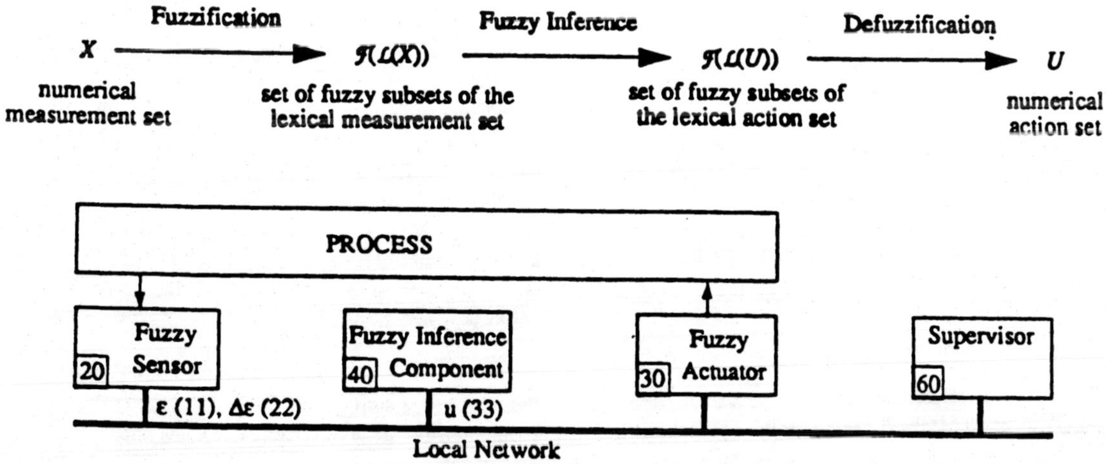

Fig. 4 : FC4 representation and its implementation.

communication format includes the address of the sender fuzzy cell and the address of the receiver fuzzy cell. Fuzzy cells are actually able to handle five types of data. A byte is used to encode the type of data to be exchanged. The encoding is 1 for double, 2 for fuzzy partition, 3 for fuzzy subset, 4 for symbol and 5 for text.

Another byte is used as to identify the variable. A number is given to each variable. In this byte, the first half byte is the variable number. The second half byte is used, if necessary, to encode a dependence on another variable. For example if the variable number 4 is a double then it leads to the following sequence :

		1	4	x	.....

Now, if the variable number 4 is a fuzzy subset which is defined with respect to the partition number 3, the sequence is :

		3	4	3	.....

The configuration of the cells is performed through the network using a specific language, called PLICAS, which is not detailed in this paper.

Now, let us develop an example. Assume that we want to implement a fuzzy controller with two inputs, the error denoted ε and the error variation denoted $\Delta\varepsilon$, and one output, the control action denoted u. This controller is based on three fuzzy components, implemented in three separate cells, as shows in Fig. 4.

The configuration of the fuzzy sensor is such that the fuzzy partitions associated with the inputs ε and $\Delta\varepsilon$ are represented in Fig. 6. Due to the encoding, the grade of membership are represented as bytes. Therefore, the grade of full membership, i.e. 1, is represented by 255.

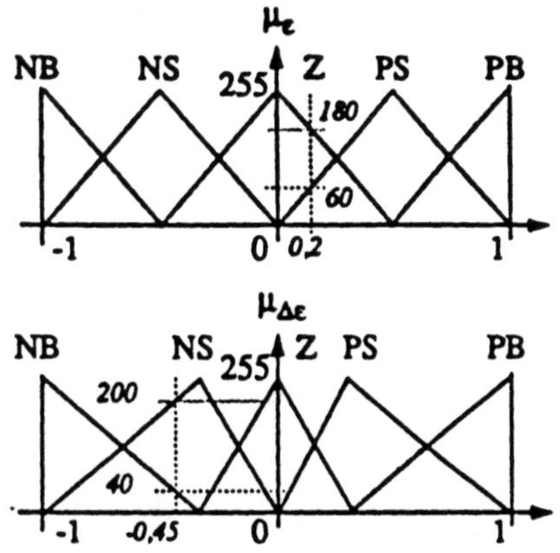

Fig. 6 : The fuzzy partitions associated with the inputs.

The configuration of the fuzzy inference component is such that a Mac Vicar-Whelan table [13], i.e. an antidiagonal rule base, is implemented as shown in Fig. 7.

Δε \ ε	NB	NS	Z	PS	PB
PB	Z	PS	PB	PB	PB
PS	NS	Z	PS	PB	PB
Z	NB	NS	Z	PS	PB
NS	NB	NB	NS	Z	PS
NB	NB	NB	NB	NS	Z

Fig. 7 : Mac Vicar-Whelan table.

Finaly, the configuration of the fuzzy actuator is such that the fuzzy partition associated with the output is given in Fig. 8.

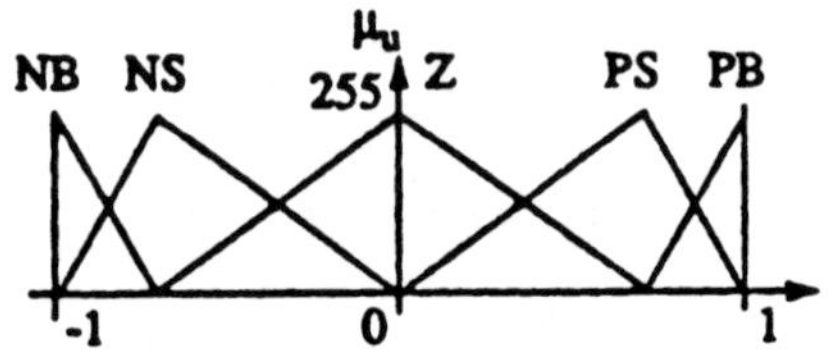

Fig. 8 : The fuzzy partition associated with the output.

In order to show the behaviour of the system, we give in the following, the data exchanged between fuzzy components when the fuzzy sensor has measured an error $\varepsilon = 0.2$ and an error variation $\Delta\varepsilon = -0.45$.

According to the configuration of the fuzzy sensor, the fuzzy description of $\varepsilon = 0.2$ is {0/NB, 0/NS, 180/Z, 60/PS, 0/PB}, while the fuzzy description of $\Delta\varepsilon = -0.45$ is {40/NB, 200/NS, 0/Z, 0/PS, 0/PB}. Thus, the fuzzy sensor (address 20) has to send to the fuzzy inference component (address 40), two information whose type are fuzzy subsets (i.e. type number is 3). The message sent by the fuzzy sensor is represented Fig. 9 where bytes are coded in hexadecimal.

40	20	3	11	0	0	B4	3C	0	3	22	28	C8	0	0	0

Fig. 9 : The message sent by the fuzzy sensor.

In this message, the first identifier 11 means that the fuzzy subset which follows is associated with the variable number 1 (here ε), and described in the fuzzy partition number 1. It means that a variable may be described in different partitions if necessary. The second identifier 22 specifies that the fuzzy subset which follows is associated with the variable number 2 (here $\Delta\varepsilon$), described in the fuzzy partition number 2.

Once the fuzzy inference component received the message, it is able to perform the inference. With two inputs and the triangular norm *min* the inference is given by Eq. (5).

$$\forall L'' \in \mathcal{L}(U),\ \mu_F(L'') = \max_{L \in \mathcal{L}(\varepsilon), L' \in \mathcal{L}(\Delta\varepsilon)} \min(\min(\mu_{\iota(\varepsilon)}(L), \mu_{\iota(\Delta\varepsilon)}(L')), \mu_\Gamma(L, L', L'')) \qquad (5)$$

The membership function $\mu_\Gamma(L, L', L'')$ is provided by the rules. For example, $\mu_\Gamma(Z,Z,Z) = 1$ but $\mu_\Gamma(Z,Z,NS) = 0$ because there is no rule that links "ε is Z and Δε is Z" to "u is NS". According to information sent by the fuzzy sensor, only four rules will be activated as shown in Fig. 7 (frame in bold). Thus , the result of the inference process is {40/NB, 180/NS, 60/Z, 0/PS, 0/PB}. Therefore, the fuzzy component inference (address 40) sends to the fuzzy actuator (address 30) one fuzzy subset associated with the variable number 3, described in the partition number 3. The message is represented in Fig. 9.

30	40	3	33	28	B4	3C	0	0

Fig. 10 : The message sent by the fuzzy inference.

As soon as this message is received by the fuzzy actuor, the symbolic defuzzification is computed. Using, for example, the smallest triangular norm in Eq. (2), (see III.2.) we obtain the membership function, represented in bold, in Fig. 8, which leads to the value u = -0.55. Finally this value is applied to the process.

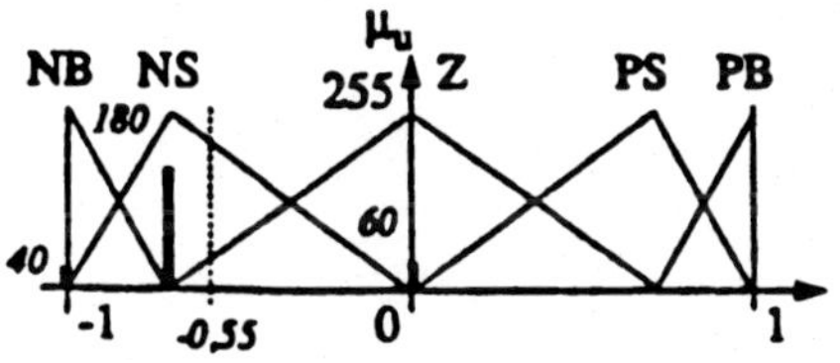

Fig. 11 : The numerical value computed by the fuzzy actuor cell.

CONCLUSION

This paper has presented the concept of fuzzy components. Such components can be easily used to realize a whole fuzzy controllers by means of connections through a local network. They can also be used to implement a specific interface, in particular the fuzzification, for usual fuzzy controllers.

Fuzzy actuators are unusal components derived from the classification of fuzzy controllers. They introduce a new representation of the control action in terms of symbols. It leads to new problems like the tuning, which has to be done at the sensor and actuator levels.

REFERENCES

[1] Togai Infra Logic, Inc., TILShell *"An Object-Based, fuzzy logic development environment"*

[2] INFORM fuzzy TECH *"fuzzy logic development system"* Development Tools Handbook, support solutions for Intel's embedded microcomputers.

[3] OMRON Corporation, *"FP-3000 Digital Fuzzy Processor"*.

[4] Togai Infra Logic, Inc., *"FC110 Digital Fuzzy Processor"*.

[5] INFORM, *"FUZZY -166 : Quick Reference Guide"*.

[6] NeuraLogix Data Sheet NLX230, *"High speed Fuzzy Logic Controller"*.

[7] Zhang J., Raczkowsky J., Mayer F., *"Development of Fuzzy Co-Processor for Real-Time Control"*, Proc. of the 1st European Congress on Fuzzy Intelligent Technologies (EUFIT 93), Aachen, Germany, Sept. 93, pp. 91-97.

[8] Foulloy L., *"Typologie des contrôleurs flous"*, Arago 14, Masson, 1994.

[9] Foulloy L., Benoit E., Mauris G., *"Applications of fuzzy sensors"*, Proc. of the European Workshop on Industrial Fuzzy Control and Applications (IFCA 93), Terrassa, Spain, April 93.

[10] Mauris G., Benoit E., Foulloy L., *"Fuzzy symbolic sensors : From concept to applications"*, Int. Journal of Measurement, in Press.

[11] Foulloy L., Galichet S., *"Fuzzy Sensor for Fuzzy Control"*, Int. Journal of Uncertainty, Fuzziness and Knowledge-based Systems, Vol. 2.1, 1994.

[12] Zadeh L.A., *"Quantitative fuzzy semantics"*, Information Sciences, Vol. 3, 1971, pp. 159-176.

[13] Mac Vicar-Whelan P.J., *"Fuzzy sets for man-machine interaction"*, Int Journal of Man-Machine Studies, 1976, No 8, pp. 687-697.

FUZZY SLIDING MODE CONTROL APPLICATION TO LARGE TIME VARYING SYSTEMS

P. Cerf, S. Boverie, F. Vernières

SIEMENS AUTOMOTIVE SA, Laboratoire MIRGAS, avenue du Mirail, BP 1149, 31036 Toulouse Cedex, FRANCE

Abstract– This paper presents an application of fuzzy control techniques for large time varying systems. This study is based on an analogy between fuzzy control and sliding mode control techniques. First, special attention will be payed to the analysis of sliding mode control properties, next a generalization of sliding mode control structure to large time varying systems is presented and finally this concept will be extended to fuzzy controllers. Some application results are shown in the last section.

Keywords: Fuzzy controllers, sliding mode, large time varying systems,process control,

1. INTRODUCTION

Fuzzy logic which appeared back in the 1960's, has just recently shown signs of strong growth. One of its prefered fields of application is process control. Fuzzy control sometimes presents an interesting alternative when the processes to be controlled are non–linear, difficult to model or are time varying. It also has the advantage of presenting a good trade–off between cost and performance.Different authors like Mac Vicar–Whelan, have proposed some formalization of the fuzzy control structures. We present in this paper an extension of the Mac Vicar–Whelan structure in order to take into account large time varying systems. This study is based on the analogy between fuzzy control and non–linear sliding mode control. The first section is dedicated to the sliding mode control. The second one presents the new control structure. Finally, we present some results obtained on an application in the last section.

2. SLIDING MODE CONTROL

2.1. SLIDING MODE CONTROL

The sliding mode control stems from the studies on relay based analogic control (Fig.1). The principles of this robust non–linear control method are quite simple and can be analyzed in the phase plane (ε , $\dot{\varepsilon}$) (Fig 2).

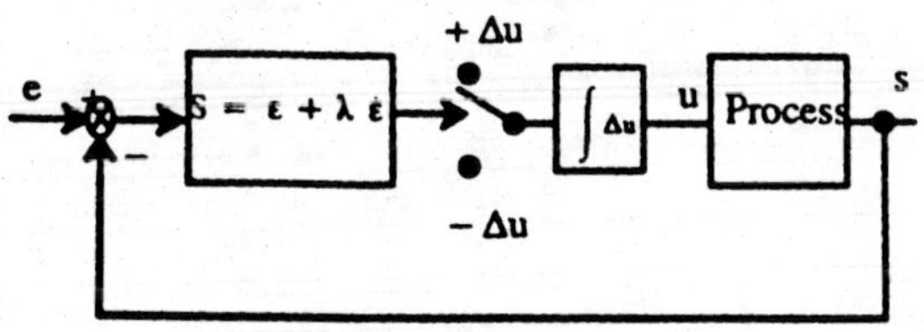

Fig.1 Sliding mode control structure

We define in the phase plane a line $S = \varepsilon + \lambda \dot{\varepsilon} = 0$ called the sliding line.

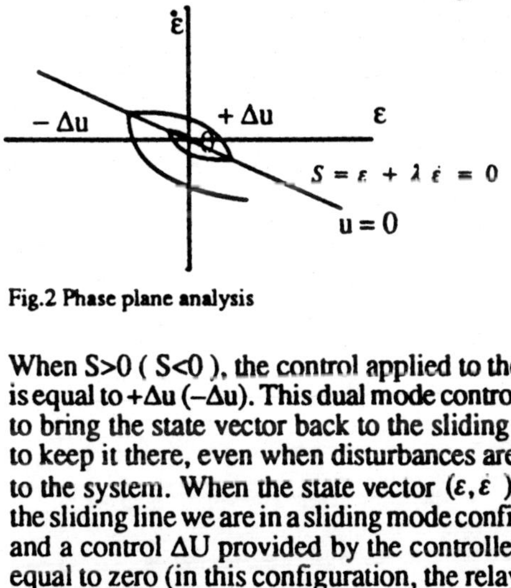

Fig.2 Phase plane analysis

When S>0 (S<0), the control applied to the system is equal to $+\Delta u$ ($-\Delta u$). This dual mode control intends to bring the state vector back to the sliding line and to keep it there, even when disturbances are applied to the system. When the state vector ($\varepsilon, \dot{\varepsilon}$) reaches the sliding line we are in a sliding mode configuration and a control ΔU provided by the controller will be equal to zero (in this configuration, the relay is in an oscillation mode). The system converges to the equilibrium ($\varepsilon = \dot{\varepsilon} = 0$) following S.

Then the equation ($S = \varepsilon + \lambda \dot{\varepsilon} = 0$) defines the behavior of a first order system; therefore, the behavior of the controlled system in the sliding mode phase is equivalent to that of a first order system.The choice of the parameter λ, which determines the time constant of the first order system could take into account the following considerations:

- λ could be determined according to the perturbation frequency we want to filter.
- λ could be determined according to the dynamic response we want to assign to the process. It is possible, in theory, to attain any dynamic response we want, if we increase the control value Δu (Fig 3).

Unfortunately, in physical processes, the values of the command are always limited and consequently λ must be chosen according to these limitations.

The determination of λ is related to the study of the open loop system behavior when we apply the maximum admissible control value.

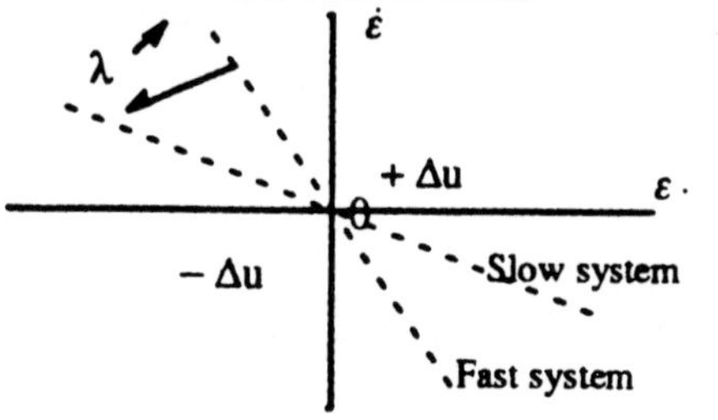

Fig.3 Influence of the process dynamic response

2.2. Generalized sliding mode control

Sliding mode control is not without problems. A rapidly varying, large amplitude control effort ± Δu applied at each S sign changing is not always imaginable (non–linearity problems, oscillations, hard constraints on the controlled process on the actuators, . . .).

With the development of digital control, it was possible to extend this concept and to consider that area around the sliding line corresponding to a smooth control. This control could be chosen proportional to the distance from the sliding line. Beyond this area, the control values are saturated (Fig4).

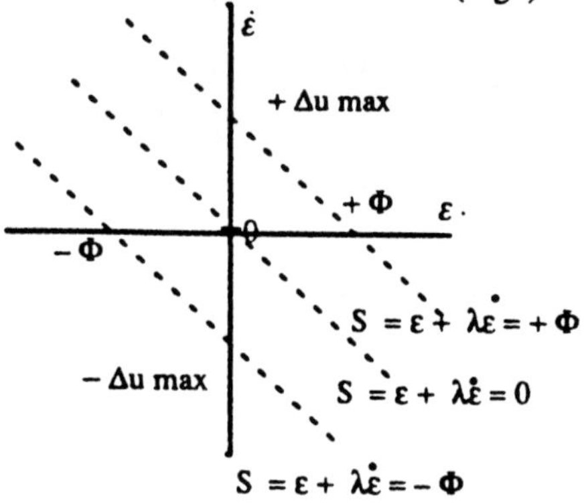

Fig.4 rotation of the sliding line for a time variant system

The control law can be expressed by (Fig 5):

If $S > \phi$ then $\Delta u = \Delta u_{max}$

If $S < -\phi$ then $\Delta u = -\Delta u_{max}$

If $-\phi < S < \phi$ then $\Delta u = K.S$

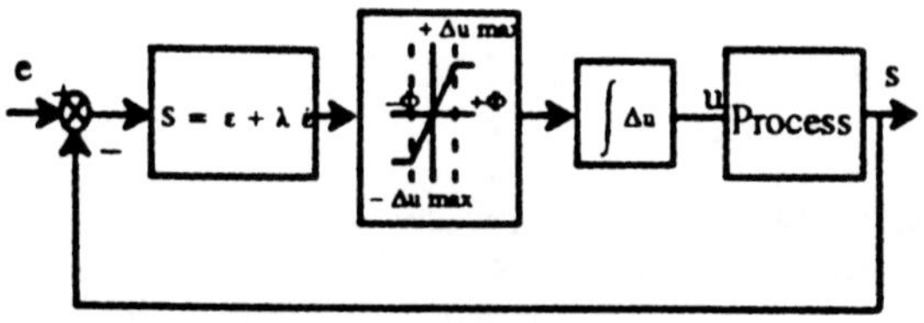

Fig.5 Generalized sliding mode controller structure

2.3. Application of the generalized sliding mode control to large time varying processes

If we consider a physical system with a time varying dynamic response, we have seen in section 2.1 that this can imply some constraints for the choice of the parameter λ. A new control structure is proposed Figure 6.

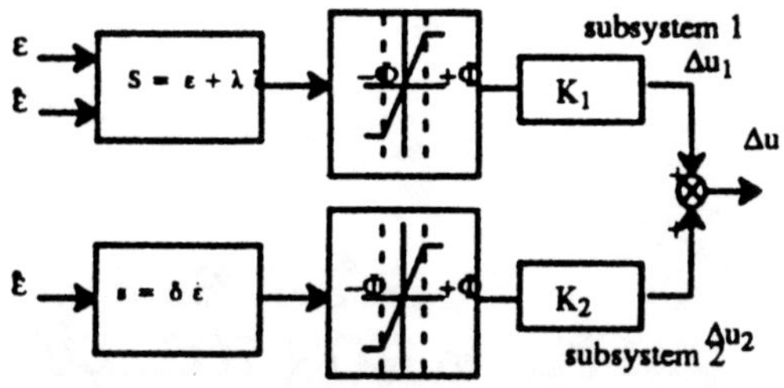

Fig.6 Representation of the sliding lines in the phase plane

For the first sub–system we can define:

$S \; (S = \varepsilon + \lambda \dot{\varepsilon} = 0)$ with

If $S > \Phi$ then $\Delta u_1 = K_1\Phi$

If $S < -\Phi$ then $\Delta u_1 = -K_1\Phi$

If $-\Phi < S < \Phi$ then $\Delta u_1 = K_1 S$

and for the second sub–system:

$s \; (s = \delta\dot{\varepsilon} = 0)$ with

If $s > \Phi$ then $\Delta u_2 = K_2\Phi$

If $s < -\Phi$ then $\Delta u_2 = -K_2\Phi$

If $-\Phi < s < \Phi$ then $\Delta u_2 = K_2 s$

Combining the output of each sub–system, in the non–saturated area we have:

$\Delta u_1 = K_1[\varepsilon + \lambda \; \dot{\varepsilon}] \qquad \Delta u_2 = K_2 \, \delta \, \dot{\varepsilon}$

$$\Delta u = \Delta u_1 + \Delta u_2 = K_1[\varepsilon + (\lambda + \frac{K_2 \delta}{K_1})\dot{\varepsilon}]$$

Then a new sliding line is defined:

$$S' = \varepsilon + \lambda' \dot{\varepsilon} = 0 \quad \text{with} \quad \lambda' = \lambda + \frac{K_2 \delta}{K_1}$$

It is possible to rotate the sliding line around the point $(\varepsilon = \dot{\varepsilon} = 0)$ by modifying the values of the gains K_i, and the equivalent dynamic response defined by the sliding line could be easily adapted to the change of the process.

When both the subsystems are saturated we can write:

$|\Delta u| = |K_1\phi \pm K_2\phi|$

In the transient areas, when only one of the two systems are saturated we have (Fig. 7):

$\Delta u = K_1[\varepsilon + \lambda \; \dot{\varepsilon}] \pm K_2\phi$

or $\Delta u = K_2 \, \delta \, \dot{\varepsilon} \pm K_1\phi$

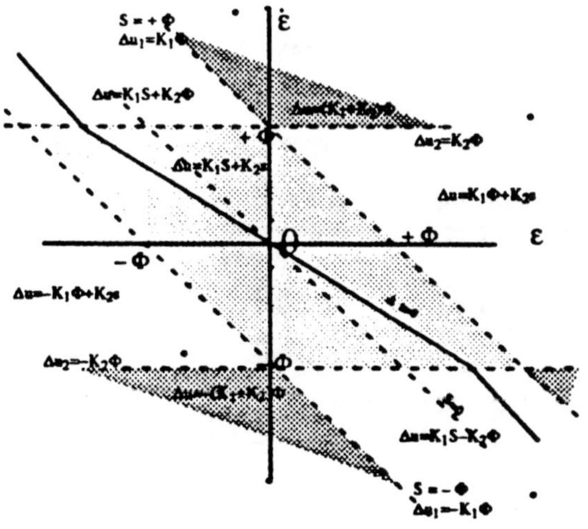

Fig.7 Rotation of the sliding line

This structure allows a global rotation of the control surface around the point ($\varepsilon = \dot{\varepsilon} = 0$). Some distortions of this control surface will appear in the transient areas. Nevertheless it is possible to choose ϕ, λ and δ in order to minimize their influences.

3. EXTENSION TO THE FUZZY CONTROL

The structure of a conventional fuzzy controller uses as inputs the error ε and the derivative of the error $\dot{\varepsilon}$. Mac Vicar–Whelan has suggested a symmetrical control structure (Fig.8) :The fuzzy control value is equal to zero on the diagonal line and positive (negative) above (below) this line; the value of this fuzzy control increases progressively to the distance from this line.

ε \ $\dot{\varepsilon}$	NB	NM	NS	ZE	PS	PM	PB
PB	ZE	PS	PM	PB	PB	PB	PB
PM	NS	ZE	PS	PM	PB	PB	PB
PS	NM	NS	ZE	PS	PM	PB	PB
ZE	NB	NM	NS	ZE	PS	PM	PB
NS	NB	NB	NM	NS	ZE	PS	PM
NM	NB	NB	NB	NM	NS	ZE	PS
NB	NB	NB	NB	NB	NM	NS	ZE

Fig.8 structure of fuzzy controller proposed by Mac Vicar–Whelan

It is possible to represent the equivalent control surface in a three dimensional space (Δu, ε , $\dot{\varepsilon}$). We can see that the control surface is non linear, depending on the choice of the modal points of ε and $\dot{\varepsilon}$ (Fig 9).

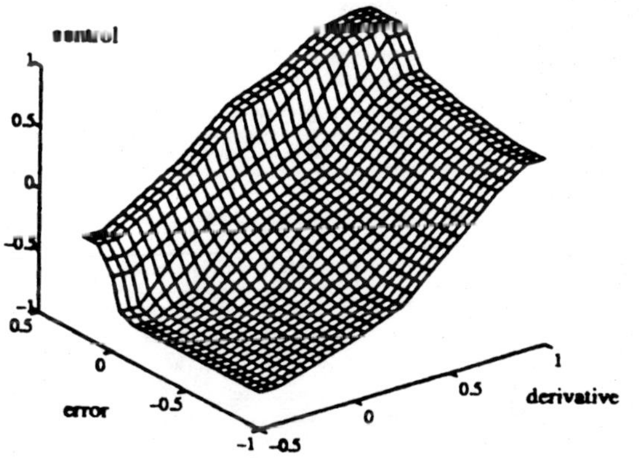

Fig.9 Fuzzy control surface in a three dimensionnal space

The intersection between this surface and the plane (ε , $\dot{\varepsilon}$) gives the zero control curve $S(\varepsilon, \dot{\varepsilon}) = 0/\Delta u = 0$ (Fig 10) which is equivalent to a sliding curve (Palm, 92).

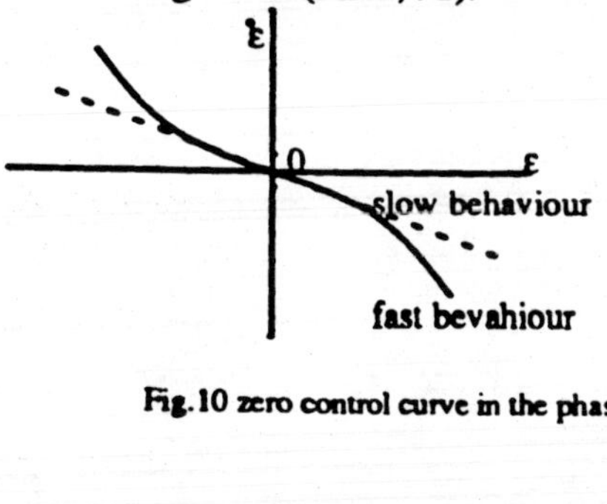

Fig.10 zero control curve in the phase plane

The control action can be determined as a non–linear function: $\Delta u = F(S(\varepsilon, \dot{\varepsilon}))$. If we examine the control surface, it is possible to define for each modal point (ε_i , $\dot{\varepsilon}_j$) a gain K ij (in the non saturated area) like $\Delta u = K_{ij} . S(\varepsilon, \dot{\varepsilon})$. S being an image of the distance to the sliding surface. Between these modal points the control values are calculated with a non–linear or a linear interpolation (Raymond *et al*, 93) (depending on the fuzzy operators: min–max, .–max) and Δu can be expressed as: $\Delta u = K(\varepsilon, \dot{\varepsilon}) S(\varepsilon, \dot{\varepsilon})$ .As for the conventional sliding modes, this sliding curve defines the behavior of the system. The control value Δu applied to the process tries to bring the state vector (ε , $\dot{\varepsilon}$) on this curve. If we consider a process with time varying response this involves some constraints in the choice of the sliding curves. In order to take into account the different dynamics, it is necessary to rotate the sliding curve around the point ($\varepsilon = \dot{\varepsilon} = 0$). To do that it is possible to consider the same structure as for the conventional sliding modes (Fig 11).

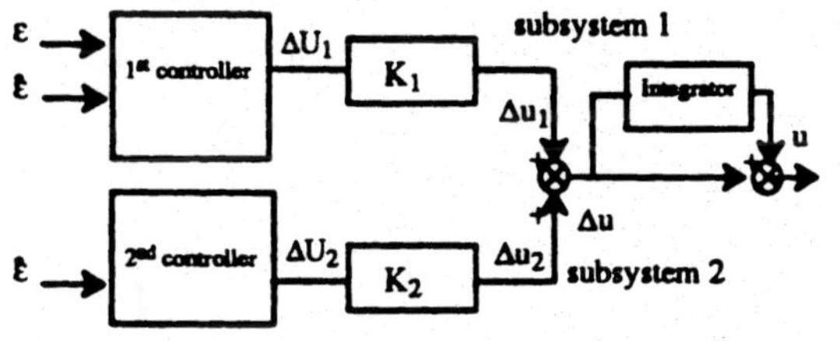

Fig.11 Fuzzy sliding mode controller structure

We have in the non–saturated area:

$$\Delta u_1 = K_1[\, K(\varepsilon, \dot{\varepsilon})\, S(\varepsilon, \dot{\varepsilon})\,] \qquad \Delta u_2 = K_2\, k(\dot{\varepsilon})_1$$

$$\Delta u = \Delta u_1 + \Delta u_2$$

$$\Delta u = K_1\, K(\varepsilon, \dot{\varepsilon})\, [S(\varepsilon, \dot{\varepsilon}) + \frac{K_2 k(\dot{\varepsilon})}{K_1 K(\varepsilon, \dot{\varepsilon})} . \dot{\varepsilon}\,]$$

$\Delta u = K_1 K(\varepsilon, \dot{\varepsilon})\, S'(\varepsilon, \dot{\varepsilon})$ and S' is the new sliding curve. In the saturated areas we have:

$|\Delta u_{max}| = |K_1 \pm K_2|\, \Delta U_{max}$ with $\Delta U_{1max} = \Delta U_{2m}$

In some cases it should be suitable to choose K_1 and K_2 as $K_1+K_2=cte$. This allows to avoid any modification of the maximum value of the control variable. Some distortions will appear between saturated and non–saturated areas, nevertheless the errors are relatively small and have no influence on the closed loop control behavior.

As for conventional sliding modes, by modifying the values of the gains K_i this structure allows a global rotation of the control surface around the point ($\varepsilon = \dot{\varepsilon} = 0$).

4. APPLICATION

This concept has been tested in real time on an automotive 4 stroke engine for controlling idle speed. Idle speed control is always considered as a difficult problem.It must be efficient every time and assuming a compromize between conflictual goals:

- for comfort, it must ensure a good stability of the engine speed and a good robustness with respect to load disturbances (power steering, lights, air conditionner, . . .),
- for consumption and pollutant emissions the regulation set point must be as low as possible.

On the other hand the features of an automotive engine are very different according to its utilization (stationary and rolling vehicle). In the first case the

process to be controlled is only the engine (engine inertia) in the second case, the process to be controlled is the overall vehicle through the mechanical coupling (vehicle inertia), which corresponds to a tremendous change of parameter values. The sensors located on the engine allow the detection of these different phases, and allow one to supervise the gains in order to rotate the sliding curve.
The results obtained show the quality of the control : the decrease of the engine speed must be smooth, without irregularities or undershoot.

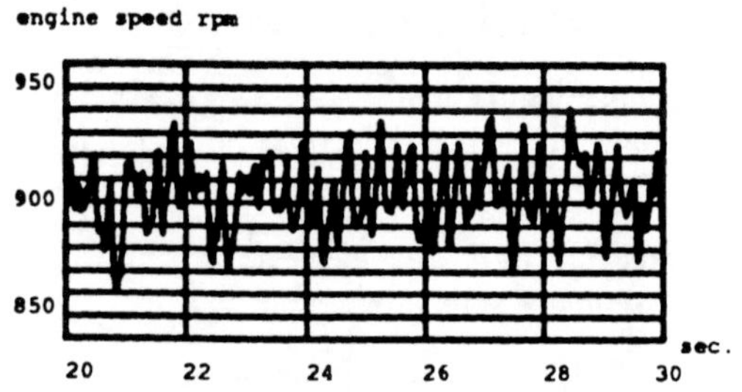

Fig.12 Idle speed regulation without load

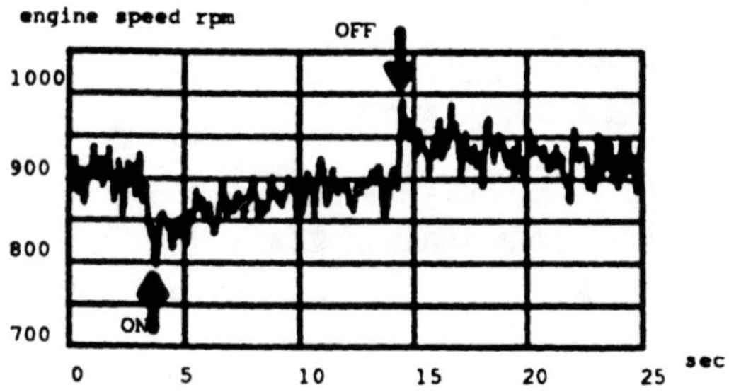

Fig.13 Idle speed regulation with load (power steering)

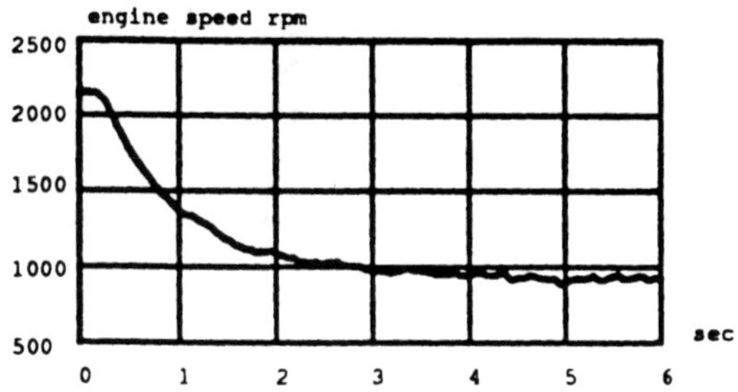

Fig.14 Reaching idle speed set point without load

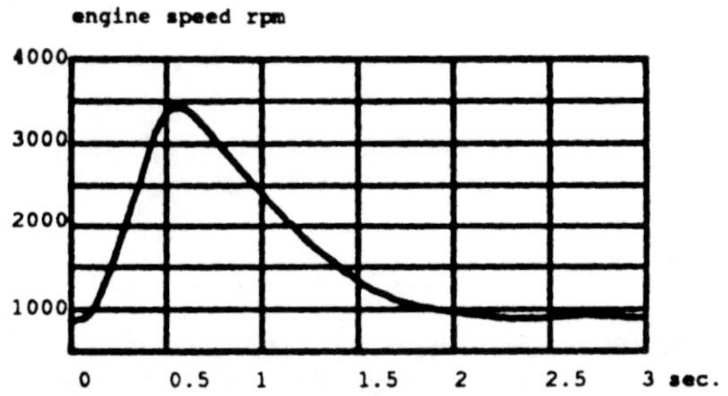

Fig.15 Reaching idle speed set point with load (power steering+ventilator)

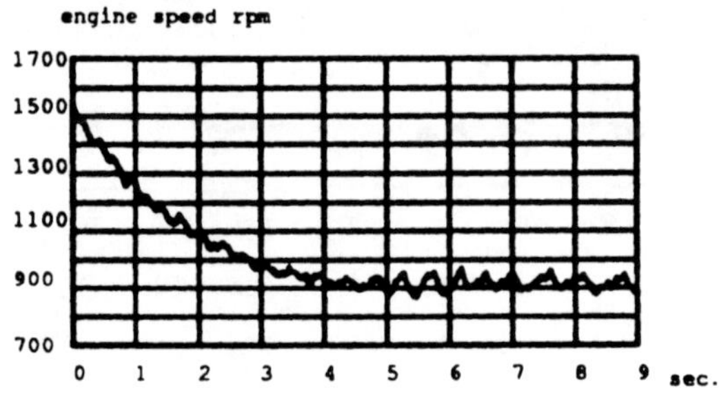

Fig.16 Reaching idle speed set point with load and with running vehicle

5. CONCLUSION

In this study, we have developed an original control structure based on an analogy between a non–linear control technique like sliding mode and the fuzzy control technique. This structure allows to take into account large dynamic variations of the processes. This algorithm has been tested succesfully in real time on an idle speed controller.

REFERENCES

C. Raymond, S.Boverie, J.M. Le Quellec, "Practical realization of fuzzy controller– comparison with conventionnal methods", Proceedings of the EUFIT symposium in Aachen, Sept 1993.

R. Palm, " Sliding mode fuzzy control" FUZZ–IEEE 92', San Diego, March 92

A FUZZY CONTROLLED PNEUMATIC GRIPPER FOR ASPARAGUS HARVESTING

G. Mattiazzo, S. Mauro, T. Raparelli, M. Velardocchia

Department of Mechanics, Politecnico di Torino
Corso Duca degli Abruzzi 24 - 10129 - Torino - Italy
Tel. +39-11-5646929 Fax +39-11-5646999

ABSTRACT. The paper presents a grasping unit for asparagus harvesting based on a commercially available pneumatic gripper. Highly reliable low-cost tactile sensors were developed to determine contact pressure during grasping. Experimental tests were conducted on all system components in order to achieve a thorough knowledge of the physical phenomena which determine system behaviour. A fuzzy control was developed on the basis of this experimental data. The paper then presents and discusses experimental data for grasping tests with different sizes of asparagus and different reference contact pressures to demonstrate the versatility and robustness of the system.

1. INTRODUCTION

Agricultural automation has become a major issue in recent years. Some of the most important applications in this extensive research area are those involving the automation of fruit and vegetable harvesting operations, which are generally time-consuming, tiring, and particularly demanding. This paper deals with asparagus harvesting. Asparagus has a very fast growth cycle and must thus be harvested selectively. To guarantee the best product quality, moreover, harvesting should be performed twice a day, and is thus costly and labor intensive.

The objective of the work described herein was to produce a low cost asparagus harvesting gripper which controls contact pressure.

A number of earlier gripper hand designs were analyzed. The literature provides several examples of anthropomorphic hands [1], [2], [3], [4], which could be used successfully in this application, though their costs are considered to be too high. There is also a fair number of simpler grasping units designed for specific operations which can provide high performance in terms of speed and reliability at lower cost [5], [6], [7].

From these observations, it was decided to base the grasping hand on a commercially available gripper, adding suitable sensors and control.

As for the gripper hand as a whole, the approach used in developing sensors was that of finding a low-cost, highly reliable solution. Consequently, the idea of using strain gauges was rejected because of the high cost involved. On the other hand, attempts to use resistive sensors or conductive rubber sensors were unsuccessful, as tests showed these components to be insufficiently reliable. A technique which is rarely used [8] but which is in fact highly effective in this type of application is the use of water-filled fingertips capable of detecting applied contact pressure. As this technique was chosen, several types of fingertip were designed and tested for serviceability and reliability. The gripper with these sensors was supplied via high frequency PWM driven digital valves installed in a layout which enables them to operate as flow proportional valves. Finally, system control was implemented using fuzzy logic. This choice was dictated by the need to manage system nonlinearities and the low development costs associated with this type of control, which result from the speed with which optimum results can be achieved [9], [10] [11]. Experimental tests demonstrated that the resulting system can handle asparagus of different sizes and at different stages of growth thanks to the robustness of both the control and the sensors.

2. SYSTEM DESCRIPTION

Double-acting pneumatic gripper with two parallel fingers was used (Figure 1).

The gripper is actuated by two pneumatic cylinders, one for opening (1) and both for closing (1-2). The axis of both cylinders is perpendicular to the motion of fingers (3). Movement is transmitted from the cylinders by means of two bell cranks (4). The fingers run on two roller bearings (5).

Two 80 mm long brackets (6) were installed on the fingers in order to provide more room for mounting the sensors and to ensure a firmer grip when harvesting asparagus.

Each fingertip consists of a small rubber bag filled with water. A schematic view of the system is shown in Figure 2. Two sensors (of which only one is shown) (1) are connected to each other by a 4 mm diameter tube (2). In addition, the sensors are connected to a pressure transducer (3) that supplies a voltage signal which is directly proportional to contact pressure.A linear amplifier must be added between the pressure transducer and the computer. The hydraulic circuit is completed by a compensator (4) used to establish a predetermined pressure in the fingertips and to permit rapid hydraulic system filling.

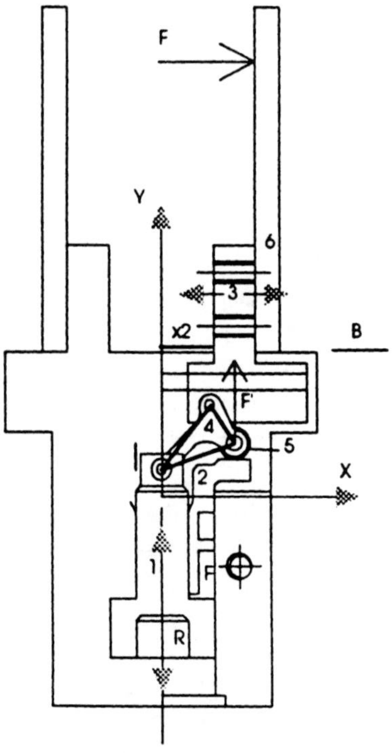

Fig. 1. Schematic view of gripper

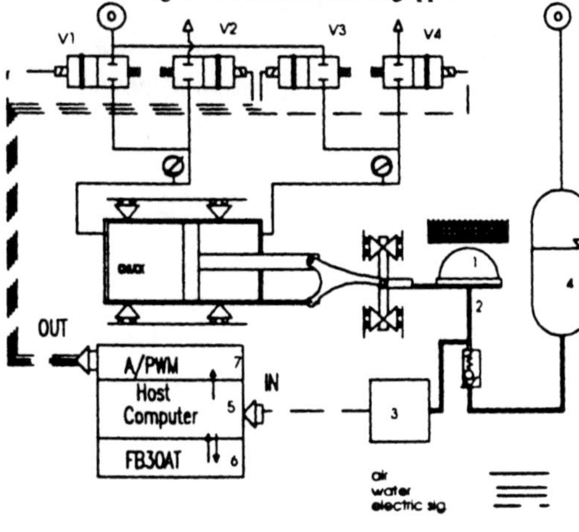

Fig. 2. Schematic view of system

The system is managed by a PC (5) equipped with an Omron FB-30AT fuzzy board (6) and the appropriate I/O boards.

Supply and discharge of both chambers are controlled by two pairs of high frequency two-position two-way valves V1, V2 and V3, V4. One valve pair controls exhaust, while the other controls supply. Thanks to their high performance, these valves can be used with PWM control as flow proportional units [12].The analog control signal is translated into a PWM signal by a dedicated board (7).

3. CLAMPING FORCES

Clamping forces must be known in order to develop a functional control. The clamping forces which can be applied to an object in relation to the latter's distance along the longitudinal axis of the fingers are affected by friction forces. With constant supply pressure, it was found that clamping force drops as the grip point moves farther away from the base B of the gripper. This can be explained by the fact that as the grip point moves farther away, the moment unloaded on the bearings rises (Figure 1), thus increasing force F and hence the corresponding friction force T which opposes motion.

The effect was evaluated through an experimental test. Clamping force versus distance from the base was measured by means of a load cell adhesive-bonded to one finger (Figure 3).

Curve 3a) illustrates how clamping force drops as distance increases for any given supply pressure. In Figure 3b), it will be noted that clamping force, which should increase linearly with supply pressure in the gripper chambers, shows a saturation phenomenon above certain pressure values.

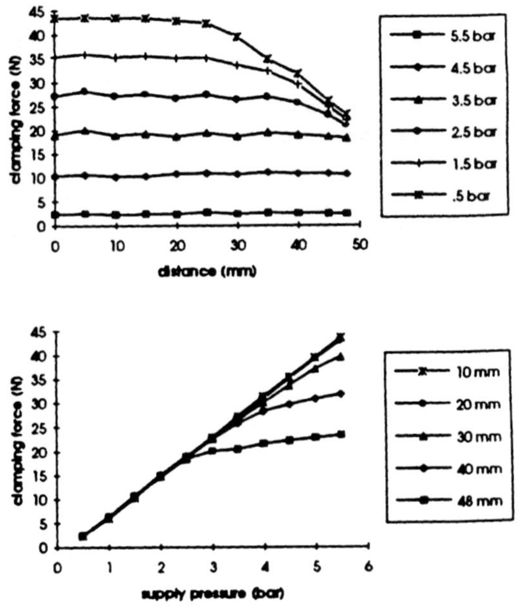

Fig. 3a), 3b). Clamping force curves

4. FINGERTIP DESIGN

A major fingertip design requirement was to ensure good watertightness and high stiffness. The fingertips consist of a metal structure and a rubber diaphragm, which acts as the sensing element to which contact forces are applied. The fingertip's overall stiffness depends on the type of metal frame and on the type of rubber. The following considerations were borne in mind at the design stage:

- the contact surfaces should be as constant as possible during the grasping operation so as to maintain a constant ratio between clamping pressure and force.

- fingertip shape and dimensions should make it possible to grasp and handle objects of varying size.
- fingertips should be very thin so as to make the most of the fingers' closing stroke.

A number of different solutions as shown in Fig. 4 were analyzed in the light of these considerations:

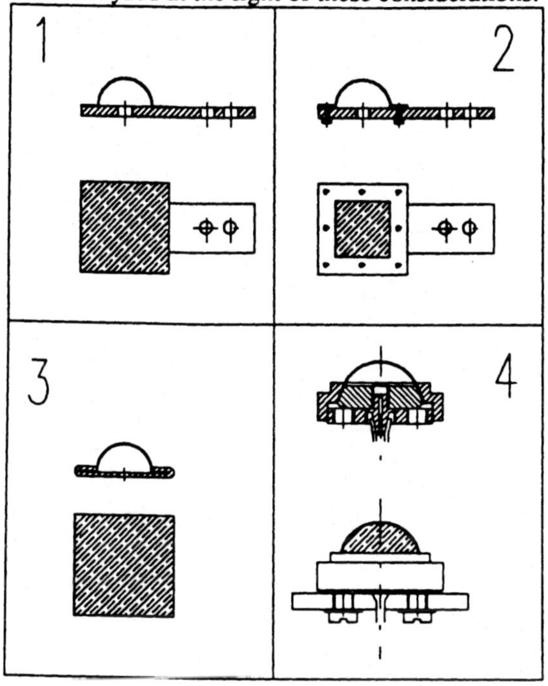

Fig. 4. Fingertips

In type 1), the rubber diaphragm is adhesive-bonded to the metal structure. As this solution shows reliability problems, models 2) and 3) were developed, in which retention is accomplished by a combination of adhesive bonding and mechanical means. However, the reliability problems were not entirely eliminated. Finally, solution 4) was developed, where the structure consists of three metal parts whose function is to clamp the rubber diaphragm between two truncated conical surfaces. A spacer was added to prevent the rubber from being damaged by relative motion between the two conical surfaces during assembly. This solution makes it possible to avoid using adhesive, and thus increases the fingertip's mechanical strength.

Two different types of rubber with the following mechanical properties were tested:

α) Tensile strength 2000 N/cm^2
Ultimate elongation 700 %

β) Tensile strength 6500 N/cm^2
Ultimate elongation 80 %

These tests where conducted to obtain a relation between the pression transducer signal (expressed in volt) and closing force developped by the gripper's actuators with different pressions in chamber R (Fig.1).

Figure 5 shows experimental results for rubber α and β when grasping an object with flat and cylindrical surfaces (a 15 mm thick square prism and a 10 mm diameters cylinder).

Clamping force depends on the difference of pressure ΔP between front (F) and rear (R) chamber pressures.

Curves are obtained with different supply pressure for chamber R.

Figure 5 shows:

a) that type α rubber exhibits a higher voltage response for any given supply pressure than type β rubber.

b) that the response curve is approximately linear when grasping a square prism, while this characteristic no longer obtains when grasping a cylinder.

The nonlinearity of the curves for grasping a cylinder is due to the increase in the contact surface resulting from diaphragm deformation. The stiffer rubber provides a curve with fewer variations in linearity, particularly when grasping a flat object.

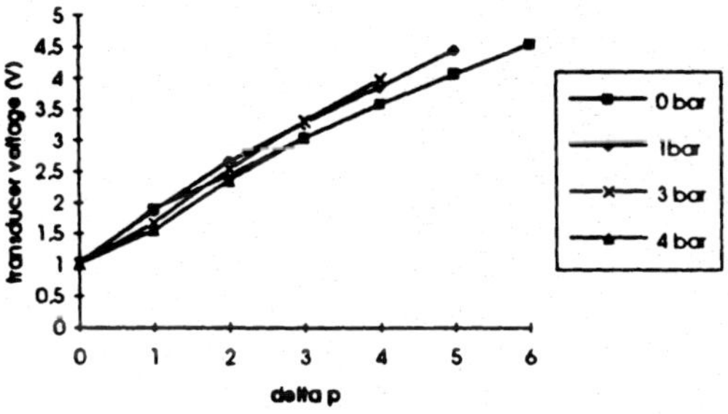

Fig. 5a). Rubber α - flat surface

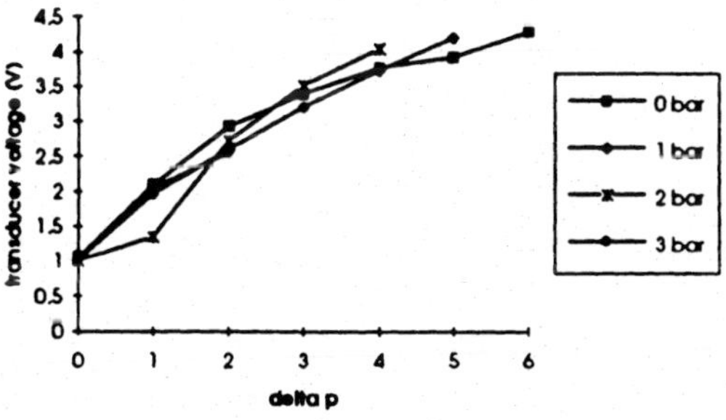

Fig. 5b). Rubber α - cylindrical surface

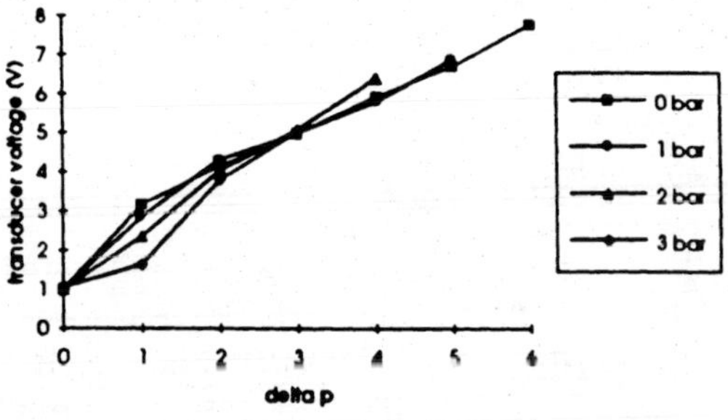

Fig. 5c). Rubber β - flat surface

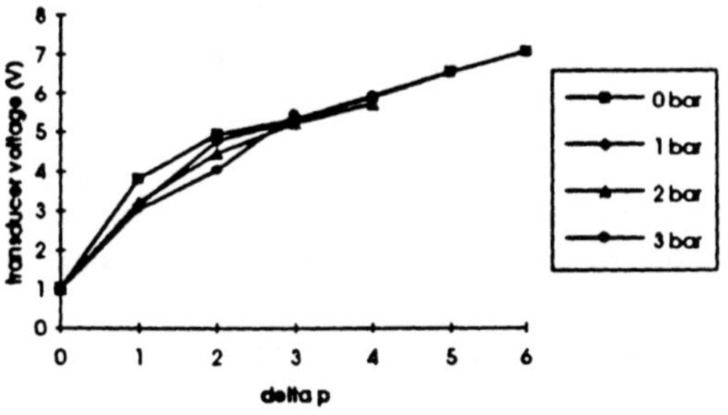

Fig. 5d). Rubber β - cylindrical surface

Fig. 5. Contact pressures

5. VALVES

Supply and discharge of both gripper chambers are controlled by two-position two-way digital poppet valves operating at high frequency ($>100Hz$). When these valves are used with PWM driving mode in the installation layout shown in Fig. 2, their behaviour is essentially that of a two-position three-way valve where flow rate is proportional to percentage duty-cycle t/T where t is the valve open time and T is the total period. However, this proportionality is not perfectly linear: though the valve response times are low, they nevertheless entail a dead band around zero and another one around the maximum flow rate. There are thus upper and lower limits to the percentage duty-cycle in which effective modulation can be accomplished.

The actual flow rate curves for a single valve were measured using a bench equipped with a compensator reservoir to ensure regular flow upstream and downstream of the valve and a flow meter installed downstream. Upstream and downstream pressures and flow rates were also measured.

Experimental results, carried out with a supply pressure of 6 bar show that the dead band around zero broadens considerably at high downstream pressures. For example with a 5 bar downstream pressure duty-cycle must be greater than 30% to open the valve.

6. DYNAMIC TESTS

Tests were conducted to measure the response times of the valves and of the complete system using both the soft and hard rubber.

The command consisted of a step signal.

For the valve, downstream pressure was measured by a transducer. A Cetop tube ensured uniform downstream flow. This type of test made it possible to determine valve delay time, which was 2 ms measured as the time lapsed between 10% of the command signal and 50% of the output signal.The dynamic performance of the fingertip was then studied.

The delay time between the solenoid command signal and transducer output voltage was measured. These magnitudes are heavily influenced by the type of rubber used. Thus, with the rubber types indicated in the paragraph on fingertip design, we have:

- type α rubber: Delay time ≈ 0.08 s
- type β rubber: Delay time ≈ 0.05 s

These results and those obtained from the static tests indicate that it is advisable to use type β rubber, which provides better response times and better sensor linearity.

7. CONTROL

The input signal is the pressure sensed by the fingertips, while output magnitudes are two voltages varying from 0 to 5 volts and converted by a suitable PWM driver for the two chambers' supply and discharge valves. In particular, a signal between 0 and 2.5 volt activates the valve connected to discharge, while the supply valve is activated by a signal in the 2.5 to 5 volt range.

This type of operation limits hardware costs, as a single channel is used to command a pair of valves. The 0 to 2.5 volt range is translated linearly into a variation in the duty-cycle of the valve connected to discharge from 100 to 0%, while the 2.5 to 5 volt range corresponds to a variation in the duty-cycle of the supply valve from 0 to 100%. Consequently, both valves are fully closed at 2.5 volt.

The fuzzy coprocesser used handles 12-bit digital input and output signals. The controller makes it possible to define the membership functions for the input magnitudes with a trapezoidal shape, while those for the processed magnitudes must be singleton.

The additional board hosting the fuzzy coprocessor is managed by software written expressly for the purpose, which superintends the dialog between the fuzzy board and the I/O board and establishes the value for the reference pressure. The cycle time of this software is around 3 ms. Duty-cycle period was selected to be of 20 ms.

The parameters summarized above were determined a priori, and were not changed during control tuning. The tuning operation thus involved only the fuzzy set.

The minimum-maximum center of gravity method was used for inferences.

The adopted rule set is as follows:

rule n°	IN0	U1	U2
1	PM	PM	NM
2	PS	PS	NS
3	ZR	ZR	ZR
4	NS	NS	PS
5	NM	NM	PM

where IN0 is the difference between the references contact pressure and the real contact pressure, U1 and U2 are the control voltages of the gripper's front and rear chamber valves respectively.

The membership functions were optimized starting from a first, perfectly symmetrical set. With this set, the system was unstable and insufficiently reactive. This poor reactivity can be attributed to the fact that the NS and PS functions for magnitudes U1 and U2 are too close to the value 2.5 V, at which both valves are closed. In fact, when rules 2 or 4 are involved with a low degree, a valve control voltage is established which corresponds to a percentage duty-cycle falling in the dead band found during the tests. To overcome this problem, the values corresponding to labels PS and NS must be farther from the central position (2.5 V). The values for U1 must be higher than the corresponding values for U2 to enable the gripper to develop the desired forces.

Supply valves of both chambers should never be openet too abruptly in order to avoid grasped objeect sliping or being damaged.

The input magnitude's membership functions were also modified. The function corresponding to label ZR was narrowed to improve system reactivity. At the same time, the range of the NM and PM function was broadened to ensure faster initial closing and to prevent excessively high peaks from being reached.

Optimum results were achieved with a slight adjustment to functions NS and PS. The corresponding membership functions are illustrated in Fig. 6.

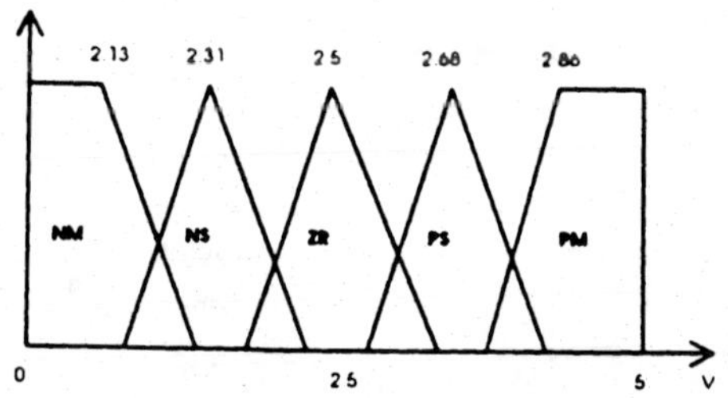

Fig. 6a). Membership functions for magnitude IN0, starting set

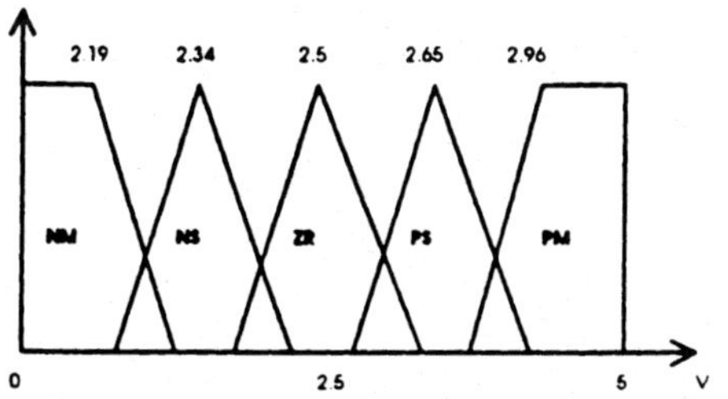

Fig. 6b). Membership functions for magnitududе IN0, optimum Set

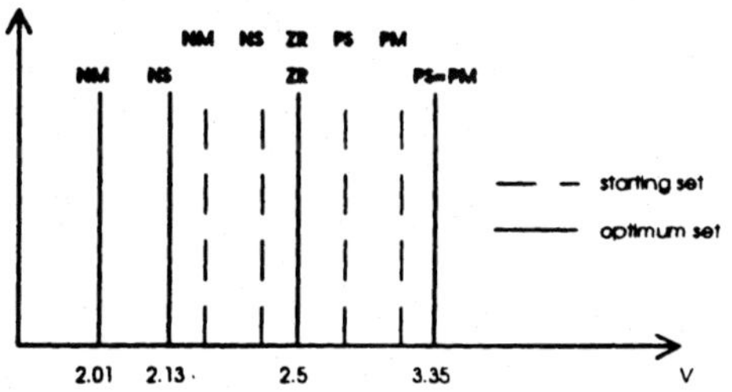

Fig. 6c). Membership functions for magnitude U1

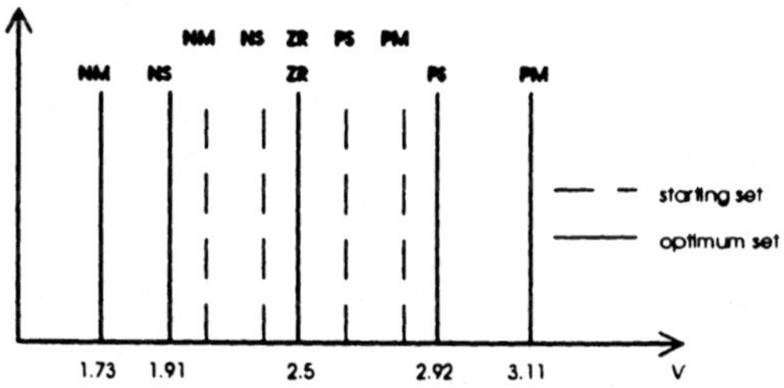

Fig. 6d). Membership functions for magnitude U2

8. GRASPING TESTS

Figure 7 shows contact pressure versus time under various operating conditions. Two stalks of asparagus of different diameter are grasped using two reference pressures corresponding to the pressures which the asparagus can withstand at different stages of growth. External disturbance was introduced by manually forcing the fingers to close. The pressure peaks on the graphs indicate when this disturbance was applied. In all cases, the system reacted immediately when the gripper was released, with an abrupt drop in contact pressure which was then recovered in a very short time so that the object did not slip. The system's prompt reaction at the moment the small stalk was removed from the soil can be observed in Fig. 7d).

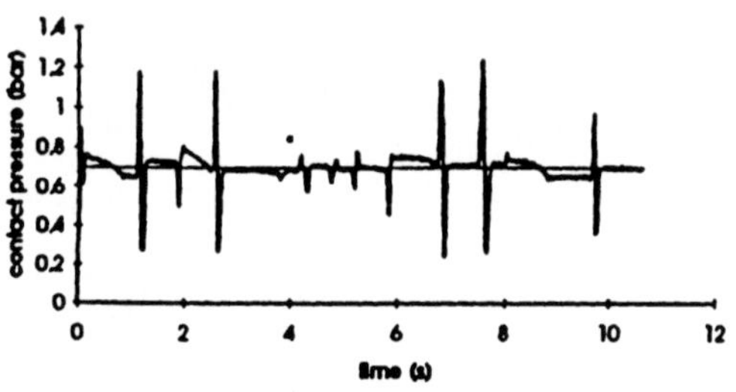

Fig. 7a). Grasping a 13 mm diameter stalk of asparagus at an advanced stage of growth (p_{rif} =0.7 bar)

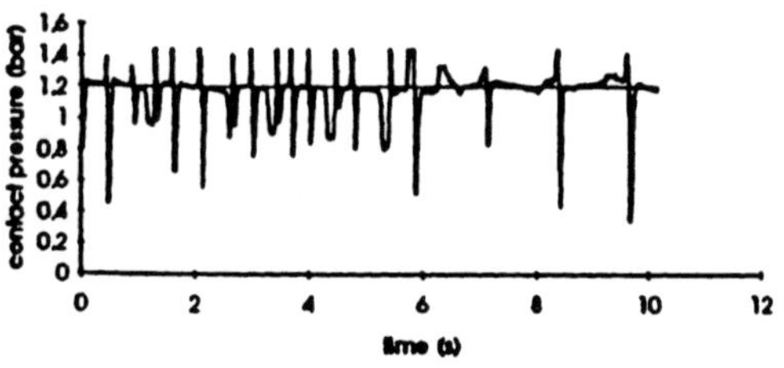

Fig. 7 b). Grasping a 13 mm diameter stalk of asparagus at an early stage of growth (p_{rif} =1.2 bar)

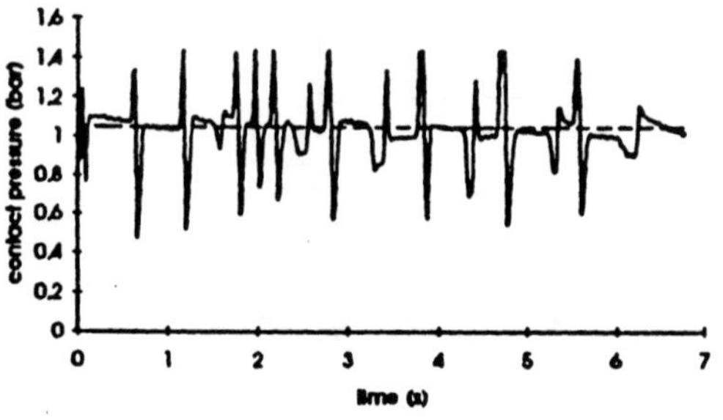

Fig. 7c). Grasping an 8 mm diameter stalk of asparagus

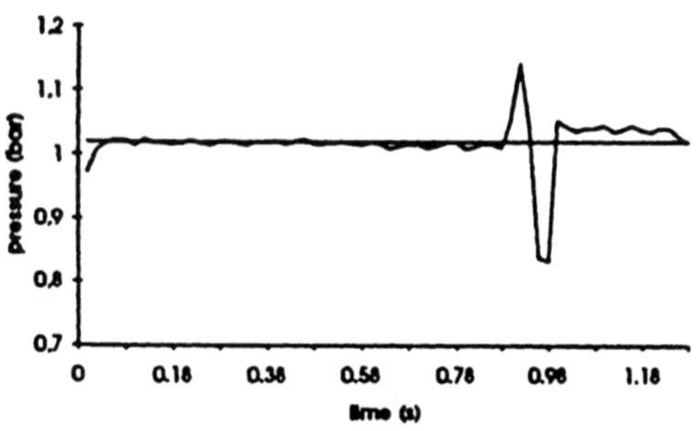

Fig. 7d). Grasping an 8 mm diameter stalk of asparagus

9. CONCLUSIONS

The gripper hand showed good performance. This result was achieved thanks to the type of tactile sensor developed and to the fuzzy control, which made it possible to overcome the problems associated with system nonlinearity. Valuable experience in the design of tactile sensors was gained in the course of the study.

The system showed itself to be extremely robust, with excellent adaptability in grasping different types of objects, even in the presence of external disturbance.

Acknowledgements

This study was carried out as part of the Italian National Research Council's Finalized Robotics Project with the contribution of SMC Italia.
The authors would like to thank Paolo Ariagno for his help.
The authors also thank ENEA for the post-graduate grant extended to S. Mauro.

10. REFERENCES

[1] Miodrag Rakic (1989). "Multifingered robot hand with selfadaptability" -*Robotics & computer intergrated Manufacturing*, 5, No. 2/3, 269-276.

[2] A. Romiti, T. Raparelli, F. Bellosta (1991). "Multi-fingered robot hand with very simple actuation and control" - *Journal of Robotics and Mechatronics*. 3 No. 5, 409-415.

[3] C. Bonivento, E. Faldella, G. Vassura (1991). "The University of Bologna robotic hand project: current state and future developments", - *Proc. of 5th International Conference on Advanced Robotics* - Pisa, 349-356.

[4] H. Maekawa, K. Yokoi, K. Tanie, M. Kaneko, N. Kimura, N. Imamura (1991). "Position/Stiffness based manipulation by three-fingered robot hand" - *Proc. of ISART*.

[5] G. Belforte, N. D'Alfio, A. Romiti (1986). "Brake drums bin picking" - *Proc. of ISATA*. Flims.

[6] Osamu Ouama (1993). "Grasping a sheet of paper by using pneumatic actuated finger" - *Fluid Power*, Tokyo, 689-693.

[7] G. Belforte, A. Romiti, M. Sorli (1986). "A sensorized manipulation system for food industry products" *Proc. of British Robot Association Conference*, Stratford upon Avon,.33-42.

[8] G. Belforte, F. Bellosta, N. D'Alfio, T. Raparelli (1987). "A sensorized system for delicate piece selection handling and packing" - *Proc. of 2nd Int Conf. on "Robotics and Factories of the Future"* , Ed. Radhoraman, Springer-Verlag, 483-488.

[9] Masafumi Uchida, Hideto Ide (1992). "The control method for the robot hand based on the fuzzy theory" *Journal of robotics and Mechatronics* 4, 262-267.

[10] G. Belforte, T. Raparelli, M. Velardocchia "Pneumatic positioner with fuzzy control"- *Proc. of IFAC - Workshop on "Motion control for Intelligent Automation"*, Perugia, 229-234.

[11] G. Belforte, T. Raparelli, M. Velardocchia (1993). "Fuzzy control of a pneumatic servo system and comparison with system using other controls" - *Proc. of IFCA* - Terrassa (Barcelona), 7-12.

[12] T. Raparelli, M. Velardocchia (1993) -"Fuzzy control of pneumatic servomechanism by on-off and PWM digital valves: a general method"- *Proc. of Int. Conf. on Automation*- Milano, 263-277.

APPLICATION USING VLSI HARDWARE REALIZATION OF SELF–LEARNING RECURSIVE FUZZY MODEL

M.F. Scheffer* and I.S. Shaw**

* *Rand Afrikaans University, Cybernetics Laboratory, Faculty of Engineering, P O Box 524, Aucklandpark 2006, Johannesburg, Republic of South Africa.*

** *Rand Afrikaans University, Faculty of Engineering, Research Group for Industrial Electronic Technology, P.O. Box 524, Aucklandpark 2006, R.S.A.*

Abstract. This paper presents a **hardware** application of a **self–learning** fuzzy logic controller on industrial gas furnace data. The fuzzy algorithm is based upon a unique probability theory approach to fuzzy identification and estimation. This method was chosen because of an **inherent filtering operation** making it **highly immune to noise**. In the past the majority of fuzzy inference processors have been designed as application–specific circuits or custom IC's. Using semi–custom programmable IC's resolves the problems of development costs, adaptability of designs and execution speed.

Keywords: Modelling ,Fuzzy Control, Fuzzy Set Theory, Prediction.

1. INTRODUCTION.

In recent years fuzzy logic and fuzzy control developed rapidly. Users of fuzzy logic have ventured ever deeper into the realm of the control of very complex, non–linear or incompletely known processes and systems. Data processing based in fuzzy logic theory requires extensive computation. The computation time of fuzzy control algorithms for real–time applications processing large amounts of data, becomes a critical issue.

Commercially available fuzzy inference processors are limited to well established fuzzy algorithms like max–min. This is a considerable limitation, especially for researchers who want to experiment with applications of alternative fuzzy algorithms. To overcome this problem and the problem of execution speed, many researchers have turned to micro–processor based fuzzy systems. These systems however, are limited in their execution speed by their essentially non–fuzzy instruction sets and often resulted in very complex parallel processor designs.

This paper presents an approach of overcoming the above mentioned limitations by using highly integrated semi–custom programmable logic arrays and thereby creating totally reconfigureable hardware designs. The theoretical basis of our approach is the design of a self–learning fuzzy controller based on the fuzzy learning and estimation method by Ridley *et al.* (1988), Shaw *et al.* (1989), Shaw *et al.*(1992), who proposed a probabilistic fuzzy relation $\overline{R}$, computed as the weighted average of the fuzzy state variables of a system. For example, for a single–input single–output self–learning fuzzy controller utilizing a moving average window, the estimated fuzzy relation,

$\bar{R}$ would be given by:

$$\hat{\bar{R}}_{Wk}(U,X,Y) = \frac{\sum_{l=k-W-1}^{N} \bar{u}_{1l} \cdot \bar{z}_{1l-1} \cdot \bar{y}_l}{\sum_{l=k-W-1}^{N} \bar{u}_1 \cdot \bar{z}_{1l-1}} \quad (1)$$

where: overbar denotes a fuzzy value, k is the observation sample number, W is the window size, $\bar{z}_{1l-1} = \Phi_\rho(\underline{y}_{l-\tau_y})$, $\bar{u}_{1l} = \Phi_\rho(\underline{u}_{l-\tau_u})$, $\bar{y}_l = \Phi_\rho(\underline{y}_l)$, Φ_ρ is the fuzzification operator producing a *1-to-ρ* mapping between crisp and fuzzy numbers, ρ is the number of referential sets in the universe of discourse, τ_u and τ_y are the integer time delay applicable to u and y, U is the universe of discourse of the system inputs u, X is the universe of discourse of the system states z, Y is the universe of discourse of the system output y.

This weighted average operation and the use of a moving-average window produces an inherent filtering operation translating into a fuzzy system, that is highly immune to noise (Postlethwaite, 1991 a,b). This noise rejection property was considered to be of benefit for industrial control applications. A system using a recursive estimator (Shaw *et al.*,1992) was implemented for this paper. See Fig. 1 for a simplified graphical representation of the single input single output recursive fuzzy estimator.

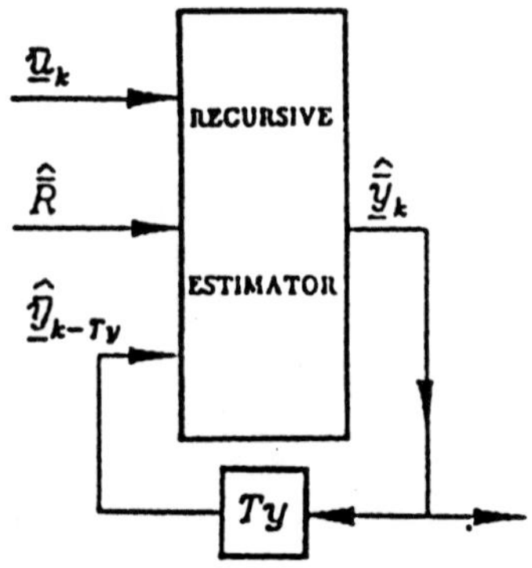

Fig 1 Recursive Fuzzy Estimator.

The fuzzy self-learning algorithm can be divided into four functional parts:

1) the **fuzzifier** [fuzzifies the system variables],
2) the self-learning fuzzy system identifier [creates the fuzzy relation],
3) the **fuzzy estimator** [estimates fuzzy output based on 2), 3) and 4)],
4) the **defuzzifier** [transforms fuzzy estimate back to discrete value by means of a *ρ-to-1* mapping of fuzzy numbers to a single crisp value].

The fuzzy process is divided into two phases the learning-phase, were the system dynamics are captured in the fuzzy relation. As can be seen in Fig. 1 in the estimation-phase the fuzzy relation is used to estimate system behavior. Further more, it is necessary to provide an initial system estimate, to initialize the recursive estimation proses.

2. DESIGN CONSIDERATIONS

To optimize the fuzzy algorithm with an emphasis on the hardware design, with specific limitations and optimal configurations, the following changes were made to the original fuzzy algorithm(Shaw *et al.*,1992).

2.1. Table-based Fuzzification

It is conventional practice to normalize incoming non-fuzzy data so as to limit the universe of discourse to acceptable values. This practice leads to the fact that the universe of discourse is known beforehand. This implies that the values within the universe of discourse could be pre-fuzzyfied and then used as a discrete look up table of fuzzy values. Thus instead of fuzzifying the incoming state variables on-line the values could be fuzzified prior to being used and then be retrieved in real-time, interpolating between the discrete values for any alternative values. This method speeds up the fuzzification process considerably, with the only prerequisite that the system variables must be known prior to operation.

2.2. Negative Values

To facilitate the simplicity of the hardware design it is sensible to get rid of all negative values in the universe of discourse. This practice simplifies the design of mathematical operators like adders, subtracters, multipliers and dividers greatly. It is accomplished by normalizing and shifting the system variables to fall within the region from 0 to 2 instead of the region from −1 to 1.

2.3. Zero–trapping

The self–learning process is unsupervised and all possible rules are interrogated, thus there would always be a large amount of rules that stays unused or are weighted as zero. As few as 20% of the rules may be weighted other than zero. Thus the learning and estimation phases are greatly accelerated by some method of trapping zeros. This implies that the estimation and learning operations are only executed for situations where all the rules used in such a situation, are weighted non–zero.

The fuzzification and defuzzification processes are not affected by this method of zero–trapping.

3. THE HARDWARE

The aforementioned fuzzy algorithm, using a recursive estimator, was implemented in hardware by using a commercially available programmable VLSI Logic Cell Array (LCA) system(Xilinx,1991). Using programmable circuitry with a cell structure has some design constraints. Most of the parallel array design methods like bit slicing or Wallace trees, can not be used because of the limitation on available programmable cells. Adders, subtracters, multipliers, dividers and other basic operators must be limited to sequential or serial–parallel designs. These types of designs are still very fast and take up very little space, leading to the fact that several could be programmed onto the same LCA to be executed in parallel.

The design is based in software and could be reconfigured quickly giving the user more freedom compared to application–specific IC's or circuits.

3.1. Description

The fuzzy hardware was designed as a PC–compatible plug–in card with separate on–board static RAM to allow easy evaluation and error checking. Although the internal bit–resolution of the VLSI fuzzy circuitry was designed to be 16–bit, the interface between the PC and the fuzzy card is a 8–bit data channel. This allows the fuzzy card to be used with any IBM PC–compatible computer. The relatively high internal bit–resolution, compared to other fuzzy processors, is required because the self–learning fuzzy algorithm is computation–intensive which contributes to problematic error accumulation. Increasing the bit resolution has proven previously (Scheffer,1991) to be the most effective solution to this problem. The fuzzy algorithm is a self–learning recursive system. This means that the fuzzy hardware can autonomously identify the plant dynamics in the learning phase and could be operated in a stand–alone fashion in the recursive estimation phase.

3.2. Comparison and Evaluation

Three commercially available fuzzy processors were chosen for comparison with the authors's design:
the **FC110 Digital Fuzzy Processor** from Togai Infra Logic Inc.(Togai,1991),
the **NLX230 Fuzzy Micro Controller** from NeuraLogix (NeuraLogix,1992) and
the **FP–3000 Digital Fuzzy Processor** from Omron (Omron,1991).

At this time the authors are not aware of commercially available self learning fuzzy hardware and also the fuzzy processors mentioned before are rule–based systems. The system from Togai does however have a software self–learning option.

An objective comparison between the four fuzzy hardware systems could be tabulated as seen in Table 1.Fuzzy hardware systems are normally evaluated on the basis of the following points:

1) the number of fuzzy rules evaluated per second,
2) the number of non–fuzzy or crisp evaluations per second, (thus including defuzzification),
3) the number of rules supported,
4) the number of external inputs,
5) the number of outputs available and
6) the internal bit resolution of calculations.

Table 1 System Comparison

System	(1)	(2)	(3)	(4)	(5)	(6)
FC110	200000	60000	800	n/a	n/a	8
NLX230	n/a	30 M	64	8	8	8
FP–3000	n/a	30709	128*3 groups	8	4	12
Author	99715	25669	125	1	1	16

Togai does not publish a limit on the number of inputs or outputs of the fuzzy processor because these values are limited by the software development system.

4. APPLICATION

The results of an application of the above mentioned fuzzy algorithm for a single input and single output system is graphically shown in Fig.2. The driving functions for this system is given by the industrial gas–furnace data of Box and Jenkins(Box *et al.*,1970). The fuzzy hardware was operated as a batch processor, and could be seen as fuzzy "co–processor" board. As was described before, the fuzzification was done by using Table–based fuzzification. A total of 256 values of 16–bit resolution was pre–fuzzified in the universe of discourse of each separate system variable. Five identical triangular referential sets were used in each universe of discourse. Furthermore a recursive estimator with zero–trapping was used to obtain an autonomous self–learning fuzzy system. Lastly centroid de–fuzzification was used.

The learning window was limited to the first 150 points of the Box–Jenkins data set. These points were re–estimated along with the rest of the remaining 146 points in the set, to be compared with the original data, as shown in Fig. 2. An execution time of 0.041 sec. and a mean–squared error, J=2.171, was obtained.

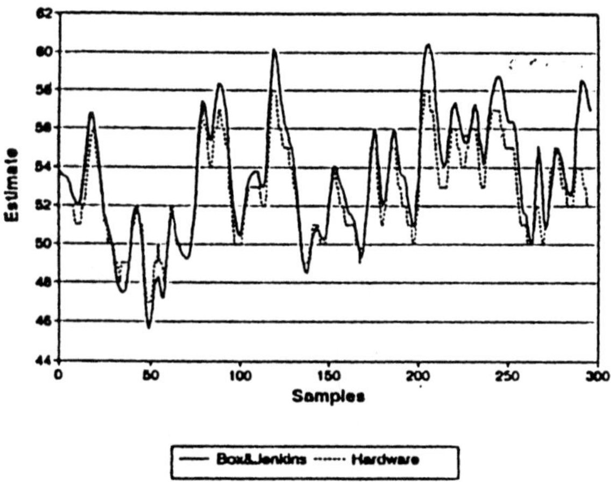

Fig. 2. Application Results

Table 2 shows the execution times and mean–squared errors, $J=\Sigma(y-y_{est})^2/N$, of:

1) a software implementation of the original algorithm(Shaw *et al.*,1992),
2) a software implementation of the adapted fuzzy algorithm with execution time optimization and table– fuzzification,
3) the fuzzy hardware system.

Table 2 Comparison with Original Algorithm

Result	Run Time(s)	Error (J)
1)	9,66	0,791
2)	1,96	1,465
3)	0,041	2,171

5. CONCLUSION

As can be seen from the results in Table 1 the fuzzy hardware system compares well with commercially available fuzzy processors. It must be kept in mind that only the authors' hardware system is self–learning. Table 2 shows that the modified algorithm has produced a considerable improvement in execution time, in comparison with the original algorithm(Shaw *et al.*,1992) . The use of LCA–circuitry limits the execution speed of the design but improves versatility and adaptability. It must also be considered that all the authors data was accumulated on a 35 MHz LCA device and that devices with clock rates of up to 260 MHz are available.(Xilinx,1991). The authors' approach of using LCA–circuits could be of great benefit to researchers who require real–time execution speeds but do not have the resources or time to develop application specific circuitry.

5. REFERENCES

Box, G.E.P., Jenkins, G.M.(1970). *Time series analysis, forecasting, and control.* Holden Day, San Francisco.

NeuraLogix (1992). · *NLX230 fuzzy micro controller data sheet.* American Neuralogix,Inc. Sanford, FL, U.S.A.

Omron (1991). *FP–3000 fuzzy processor data sheet.* Omron Corporation, Fuzzy Technology Center, Kyoto, Japan.

Postlethwaite, B. (1991a). *Probabilistic fuzzy Model in a Noisy Environment.* Internal publication, Department of Chemical and Process

Engineering, University of Strathclyde, Glasgow, Scotland.

Postlethwaite, B. (1991b). *Empirical Comparison of Methods of Fuzzy Relational Identification.* IEE Proceedings–D, Vol.138, No.3.

Ridley, J.N., Shaw, I.S., Krüger, J.J. (1988). *Probabilistic fuzzy model for dynamic systems.* Electronics Letters, Vol.24, 14, pp. 890–892.

Scheffer, M.F. (1991). *Die bestudering van 'n Wasige Beheermodel wat gebasseer is op 'n Texas Instruments Mikro –verwerker vir toepassing op Industiele Gasoond–data.* Internal publication Rand Afrikaans University.

Shaw, I.S., Krüger, J.J. (1989). *New approach to fuzzy learning in dynamic systems.* Electronics Letters, Vol.25, 12, pp. 796–797.

Shaw, I.S., Krüger, J.J. (1992). *New fuzzy learning model with recursive estimation for dynamic systems.*
Fuzzy Sets and Systems, Vol.48, 2, pp.217–229.

Togai (1991). *FC110 fuzzy processor data sheet.*, Togai InfraLogic, Inc., Irvine, CA.,U.S.A.

Xilinx (1991). *Programmable Gate Array Data Book.* XILINX San Jose, California.

FERROPIEZOELECTRIC TACTILE SENSOR ARRAY

V.TODOROVA, S.MILCHEV

Higher Institute of Mechanical and Electrical Engineering,4,H. Dimiter St., 5300 Gabrovo, Bulgaria

Abstract. A ferroelectric array sensor of a functional type for primary processing of tactile information, representing a homogeneous field of sensors in an active PZT ceramic substrate, has been developed. The results, obtained by physical-mathematical simulation of certain array sensor element, which is a non resonance piezoelectrical transformer with a running bulk acoustic wave (NRPT with RBAW), taking into accound its interaction with the homogeneous common ferroelectric medium are presented. Tactile arrays are designed for using for identification of tactile images to be applied to intelligent robots and other specialized means of information inputting.

Key Words. Delay, Electrical Sensing Devices, Identification, Image Processing, Mathematical Models, Robotics, Sensors

1. INTRODUCTION

Tactile transducers are a perspective trend in the field of modern information collection and processing systems. A wide range of principles, materials and technologies can be used (Skopalic et al. 1984, Yardley and Backer 1986, Todorova et al. 1987). The application of electromechanical type of sensors, i.e. transducers of mechanical action into an electrical signal, is predetermined by tactile principle of information inputting.

The results of the research and development of a functional piezoelectric tactile array sensor have been presented in this paper. The object of study is an array of non resonance piezoelectrical transformers (NPRT) with a running bulc acoustic wave (RBAW). The NRPT differ from the resonance piezoelectric transformers (RPT), studied by Meson (1966), Lavrinenco (1975) and Kartashov and Marchenko (1978), in the type of fixing and excitation.

NRPT operate normally in a pulse excitation mode. Square pulse of electric voltage causes non stationary acoustic transverse oscillations **z(t)** in a limited volume of the NRPT active medium at the expense of the reverse piezoeffect. At each moment of time, a volume of medium different in location is activated due to the displacement of the excited RBAW at a constant velocity v_{ac}. That eliminates energy accumulation in the active medium and enables us to consider RBAW as a tactile effect information carrier by comparing the delay time τ ($t_i = t - \tau$) and the RBAW propagation co-ordinates **r** in a certain point or area of the piezowafer:

$$\tau = r / v_{ac} . \quad (1)$$

Electromechanical processes in NRPT with pulse exciting RBAW were presented by Golubenko et al. (1984) and Zavadski (1989) by means of one- and two-dimensional models.
NRPT require rigid fixing when assembled. It allows homogeneous array fields of rigidly fixed to each other and to the base NRPT to be created in a PZT ceramic substrate. The NRPT are realized by **n** output electrodes.

2. MATHEMATICAL SIMULATION AND CHOOSING OF CONSTRUCTION VARIANTS OF TACTILE ARRAYS

The theoretical consideration of the one-dimensional model is sufficient for choosing the construction variants of the array sensor. The theory is reduced to the following formulations:
1) description of the electromechanical process of excitation and RBAW formation in the excitation section zone of the NRPT is carried out according to Kolski (1953) by a differential equation:

$$\ddot{z}(t)+2n\dot{z}(t)+\omega_0^2(z+\gamma z^3)=C_1 A_I U_I(t)/h, \quad (2)$$

where: $z(t)$ is the amplitude of the excited acoustic transverse oscillations; **n** is the internal friction coefficient; $\omega_0^2(z+\gamma z^3)$ is the symmetric elastic characteristic of an anisotropy solid; C_1 is a general coefficient of the piezoelectric material; A_I and h are design parameters of NRPT - the exciting electrode area and the piezowafer thickness, respectively; $U_I(t)$ is the exciting electric pulse;

2) description of the RBAW propagation along the wafer, taking into account the losses:

$$Z(t-\tau,x)=z(t)e^{-\mu x}, \quad (3)$$

where μ is the loss coefficient;

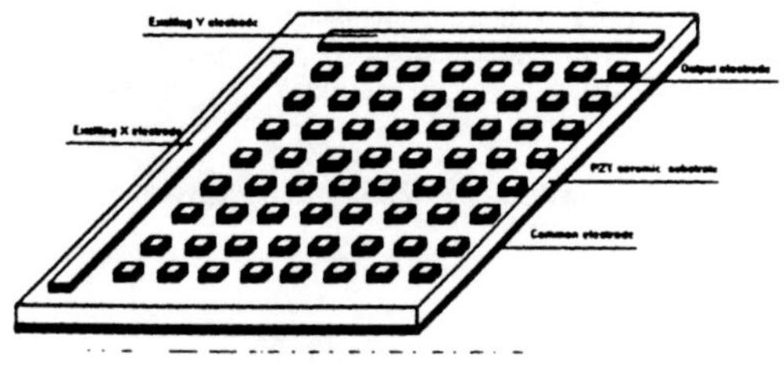

Fig. 1. An array construction with exciting in X and Y directions

3) description of the output electric signal received from the corresponding activated output electrode at the expense of the direct piezoeffect:

$$U_o(t-\tau,x)=C_2Z(t-\tau,x), \quad (4)$$

where C_2 is a general coefficient of the piezoelectric material.

The mathematical simulation of the processes in NRPT with RBAW has been carried out by means of program package „TUTSIM“. The electrophysical parameters of the piezoelectric materials, the exciting pulse parameters and the design parameters of the array related to the output signal shape and amplitude have been optimized. In accordance with the simulation results, two constructions of tactile array sensors have been designed, shown in Figures 1 and 2.

Standard piezoceramic material of the type PZT ПКМ - O61 has been selected with wafer thickness of h = 0,35 mm and dimensions of the array field of output electrodes - (8 x 8) elements. The optimum parameters of exciting square pulse related to the selected design are: amplitude U = 20 V; pulse duration - 0,7 μs; repetition period over 2 μs; slope steepness - below 50 ns. In the design variant of the array in Fig. 1, two exciting electrodes forming RBAW in X and Y directions have been used and pulses sent in these directions have opposite polarity. The array with sectional exciting electrode of Fig. 2 is exciting in X direction only.

The arrays shown in Fig. 1 are applicable to tactile effect registration concentrated in a region commensurable with the area of a single sensitive element, i.e. for point effects. The accuracy obtained by such tactile array is defined within ± one discrete. But when recognizing objects with tactile images with outlines or arbitrary area, the deviation from the real object outlines might reach a considerable value. This effect is a consequence of the obtained „hidden zones“. For example, when two or more adjacent sensors, diagonally activated, are available, they immediately form an area of x^2 points, which prevents the ambiguous identification of the touched points causing the response. This effect is a result of the excitation and propagation of RBAW separately in X and Y directions, neglecting their mutual influence.

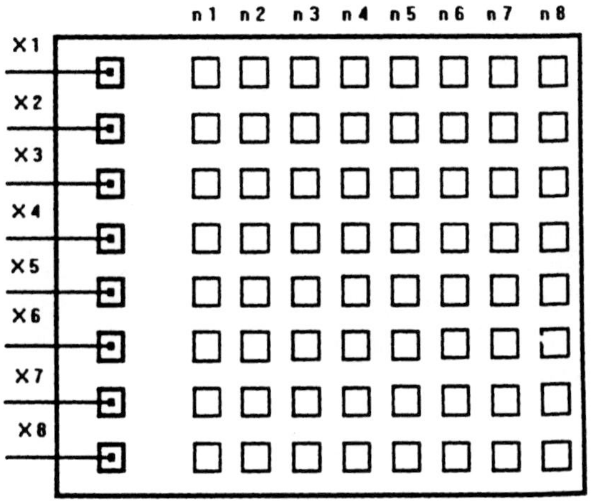

Fig. 2. Sectional tactile array

This problem is solved by the construction of Fig. 2. It is a row-aray with one exciting and **n** output electrodes which is technologically multiplied in rows - **m** in number. The use of such array is more justified from the point of view of the reliability of the one-dimensional mathematical model of RBAW processes in NRPT since only under certain limiting conditions concerning the electrode width, it could be assumed, that RBAW propagates only in one direction and those conditions are:

$$b_I \leq \lambda_o / 2, \quad (5)$$

$$b_o \leq \lambda_o / 2, \quad (6)$$

where b_I , b_O are the exciting and output electrode widths, respectively; λ_o is the length of excited RBAW, $\lambda_o = v_{ac} / f_0$. For the selected ceramics and design of tactile array $\lambda_o = 2{,}4.10^{-3}$ m. Therefore if b_I and b_O are less than 1,2 mm, the construction shown in Fig. 2 guarantees the reliability of the one-dimentional mathematical model of the processes in NRPT as well.

This construction has a shortcoming compared to the one shown in Fig. 1 because the sectional tactile array has (m - 2) more terminals and their number will grow when the dimensions of the array are increased in the Y direction whereas tactile array in Fig. 1 has a fixed number of terminals regardless of its dimensions. This is a technological problem which reflect on the array cost since a package with more than four pins has to be used.

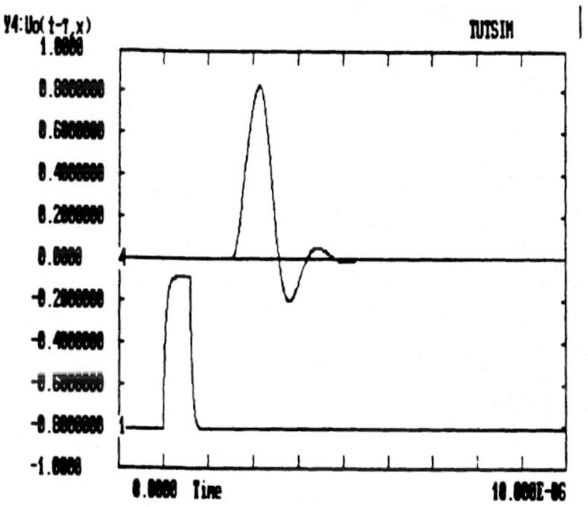

Fig. 3. Mathematical simulation data of the processes in the NRPT with RBAW

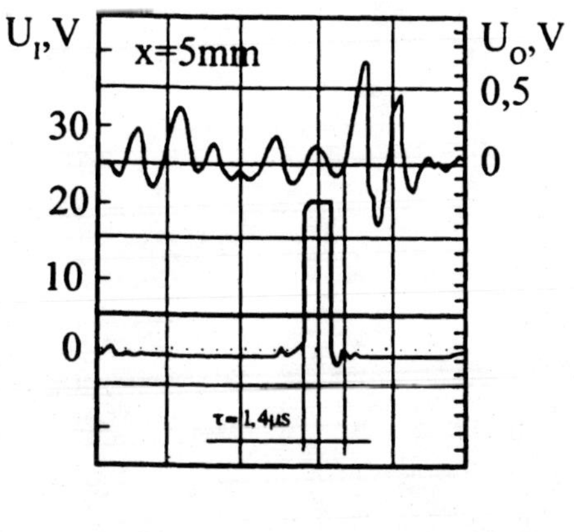

Fig. 4. The experimental oscillogram of the exciting and output electric sign

However, this construction has an advantage: the exact co-ordinates of each activated sensitive element can be obtained irrespective of its location in relation to the rest of the activated elements. The number of the scans is m - 2 and the information word has a length of n bits and is grid copy of the tactile image of one row with accuracy within ±one discrete.

3. EXPERIMENTAL RESULTS

The variants of tactile arrays showw in Fig. 1 and Fig. 2 have been developed by means of a thickfilm technology. The electrode system has been formed by offset printing of Ag paste followed by baking. After polarization of the NRPT exciting and output sections and assembling into a package, the arrays have been experimentally tested.

Fig. 4 shows an oscillogram of the exciting and output electric signals for an activated output electrode of the array shown in Fig. 2. As a comparison, Fig. 3 shows the mathematical simulation result from a row-array of the same construction. Table 1 shows a comparison of the signal time-delays obtained from the mathematical simulation and experiment for the row-array with 8 output electrodes. A comparatively satisfactory compliance of the results (within ±1%) has been observed.

Table 1. The signal time-delays

Results from mathematical simulation							
τ_1,μs	τ_2,μs	τ_3,μs	τ_4,μs	τ_5,μs	τ_6,μs	τ_7,μs	τ_8,μs
1.56	2.19	2.81	3.44	4.06	4.69	5.31	5.94
Experimental results							
1.40	2.20	2.90	3.50	4.20	4.80	5.40	5.90

4. CONCLUSION

According to the autors, the theoretical and experimental results obtained cofirm the opportunity of practical application of the developed constructions of tactile arrays. Moreover, they can be used for solving both direct (inputting) and the reverse (recognition) problems for tactile data processing. For example, tactile fields with dimensions (p x q) arrays can be developed based on the construction shown in Fig. 1. They might be applied to the inputting information systems. Unlike them, the construction shown in Fig. 2 can be used for tactile image recognition which makes them of practical interest for modern robots and other devices of the same type.

Taking into account the operating principle of the discussed constructions of ferropiezoelectric tactile arrays the problems of secondary processing of tactile information can be definitely formulated.

5. ACKNOWLEDGEMENTS

The authors are grateful to V. Zavadski from Kiev Polytechnic (Ukraine) and R. Radev from HIMEE - Gabrovo for their help in the physical and mathematical simulation of the electromechanical processes in NRPT with RBAW.

5. REFERENCES

Голубенко,А.В.,В.А.Завадский,Ю.П.Зайка, А.В.Згурский (1984). Электромеханические процессы в нерезонансных пьезотрансформаторах с бегущей объемной акустической волной. In: *Диэлектрическая элементная база и ее применение* pp. 16-31. Техника, Киев. [Golubenco A.V., V.A. Zavadski, U.P. Zaika, A.V. Zgurski (1984). Electromechanical processes in non resonanse piezotransformers with a running bulk acoustic wave. In: *Dielectric Element Basis and its Application*. pp. 16-31. Tehnika, Kiev.]

Карташов А.И., Н.Б.Марченко (1978). *Пьезоэлектрические трансформаторы тока.* Техніка, Киев. [Kartashov A.I. and N.B.Marchenko (1978). *Piezoelectric Current Transformers.* Tehnika, Kiev.].

Кольский Г. (1953). *Волны напряжений в твердых телах.* Издательство иностр. лит., Москва. [Kolski G. (1953). *Waves of Strain in Solids.* Publishing House of Foreign Literature, Moscow].

Лавриненко В.В. (1975). *Пьезоэлектрические трансформаторы.* Энергия, Москва. [Lavrinenko V.V. (1975). *Piezoelectric Transformers.* Energy, Moskow].

Мэзон У. (1966). *Физическая акустика. Методы и приборы ультразвуковых исследований. Т.1, Ч.А.* Мир, Москва. [Mason W. (1966). *Physical Acoustic. Methods and Devices of Supersonic Investigations.* V.1, P. A. World, Mockow].

Skopalic E., V. Todorova, P. Todorov (1984). Bulgarian tactile transducers and tactile arrays. *National Youth School, St. Zagora.* Dep. in ZINTI N º ND 1265/85.

Todorova V.D., S.M. Hristov, P.J. Todorov (1987). Ferropiezoelectric tactile sensors and arrays. *International Conference ROBCON-4,* Sofia, pp. 218-228.

Завадский В.А. (1989). Проектирование сегнетопьезоэлектрических функциональных устройств. В: *Арифметика, принципы, организация, диагностика и формализованное проектирование вычислительных структур и устройств.* Вища школа, Киев, 289-340. [Zavadski V.A. Design of ferropiezoelectric functional devices. In: *Arithmetic, Principles, Organization, Diagnostic and Fomalistic Design of Computing Structures and Devices.* Higher School, Kiev. pp. 289-340.

Yardley A.M.M. and K.D. Baker (1986). Tactile Sensors for Robotics: A Review. *World Yearb. Rob. and Dev.,* London. pp. 47-83.

ULTRASONIC FLOW MEASURE FOR WATER AND WASTEWATER USING OPEN CHANNEL WEIR

E.M. Sroczan * and A. Urbaniak**

* *Institute of Power Engineering*
** *Institute of Computing Science*
Poznań University of Technology, ul. Piotrowo 3A, Pl- 60-965 Poznań, Poland

Abstract. The intelligent measurement systems are usually based on the microcomputer circuit. The hardware and software allow to realize various conversions of characteristics and their fast calculations. A method of flow measuring based on ultrasonic devices and on the given known characteristic of open channel weir is presented in the paper. The way of obtaining of characteristics, which is convenient to use in the industrial control systems is presented. Three main approaches are analyzed the modelling of weir characteristic, the discrete modelling and the calculations on-line. The negative influences of outside factors which can affect the measuring are discussed.

Key Words. Flowmeters, industrial control, ultrasonic transducers, water pollution, water supply.

1. INTRODUCTION

A development of the intelligent measurement systems based on the single-chip microprocessors allows for effective application of measurement methods useful for environmental protection. These methods were known earlier but due to complexity of signal processing there were no possibility for their use.

In the continuous measurements of process variables physical parameters can be represented by signals that should be convenient for further processing. This processing is usually made by digital technics of signal processing. There exist now new hardware and software tools for various conversions using projection of transient functions or very quick executions of arithmetic and logical operations (Olsson and Piani, 1990; Urbaniak, 1990).

Control methods of technological processes especially in the field of water and wastewater treatment often demands the solution of complicated hydraulic tasks. The most complicated tasks from this set are problems concerning the flow measurements without invasion to the stream of the medium especially in the open channel. In this paper the method of stream flow measurement using the weir characteristic and ultrasonic devices is presented.

2. MODELING OF WEIR CHARACTERISTIC

The unsinked weir was used for both stream flow measuring and total quantity of the medium, with known flow characteristic as a function of the velocity of flow which is precisely described by Majcherek (1990). There are two main difficulties which designer must solve. First one is the effective measurement of the medium level and the second is time-consuming calculation of flow following given complicated mathematical formulae.

The mathematical formulae describing the values of liquid level, the velocity of flow, the intensity of flow and integrated flow of the medium are usually strong nonlinear and implicit.

Additional, in many real situations the media are chemically aggressive and hence the measurement have to be made in the noninvasive way. For this purpose, the technology using the ultrasonic valves is very useful. Generally speaking, the measurements of physical variables based on electrical methods are realized in the following scheme.

The measured physical variables is converted by the indicator to another intermediary magnitude which is convenient for further processing. The next step is based on the conversion of intermediary magnitude to electrical, pneumatic or hydraulic signals convenient for transmission and processing.

The following conditions have to be fulfilled for obtained the correct values of measures:

- the method of direct measurement of physical variable closely connected with analyzed variable must be known,
- the function describing the dependence of the measured quantity of physical variable and analyzed one is known.

There are three various approaches to obtaining the value of process variables based on the different measurable variables:

1° Modeling of the characteristic describing the analyzed variable as a function of measured variable, using the dedicated analog model.

2° The mapping of the transient characteristic: measured variable - searched variable, using the finite number of points with linear or polynomial approximation between given points of characteristic.

3° The on-line calculation of the analyzed value based on the actual measured value of process variable.

Fig.1 presents the scheme diagram of the flow measurement system with the use of conversion of searched process variable to electrical intermediary variable.

As a measurement sensor the special construction of the measurement channel is used with the given length and determined profile which guarantee the stream flow stabilization of the medium. The system transmitter - receiver of ultrasonic signal is placed in the chosen point of a channel. The signal generated by the transmitter is reflected from the liquid surface and the level of surface liquid-air is determined. The travel time of the ultrasonic signal is then processed by the electronic system to the intermediary variable.

The system gives the linear dependence between frequency and the level but further processing for flow value and flow intensity value is needed to obtain the voltage or current signal directly representing these both values. This task is realized by final converter.

The method of signal conversion, modeling of complex weir characteristics and processing of ultrasonic signal are presented below.

2.1. Analog model of weir characteristic

The formulae describing the dependence between flow value and measured value i.e. the level of stream in the channel are very complex (Majcherek, 1990). After some transformations the following formulae connecting flow and level can be obtained:

$$Q = \frac{8}{15} \mu \, tg\alpha \, h^2 \sqrt{2gh} \qquad (1)$$

where

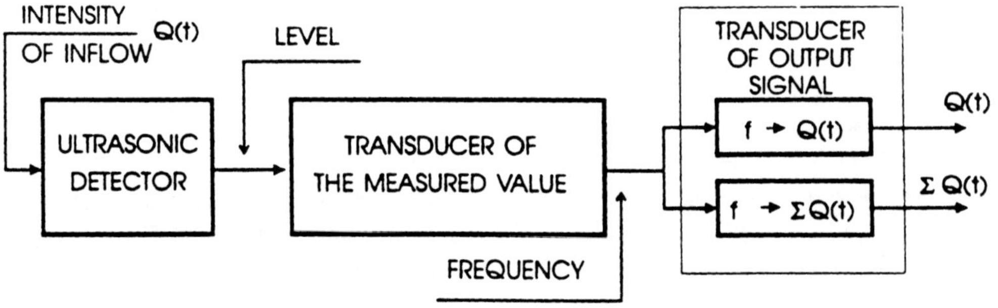

Fig.1. The structure of the ultra sound device applied to the measuring the stream level at the unsinked weir.

$$\mu = 0.575(1+ 0.01015\ ctg\alpha)(0.0328\ h)^{-\beta}$$

$$\beta = -0.02\ (ctg\alpha)^{0.75}$$

and

α - angle of edge inclination of triangular weir,

Q - intensity of liquid flow through the weir,

h - high of flow stream through the weir

The characteristic which is obtained from the formula (1) and the values of parameters describing the the weir, are given by:

$$\ln Q = \ln 0.790388 + 2.5 \ln h - 0.030196 (\ln 0.0328 + \ln h) \qquad (2)$$

The example curve is presented in Fig.2.

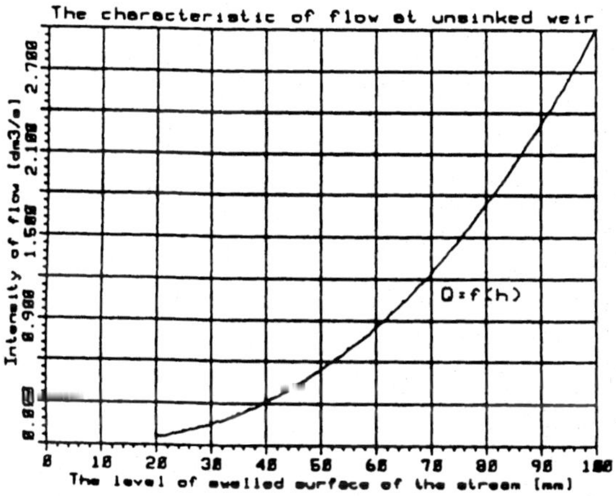

Fig.2. The calculated characteristic of flow at the unsinked weir.

Microcomputer hardware allows for effective generation of function tables, graphic presentation of the searched dependencies and the approximation formulae with the given accuracy. The electronic-designer has to model the circuit with the inverse characteristic and with maximum possible accuracy. This way is certainly complicated for complex characteristics but construction of inverted characteristic allows immediately to obtain the result as a flow intensity value. The integration of the signal gives the sum of the flow for a assumed time period.

The very important problem which must be solved is the method of compensation disturbances influencing modeling system and connected with the loss of primary accuracy.

2.2. Discrete characteristic projection

The analyzed relation can be represented in the table form describing the discrete values of characteristic with the given accuracy. The flow values are obtained from the measured values (levels) by means of linear or polynomial interpolation. For realization of this projection system containing the appropriate memory and ability to make simple arithmetic operations in short time are needed. The majority of professional logic, microprocessor controllers have these possibilities.

This approach does not demand for execution of any time-consuming calculations and the final value is obtained from simple interpolation formulae.

2.3. On-line flow determination

In the case of multifunctional microcomputer system with large computing possibilities, current value of stream flow according to relation (1) and (2), may be determined.

It is particularly important in case when accuracy of flow measurements is necessary. The nowadays construction, for industrial application, the big computer control systems contain the software for processing and data acquisition. These systems allow for modeling of measurement sensor characteristic and recalculation the measurement signal to chosen technical units.

3. THE ULTRA SOUND WAVES APPLIED TO MEASURE THE LIQUID SWELLING WITH THE UNSINKED WEIR.

The ultra sound waves are mechanic oscillations of range from 16 kHz to 10 GHz which propagate in elastic media in form of longitudinal or lateral wave or both. In case of applying the ultra sound to measure a distance between a source and an obstacle - in this instance - the level of the border between a liquid and air or air and solid, only the waves which propagate in the direction compatible with the direction of a measured distance should be taken into consideration. State of the analyzed medium is described with the following parameters: pressure of acoustic wave, propagation speed of ultrasonic waves, density of medium and its temperature (Matuschek, 1985). The influence of these parameters on the measured level of liquid stream at the weir is discussed below.

3.1. Influence of parameters of an acoustic field on the application of ultra sonic device to measure the weirs .

The acoustic pressure and wave velocity as a function of time are described as follow:

$$p = p_m \cos \omega\left(t - \frac{x}{c}\right) \quad (4)$$

$$V = V_m \cos \omega\left(t - \frac{x}{c}\right) \quad (5)$$

where:

p_m, V_m - amplitude of the acoustic pressure and velocity, respectively;

$\omega = 2\Pi f$ - angular frequency;

f - frequency of oscillation;

x - coordinates of three-dimensional space;

c - velocity of wave propagation.

However for purposes of describing the acoustic field it is necessary to consider the volume of acoustic pressure which involves the acoustic impedance of considered medium:

$$p = v\, z \quad (6)$$

and

$$z = \rho\, c \qquad \text{or} \qquad z = \sqrt{\rho\, E}$$

where:

z - acoustic impedance;

ρ - density of considered medium;

E - modulus of volume compressibility of medium.

It is important that between liquid and gases the difference of acoustic impedance approximates about 10^3units. The parameters of medium, mentioned above, can be involved in form of an equation describing the volume of flat longitudinal wave:

$$I = \frac{1}{2} A^2\omega^2 \rho\, c \quad \text{or} \quad I = \frac{p_m^2}{2\,\rho\, c} \quad (7)$$

The definition of characteristic of acoustic field for two fields - the near and the long distance - is important in case of using the ultra sound to measure the distance between source and obstacle - the border of considered media. In case of free acoustic field the level of wave intensity decreases depending on distance from source according to equation:

$$L_n = L_o - 20 \lg \frac{r_n}{r_o} \quad (8)$$

where:

L_o, L_n- the levels of the intensity of the acoustic wave in the distance of r_o and r_n from the source.

Measuring of the level of the intensity could not be applied to calculate the distance between the source and acoustic obstacle, because the distance L_n is involved in parameters of discussed medium in which the ultra sound wave propagates. It is possible to compensate the effect of these parameters by investigating the phenomena of reflexing ultra-acoustic waves at the border of two media. The reflexed part of wave energy, which is described by the rate of reflex is calculated as follow:

$$R = \frac{I_r}{I_o} + \left(\frac{z_1 - z_2}{z_1 + z_2}\right)^2 \quad (9)$$

where:

I_o, I_r - the intensity of the direct and the reflexed waves, respectively.

The phenomenon of Reyleigh dissipation concomitants the propagation of ultra-acoustic waves for the structures comparable with the length of the wave. Due to this phenomenon it is necessary to take into account the frequency of propagated wave:

$$\alpha_R(f) = k_R f^4 \tag{10}$$

where:

α_R - the value of Reyleigh dissipation factor;

k_R - the constant Reyleigh's value;

f - frequency of ultra sound waves.

Resuming, the attenuation of ultra sound wave can be described as follow:

$$p = p_o e^{-ax} \sin(\omega t - kx) \tag{11}$$

where $k = \frac{2\Pi}{\lambda}$.

The proper application of ultra sound technic to continuous measuring of the level of liquid surface swelled with the unsinked weir should take into consideration the following environmental parameters:

- the humidity of the air and concomited dissipation and attenuation of the ultra sound wave;
- the temperature of media which determines the value of velocity of the wave propagation;
- the angle of incidence of the wave on the media border: air and liquid;

More over the ultra sound wave frequency depends on attenuation and dissipation and the resonance parameters of ultra sound transducers.

The negative influence of change of the above mentioned parameters on the result of the measure is reduced by correction of measured signals in electronic circuit or partially reduced by the careful selection of the measuring method. For practical purposes the back time of the reflexed wave of the media border, air and the surface of liquid swelled with the weir is measured. Other variables are reduced by electronic circuits of the measuring device. They are also compensated by specialized integrated circuit "ASIC" named "Echo-processor" (Cieślewicz, 1992).

3.2. Function realized by "Echo-processor"

The "Echo-processor", VLSI integrated circuit, realizes the following functions:

- controls processes of generation, emission and receiving ultra sound waves by the transmitter and receiver;
- eliminates parasite and reflexed signals;
- processes measured results to the form demanded by arithmetic processor of of the measuring device;
- stores the data in the period of a single measure cycle;
- monitors and detects the values out of range of possible time of propagation of forward and back ultra sound wave;
- accomodates the processor circuit signal to any type of ultra sound converter.

The pattern frequency generator, built in the "Echo-processor", enables the choice of the measured distances.

4. SIMULATION OF MEASURING PROCESS OF SWELLING AT UNSINKED WEIR

The scheme of the measuring stand over the surface of considered weir is shown in Fig.3. The ultra sound wave used to

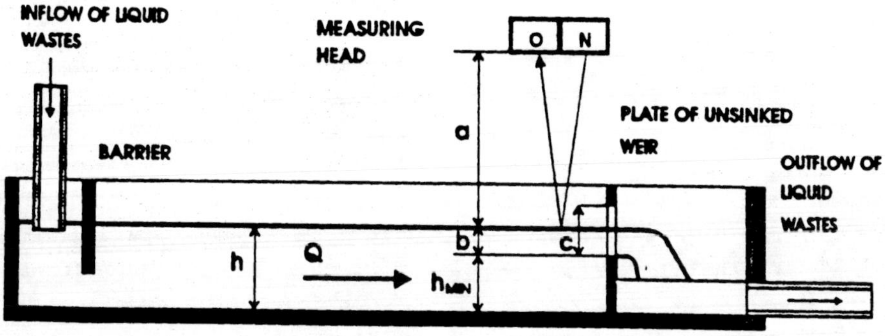

Fig.3. The installation of measuring device over the open channel weir.

measure the level of swelled stream in the cross channel of the weir propagates with frequency f. The frequency has been chosen to minimaze the influence of the changes of environmental parameters (especially air) and acoustic resonance of ultra sound transmitter and receiver on the result of measure. The obtained binary electric signal corresponds with a swell of the stream in the channel. This signal is converted to the signal of the intensity of liquid flow.

The inverted characteristic of assembly: weir - ultra sound detector - converter is used to calculate the value of the liquid flow. The measured time of the echo of ultra sound wave is the basis to obtain the searched value of flow. The acoustic wave reflexed from border surface air-liquid is detected by a receiver of ultra sound head. The analogue signal from the receiver is converted by "Echo-processor" into binary one which describes the value of the distance a_t - between the surface of liquid and the level of ultra sound head. The expected value h_t of swelling surface of stream of liquid is obtained as from:

$$h_t = K - a_t \tag{12}$$

where:

K - constant value involving the mounting level of measure assembly over the stream surface;

$a_t = f(\Delta t, v)$ - current value of the swell level measured over the liquid surface (the velocity of acoustic wave propagation included).

The value of the level h_t allows to calculate the flow through the considered weir channel by using the equation (3) or obtained with characteristic showed in the Fig.2.

5. FINAL REMARKS

The technical solution was developed for flow measurement that has the many advantages and the most important are:

- the possibility of continuous flow and flow intensity measurement,
- the non-invasion in the medium flow.
- no moving mechanical element of the sensor,
- signal easy to conversion to final value,
- the possibility of measurement in the gravitation installation for any fill of the canal.

The above mentioned advantages are very useful for technologist-designers and for control-designers dealing with physical, chemical and biological measurements in the wastewater treatment processes.

The accuracy of flow measurement is basically the same for evaluation of water and wastewater management, especially in case of water deficit and the active tariff of water and wastewater charge.

The developed device gives the possibility of registration and reporting of measured variables in the arbitrary chosen time period. It is really possible thanks to connection with programmable control system.

6. REFERENCES

Cieślewicz, J. (1992). *Ultrasonic System for Remote Robot Control*, Master Degree, Chair of Control, Robotics and Comp. Sci., Technical University, Poznań (in Polish).

Majcherek, H. (1990). *Mathematical modelling of channel with a given characteristic of flow velocity*, Technical University Edition, Dissertation no. 240, Poznań (in Polish).

Matuschek, J. (1985), *Ultra sound technics*, WNT Warszawa (in Polish).

Olsson, G., Piani, G. (1990). *Computer System for Automation and Control*, Prentice Hall, London.

Urbaniak, A. (1990). *Automation in Environmental Engineering*, Technical University Edition no.1643, Poznań

Reference Manual: *Specialized Integrated Circuit YO0007 for Ultrasonic Remote Control "Echo-Prcesor"*, ATMOS - Warszawa

AN APPLICATION OF IMAGE SENSING AND ANALYZES IN OPHTHALMOLOGY.

J. VALENČÍK, J. HOSCHEK

SLOVAK ACADEMY OF SCIENCES, Institute of Control Theory and Robotics, Severná 5, 974 00 Banská Bystrica, SLOVAKIA

Abstract. An application for image sensing and image analysis in medicine (ophthalmology) is described. A new approach to corneal endothelium microscopy image grabbing and analysis by means of the CCD camera; frame grabber and PC improves operability and quality of this examination. Some specialized algorithms for image filtering and cellular walls detection are also described and examples of their application are shown.

Key Words. Image processing; medicine; medical information processing, computer application

1. INTRODUCTION

Much research work has been done in the development of image sensing technology and the processing techniques widely applied in robotics. The use of these sensing techniques in other sciences brings a new quality to these areas. This paper deals with a specific application of image sensing for diagnostics in ophthalmology.

In the daily praxis of ophthalmology there are a large number of visual examinations of patient's eyes by means of special microscopes. The physician subjectively observes qualitative and quantitative parameters of an eye's image (or its part) and makes decision about eye's condition. There are a several devices for such medical diagnostics, where there is a possibility to take a photo of an observed object in order to store it in the archive or to make following analysis of the image. There seems to be a very suitable place for the application of sophisticated technical devices - like CCD cameras, frame grabbers and computers - used for image sensing, analysis of its parameters and storage. A typical example is in endothelial cornea microscopy where the analysis of quantitative parameters (cells' area density) is a crucial part of the examination.

Corneal endothelium (see Fig. 1) is a one layer tissue created from cells mainly of hexagonal shape with an average size of 9 μm. The endothelium plays an important role in water and ion transport as well as balance.
Endothelial cornea microscopy is a modern method for diagnostics of the actual state of cornea endothelium's tissue and its main advantage is in the early detection of cell changes. A main disadvantage of present evaluation methods used in endothelial microscopy is the time delay combined in many cases with unsatisfactory results. This produces a decreasing quality and operability in such diagnostic techniques. The main reason for this state is the use of old-fashioned technical equipment and dated methods for evaluation.

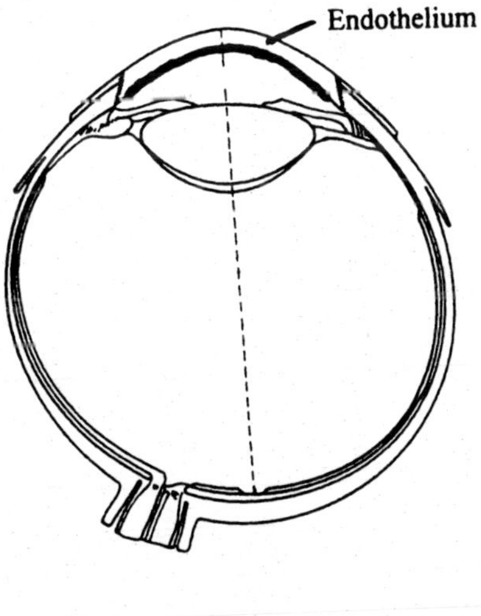

Fig. 1. Cross section of the human eye

First experiments with a CCD camera and frame grabber for endothelium's tissue image acquisition were made in cooperation with Ophthalmologic Clinic in 1992 with the promising results, and from this time work on cornea endothelium image sensing and evaluation continues.

2. HARDWARE ARRANGEMENT

First experiments in image acquisition started with RODENSTOCK's endothelial microscope and visual system (VS) with 256 x 256 pixel resolution and 6-bit A/D-converter. The principle scheme for hardware arrangement is shown in Fig.2. Although connection of CCD camera to the endothelial microscope was successfully tested and first images were captured to PC; the quality of the obtained image was very poor and it appeared insufficient for further evaluation. It was because of the poor quality of observed image from the microscope itself and the low spatial and intensity resolution of the visual system. Due to technical problems with the above mentioned microscope, further experiments are being made with the new KEELER KONAN SP-5500 endothelial microscope. Also a new visual card with 512 x 512 pixels and 8-bit A/D converter connected to PHILIPS B-W camera with 604 x 588 pixels and 0.03 lux sensitivity are used.

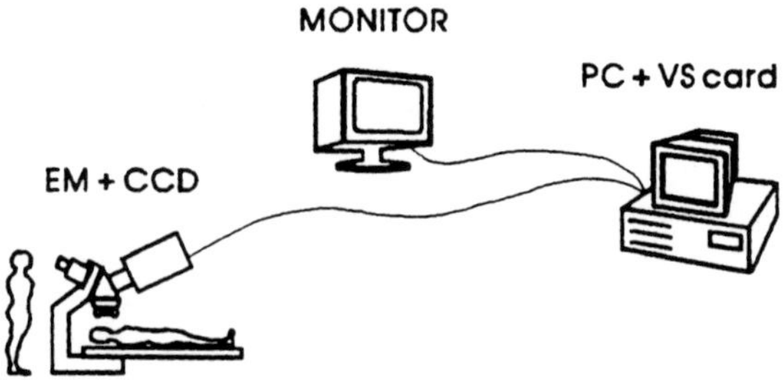

Fig. 2. Hardware arrangement

An analog video-recorder for storage of continual sections of endothelium observation is used. In this case the images are taken to PC from video-recordings.

3. IMAGE PROCESSING

The application software is written in C++ language and at the present time enables MONITORING of the image observed by microscope on a VS monitor; ACQUISITION of up to 4 images into VS memory; SAVE/LOAD images to/from PC's hard-disk; and image PROCESSING.

The main problem in endothelial microscopy is the relatively poor image quality (see Fig. 3), that is caused by the difficulty of examination method itself. That's why the mechanical application of standard algorithms for image filtering and enhancement did not lead to success. Some specific modifications of the standard algorithms based, on expected image properties, were developed and tested.

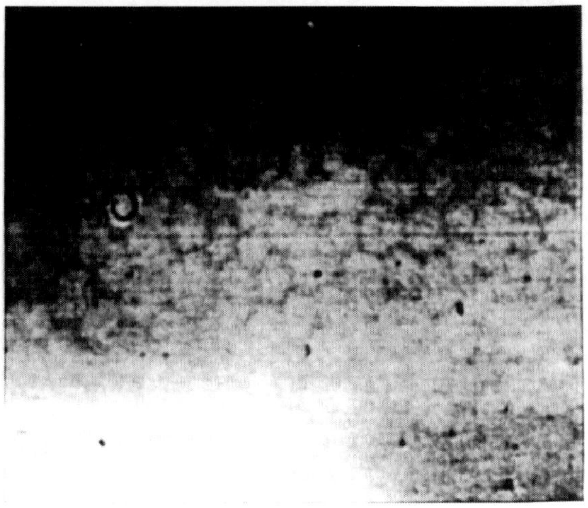

Fig. 3. Image of corneal endothelium

3.1. Image filtering

The first step in image processing is the separation and removal of those parts of the image structures that are apparently not parts of the observed tissue. They are mainly caused by dust particles on the optics and CCD chip, and their presence in image increases errors in further processing. Removal of these structures is based on the fact, that the gradient of intensity on the border of the image for such particles is very high and the absolute value of the intensity step between their pixels and neighborhood pixels is higher than any intensity step between the cellular wall and inside the cell.
Fig. 4 shows the intensity profile for a cross-section of a partial image with such a structure.

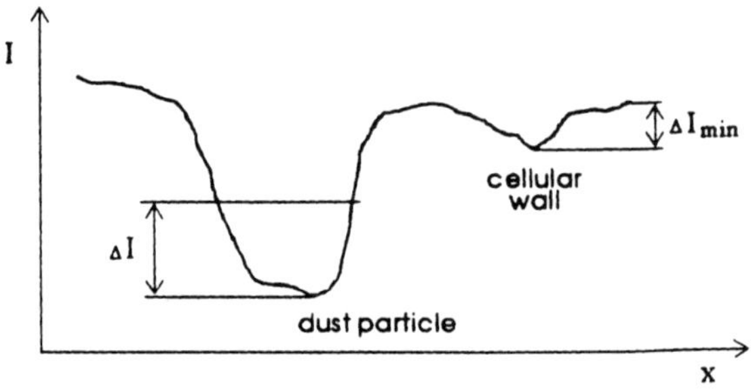

Fig. 4. Intensity profile of image with disturbance structure

Consider a rectangular mask of n x n = N pixels moving around the image (see Fig. 5).
For each pixel the difference of intensity between the central pixel I_C and the intensities of others pixels in mask I_j are computed.
For pixels where

$$|I_C - I_j| > \Delta I, \qquad (1)$$

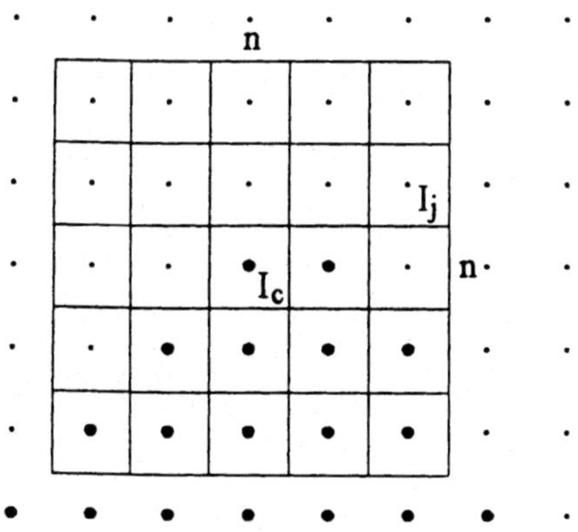

Fig. 5. Mask for disturbance structures filtering

where ΔI is a chosen parameter, is the sum of intensities S,

$$S = \sum_{J=1}^{m} I_J, \tag{2}$$

and the number m of such pixels counted.
If $m < N \cdot k$, where $k \in \langle 0, 1 \rangle$ is a chosen parameter, then the intensity of the central pixel I_C is replaced with I_S - the average intensity of pixels for that (1) is satisfied:

$$I_S = \frac{S}{m} \tag{3}$$

Calculation for the next pixel in an image is done with the intensity values of its neighbors modified in previous steps. This ensures, that described mask algorithm works in an erosive way and the efficiency of removing disturbance structures does not depend on their size. The image below these structures is computed from the values of 'valid' pixels in their neighborhood. Fig. 6 shows the image from Fig. 3, after an application of the above described algorithm with $n = 11$, $\Delta I = 20$, and $k = 0.3$.

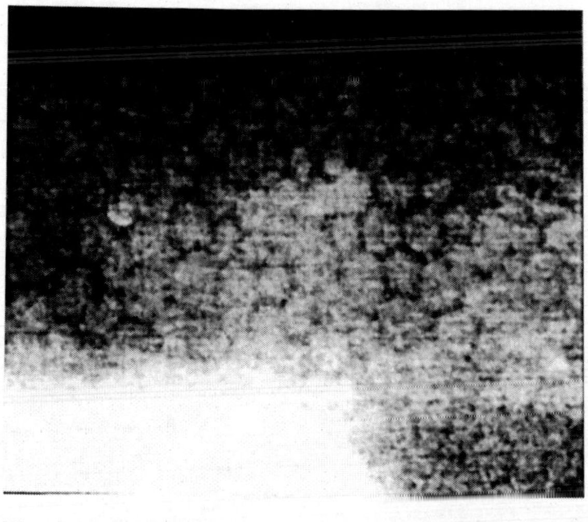

Fig. 6. Image from Fig.3 after disturbances filtering

3.2. Cellular walls detection.

The following algorithm for cellular walls detection is used:
A circle area with R - points radius is defined around each pixel (see Fig. 7). For each possible direction of the circle diameter in the pixel's net the average intensity I_A is computed. For the direction D_M with a minimal average intensity I_M, is an average intensity I_P in the perpendicular direction D_P

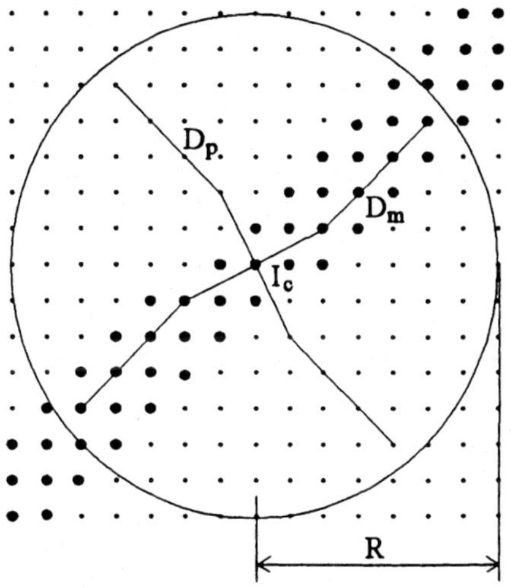

Fig. 7. Principle of cellular wall recognition algorithm

and their quotient P,

$$P = \frac{I_A}{I_P}. \tag{4}$$

is computed. If $P < t$, where t is a parameter, the point is considered for further evaluation.

In the next step the distinction r of the center of gravity for intensities from the considered point in direction D_P is computed.

If the eccentricity of the center of gravity E_G is

$$E_G = \frac{r}{R} < e, \tag{5}$$

where e is a parameter, $e \in \langle 0, 1 \rangle$, the pixel is considered to be lying on the cellular wall.

Fig. 8 shows the image from Fig. 6 after application of the above described algorithms. Radius R of the circle mask was 15 pixels and values for parameters t and e were 0.95 and 0.03 respectively.

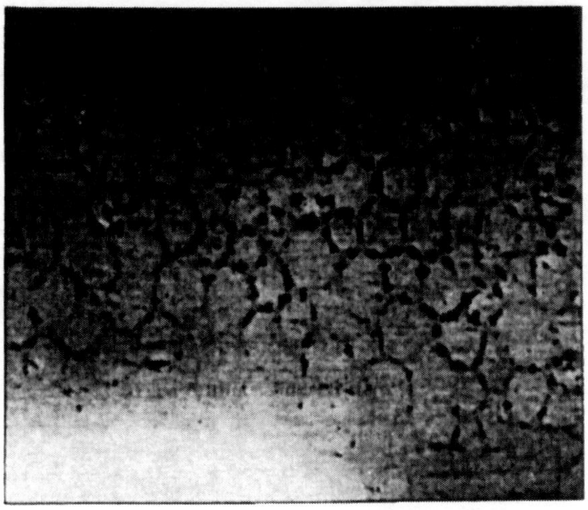

Fig. 8. Image from Fig. 6 after detection of walls

4. CONCLUSION

The described algorithms assure the first steps in sensing and automated processing of endothelium tissue's image. Values of parameters n, ΔI, k, R, t, e depend on specific properties of the image of endothelium - mainly the width of cellular walls and the difference in intensity between the walls and inside the cells. After experimental optimalization of these parameters, their optimal values for different images vary only a little. However, some corrections done respecting the processed image quality could improve the efficiency of algorithms.

Further processing of image consists of removing small areas, that are expected not to be part of the walls. This is done on the basis of their size and geometry. Two methods are tested for cell counting. The first one is based on building up a polygonal cellular net from parts of walls by computing intersections of the lines that are interpolated through the centers of neighbour walls. Counting the cells then can be done by a standard algorithm.

A second method is based on putting a dense net onto the image and checking the mutual visibility of neighbouring points in this net. If the connecting line of the neighbour point does not cross any part of wall, the pairs of points are replaced step by step with one point. Finally only one point leaves in each cell. Because the cellular wall structure is not closed in this case; there can be pairs of such points that see each other but lie in different cells. This can be eliminated by varying the position of points and repeating the checking more times before their replacement.

These algorithms are being tested at present time together with experiments on increasing the quality of the image by averaging more images, snapped in a very short time interval.

The goal is to develop a complete system that would help physicians to improve the operability and quality of endothelial microscopy, as well as enable storage of the text and image information obtained in various examinations in the computer database.

5. REFERENCES

Gonzalez, R.G., P. Wintz (1987). *Digital image processing*, 2nd ed., Addison-Wesley Publ. Comp., California

Jaroslavski L.P., Bajla I. (1989). *Methods and Systems of Digital Image Processing*, Alfa, Bratislava, (in slovak language)

Krejci L., Rezek P. (1983). Specular Mikroscopy of the Corneal Endothelium. Presence and Future., *Cs. Oftal.*, 39, No. 5, pp. 344-349, (in czech language)

Hoskovcova-Krejcova H., Hoskovec P., Krejci L., Rezek P. (1985). Classification of the Endothelial Lesions in Congenital Glaucoma by Means of the Computer Analysis, *Cs. Oftal.*, 41, No. 4, pp. 252-257, (in czech language)

IDENTIFICATION TECHNIQUES FOR MULTISENSOR SYSTEMS

PHAM HUY THOA

Computer and Automation Research Institute, Hungarian Academy of Sciences
H-1111 Budapest, Kende u. 13-17, Hungary
On leave from Institute of Information Technology, Nghia Do, Tu Liem, Hanoi, Vietnam

Abstract. A system parameter estimator for correlated noise systems with multisensors is presented in this paper. The system is described by a stochastic difference equation, while the integrated values of the output are observed by different sensors at discrete times. The estimator is developed on the basis of the variational method and the conjugate equation theory. The estimate obtained by the proposed method is identical with the optimal solution of the direct canonical minimization. It is shown that the proposed method can be considered as one of the identification methods using template functions.

Key Words. System identification; parameter estimation; multisensors; template functions; conjugate equation; variational method.

1. INTRODUCTION

In many practical control applications it is necessary to have a high reliability of the measuring system to be used. The multisensor system will help to improve the safety of the sensing system compared to that of a single sensor system. Two different structures of multisensor systems have been presented in (Tränkler, 1992). In a homogeneous redundant structure several sensors of the same kind are used. The diversity structure contains sensors with different physical principles for the same measuring value. Applications of such a multisensor system can be found e.g. in a gas-monitoring system or in an intelligent microphone system (Takahashi and Yamasaki, 1990).

The aim of this paper is to investigate identification techniques for dynamical systems with such a class of sensing systems. There are many methods for system parameter identification proposed in the literature (Åström and Eykhoff, 1973; Keviczky *et al.*, 1977; Ljung and Söderström, 1983; Isermann, 1988). A system parameter estimation method based on the variational method and the conjugate equation theory has been developed in (Thoa *et al.*, 1991a; 1991b; Thoa, 1991; Thoa and Son, 1991). It is assumed in this method that the system measurements are given by the measurement functionals of the integral form. That is to say, the system is described by a stochastic difference equation, while the integrated values of the output are observed by different sensors at discrete times. Such a measuring problem can be found in the measurement of the concentration of pollutants in environment systems (Fujishige, 1975). In this paper, the variational and conjugate equation method will be developed for correlated noise systems with multisensors. It will be shown that the proposed method may be considered as one of the identification methods using template functions presented in (Eykhoff, 1980; 1982; Eykhoff *et al.*, 1981).

The paper is organized as follows. After the problem statement in section 2, the method of the direct canonical minimization is presented briefly in section 3. The variational and conjugate equation method is developed in section 4 for systems with correlated noises. The relationship between the proposed method and the methods using template functions is derived in section 5. Some conclusions are given in section 6.

2. PROBLEM STATEMENT

Let us consider the dynamic time invariant single-input / single output system as shown in Fig.1.

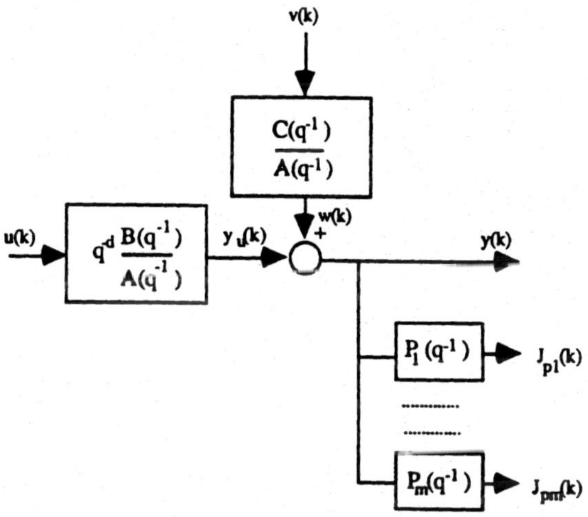

Fig.1. Block diagram of the system

and described by

$$A(q^{-1})y(k) = B(q^{-1})u(k-d) + C(q^{-1})v(k) \quad (1)$$

where u(k) and y(k) denote the system input and output, respectively; q^{-1} is the backward time shift operator; d is the discrete deadtime; $A(q^{-1})$, $B(q^{-1})$ and $C(q^{-1})$ are polynominals of the form:

$$A(q^{-1}) = 1 + a_1q^{-1} + a_2q^{-2} + \dots + a_{na}q^{-na}$$

$$B(q^{-1}) = b_0 + b_1q^{-1} + b_2q^{-2} + \dots + b_{nb}q^{-nb}$$

$$C(q^{-1}) = 1 + c_1q^{-1} + c_2q^{-2} + \dots + c_{nc}q^{-nc}$$

and v(k) is the white noise.

Let us denote the system parameter vector by $\Theta = [a_1, \dots, a_{na}, b_0, \dots, b_{nb}, c_1, \dots, c_{nc}]^T$.

The measurements given by m different sensors can be descibed by the measurement functionals of the form:

$$J_{p1} = P_1(q^{-1})y(k)$$

$$\vdots \qquad (2)$$

$$J_{pm} = P_m(q^{-1})y(k)$$

with

$$P_j(q^{-1}) = p_{j,n_{pj}} + p_{j,n_{pj}-1}q^{-1} + \dots + p_{j,1}q^{-n_{pj}+1};\ n_{pj} \le N$$

where N denotes the measurement interval and $j = 1, \dots, m$.

Let us denote the observation vector by

$$j = \left[J_{p1}, \dots, J_{pm}\right]^T$$

and consider the case of $n_{pj} = N$, $j = 1, \dots, m$. In this case, the measurement functionals J_{pj} can be written in the following form:

$$J_{pj} = p_j^T y \qquad (3.a)$$

where

$$p_j^T = \left[p_{j,1}, \dots, p_{j,N}\right]$$

$$y = \left[y(1), \dots, y(N)\right]^T$$

or in the matrix / vector form:

$$j = P^T y \qquad (3.b)$$

with

$$P^T = \left[p_1^T, \dots, p_m^T\right]^T$$

Note that in this case P plays the role of template function matrix (see section 5). Here, P is assumed to be known and regular, i.e. P is invertible.

The identification problem is to determine the unknown system parameter vector Θ by using the measurement functionals $j = \left[J_{p1}, \dots, J_{pm}\right]^T$.

3. THE DIRECT CANONICAL MINIMIZATION METHOD

For $k = 1, \dots, N$, the system described by Eq.(1) may be written in the form:

$$Ay = Bu + Cv \qquad (4)$$

where

$$A = I + a_1S + \dots + a_{na}S^{na} = I + \tilde{A}$$

$$B = b_0I + b_1S + \dots + b_{nb}S^{nb}$$

$$C = I + c_1S + \dots + c_{nc}S^{nc} = I + \tilde{C}$$

Here, $S^{(\bullet)}$ denotes the TOEPLITZ shift matrix (Keviczky *et al.*, 1977).

Assume that y is measurable, we may estimate the noise v by

$$\hat{v} = y - \hat{y} \qquad (5)$$

with

$$\hat{y} = -\hat{\tilde{A}}y + \hat{B}u + \hat{\tilde{C}}\hat{v} = F(y,u,\hat{v})\hat{\Theta} \qquad (6)$$

where

$\hat{y}$ is the estimated output; $\hat{\tilde{A}} = \hat{A} - I$; $\hat{\tilde{C}} = \hat{C} - I$,

$\hat{A}$, $\hat{B}$ and $\hat{C}$ are estimated values of A, B and C, respectively.

From Eq.(5) and Eq.(6) it follows that

$$\hat{A}y = \hat{B}u + \hat{C}\hat{v} \qquad (7)$$

or

$$\hat{v} = \hat{C}^{-1}\hat{A}y - \hat{C}^{-1}\hat{B}u \qquad (8)$$

Let us define the output error by

$$\varepsilon = y - \hat{y} = \hat{C}^{-1}\hat{A}y - \hat{C}^{-1}\hat{B}u = \hat{C}^{-1}(\hat{A}y - \hat{B}u) \qquad (9)$$

and use the criterion

$$Q = \frac{1}{2}\varepsilon^T\varepsilon \qquad (10)$$

Then, we obtain

$$\frac{dQ}{d\hat{\Theta}} = \frac{d\varepsilon^T}{d\hat{\Theta}}\varepsilon \qquad (11)$$

where $\hat{\Theta} = \left[\hat{a}_1, \dots, \hat{a}_{na}, \hat{b}_0, \dots, \hat{b}_{nb}, \hat{c}_1, \dots, \hat{c}_{nc}\right]^T$

Differentiating ε with respect to the parameters leads to

$$\frac{d\varepsilon}{d\hat{a}_i} = \frac{\partial}{\partial\hat{a}_i}\hat{C}^{-1}(\hat{A}y - \hat{B}u) = \hat{C}^{-1}\frac{\partial\hat{A}}{\partial\hat{a}_i}y = \qquad (12.a)$$
$$= \hat{C}^{-1}S^iy = S^iy^F$$

$$\frac{d\varepsilon}{d\hat{b}_i} = \frac{\partial}{\partial\hat{b}_i}\hat{C}^{-1}(\hat{A}y - \hat{B}u) = -\hat{C}^{-1}\frac{\partial\hat{B}}{\partial\hat{b}_i}u = \qquad (12.b)$$
$$= -\hat{C}^{-1}S^iu = -S^iu^F$$

$$\frac{d\varepsilon}{d\hat{c}_i} = \frac{\partial}{\partial\hat{c}_i}\hat{C}^{-1}(\hat{A}y - \hat{B}u) = -\hat{C}^{-1}\frac{\partial\hat{C}}{\partial\hat{c}_i}\hat{C}^{-1}(\hat{A}y - \hat{B}u) =$$
$$= -\hat{C}^{-1}S^i\hat{v} = -S^i\hat{v}^F \qquad (12.c)$$

where $y^F = \hat{C}^{-1}y$, $u^F = \hat{C}^{-1}u$, $\hat{v}^F = \hat{C}^{-1}\hat{v}$

It follows that

$$\frac{d\varepsilon^T}{d\hat{\Theta}} = -F^T(y^F,u^F,\hat{v}^F) = \qquad (13)$$
$$-\left[-Sy^F, \dots, -S^{na}y^F, Iu^F, Su^F, \dots, S^{nb}u^F, S\hat{v}^F, \dots, S^{nc}\hat{v}^F\right]^T$$

and

$$\frac{dQ}{d\hat{\Theta}} = -F^T(y^F, u^F, \hat{v}^F)\varepsilon \qquad (14)$$

The system parameter vector Θ may be determined iteratively by

$$\hat{\Theta}_{i+1} = \hat{\Theta}_i - H^{-1}_{\hat{\Theta}_i}\left[\frac{dQ}{d\hat{\Theta}}\right]_{\hat{\Theta}_i} \qquad (15.a)$$

where $H_{\hat{\Theta}_i}$ denotes the Hessian matrix:

$$H_{\hat{\Theta}_i} = \left[\frac{d^2Q}{d\hat{\Theta}d\hat{\Theta}^T}\right]_{\hat{\Theta}_i} = \left[\frac{d\varepsilon^T}{d\hat{\Theta}}\frac{d\varepsilon}{d\hat{\Theta}^T}\right]_{\hat{\Theta}_i} + \left[\frac{d}{d\hat{\Theta}}\varepsilon_i^T\left(\frac{d\varepsilon^T}{d\hat{\Theta}}\right)\right]_{\hat{\Theta}_i} \qquad (15.b)$$

or

$$\hat{\Theta}_{i+1} = \hat{\Theta}_i + \left[F^T_{\hat{\Theta}_i}(\bullet)F_{\hat{\Theta}_i}(\bullet) - \frac{d}{d\hat{\Theta}}\varepsilon_i^T F^T_{\hat{\Theta}_i}(\bullet)\right]^{-1} F^T_{\hat{\Theta}_i}(\bullet)\varepsilon_i \qquad (16)$$

where $F^T_{\hat{\Theta}_i}(\bullet) \hat{=} F^T_{\hat{\Theta}_i}(y^F,u^F,\hat{v}^F)$,

$\varepsilon_i = y - \hat{y}_i = y - F(y,u,\hat{v})\hat{\Theta}_i$.

4. THE VARIATIONAL AND CONJUGATE EQUATION METHOD

The variational and conjugate equation method presented in (Thoa *et al.*, 1991a, 1991b) will be applied in this section to derive the parameter estimators for the system to be considered.

From Eq.(4) it follows that the noise process is described by

$$Cv = Ay - Bu \qquad (17)$$

Corresponding to Eq.(17), the following estimated noise model is used:

$$\hat{C}_i\hat{v} = \hat{A}_i y - \hat{B}_i u \qquad (18)$$

where $\hat{v}$ is the estimate of noise v.

The noise v may be estimated by

$$\hat{v} = y - \hat{y} \qquad (19)$$

with

$$\hat{y} = -\hat{\bar{A}}_i y + \hat{B}_i u + \hat{\bar{C}}_i \hat{v} = F(y,u,\hat{v})\hat{\Theta}_i \qquad (20)$$

where $\hat{y}$ is the estimated output, $\hat{\bar{A}}_i = \hat{A}_i - I$, $\hat{\bar{C}}_i = \hat{C}_i - I$ and $\hat{A}_i$, $\hat{B}_i$, $\hat{C}_i$ are the estimated values of A, B, C at the i-th step, respectively.

Let us define

$$L' \triangleq C,\ \varphi' \triangleq v,\ q' = Ay - Bu$$
$$L \triangleq \hat{C}_i,\ \varphi \triangleq \hat{v},\ q = \hat{A}_i y - \hat{B}_i u \qquad (21.a)$$

In the scalar form, this means that

$$L' \triangleq C(q^{-1}),\ \varphi' \triangleq v(k),\ q' \triangleq A(q^{-1})y(k) - B(q^{-1})u(k-d)$$
$$L \triangleq \hat{C}_i(q^{-1}),\ \varphi \triangleq \hat{v}(k),\ q \triangleq \hat{A}_i(q^{-1})y(k) - \hat{B}_i(q^{-1})u(k-d) \qquad (21.b)$$

It follows then

$$\delta L = \delta C_i = C - \hat{C}_i = \left[\delta c_1^i S^1 + \dots + \delta c_{nc}^i S^{nc}\right]$$

$$\delta q = \delta A_i y - \delta B_i u = $$
$$= \left[S^1 y, \dots, S^{na} y\right]\delta a_i - \left[Iu, S^1 u, \dots, S^{nb} u\right]\delta b_i$$

with

$$\delta a_i = \left[\delta a_1^i, \dots, \delta a_{na}^i\right]^T,\ \delta b_i = \left[\delta b_0^i, \dots, \delta b_{nb}^i\right]^T$$
$$\delta c_i = \left[\delta c_1^i, \dots, \delta c_{nc}^i\right]^T,\ \delta\Theta = \Theta - \hat{\Theta}_i = \left[\delta a_i^T, \delta b_i^T, \delta c_i^T\right]^T$$

Let us define the noise measurement functionals by

$$j_v = P^T v$$
$$\hat{j}_v = P^T \hat{v} \qquad (22)$$

and introduce the conjugate equations corresponding to Eq.(18):

$$\hat{C}_i^T v_j^* = p_j,\ j = 1, \dots, m \qquad (23.a)$$

or

$$\hat{C}_i^T V^* = P \qquad (23.b)$$

and corresponding to Eq.(17):

$$C^T v_j^* = p_j,\ j = 1, \dots, m \qquad (24.a)$$

or

$$C^T V^* = P \qquad (24.b)$$

Note that in the scalar form, the first conjugate equation is

$$\hat{C}_i(q^{+1})v_j^*(k) = p_j(k),\ j = 1, \dots, m \qquad (25.a)$$

where $\hat{C}_i(q^{+1})$ is the conjugate operator of $\hat{C}_i(q^{-1})$ described by

$$\hat{C}_i(q^{+1}) = 1 + \hat{c}_1^i q + \dots + \hat{c}_{nc}^i q^{nc} \qquad (25.b)$$

and the later conjugate equation has the form:

$$C(q^{+1})v_j^*(k) = p_j(k),\ j = 1, \dots, m \qquad (26.a)$$

where

$$C(q^{+1}) = 1 + c_1 q + \dots + c_{nc} q^{nc} \qquad (26.b)$$

The following theorem has been proved in (Thoa *et al.*, 1991a):

<u>Theorem 1</u>. Let φ be the solution of Eq.(18) and φ_p^* (= v_j^*) be the solution of the conjugate equation (23) or (24). Then, the following relation holds for the variations δJ_p, δL and δq:

$$\delta J_p = -< \varphi_p^*, \delta L\, \varphi > + < \varphi_p^*, \delta q > \qquad (27)$$

where $\delta J_p = J_p' - J_p$; $\delta L = L' - L$; $\delta q = q' - q$ and $<\cdot,\cdot>$ denotes the inner product.

Using Theorem 1, one obtains

$$\delta J_{vj} = -< v_j^{*T}, \delta C_i \hat{v} > + < v_j^{*T}, \delta A_i y > - < v_j^{*T}, \delta B_i u > \qquad (28)$$

$$\text{or } \delta j_v = j_v - \hat{j}_v = P^T v - P^T \hat{v} = V^{*T}\left[S^1 y, \dots, S^{na} y\right]\delta a_i -$$
$$- V^{*T}\left[Iu, S^1 u, \dots, S^{nb} u\right]\delta b_i - V^{*T}\left[S^1\hat{v}, \dots, S^{nc}\hat{v}\right]\delta c_i \qquad (29)$$

Solving the first conjugate equation (23.b) for V^{*T}, we obtain

$$V_1^{*T} = P^T \hat{C}_i^{-1} \qquad (30.a)$$

and from the second conjugate equation (24.b) it follows that

$$V_2^{*T} = P^T C^{-1} \qquad (30.b)$$

Corresponding to the first conjugate equation, we have

$$P^T v - P^T(y - \hat{y}) = -P^T F_{\hat{\Theta}_i}(y^F, u^F, \hat{v}^F)\delta\Theta_i$$
$$P^T v - \delta j = -P^T F_{\hat{\Theta}_i}(y^F, u^F, \hat{v}^F)\delta\Theta_i$$

$$\delta j = P^T F_{\hat{\Theta}_i}(y^F, u^F, \hat{v}^F)\delta\Theta_i + P^T v \qquad (31.a)$$

$$P^{T^{-1}}\delta j = F_{\hat{\Theta}_i}(y^F, u^F, \hat{v}^F)\delta\Theta_i + v \qquad (31.b)$$

where

$$F_{\hat{\Theta}_i}(y^F, u^F, \hat{v}^F) =$$
$$= \left[-Sy^F, \dots, -S^{na}y^F, Iu^F, Su^F, \dots, S^{nb}u^F, S\hat{v}^F, \dots, S^{nc}\hat{v}^F\right]$$
$$\delta\Theta = \Theta - \hat{\Theta}_i = \left[\delta a_i^T, \delta b_i^T, \delta c_i^T\right]^T;\ \delta a_i = \left[\delta a_1^i, \dots, \delta a_{na}^i\right]^T,$$
$$\delta b_i = \left[\delta b_0^i, \dots, \delta b_{nb}^i\right]^T,\ \delta c_i = \left[\delta c_1^i, \dots, \delta c_{nc}^i\right]^T$$
$$y^F = \hat{C}_i^{-1}y,\ u^F = \hat{C}_i^{-1}u,\ \hat{v}^F = \hat{C}_i^{-1}\hat{v}$$

By the same way, one obtains corresponding to the second conjugate equation

$$\delta j = P^T F_\Theta(y^F,u^F,\hat{v}^F)\delta\Theta_i + P^T v \qquad (32.a)$$

$$P^{T^{-1}}\delta j = F_\Theta(y^F,u^F,\hat{v}^F)\delta\Theta_i + v \qquad (32.b)$$

where

$$F_\Theta(y^F,u^F,\hat{v}^F) =$$

$$= \left[-Sy^F, \ldots, -S^{na}y^F, Iu^F, Su^F, \ldots, S^{nb}u^F, S\hat{v}^F, \ldots, S^{nc}\hat{v}^F\right]$$

$$y^F = C^{-1}y,\ u^F = C^{-1}u,\ \hat{v}^F = C^{-1}\hat{v}$$

For some fix model parameters $\hat{\Theta}_i$, $P^{T^{-1}}\delta j$ and $P^T F_{\hat{\Theta}_i}(y^F,u^F,\hat{v}^F)$ in Eq.(31) are independent of $\hat{\Theta}$. Under the condition that $F^T_{\hat{\Theta}_i}(*)F_{\hat{\Theta}_i}(*)$ is not singular, i.e. $\det\left[F^T_{\hat{\Theta}_i}(*)F_{\hat{\Theta}_i}(*)\right] \neq 0$, Eq.(31) can be solved for $\hat{\Theta}$ by using the criterion

$$Q_1 = \frac{1}{2}e_1^T e_1 \qquad (33)$$

with

$$e_1 = \varepsilon_i - F_{\hat{\Theta}_i}(y^F,u^F,\hat{v}^F)(\hat{\Theta} - \hat{\Theta}_i)$$

where $\varepsilon_i = P^{T^{-1}}\delta j$.

The solution is given by

$$\delta\hat{\Theta}_i = \hat{\Theta} - \hat{\Theta}_i = \left[F^T_{\hat{\Theta}_i}(*)F_{\hat{\Theta}_i}(*)\right]^{-1} F^T_{\hat{\Theta}_i}(*)\varepsilon_i \qquad (34)$$

where $F^T_{\hat{\Theta}_i}(*) \triangleq F^T_{\hat{\Theta}_i}(y^F,u^F,\hat{v}^F)$,

$$\varepsilon_i = y - \hat{y}_i = y - F(y,u,\hat{v})\hat{\Theta}_i .$$

The system parameter vector is determined by

$$\hat{\Theta} = \hat{\Theta}_i + \left[F^T_{\hat{\Theta}_i}(*)F_{\hat{\Theta}_i}(*)\right]^{-1} F^T_{\hat{\Theta}_i}(*)\varepsilon_i \qquad (35)$$

The above procedure can be repeated if we put $\hat{\Theta}_{i+1} = \hat{\Theta}$ obtained by Eq.(35) in the next iteration step :

$$\hat{\Theta}_{i+1} = \hat{\Theta}_i + \left[F^T_{\hat{\Theta}_i}(*)F_{\hat{\Theta}_i}(*)\right]^{-1} F^T_{\hat{\Theta}_i}(*)\varepsilon_i \qquad (36)$$

where

$$\varepsilon_i = P^{T^{-1}}\delta j = P^{T^{-1}}j - F(y,u,\hat{v})\hat{\Theta}_i =$$

$$= P^{T^{-1}}j - F(P^{T^{-1}}j,u,\hat{v})\hat{\Theta}_i$$

Comment. The solution (36) is a simplified solution of the direct canonical minimization method given by Eq.(16) where the second order term of the Hessian matrix is neglected. This neglection is allowed, however, in many applications. Therefore, the solution (36) is applicable for iterative determination of system parameters in correlated noise case.

To obtain more exact system parameter estimates, one needs to solve Eq.(32) for $\delta\Theta_i$. In this case the dependence of $F_{\hat{\Theta}}(y^F,u^F,\hat{v}^F)$ on $\hat{\Theta}$ should be considered. To find the minimum of

$$Q_2 = \frac{1}{2}e_2^T e_2 \qquad (37)$$

with $e_2 = \varepsilon_i - F_{\hat{\Theta}}(y^F,u^F,\hat{v}^F)(\hat{\Theta} - \hat{\Theta}_i)$

we set

$$\frac{dQ_2}{d\hat{\Theta}} = \frac{1}{2}\frac{de_2^T e_2}{d\hat{\Theta}} = 0 \qquad (38.a)$$

or

$$-F^T_{\hat{\Theta}}(*)\varepsilon_i - \frac{d\varepsilon_i^T F_{\hat{\Theta}}(*)}{d\hat{\Theta}}(\hat{\Theta} - \hat{\Theta}_i) + M(\hat{\Theta} - \hat{\Theta}_i) + \qquad (38.b)$$

$$+ \frac{1}{2}\left[\frac{\partial m_1^T}{\partial\hat{\Theta}}(\hat{\Theta} - \hat{\Theta}_i), \ldots, \frac{\partial m_n^T}{\partial\hat{\Theta}}(\hat{\Theta} - \hat{\Theta}_i)\right](\hat{\Theta} - \hat{\Theta}_i) = 0$$

where $M = F^T_{\hat{\Theta}}(*)F_{\hat{\Theta}}(*)$ and $m_1, \ldots, m_n$ are the column vectors of M.

Neglecting the last second order term of Eq.(38.b), the following solution is obtained

$$\hat{\Theta} = \hat{\Theta}_i + \left[F^T_{\hat{\Theta}}(*)F_{\hat{\Theta}}(*) - \frac{d\varepsilon_i^T F_{\hat{\Theta}}(*)}{d\hat{\Theta}}\right]^{-1} F^T_{\hat{\Theta}}(*)\varepsilon_i \qquad (39)$$

Computing the last term of Eq.(39) at $\hat{\Theta} = \hat{\Theta}_i$ and denoting the estimated result by $\hat{\Theta}_{i+1}$, one obtains the following iterative estimation procedure:

$$\hat{\Theta}_{i+1} = \hat{\Theta}_i + \left[F^T_{\hat{\Theta}_i}(*)F_{\hat{\Theta}_i}(*) - \frac{d}{d\hat{\Theta}}\varepsilon_i^T F^T_{\hat{\Theta}_i}(*)\right]^{-1} F^T_{\hat{\Theta}_i}(*)\varepsilon_i \qquad (40)$$

Comments.

-The parameter estimate (40) obtained by the proposed identification method is identical with the solution of the direct canonical minimization (16). This means that the solution (40) is an optimal solution of the identification problem in the correlated noise case.

-In the proposed method it is necessary to integrate the model equation forward and n conjugate equations backward in time to obtain the gradient of the lost function with respect to the parameters given by Eq.(14). The computation cost is therefore equal to that of the direct canonical minimization.

5. RELATIONSHIP TO THE TEMPLATE FUNCTION METHOD

Due to process disturbances or unobservable inputs and due to measurement errors, the template function method is based on observation and data processing over an interval of time.
In a rather general way this can be described through integral operators working over such a time interval on the output signal y(t) (Eykhoff,1980).
The time functions $p_j(t)$ are called "template functions" and they distinguish the operators that may be

- continuous in time :

$$J_j[y(t)] \triangleq \langle y, p_j \rangle = \int_0^T y(t)p_j(t)\,dt \qquad (41)$$

- discrete in time :

$$J_j[y(t)] \triangleq \langle y, p_j \rangle = \sum_{i=0}^{k} y(iT_s)p_j(iT_s) = p_j^T y \qquad (42)$$

with the proper definition of the corresponding vectors.

The notation $\langle\cdot\cdot\rangle$ refers to the notion of scalar product in function (abstract) spaces (e.g. E_k, E_∞, L_2) which may imply both discrete (finite and infinitely dimensional, E_k viz. E_∞) and continuous (L_2) cases.
Note that Eq.(42) is identical with Eq.(3) describing the measurement functionals provided by the sensors.

The use of functional operators and of the template functions in parameter estimation can be illustrated as follows:

The system (1) with unknown parameters Θ may be described by

$$y(k) = F(y,u)\Theta + w(k) \quad (43)$$

where w(k) is a colored noise.

Let us denote the estimates of unknown process parameters Θ by $\hat{\Theta}$, then

$$y(k) - F(y,u)\hat{\Theta} = \varepsilon(k) \quad (44)$$

Using m operators that are different in the sense that the corresponding template functions are linearly independent, i.e. that they span an m-dimensional space

$$j[\;] = [\; J_1[\;], \ldots, J_m[\;]\;] \quad (45)$$

and the projection theorem (Eykhoff, 1982):

$$J[\varepsilon(k)] = \langle p, \varepsilon \rangle = 0 \quad (46)$$

leads to

$$J[F(y,u)]\hat{\Theta} = j[y(k)] \quad (47)$$

Note that the error is linear in the parameters and if the parameters are assumed to be constant in time, the use of such a set of operators leads to a set of linear equations from which the unknown process parameters Θ can be found.

Let P play the role of the template functions. P is a matrix of the same dimensions as F. Matrix P is called the "template function matrix". Along the line given before, the template function estimator can be written as:

$$\hat{\Theta} = \left[\; P^T F(y,u) \;\right]^{-1} P^T y \quad (48)$$

provided of course that the matrix $P^T F$ is invertible.

The substitution of expression (43) for the process output leads to

$$\hat{\Theta} = \Theta + \left[P^T F(y,u)\right]^{-1} P^T w \quad (49)$$

from which statistical properties like (asymptotic) bias and (asymptotic) covariance can be found. From Eq.(48) different parameter estimators can be obtained by making particular choices of the template function matrix P (Eykhoff and co-workers, 1981; Eykhoff, 1982).

The application of the variational and conjugate equation method presented in section 4 for the system (43) leads to the following relationship:

$$\delta j = P^T F(y^F, u^F)\delta\Theta + P^T w^F \quad (50)$$

where

$$F(y^F, u^F) = \left[-S^1 y^F, \ldots, -S^{na} y^F, Iu^F, S^1 u^F, \ldots, S^{nb} u^F\right]$$

$$y^F = \hat{A}^{-1} y,\; u^F = \hat{A}^{-1} u,\; w^F = \hat{A}^{-1} w,\; \delta\Theta = \Theta - \hat{\Theta}$$

Under certain conditions P can be chosen such that (see Eq.(46))

$$\delta j = P^T \varepsilon = 0 \quad (51)$$

It follows, then, from Eq.(50) and Eq.(51) that

$$\hat{\Theta} = \Theta + \left[P^T F(y^F, u^F)\right]^{-1} P^T w^F \quad (52)$$

or

$$\hat{\Theta} = \left[\; P^T F(y^F, u^F) \;\right]^{-1} P^T y^F \quad (53)$$

Note that the estimator (53) derived from the proposed method has a similar form as the template function estimator given by Eq.(48). In this case P given in Eq.(3) plays the role of a deterministic template function matrix. By particular choices of the template function matrix P one can obtain asymptotically unbiased estimators such as the instrumental variable estimators.

6. CONCLUSION

In this paper, a system parameter estimator for correlated noise systems with multisensors has been presented. The estimator has been developed on the basis of the variational method and the conjugate equation theory. The estimate obtained is identical with the optimal solution of the direct canonical minimization. It has been shown that the proposed method is one of the identification methods using template functions. By making particular choices of the template function matrix P, different parameter estimators can be obtained. The presented estimation procedure is called off-line estimator. Due to the linearity in the parameters, this estimator can easily be brought into a recursive form.

7. ACKNOWLEDGMENT

The author wishes to express his sincere gratitude to Prof. Dr. L. Keviczky for his numerous suggestions and encouraging support through this work.

8. REFERENCES

Åström, K.J. and P. Eykhoff (1971). System identification - a survey. *Automatica*, 7, 123.

Eykhoff, P. (1980). System identification: approach to a coherent picture through template functions. *Electronics Letters*, 16, pp. 502-504.

Eykhoff, P., A.J.W. Van den Boom and A.A. Van Rede (1981). System identification methods - unification and information development using template functions. *Preprints/ Proc. 8th IFAC World Congress*. Kyoto, Aug. 1981.

Eykhoff, P. (1982). On the coherence among the multitude of system identification methods. *Proc.6th IFAC Symp. on Identification and System Parameter Estimation*, Washington DC, pp. 737-748.

Fujishige, S. (1975). Minimum variance estimation for a linear continuous-discrete system with noisy state-integral obsservation. *IEEE Trans. on AC*, pp. 139-140.

Isermann, R. (1988). *Identifikation dynamischer Systeme*, Springer-Verlag, Berlin.

Keviczky, L., C. Bányász and M. Hilger (1977). *Folyamatidentifikáció,Folyamatirányítási Kutatások a Szilikátiparban*, SZIKKTI, Budapest.

Ljung, L. and T. Söderström (1983). *Theory and Practice of Recursive Identification*, MTI Press, Cambridge.

Takahashi, K. and H. Yamasaki (1990). Self-adapting multiple microphone system. *Sensors and Actuators*, A21-A23, pp. 610-614.

Thoa, P.H., N.T. Loan and H.H. Son (1991a). Self-tuning adaptive control based on a new parameter estimation method. *Proc. IFAC Symp. on Intelligent Tuning and Adaptive Control*, Singapore, pp. 345-350.

Thoa, P.H., H.H. Son and N.T. Loan (1991b). System parameter identification based on variational method and optimal filtering theory. *Preprints 9th IFAC/IFORS Symp. on Identification and Parameter Estimation*, Budapest, Hungary, pp. 1465-1470.

Thoa, P.H.and H.H. Son (1991). Decentralized adaptive control of robotic manipulators. *Preprints IFAC/IFIP/ IMACS Symp. on Robot Control*, Vienna, Austria, pp. 223-228.

Thoa, P.H. (1991). Microcomputer-based adaptive control of DC motor actuators. *Proc. ChemAsia91 / Instrument Asia91 International Conference*, Singapore, pp. B21-B25.

Tränkler, H.R. (1992). Information processing in sensing devices and microsystems. *Preprints IFAC Symp. on Intelligent Components and Instruments for Control Applications*, Málaga, Spain, pp. 3-14.

SIMULTANEOUS DETECTION, LOCATION AND IDENTIFICATION OF FAULTS FOR DYNAMIC SYSTEMS FROM ROBUST PARAMETERS IDENTIFICATON

G. BLOCH, M. OULADSINE and P. THOMAS

Centre de Recherche en Automatique de Nancy, CNRS URA D821
ESSTIN, Rue Jean Lamour, 54500 VANDOEUVRE, FRANCE
Tel: (33) 83 50 33 44; Fax: (33) 83 54 21 73; E-mail : bloch@esstin.u-nancy.fr

Abstract. A procedure achieving simultaneously the detection of failures, their location and their identification is presented. The considered systems are MISO systems represented by ARX models. The a priori knowledge on the faults which can occur is particularly used. The faults modeled here are outliers, biases or drifts and can act upon the output, the inputs or even the noise. So models of the different faults, taking into account their dynamics and their location, are builded and then reduced to an unique form. The magnitude of the failure can be then estimated from the prediction error sequence simply by a least squares procedure. The nature and the location of the faults being actually not known, statistical tests for significance of the estimated parameters corresponding to the different faults are performed. Because the fault occurence time is also unknown, the previous estimation is made on a moving window. Moreover a robust recursive algorithm provides the fault diagnosis procedure with the parameters of the model.

Key Words. Failure (fault) detection, failure isolation, failure identification, analytical redundancy, difference equation, discrete systems, robust recursive identification.

1. INTRODUCTION

Fault detection and isolation (FDI) using analytical redundancy methods are currently the subject of extensive research and numerous survey papers can be found. See for instance Patton and Chen (1991). Although several works are based on the representation of systems by transfer functions (Mironovsky, 1980), most of the time the methods are based on a state representation of systems and assume that the nature of the possible faults is not a priori known. So the different steps are clearly separated: detection, isolation and if possible estimation of the fault magnitude.

On the contrary, a procedure is presented here achieving simultaneously the detection of faults, their location and their identification. The main idea is to make an intensive use of the knowledge on the faults which are supposed potentially occur on the system. The procedure can be related to the intervention analysis, provided by Box and Tiao (1975) but for time series modeling, and to further works aiming at handling outliers in identification (Tsay, 1988; Kobi et al., 1993).

In order to achieve the detection, location and identification of the faults, the model of the considered system must be known. Nevertheless the case of unknown model parameters is actually the most frequent. So the parameters of the model must be on-line identified, by a recursive procedure upon which the events to be detected do not act. A robust identification algorithm, due to Puthenpura and Sinha (1990), is then used to provide the fault detection, location and identification procedure with parameters of the model.

In the first part of this paper, models of the different faults, taking into account their dynamics and their application points (locations), are builded and then reduced to an unique form.

In the second part, the procedure of fault detection, location and estimation is presented. Different assumptions are considered. With the first one, the type, the location, the occurence time of the fault and the system model parameters are assumed to be known and the estimation of the fault magnitude is performed simply by least squares. With the second, only the fault occurence time and the system model parameters are known and statistical tests for significance of the estimated parameters corresponding to the different faults are performed. In the third place, the fault occurence time is also unknown and the previous estimations and tests are carried out on a moving window. Finally, the parameters of the model of the system are assumed to be unknown and are on-line identified allowing the complete diagnosis procedure wich can be implemented in real time.

A simulation example is finally shown, illustrating the efficiency of the strategy.

2. THE MODELS

A multi-input single output system (MISO), represented by a discrete ARX model, is considered :

$$A(q)\,y(t) = \sum_{j=1}^{p} B_j(q)\,u_j(t-nd_j) + e(t) \qquad (1)$$

where : $A(q) = 1 + a_1 q^{-1} + \ldots + u_{na} q^{-na}$,

$B_j(q) = b_{j1} + b_{j2} q^{-1} + \ldots + b_{jnb_j} q^{-nb_j+1}$, j=1, ..., p,

are polynomials in the delay operator q^{-1}, such as $q^{-1} y(t) = y(t-1)$, of orders na and nb_j, j=1, ...p respectively, y(t) is the output of the system, $u_j(t)$, j=1, ..., p, are the p inputs of the systems, nd_j, j=1, ..., p, are the numbers of delays from inputs to output and $\{e(t)\}$ is without fault a sequence of independant and normally distributed variates (white noise) with mean zero and variance σ_ε^2. In model (1), it is assumed that all of the zeros of the polynomials are outside the unit circle.

When various disturbances act on the inputs and the output, they are inobservable and the observed variables can be written as :

$$\begin{aligned} Y(t) &= y(t) + f_y(t) \\ U_j(t) &= u_j(t) + f_{u_j}(t) \quad , \quad j = 1, \ldots, p \end{aligned} \qquad (2)$$

where $f_y(t)$ and $f_{u_j}(t)$, j=1, ..., p are parametric functions representing exogeneous deterministic disturbances, for instance measurement errors, such as outliers or biases. Note that for the inputs, $f_{u_j}(t)$ can also represent, apart from the sign, a disturbance of actuator. The real input $u_j(t)$ of the system is then linked to the measured input $U_j(t)$ by:

$$u_j(t) = U_j(t) + f_{u_j}(t)$$

A change in the system itself can be to some extend represented by a change of the noise e(t), expressed as:

$$E(t) = e(t) + f_e(t) \qquad (3)$$

where $\{e(t)\}$ is always white noise with mean zero and variance σ_ε^2.

The function $f_e(t)$ can be deterministic, or stochastic as presented by Tsay (1988) leading to tests of variance change. The deterministic case is only considered here.

The disturbed system can be represented by the diagram of figure 1.

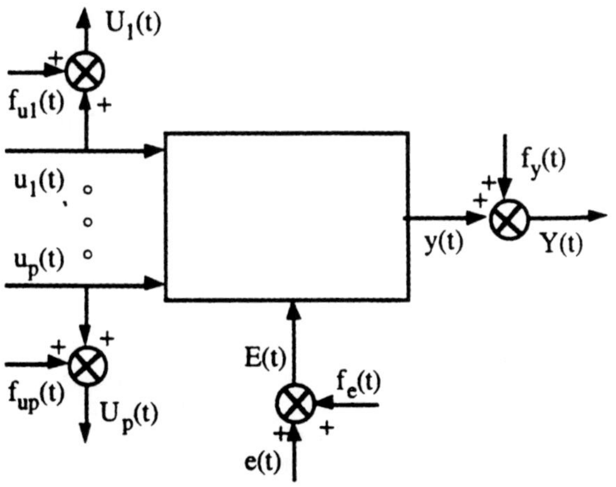

Fig. 1. The system and its disturbances

Whatever its point of application (location) may be, on the output, on the inputs or on the equation noise, it is assumed that the disturbance function (respectively $f_y(t)$, $f_{u_j}(t)$ and $f_e(t)$), noted simply f(t), can be written as :

$$f(t) = \omega \frac{\Omega_D(q)}{\Delta_D(q)} \xi_t^d = \omega\, P_D(q)\, \xi_t^d \qquad (4)$$

where ξ_t^d is an indicator variable signifying the occurence of a disturbance at the time point d ($\xi_t^d = 1$ if t=d, $\xi_t^d = 0$ if t≠d), $\Omega_D(q)$ and $\Delta_D(q)$ are polynomials representing the own dynamics of the disturbance and ω is a constant denoting its initial impact.

The main disturbance shapes wich can be met in practise are the following. If $P_D(q) = \frac{\Omega_D(q)}{\Delta_D(q)} = 1$, f(t) represents an outlier of magnitude ω, at time d. If $P_D(q) = \frac{\Omega_D(q)}{\Delta_D(q)} = \frac{1}{1-q^{-1}}$, f(t) represents a level change of magnitude ω, at time d and renders a bias of the considered sensor of magnitude ω. If $P_D(q) = \frac{\Omega_D(q)}{\Delta_D(q)} = \frac{1}{(1-q^{-1})^2}$, f(t) represents a drift of slope ω from instant d.

The different models can be summarized as follows:

$$A(q)\,y(t) = \sum_{j=1}^{p} B_j(q)\,u_j(t-nd_j) + e(t) \qquad (5)$$

$$Y(t) = y(t) + f_y(t) \tag{6a}$$

$$U_j(t) = u_j(t) + f_{u_j}(t) \quad , \quad j = 1, ..., p \tag{6b}$$

$$E(t) = e(t) + f_e(t) \tag{6c}$$

$$f(t) = \omega \, P_D(q) \, \xi_t^d \tag{7}$$

With equation (5), the prediction error e(t) can be calculated from the measurements assumed without faults as:

$$e(t) = A(q)\, y(t) - \sum_{j=1}^{p} B_j(q)\, u_j(t - nd_j) \tag{8}$$

The variable Z(t) can also be builded from the measurements:

$$Z(t) = A(q)\, Y(t) - \sum_{j=1}^{p} B_j(q)\, U_j(t - nd_j) \tag{9}$$

If a failure occurs on y(t) only, the previous equation becomes with eq. (6a):

$$Z(t) = A(q)\, y(t) - \sum_{j=1}^{p} B_j(q)\, u_j(t - nd_j) + A(q)\, f_y(t)$$

If a failure occurs on $u_k(t)$ only, Z(t) becomes now with eq. (6b):

$$Z(t) = A(q)\, y(t) - \sum_{j=1}^{p} B_j(q)\, u_j(t - nd_j) - B_k(q)\, f_{u_k}(t)$$

Moreover if a change occurs on the noise e(t), it comes simply by eq. (6c):

$$Z(t) = A(q)\, y(t) - \sum_{j=1}^{p} B_j(q)\, u_j(t - nd_j) + f_e(t)$$

The three preceding versions of equation (9) can be reduced in an unique form and combined with equation (8):

$$Z(t) = e(t) + F(t) \tag{10a}$$

where:

$$F(t) = P_L(q)\, f(t) = P_L(q)\, P_D(q)\, \xi_t^d \, \omega \tag{10b}$$

and where $P_L(q) = A(q)$, $P_L(q) = -B_k(q)$ or $P_L(q) = 1$ if a failure or a change occurs on the output y(t), on the kth input $u_k(t)$ or on the noise e(t), respectively.

3. THE DETECTION-LOCATION-ESTIMATION PROCEDURE

3.1. Estimation of the disturbance magnitude

For n time points, the previous equation (10) can be expressed as :

$$\underline{Z} = \underline{H}\, \omega + \underline{E} \tag{11}$$

where :

$\underline{Z}$=[z(1) z(2) ... z(d) z(d+1) z(n)]^T

$\underline{H}$=[0 0 ... h(d) h(d+1) h(n)]^T

$\underline{E}$=[e(1) e(2) ... e(d) e(d+1) e(n)]^T

and $\underline{H}$ is the impulse response at time d of the transfer function $P_L(q)\, P_D(q)$.

With these notations, the magnitude of the disturbance is simply given by least-squares estimation:

$$\hat{\omega} = (\underline{H}^T \underline{H})^{-1} \underline{H}^T \underline{Z} \tag{12}$$

Note that the previous calculation assumes that the fault occurence time d, the fault location (and the corresponding polynomial $P_L(q)$), the fault type (and the corresponding polynomial $P_D(q)$) and moreover the model parameters are known.

3.2. Known occurence time and model parameters

If the dynamics and the location of the failure are unknown, N vectors $\underline{H}_i$, i=1, ..., N, can be builded, each one representing a particular disturbance. So $N = n_L . n_D$, where n_L is the number of possible application points (n_L = p+2) and n_D is the number of possible shapes of the fault.

The localisation and the determination of the disturbance dynamics can be made by classical tests related to the linear model. The following variable can be for instance calculated:

$$t_i = \frac{\hat{\omega}_i}{\hat{\sigma}_i} \quad i = 1, ..., N \tag{13}$$

which, without fault, i.e. under the hypothesis of nullity of the fault magnitude ω_i, follows a Student distribution with n-1 degrees of freedom, with the classical hypotheses of normality for vector $\underline{E}$. $\hat{\sigma}_i$ is the standart deviation of the coefficient $\hat{\omega}_i$ given by:

$$\hat{\sigma}_i^2 = \frac{\hat{e}_i^T \hat{e}_i}{n-1} (\underline{H}_i^T \underline{H}_i)^{-1} \quad i = 1, ..., N \tag{14}$$

where $\hat{\underline{e}}_i = \underline{Z} - \underline{H}_i\, \hat{\omega}_i$.

Without faults, for each possible fault, i.e. for i=1, ..., N:

$$|t_i| < t_{1-\alpha/2}$$

where $t_{1-\alpha/2}$ is the value of a Student variable with n-1 degrees of freedom corresponding to a distribution function equal 1-α/2, with α a confidence level fixed a priori. The index for which $\max_i \{|t_i| \,/\, |t_i| > t_{1-\alpha/2}\}$ provides both the location and the type of the fault. The corresponding coefficient gives its magnitude.

3.3. Known model parameters

For on line operation, the determination of the occurence time can be made by processing on a moving window. At each time point t, i.e. on each moving window of n=2m+1 points, a vector $\underline{Z}(t) = [z(t-m)\dots z(t-1)\, z(t)\, z(t+1)\dots z(t+m)]^T$ is formed and fault magnitudes $\hat{\omega}_i$, $i = 1,\dots,N$ are calculated from vectors $\underline{Z}(t)$ and $\underline{H}_i = [0\dots 0\; h(d=m+1)\; h(m+2)\dots h(2m+1)]^T$ corresponding to the different faults. Then $\hat{\sigma}_i^2(t)$ and $t_i(t)$, $i = 1,\dots,N$ are calculated by :

$$\hat{\sigma}_i^2(t) = \frac{(\underline{Z}(t) - \underline{H}_i\hat{\omega}_i)^T(\underline{Z}(t) - \underline{H}_i\hat{\omega}_i)}{n-1}(\underline{H}_i^T\underline{H}_i)^{-1} \quad i = 1,\dots,N$$

$$t_i(t) = \frac{\hat{\omega}_i(t)}{\hat{\sigma}_i(t)} \quad i = 1,\dots,N$$

Moreover, the maximum of $|t_i(t)|, i = 1,\dots,N,$ is calculated and compared to a threshold, allowing the detection at time t. As shown in the example, the final location and identification are achieved by considering the $t_i(t)$, $i = 1,\dots,N$, for the next 2m+1 points after the detection point t_D. The indexes i_M and t_M for wich $\max\limits_{\substack{i=1,\dots,N,\\ t=t_D,\dots,t_D+2m}} (|t_i(t)|)$ is reached give for i_M the location and the type of the fault and for t_M its occurence time. Its magnitude is given by $\hat{\omega}_{i_M}(t_M)$.

3.4. Robust identification

If the parameters of the model (5) are unknown, they can be estimated by a recursive procedure. But the disturbances which act upon the measured variables must not influence the parameters identification. So a robust procedure must be carried out. The algorithm, due to Puthenpura and Sinha (1990), is used in the following. The scheme is a weighted least squares algorithm with particular weights and is very similar to the robust Kalman filter obtained by Masreliez and Martin (1977).

It is based on the modelling of outliers due to Huber (1964) and considers that the measurement noise $\varepsilon(t)$ which contamines the noise free output is from the family D_μ, defined by:

$$D_\mu = \{D \mid D = (1-\mu)G + \mu H,\; 0 \le \mu \le 1\} \quad (15)$$

where G is the usual normal distribution, H an arbitrary symmetric long-tailed distribution and μ the probability of occurring large errors $(0 \le \mu \le 1)$. In fact H is assumed to be also normal, but with a larger variance compared to G:

$$\varepsilon(t) \sim (1-\mu)\, N(0,\sigma_1^2) + \mu\, N(0,\sigma_2^2) \quad (16)$$

where N represents a normal distribution with $\sigma_2^2 > \sigma_1^2$.

The probability μ of occurring large errors being unknown, the preceding model is replaced by:

$$\varepsilon(t) \sim [1-\delta(t)]N(0,\sigma_1^2) + \delta(t)N(0,\sigma_2^2) \quad (17)$$

$\delta(t) = 0$ for $|e(t)| \le M$ and $\delta(t) = 1$ for $|e(t)| > M$ where e(k) is the prediction error and M a preassigned bound which can be taken as $3\sigma_1$ (Aström, 1980).

Model (5) is expressed by a linear regression:

$$y(t) = \phi^T(t)\,\theta + \varepsilon(t)$$

where θ is the vector of the parameters, $\phi(t)$ is the regression vector. So the parameter vector θ is recursively estimated by:

$$\theta(t) = \theta(t-1) + \frac{P(t-1)\,\phi(t)\,e(t)}{[(1-\delta(t))\sigma_1^2 + \delta(t)\sigma_2^2] + \phi^T(t)P(t-1)\phi(t)}$$

(18)

$$P(t) = P(t-1) - \frac{P(t-1)\,\phi(t)\,\phi^T(t)\,P(t-1)}{[(1-\delta(t))\sigma_1^2 + \delta(t)\sigma_2^2] + \phi^T(t)P(t-1)\phi(t)}$$

Moreover, the variances σ_1^2 and σ_2^2 can be updated as:

$$\sigma_1^2(t) = \sigma_1^2(t-1) + \frac{1}{(t-\tau)}(e^2(t) - \sigma_1^2(t-1)) \quad \text{for } |e(t)| \le 3\sigma_1(t-1) \quad (19a)$$

$$\sigma_1^2(t) = \sigma_1^2(t-1) \quad \text{otherwise}$$

$$\sigma_2^2(t) = \sigma_2^2(t-1) + \frac{1}{\tau}(e^2(t) - \sigma_2^2(t-1)) \quad \text{for } |e(t)| > 3\sigma_1(t-1) \quad (19b)$$

$$\sigma_2^2(t) = \sigma_2^2(t-1) \quad \text{otherwise}$$

with τ=0, for t=1 and $\tau = \tau+1$ whenever $|e(t)| > 3\sigma_1(t-1)$. $\sigma_2^2(0)$ can be chosen as $\sigma_2^2(0) = 3\sigma_1^2(0)$.

The so estimated parameters are provided to the calculations described in the previous section for the fault detection-location-identification procedure. At each time point t, the vectors $\underline{H}_i$, becoming $\underline{H}_i(t)$, are calculated from the parameters estimated at the same time t.

4. EXAMPLE

A MISO system is considered, described by the following discrete ARX model:

$$A(q)\,y(t) = \sum_{j=1}^{2} B_j(q)\,u_j(t - nd_j) + e(t)$$

where $A(q) = 1 - 1.5q^{-1} + 0.7q^{-2}$

$B_1(q) = 1 + 0.5q^{-1}$ and $nd_1 = 1$

$B_2(q) = 0.5 - 0.9\,q^{-1}$ and $nd_2 = 1$

$\{e(t)\}$ is a gaussian noise such as $e \sim N(0, \sigma_e = 0.1)$. The system is excited by sequences of random length steps and produces the following output :

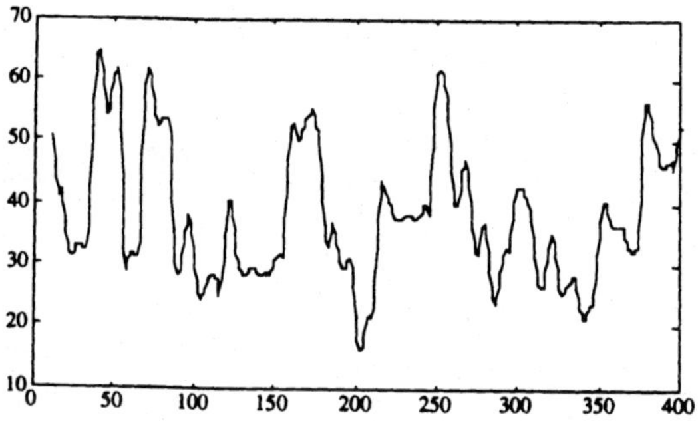

Fig. 2. The system output

At time point 300, a disturbance is added to the measured input u_1. The disturbance function $f_{u_1}(t)$ represents a level change of magnitude $\omega = 5\sigma_e$ and renders a bias of the considered sensor. The real input u_1 and the corresponding measured and biased input are shown figure 3.

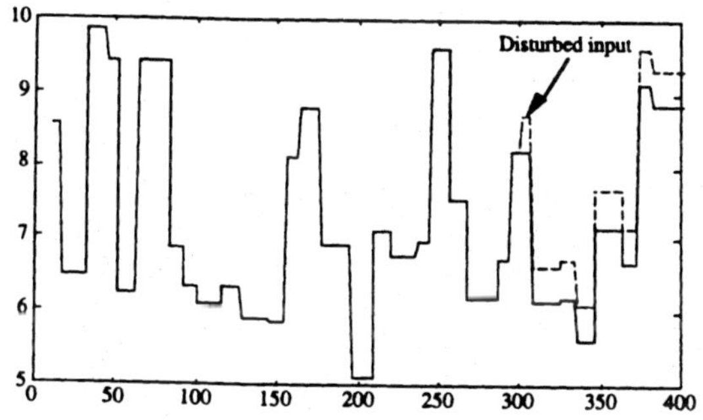

Fig. 3. Input u_1

Consequently, this disturbance acts on the prediction error sequence z(t) calculated from the measurements, shown in figure 4 :

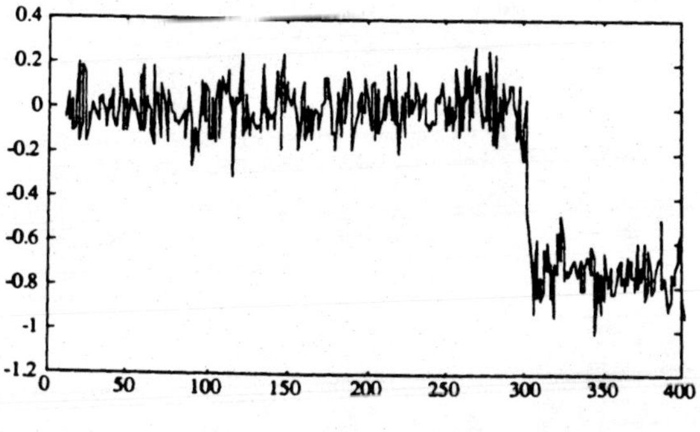

Fig. 4. Residuals z(t)

It is supposed that n_D=3 shapes (impulse, step and drift) and n_L=4 application points (u_1,u_2, e and y) of a failure can be met ($N = n_D . n_L = 12$ types of fault).

The previously described procedure is then carried out with a moving window of n=2m+1=21 points.

Figure 5 shows the maximum, for each moving window, of the N variables $|t_i(t)|$ values.

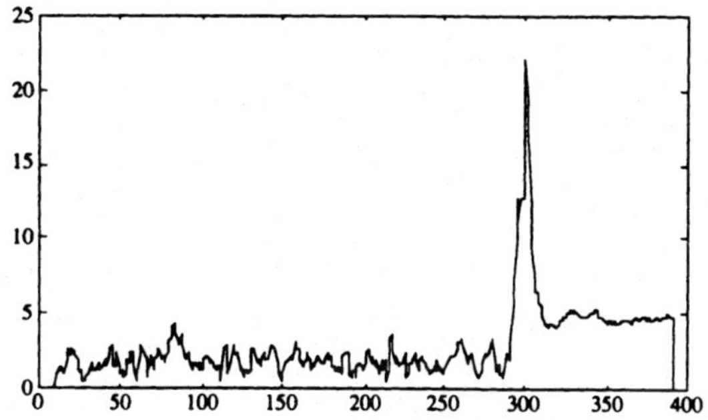

Fig. 5. Maximum of $|t_i(t)|$ versus time

Notice that a threshold strictly chosen as $t_{1-\alpha/2} \simeq 2$, with $\alpha = 0.05$ classicaly and in this case n-1=20 degrees of freedom, can lead to many false alarms because the slightest disturbance in the residual can be explain by one of the different shapes $\underline{H}_i(t)$. So the threshold must be tuned to a higher value for providing the classical compromise between false alarms and miss.

To illustrate the procedure, figure 6 focuses on the closeness of the disturbance occuring instant equal to 300.

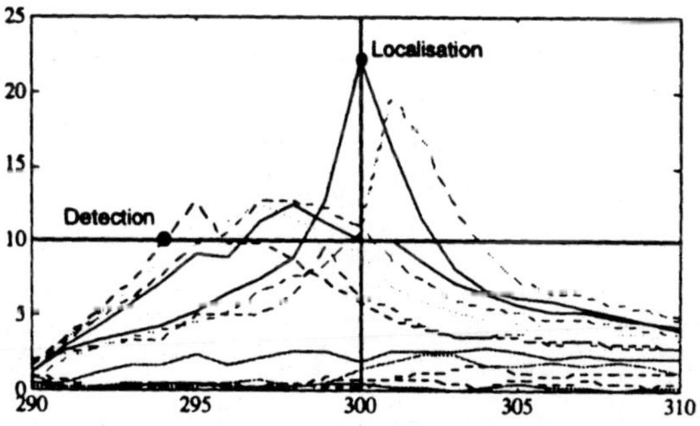

Fig. 6. $|t_i(t)|$i=1,...,12 versus time

The N variables $|t_i(t)|, i = 1, \ldots, N$, for t=290,..., 310, are shown. Detection takes place at time point 294 and the location and the type determination of the fault are achieved exactly at time point 300. The level change type and the location on u_1 are founded. The corresponding estimation of this level change is $\hat{\omega} = 0.5189$ ($\omega = 5\sigma_e = 0.5$).

The following figures show the values of $\underline{Z}(t)$ for the moving window centered at t=300, and the shapes $\underline{H}_i(t)$ corresponding for figure 7 to an outlier (impulse response) and to a bias (step response) on input u_1 and for figure 8 to biases (step responses) on inputs u_1 and u_2.

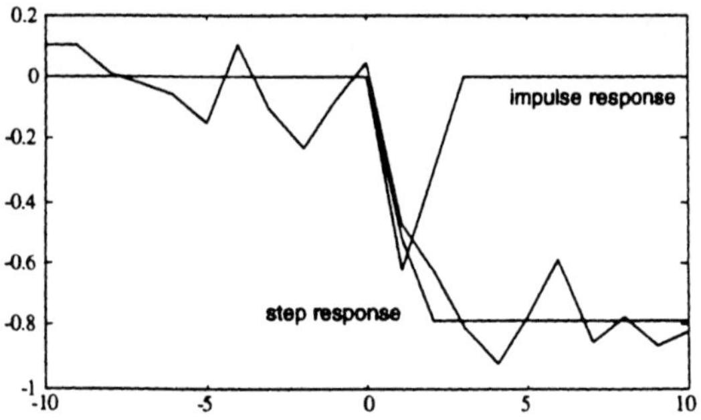

Fig. 7. Residual and 2 shapes $\underline{H}_i(t)$

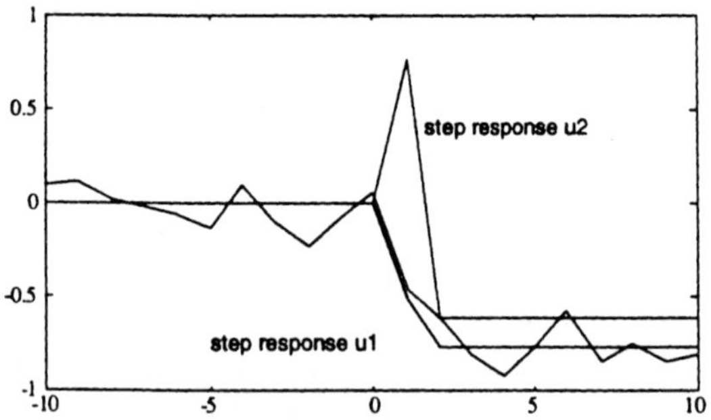

Fig. 8. Residual and 2 shapes $\underline{H}_i(t)$

Figures 7 and 8 highlight the basic principle of the procedure: to fit on a moving window shapes characterizing the different faults (type and location) on the transient period of the residual.

The last figure shows the evolution of the parameters by recursive identification. Notice that the estimated parameters are weakly influenced by the bias on input 1, beginning at time point 300.

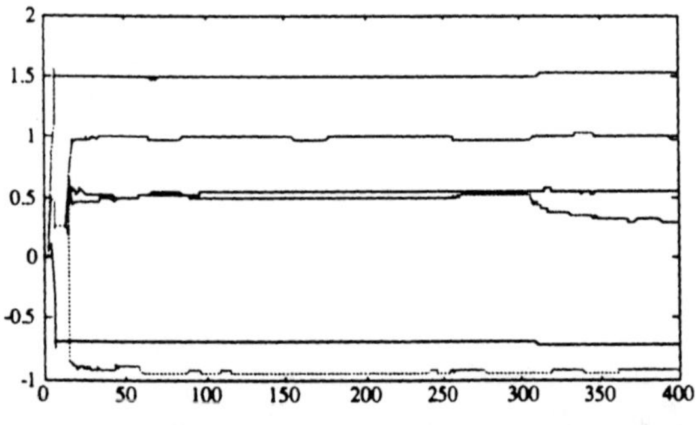

Fig. 9. Parameters during recursive identification

5. CONCLUSION

The proposed method seems an attractive one by its simplicity and its unified character. The robustness of the detection can be improved by using adaptative thresholds wich are varied according to the level of the different t_i variables on training periods without faults. On the other hand, the proposed method is based on the discriminant character of the different shapes. So if some polynomials of the model are close, the location of the fault will be problematic. Extensions of the proposed method can be considered concerning for instance the stochastic case, the link with state space formulation.

6. REFERENCES

Aström, K.J. (1980). Maximum likelihood and prediction error methods. *Automatica*, **16**, 551-574.

Box, G. E. P. and G. C. Tiao (1975). Intervention analysis with applications to economic and environmental applications. *J. Amer. Statist. Assoc.*, **70**, 70-79.

Huber, P.J. (1964). Robust estimation of a location parameter. *Ann. Math. Stat.*, **35**, 73-101.

Kobi, A., M. Ouladsine and G. Bloch (1993). Outliers detection, location and identification in time series. *Proc. IMACS/IFAC Sec. Int. Symp. on Mathematical and Intelligent Models in System Simulation MIM-S²'93*, 275-280, Brussels, Belgium, april 12-16.

Kobi A., M. Ouladsine and J. Ragot (1993). Fault detection in ARX process. *Proc. IEEE Int. Conf. on Systems, Man and Cybernetics*, **3**, 394-399, Le Touquet, France, october 17-20.

Masreliez, C.J., and R.D. Martin (1977). Robust bayesian estimation for the linear model and robustifying the Kalman filter. *IEEE Trans. A.C.*, **22**, 361-371.

Mironovski, L. A. (1980). Functional diagnosis of linear dynamic systems - A survey. *Automation and Remote Control*, **41**, 1122-1143.

Patton, R. J., and J. Chen (1991). A review of parity space approaches to fault diagnosis. *Proc. IFAC Symposium SAFEPROCESS'91, Baden-Baden*, september 10-13.

Puthempura, S., and N.K. Sinha (1990). A robust recursive identification method. *Control-Theory and Advanced Technology*, **6**, 683-695.

Tsay, R.S. (1988). Outliers, level shifts, and variance changes in time series. *J. of Forecasting*, **7**, 1-20.

SELF-DIAGNOSIS AND REDUNDANCY FOR ELECTRONIC CONTROL OF DIESEL ENGINE

R.KLAUS , A.URBANIAK
Institute of Computing Science, Technical University of Poznan, ul.Piotrowo 3A, 60-965 Poznań, Poland

Abstract. Redundancy and self-diagnosis applied together with mechatronics make the Diesel engines more reliable and safe The paper shows particular issues referred to redundancy in electronic controllers of Diesel engine. The system was prepared to be applied in the excavator.

Key Words. Redundancy; automating testing; computer control; diesel engine control; mechatronics

1. INTRODUCTION

The compression-ignition engines as astatic objects require the controllers to be applied (Klaus, 1992). Hence, improving quality of engines' running and avoidance of damage is a strategic purpose of controllers application.

The conventional controllers applied in the Diesel engines make work possible only in a limited range of parameters. This is the reason for many defects of the engines, such as the increased smoke level, noise, bad dynamics (Ochocki et al. , 1993)

At the present stage of internal-combustion engines development mechanical governors of rotation speed are more and more frequently replaced by controllers which make possible numerical control of engines. The controllers which don't have the mechanical governors' defects appeared together with progress in application of microprocessors.

The digital controller's task is to work out appropriate setting of control rod of the injection pump and injection advance angle which depends on the present engine's working point.

In microprocessor controllers it is very easy to change the control structure, dynamic characteristic and working points in relation to the required criterion.
Usually the following criteria of control are considered:
- maximisation of the power rated,
- minimisation of the smoke level,
- minimisation of the fuel consumption (Klaus, 1992).

2. CONSTRUCTION OF CONTROL SYSTEM

2.1. Flexibility of solution

The system shown in the paper is prepared to control dose of fuel and injection advance angle by the conventional injection pump without the centrifugal governor
The diesel engine Andoria S400/K2/9 has been subjected to an examination with the conventional controller replaced by microprocessor one.

The modular construction of the whole system has been applied here with the separation of the injection pump subsystem and the digital control subsystem and injection advance angel. The advantage of such a construction is the possibility of applying it in the new engines as well as in the already exploited ones. In the exploited machines it is very easy to replace the old control system with the new one. It requires only dismantling the injection pump with the centrifugal governor and then installing the new one together with an equipment.

Moreover, the modular construction guarantees system's flexibility and make the service time much shorter. Every module can be developed improved or replaced independently, obviously with regard to the rules of co-operation. The modular construction of the system is shown in the Fig. 1.

The control system of the engine make also possible the co-operation with the master microcomputer of the whole machine, such as excavator.

2.2. The selected elements of construction

In the 3rd module (see Fig. 1.) we can find the injection pump and the equipment.
The equipment of the injection pump consists of:
- induction sensor for the measurement of the speed and the injection advance angle,
- hydraulic mechanism changing the injection advance angel,
- step motor changing the position of control rod,
- rate generator and special system protecting the engine from exceeding the limit velocity.

The 2nd module is composed of interface protection and redundancy systems and microcontroller.

In this design the microcontroller NEC 78310 is applied. It is composed of high performance 16-bit CPU, a multi-function pulse input/output facility, macro service facility which acts like 8 DMA channels, and 8-level priority interrupt controller witch includes high speed context switching.

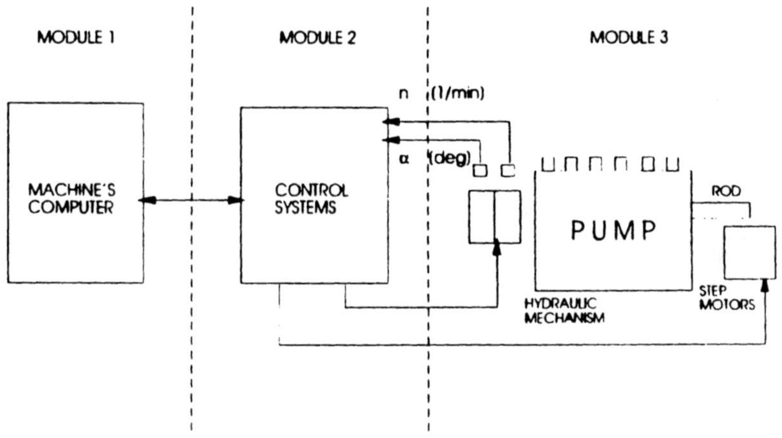

Fig.1. System blok diagram

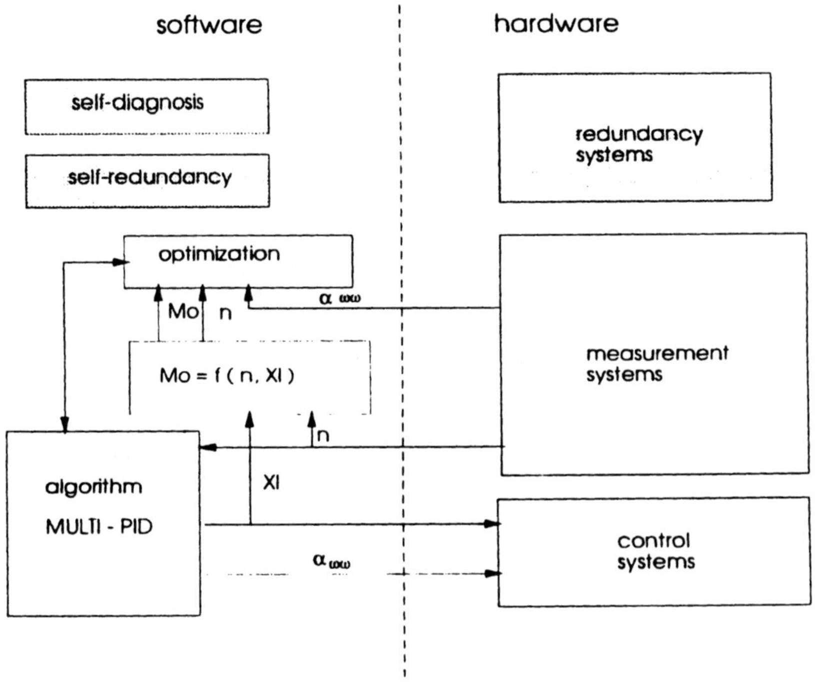

Fig.2.Partition of the system function

In the simplest version control algorithm operates on the principle of conventional PID governors selected for various parameters of engine operation. The parameters of this MULTI-PID controller change also according to the accepted criterion of control (Olsson, 1990).

In the 2nd module the microprocessor technology is applied. This makes possible the software realisation of some of the system's functions.

Moreover the system is more flexible and easy to adapt to different engines and energy receivers. Demands made on controller depend on the way of energy reception and static and dynamic characteristics of different engines.

Owing to the using of the hardware and software it is possible to apply the system to different machines without any hardware changes, only the program is adapted to the current requirements. Consequently, it decreases costs and makes the stage of the system adjustments shorter. Division of function into the software and hardware is shown in the Fig.2.

3. SELF-DIAGNOSIS AND REDUNDANCY

3.1. Overspeeding of the Diesel engine

The diesel engines in dynamic conditions behave like the astatic objects. This means that self-stabilisation is reached beyond the admissible running speeds (see Fig.3.). If these speeds are exceeded, the engine is in danger of damage. That's way one of the most important task of the control system is to prevent the engine's overspeeding and to guarantee safe stoppage in different states of damage.

Considerable number of analog and digital components together with additional power supplier make electronic controllers less reliable than mechanical ones. Hence, the machine is at risk of overspeeding and damage.
The most common damages are:
- the electric power supplier damage,
- the sensor damage,
- the damage to the actuator,
- the damage to the microprocessor controller.

All these damages can result in the loss of control over the engine.

There is an example in Fig.3. of the Diesel engine's rotation speed characteristic while without control. The safety improvement can be archived by application of the redundancy and proper strategy of the system designing.

3.2. Redundancy

Redundancy is using additional system's components that keeps the system running even of the defects are detected and signalled. In the presented system, the redundancy encompasses:
- programs and additional diagnostic systems used for fast search of defects,
- the master computer of the working machine, controlling the Diesel engine's control system,
- systems of mechatronics preventing the overspeeding of the engine.

An approach based on mechatronics enables the effective union of mechanics, electronics and software, so the machines and systems can be more flexible, more adaptable, safer and more reliable.

Application of self-diagnosis algorithms enables the use of proper redundancy systems of mechatronics and prevents the overspeeding of the engine and consequently its damage in the hard real time environment. During the starting of the engine the states of actuators, measuring and protection elements are tested. The engine will start action if the results of the tests are positive.

The correct work of the microprocessor system is controlled by the built-in watchdog timer.

A detection of the microprocessor system's break-down causes the turning on of electrohydraulic systems switching the engine off. The master computer of the excavator controls also the engine's microprocessor system. It keeps the continuous contact with the engine through the interface RS 232 and controls the state of engine's action.

There are various elements protecting against exceeding of the speed limit applied according to the kind of the control system's construction. The tachogenerator with comparator has been applied in the basic version. It triggers the action of mechanical elements, cutting off the fuel inflow to engine's cylinders, when ever the value of the speed limit is exceeded (Klaus, 1992). In the next versions of the system the tachogenerator has been replaced by the system of counters with a standard generator and the mechanical elements have been replaced by an electrohydraulic system.

The supervision over the whole running system takes the operator of the excavator. He can switch off the electronic control system any time. In case of electronic system damage the operator has a chance to switch on a manual control. The block diagram of protection hierarchy is shown in Fig.4.

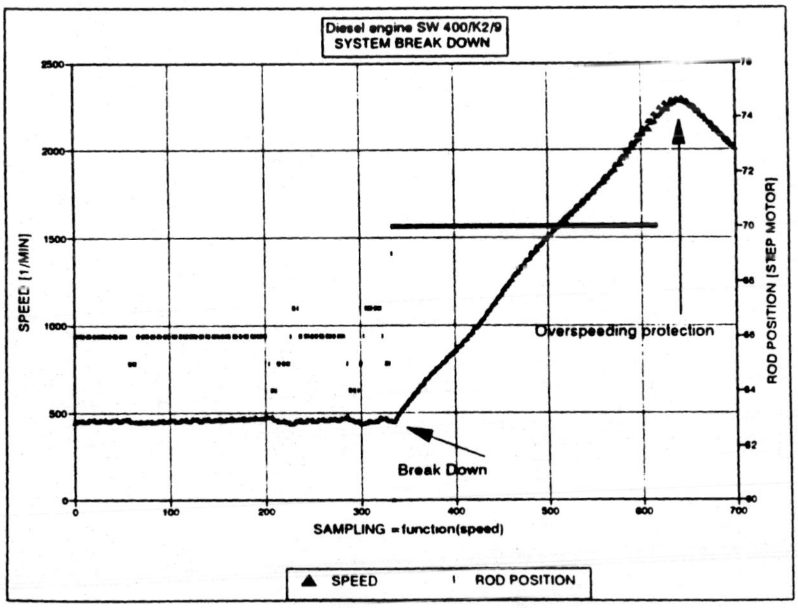

Fig.3. System break down

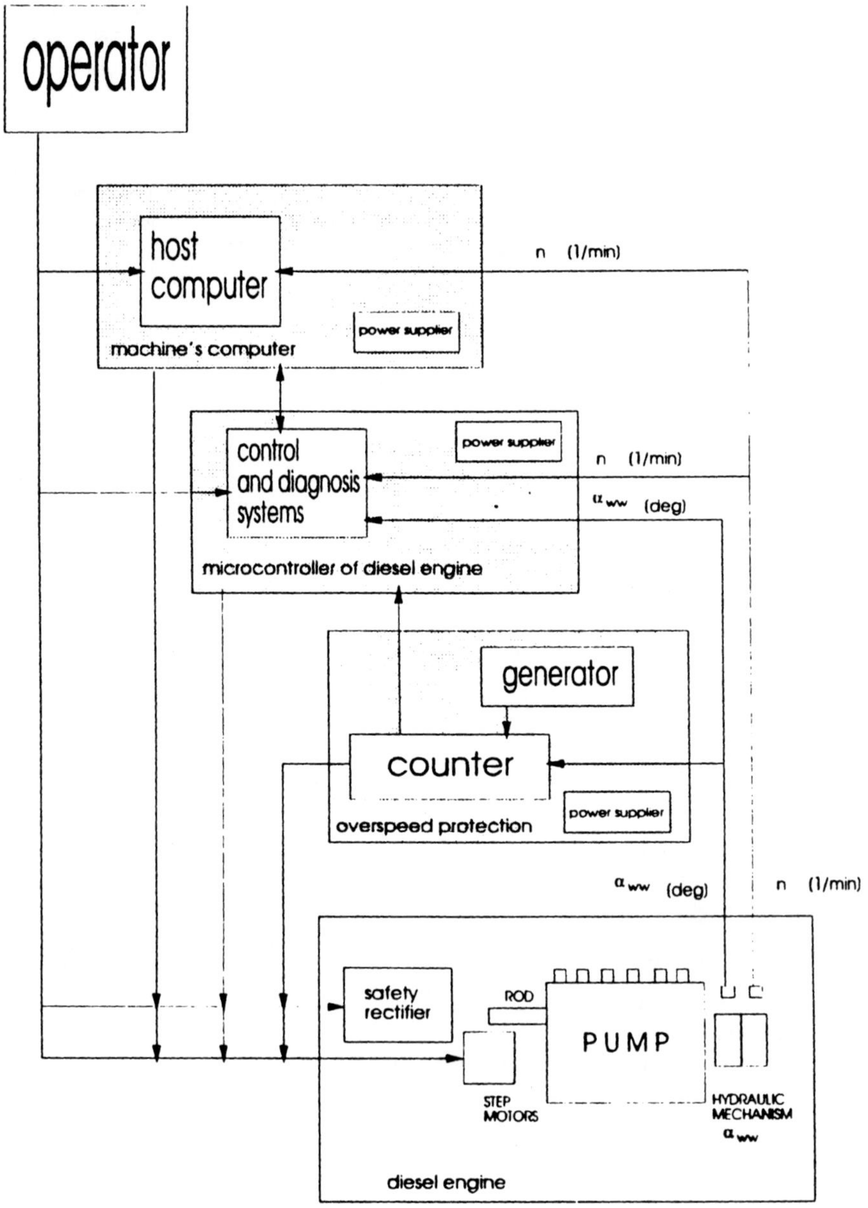

Fig.4. Protection systems blok diagram

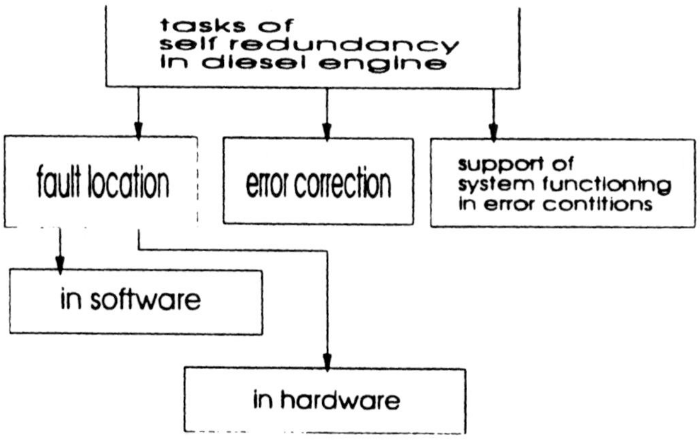

Fig.5.Tasks of self-redundancy

There are three main elements controlling the proper run of the operation contained in the test programmes:

- the control over the running of a given process: checking if the estimated values are feasible, which means be examine the internal discrepancy in the results of the program execution
- the control of correctness of the order of the running processes checking if the modules of the program are executed in the proper way. This correctness is checked by using the control word which is transmitted to another module of the program. The next module can be executed if the value connected with a certain control word is recognised as an admissible one.
- the data control and checking if the data are read from the current file and if the determined, limited values are stored.

The application of the software redundancy in the presented system assures its running even if some damages had been detected before.

If the engine works correctly the master computer of the excavator sends information about the real load moment to the microcontroller of the Diesel engine. This allows choosing the optimal parameters (Klaus *et al.* 1992).

In case of a damage of the transmission or the master computer of the excavator, the microcontroller of the engine triggers the redundancy software and estimates an approximate value of the load moment on the basis of speed and the rod position (see Fig.2.).

If it's working correctly the control system controls a fuel doze as well as the injection advance angle. Both, the engine's speed, and the injection advance angle are estimated by the independent sensors . In case of damage of any element of the speedometer channel the engine's speed is estimated using of the software redundancy from the injection advance angle channel. The operator is being informed about all the defects.

Some elements applied in the system can double their functions, e.g. the limited speed is detected in the hierarchical way during the standard running. The software protection which controls the step motor until the moment of the fuel doze cutting off should work as the first element. If it doesn't work and the speed still increases then the feed of the step-motor is switched off and the mechanical system moves the rod until the fuel doze is cut off. If this protection should fail the counter-standard generator system is triggered setting the electrohydraulic system to work. Obviously the operator of the excavator can switch the machine off, too. The task of the applied redundancy software's are shown in Fig.5.

4. CONCLUSIONS

The specific construction of the system of microcomputer control of the fuel doze and the injection advance angle presented in this paper gives the following advantages:

- easy and fast assembly and disassembly even in difficult conditions,
- possibly of modernisation and easy service of every particular module,
- easy adaptability of the system to many different devices (excavators, loaders, bulldozers, generators).

The main advantage of the developed protection system is making the engine's running more safely.

Main disadvantages of the redundancy application are: extended time of the program's run because of the connected additional redundant programs' execution and possibility to give rise to new defects' by the additional software what creates new sources of damages.

5. REFERENCES

Klaus, R. (1992). *Control Systems for an Diesel Engine with Application of PID Controllers*, Proc. of the Conference AUTOMA-SIL'92, Technical University of Poznań, pp. 1-10.

Klaus, R., Urbaniak, A., Sobkowiak, R. (1993). *Research Strategy of Algorithms for Digital Control of Diesel Engines*, Proc. of the Conference "Alternative Fuels" IBMER Warszawa, pp.21-24 (in Polish).

Ochocki, Wl. Klaus, R. Rybarczyk, P., (1993). *Selected Results of Numerically and Conventionally Controlled Diesel Engine Research*, Proc. of the Conference "Engines Control", Stawiska pp.68-76.

Olsson, G., Piani, G., (1990). *Computer System for Automation and Control*, Prentice Hall, London.

COMPUTER-AIDED TUNING AND DIAGNOSING BASED ON PERSONAL COMPUTERS USAGE

I.B. YADYKIN* and E.K. KORNOUSHENKO*

*Institute of Control Sciences, Profsouyuznaya, 65, 117806, Moscow, Russia

Abstract. Wide-range usage of the advanced controllers are changing the views concerning technical service system assignment (a tuning, diagnosing, reconfigurating). These service tasks could be solved by means of reference model principle and expert systems in Computer-Aided Tuning and Diagnosing System (CATDS). This paper treates the design principles, architectural features and ways of CATDS realization on PC. The main requirements which must be fulfilled by modern CATDS are specified and the approach to design of CATDS depending on their applications and "intelligence" level is proposed.

Key words. Automatized controller tuning, redundancy, intelligent machines, signal processing, failure detection.

1. INTRODUCTION

Now the modern personal computers provide a good opportunity to integrate control and technical service functions in intelligent distributed control systems (Astrom, 1987; Keviczky *et al.*, 1978; Navarro *et al.*, 1990; Tsypkin and Avedîan, 1985). It is obvious that computational procedures designing by means of using parallel processing and background processing allows to gain the efficiency of an application of such systems. Since hardware and software of modern distributed control systems get more sophisticated, the significance of maintenance operation grows. Controller tuning is the complex, difficult and expensive process even in case of PID controllers, because it requires high skill and deep knowledge. There is an objective trend towards increasing the range of industrial controllers tuning and developing more stringent requirements to the tuning quality. That is favored by the trend towards the optimization of the process characteristics, multiple modes of technological rules, introduction of flexible manufacturing systems, and more strict technological standards due to the ecological safety requirements.

Modern control systems with reinforced requirements for their functioning (such as CATDS) are designed using sufficiently exact mathematical models of their elements. It is due to the availability of such models that these systems may be diagnosed using subtle analytical methods with considering latest results from control theory and system analysis. Analytical diagnostic methods are based on the use of analytical redundancy contained in mathematical representation of the system under supervision (the hardware redundancy is awkward and in many cases expensive). The use of analytical redundancy for diagnosing has been considered by many authors and various diagnostic approaches have been proposed (see Frank (1990) for survey). On-line fault detection and isolation in a large-scale control system is a very hard problem from the mathematical point of view.

Modern control systems are characterized by some peculiarities making difficult the implementation of the known methods for on-line fault detection and isolation:

- the ramified structure with a large number of units and interconnections and with a small number of accessible points to be used for sampling information on system functioning;
- the presence of strongly nonlinear units as well as nonstationary and (in some cases) logical units;
- the presence of both unobservable disturbances and unit approximate models of units.

The direct implementation of known diagnostic methods for such system may have a small effect in practical situations. Therefore, developing new methods suitable to diagnose systems with above mentioned peculiarities is a very urgent problem.

The creation of CATDS would not be possible, but for power computational hardware and software due to PC development. The commercialization of PC has permitted a convenient interface with an operator, namely, a tuner on the basis of computer graphics. It also allows the data archiving for tuning and provides the utilization of a simulation modelling methods for better tuning forecasting. The main fields of CATDS applications are shown in Fig. 1.

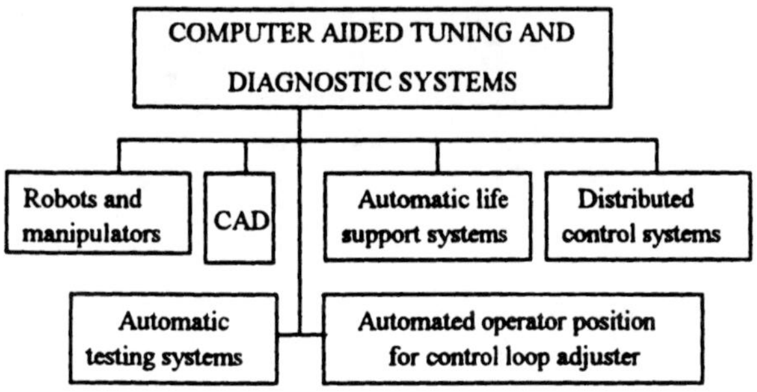

Fig. 1. The main fields of CATDS application

2. PRINCIPLES OF CATDS DESIGN

In case of a manual controller tuning under varying conditions the tuner seeks to use the variation of controller tuning to provide similarity between the closed loop tuning response and some response realizing the desired performance of the control system. If the mathematical models of the plant and other control loop elements are known, then it is possible to obtain full mathematical description of the closed control loop in the time or frequency domain corresponding to the transition process. The obtained description corresponds to an implicit reference model (IRM), and the set of parameters of that description determines assigned IRM (Yadykin, 1989). In the case of an automatic tuning the computer varies these parameters in order that to approximate the closed loop operator to the reference model operator.

In Computer-Aided Tuning System (CATS) the tuner is informed about preferable tunings and it makes a final decision. Therefore, the problem of controller tuning is the basic one among the problems of the parametric adaptive control with an implicit or explicit reference model. The solution of this problem is well known in the adaptive control theory. The solution s of this problem may be obtained on the basis of a direct or indirect adaptive control using search or nonsearch methods (Astrom, 1987; Goodwin and Sin, 1984; Wiemer *et al.*, 1989; Krasovsky, 1987; Rotach *et al.*, 1984). In most cases existing CATS are based on the above principle, excluding expert and neural network CATS (Astrom, 1987). The indirect control principle is realized with the use of parametric identification procedures to determine the mathematical model of the plant. It is usually assumed that the structural identification stage for the required model construction is executed by some method (Goodwin and Sin, 1984). After the determination of the mathematical model of the plant the controller parameters can be evaluated by the following methods:

1. Euristic procedures of the Nickols-Ziegler rule type (Ziegler and Nickols, 1942).
2. A direct synthesis of a controller by means of CATS (Jamshidi and Herget, 1985).
3. Adaptive procedures for the control loop adjustment with the reference model (Horn, 1988; Yadykin, 1989).
4. Neural network procedures (Astrom, 1987).
5. Fuzzy procedures (Kwok *et al.*, 1990)

3. THE MAIN REQUIREMENTS FOR CATDS

Let us specify the main requirements to be satisfied with CATDS:

- CATDS must operate, as a rule, in a real time and have a short tuning time for each variant of tuning;
- the requirements to the tuner-operator skill must be minimal;
- CATDS must be friendly with a tuner, i.e., propose to him alternative solutions, assist him in seeking the optimal tuning variant with taking into account his personal preferences;
- CATDS is to be preferably used in the mode of the normal plant operation with a controller, however it should permit to use small-amplitude test signals, as well as to solve the tuning problems by stages;
- CATDS is to be preferably used to determine the degree of tuning of a control loop;
- CATDS must allow the archiving of tuning variants with common database for diagnostics and tuning;
- CATDS must allow to comparison of various variants of tuning in a convenient (graphical) form.

Let us consider the main requirements for the software package of the workstation on the PC base. It is preferable to use multiple windows into process simultaneously. Each window should be separate, might be animated and should correspond to some part of the control loop. The software package must support real time/historical trend of process and X-Y plots on line changing span, tag, scales and display rates. It should support the Windows Data Dynamic Exchange client/server interface, dynamically links real time data to the program such as spreadsheets, databases and statistical packages. It is preferable to use AUTO-CAD^R like windows display builder. It let user to create process diagrams, dynamic attributes and key macro functions. The software package must provide on-screen alarm summary, customized alarm groups, alarm log storage. The conditional historic classification of CATDSs is shown in Fig. 2. Many of the above requirements are inherent in modern intelligent control systems as well, moreover, some important CATDS function can be realized in them as embedded firmware subsystems.

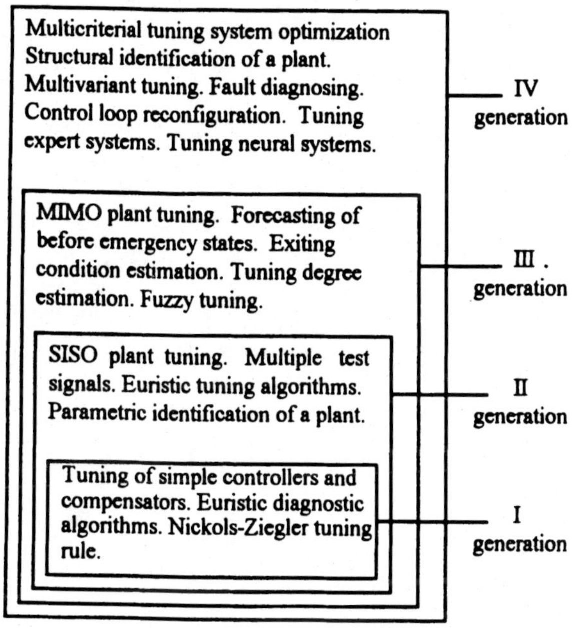

Fig. 2. Four generations of CATDS

4. FAULT DIAGNOSIS FOR CATDS

The computer pertaining to CATDS introduces the opportunities to diagnose the control system as a whole as well as its elements using information to be treated during the control process. It is obvious that to attain and to maintain over time the required performance of industrial control system is undoubtedly a very important practical problem and its complete solution affects economic efficiency of system being considered. As far as the system's accident has occured, the expenditures on recovering of system performance may essentially exceed the ones needed for timely diagnosing.

It should be noted that diagnosing process has some peculiarities one has to take into account before developing diagnostic procedures for CATDS. These peculiarities are as follows.

1. A fault having appeared in control system may change a system functioning regime whence it follows that unit models (linearized, less order, etc.) having been used for control purposes are unapplicable for diagnostics. Thus, effective diagnostic algorithms should be based on unit diagnostic models (may be nonlinear) tolerated to large deviations of their parameters.

2. The problem of diagnosing control system is in a certain sense secondary being the main problem to provide the control quality required. Hence, a minimal number of additional accessible points needed for diagnostic purposes only should be required.

3. Diagnostic inferences should be developed very quickly having prevented a system to transfer into an accident state.

The known diagnostic methods may be bad suitable for diagnosing CATDS due to a disregard of the requirements 1-2. The approach proposed in Kornoushenko and Gerasimov (1990) seems to be more suitable in this case. This approach is based on a feedforward and/or backward recalculation of observable signals of the control system to inputs and outputs of its internal units and an analysis of compatibility of unit input-output representations obtained with their diagnostic models. After drawing the observable input and output signals of CATDS to inputs and outputs of some internal unit in CATDS, we have the estimated input-output representations of this unit. Using such presentation, ordinary identification to estimate the unit diagnostic model parameters is to be done. The diagnostic model of a unit is its extended parameterized mathematical model such that the admissible fault appearence implies the deviations of some coordinates in its parameter vector. The deviations may be either constant or varying over time. If CATDS is faulty, some unit representations obtained will be "incorrect" due to the fact that some observable signals were drawn along the ways contained the faulty units. In these situations the least squares solution residual norm for such units will be much more in comparison with this norm in fault free case. Obviously, the parameter estimates obtained for these units will have no common with their true parameter values. The diagnostic models of the units in CATDS are assumed to be the linear and bilinear either stationary or nonstationary systems, nonlinear static systems, logical systems, etc. Having been developed the applicable deconvolution procedures for these systems (Kornoushenko, 1992; Kornoushenko and Gerasimov, 1994) and keeping in mind that the corresponding modelling procedures are obvious, fault detection and isolation in CATDS to be composed of these "basic" units is no difficult problem. Really, using the observable signal drawing, one is able to turn an unobservable behavior of internal units into the "observable" one.

The defect of an approach considered is the same as in any parameter estimation method: its performances deteriorate with occurring uncertainties in the unit mathematical models. In other words, it has no robust properties. It is the price that we have to paid for possibility to diagnose large-scale control systems (such as CATDS) with sufficient completeness.

5. PURPOSE AND STRUCTURE OF CATS ADACOM

CATS ADACOM has been developed in Institute of Control Sciences and the firm "Tunex" Ltd to automatize the tuning of industrial control systems and in CAD DSC problems. The considered CAD allows:

- specification of requirements for obtaining the desired dynamic characteristics of a tuned controller;
- centralized control of tuning accuracy indexes of control loops;
- determination, mapping and registration of the dynamical characteristics of control plant;
- generation of test signals; multivariant tuning of PID-controller parameters in order to use them as recommendation for an operator-tuner;
- automatic tuning of microprocessor of the PID laws family having a supervisor parametric control of the tuning parameters;
- interaction with operator-tuner.

The system is realized on the basis of the ADACOM application software package for IBM PC which includes the following program modules: identification of the plant structure, parameters and disturbances; evaluation of multivariant tuning controller parameters; estimation of a tuning degree; assignment of the control loop reference model; dispatcher; expert system; data archive; computer graphics; support programs (see Fig. 3). ADACOM application software package: 3000 operators, 356 Kb RAM, MS DOS, written in Fortran-77 and Pascal.

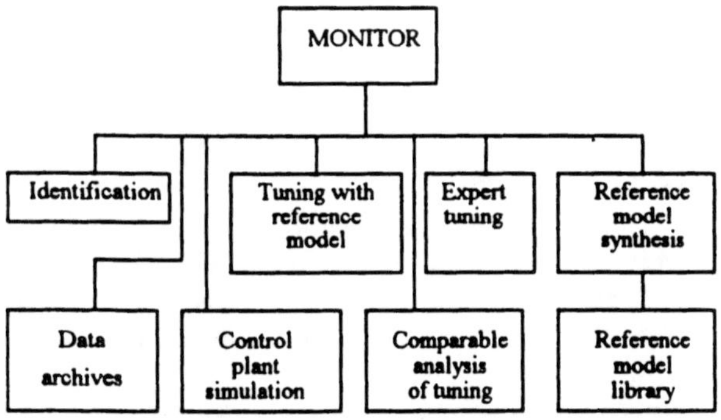

Fig. 3. Structural scheme of ADACOM

6. CONCLUSION

The advent of PC and PC-based workstation in the field of industrial automation has permitted the development of automated tuning and diagnosing systems for industrial controllers. Such systems can be installed on lap-top PC as a part of standard subsystems of tuning degree monitoring and diagnostics for operator and process engineer station in modern distributed control systems. CATS based on ADACOM solves the controller tuning problems via the principle of adaptive control with reference model. The utilization of the MS DOS includes the entry of the plant I/O data into PC and determination of controller tuning by some methods depending of preference of the operator-tuner. The CATS ADACOM operation experience has shown its high efficiency especially for complex controller tuning for industrial MIMO plants.

7. REFERENCES

Astrom, K.J. (1987). Adaptive feedback control. *Proc.IEEE*, **75**, 185-217.

Frank, P.M. (1990). Fault diagnosis in dynamic systems using analytical and knowledge based redundancy - A survey and some new results. *Automatica*, **26**, 459-474.

Goodwin, G.C., and K.S. Sin (1984). *Adaptive filtering prediction and control.* Prentice-Hall, Englewwood Cliffs.

Horn, B.C. (1988). Optimal self-tuning of PID-controllers. In: *Proc. ACC*, pp. 2362-2367. Atlanta.

Jamshidi, M., and C.J. Herget (Eds.) (1985). *Computer-aided control systems engineering.* North Holland, Amsterdam-New York-Oxford.

Keviczky, L., J. Hetthessy, M. Hilger, and J. Kolostory (1978). Self-tuning adaptive control of cement raw material blending. *Automatica*, **14**, 1-8.

Kornoushenko, E.K. (1992). Input signal recovery from stored data in discrete nonstationary linear systems. *Autom. Remote Control*, **53**, 1852-1862.

Kornoushenko, E.K., and V.V. Gerasimov (1990). Diagnostics of large-scale dynamic systems on the basis of finite records of their behavior. In: *Proc. 7-th Intern. Conf. on Technical Diagnostics*, pp. 477-484. Helsinki.

Kornoushenko, E.K., and V.V. Gerasimov (1994). The least squares deconvolution procedures for both linear and bilinear discrete-time nonstationary systems. To be published in *Autom. Remote Control*, **55**.

Krasovsky, A.A. (Ed.) (1987). *Reference book on automatic control theory.* Nauka, Moscow [in Russian].

Kwok, D.P., P. Tam, C.K. Li, and P. Wang (1990). Linguistic PID controllers. In: *Proc. 11-th IFAC World Congress*, Vol. 9, pp. 192-197. Pergamon Press, London.

Navarro J.L., P. Albertos, M. Martinez, and F. Morant (1990). Intelligent industrial control. In: *Proc. 11-th IFAC World Congress*, Vol. 10, pp. 87- 92. Pergamon Press, London.

Rotach, V.Ya., V.F. Kuzishev, A.S. Klujev et al. (1984). *Automation of the tuning of control systems.* Energoatomizdat, Moscow [in Russian].

Tsypkin, Ya.Z., and A.D. Avedyan (1985). Discrete adaptive control systems for deterministic plants. In: *Results of Sciences. Cybernetics*, pp. 45-77. VINITI publ., Moscow [in Russian].

Wiemer, P., G. Olejua-Torres, and H. Unbehauen (1988). A robust adaptive controller for systems with arbitrary zeros. In: *Proc. Intern. Conf. Control-88*, pp. 598- 603. Oxford.

Yadykin, I.B. (1989). Principles of construction, architecture and software of the automatic tuning systems of industrial controllers. In: *Computer Facilities. Systems. Control*, pp. 25-36. Int. Center of Science and techn. inf. and Inf. Center on Tech. transfer., Moscow-Sofia.

Yadykin, I.B., and E.K. Kornoushenko (1990). Integrated Intelligent Control Systems - Conceptual Approach. In: *Proc. 11-th IFAC World Congress*, Vol. 8, pp. 152-157. Pergamon Press, London.

Ziegler, J.G., and N.B. Nichols (1942). Optimum settings for automatic controllers. *Trans. ASME*, **64**, 759-768.

OBJECT RECOGNITION BY NEURAL NETWORK

Z. NAGY

Technical University of Budapest
Department of Production Engineering
1521 Budapest, P.O.Box 91, Hungary

Abstract. This paper introduces a new, automated method to adjust the recognition parameters of a second generation vision module (with statistical pattern recognition algorithm) by neural network and its application in an assembly cell to identify workpieces and their quality.

Key Words. Industrial robots; Flexible manufacturing; Failure detection; Identification; Neural nets; Pattern recognition

1. INTRODUCTION

The realisation of flexible automation of a robotic assembly cell requires different kinds of sensors and advanced supervision algorithm to cover mistakes that occur most frequently (Alpek, 1990; Alpek, 1991; Alpek, 1993; Nagy, 1992; Szalay, 1993). For visual testing vision module or modules are applied typically. By a vision module parts can be recognised, but the adjusting of the recognition parameters is very difficult in case of a second generation vision module (with statistical pattern recognition algorithm) (Solt, 1990; Nagy, 1992).

This paper introduces the development of an automated pattern recognition system. The system includes a common second generation pattern recognition module, which records the pictures and calculates the most important parameters of the pictures. Through the description of the standard usage of a second generation pattern recognition system, the difficulties of its application are presented.

The suitability of the application of a neural network to tune the pattern recognition system is suggested. In this case the parameters of the picture (coming from the pattern recognition module) become the input values of the neural network, which is trained to be able to select the good workpieces from the bad ones. The paper describes the assembly cell, where this automated subsystem was applied and the visual inspection tasks, which were executed by this subsystem.

2. THE PATTERN RECOGNITION PROCESS AND ITS AUTOMATION

First the statistical pattern recognition algorithm will be studied, then its possible automation by neural network will be introduced.

2.1. The Statistical Pattern Recognition Algorithm

The vision module calculates the most important parameters (see Appendix) of the current picture. The parameters are called p_{ij}, where i is the number of the model and the j is the number of the recognition parameter of the ith model.

During the teaching process the selected values are multiplied by weighting factors (w_{ij}). These factors must be determined by an expert in advance. The result of these multiplications is summed and the product (a single number) becomes the attribute value of the ith part (ith model) on the current picture.

During the recognition process the same steps are executed on the present picture. The result of this process isn't stored, but it is compared with the attribute value of the models. If the difference between the attribute value of the models and the current part exceeds a deviation limit, which was tuned previously by the expert too, the current part is different from the models, otherwise they are the same. The vision module tests all the models that are stored in the memory, subsequently.

Although the set of recognition parameters and their weighting factors adjusted by the expert is able to check the failures occur most frequently, but not all the failures can be managed. For example to check the number of the holes on the picture of a workpiece is easier than to check i.e. the diameter of each hole or to recognise any smaller failure or deformation on the same picture.

With this method a lot of errors, which occur most frequently during the assembly process, can be identified, however the complexity of the tuning process is increasing with the number of possible errors. So after a while the set of weighting factor becomes vast and puzzling.
This is why the application of an Artificial Intelligence method, the Neural Network seems advantageous:

- by a NN the tuning process could be automated easily, i.e. the tuning procedure doesn't require the existence of an expert and it needs less time
- the fineness of the tuning can be increased without bigger effort
- in case of NNs the continuous self-learning can be guarantied, i.e. the number of covered errors can be increased

2.2. The Automation Of The Adjusting Process

The process described previously can be replaced by NN (Nagy, 1993). A NN is able to tune itself and, as a result of it, after a training process the NN is able to command a system through its output according to its input from the system.

The elementary object of a NN is the neuron. The neuron has inputs that are multiplied by weighting factors and has an output value that is calculated from the sum of the multiplied values. During the training the weighting factors are modified automatically according to the difference between the desired and the real output of the NN for the input values.

The similarity of these two processes is obviously, this is why the NN seemed to be proper to automate the adjusting process of recognition parameters.

3. THE AUTOMATED PATTERN RECOGNITION SUBSYSTEM AND ITS APPLICATION IN THE ASSEMBLY CELL

First the automated pattern recognition subsystem will be described in details, then the assembly cell including this subsystem and those steps of the assembly process in which the subsystem has role. Finally the results achieved by the application of the automated pattern recognition subsystem will be presented.

3.1. The Automated Pattern Recognition Subsystem

The automated pattern recognition subsystem includes the following parts:
- a Mikromat M8001 pattern recognition module (MTA SZTAKI, Budapest, Hungary (Solt, 1990)) with a CCD camera
- an IBM PC (the cell controller) with a NN program

The recognition process was controlled by the IBM PC. It sent a command to the vision module to record a picture. The picture gathered from the scene by the CCD camera is proceed by the M8001 pattern recognition module. The module calculated the most important attributes of the recorded picture and sent them to the IBM PC through serial line. With these values as input values the NN (12-6-3-1) could be trained. Using the trained NN, the IBM PC was able to identify the workpieces.

The training means that the picture of a good part is taught, i.e. at the input of the NN the parameters of workpieces (good and bad) are fed and the weighting factors are modified to decrease the difference between the desired and the real output values of the NN. With these values (training set) the weighting factors of the NN are tuned.

During the execution phase these tuned NN was applied. In this case the NN received the input values from the vision module through serial line and it gave its result to the cell control program, which was able to determine the following step of the assembly process. This solution has advantage in its automation and self-learning ability.

3.2. The Application Of The Automated Pattern Recognition Subsystem In The Assembly Cell

The experimental subsystem was part of the assembly cell (Fig. 1.) of the CIM Pilot System at TU Budapest. The assembly cell included the following components:

- a PUMA 760 robot with a KAWASAKI VAL-I controller
- a supervised gripper
- a six-component force/torque sensor (Ledeczi, 1993)
- two vision modules (with a fixed and a mobile cameras) (Nagy, 1992)
- a palette with work pieces (house, ball and cover)

- a supervised fixture (assembly place)
- cell controller (IBM PC) with NN program

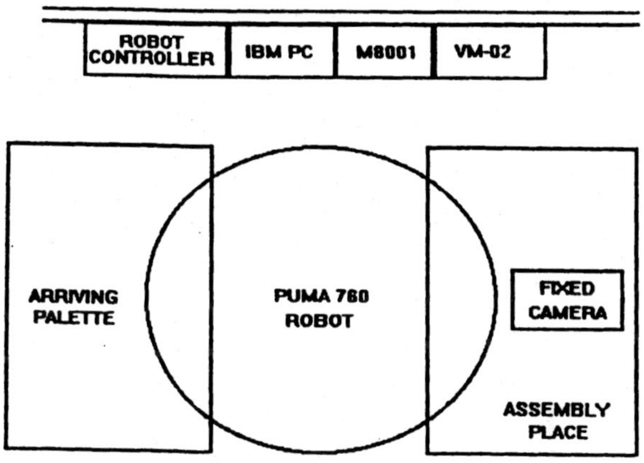

Fig. 1.: Layout of the assembly cell

In this cell the flexible automated assembly of ball taps (Fig. 2.) was realised. The parts were checked, the execution process was monitored by the sensors and their signals were handled by the supervision algorithm based on neural network that caused the system to be able to cover the mistakes, which occurred most frequently during the assembly process.

The assembly sequence of the ball tap and the complete supervising system are presented in more articles (Alpek, 1991; Alpek, 1992; Nagy, 1992; Alpek, 1993), here only the quality checks will be detailed.

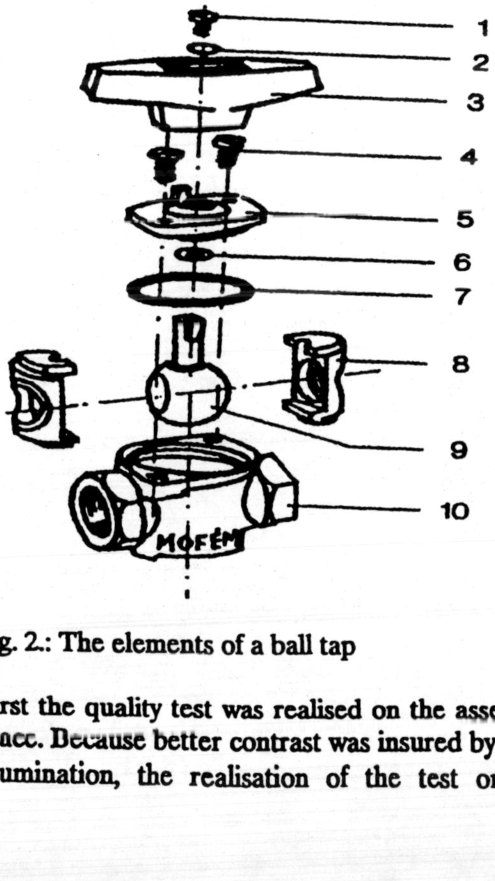

Fig. 2.: The elements of a ball tap

First the quality test was realised on the assembly place. Because better contrast was insured by back illumination, the realisation of the test on the assembly place was easier (see Fig. 3.b.), here only the contour of a workpiece could be checked, but not the inside.

Later the quality of the house was verified at the beginning of the assembly process, when the house was on the palette. Here the house was in front illumination and the picture (Fig. 3.a.) was received from the mobile camera.

A.Pictures of the ball tap in front illumination

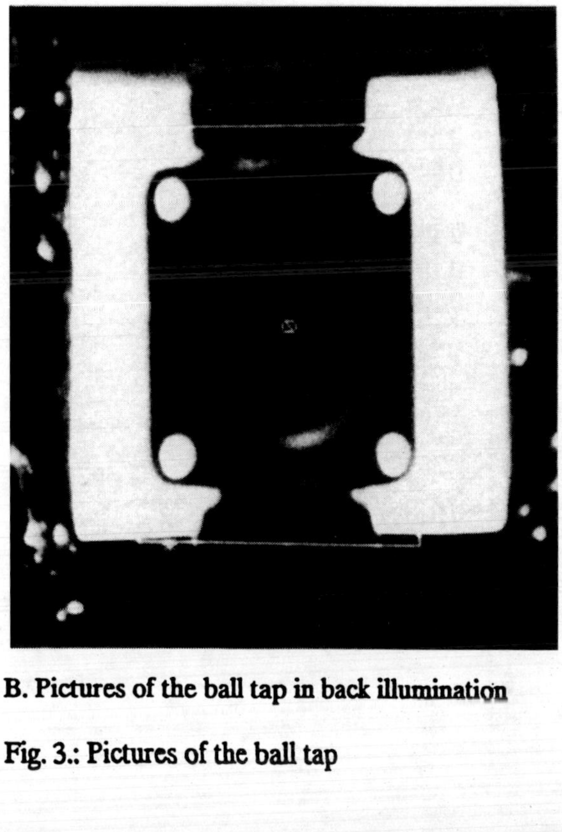

B. Pictures of the ball tap in back illumination

Fig. 3.: Pictures of the ball tap

The realisation of this test required a little more attention than the other on the assembly place, because of the illumination. With this test more details were observed within the shape of the house, e.g. the existence of the O ring inside the house.

3.3. The Results

The result of the training can be seen on Fig. 4.a. (in case of front illumination) and on Fig. 4.b. (in case of back illumination). These results proof also that the adjusting process in case of front illumination (60 steps) is more difficult than in case of back illumination (15 steps).

To test the biggest accuracy attainable by the automated recognition system, workpiece with smaller holes was tested. The hole was one mm smaller in diameter (e.g. the hole wasn't threaded). The test was executed in back illumination. The results can be seen on Fig. 4.c.

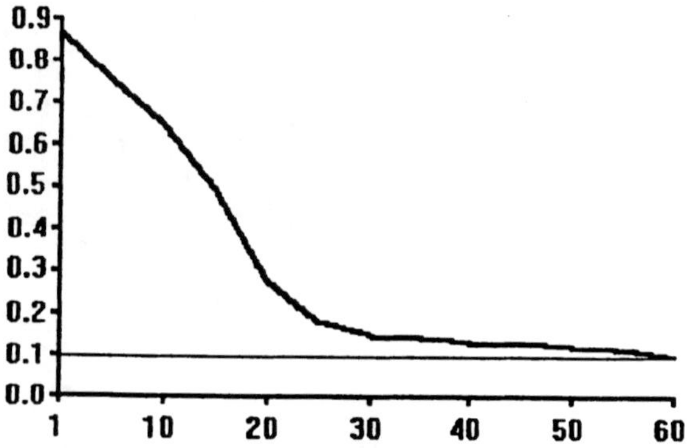

Cycle	Max error	RMS error
1	0.894	0.304
2	0.827	0.293
3	0.802	0.286
:		
58	0.102	0.050
59	0.101	0.049
60	0.100	0.049

A. in case of front illumination

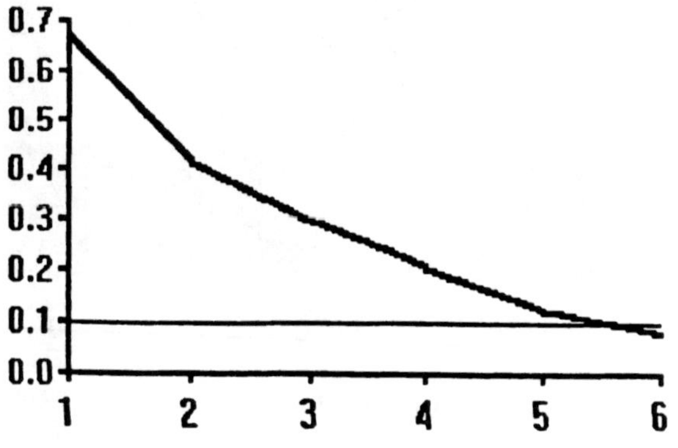

Cycle	Max error	RMS error
1	0.515	0.381
2	0.594	0.277
3	0.706	0.317
:		
13	0.122	0.065
14	0.102	0.058
15	0.093	0.053

B. in case of back illumination

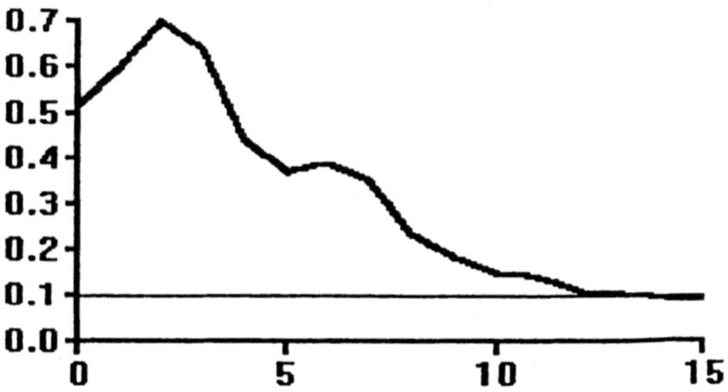

Cycle	Max error	RMS error
1	0.675	.350
2	0.416	.254
3	0.302	.162
4	0.219	.111
5	0.125	.075
6	0.083	.057

C. in case of smaller hole

Fig. 4. The results of the training

4. CONCLUSION

The result of the application proofed the ability of the NN for automatic adjusting of recognition system based on a second generation vision module with statistical pattern recognition algorithm. The application of the automated recognition system in the assembly cell confirms the advantages of this technique:

- the tuning of the recognition system could be automated
- the tuning time of recognition system could be reduced
- the fineness of the automated system became better
- the safety and reliability of the automated system became better in selection of good and bad parts
- the difficulty of tasks won't increase with the increasing number of pattern
- the handling of the automated system requires experiment neither on vision nor on AI

5. REFERENCES

Alpek, F., Berkes, O., Szelig, K., and Nagy, Z. (1990). *Some Element of The Automatic Supervision in a CIM Pilot-System*, AC'90 CIRP Symposium, Rydzyna (Poland)

Alpek, F., Berkes, O., Nagy, Z., and Szelig, K. (1991). The Automatic Inspection in a CIM System, *Computers in Industry*, **17**, 156-167.

Alpek, F., Nagy, Z., Szalay, T. and Szelig, K. (1991). Uberwachungs-probleme der Montagezelle im CIM-Pilotsystem der TU Budapest, *e&i Automatisierungs- und Messtechnik*, **110**, 127-134.

Alpek, F., Jerzsabek, P., Nagy, Z., Szalay, T., Szelig, K., and Sallay, P. (1992). *Concept and realisation of supervision tasks in assembly cell of CIM Pilot System of TU Budapest*, Mechatroinfo'92, MATE, Budapest (Hungary), pp. 245-255 (in Hungarian)

Alpek, F., Nagy, Z., Sallay, P., Szalay, T., Szelig, K., and Toth, K. (1993). *Multi-sensor technique for increasing intelligence of assembly robots*, 2nd International Workshop on Robotics in Alpe-Adria Region, Krems (Austria), pp. Su.5.3-1..8.

Alpek, F., Nagy, Z., Sallay, P., Szalay, T., Szelig, K., and Toth, K. (1993) *Multisensor System - Supervision problems and possible solutions in CIM*, Automation and Metrology: Man, Technology, Environment: 4th International DAAAM Symposium, Brno (Czech)

Ledeczi, T. (1993). *MiniForce User's Guide*, MTA SZTAKI

Nagy, Z., Toth, K., Gladkova, I. (1992). *Vision module applications in Assembly Cells*, Automation'92 Conference with International Participation, MATE, Budapest (Hungary)

Nagy, Z., Alpek, F., and Toth, K. (1992). *Increase of Intelligence of Assembly Robots with Force- and Torque Sensor and Vision Module*, 3rd International DAAAM Symposium on Flexible and Intelligent Automation, Budapest (Hungary), pp. 125-126.

Nagy, Z., and Szalay, T. (1993). *Control of Multisensor Systems by Neural Networks*, 30th MATADOR Conference, Manchester (UK), pp. 355-360.

Solt, P., and Banki, Zs. (1990). *M8001 User's Guide*, MTA SZTAKI

Szalay, T., and Nagy, Z. (1993). *Increasing the Intelligence of Industrial Robots Using Sensors*, 24th International Symposium on Industrial Robots, Tokyo (Japan)

6. APPENDIX

List of recognition parameters:

1. x coordinate of the mass centre point of the object
2. y coordinate of the mass centre point of the object
3. the smallest x coordinate of the object
4. the smallest y coordinate of the object
5. the biggest x coordinate of the object
6. the biggest y coordinate of the object
7. the x coordinate of the furthest point of the object
8. the y coordinate of the furthest point of the object
9. area of the object
10. perimeter of the object
11. shape factor of the object (= perimeter^2/area)
12. rotation angle
13. 1st order torque to the lower x axis of the window
14. 1st order torque to the left y axis of the window
15. 2nd order torque to the x axis passing through the mass centre
16. 2nd order torque to the y axis passing through the mass centre
17. 2nd order torque to the x and y axes passing through the mass centre
18. square of distance of the furthest point from the mass centre
19. average of square of distance between the mass centre and the boundary points

NEURAL ADAPTIVE CONTROL OF A BIOREACTOR

R. E. LOKE, G. CEMBRANO*

**Instituto de Cibernética, Universidad Politécnica de Cataluña, Consejo Superior de Investigaciones Científicas, Diagonal 647, Barcleona 08028, SPAIN*

Abstract. This paper describes the application of two different adaptive control schemes based on neural networks to the control of a bioreactor. The results are a useful assessment of the adequacy of neural networks for identification and control of nonlinear systems.

Key Words. System identification, Adaptive Control, Neural nets, Nonlinear Systems

1. INTRODUCTION

The use of connectionist techniques is currently considered a promising approach to the adaptive identification and control of certain classes of non-linear systems, since they have been proven capable of efficiently ***approximating broad classes of non-linear functions***, as reported in Funahashi (1989) and Girosi(1989). One of the most relevant features of neural networks is their ***adaptivity*** to a changing environment, which allows them to emulate time-varying behaviours of nonlinear plants and of the adaptive controllers required to effectively control them. Additionally, they have shown great potential in terms of their ***robustness*** with respect to noise, and ***fault-tolerance*** with respect to local disablement of neurons, much like the qualities observed in their biological counterparts which, to some extent, they were designed to imitate. Some relevant studies on the application of neural networks to control are presented , e. g., by Narendra and Pathasarathy (1990), Cembrano and Wells (1992) and Hunt *et al.* (1992). A number of interesting applications of neural identification and control of dynamic systems have been reported lately, especially in the fields of motion and navigation control. These fields are also the object of important on-going research projects. In robotics, the main goal has been to train robot arms to perform certain tasks in partially unknown environments, where the robot must take motion decisions on-line, based on sensory information, as shown e.g. in (Kawato, 1990; Jordan, 1991; Martinetz, 1990 and Millán, 1992) Another interesting illustrative example is the truck backer-upper developed by Nguyen and Widrow and coworkers (Nguyen, 1990).

Several other control problems are currently the object of research and, since neural control is still in its infancy, no general methodologies or approaches exist. In an attempt to challenge experimenters in the field to compare the performance of different approaches with a common test-bed, a collection of control problems was presented by Anderson and Miller (1990). The problems include balancing a pole, steering a ship, landing an aircraft and controlling bacterial growth in a bioreactor. The latter is the object of the work described in the paper.

The outline of the paper is as follows. In the next section, the nonlinear system is described, and its behaviour in different operating conditions is shown. The third section illustrates the design process of a neural identifier for the dynamic system. Subsequently, in the fourth section, two types of neural controllers are used to control the system in different regions of the state space. In the first case, the neural controller is based on forward modelling (Jordan, 1991), and in the second, on iterative inversion of neural networks (Hoskins, 1992). Finally, the conclusions of the work and some directions for future research are given.

2. SYSTEM ANALYSIS

The bioreactor dealt with in this work is a constant volume, isothermic, continuous flow, stirred tank reactor. It contains water, biological cells and nutrient. Water containing nutrient is continuously added and stirred into the contents of the tank. The volume of solution in the tank is maintained at a constant level by removing tank contents at a rate equal to the incoming rate. This flow rate w is the variable with which the bioreactor may be controlled. The state of the bioreactor can be expressed by the amount of cells C, and the amount of nutrient N in the tank at this time.

The differential equations modelling the dynamic behaviour of the system in a state-space representation are as follows:

$$\frac{dC}{dt} = -Cw + C(1-N)e^{N/\gamma} \quad (1)$$

$$\frac{dN}{dt} = -Nw + C(1-N)e^{N/\gamma}\frac{1+\beta}{1+\beta-N} \quad (2)$$

in which β and γ are constants determining the rate of cell growth and nutrient consumption. C and N are, respectively, dimensionless cell mass and subtrate concentrations. The defined ranges for C and N are [0, 1]; the range for w is [0, 2].

The dynamics of the system are highly nonlinear, as shown in the differential equations. Additionally, the presence of unstable behaviour in some operating conditions, as well as limit cycles, make the control problems challenging.

3. IDENTIFIER DESIGN WITH FEEDFORWARD BACKPROPAGATION NETS

The neural identifier was implemented with feedforward neural networks. The input layer contains the current system states and controls, as well as tapped delays thereof; the output layer yields the next state. The networks were trained with the standard backpropagation training algoritm, see Werbos (1974) and Rumelhart (1986), and the activation functions used in the hidden and output layers is the sigmoid. The training examples were generated by means of a simulator coded in ACSL language (Mitchell, 1991), which integrate the differential equations (1) and (2).

3.1. *Network Structure*

Number of Delays. For feedforward networks, as well as in other identification methods, it is common to attempt to reconstruct the state of a dynamical system using past measurements of the input and the output, so that the output of the system at time k+1 is expressed in terms of inputs at previous times k, k-1, k-2, ..., k-p, and outputs at previous times k, k-1, k-2, ..., k-q. The significant p and q can be chosen by comparing the identifier output errors, which result from partially trained networks with different delays as inputs. When the output error decreases after some additional delay is added, this delay can be considered as relevant for the representation of the system.

An appropriate number of delays for this specific system was determined through a comprehensive set of tests. The best experimental results were obtained with 3 delays in the states. The results of adding C_k-C_{k-1} and N_k-N_{k-1} and the identifier errors on the identifier performance were studied. Since the inclusion of these new nodes did not significantly improve the modelling results, they were not incorporated in the final identifier

Number of Hidden Nodes. It has been proven that a single hidden layer with hidden nodes can perform any nonlinear function mapping to any desired degree of accuracy by Girosi (1989). In choosing the number of hidden nodes, a compromise between two concepts must be found. Although it is possible for a network to represent exactly each example of a training set by adding sufficient hidden nodes to a hidden layer, this kind of "learning" must be avoided, (see Le Cun, 89 and Denker, 1987). In such a case, network performance will be limited because important input features are not learned, but exactly copied. Another negative effect of choosing too many hidden nodes is an increase in computation time. On the other hand, if the number of hidden nodes is too small, a network will not be able to converge during training. Automatic sizing techniques have recently been developed to overcome these problems (Fahlman, 1990; Sanner and Slotine, 1991 a, b), but for a reletively low number of necessary nodes, the experimental design was considered appropriate. In an experiment the identifier errors for networks with different numbers of hidden nodes was determined.

3.2. *Training*

The system was simulated in ACSL during 50 seconds by numerically approximating the equations of growth using the Runge-Kutta-4 method with a time step of 0.001 seconds. The sampling interval was determined with the sampling theorem of Nyquist or Shannon. A sampling interval of 0.5 seconds was chosen. With this sampling interval the continuous time signals for C and N were discretized. This discretized data was used for the neural network training.

Hence, the identifier network was trained with dynamic trajectories obtained from the ACSL simulator for combinations of control strategies and initial states which span the operating ranges of the system efficiently. The simulated state trajectories led to a total amount of 297 training examples. During the training process the examples were presented in random order to the neural network. The values for the learning rate and momentum terms in the backpropagation learning rule used were 0.15 and 0.4 for the output layer, and 0.3 and 0.4 for the hidden layer.

The identifier performance during the training process was tested applying various step for the input w. The training process was stopped when the error made by the network on this test control sequence stopped decreasing. Thus, overtraining was avoided.

3.3. *Results*

Figure 1 shows the identification errors in C_{k+1} and N_{k+1} for a completely trained feedforward neural network (after 500000 training iterations), when a test control sequence consisting of various steps in the system input w_k is applied

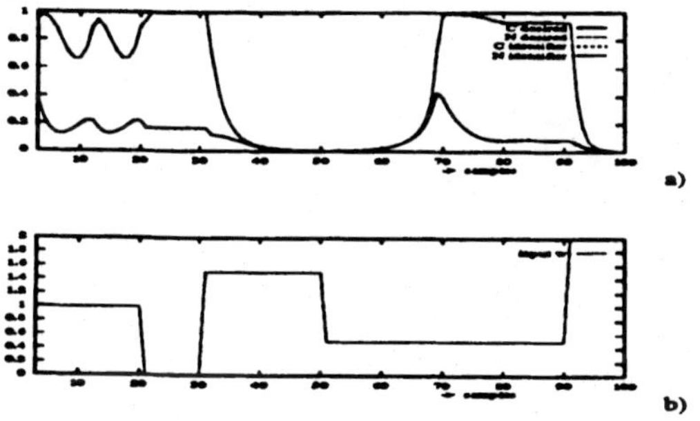

Fig. 1. a) Identification errors of a feedforward network with a piecewise constant input b) Input signal

4. NEURAL CONTROLLER

4.1. *Control Problems*

The control objective studied in this work is to achieve and maintain a desired cell concentration C^* and a desired nutrient concentration N^* constant, by altering the flow rate w. Different control problems can be defined, since to each set of values for w, C^* and N^* belong different system dynamics.

Three different control problems have been considered, following the setup of Anderson and Miller (1990). These correspond to different operating conditions, where the system behaves stably or unstably. In the first problem, the bioreactor is started in a region of the state space from which a stable state is easily achieved, namely around a spiral attractor. Then, control commands must be generated, which bring the system in the state belonging to the center of the spiral attractor. In the second and third problem the desired states are, respectively, the center of a spiral source, and a state around the center of a spiral attractor. The fact that the desired states for the latter two problems are unstable makes their control much harder. In the first problem, the stable spiral attractor the pair (0.1207, 0.8801), ($C = 0.1207, N = 0.8801$), is the desired system state. For the second problem, the desired concentrations for C and N are in the center of the spiral source; (C^*, N^*) is (0.2107, 0.7226). In the third problem, like in the first one, the phase-plane diagram shows a stable spiral attractor (0.1237, 0.8760). However, in this case, the desired state is situated in the area of attraction of this spiral source; (C^*, N^*) is (0.1737, 0.8760).

4.2. *Control Schemes*

Forward Modelling. This method requires the use of two feedforward neural networks, one to model the forward dynamics of the plant, and a second for the controller (Jordan, 1991). The earlier constructed neural network identifier can be used as a forward model of the plant. Its main function is to replace the real plant as a neural model, so that it is possible to backpropagate the output errors of the plant through its layers. Hence, the error in the output of the controller network (the control command w_k) can be calculated and the neural controller can be trained. The controller network is placed in series with the neural identifer model and the real plant; see Fig. 2 for the complete control scheme. The reference model defines the desired system output at time k+1.

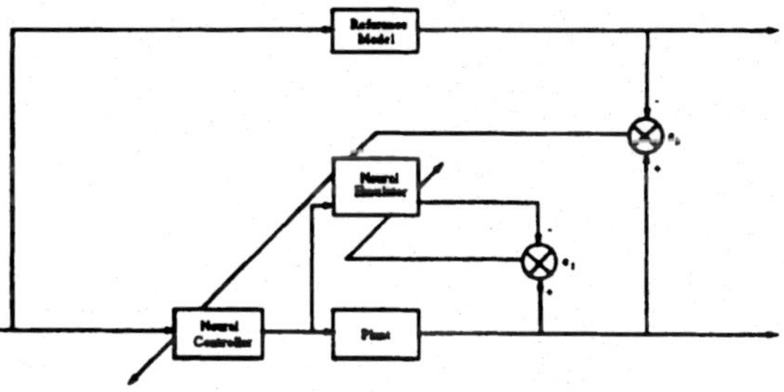

Fig. 2. Forward modelling control structure

The structure of the neural controller was determined in a series of experiments similar to the ones used for the neural network identifier. The inputs to the network are the actual cell and nutrient concentrations, C_k and N_k, the command input applied at time k-1, w_{k-1}, and the desired concentrations C^*_{k+1} and N^*_{k+1}. The further input to the network consists of significant delays for C_k, N_k and w_{k-1}. The relevant delays found are C_{k-1}, C_{k-2} and w_{k-2}. The network consists therefore of one input layer with 8 neurons, one hidden layer of 16 hidden nodes and one output layer of 1 neuron. The output neuron represents the next control command w_k, which leads to a new system state when fed into the system.

The controller was trained for the three problems defined above. In each training iteration, the inputs to the controller C_k, N_k, w_{k-1} and its delays, were randomly chosen with a uniform distribution in a ten percent range of their desired values. With these inputs, the controller network produces a control command w_k, which is then used in the real plant. The output of the plant is then compared to its desired values C^*_{k+1} and N^*_{k+1}. In order to update the controller an error in the control command is required. This is obtained by backpropagating the error between the actual and desired plant outputs at timestep k+1 with the neural network identifier. During its training, the neural identifier was considered as a completely learned network and its network weights were not changed.

The complete training process of the controller consisted of two stages. Firstly, in order to obtain a good initial weight set, the controller was trained off-line as desribed above. Thereafter, in order to perform adaptive control, the controller was trained on-line.

Iterative Inversion of Neural Networks. This method, proposed by Hoskins *et al.* (1992), is another form of adaptive control. Iterative inversion is, like forward modelling, also based on the use of the backpropagation technique. Whereas in the forward modelling control scheme two networks are used, in this method just one neural network is required, the neural identifier, which was trained as described in section three. In the overall control scheme (see Fig. 3) three models are used. The *performance model* specifies the desired dynamic response. The *convergence model* determines the desired dynamics of the error between the real plant and the performance model.

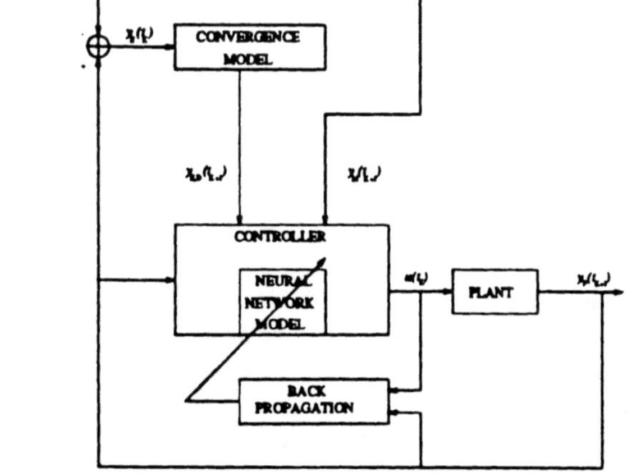

Fig. 3. Structure of the iterative inversion scheme (Hoskins *et al.* 1992).

In the iterative inversion method, the output errors are backpropagated through the identifier network, without affecting its weights, but leading to an error in the input space. Thus, iterative inversion of neural networks proceeds by a gradient descent search of the network input space, and control commands can be generated on-line.

A first advantage of this method, as compared to forward modelling, is that a good accuracy with respect to the reference model is obtained. Secondly, the method has been reported to have good stability and convergence properties. Moreover, it is possible to specify the desired error dynamics, which can lead to a smoother response. Finally, it is possible to obtain fast adaptation to the reference model. The desired next output error, $y_{e,d}(t_{k+1})$, is calculated by applying a linear convergence model to the real plant error $y_e(t_k)$. For the bioreactor, the following Lyapunov-like function was used to define a global error landscape for the error in C and N:

$$v(y_e) = y_{ce}^2 + y_{ne}^2, \tag{3}$$

where y_{ce} and y_{ne} are the components of the two dimensional output error vector y_e of the system. This Lyapunov function is applied to the desired next output error in order to find an evaluation for the next desired state. The constructed identifier (see section 3) was used as *neural network model.*

The controller is primarily based on the convergence and the neural network models. The convergence model defines the error dynamics which must be followed. The neural network model is iteratively inverted to find new candidate control commands. The control selection process is illustrated in Fig. ??. The control command w_k is selected by an iterative search to minimize the cost function:

$$P(y_{e,p}, y_{e,d}) = \frac{(v(y_{e,p}) - v(y_{e,d}))^2}{v(y_{e,p})} \tag{4}$$

where $y_{e,d}$ is the desired next output error (obtained from the convergence model), and $y_{e,p}$ is the predicted next output error (obtained from the neural network model and a candidate control input). The update equation for the candidate control used in Hoskins *et al.*(1992), based on simple gradient descent, was changed into a polynomial interpolation of P based on P and its two derivatives. This resulted in a significant speed-up in the rate of convergence of the search process.

In order to perform polynomial interpolation, P, $\frac{\partial P}{\partial u}$ and $\frac{\partial^2 P}{\partial^2 u}$ must be determined for the candidate control input. P is calculated as specified in (4). Its deriviative is calculated as follows:

$$\frac{\partial P}{\partial u} = \frac{\partial y_{e,p}}{\partial u} \frac{\partial P}{\partial y_{e,p}}, \tag{5}$$

where $\frac{\partial y_{e,p}}{\partial u}$ is obtained explicitly from the neural network using the mechanics of error backpropagation, and $\frac{\partial P}{\partial y_{e,p}}$ is found analytically. The second deriviative of P is obtained using a small finite step in u. Thus, P and $\frac{\partial P}{\partial u}$ are calculated for a $u + \Delta$, where Δ is a small finite constant. The situation in which the interpolation method finds a hill-parabola must be handled especially, since in such a case the minimum for P is not located at the point where the deriviative of the parabola is zero. To deal with these cases, a simple gradient descent was used to provide a new candidate input. The polynomial interpolation search was started with the best candidate control input found for the previous time point as the first candidate. The search process was stopped when the predicted cost P was below a threshold, or when a maximum number of search iterations was reached.

4.3. *Results*

The results obtained for the different control problems with both types of neural controllers are illustrated in the following figures. Figure 4 shows the control errors in C_{k+1} and N_{k+1} and the generated control commands for the first control problem with the use of both forward modelling (FM) and iterative inversion (II). In the latter method, the value used for the linear convergence model was 0.25, the maximum number of iterations was 20, and the gain used for the gradient descent when a hill-parabola is encountered was 0.005. For this control problem, the desired state is stable, namely the source of a spiral attractor (C^*=0.1207, N^*=0.8801, w=0.75). Whereas the command generated with the forward modelling controller keeps oscillating, the iterative inversion command stabilizes quickly around the desired value for w.

The control errors in C_{k+1} and N_{k+1} and the generated control commands for the second control problem can be seen in Fig. 5. In this problem the desired state is unstable, namely the center of a spiral source (C^*=0.2107, N^*=0.7226, w=1.25). Although the iterative inversion method settles its control command rapidly close to its desired value, some oscillations occur in the system output. After 50 samples, the forward controller stabilizes its control command practically on the required value. In the iterative inversion method, the value used for the convergence model was 0.75.

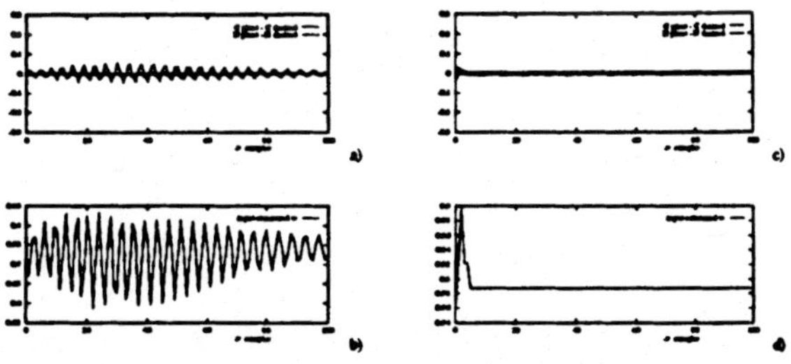

Fig. 4. Results for the first control problem: a) Errors with respect to reference using FM, b) Control command using FM, c) Errors with respect to reference using II, d) Control command using II .

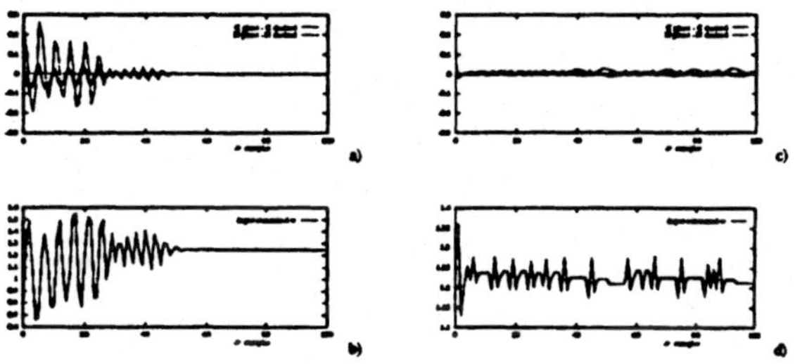

Fig. 5. Results for the second control problem: a) Errors with respect to reference using FM, b) Control command using FM, c) Errors with respect to reference using II, d) Control command using II .

Finally, Fig. 6 shows the performance of both controllers for the third control problem. The value for the linear convergence model in the iterative inversion method was 0.9. The desired state is off the center of a spiral attractor (C^*=0.1737, N^*=0.8760, w=1/1.3), which makes the control problem unstable. Both controllers maintain a significantly big error for the amount of cells C, and they do not manage to bring the system to its desired state. However, they bring the system to the closest stable state, and achieve a stationary behaviour.

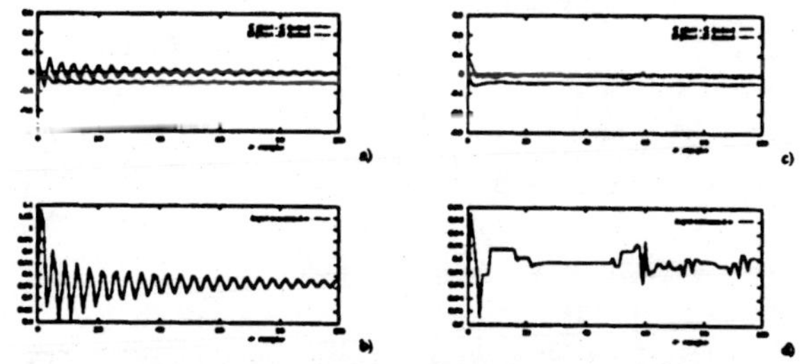

Fig. 6. Results for the third control problem: a) Errors with respect to reference using FM, b) Control command using FM, c) Errors with respect to reference using II, d) Control command using II .

The performance of the controllers was also tested adding a five-percent gaussian white noise to the control command or the output. The results show that both control methods can cope well with this noise in sensors and actuators. Also, five percent gaussian white noise was added to the system parameters β and γ during the control process. For this amount of noise, both control methods were robust.

5. CONCLUSIONS AND FUTURE WORK

The described bioreactor is an interesting problem for neural network controllers, because the nonlinearity in the system dynamics leads to different stability behaviours of the system in different operating conditions. In order to cope with the nonlinearity of the system, a feedforward network with nonlinear activation functions in the hidden and output layer was used for the system identification. A forward modelling controller and a controller based on iterative inversion of neural networks have been used to adaptively control the system in different operating conditions. For the three control problems, the generated control command with the forward modelling approach keeps oscillating a long time, whereas the command calculated with the iterative inversion method settles down rapidly The forward modelling method achieves better output accuracy after a big number of training samples.

The results obtained in this study constitute a useful assessment of the adequacy of neural networks for nonlinear identification and control of dynamical systems. Furthermore, a comparison of two valid neural adaptive control structures is provided. Ongoing research in this field at Instituto de Cibernética is aimed at the design of neural adaptive control schemes and stability analysis thereof, using feedforward and recurrent networks.

Acknowledgements

Research in neural networks and their applications to process control at the Instituto de Cibernética is partially supported by the research grant CICYT-TIC91-0423 of the Spanish Science and Technology Council and by the European Community, under the ESPRIT project No. 6715 "Robot Control based on Neural Network Systems. The work of R.E. Loke is partially supported by the Delft University of Technology in The Netherlands, under the supervision of Prof. dr. E.J.H. Kerckhoffs.

REFERENCES

Anderson C. W., W. T. Miller (1990) "Challenging Control Problems" *Neural Networks for Control, T. Miller, R.S. Sutton, P. J. Werbos, editors*, MIT Press.

Barto A. G. (1989) "Connectionist Learning for Control: An Overview", COINS Technical Report (89-89, Dept. of Computer Science. Univ. of Massachusetts.

Cembrano G., G. Wells (1992) "Neural Networks for Control", *Application of artificial intelligence in process control, L. Boullart, A. Krijgsman, R. A. Vingerhoeds, editors*, Pergamon Press, pp 388-402.

Denker J., D. Schwartz, B. Wittner, S.A. Solla, R. Howard, L. Jackel, and J. Hopfield (1987) "Large automatic learning, rule extraction and generalization", *Compl. Syst. 1*, 877-922.

Fahlman S. E., C. Lebiere (1990) " The cascade-correlation learning architecture, *School of Computer Science Report CMU-CS-90-100*, Pittsburgh PA. Carnegie Mellon University.

Funahashi K. (1989) "On the approximate realization of continuous mappings by neural networks", *Neural networks*, Vol. 2, pp. 183-192.

Girosi F., T. Poggio, (1989) "Networks and the best approximation property", *Artificial Intelligence Lab. Memo*, No. 1164, MIT, Cambridge, MA, USA.

Hoskins D. A., Hwang J. N., Vagners J., (1992) "Iterative Inversion of Neural Networks and Its Application to Adaptive Control" *IEEE Trans. on Neural Networks*, Vol. 3, No. 2.

Hunt K. J., Sbarbaro D., Zbikowski, R. and Gawthrop, P. J. (1992), "Neural Networks for Control Systems - A Survey." *AUTOMATICA*, Vol. 28, No. 6.

Jordan, M. L., Rumelhart, D. E. (1991), "Forward Models: Supervised Learning with a Distal Teacher", Occasional Paper 40, MIT Center for Cognitive Science.

Kawato M. (1990) "Computational Schemes and Neural Network Models for Formation and Control of a Multijoint Arm Trajectory" , *Neural Networks for Control, T. Miller, R.S. Sutton, P. J. Werbos, editors*, MIT Press.

Le Cun Y. (1989) "Generalization and network design strategies", in *Connectionism in Perspective*, eds. R. Pfeiger, Z. Schreter, F. Fogelman and L. Steels. Elsevier, Switzerland.

Maren, A., Harston, C. and Pap, R. (1990) *"Handbook of neural computer applications"*, Academic Press.

Martinetz, T. M., Ritter H. J. and Schulten K. J.(1990), "Three-Dimensional Neural Net for Learning Visuomotor Coordination of a Robot Arm" *IEEE Trans. on Neural Networks*, Vol. 1, No. 1.

Mel, B. (1990) "Vision-Based Robot Motion Planning", *Neural Networks for Control, T. Miller, R.S. Sutton, P. J. Werbos, editors*, MIT Press.

Millán, J. del R., and Torras, C. (1992) "A Reinforcement Connectionist Approach to Robot Path Finding in Non-maze-like Environments", *Machine Learning* vol. 198 No. 3 and 4.

Mitchell, and Gauthier (1991), "Advanced Continuous Simulation Language", Mitchell and Gauthier Associates.

Narendra, K. S. and K. Parthasarathy (1990) "Identification and Control of Dynamical Systems Using Neural Networks", *IEEE Transactions on Neural Networks*, Vol. 1, No. 1.

Nguyen, D. H. and Widrow, B. (1990) "Neural Networks for Self-Learning Control Systems", *IEEE Control Systems Magazine*.

Rumelhart (1986) Rumelhart, D. E. and McClelland, J. L., "Parallel Distributed Processing: Explorations in the Microstructure of Cognition", Vols. I, II and III, MIT Press, 1986 and 1987.

Sanner, R. M., and Slotine, J.-J. E. (1991 a), "Gaussian Networks for Direct Adaptive Control", *NSL Report*, No. 910503, May 1991.

Sanner, R. M., and Slotine, J.-J. E.(1991 b), "Stable Adaptive Control and Recursive Identification Using Radial Gaussian Networks", *NSL Report*, No. 910901.

Werbos, P. J. (1974) "Beyond Regression: New Tools for Prediction and Analysis in the Behavioral Sciences", Ph.D. Thesis, Applied Mathematics, Harvard University.

GENETIC ALGORITHMS IN INTELLIGENT CONTROL SYSTEMS DESIGN

C. BUIU*, I. DUMITRACHE*

*Polytechnic University of Bucharest, Department of Intelligent Control and Bioengineering, Spl. Independentei 313, 77206 Bucharest, Romania

Abstract. New methods for designing and analyzing intelligent control systems are required. Methods for using genetic algorithms for evolving neurocontrollers are reviewed. Some architectures for integrating genetic algorithms with fuzzy logic controllers are presented and discussed. A new application of a genetic algorithm for evolving the algebraic model of a fuzzy controller is proposed. Genetic algorithms are shown to be able to deduce the algebraic model of a simple fuzzy controller used for controlling a servo-system. The genetic algorithm is further used to tune the coefficients of the deduced algebraic model.

Key Words. Control systems; neural nets; expert systems; fuzzy control; process models; genetic algorithms.

1. INTRODUCTION

Taking into account the natural synergism that exists between different intelligent techniques, such as neural nets, fuzzy logic, expert systems, genetic algorithms, there has been much interest lately in developing and analyzing hybrid systems. Such hybrid systems are using the complementarity of these techniques in order to maximize their advantages, while minimizing their drawbacks (Dumitrache *et al.*, 1993b).

Genetic algorithms (GAs) are stochastic adaptive algorithms whose search method is based on simulation of natural genetic inheritance and Darwinian striving for survival. GAs have been quite successfully applied to optimization problems like wire routing, scheduling, game playing, cognitive modeling, the transportation problem, the traveling salesman problem, adaptive and optimal control, fuzzy control problems, pattern recognition, neural networks, machine learning (Dumitrache and Buiu, 1994).

The purpose of this paper is twofold. It tries to review the applications of genetic algorithms in developing intelligent control systems, and to propose new ideas for using this kind of algorithms in intelliegnt control systems.

The paper is organized as follows. Section II gives an overview of genetic algorithms, and section III briefly describes the concept of intelligent control, and some applications of GAs in designing intelligent control systems. Section IV presents the experimental framework and some results obtained by using GAs for obtaining the algebraic model of a fuzzy logic controller and for tuning the obtained model. Section V gives some conclusions and directions for further research.

2. GENETIC ALGORITHMS

Genetic algorithms (GAs) belong to the class of adaptive algorithms whose search methods are based on simulation of natural genetics and survival pressure. They are very different from random algorithms as they combine elements of direct and stochastic searches. For hard optimization problems they are superior to hill-climbing methods by providing simultaneous both exploitation of the best solutions and exploration of the search space. This property also makes them more robust than existing directed search methods. Another important property of such genetic based search methods is their domain-independence.

A GA performs a multi-directional search by maintaining a population of potential solutions and encourages information formation and exchange between these directions. This population undergoes a simulated evolution : at each generation, the relatively "good" solutions reproduce, while the relatively "bad" solutions die. To evaluate the quality of a solution, a special function is used which plays the role of the environment. The structure of a simple GA is shown in Fig 1.

```
procedure genetic algorithm
   begin
      t=0
      initialize P(t)
      evaluate P(t)
      while ( not termination-condition ) do
       begin
          t=t+1
          select P(t) from P(t-1)
          recombine P(t)
          evaluate P(t)
       end
   end
```

Fig. 1. A simple genetic algorithm

A genetic algorithm run is an iterative simulation with a global population $P(t) = \{x_1(t), x_2(t),...,x_n(t)\}$, where $x_i(t)$ is a feasible solution, t is an iteration number, and n is size of the population. The algorithm operates as follows : in each generation it evaluates each chromosome (using the evaluation function on the decoded sequences of variables), select a new population according to the probability distribution based on fitness values, and recombine the chromosomes in the new population by using genetic operators.

After a number of generations, when no further improvement is observed, the best chromosome represents an optimal solution.

A genetic algorithm to solve a problem must have five components: a genetic representation of the potential solutions of the problem; a way to create an initial population of solutions; an evaluation function that plays the role of the environment, rating solutions in terms of their "fitness"; genetic operators that alter the composition of children during reproduction , and values for the parameters that the GA uses (population size, probabilities of applying genetic operators).

3. GENETIC ALGORITHMS AND INTELLIGENT CONTROL

3.1. *Intelligent control*

In conventional control, the goals of the controller are fixed and defined by the designer. This approach is unacceptable for real-life situations, because such complex systems require some forms of autonomy. These complex, ill-defined processes are characterized by : poor mathematical models, high noise levels, stringent perfromance requirements, presence of nonlinearities, distributed sensors and decision makers.

An intelligent control systems (ICS) has the ability to learn and to make decisions. These characteristics enable an ICS to adaptively and predictively reason about dynamic environments. They have the ability to react in a successful manner to a variety of uncertainties.

With rapid development of fuzzy logic, knowledge engineering and artificial neural networks, the using of these techniques for developing ICSs has become an attractive new direction. However, an ICS is a very complex system, highly nonlinear and therefore is difficult to be designed and analyzed.

3.2. *GAs for evolving neurocontrollers*

The neural net (NN) based approach is a good tool in order to tackle the control of processes without an explicit analytical model.The property of NNs for building an adaptive model of a complex plant allows the use of this technique for solving control problems which have difficult solutions using traditional techniques.

The neurocontroller is included in a closed loop and responds straightforwardly to the process observed behavior by supplying control parameters.The task of the neurocontroller is to learn to supply the appropiate control parameters for desired targets given as input.

The interest in developing more powerful artificial hybrid systems has lead to the study of evolutionary neural net (ENN) (Yao, 1993). An ENN has the meaning of an artificial neural network developed using evolutionary computing techniques.

One can distinguish three possible ways of using GAs for evolving NNs : the evolution of connection weights, the evolution of the topology of the neural network, and the evolution of learning rules.

The most researches were done until now in the first two directions. The obtained results have shown that using a GA as a replacement for back-propagation does not seem competitive with the best gradient method, e.g.quickprop . But GAs seem to be an useful learning method when we deal with discontinous optimality criteria or discontinous node transfer functions.

Some attempts have been made to adaptively adjust standard training algorithms' parameters using an evolutionary approach. For example, such parameters may be the learning rate and the momentum of a standard back-propagation algorithm. A more difficult approach is to evolve learning rules.

3.3. *GAs for tuning fuzzy controllers*

Fuzzy logic is a innovative technology with a broad range of applications. Reduced cost, small size and reliability are key advantages of fuzzy logic control systems. They also provide the simplest solution and the easiest operation. A fuzzy logic controller is very robust and superior in performance to a conventional PID controller. It can be implemented with very little silicon surface. Special fuzzy logic chips are available, and they can easily be interfaced to sensors and actuators.

Fuzzy controllers are typically highly non-linear systems. Therefore, a fuzzy controller is difficult to understand and analyse. The key problem with fuzzy logic controllers is that they are difficult to calibrate.

Some approaches of this problem using GAs (Dumitrache *et al.*, 1993 and Janikow *et al*, 1994) make the assumption that the mathematical model of the process is available, or it could be obtain by using some classic identification algorithm. In the proposed fuzzy representation, the knowledge is distributed among three different levels : symbolic rules, numerical weights, and fuzzy linguistic definitions. In order to obtain an optimal fuzzy controller with respect to some performance index, the search must be guided to all three levels simultaneously. The GA is used for both extracting the necessary knowledge and tuning the existing knowledge of a fuzzy logic controller.

The genetic operators utilized there include random mutation, dynamic mutation designed for better local exploration, single crossover on whole genes, and arithmetical crossover averaging selected genes. For evaluation, there are used different objective functions, reflecting various criteria. Presented results indicate that the modifications made by the GA could lead to optimizing the fuzzy controller both in terms of simplifying the symbolic structure (and thus increasing its comprehensibility) and optimizing the structure using the numerical weights.

What to do when the process model is not available? In order to cope with this situation an architecture of a hybrid geno-fuzzy control system is presented in Fig. 2. The function of process identification is performed on-line by a GA, called GAI (GA Identifier), which is providing each time step Θ, an estimate of the process model, by minimizing a performance index V. This model is used by another GA, called GAT (GA Tuner) which is performing the adaptation of the fuzzy logic controller (FLC), with respect to a performance index, *J*. The adaptation is performed simultaneously at all three levels of knowledge described above. The GAT provides Π, the set of tuned controller parameters (numerical weights, fuzzy definitions, fuzzy rules).

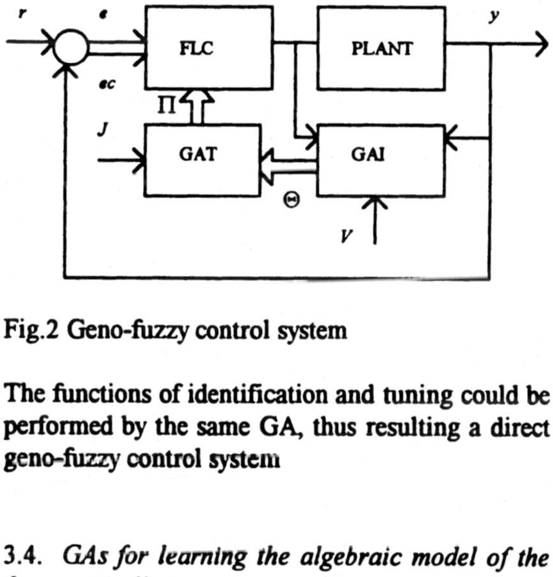

Fig.2 Geno-fuzzy control system

The functions of identification and tuning could be performed by the same GA, thus resulting a direct geno-fuzzy control system

3.4. *GAs for learning the algebraic model of the fuzzy controller*

The big numbers of parameters to be adaptively modified is a drawback of the previous approach. So, this problem has lead the authors to the idea of aggregating these parameters into a smaller number of parameters. This would have as an effect an easier tuning procedure and would reduce the time required for computing the inputs to the process.

The idea was to use for this purpose the algebraic model of the fuzzy controller, introduced in (Braae, 1979). Suppose that the defuzzified output of the fuzzy controller is δ, where δ is in [-1,1]. If the inputs to the fuzzy controller are r_1, r_2, ... , the functional relationship between the defuzzified output δ and the inputs, $\delta = F(r_1, r_2, \ldots)$ is called the algebraic model of the fuzzy controller. For example, if the inputs of the fuzzy controller are, as in the most actual cases, the error e and the change of error, ec, the output of the controller would be $\delta = F(e, ec)$. If F it is known, and this function is not very complicated, it may be used for computing the process input, and fire no fuzzy rules. Also, this algebraic model would aggregate all the fuzzy controller's parameters into a much smaller number of numeric coefficients, thus making the adaptation process more faster.

In (Buckley, 1992) there are reported some situations when it was possible to construct F. This was possible when were used a small number of rules, and special logics to evaluate the fuzzy control rules. There are some important cases in what regards the structure of F. If $\delta = \alpha * e + \beta * ec$ and α and β are constants, then F is linear. If F has the same structure as above, but α and β are functions of e and/or ec, then the fuzzy controller is a nonlinear PI controller. There were reported many experiments on the sensitivity of δ with respect to various parameters, such as the defuzzifier, the fuzzy logic, the scaling constants.

An interesting opportunity to determine the structure of the function F is to use a GA for this purpose, and to further use the algebraic model (AM) of the fuzzy controller for tuning purposes.

An architecture which is using this approach is presented in Fig.3. One can observe the GAT and GAI blocks which have the same purposes as in Fig.2. There exists GAM (GA Modeller) which receives as input the look-up table Λ of the fuzzy logic controller (FLC), which is known to perform good on the respective class of processes.

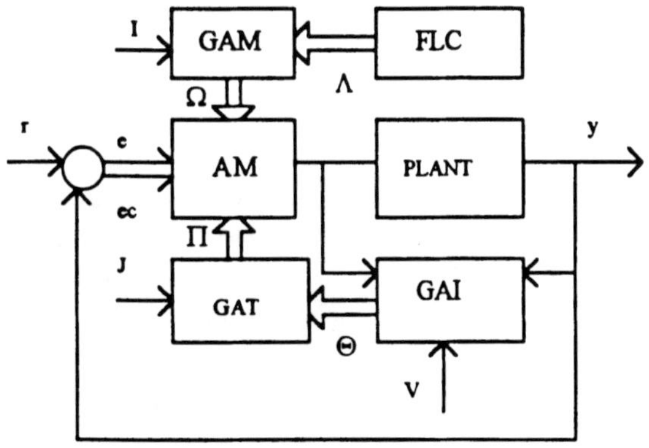

Fig.3 Algebraic model based geno-fuzzy control system

The GAM performs off-line and provides Ω, an AM of the fuzzy controller, by minimizing some performance index I. This AM is further used for controlling the process, and for tuning purposes, using GAI and GAT.

4. EXPERIMENTAL FRAMEWORK

4.1. *Test problem*

The experiments consisted in testing the algebraic model based geno-fuzzy control system on a servo-system. The task to be performed was to rotate the shaft of a servomotor to a set point. The fuzzy controller has two inputs, error e and error change ec, and one defuzzified output. The error e is defined as the difference between the setpoint r and the output y from the shaft encoder : $e = r - y$. The error change is defined as follows : $ec = e_1 - e_2$ where e_1 is the error at the time step t_1, and e_2 is the error at the time step t_2. The input to the process is called u, and it is a voltage output, defined as $u(t) = u(t-\Delta) + \delta(t)$, at times $t = \Delta, 2\Delta, \ldots,$ where Δ is the sampling time.

For the linguistic values, there are used the same fuzzy sets for error, error change, and control change specified in [-1,1]. They are presented in Fig. 4 . They have the following meaning : SN = "small negative", ZE = "zero", and SP = "small positive".

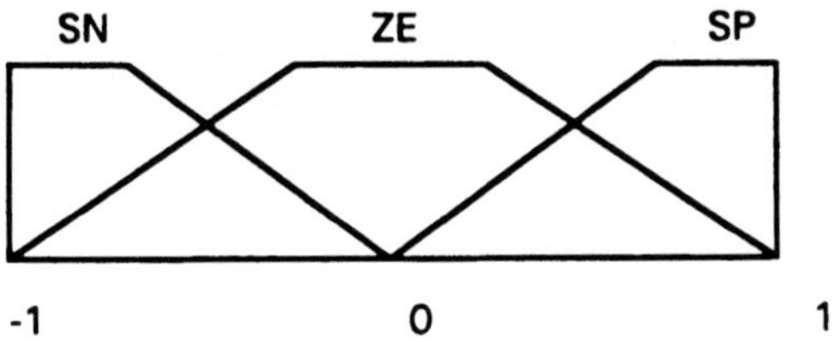

Fig.4 Linguistic values and their fuzzy sets

In Table 1 it is shown a simple fuzzy controller which contains three fuzzy control rules,relating the output δ to the inputs e and ec. This controller is known to perform good and it is further denoted by FC1.

Table 1. Human expert generated rules for FC1

Error	Change in error		ec
e	SN	ZE	SP
SN		SN	
ZE		ZE	
SP		SP	

First, the corresponding look-up table for FC1 was created, by uniformly quantizing the universse of discourse for e and ec, which is the interval [-1,1], and computing the corrresponding output, using

Zadeh logic for inference and the center of gravity method for defuzzifier.

4.2. *Genetic representation*

The following model was fitted to data obtained by creating the look-up table of the fuzzy controller.

$$\delta = \frac{g_1 + g_2 * e + g_3 * ec}{g_4 + g_5 * e + g_6 * ec} + \frac{g_7 + g_8 * e + g_9 * ec}{g_{10} + g_{11} * e + g_{12} * ec} + \frac{g_{13} * e * ec}{g_{14} + g_{15} * e + g_{16} * ec} \quad (1)$$

The evaluation function I, used for rating the potential solutions was choosed to be a quadratic index :

$$I = (1/P) \sum_{p=1}^{P} (r_p - m_p)^2 \quad (2)$$

where p is the *p*th entry of the look-up table, p=1..P, m_p is the output of the potential algebraic model (chromosome) for *e* and *ec* corresponding to this entry and r_p is the output from the *p*th entry of the look-up table. The best algebraic model should have the smallest I. So, in this case the GA operates for minimizing I.

4.3. *Genetic Operators*

All the operators used in the reported experiments are based on floating point representations of the chromosomes (Michalewicz and Janikow, 1991).

Uniform mutation is an unary operator and selects a random component (gene) of a selected potential solution (chromosome), which is also selected randomly with a probability equal to the uniform mutation rate. The selected gene is modified randomly.

Dynamic mutation is an unary operator which is responsible for the fine tuning capabilities of the system. For a parent x, if the element x_k is selected for dynamic mutation, the result is $x' = (x_1, \ldots, x'_k, \ldots, x_n)$, where $x'_k = x_k + D(t, u_k - x_k)$, if a random digit is 0, and $x'_k = x_k + D(t, x_k - l_k)$, if a random digit is 1, and l_k and u_k are the lower and upper limits of x_k. The function D(t,y) returns a value in the range [0,y] such that the probability of D(t,y) being close to 0 increases as t increases. (t is the generation number). This property causes this operator to search the space very locally when t is big. An example of D is :

$$D(t,y) = yr(1 - t/T)^b \quad (3)$$

where r is a random number in [0,1], T is the maximal generation number, and b is a system parameter (usually b=6).

Simple crossover is the classic crossover operator which is operating by swapping corresponding segments of two parents and produce two offsprings.

Whole arithmetical crossover is defined as a linear convex combination of two parents, i.e. if x_1 and x_2 are chosen to be the parents, the resulting offsprings are $x'_1 = ax_1 + (1-a)x_2$, and $x'_2 = ax_1 + (1-a)x_2$, where a is a random number in [0,1].

4.4. *Experimental results*

The step response of the servo-system with FC1 is presented in Fig.5. For FC1, the obtained algebraic model is denoted by AM1, and is as follows.

$$\delta = \frac{0.0326 + 3.7279 * e - 0.0025 * ec}{4.9249 + 1.6736 * e + 0.0054 * ec} + \frac{-0.0015 + 0.9076 * e + 0.0099 * ec}{4.4573 + 0.201 * e + 2.8231 * ec} + \frac{1.0359 * e * ec}{4.7052 + 0.0052 * e + 4.0244 * ec} \quad (4)$$

and the corresponding value of the performance index was I=0.019756. The step response of the servo-system controlled with AM1 is presented in Fig.6. This model was further tuned by a GA which minimized the following performance index.

$$E = \sum e_t^2 \quad (5)$$

where e_t is the error at time *t*, and is as follows:

$$\delta = \frac{0.04696 + 9.9999 * e + 2.399 * ec}{3.3101 - 9.9999 * e + 0.0054 * ec} + \frac{-0.0015 + 1.0738 * e + 9.9999 * ec}{1.4139 - 0.8834 * e + 0.5255 * ec} + \frac{5.4436 * e * ec}{0.2083 + 3.1907 * e + 8.3747 * ec} \quad (6)$$

The performance index for AM1 is E=12.200. The tuned AM1 assures E=2.404019, and the step response of the control systems with this tuned algebraic model is presented in Fig.7.

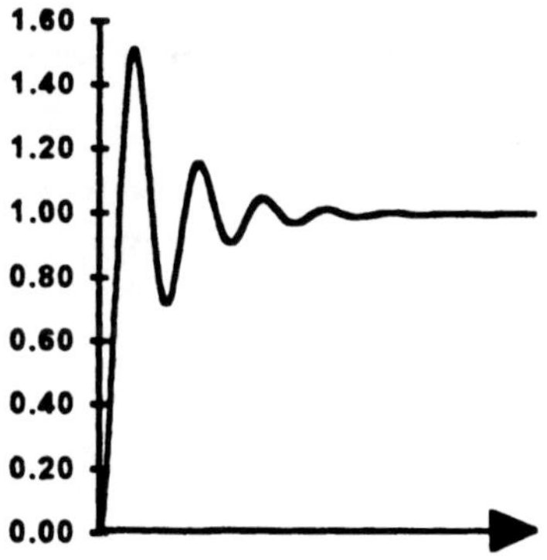

Fig. 5. Step response with FC1

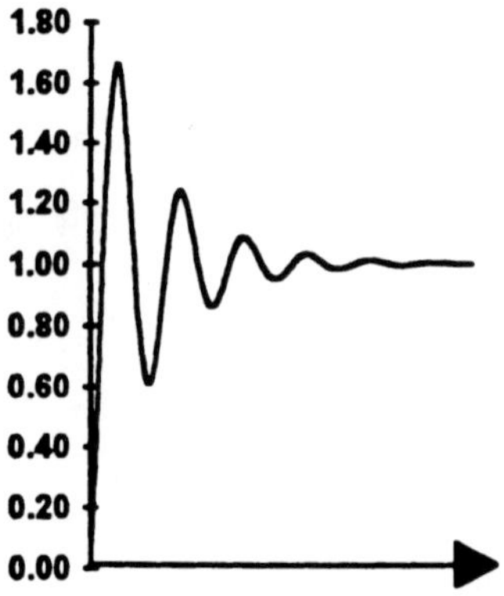

Fig.6. Step response with AM1

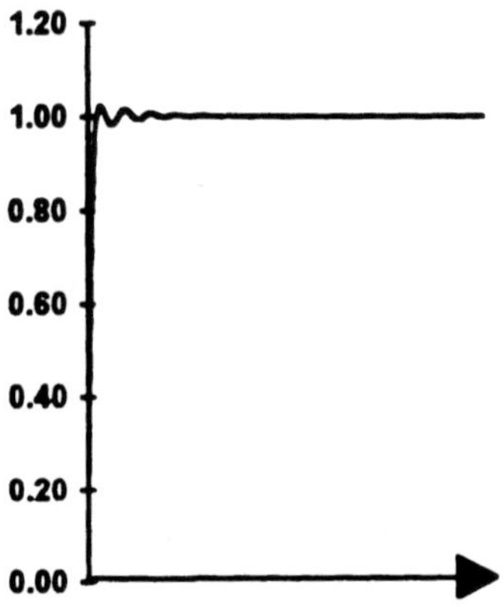

Fig.7. Step response with tuned AM1

For both experiments, the population size was 40, the uniform mutation rate was chosen to be 0.1, the dynamic mutation rate was set to 0.15, the simple crossover probability was 0.6, and the arithmetic crossover rate was 0.2. The initial population of solutions was created randomly.

5. CONCLUSIONS

This paper presents some ideas for using genetic algorithms in the design of intelligent control systems, and especially of fuzzy control systems. Genetic algorithms are attractive approaches for designing intelligent controllers, such as neurocontrollers, and for analyzing and tuning fuzzy logic controllers. A new application of GAs for evolving the algebraic model of a fuzzy logic controller was introduced. The presented results indicate that this approach lead to interesting and useful results.

Further experiments will consist of testing different coding methods for the potential algebraic models (chromosomes), and implementing domain-specific genetic operators, all these done for more complex sets of rules. A challenging domain of research would be the possibility of evolving algebraic multi-models, i.e. different algebraic models of the fuzzy controller, evolved for different domains in the input space of the fuzzy controller. This would bring the necessity of providing an intelligent supervisor of the fuzzy controller, whose purpose would be to select different algebraic models. The same approach could be applied to evolving mathematical models of the processes described by fuzzy relations relating the process inputs to the process outputs.

6. REFERENCES

Braae, M. and Rutherford, D.A. (1979). Theoretical and linguistic aspects of the fuzzy logic controller.*Automatica* 15, 553-577.

Dumitrache, I., Janikow, C. Z., and Buiu, C. (1993a). Tuning a fuzzy logic controller using genetic algorithms. *Proceedings of the 9th International Conference on Control Systems and Computer Science*, CSCS9, Bucharest, 450-461, May 25-28.

Dumitrache, I., Calcev, G., Teta, P., Buiu, C., and R. Constantinescu (1993b). *Intelligent Process Control Techniques*, Polytechnic University of Bucharest Press, Bucharest.

Dumitrache, I., Buiu, C. (1994). Genetic algorithms in process identification and control, in press.

Janikow, C. Z. , Buiu, C. and Dumitrache, I (1994). Learning a fuzzy controller by genetic algorithms, in press.

Michalewicz, Z., Janikow, C.Z. (1991). Genetic algorithms for numerical optimization. *Statistics and Computing* (1), 75-91.

Yao, X. (1993). A review of evolutionary artificial neural networks. *International Journal of Intelligent Systems*, vol.8, 539-567.

THE ROUTING PROBLEM IN TRAFFIC CONTROL USING GENETIC ALGORITHMS

F.J. MARIN*, F.J. GONZALEZ* and F. SANDOVAL**

*Universidad de Málaga, Departamento de Arquitectura y Tecnología de Computadores y Electrónica, Plaza El Ejido s/n, 29013 Málaga, SPAIN
**Universidad de Málaga, Departamento de Tecnología Electrónica, Plaza El Ejido s/n, 29013 Málaga, SPAIN

Abstract. This paper proposes the use of evolutionary techniques for optimization in the routing problem. This problem consists in finding the minimum connection cost between a source node and a destination node of an arbitrary network. It is necessary a strong optimization for working in real time. One advantage that presents the evolutionary algorithms is a good performance to resolve adaptive dynamic problems. The simulations show that this algorithm requires less computation time than the traditional techniques and provides a better performance than artificial neural networks.

Key Words. Traffic control, Optimization, Dynamic response, Adaptive control, Routing problem, Genetic algorithms.

1. INTRODUCTION

The routing problem is at the moment a task largely investigated because its application to centralized or decentralized guided traffic in cities, to telecommunication networks in Packet Switching Networks, and to network computers. In general, the routing problem consists of finding the shortest path between any source node, s, and any destination node, **d**, of a network, **G**. This network **G={N,A}**, is composed of a finite set, not empty, of nodes **N** and a collection of orderly pairs of different nodes, **A**, belonging to **N**. In the network, each pair of nodes (i,j) is called arc or edge. Each arc has a weight, w, which is integer, positive and independent of the other weights. In general, for any two nodes (i,j), it applies $w_{ij} \neq w_{ji}$. The solution to the routing problem is an orderly list of nodes $< n_\pi(1) \ldots n_\pi(q) >$ which minimizes:

$$\sum_{p=1}^{q-1} W_{p,p+1} \qquad (1)$$

The application of the routing problem to traffic in cities implies: the arcs represent the streets of the city; the nodes represent the intersections of the streets; and the weight of the arc (i,j) depends on the action of parameters such as: the length to be run, the time estimated, the saturation in the arc, the number of vehicles in the route, etc. This is a classical problem of optimization with combinatorial complexity. In a n-nodes fully connected network there are (n-2)! possible paths from any source node to any destination node.

The way the routing problem is resolved has to be strongly optimized to be able to work in real time. Traditional algorithms (Brassard and Bratley, 1988) do not resolve the long computational time needed, and new techniques have been introduced. Cavalieri *et al.* (1993) use a Hopfield neural network, but the energy function presents many local minima, doing the not optimal paths rate to increase significantly. Minot (1993) uses a cellular automata for an adaptive routing, but when the networks are more complexes, the convergence speed decreases quicker than the convergence speed of the classical algorithms.

Last years a new technique, based on genetic algorithms (GAs), has emerged as a suitable optimization way to obtain the shortest paths in routing problem. In this paper it is shown how to implement it. The paper is made up by the following sections. Section 2 gives basic notions of the genetic algorithms. Section 3 shows the solution selected for the codification of the problem, the genetic operators, a new operator called Unblock, and the values of the parameters used in the genetic algorithms. Section 4 presents the network target of the experiments and the results obtained considering one or several vehicles in the network. Finally, section 5 discusses the relevant problems with the real use of genetic algorithms and suggests further investigations.

2. GENETIC ALGORITHMS

The genetic algorithms are evolutive algorithms of search based on natural selection mechanisms. They were first described by Holand (1975) and further developed by Goldberg (1989). The way they work is very simple: it starts with a randomly generated initial population of individuals, which are candidates for the solution of a specific problem. Each individual is composed of a string (e.g., 0 or 1) which is evaluated through a function (fitness function) that gives the approximated global solution of the problem. The better individuals (better fitness values) are selected (selection operator) as parents, the better they develop a new population. To accelerate the search it is necessary to apply other genetic operators. The most widely used genetic operators are the crossover and mutation operators. The crossover operator exchanges substrings between two individuals (see Fig. 1); once two individuals are selected, a position (the crossover position) is randomly obtained; from this position the exchanges between the subchains takes place to yield two new sons which replace to the parents.

The mutation operator randomly changes the value of one field of the individual (see Fig. 2). This operator first randomly selects a field of the individual, and then changes its value.

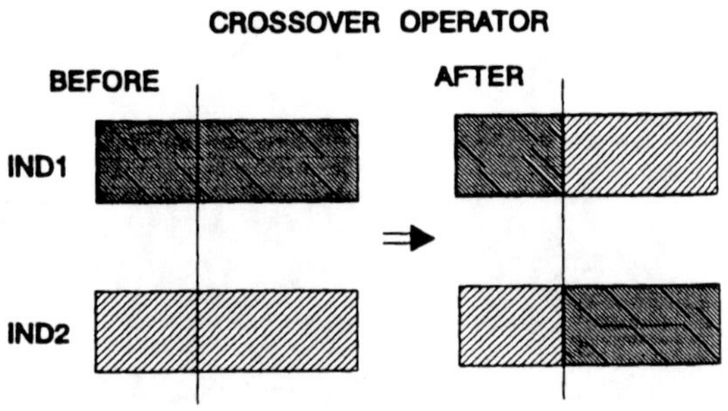

Fig. 1. Crossover operator.

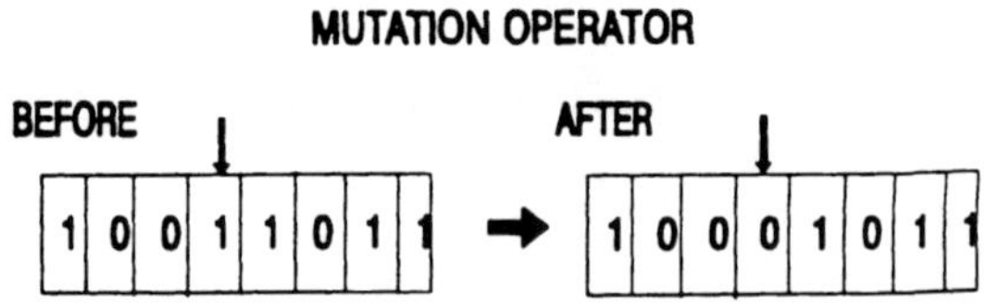

Fig. 2. Mutation operator.

This process is iterated over many time steps, each called generation. After several generations, the result is often highly fitted individuals in the population.

The process developed above may be stated in pseudocode as follows:

```
{ GetParameters()
  PerformRequiredInitializations();
  do
    {
      if (FirstGeneration)
              CreatePopulation();
      else {
              Selection();
              Crossover();
              Mutation();
           }
      EvaluatePopulation();
      Statistic();
      ReportResults();
    } while (!EndOfJob)
}
```

3. METHODOLOGY

The most important step in the application of GAs to any problem is the choice of a good codification. The speed and the correct running are closely dependent on an appropriate codification. In the routing problem, it has been used a codification through a diploide structure (two strings) (see Fig. 3). The first string has all the nodes of the network in any order except the first and the last field, which correspond to the source and destination nodes, respectively. The second string has the same number of fields than the first string and it is formed by binary digits 0 and 1, where 0 means than this node does not belong to the path, whereas 1 does.

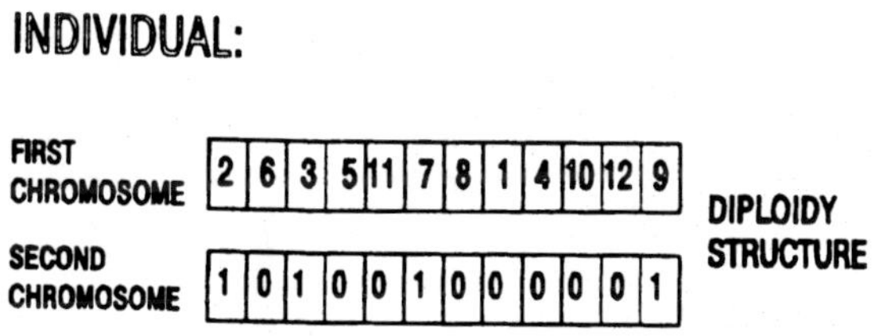

Fig. 3. Codification of the problem. Optimal path from node 2 to node 9.

The fitness function is coincident with the relation (1). The selected operator is the Baker's universal stochastic sampling (Baker, 1987) where the number of descendants of any individual is function of the performance. In this case, the individuals with lower fitness value, those that minimize (1), can have a larger number of descendants in the next generation.

The solution of the routing problem strongly depends on how the nodes are ordered in the network codification. For instance, if nodes 3 and 5 are generated in order, we never go from node 5 to node 3; therefore, we need a crossover operator to generate good ordering, and, in this way, avoiding the destruction of sets of nodes that are essentials for finding the optimal solution of the problem. On the other hand, it is necessary that in the crossover of two individuals do not appear repetitions or omissions of nodes. We have used a variant of the partially mapped crossover (PMX) (Goldberg *et al.*, 1985), which operates with a certain crossover rate. The mutation operator is only concerned with the second string of the diploide structure. This operator changes its binary value with a certain mutation rate. We have implemented a well-known operator: the permutation operator; it randomly exchanges the positions of two nodes of the first string, selected according to a certain permutation rate.

The experiments carried out with the above operators have shown that the population converged with a relative frequency towards not valid routes. To overcome this problem, for the first time it is introduced a new operator called unblock operator; it regenerates every certain number of generations those individuals that may perform the genetic algorithm operates unsuitably. This regeneration is only applied to those individuals that oblige to run one or two not available paths; the unblock operator sets to zero the fields of the second string corresponding to those nodes that do not take part in a valid path. The action of the unblock operator to regenerate all the individuals that perform not valid paths take us to a random search of the solution; therefore, these parameters must carefully be selected. In this way, it is avoided the genetic algorithm falls down in not valid routes.

4. EXPERIMENTS AND RESULTS

The network under study is composed of 12 nodes (see Fig. 4) with bidirectional weights and incompletely connected. The connection rate (number of arcs of the network over the total number of possible arcs) is very small.

The dimension of the population is always of 100 individuals, and 100 generations have been performed. The best results have been obtained with the following parameters: crossover rate = 0.7; mutation

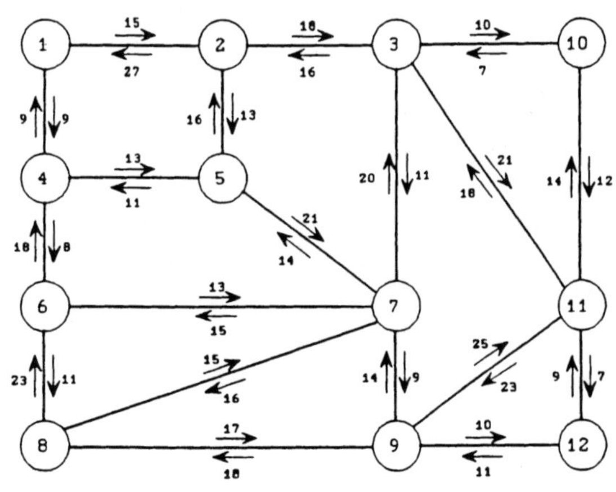

Fig. 4. Network.

rate = 0.3; permutation rate = 0.6 and the unblock operator works every five generations. All the data have been averaged over ten runs (presentation frequency).

In the first experiment the source node is the node 2 and the algorithm searches the shortest path with any other node of the network. Only one vehicle is in the route and therefore there are not modifications of the initial weights of the network (see Fig. 5). From among the 110 runs performed, 82 get the optimal path, 28 find a valid path but not optimal, and just one fails finding a not valid path.

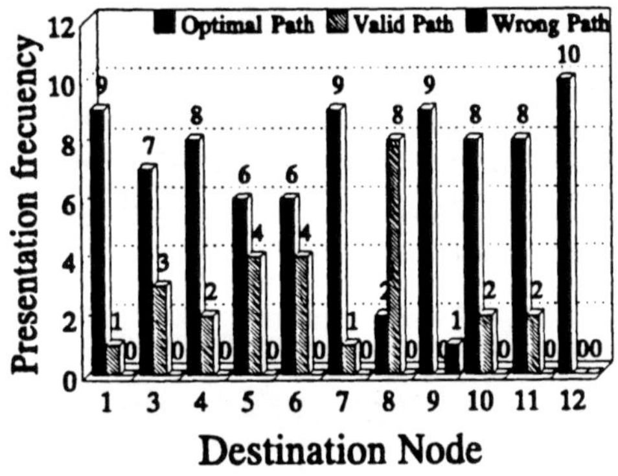

Fig. 5. Optimal,valid and wrong paths from node 2 to all the other nodes.

In the second experiment the path to be run is fixed (from node 2 to node 9), but there are five vehicles in the network. Therefore, the weights of the network are modified along the time, and the algorithm must work in a dynamic way to achieve an adaptive control. (see Fig. 6). In this case, all the runs performed (50) get a valid path, being 41 which find an optimal path.

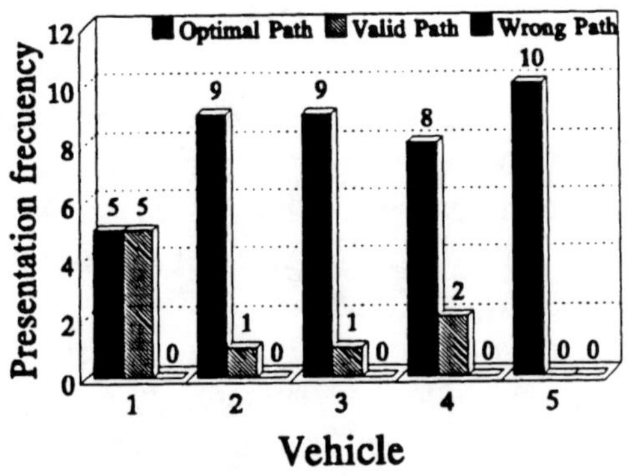

Fig. 6. Optimal, valid and wrong paths from node 2 to node 9 with five vehicles.

In the third experiment the paths are randomly chosen and there is just one vehicle in the route (see Fig. 7). Results are even better: Most of the runs (93%) find the optimal path, and the rest 7% get a valid path.

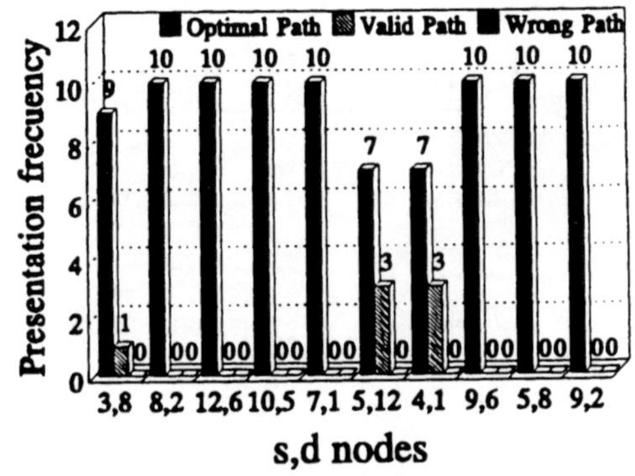

Fig. 7. Optimal, valid and wrong paths for random s and d nodes with 1 vehicle.

In the last experiment the paths are randomly chosen and there are five vehicles in the route (see Fig. 8). Here again, the amount of runs finding the optimal path is greater than 90%.

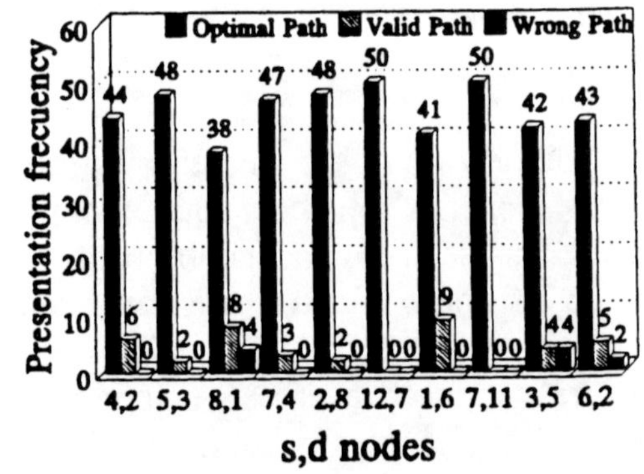

Fig. 8. Optimal, valid and wrong paths for random s and d nodes with five vehicles.

Independently of the pairs (s,d) selected and the amount of vehicles in the network, the rate of optimal paths over the valid or wrong paths is very high. This success is enhanced if it is considered that a valid path presents a fitness value very near to that of the optimal path.

5. CONCLUSIONS

The experiments show how using evolutive algorithms is possible to reduce the computational time, carrying out a fewer number of evaluations and reducing the number of generations without losing precision, i.e., GAs use less computational time than traditional algorithms (Brassard and Bartley, 1988).

However the network has two nodes more than the network used by Cavaliers *et al.* (1993), GAs have obtained better results. In the best experiment by Cavalieri *et Al*, the neural network only find the optimal path in the 68% of runs.

Presently we are investigating new codification methods to reduce the large search space of the GA. In this codification the GA must explore a very large search space. There are 10! possible paths because of the codification of the first string of the individual, and, because of the second string, there are 2^{10} possible ways of encoding the first string; therefore, we have $10! \times 2^{10} = 3{,}7 \times 10^9$ total possible paths.

The unblock operator must be improved when it is used in networks with larger dimensions. We are checking other order crossover operator and studding the use of some advanced operators.

The GA has been put in parallel over a LAN-based message passing computer architectures, using parallel virtual machine - public domain software (PVM 3.1) - as software integration tool (Marín *et al.*, 1994). A good efficiency and computational speed up have been obtained.

6. REFERENCES

Baker, J.E. (1987). Reducing Bias and Inefficiency in the Selection Algorithm. In: *Proceedings of the second International Conference on Genetic Algorithms and Their Applications* (J.J. Grefenstete, Ed.) Lawrence Erlbaum Associattes.

Brassard, G. and Bratley, P. *Algorithmic. Theory and Practice.* Prentice Hall International.

Cavalieri, S., A. Di Stefano, O. Mirabella (1993). Hopfield Neural Network for Routing. In: *New Trends in Neural Computation* (J. Mira, J. Cabestany, A. Prieto, Eds.), pp 583-592. Springer-Verlag.

Goldberg, D.E. (1989). *Genetic Algorithms in Search, Optimization and Machine Learning*. Addison-Wesley Reading, MA.

Holland, J.H. (1975). *Adaptation in Natural and Artificial Systems.* MIT Press, Cambridge, MA.

Marin, F.J., O. Trelles, F. Sandoval (1994). Genetic Algorithms on LAN- message passing architectures using PVM and Its Applications to Routing Problem. Submitted for publication.

Minot, J. (1993). Adaptive Routing using Cellular Automata. In: *New Trends in Neural Computation* (J. Mira, J. Cabestany, A. Prieto, Eds.), pp 599-604. Springer-Verlag.

COMBINING NEURAL AND FUZZY TECHNIQUES IN MONITORING AND CONTROL OF MANUFACTURING PROCESSES

L. Monostori

Computer and Automation Research Institute, Hungarian Academy of Sciences
H-1518 Budapest, POB 63, Hungary

Abstract. Real-time nature, uncertainty handling and learning ability are essential requirements for knowledge representation and processing techniques to be applied at lower levels of intelligent manufacturing systems. The paper demonstrates and compares the applicability of neural networks and fuzzy techniques for monitoring and control of manufacturing processes. Combined use of neural and fuzzy techniques is illustrated, and the integration of such a hybrid system in an intelligent manufacturing environment is investigated.

Key Words. Intelligent manufacturing; manufacturing processes; monitoring; artificial intelligence; pattern recognition; learning systems; neural nets; fuzzy systems

1. INTRODUCTION

In *intelligent manufacturing systems*, within certain limits, unprecedented, unforeseen situations and problems are expected to be solved even on the basis of incomplete and imprecise information, as was stated by Hatvany (1983).

In the last few years *artificial neural networks (ANNs)* or *connectionist systems* came into the limelight and divided to some extent the AI community (Minsky and Papert, 1969; Rummelhart and McClelland, 1986). As a subsymbolic approach, connectionism has some true benefits over symbolicism exactly in those fields (real-time feature, quantitiveness, uncertainty handling, effective learning ability), which are considered as fundamental requirements for intelligent manufacturing.

Investigations confirmed (Chryssolouris and Domroese, 1988; Rangwala and Dornfeld, 1989; Dornfeld, 1990; Monostori and Nacsa, 1990) that - similarly to our present conception of biological systems - ANN techniques seem to be a viable solution for the lower level of intelligent, hierarchical control and monitoring systems. Since the higher levels of the control/monitoring hierarchy require mostly symbolic knowledge representation and processing, the integration of symbolic and subsymbolic methods can be predicted (Barschdorff and Monostori, 1991; Monostori and Barschdorff, 1992).

The paper summarizes the most important results of previous works to which the author contributed on the application of artificial neural networks in the lower levels of intelligent manufacturing systems (Monostori and Nacsa, 1990; Barschdorff *et al.*, 1993a, 1993b; Monostori *et al.*, 1992; Monostori, 1993).

The fundamental aim of the paper is to demonstrate and compare the applicability of neural networks and fuzzy techniques for monitoring and control of manufacturing processes. Combined use of the neural and fuzzy techniques is illustrated, and the integration of such a hybrid system in an intelligent manufacturing environment is investigated.

2. FUZZY SYSTEMS

2.1. Fuzzy Inference Systems

Fuzzy conditional statements or *fuzzy if-then rules* are expressions in the form of *IF A is A1 THEN B is B1*, where A and B are *linguistic variables*, labels of *fuzzy sets*, A1 and B1 are *linguistic values* characterized by *membership functions, (MBFs)* (Zadeh, 1965).

Fuzzy inference systems or - when they are used as controllers - *fuzzy logic controllers, (FLCs)* are basically composed of five functional blocks (Fig. 1, without feedback):

- a *rule base* containing fuzzy if-then rules;
- a *data base* where MBFs of fuzzy sets used in the fuzzy rules are defined;
- a *decision making unit* or *inference engine* which performs the inference operations on the rules;
- a *fuzzification interface* or *fuzzifier* which transforms the crisp data into degrees of match with linguistic values;
- a *defuzzification interface* or *defuzzifier* which transforms the fuzzy results of the inference into crisp output(s).

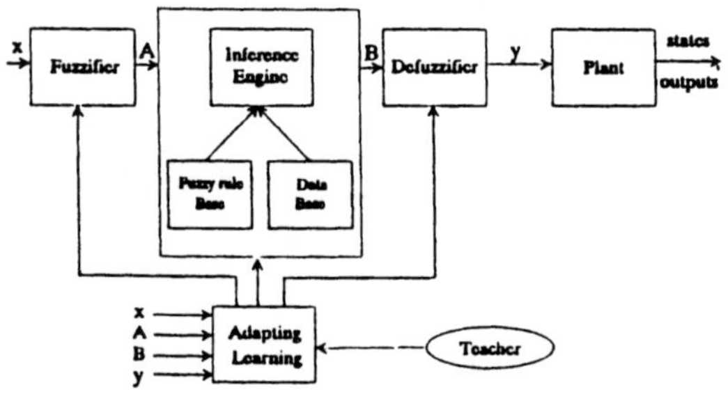

Fig. 1. General model of a fuzzy logic controller

The FLC of Fig. 1 is equipped with a *learning adapting block* which is to find suitable fuzzy logic rules and proper shapes and overlaps of MBFs by learning the desired output or regarding to an external reinforcement signal.

2.2. Fuzzy Associative Map (FAM)

Fuzzy associations between fuzzy properties (e.g. HIGH, SEVERE) can be entered in a *FAM-rule matrix* (FAM: fuzzy associative map or memory), Fig. 2. The system maps input fuzzy sets A to output fuzzy sets B. Separate FAM rules are stored separately and for each input they are fired in parallel. In the fundamental case the w weights are set to 1 or 0.

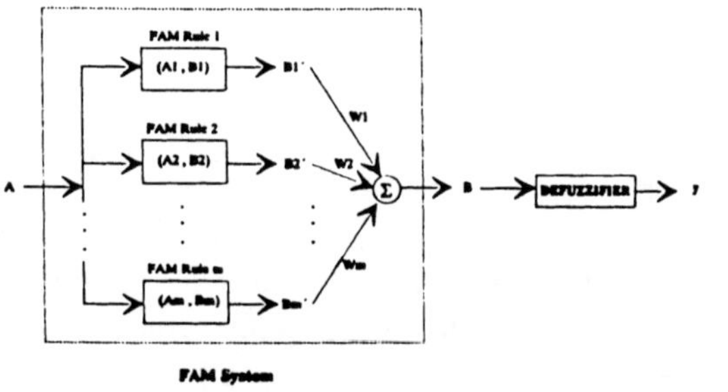

Fig. 2. FAM based fuzzy system architecture for one input and one output variable (Kosko, 1992)

2.3. Artificial neural networks versus fuzzy systems

Fuzzy systems (Zadeh, 1965; Negoita, 1985; Zimmermann, 1987) lie in some sense between traditional AI systems (knowledge based expert systems, KBESs) using structured knowledge representation in a symbolic manner and neural networks which cannot directly encode structured knowledge (Fig. 3). The fuzzy approach combines the purely numerical approaches of neural networks and mathematical modelling with the symbolic, structure-rich approaches of AI.

	Symbolic	**Numerical**
Structured	KBES	Fuzzy system
Unstructured	-	ANN

Fig. 3. Taxonomy of model-free estimators (Kosko, 1992)

Neural and fuzzy systems encode sampled information in a parallel-distributed framework. Both frameworks are numerical. Besides learning ability, model-free estimation is, after all, the central computational advantage of neural networks. The cost is system inscrutability. In contrast to the neural approach, knowledge representation in fuzzy systems is comprehensive for users. Using fuzzy systems, the user can determine which FAM rules contributed how much membership activation to a concluded output. FAM-structured knowledge can be added or deleted without disturbing stored knowledge.

2.4. Neuro-fuzzy systems

Fuzzy systems are considered to be a natural link between symbolic and subsymbolic approaches. On the one hand they can work in real-time circumstances and handle uncertainties as artificial neural networks, on the other hand they can manage both symbolic and numerical information. However, fuzzy systems usually do not incorporate automatic learning ability.

It seems, best performance can prospectively be obtained by combining neural and fuzzy approaches, integrating their benefits. The resulting *neuro-fuzzy system*, is a *hybrid system*, where the system architecture remains fuzzy, but using neural learning techniques, it can be trained automatically.

Training procedures originating from the ANN world, can be used on two levels. First, unsupervised, competitive techniques (Rummelhart and McClelland, 1986) can generate FAM rules from training data. At this stage, the input-output space is clustered with neural network techniques, and each cluster estimates a FAM rule. The experts need not articulate these rules, only generate input-output data for learning. Secondly, a further training procedure (e.g. back propagation) modifies the w weights in the system (Fig. 2).

Therefore, the neuro-fuzzy approach can comply with the requirements enumerated earlier in this paper, i.e. *real-time nature, uncertainty handling and learning ability*, moreover, *expert knowledge* can easily be incorporated in these systems.

3. EXPERIMENTS, APPLIED TECHNIQUES

3.1. Cutting experiments, preprocessing of signals

The main goal of this paper is to investigate the applicability of the neuro-fuzzy technique in monitoring and control of manufacturing processes and compare the results with the pure ANN approach.

The data from *turning experiments* made in the Institute of Manufacturing and Production Engineering, University of Kaiserslautern, FRG which were formerly analyzed by ANN techniques (Barschdorff *et al.*, 1993a, 1993b) were used.

Numerous (100 - 150) statistical and spectral features were computed from the measured force and vibration signals. Because of the great number of features considered and in order to enhance the learning, estimation and classification performances of the modelling and monitoring procedures, SFS (sequential forward search) *feature selection* was accomplished for every group of investigations using the criterion:

$$J = \frac{trace(S_b)}{trace(S_w)} \qquad (1)$$

where S_b is the between class and S_w is the within class scatter matrix. Through this feature selection the relative importance of measured signals in wear estimation/classification can also be evaluated.

3.2. The *fuzzy*TECH development system

Preprocessing of the measured signals, automatic feature selection and ANN based learning, classification and modelling were fulfilled by software systems written in C and C++ (Monostori and Bothe, 1992). For fuzzy investigations the WINDOWS-version of the *fuzzy*TECH development system (INFORM GmbH, Aachen) was applied. Its main features are the following (*fuzzy*TECH 3.1 Explorer, 1993):

- graphical development environment,
- arbitrary curved and linear membership functions,
- standard MAX-MIN/MAX-DOT and advanced FAM inference methods with compensatory operators,
- individual rule weighting,
- rule blocks,
- various defuzzification methods (Center-of-Maximum, Center-of-Area, Mean-of-Maximum),
- off-line simulation with graphic tools,
- real-time code generation, etc.

Through the Neuro-Fuzzy Module of the system degree of support (DoS) for rules and membership functions can be trained automatically using a supervised learning technique which combines the idea of competitive learning and error back propagation (*fuzzy*TECH 3.1 NeuroFuzzy Module, 1993).

4. NEURO-FUZZY MONITORING OF CUTTING TOOLS

The main results of the investigations mentioned in 3.1 were the following:

- multisensor integration through ANNs,
- classification of wear states of cutting tools,
- estimation of the flank wear,
- incorporation of the cutting parameters into the learning and classification phases (Barschdorff *et al.*, 1993a, 1993b),
- inverse modelling of the cutting process through neural networks,
- application of inverse models for tool monitoring (Monostori *et al.*, 1992; Monostori, 1993).

The investigations and comparisons described in this paper refer to state classification of turning tools below constant process parameters (cutting speed, feed rate, depth of cut).

In these experiments patterns determined from signals collected between 0 and 0.2 mm flank wear were assumed to be generated due to cutting with a fresh tool. Patterns associated with a flank wear level greater than 0.5 mm were treated as belonging to worn tool category (two-class problem).

The right side of Fig. 4 shows the performance of a 6*6*2 (6 input elements, one hidden layer with 6 nodes, two output elements) back propagation (BP) network during the whole tool life cycle with patterns not incorporated in the learning set (Barschdorff *et al.*, 1993a, 1993b). Using 6 features of force components, the two network outputs produce relatively smooth signals during the life cycle remarkably in the middle period (flank wear between 0.2 and 0.5 mm) too, the period which was not represented at all in the learning phase. (Target elements of 0.1 and 0.9 were given to get faster convergence.) The network classifies the tool state into worn category, when output 1 is greater than output 0. (Using test data, high target values (0.9) were given only for flank wear > 0.5 mm.)

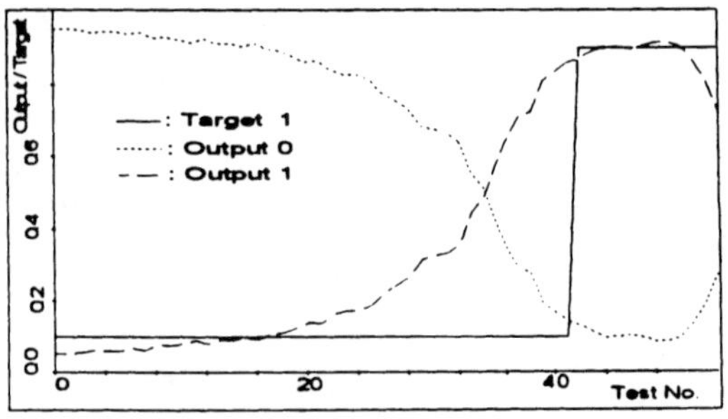

Fig. 4. Classification result with a 6*6*2 BP network, tool life cycle, target for class 1 (worn), outputs for class 0 (sharp) and for class 1 (worn)

To solve the same problem with the neuro-fuzzy approach, the fuzzy development system was set up as follows (Fig. 5).

The six force features were fuzzified using linear MBFs. In the investigations described here, only two MBFs were attached to every linguistic variable. Taking into account all variations, 128 (2^7) FAM rules were generated and their DoS were initiated with 0.5. MIN and PROD operators were chosen as composition and aggregation operators respectively. Center of Maximum (CoM) defuzzification method was selected.

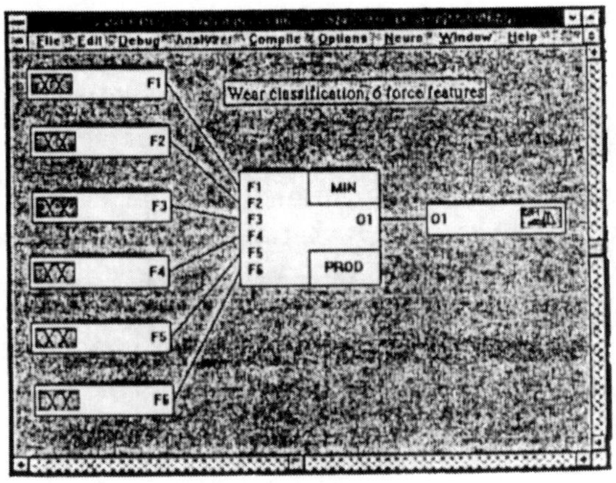

Fig. 5. Structure of the fuzzy system for state classification of cutting tools with 6 force features

Using the neuro-fuzzy module of *fuzzy*TECH, similarly to the ANN approach, very fast convergence was experienced. The most important rules, their learnt Degrees of Support (DoS), and the learnt MBFs for feature 4 are illustrated in Fig. 6.

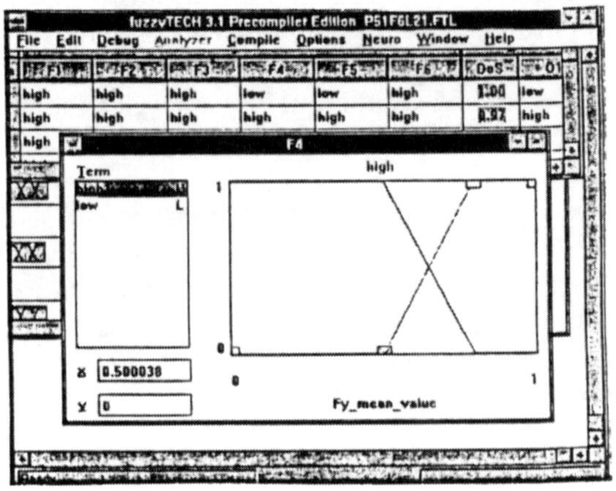

Fig. 6. Learnt MBFs and fuzzy rules with corresponding DoS values

The first rule on Fig. 6 can be interpreted as follows: *IF F1 is high and F2 is high and F3 is high and f4 is low and f5 is low and f6 is high THEN O1 is low, with a DoS value of 1.00.*

One of the most important questions is, how do the trained neuro fuzzy system perform through the whole life cycle of a tool (Fig. 7). (For the sake of easier comparison with Fig. 4, this figure illustrates the results, if the system had two outputs, one for the sharp, one for the worn state.)

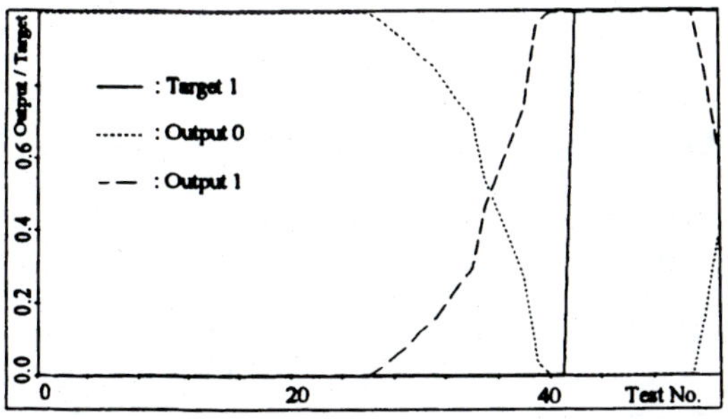

Fig. 7. Classification result with the neuro-fuzzy approach, tool life cycle, target for class 1 (worn), outputs for class 0 (sharp) and for class 1 (worn)

Comparing the results gained by BP networks (Fig. 4) and neuro-fuzzy technique (Fig. 7), one can conclude that the two approaches resulted in similar performance. On the one hand, classification with the trained neuro-fuzzy system seems to be more unambiguous, on the other hand, outputs of the BP networks follow more precisely the gradual increase of flank wear. Both of these phenomena can be attributed to the small number of applied MBFs.

5. CONCLUSIONS

The fundamental aim of the paper was to focus on some requirements of knowledge representation and processing techniques to be used in intelligent manufacturing. Artificial neural networks and fuzzy systems seem to be able to comply with these requirements. Having summarized the most important results of ANN applications in modelling and monitoring of manufacturing processes, the neuro-fuzzy approach was outlined and compared with previous results gained by ANN techniques.

The first results seem to be promising. Having limited the investigations to the state classification of cutting tools below constant operating parameters, the neuro-fuzzy technique presented in the paper showed comparable performance to the BP solution, with the additional benefits of *explicit knowledge representation* which is easily *modifiable* and incorporates a kind of *explanation facility*.

Further research was initiated to incorporate cutting parameters (cutting feed, feed rate, depth of cut) into the neuro-fuzzy approach, and to apply this technique for the estimation of the tool wear and for the inverse modelling of cutting processes.

In these further investigations, among others the following main issues are addressed:

- incorporation of more features (including process parameters),
- application of a greater number of MBFs which are not necessarily linear, and in connection with these,
- keeping the initial number of rules on a reasonable level by unsupervised fuzzy clustering,
- development of more complex structures using rule blocks.

As a conclusion, neuro-fuzzy systems as hybrid AI systems can comply with the fundamental requirements of intelligent manufacturing, i.e. *real-time nature, learning ability, handling of uncertainties and managing both symbolic and numerical information*. It is expected that the combination of neural and fuzzy approaches can provide sufficient framework for the solution of numerous problems in manufacturing, and can prospectively contribute to the future realization of intelligent manufacturing systems.

6. ACKNOWLEDGEMENTS

This work was partially supported by the *National Research Foundation*, Hungary, Grant No. T 014514 (Fundamental research for intelligent manufacturing). The cutting experiments were accomplished at the University of Kaiserslautern, 1992 during the stay of László Monostori at the Institute of Electrical Measurement, University of Paderborn with a Research Scholarship from the Humboldt-Foundation in 1990-92. The investigations of data with ANN techniques proceeded there (Barschdorff *et al.*, 1993a, 1993b). The author expresses his gratitude to the heads of corresponding Institutes, *Prof. G. Warnecke* and *Prof. D. Barschdorff*.

Thanks to the *PHARE ACCORD Programme* of the *European Union*, Grant No.: H 9112-0216 (Hybrid knowledge processing in CIM) which allows the continuation of the above scientific cooperation between the two German Institutions and the Computer and Automation Institute of the Hungarian Academy of Sciences.

7. REFERENCES

von Altrock, C. (1993). Fuzzy logic, Band 1, Technologie, *Oldenburg*, München, Wien

Barschdorff, D. (1992). Artificial intelligence: Diagnostic expert systems, fuzzy logic and neural networks, Proc. of the 8th *International IMEKO Symposium on Technical Diagnostics*, Sept. 23-25, Dresden, Germany, pp. 46-55.

Barschdorff, D., and L. Monostori (1991). Symbolicism and connectionism: rivals or allies in intelligent manufacturing?, Proc. of *The 23rd CIRP International Seminar on Manufacturing Systems*, 6-7 June, Nancy, France, Section 4, pp. 14-28.

Barschdorff, D., L. Monostori, T. Kottenstede, G. Warnecke, and M. Müller (1993a). Wear estimation and state classification of cutting tools in turning via artificial neural networks, Proc. of *Tooldiag'93, International Conference on Fault Diagnosis*, April 5-7, Toulouse, France, pp. 669-677.

Barschdorff, D., L. Monostori, T. Kottenstede, G. Warnecke, and M. Müller (1993b). Cutting tool monitoring in turning under varying cutting conditions via artificial neural networks, Proc. of *The sixth International Conference on Industrial & Engineering Applications of Artificial Intelligence & Expert Systems*, June 1-4, Edinburgh, Scotland, Gordon and Breach Science Publishers, pp. 353-359.

Chryssolouris, G, and M. Domroese (1988). Sensor integration for tool wear estimation in machining, Proc. of *The Winter Annual Meeting of The ASME on Sensors and Controls for Manufacturing*, Chicago, Illinois, Nov. 27 - Dec. 2, pp. 115-123.

Dornfeld, D.A. (1990). Unconventional sensors and signal conditioning for automatic supervision, Proc. of the *AC'90, III. CIRP International Conference on Automatic Supervision, Monitoring and Adaptive Control in Manufacturing*, 3-5. Sept, Rydzyna, Poland, pp. 197-233.

Erdélyi, F. (1993). Control of machine tools and discrete manufacturing processes, *Dissertation for the C.Sc. degree of the Hungarian Academy of Sciences*, Miskolc, Hungary

Goebel, K.F., and P.K. Wright (1993). Monitoring and diagnosing manufacturing processes using a hybrid architecture with neural networks and fuzzy logic, Proc. of *EUFIT'93, First European Congress on Fuzzy and Intelligent Technologies*, Sept. 7-10, Aachen, Germany, pp. 574-580.

Hatvany, J. (1983). The efficient use of deficient information, *Annals of the CIRP*, **32**, No. 1, pp. 423-425.

Jyh-Shing, R.J. (1993). ANFIS: Adaptive-network-based fuzzy inference system, *IEEE Trans. on Systems, Man, and Cybernetics*, **23**, No. 3, May/June, pp. 665-685.

Kosko, B. (1992). Neural networks and fuzzy systems; A dynamical systems approach to machine intelligence, *Prentice Hall, Englewood Cliffs*, New Jersey

Lin, C-H., and C.S.G. Lee (1991). Neural-network-based fuzzy logic control and decision system, *IEEE Trans. on Computers*, **40**, No. 12, December, pp. 1320-1336.

Minsky, M., and S. Papert (1969). Percetrons: an introduction computational geometry, *MIT Press*, Cambridge

Monostori, L., and J. Nacsa (1990). On the application of neural nets in real-time monitoring of machining processes, Preprints of the *22nd CIRP International Seminar on Manufacturing Systems*, 11-12, June, Enschede, the Netherlands, pp. 6A/15-27.

Monostori, L., and A. Bothe (1992). Convergence behaviour of connectionist models in large scale diagnostic problems, in *Belli, F. and Rademacher, F.J.: Industrial & Engineering Applications of Artificial Intelligence and Expert Systems, Proc. of the 5th International Conference, IEA/AIE-92, Paderborn, Germany, Lecture Notes in Artificial Intelligence, 604, Springer*, 1992, pp. 113-122.

Monostori, L., S. Markos, S. Krämer, J. Nacsa, and G. Szöllősi (1992). Neural networks for modelling and diagnosis of manufacturing processes; Milling, Proc. of the 8th *International IMEKO Symposium on Technical Diagnostics*, Sept. 23-25, Dresden, Germany, pp. 502-512.

Monostori, L., and D. Barschdorff (1992). Artificial neural networks in intelligent manufacturing, *Robotics and Computer-Integrated Manufacturing*, Pergamon Press, Vol. 9, No. 6, pp. 421-437.

Monostori, L. (1993). A step towards intelligent manufacturing: Modelling and monitoring of manufacturing processes through artificial neural networks, *Annals of the CIRP*, **42**, No. 1, pp. 485-488.

Negoita, C.V. (1985). Expert systems and Fuzzy systems, *The Benjamin/Cummings Publishing Company, Inc.*, Menlo Park, California

N.N. (1993). *fuzzy*TECH 3.1 Explorer, Manual and Reference Book, *INFORM GmbH Aachen*

N.N. (1993). *fuzzy*TECH 3.1 NeuroFuzzy Module, Reference Manual, *INFORM GmbH Aachen*

Rangwala, S.S., and D.A. Dornfeld (1989). Learning and optimatization of machining operations using computing abilities of neural networks, *IEEE Transactions on Systems, Man, and Cybernetics*, **19**, No. 2, March/April, pp. 299-314.

Rummelhart, D.E., and J.L. McClelland (1986). Parallel distributed processing, *MIT Press*, Cambridge

Wang, L-X., and J.M. Mendel (1992). Generating fuzzy rules by learning from examples, *IEEE Trans. on Systems, Man, and Cybernetics*, **22**, No. 6, Nov./Dec., pp. 1414-1427.

Warnecke, G., A. Jenewein, and A. Reinfelder (1990). Tool Monitoring based on Process Identification, *Monitoring and Control for Manufacturing Processes, ASME Winter Annual Meeting*, Dallas, Texas, Nov. 25-30.

Yager, R.R. (1992). Implementing fuzzy logic controllers using a neural network framework, *Fuzzy Sets and Systems*, **48**, pp. 53-64.

Zadeh, L.A. (1965). Fuzzy sets, *Information and Control*, **8**, pp. 338-353.

Zimmermann, H.J. (1987). Fuzzy sets, decision making, and expert systems, *Kluwer Academic Publishers, Boston, Dordrecht*, Lanchester

AN OBJECT-ORIENTED SPECIFICATION OF A CONTROL LOGIC FOR AN FMS

JAROSLAV FOGEL, MIROSLAV KOCIAN, JOLANA SEBESTYÉNOVÁ

Institute of Control Theory and Robotics, Slovak Academy of Sciences, Bratislava, Slovakia

Abstract. The paper concerns with the application of object-oriented modelling to the development of real-time control software for FMS. Object- oriented model integrates the features of object-oriented database, features of real-time system and features of artificial intelligence aspects. The goal is to develop intelligent decision making logic for automatic creation of the control actions in both the standard and the nonstandard (error) situations. The knowledge is represented by the rules and the constraints. Three type of rules are described in the paper: process rules (create the dynamic model of the devices), production plan rules (define the production sequence of the technological operations), and expert knowledge rules (create reasoning model for solving the fault situations). The cooperation among these rules is also discussed in the paper.

Key Words. Flexible manufacturing; knowledge engineering; object-oriented database; production control; modelling

1. INTRODUCTION

Last years, the scientific community's research focuses on the new perspective modelling tools for DEDSs, which may be used in FMSs, with a profit of the advantage of the OOP. FMSs are large and complex systems with a modular and hierarchical structure, and with a requirement for a bottom - up design or analysis, i.e., from the basic autonomous modules of the production equipment up-to the complex modules of the workshops on the higher level. The OOP is very suitable for these purposes. Besides the complexity, the basic characteristics of FMSs are: a parallelism, an asynchronous occurrence of the discrete events and a cyclic repetition (Buzacott and Yao, 1986). Besides the mentioned modularity and the hierarchical structure, the OOP enables the modelling of the static and dynamic relations between the state variables of the objects. The state variable expresses the abstract notion of a state and its meaning is given on the sublevel of the specialized objects (Attoui and Schneider, 1991).
Generally, the main problems of the paper are related to effective representation of object and data relationships in application area, to suitable specification of required goal (control) and possible dynamics of active objects (devices). These problems are interrelated but for better understanding they will be considered relative separately and then approach as a whole will be presented.

Effective representation of (physical) objects and data consists in making abstraction of them into suitable defined classes. There is also need for defining more abstract classes which are not directly related to physical objects but they reflect abstract notions in application area. Some of them are mentioned later, during explanation of main ideas of the paper.

2. OBJECT-ORIENTED DATABASE

Object-oriented database management systems (OODBMS) represent one of the most promising evolution in the database area. The design of an OODBMS implies the integration of database technology with object-oriented technology. It is needed when to share complex information, like in such applications as expert/knowledge based systems, CIM, decision support information systems.

During analysis and in low-level design, objects are chosen as basic run-time entities in the object-oriented system. A set of procedures and functions working on them is joined to each object. This means that the object comprises data and behaviour. The objects are organized into classes. Inheritance is an important reusability mechanism, and so it avoids code redundancy. It is a powerful modelling tool, because classes can be refined in several steps. It supports a concise and

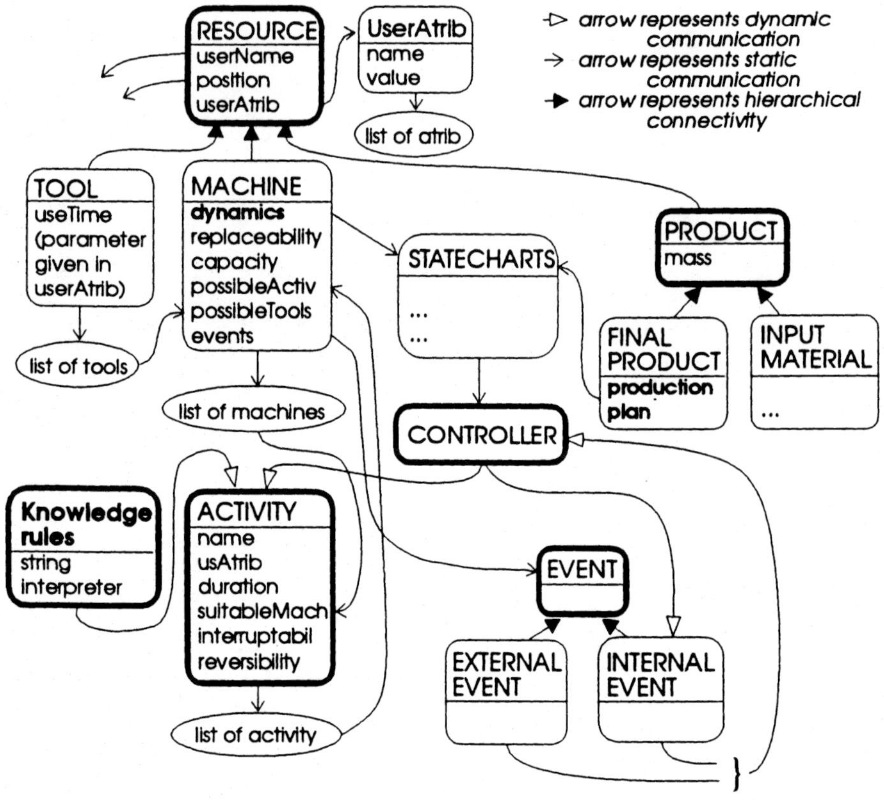

Fig. 1: OODB schema (restricted)

better organized representation of the database schema.
OODBMS provides library of classes, which are application-specific data types. The components of an FMS, that have to be described in an ODB, fall into two basic groups: the resources and the activities. The resources consist of the active and passive equipment (e.g., machines, their tools, ...), as well as all material entering the production process, and the products (material flow and output).
Objects contain data, member function, knowledge rules, constraints. About class hierarchy and class definitions (attributes and member functions) see (Fogel and Sebestyénová, 1993). The user can extend the set of types (for now only in restricted manner, but in further this facilities will be generalized in schema definition language). Fig.1 shows the knowledge distribution in DB.

Objects often represent real-world entities and so they must be able to mirror evolution of real-world entities. Object evolution may imply that an object changes class. Object migration may be dangerous for DB integrity. This must be taken into care when production plan of the final product is broken into subgoals (for the cyclic repetition).
Objects may be accessed by pointers or by external names (user-defined names provide users with semantically meaningful handlers to objects) or associative access through query language (query language is crucial when dealing with a large number of objects).
Database programming language has to support persistence of objects. Object deletion is not based on garbage collection performed by the system, but objects are explicitly deleted by users. Furthermore it is possible to change parameters of object and in this cases the system recreates the object.
OODBMS provides some lists (e.g., list of objects which represent the workshop production in it has to be controlled, list of activities which may be carried out on the machine, list of machines which are able to do some activity, ...).
These lists of objects are stored. The user (technologist) may read and recreate them by the requirements and store the recreated workshop definition.

3. RULE GROUPS

3.1. The production plan and the dynamic resource model

The two basic groups of rules are the production

plan (or task) and the process knowledge rules (or model of possible dynamics of resource).
The production plan reflects the goal of manufacturing process and it represents the required sequence of operations on input materials or semiproducts to produce the desired final product. It enable to concentrate on the qualitative or quantitative state variables of materials under processing and required operations on it without detail consideration of resource (device) state variables.

The model of possible dynamics of the resource is the dynamic model of the state changes of it. This model contains state variables of the resource and their changes as the results of starting and ending of resource technological operations, or as the results of such events like failures of the resource. The model can be relatively more conservative part of the design specification of the manufacturing process because the model is not directly coupled to specific final products.

The operation in the production plan has directly or indirectly (in the database) specified the context of the operation, i.e., the resources that can carry out the operation. If the operation is required in the production plan (it has fulfilled the starting conditions related to the state of materials under processing) then it is necessary to look into dynamic models of devices in the context of the operation. There are the conditions that have to be fulfilled before the start of the operation is possible. As the result, the operation can start, if it is required and possible.

The above presented idea is only the base for the additional consideration. There are many additional features, problems and alternatives, where to place, when and how to use certain kind of knowledge in the database.

The parametrization.
If considering reusability of the dynamic resource model, then it is an advantage to use parameters in this model, for which production plan can give actual values.
This is one of ways how to effectively manage the complex state/transition model of the manufacturing process. For instance, the dynamic model of the drilling machine can contain variable standing for the parameter of drilling tool. The required parameter of the drilling tool from the production plan is passed to the dynamic model of the drilling machine where it replaces the related variable.

The complementation of necessary operations.
The production plan can contain only the main operations through the input materials have to pass in order to create the desired final products. In the case that starting conditions for the operation are not valid, it is possible to look into the dynamic resource models with the goal to find such operations (using backward chaining) that can make them valid.
But there is also the requirement to consider the case when nonvalid starting conditions are postconditions of previously started operations, then no new operations are required and the system simply have to wait. Knowledge of possible operation postconditions in the database is useful for recognizing this case.

The context.
Each part of the database contains fully specified primary information specific to this part and explicit or implicit specifications of the other parts of the database that contain additional information that have to be taken into account together; these additional information together with the interpretation of their connection to the primary information represent the context of the primary information.

The set of production plans.
The goal of the manufacturing process can be represented by the set of production plans (the set of final products). The inputs into the production plan are input materials of the process and/or the produced outputs (the final products) from the other production plans. By this way the global final product is represented by the tree structure whose branches are production plans.
The production plan represents single pass of the input materials through manufacturing operations to produce the desired final products but in addition, there is required cyclic production of the final products from batches of the input materials.

The resulting control Statechart model.
Both, the set of production plans and the set of dynamic models of devices are expressed in terms of the Statechart formalism (Harel *et al.*, 1987). These two sets are combined to the resulting Statechart model reflecting how possible operations of devices are used in order to produce the desired final products. The resulting Statechart model is the parallel model.
The parallel parts can reflect the working cycles of devices or the concurrent existence of the objects of the same product class on the different level of processing.
In any case the resulting Statechart model is produced by combining the states, the transitions and labels from the production plans and the dynamic models of devices. During this process there is the need to divide the production plans into parallel parts in order to respect the concurrent existence of the object of the same product class on the different level of the processing.
The resulting off-line produced parallel Statechart model can be used during control of the manufacturing process. There are some problems

and alternatives in the producing and using the resulting Statechart model.

During creation of the resulting Statechart model, the conditions that stand for the cyclic repetition have to be added. It must be emphasized that, the process of the creation of the resulting Statechart as well as its use for the control is more complex that was indicated above.

The process of its creation is more complex because of the above mentioned features of the approach (the parametrization, the context, the complementation of operations) have to be taken into account in the whole and other groups of rules mentioned later in the paper have to participate in the process of the creation of the resulting Statechart model (e.g., push rules).
During control, the complexity is caused by the fact that the resulting Statechart model is used in the combination with information originally specified in the database because some knowledge and some related problems have to be taken into account on-line. Some problems cannot be solved off-line or it is not suitable to solve them so because it is not practical take into account all the possible states or situations in the process off-line.

3.2. The coordination and synchronization problems

The starting conditions for the operations in the production plans and in the dynamic models of devices are related basically to state variables of materials under processing and state variables of devices as individuals. But there is necessary to use properly the synchronization variables in order to solve problems like the mutual exclusion, the deadlock avoidance and respecting capacity constraints.
These problems are caused by the concurrent existence of the objects of the same product class (with the same production plan) on the different level of processing, the concurrent activities of devices and using of devices as shared resources.

The notion of the context is useful for **the problem of capacity constraints** and **the mutual exclusion requirements**. The information on the capacity constraints and the mutual requirements of the resource is placed in the database as the attribute of the resource, but it is used whenever the constrained resource is in the context of the operations in the dynamic models of the other devices.
It have to be recognized that, the constrained resource is in the context of that operations, and test and update of its state variables (as well as proper use of synchronization tools like semaphores) are added to the preconditions and postconditions of that operations.

Generally, the problem of resources allocation is more complex because there are also problems of the optimization of the resources allocation.

The theoretical framework of **the deadlock avoidance** is more complex. It is possible to identify off-line causes of deadlock situations and to use synchronization variables in order to avoid deadlocks, because possible behaviour sequences are known in advance in the manufacturing process. But in some cases many synchronization variables have to be used, because all possible sequences are taken into account.
Also it is possible to combine off-line analysis of the system behaviour and to use on-line allocator in the following sense.
When the resource allocation is requested, the information on the future requests before releasing the requested resource is passed to the allocator. The allocator have to consider previous requests and to enable or deny the current request. Generally, the coordination problems especially deadlocks need special attention and some problems are subject of the future research.

3.3 The other groups of rules

The push rules.
There are terms like push, pull and just-in-time control strategies. In the push control, the operation is started each time when the operation is possible and in the future the outputs of it will be required. In the pull control, the operation is started when the outputs of it are required and the operation is possible. The just-in-time control can be seen as a combination of the push and pull policies.
From this point of view the production plan is more related to the pull control.
If the preconditions of an operation are not valid, their validity is pulled into the process by carrying out some other operations. It takes some time to carry out the operations that make these preconditions valid. As a result, it is suitable to enable the push strategy of the control.
The push rules are considered to be the special separate group of rules. The form of individual rules (IF-THEN or EVENT-CONDITION-ACTION) can be used with the following interpretation - if the preconditions are valid then operations on the right side are required to carry out.
From this point of view, the push rules are also related to the production plan, because they express certain goal of the process but in the different sense than the production plan. From the implementation point of view, they are related to system devices and they have to be coupled with events and conditions on their left sides in order to make the process of finding the rules with valid

preconditions more effective.

The nonstandard situation rules (knowledge rules for failure processing).
The following groups of rules are connected to nonstandard situations (behaviour) in the manufacturing process. Generally, nonstandard situation rules are nonstandard in the sense that they are used only under some special conditions. While the behaviour of the manufacturing process is standard, there is no need to use information related to the nonstandard behaviour and so there is an advantage to specify this information separately. The problem is where to place, when and how to use this information.

The failure processing is a kind of nonstandard situations. The failure processing (detection, analysis and recovery) can be explicitly stated as a part of the production plans and the dynamic resource models. But for more structured view of the underlying manufacturing process, it can be convenient to specify separately normal control goals and the failure processing.
In the presented framework the failure processing is used as an example of nonstandard situations in the following way. First, the part of the dynamic resource model is related to the failure processing. This is an integral part of the model expressed in the Statechart formalism but it is used only in the case of failures.

Next, the new group of knowledge rules expressed in the form of individual rules is attached to the resources. The form of individual (but related) rules can be considered for this purpose more convenient because of certain declarative nature of these rules. From the set of rules, the set of state update and/or recovery sequences can be derived, even such sequences that are not directly foreseen by the designer of the rule set (base).
These groups of rules are used in the case of explicit demands (through a special keyword or after timeouts) or in the case of implicit demands.

In the latter case, they are used when some desirable substate of the production plan cannot be reached (conditions are not valid), and operation that can make them valid are not possible or they are not found in the part of dynamic resource models that is related to standard behaviour of manufacturing process.

The additional useful characteristics.
In addition, there are many special characteristics of objects (classes) in manufacturing systems that can add useful information for the state space consideration. Some of them (e.g., reversibility of an operation, interruptibility of an operation) can be used in the sense that they add some new states and transitions into the state space of the underlying manufacturing process.
The specification of transitions in the state space that is related to technological operations is simplified by the identification of common properties of technological operations of the same kind.

4. EXAMPLES

The functionality of the rules in the production plan together with the rules describing resource dynamics can be illustrated in the following three examples.

Example 1:
Some postconditions of the technological operations scheduling from the production plan as well as some preconditions can be stated by the system means automatically, e.g., the execution of the operation get_item from the buffer has to set the new value of the variable buffer_capacity and in the case of empty buffer the operation putting of new item into the buffer has to be started. This can be reached introducing the following transitions into the state/transition dynamic model of the buffer.
If the state of the buffer is nonempty the event starting of get_item operation causes call the decrementing function from the class buffer automatically. In the case of empty buffer (buffer_capacity = 0) it results in the triggering the transition to the state empty_buffer with the simultaneous start of the operation to fulfil it, e.g., moving operation of the transporter system carrying items. This event causes again the change of the state_variable of the transporter system from the idle_state to busy_state automatically. After the external signal about finish of put_item operation arises, the incrementing function from the class buffer is called automatically and the transition to nonempty_buffer is triggered. This same event causes triggering the transition to idle_state in the transporter dynamic model too.
In the case that buffer is shared by several devices, the situation is more complicated. The problem of mutual exclusion in the common space must be solved (see example 3), because the object buffer can receive concurrent requests from the clients nondeterministically and thus the consistency of the object state must be controlled.

Example 2:
During milling operation the external signal broken_tool entered into the system. In the dynamic model of the milling machine this event causes triggering the transition from active_state to faulty_state with simultaneous invoking the operation tool_change. We suppose in this example, that this operation can be performed by the robot. In the state/transition dynamic model of the robot, the transition from the idle_state to the repair_state is triggered as a result of the

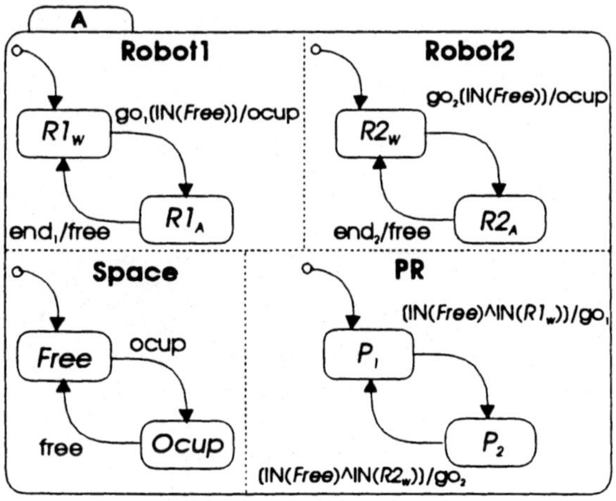

Fig. 2: Mutual exclusion

event start of tool_change operation. In the case of working robot (this is the chance because this event occurs nondeterministically), the event will enter into the queue of waiting messages that is declared in the class robot. The items can be ordered in the queue by some strategy, e.g., FIFO, priority, etc. In the last case, the priority of the event must be declared in the class event.
After finish the current operation, the robot will try to fulfil the waiting demands, preferable the first message in the queue.

Example 3:
Two robots perform operations in the common region (collision space). It is necessary to secure their mutual exclusion in this shared space. The solution of this problem is given in Fig.2, where Robot1, Robot2, and Space are state/transition models of two robots and collision space respectively, and PR is the production plan. Concurrent activities of these models are expressed in the Statecharts syntax as orthogonal (AND) state A decomposed into four substates (OR states) Robot1, Robot2, Space and PR.
Where $R1_W$, $R2_W$ are the waiting states of the robots R1, R2 respectively,
$R1_A$, $R2_A$ are the active states of the robots R1, R2 respectively,
The transitions in the diagram are labelled by e[c]/a, where e is an event c is a condition and a is an action,
IN(*state*) is the condition of the model being in the given state,
go_1, go_2 are the actions of starting the moving operation of the robots R1, R2 respectively,
end_1, end_2 are the external signals about finishing of the movement operation for the robots R1, R2 respectively,
occup and free are events for locking and unlocking shared space respectively.

In the given examples, it was illustrated how the system handles the retrieval and routing of events, and also supplies default processing of ones automatically. The implementation of this ability corresponds to user-defined messages in Object Windows, and it contains the definitions of the response functions on the events, which are unique to given application in the corresponding object's classes. What must be specified for the given application, it is the declaration of events in the corresponding classes and their connections to the response functions in these classes. The semantics of the object processing events is similar as the active objects in (Attoui and Schneider, 1991).

5. CONCLUSION

In this paper the object-oriented database that integrates data modelling, artificial intelligence and real-time control aspects is presented. The database design is problem oriented to control of FMS and it is determined for the control synthesis and analysis in the stage of simulation.

The authors are grateful to the Slovak Grant agency for science (grant No. 2/999468/93) for partial supporting of this work.

6. REFERENCES

Harel, D., A. Pnueli, J.P. Schmidt, R. Sherman (1987). On the formal semantics of statecharts, *Proc. 2nd IEEE Symp. Logic in Computer Science*, New York, pp. 54-64.
Fogel, J., J. Sebestyénová (1993). An object-oriented concept for real-time control of an FMS, *Control Engineering Practice*, Vol. 1, Number 4, pp. 691-698.
Attoui, A., M. Schneider (1991). MO2: An object-oriented model for database and real-time applications, *East EurOOPe'91 Proceedings*, Bratislava, pp. 105-115.
Buzacott, J. A., D. D. Yao (1986). Flexible manufacturing systems: A review of analytical models, *Management Science*, Vol. 32, Number 7, pp. 890-905.

FACTORY COMMUNICATIONS FOR AUTOMATING A CIM CENTER EXPERIMENTAL PLANT

J.M.FUERTES*, J.OJEDA** and E.ARTIAGA**

*Departament d'Enginyeria de Sistemes, Automàtica i Informàtica Industrial, Universitat Politècnica de Catalunya, Pau Gargallo 5, 08071 Barcelona, Spain.
**Centre CIM (UPC/ICT) Universitat Politècnica de Catalunya and Institut Català de Tecnologia, Adolf Florensa s/n, 08028 Barcelona, Spain.

Abstract. Development of networking solutions in a factory needs not only specific links to connect individual machines but a framework to organize and encompass architecture, structure and protocols of the communication system. The CIM Center is an experimental pilot plant where this, among other Computer Integrated Manufacturing (CIM) concepts, are used and demonstrated. This paper presents the results obtained at the CIM Center (UPC/ICT) when using Information Technologies for Communications at the factory level and how they are impacting plant flexibility and productivity.

Key Words. Factory information technologies, computer integrated manufacturing, factory automation, integrated plant control, industrial local area networks, communication control applications, manufacturing message specification, systems enabler platforms.

1. INTRODUCTION

The conceptual Computer Integrated Manufacturing (CIM) architecture model integrates management, engineering, and planning systems with the devices that control the physical operations on the production lines. To create a fully automated manufacturing environment, it is required having communication links between all the machines, devices, and computers used in the factory. Manufacturing plants (the CIM Center for example) usually have a variety of workcells, each composed of machines and robots from different vendors. Corporate computers and engineering workstations in such an environment have to transmit orders, programs and supervisory data to factory devices, robots, NC machines, and workcell controllers. But no single vendor actually supply all the factory devices, and the proprietary vendor components can't easily communicate with other vendor equipments. Establishing the communication links environment is not an easy task, because the machines have to be well conditioned for powerful communication tasks, and such tasks are not usually well organised enough in the majority of the present industrial equipment.

An experimental pilot plant for Computer Integrated Manufacturing (CIM Center, UPC/ICT) is in development at Barcelona as a collaborative effort between the Polytechnic University of Catalonia and the Technology Institute of Catalonia, Ferré (1991). The CIM Center is a pilot manufacturing facility where automation manufacturing technologies can be tested in a real setting, with relevant industrial manufacturing machines and equipment. The center is open to all technologies, systems and products included in the CIM concept.

The center provides:

- Services related to production and automatization techniques.
- Support for research and development activities.
- Training of professionals and students in the fields of CIM.
- Product exposition and application demonstrations.
- Development, testing of products and applications.

The CIM Center is organized into three main areas: Management area, Engineering area, and Production area (Fig. 1). The Management area is responsible for the global center management and direction tasks, and has the support of MAPICS-DB software package and some work-site office terminals. The tasks that are carried out in the engineering area are mainly concerned to CAD, CAE and CAM applications, by using software packages (CATIA, AUTOCAD, NC-POLARIS, ORTHOCAM, TEBIS, ME30, ANSYS.) on advanced workstations. The production area is organized into a set of production cells, each one concerning some specific operation of the plant: material storage and moving, manipulation and manufacturing, and measuring.

The communication system is a key element into the CIM architecture. Through it, all the elements, machines, devices, computers, operators, and users can exchange plant and center information by means of a distributed management information storage and retrieval system.

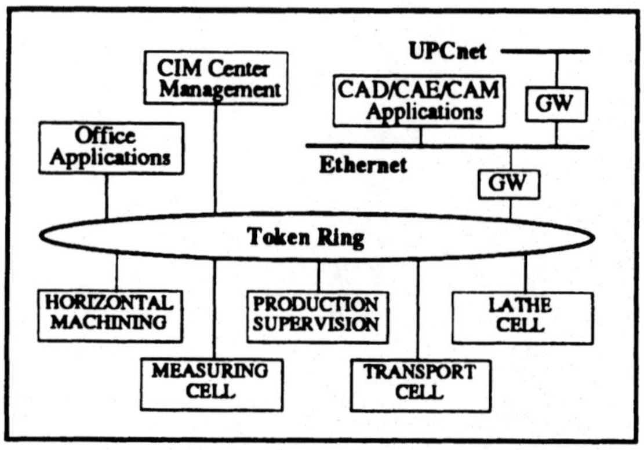

Fig.1 CIM center systems architecture

During the center initial installation phase, it was organized a communication system divided into three areas with different networking solutions: A HP HUB Repeater was used to link the external UPC-Net University 10-base-5 Ethernet network to the CIM Center internal 10-base-2 Ethernet network, with TCP/IP support; this last section interconnects Unix based engineering workstations for CAD/CAE/CAM applications. Through a Router on a Unix based computer (IBM RISC 6000), equipped with two communications adapters and routing software, the Ethernet section interconnects with a IEEE 802.5 Token Ring section, which provides communication to the center and to the plant management computers.

Within the management area, there is the center management mainframe (an IBM AS/400), linked to personal computers through the APPC (Advanced Program to Program Communication) protocol. The other computers of this area (DOS based and OS/2 based personal computers), such as those used for application development and for managing production databases, use APPC and NetBios communication protocols while TCP/IP services allow connection to the Unix based systems of the engineering network.

The production plant area communications were based on a star configuration, Fuertes (1992). The control computer (an IBM PS/2 80) used a Real-Time Interface Coprocessor RIC-ARTIC) to handle up to eight serial lines. The different plant machines were connected to the control computer via RS-232 or RS-422 point-to-point communication links. The control computer was connected to the Token Ring network, for communicating the center management computer with plant devices, (Fig.2).

The solution which has been described was an easy way to start the operation of the CIM Center production plant. Nevertheless, it was clear that this couldn't be the final solution. The production plant area communications have been sensitive to several facts, being the most relevant the difficulties to link machines and computers in a way not well enough established, normalised, or with poor connection interfaces. The star configuration cannot be always expanded with new devices as the number of ports handled by the RIC card is limited. Besides, the concentration of all the control software in a single computer could easily overload it, putting response time requirements in jeopardy. Finally, it's easy to see that this was not a very reliable model, because a system crash in the control computer would be affecting the whole production plant.

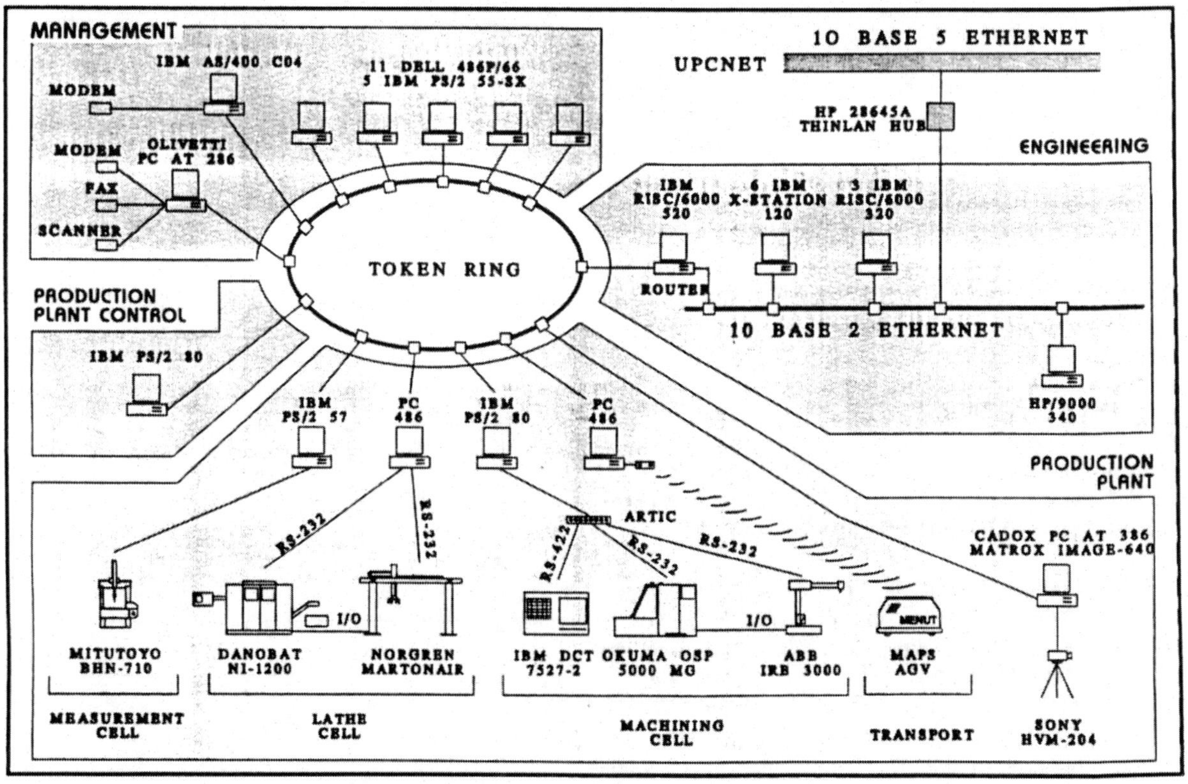

Fig.2 Communications systems structure of the CIM Center

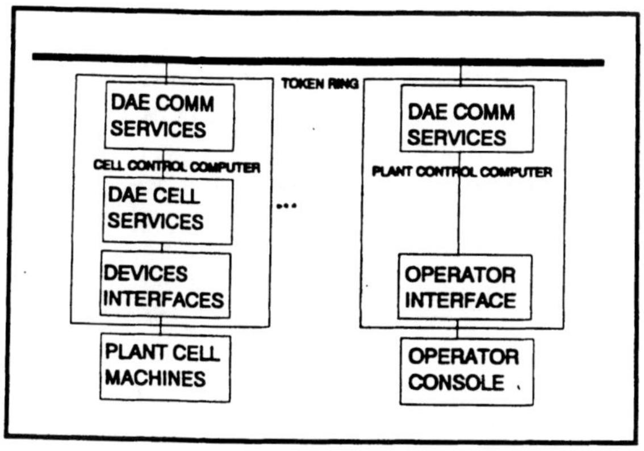

Fig.3 Plant communications architecture

So, as the CIM Center was growing and the number of machines was increasing, many reasons made the plant communications system evolve to a new structure. Some modifications affected the token ring network. Several *logical* networks (with various protocols) are now running over the physical Token Ring. The office network uses Novell Netware (based on ODI - Open Data-link Interface) to connect PCs to a server, which provides access to commonly used application programs. One of the office computers runs SCO (System V Unix), providing remote access (via modem) to relational databases at the ICT sites.

2. PLANT COMMUNICATION SYSTEM

2.1 Communications System Architecture

The actual communications system architecture is the result of the evolution of the previous one. The idea was not to dispose of existing material (hardware and software) but to use it in a new way.

The most important change affected the production plant network. The new communications system distributes the weight of the plant control among several computers, using a modular configuration which allows the easy growing of plant equipment (Fig.3). The production plant has its own *logical* network integrated into the global Token Ring network. There is a computer controlling each *work cell* in the plant (a *work cell* is a collection of plant devices that make related manufacturing operations).

Furthermore, there is one computer for production plant control, doing tasks which affect several cells. The different plant devices are connected to their corresponding control computer through conventional serial ports, RICs or vendor specific communications media, depending on the available options.

Using a different control computer for each work cell has advantages: Cells can work in autonomous mode if needed. This allows retaining control (or, at least, doing a 'smooth' stop) of the plant devices when the communications system fails. Moreover, it is now possible to *logically* disconnect certain cell from the plant network for trouble-shooting or testing new configurations, without affecting the rest of the plant while keeping the full functionality provided by computer software. Additionally, the cell can be provided with a higher degree of *intelligence*. In this way the ability of each cell for doing complex jobs is increased and lets the global control system be simpler by distributing the control tasks.

2.2 Communications System Functionality

The normal operation process of the plant system is as follows: the center management programs produce a Production Command list, which is validated by the responsible for the management area. The information is transmitted to the production plant control computer, which distributes it to the cell controllers.

During the production process, the different cell computers and the plant control computer are mutually communicating. The communication procedures for the integration of the plant cells is mainly organized around the Distributed Automation Edition (DAE) system enabler. DAE provides a set of programming interfaces and services which isolate application programs from the differences between computers, operating systems, networks, devices, machines, and communication protocols, on which the applications run. DAE services provide system function for transforming data, program to program link and synchronization, device, database or operator data transfer without regard to the physical location of the peers. As an example, if a cell controller is sending a request for the AGV, the message is sent to a *logical device* (in fact, a message queue) and the requester is not concerned about which system is reading the information, whether it is the transport controller or the plant control system (which will be sending another message to the real transport controller).

All the relevant information related to the manufacturing process (production data and different events which may occur during de process) are saved in local files and transmitted, at the same time, to the plant control computer. This fact allows to supervise, in real time, the whole plant status. Finally, the plant control computer provides data to the management computers, which obtain more elaborated data (such as statistics), and feeds-back the planning system. This last task is not considered as time-critical, and may be done with less priority for not disturbing the real time processes. Results are stored in databases and, as required, programs are transmitted to cell control computers at the production network. A copy of the most commonly used programs may be stored locally into the cell computer memory.

2.3 Communications System Structure

The system is structured in two main levels: at the bottom of the structure there are the individual cells and above the cells there are the plant control and management systems.

The plant machines are at the lowest level of the structure. At the CIM Center there are devices from different vendors using proprietary communication drivers. PLCs and control devices for these machines are connected to control computers via RS-232 and RS-422 lines and, in some cases, using special communication systems provided by the vendor.

Serial communication lines arrive to the computers either through conventional computer serial ports (COM*x*) or serial multiport cards or using real time communications coprocessors (in the CIM Center we are currently using a RIC Multiport/2). The choice of one of those methods depends on several factors, such as the amount of data to be interchanged with the machine, communication costs (i.e. costs of data packetizing, and of managing a complex protocol) or, simply, the availability of the proper software or hardware. For example, when using RIC-ARTIC cards, a machine specific CPP (Communications Protocol Program) is required; CPPs are device-specific routines to communicate them with general communication services and, in particular, to DAE services.

When information reaches the cell control computer, it is necessary to make it accessible by the whole production system. This is done by means of a System Enabler software package. This software offers common interface services to integrate different systems and resources. Furthermore, communication services offered by the system make possible the interconnection in a single network of computers using different protocols (APPC, NetBios, TCP/IP) and running different operating systems (DOS, OS/2, Unix). The way information is communicated to the system enabler depends on the special configuration of each cell.

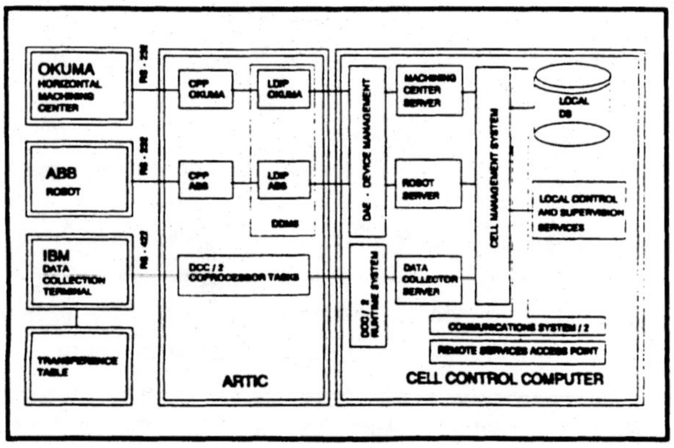

Fig.4 Horizontal machining cell internal structure

3. PRODUCTION PLANT CELLS

The cell structured production plant in the CIM Center is composed by: an horizontal machining cell; a lathe cell; a measuring center cell; and a transport cell. An automated storage area cell is actually in the installation phase.

The horizontal machining cell consists of:

- OKUMA horizontal machining center, with four axis, automatic change of tools and double pallet.
- ABB IRB-3000 six axis robot.
- IBM DCT 7527 Model 002, a data collector terminal with touchscreen, magnetic badge slot reader and a DIDO port
- A machining transference table.

The horizontal machining center and the ABB robot (Fig.4) are communicating with the RIC via RS-232 lines. Two special tasks running in the coprocessor (the CPPs) manage all the communication aspects (protocol, error handling, retrying). Those programs communicate with their respective LDIPs (Logical Device Interface Procedures), which are integrated in DAE DDMS (Device Data Management Support). LDIPs provide a logical device interface which allows performing common operations without being concerned about the real physical device attached to the computer. This means that the same command can be send to start both the machining center and the robot, since the LDIP performs the proper command translation. In case of changing one of the machines, only CPP and LDIP should be replaced, and system could go on working with minimal changes (just those related with device specific commands).

The transference table is connected to the Data Collection Terminal (DCT) through a DIDO (Digital Input, Digital Output) port, while DCT is linked to RIC via RS-422 serial line. The DCT is controlled by proprietary software (IBM DCC, Data Collection Control). This software provides control and supervision facilities for the DCT.

In the system unit, each device is controlled by a *server*, which can manage concurrent access to devices and offer higher level interfaces. Furthermore, they are used to link local management systems (DDMS, DCC/2) with system enabler communication services, allowing remote access to devices (those links are not explicitly shown in the figure for clearness).

The cell management is above the device servers. This is the main process in the control computer, which receives orders from the plant control computer, and redistributes them to the different cell machines. The process communicates with device servers via DAE API. Other tasks made by the program involve coordination (and synchronization) of the several devices, maintenance of the local database, logging and alert handling. Human operators may use *Local Control and Supervision Services* to access the cell management system. This program allows to perform control tasks and real time monitoring within the cell control computer. Such program uses the same interface to communicate with the cell management system that remote control services (e.g. those used by the plant control computer) use.

The lathe cell has the following elements:

- DANOBAT lathe center with five axis, double tower, four motorized tools, tool wear control, and an automated bars feeding system. The NC is a SINUMERIK 850-T, and the communications PLC is SIMATIC 130 W/b, both from Siemens.
- NORGREN-MARTONAIR linear portal robot with a local pallet store and with a Siemens SIROTEC ACR control.
- Transference table equipment.

The internal structure is very similar to that one described for the machining cell. However, some differences must be pointed. First, this cell don't use any coprocessor (because the required software is not available for those plant machines). Serial lines arriving from plant devices are connected directly to computer serial ports. This configuration implies that device servers have extra code to handle protocols and possible errors. Accessing ports is not a problem since OS/2 offers complete support. In this cell, digital inputs and outputs are used for direct communication between the lathe center and the portal robot.

The measuring cell (Fig.4) is composed by

- MITUTOYO BHN-710 Measuring Center

This cell is structured in a different way. Communication between device and control computer is carried out through special communication card provided by vendor. Control and supervision system is implemented in a single program that runs over DOS. The program is structured in different parts: device server, user interface and communications subsystem. Integration in the production plant network is carried out through the system enabler (DAE/DOS) which provides communication services.

The transport cell is composed by

- MAPS Automatically Guided Vehicle (AGV).

The control program provided by MAPS runs over DOS. This program will be connected to DAE/DOS services to join the rest of the network.

4. PLANT CONTROL SYSTEM

There is a plant control computer coordinating the different cells. This computer has the following functions: Coordination of the different cell operations, distributing the commands to the different cells in the plant; Knowledge, at any time, of the status of the different production commands, for decision taking. Supervision of the plant global status, taking the necessary decisions in case of alert. Providing communications support to the cell computers which don't have full communications capabilities (e.g. those running DOS).

The main process in the control plant computer is the Plant Supervision and Control System. This system is connected to the management and planning systems and with cell control computers through a system enabler. There is a subservient process to help in this task (Cell Support Procedures), needed to get a higher degree of transparency, since some communication services are not provided for DOS. The system enabler does the necessary tasks related with networking and protocol managing.

Running above the system enabler, there is an application enabler (IBM Plantworks). This system provides support for designing and executing control and monitoring applications. Plantworks is fully compatible with DAE and it has been used for the operator interface at the plant control computer. The information required by the supervision and control system to run is stored in local databases. This information includes a list of production commands, log files and performance data. Data available to the supervision and control system can be accessed or maintained by the management and planning systems.

The control and supervision applications at the center are developed in a generic form, creating client-server associations, and using an Application Program Interface (API), Brill (1991), to link with the present plant systems, but maintaining an open conceptual framework for facilitating future system growth or machines and controllers modifications.

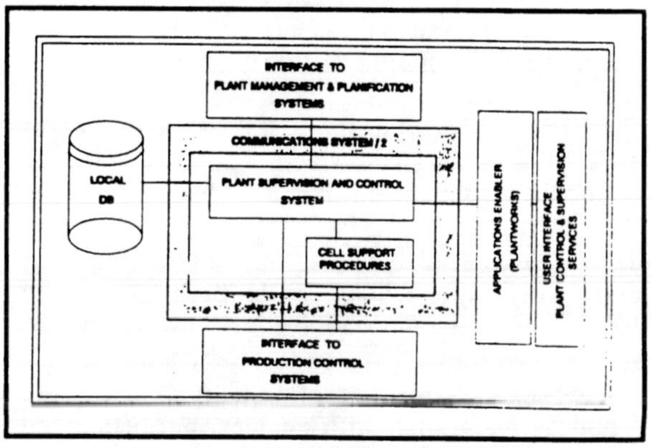

Fig.5 Plant control computer structure.

The control application programming style is based on the MMS (Manufacturing Message Specification), initially developed into the MAP (Manufacturing Automation Protocol), MAP (1988). MMS acts like a factory automation language for exchanging messages and instructions between plant devices. MMS applications development recommend and facilitate a disciplined, structured and well documented systems development style.

DAE allows the use of MMS style programation for manipulating and using remote information. The plant applications are based on the following methodology: Definition of sets of objects, sets of services for manipulating those objects, and packetizing them into VMDs (Virtual Manufacturing Devices), that serve as models to represent the behaviour of real devices. It is the local device or machine (controller) responsibility the mapping of the corresponding VMD to the real plant device, Fuertes(1993). Standardization efforts are inducing device manufacturers to adopt common interface standards.

5. CONCLUSIONS

The CIM Center gives a solution for the integration of all the plant devices to a common framework that is giving full operation and flexibility. The paper covers the elements to implement factory advanced automation strategies and to allow the application of intelligent functions at different levels of supervisory control of the plant. But even when the solution presented here is a working solution, the current tasks in this area is being oriented to define and use standard common services for all the levels of the factory automation system. Such normalizations are actually a *hot potato* and the most advanced automation centers, factory automation manufacturers, and normalization offices are (slowly but positively) working for obtaining long term commitments and products conforming the standards.

Besides that, the link of the CIM Center with the University help the academic and practice formation of the future industrial and informatic engineers in such especialities as Industrial Automation based on real and high level systems. Although it is installed conventional vendor (not normalized) products to obtain connectivity, we are working to futurely include in our system the emerging standards, while keeping operational the previous solutions. At the ESAII department there are some applied works specifically oriented to apply, test and conform normalized solutions and products at a laboratory level, before their installation within a production line.

The products available today form a consistent base upon which applications can be built. But in the future, integration of the global enterprise is going to acquire two forms: more connectivity with systems and devices from different vendors, and a more seamless integration with engineering, office and non-manufacturing operations, while more emphasis will be given to adding application solutions within the architecture.

The relevance of the paper is not only on showing an applied project but on providing a common platform between academic researchers and industrial colleagues where to experiment and discuss the advances in the field of Flexible Automation and Factory Information Technologies using advanced automation solutions, integrated with classical industry systems and making special emphasis on the communication system, where actually there is a large interest for obtaining normalized solutions. The study of the possibilities and limitations of the present system allows to conclude with proposals for future developments, both theoretical and applied, in order to incorporate different communication systems, possibilities and products and to link them with the most advanced solutions for industrial automation.

Acknowledgment. This research work has received fundings from the Comisió Interdepartamental per la Recerca i la Innovació Tecnològica CIRIT, Direcció General d'Universitats, from the CICYT ref. 89-0278 and had support from CERCA, Col.lectiu d'Estudis i Recerca en Control i Automàtica

5. REFERENCES

Brill, M., Gramm, U., (1991). MAP Application Services for the Manufacturing Industry. *Computer Networks and ISDN Systems*, 21, 357-380.

Ferré, R., Ojeda, J., (1991). Concepció i Posta a Punt d'un Centre CIM. *Jornadas APLITEC'91 de Automatización y Productividad*, 229-245, Sant Cugat del Vallés, Barcelona, Noviembre.

Fuertes, J., Gine, G., Romero, G. (1993). Quasi-MMS Communication for Supervising and Control an Industrial Process. *4th. Int. Conf. on Advances in Communication and Control COMCON-4*, Rodes, Greece, June.

Fuertes, J.M., Ojeda, J., Lomascolo, R. (1992). Industrial Communications: The CIM Center Pilot Plant (UPC/ICT). *IFAC Symposium on Intelligent Components and Instruments for Control Applications, SICICA'92*. Banalmádena, Málaga, 769-774, May.

Fuertes, J., Alcober, X. (1991). Comunicaciones en entornos CIM. *Revista CIM*, 34-45, Octubre.

Fuertes, J., Lomascolo, R., Mendoza, C., Cruz, J., (1991). Las Comunicaciones Industriales en el Proceso de Automatización. *J. APLITEC'91 de Automatización y Productividad*, 249-269, Sant Cugat del Vallés, Barcelona, Noviembre.

MAP/TOP 3.0 (1988). Manufacturing Automation Protocol / Technical and Office Protocol Specifications, *General Motors / Boeing Co*, August.

Pimentel, J.R. (1990). *Communication Networks for Manufacturing*. Prentice Hall, Englewood Cliffs.

DYNAMIC SENSORS DISTORTION COMPENSATION BY MEANS OF INPUT ESTIMATION ALGORITHMS

L. M. LYUBCHIK

Kharkov Polytechnic University,
Department of Engineering Cybernetics,
21 Frunze str., 310002 Kharkov, UKRAINE

Abstract. A method of distortion compensation for dynamic sensors is presented. The problem is reduced to the dynamic system input signal estimation under uncertain concern the signal model. The inverse system approach to the arbitrary input signal recurrent estimation algorithms development is investigated. The inverse systems design method is proposed based on the theory of dynamic observers for the systems with unknown inputs. The design procedure is generalized for the random noisy measurements using the parametric optimization technique. As a result a mean-square error optimal input estimation algorithm is received in the form of digital inverse filter. The conditions of steady-state filter existence are also established and the inverse filter regularization method is proposed.

Key Words. Digital filters; error compensation; inverse systems; sensors; optimal filtering; observers; least-squares estimation.

1 INTRODUCTION

The main line of measurements efficiency improvement in process control systems is the intelligent microprocessors based sensors development including the adaptive signal processing algorithms. One of the most important problem in the field of nonstationary processes measurements technique development is the problem of dynamic transducers distortion compensation. Dynamic distortion depending from both measured signal spectrum and sensor's frequency response may essentially decrease control systems performance. There are a lot of practical approaches based on the different methods of digital filters design but the problem is yet of a great current interest.

The distortion compensation problem is closely connected with the problem of estimating the dynamic system input signals from noisy measuring output response. The common approach to the solution of such inverse problem consists in their reduction to the convolution integral equation with subsequent application a discrete-time approximation methods of the convolution integral and numerical solving the over-determined system of simultaneous linear algebraic equations with the help of some type of regularization methods (Mendel, 1983). This method cannot be directly applied to the real-time estimation.

Another approach using various state-space models and the Kalman filtering technique has also been considered (Snyder, 1970). The main difficulties is in the fact that this method require sufficiently complete *a priori* information concern the estimated signals in the form of a parameterized equations of some type of dynamic system in order to its adequate model construction. In practice, however, such information is usually absent. Moreover, if the essentially nonstationary signals are to be estimated the number of the estimated parameters determined by the complexity of the signal's forming filter is increased. As a result because the input estimating problem is ill-posed, the higher input signals complexity, the lower the estimations accuracy in the presence of measurement noise.

The possibility of input signals reconstruction without it's exact model using is connected with the dynamic systems inversion approach based on the state-space representation of the inverse system. See, for example, Silverman (1969) and Singh (1982). The method of inverse dynamic systems application for input signals reconstruction of distributed parameter system was developed by Borukhov and Kolesnikov (1988), the discrete inverse systems suitable for real-time algorithms design was investigated by Yaaksoo (1980). If the invertibility conditions (Silverman, 1969) takes place, the inverse system parameters are strictly determined by the parameters of the initial one. In certain cases the inverse system may be unstable or has an unsatisfactory performance. The spectral factorization and a polynomial equation method to the inverse system design is proposed by Sternad and Ahlen (1988).

In this paper the inverse system state-space approach to the arbitrary input signal recurrent estimation algorithms development is investigated. The inverse systems design method is proposed based on the theory of dynamic observers for the systems with unknown inputs (Kurek 1983, Hou and Muller, 1992). The design procedure for general structured unknown-input observer presented by Shao-Kung Chang and Pau-Lo Hsu (1993) is generalized for the random noisy measurements using the parametric optimization technique. As a result a mean-square error optimal input estimation algorithm is received in the form of digital inverse filters. The conditions of steady-state filter existence are also established and the inverse filter regularization method is proposed.

2 PROBLEM STATEMENT

Consider the linear state-space discrete-time dynamic sensor model

$$\begin{aligned} x(k+1) &= Ax(k) + Bu(k) + w(k), \\ y(k) &= Cx(k) + v(k), \end{aligned} \qquad (1)$$

where $x(k) \in \mathbf{R}^n$ - state vector,$\mathbf{R}^n$ - n-dimensional vector space, $u(k) \in \mathbf{R}^m$ unknown input signal to be estimated, $y(k) \in \mathbf{R}^p$ - output measured signal, $w(k)$ and $v(k)$ - random disturbances represents unmodelled dynamic and measurement noise respectively. It will be assumed that $rank\ B = m$, $rank\ C = p$, the number of outputs is greater than or equal to the number of unknown inputs $m \leq p$ and the simplest type of invertibility conditions takes place i.e. $rank(CB) = m$.

The random disturbances are represented by zero mean uncorrelated while noises with statistics

$$\begin{aligned} E\{w(k)\} &= 0, \qquad E\{w(i)w^T(j)\} = Q\delta_{ij}, \\ E\{v(k)\} &= 0, \qquad E\{v(i)v^T(j)\} = R\delta_{ij}, \end{aligned} \qquad (2)$$

and the initial conditions are

$$E\{x(0)\} = \bar{x}_0, \qquad Var\{x(0)\} = P_0. \qquad (3)$$

The initial conditions are not correlated with the noises.

The *a priori* information concern bounded input signal $u(k)$ will also be represented in statistical form

$$E\{u(k)\} = 0, \qquad Var\{u(k)\} = S_0. \qquad (4)$$

It should be marked that the *a priori* features received from the upper bounds of unknown signal components tolerations may be very rough because they are used only for the estimating algorithms regularization purposes.

The main objective of input estimation problem is to obtain the estimate $\hat{u}(k)$ using an inverse filters of the form

$$\begin{aligned} \hat{x}(k+1) &= \tilde{A}(k)\hat{x}(k) + \tilde{B}_0(k)y(k) + \tilde{B}_1(k)y(k+1), \\ \hat{u}(k) &= \tilde{C}(k)\hat{x}(k) + \tilde{D}_0(k)y(k) + \tilde{D}_1(k)y(k+1), \end{aligned} \qquad (5)$$

where $\hat{x}(k)$ - state vector of the inverse filter. The objective is to pick the matrix parameters of (5) so that to minimize the value of the performance value

$$\begin{aligned} J_u(k+1) &= E\{e_u^T(k+1)e_u(k+1)\}, \\ e_u(k) &= u(k) - \hat{u}(k). \end{aligned} \qquad (6)$$

It will be shown that the estimate $\hat{u}(k)$ may be obtained only with one step delay with respect to the output measurements.

The problem seems to be rather complicated and the proposed method consists of the three stages:
i) inverse filters structural synthesis i.e. the dynamic observer design for the system (1) with unknown input $u(k)$;
ii) observer matrix parameters optimization with respect to performance index
$J_x(k+1) = E\{e_x^T(k+1)e_x(k+1)\}$,
$e_x(k) = x(k) - \hat{x}(k)$;
iii) input signal mean-square error optimal estimating using the observer state $\hat{x}(k)$ and *a priori* information (5).

Because the structural synthesis problem of unknown-input observer has a nonunique solution if the solvability condition takes place (Kurek, 1983), the possibility of inverse filter optimization has arisen.

3 INPUT ESTIMATION ALGORITHM

The dynamic state observer for systems (1)

$$\begin{aligned} \bar{x}(k+1) &= F\bar{x}(k) + Gy(k), \\ \hat{x}(k) &= \bar{x}(k) + Hy(k), \qquad \bar{x}(0) = \bar{x}_0 \end{aligned} \tag{7}$$

will be referred to as an invariant observer if the state estimation error $e_x(k)$ is independent from the input signal $u(k)$ and state vector $x(k)$. From (1), (7) it follows that the difference equation for the estimation error $e_x(k) = (I_n - HC)x(k) - \bar{x}(k) - Hv(k)$ may be written as

$$\begin{aligned} e_x(k+1) &= Fe_x(k) + [(I_n - HC)A - F(I_n - \\ &HC) - GC]x(k) + (B - HCB)u(k) + \\ &(I_n - HC)w(k) + (FH - G)v(k) - Hv(k+1), \end{aligned} \tag{8}$$

where I_n – $(n \times n)$ identity matrix.

Let us pick the matrix parameters of (7) from the invariance condition

$$\begin{aligned} &(I_n - HC)A - F(I_n - HC) = GC, \\ &B = HCB. \end{aligned} \tag{9}$$

If $rank(CB) = m, \ m \leq p$ the linear matrix algebraic equation (9) has a solution

$$\begin{aligned} F &= \Pi A - G_0 C, \\ G &= \Pi AH + G_0 \Omega, \\ H &= B(CB)^+, \end{aligned} \tag{10}$$

where $\Pi = (I_n - HC) = I_n - B(CB)^+C$, $\Omega = I_p - (CB)(CB)^+$ - the projection matrices, $^+$ - denotes the Moor-Penrose pseudoinversion (Albert, 1972) and G_0 is (n x p) arbitrary matrix. In such a case the estimation error is given by the autonomous equation

$$\begin{aligned} e_x &= Fe_x(k) - \xi(k+1), \\ \xi(k+1) &= G_0 v(k) + Hv(k+1) - \Pi w(k). \end{aligned} \tag{11}$$

When the observer (7) initial conditions are $\bar{x}_0 = \Pi \bar{x}_0$ the estimation error is unbiased and their dynamic properties are determined by the tuning matrix G_0. If the observability conditions take place

$$rank[C^T | (\Pi A)^T C^T | ... | ((\Pi A)^T)^{n-1} C^T] = n, \tag{12}$$

the arbitrary matrix G_0 can be chosen by pole placement methods so that the dynamic matrix F has a predetermined eigenvalues.

Choosing the input signal estimate in the form

$$\hat{u}(k) = B^+_{S_0^*}[\hat{x}(k+1) - A\hat{x}(k)], \tag{13}$$

where $B^+_{S_0^*} = (B^T S_0^* B)^{-1} B^T S_0^*$, S_0^* - an arbitrary positive definite weighting matrix, one can obtain that the input estimation error is given by equation

$$\begin{aligned} e_u(k) &= B^+_{S_0^*}(F - A)e_x(k) - B^+_{S_0^*} G_0 v(k) - \\ &(CB)^+ v(k+1) + (CB)^+ Cw(k) \end{aligned} \tag{14}$$

As a result the input estimation algorithm may be represented with the help of (7), (10), (13) in the form of the inverse filter

$$\begin{aligned} \hat{x}(k+1) &= F\hat{x}(k) + G_0 y(k) + Hy(k+1), \\ \hat{u}(k) &= B^+_{S_0^*}(F - A)\hat{x}(k) + B^+_{S_0^*} G_0 y(k) - \\ &(CB)^+ y(k+1), \\ \hat{x}(0) &= \Pi \bar{x}_0 + y(0) \end{aligned} \tag{15}$$

Equations (15) may be rewritten as

$$\begin{aligned} \hat{x}(k+1) &= \Pi A\hat{x}(k) + G^0[y(k) - C\hat{x}(k)] + \\ &B(CB)^+ y(k+1), \\ \hat{u}(k) &= (CB)^+[y(k+1) - CA\hat{x}(k)] + \\ &B^+_{S_0^*} G_0[y(k) - C\hat{x}(k)]. \end{aligned} \tag{16}$$

One can see, that the equations (15) are equivalent to the discrete inverse system (Jaaksoo, 1980) with additional state feedback loop with gain matrix G_0. In so doing, the input estimation accuracy is determined by the arbitrary matrix G_0, S_0^* choosing.

4 INVERSE FILTER OPTIMIZATION

A parametric optimization approach to the invariant observer design reduce to the matrix G_0 selection using the performance index $J_x(k+1)$:

$$G_0^*(k) = arg \min \ tr \ P_e(k+1), \tag{17}$$

where state error estimation covariance matrix $Pe(k)$ may be obtained by direct calculation from (11)

$$P_e(k+1) = FP_e(k)F^T - FP_{e\xi}(k,k+1) - P_{\xi e}(k+1,k)F^T + P_{\xi\xi}(k+1),$$
$$P_{e\xi}(k,k+1) = E\{e_x(k)\xi^T(n+1) = -HRG_0^T,$$
$$P_{\xi\xi} = E\{\xi(k+1)\xi^T(k+1)\} = G_0RG_0^T + HRH^T + \Pi Q \Pi^T. \tag{18}$$

It follows from (18) that $P_e(k)$ fulfils the propagation equation

$$P_e(k+1) = FP_e(k)F^T + FHRG_0^T + G_0RH^TF^T + G_0RG_0^T + HRH^T + \Pi Q\Pi^T \tag{19}$$

Using (19) the performance measure may be written as

$$J_x(k+1) = tr[G_0W_0(k)G_0^T - G_0W_1^T(k) - W_1(k)G_0^T + W_2(k)],$$
$$W_0(k) = CP_e(k)C^T - CHR - RH^TC^T + R,$$
$$W_1(k) = \Pi A[P_e(k)C^T - HR],$$
$$W_2(k) = \Pi AP_e(k)A^T\Pi^T + HRH^T + \Pi Q\Pi^T. \tag{20}$$

One can set the gradient of $J_x(n+1)$ with respect to G_0 equal to 0 and obtain the optimal matrix G_0^* by normal equations solving as $G_0^*(k) = W_1(k)W_0^+(k)$. Finally the optimal invariant observer equations taking into account (20) can be represented in the form

$$\hat{x}(k+1) = F^*(k)\hat{x}(k) + G_0^*(k)y(k) + B(CB)^+y(k+1),$$
$$F^*(k) = \Pi A - G_0^*(k)C, \qquad G_0^* = W_1(k)W_0^+(k),$$
$$P_e(k+1) = \Pi AP_e(k)A^T\Pi^T - \Pi A[P_e(k)CT - HR]W_0^+(k)[CP_e(k) - RH^T]A^T\Pi^T + HRH^T + \Pi Q\Pi^T. \tag{21}$$

The observer (21) is initialized to be initially unbiased, i. e. $\hat{x}(0) = \Pi\bar{x}(0) + y(0)$, $P_e(0) = \Pi P_0\Pi + HRH^T$. In such a case the optimal value of performance index becomes

$$J_x^*(k+1) = tr\ [W_2(k) - W_1^T(k)W_0^+(k)W_1(k)] \tag{22}$$

For input signal estimation the regression equation is used

$$Bu(k) = \hat{x}(k+1) - A\hat{x}(k) + \eta(k), \tag{23}$$

where $\eta(k) = (F-A)e_x(k) - \xi(k+1) - w(k)$ is the equivalent random disturbance conserved with the input estimation. From (11), follow that

$$E\{\eta(k)\} = 0,$$
$$E\{\eta(k)\eta^T(k)\} = S(k),$$
$$S(k) = G_0^*(k)W_0(k)(G_0^*(k))^T - G_0^*(k)\tilde{W}_1^T(k) - \tilde{W}_1(k)(G_0^*(k))^T + \tilde{W}_2(k),$$
$$\tilde{W}_1(k) = (\Pi - I_n)A[P_e(k)C^T - HR],$$
$$\tilde{W}_2(k) = (\Pi - I_n)AP_e(k)A^T(\Pi - I_n)^T + HRH^T + (\Pi - I_n)Q(\Pi - I_n)^T. \tag{24}$$

As it follows from (24) the nonnegative covariance matrix $S(k)$ is singular and for input reconstruction it is necessary to use a general Gauss-Markov estimator (Albert, 1972):

$$\hat{u}(k) = B^*(k)[\hat{x}(k+1) - A\hat{x}(k)],$$
$$B^*(k) = B^+\{I_m - P_\eta^{\frac{1}{2}}(k)[\Omega_B P_\eta^{\frac{1}{2}}(k)]^+\}, \tag{25}$$
$$\Omega_B = I_n - BB^+.$$

The equations (21), (25) are forming the recurrent input estimation algorithm in the form of the optimal inverse filter.

5 STEADY-STATE REGULARIZED FILTER

The possibility of practical utilization of the inverse filter is intimately relocated to the asymptotic behavior of the covariance matrix $P_e(k)$. If a steady-state covariance matrix $\bar{P}_e = \lim_{k\to\infty} P_e(k)$, $tr\ \bar{P}_e < \infty$ is exist, the inverse filter become stationary.

The covariance matrix can be represented as $P_e(k) = P(k) + \Psi$, where the nonnegative matrix Ψ satisfies the equation $\Psi C^T = HR$ and a symmetric matrix $P(k)$ by virtue of (19) fulfils the Riccati equation

$$P(k+1) = \Pi AP(k)A^T\Pi^T - \Pi AP(k)C^T[C^TP(k)C + \bar{R}]^+CP(k)A^T\Pi^T + \bar{Q},$$
$$\bar{R} = \tfrac{1}{2}(\Omega R + R\Omega^T),$$
$$\bar{Q} = \Pi A\Psi A^T\Pi^T - \Psi + \Pi Q\Pi^T + HRH^T \tag{26}$$

Using the well-known condition of Kalman filters asymptotic stability (see Liptser, Shiryaev, 1974), one can received the result:

Steady-state inverse filter existence criterion

Let the following conditions take place:

1. there exists nonnegative definite symmetric matrix Ψ satisfies the equation $\Psi C^T = HR$,
2. the pair $\Pi A, C$ is observable, i.e. $rank[C^T|(A^T\Pi^T)C^T|...|(A^T\Pi^T)^{n-1}C^T] = n$,
3. the pair $\Pi A, \bar{Q}$ is controllable, i.e. $rank[\bar{Q}^{\frac{1}{2}}|(\Pi A)\bar{Q}^{\frac{1}{2}}|...|(\Pi A)^{n-1}\bar{Q}^{\frac{1}{2}}] = n$,
4. the matrices $\bar{R}$ and $\bar{Q}$ are positive definite.

Then there exists a unique positive definite solution $\bar{P}$ of algebraic equation

$$\bar{P} = \Pi A\bar{P}A^T\Pi^T - \Pi A\bar{P}C^T(C\bar{P}C^T + \bar{R})^{-1}C\bar{P}A^T\Pi^T + \bar{Q}. \tag{27}$$

and $\lim_{k\to\infty} P(k) = \bar{P}$, $\quad tr\ \bar{P} < \infty$.

It is obvious that in such a case the inverse filter (21) is asymptotically stable and steady-state filter there exists with gain matrix G_0^* determined by $\bar{P}_\varepsilon = \Psi + \bar{P}$.

By directly checking it can be established that the above conditions for inverse filter (21) are certainly violated.

In order to obtain the feasible input estimation algorithm it is possible to use the approximate invariance condition $B - HCB = \varepsilon B$, where $0 < \varepsilon < 1$ is a regularization parameter. Consider the stationary regularized inverse filter

$$\begin{aligned}\hat{x}(k+1) &= F_\varepsilon\hat{x}(k) + G_0^*(\varepsilon)y(k) + H_\varepsilon y(k+1),\\ \hat{u}(k) &= B_0^*(\varepsilon)[(F_\varepsilon - A)\hat{x}(k) + G_0^*(\varepsilon)y(k) + H_\varepsilon y(k+1)],\end{aligned} \tag{28}$$

where $F_\varepsilon = \Pi_\varepsilon A - G_0^*(\varepsilon)C$, $\quad \Pi_\varepsilon = I_n - H_\varepsilon C$, $\quad H_\varepsilon = (1-\varepsilon)B(CB)^+$

Picked matrices G_0^*, B_0^* are defined by the expressions

$$\begin{aligned}G_0^*(\varepsilon) &= \Pi_\varepsilon A\bar{P}_\varepsilon C^T(C\bar{P}_\varepsilon C^T + \bar{R}_\varepsilon)^{-1},\\ B_0^*(\varepsilon) &= [B_\varepsilon^T S_0^*(\varepsilon)B_\varepsilon]^{-1}B_\varepsilon^T S_0^*(\varepsilon),\\ S_0^*(\varepsilon) &= (\varepsilon^2 I_n + \bar{S}_\varepsilon)^{-1}.\end{aligned} \tag{29}$$

where $B_\varepsilon = (1-\varepsilon)B$, and $\bar{P}_\varepsilon, \Psi_\varepsilon$ are the positive definite solutions of the algebraic equations

$$\begin{aligned}\bar{P} &= \Pi_\varepsilon A\bar{P}A^T\Pi_\varepsilon^T - \Pi_\varepsilon A\bar{P}C^T(CPC^T + \bar{R}_\varepsilon)^{-1}C\bar{P}A^T\Pi_\varepsilon^T + \bar{Q}_\varepsilon,\\ \Psi C^T &= H_\varepsilon R,\\ \bar{R}_\varepsilon &= \tfrac{1}{2}[(I_p - CH_\varepsilon)R + R(I_p - CH_\varepsilon)^T],\\ \bar{Q}_\varepsilon &= \Pi_\varepsilon A\Psi A^T\Pi_\varepsilon - \Psi + \Pi_\varepsilon Q\Pi_\varepsilon^T + H_\varepsilon RH_\varepsilon^T + \varepsilon^2 BS_0B^T,\end{aligned} \tag{30}$$

and the stationary covariance $\bar{S}_\varepsilon$ is given by

$$\begin{aligned}\bar{S}_\varepsilon &= \Pi_\varepsilon\Phi(\varepsilon) + \Phi_1(\varepsilon)\Pi_\varepsilon - \Pi_\varepsilon\Phi_1(\varepsilon)\Pi_\varepsilon^T + H_\varepsilon\Phi_2(\varepsilon)H_\varepsilon^T + \varepsilon^2 BS_0^T B,\\ \Phi_1(\varepsilon) &= A\bar{P}_\varepsilon C^T(C\bar{P}_\varepsilon C^T + \bar{R}_\varepsilon)^{-1}C\bar{P}_\varepsilon A^T,\\ \Phi_2(\varepsilon) &= CA\bar{P}_\varepsilon A^T C^T + CA\Psi_\varepsilon A^T C^T + CQC^T + R.\end{aligned} \tag{31}$$

Since $\bar{R}_{\varepsilon=1} = R, \bar{Q}_{\varepsilon=1} = Q + BS_0B^T$, $\Pi_{\varepsilon=1} = I$ and $\det\Pi_\varepsilon \neq 0$ there exist $\varepsilon^* < 1$, so that the steady-state inverse filters existence conditions are fulfilled if the system is observable.

The optimal value of the tuning regulazation parameter ε may be obtained using the steady-state performance index

$$\bar{J}_u(\varepsilon) = B_0^*(\varepsilon)\bar{S}_\varepsilon[B_0^*(\varepsilon)]^T. \tag{32}$$

Another approach using the deterministic type of *a priori* information may be based on the trade-off between the steady-state mean square error and bias induced by the unknown input signal.

6 CONCLUSIONS

An input estimation algorithm has been presented in this paper in order to dynamic distortion compensation. The algorithm ensures recurrent step-by-step input signal estimation without signal model specifying necessity. The proposed regularization procedure leads to the stable computational scheme which can be realized by the intelligent sensor's software.

7 REFERENCES

Albert A., (1972). *Regression and the Moor - Penrose Pseudoinverse.* Academic Press, New-York and London.

Borukhov V. T., and P.M. Kolesnikov (1988). Methods of inverse dynamic systems and its application for recovering internal heat sources. *Int. J. Heat Mass Transfer., 31,* 1549 - 1556.

Hou, M. and P.C. Muller, (1992). Design of observers for linear systems with unknown inputs. *IEEE Trans. Automat. Contr., 37,* 871 - 875.

Jaaksoo, Yu. I., (1980). To the discrete inverse systems theory. *Izv. AN SSSR, Technicheskaya Kibernetica* (in Russian), *N 5,* 165 - 168.

Kurek, J. (1982). Observation of the state vector of linear multivariable systems with unknown input. *Int. J. Control, 36,* 511 - 515.

Liptser, R.Sh. and A. N. Shiryaev, (1974). *Statistics of Random Processes.* Nauka, Moscow (in Russian).

Mendel, J.M., (1983). *Optimal Seismic Deconvolution. An Estimation-based Approach.* Academic Press, New-York and London.

Shao-Kung Chang, and Pau-Lo Hsu, (1993). State estimation using general structured observers for linear systems with unknown input. *Proc. 2-nd European Control Conference, 3,* 1794 - 1799.

Silverman, L.M., (1969). Inversion of multivariable linear systems. *IEEE Trans. Automat. Contr. 14,* 270 - 276

Singh, S. N., (1982). Invertibility of Multivariable Parabolic distributed Parameter Systems. *IEEE Trans. Automatic Control, 27* 276 - 279.

Snyder, D., (1973). *The State-variable Approach to Continuous Estimation with Application to Communication Theory.* The MIT Press, Cambridge, Massachusetts.

Sternad, M. and A. Ahlen. (1988). Corresponds between input estimation and feedforward control. *Int. J. Control, 48,* 1723 - 1734.

EXTRACTING PERIODIC COMPONENTS FROM MEASURED OSCILLATORY SIGNALS

Hengjun Zhu and D. P. Atherton*

**School of Engineering, University of Sussex, Brighton BN1 9QT, UK*

Abstract. Oscillatory signals, such as those from a turbine flowmeter or pressure signal representing a pulsatile flow, may contain useful information regarding the status of the instrument and/or that of the associated process plant. As the signals are dominated by the oscillatory components, other useful information may be dramatically overshadowed. In this paper, an algorithm for extracting periodic components from oscillatory signals, whose periods are not necessarily an integer multiple of the sampling interval, is described. By applying this algorithm, periodic components can be extracted satisfactorily so that useful information in the residual may be highlighted for condition monitoring.

Key Words. Digital signal processing; condition monitoring; process measurement; adaptive filter; change detection.

1. INTRODUCTION

In the field of process measurement turbine flowmeters generate oscillatory signals when the turbine blades pass the electro-magnetic sensors. Even transmitters which generate an analogue output, such as pressure transmitters with reasonably wide frequency bandwidth, can also have a significant oscillatory component when the flow is pulsatile. Both the periodic components and other components in the measurement signals from the process may contain useful information regarding the status of the instrument and/or that of the associated process plant, but such features may be difficult to detect in the total signal.

In the above cases, the magnitude of the oscillatory components is much higher than that of the other components in the signal; therefore the other components cannot be examined properly either in the time domain or frequency domain. Consequently, an algorithm is required to extract the periodicity from the oscillatory signals so that useful information can be highlighted for the purpose of condition monitoring.

In the field of time series analysis[1], periodic components are normally extracted under the assumption that the period is constant and an integer multiple of the sampling interval. In practice, however, the measured oscillatory signals in a flow system never have a precise and constant period which can be made an integer multiple of the sampling interval, because the flow is not ideally constant. Rotating or reciprocating machines do not operate at a constant speed and cause pulsations either in flow rate or pressure and instrument signal generating mechanisms, such as the rotor of a turbine flowmeter, are not perfectly symmetric. Because of these random variations, it is difficult, when carrying out data acquisition, to set the sampling interval adaptively so that the periods of the measured signals are integer multiples of the sampling interval.

To remove components in a certain frequency band, it is efficient and convenient to use a digital filter[2], but there are two major difficulties in using digital filters to extract periodic components in the aforementioned situation.

- It is not easy to design a filter with a bandwidth so narrow that the periodic component can be removed without affecting any neighbouring components. When the periodic component is not sinusoidal, several narrow band filters have to be applied to remove significant harmonics, thus the chance of either removing or affecting other components present is increased.
- As the flow rate and pulsation frequency are not constant, the frequencies of the dominant periodic components are always shifting around their average values, and to track this shift complicated adaptive filters are required. This not only causes further difficulty in filter design, but also enlarges the possibility of affecting further other components.

In this paper, an algorithm for extracting periodic components from oscillatory signals whose periods are not necessarily an integer multiple of the sampling interval is described. By applying this algorithm together with another algorithm, which unifies the periods of the oscillatory signal[3], perturbed periodic components can be extracted satisfactorily from the original signal.

The algorithm is described concisely in Section 2 and its effectiveness is demonstrated in Section 3 by applying it to a synthesized test signal. In Section 4, measured signals are processed using the algorithm and it is seen from these results how useful information can be highlighted to indicate the status of either the process plant or the measurement instrument itself. A brief conclusion is given in Section 5.

2. ALGORITHM

2.1. *Test signal*

Many typical signals may be considered to contain additively several components such as:

- The average value of the measurand,
- A periodic component,
- A trend component,
- A stochastic component.

To clearly describe the algorithm, a test signal of the form

$$y(t) = 3 + \sin(2\pi f t + 0.8) - 0.8\sin(6\pi f t + 0.5) + 0.2\mathcal{N}(0,1) + 0.3\cos 15t \quad (1)$$

is assumed, where $\mathcal{N}(0,1)$ denotes Gaussian noise with zero mean and a variance of unity. The test signal possesses a mean value of 3 units and its periodic components consist of a sinusoidal wave at a frequency $f = 147$ Hz and amplitude 0.3, together with its third harmonic with an amplitude level of 0.8. A stochastic noise of level 0.2 is superimposed on the signal and the term $0.3\cos 15t$ represents a low frequency trend. The time series plot of the test signal is shown in Fig.1.

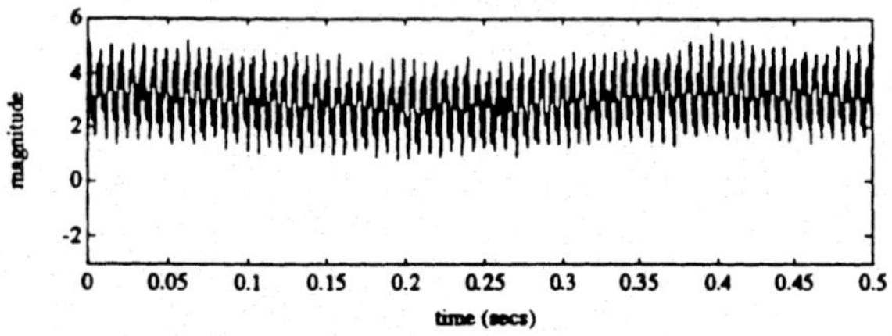

Fig. 1. Synthesised test signal

2.2. *Period determination*

In order to extract a periodic component from the other components of the signal, it is necessary to determine the time duration of every complete cycle of the periodic component under the assumption that the periods are almost equal. When the signal has a noticeable frequency shift, an algorithm for unifying the periods can always be applied to remove the shift[3]. Briefly, to unify the periods, this algorithm converts the data of the signal from the ordinary time domain to a modified time domain under the assumption that the periods in the modified time domain are constant. The conversion is conducted under the assumption that the signal crosses the zero value where its phase angle is $k\pi$, $k = 1, 2, \ldots$, in both the actual and the modified time domains.

Period determination can be done by one of two methods.

- Use the algorithm introduced in [4] to compute the duration of zero-crossing intervals of the signal after its mean value is removed. This method is suited for signals with a steady mean value but unequal periods. The zero-crossing period data is needed for period unification.
- Compute the normalised auto-correlation function (NACF) of the signal. The time lag of the maximum stationary point of the NACF is the average duration of periods. Because the NACF which results from sampled data is discrete, the stationary points have to be computed via interpolation.

A plot of the NACF of the test signal to illustrate the period determination, is shown in Fig.2.

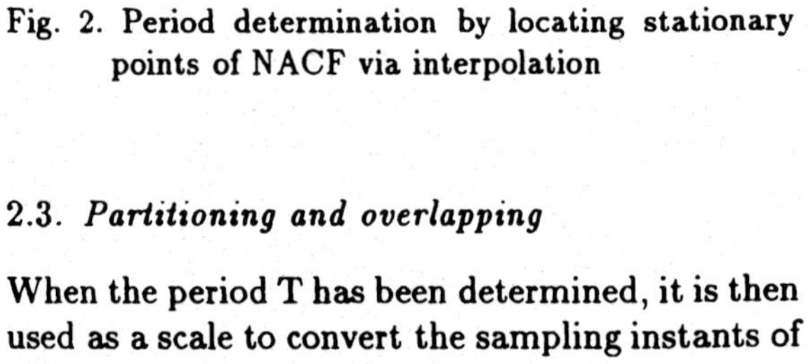

Fig. 2. Period determination by locating stationary points of NACF via interpolation

2.3. *Partitioning and overlapping*

When the period T has been determined, it is then used as a scale to convert the sampling instants of the data set into phase angles according to

$$\theta_i = 2\pi \frac{iT_s}{T} \quad (2)$$

Then the data set of the sampled signal can be expressed by $y_i = y(\theta_i)$. Since a periodic function satisfies

$$y(\theta) = y(\theta + 2n\pi) \quad n = 0, \pm 1, \pm 2, \ldots \quad (3)$$

the signal can be converted to

$$y(\theta_i) = y(\text{mod}[\theta_i, 2\pi]) \tag{4}$$

The significance of this conversion is to partition the data set with the period T and translate all the data points in the second, third and subsequent cycles to the first one as shown in Fig.3(a), where '+' stands for the data points in the second cycle and 'o' the points in the third cycle, *etc.* Every point so moved has its own position in the first cycle which then becomes the reference cycle. When the data set is sorted according to the magnitude of the phase $\varphi_i = \text{mod}[\theta_i, 2\pi]$, a new series is formed as

$$y_i = y(\varphi_i) \qquad 0 \le \varphi < 2\pi \tag{5}$$

and the plotting of the re-sorted data points is shown in Fig.3(b). For an arbitrary signal containing a periodic component, if the periodic component is not too low and its frequency is fairly constant, its waveform is easy to recognize from the plotting of the re-sorted data points.

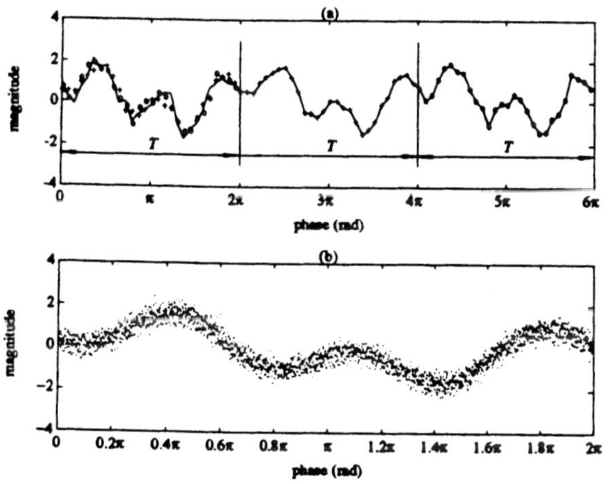

Fig. 3. Sorting data points; (a) partitioning and overlapping; (b) resultant reference cycle

2.4. *Extracting periodicity*

Because the set $y(\varphi_i)$ has various intervals between its re-sorted data points, the conventional moving averaging algorithm, using least square curve-fitting on equally spaced samples[1], is not suitable for this case. Hence, a moving quadratic polynomial curve-fitting algorithm based on the least squares principle and non-constant sampling intervals has been developed to extract the periodic component from the reference cycle containing all the data points translated from other cycles. The curve-fitting algorithm calculates a set of quadratic polynomial coefficients and interpolates the value of one point which is located at the centre of the curve-fitting base. To work out the size of the curve-fitting base, an integer m is defined by

$$m = \text{int}\left(\frac{NT_s}{2T}\right) \tag{6}$$

where N is the total number of data points, T_s denotes the sampling interval and the function int means to round the quotient to the nearest integer which is greater than the quotient. Therefore, m is approximately half the number of the periods within the time duration of N samples. Then $2m+1$ data points in the reference cycle are used as the curve-fitting base so that each cycle in the original signal set contributes at least one data point to the curve-fitting base. The periodic component extracted from the test signal with the moving quadratic curve-fitting method is plotted in Fig.4 together with the analytical periodic component defined by

$$y_p = \sin(2\pi 147t + 0.8) \\ -0.8\sin(6\pi 147t + 0.5) \tag{7}$$

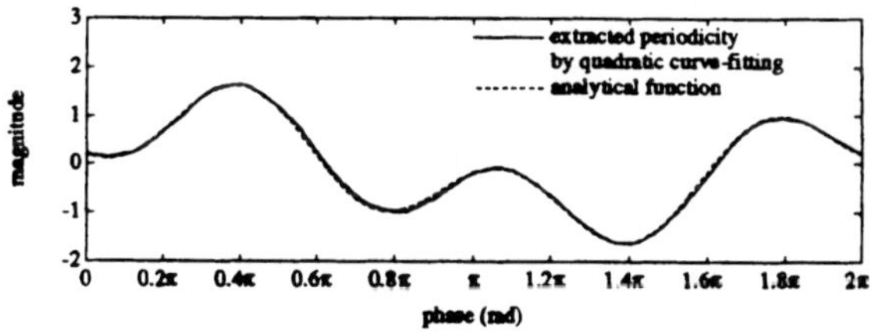

Fig. 4. Comparison of the extracted periodic components to the analytical periodicity

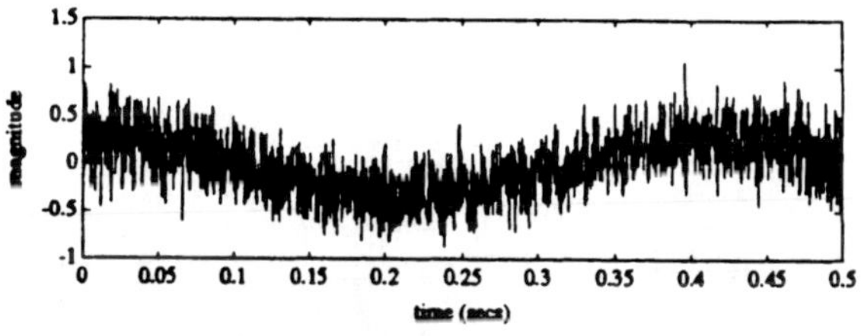

Fig. 5. Residual component after the periodicity is removed

It is evident that the extracted curve is very close to the analytical periodicity, but a certain amount of loss in amplitude occurs in the sections where the value of the curvature is relatively large. Compared to the level of the noise and trend of the original test signal, the curve extracted is seen to be a very satisfactory result.

By examining the computed points on the interpolated curve, it is evident that the value of this computed point is the contribution of its original sample to the periodicity and the difference between the measured point and the computed point is the residual component. The residual component of the test signal is shown in Fig.5.

3. APPLICATIONS

3.1. *Application to synthesized signal*

The algorithm has been applied to a signal synthesized as

$$y = y_1 + y_2 \qquad (8)$$

where

$$y_1 = \sin 2\pi f_1 t$$
$$\text{with } f_1 = 147 \text{ Hz}$$
$$y_2 = \begin{cases} 0.01 & \text{when } t = \dfrac{n}{f_2} \\ 0 & \text{otherwise} \end{cases}$$
$$\text{with } f_2 = 95 \text{ Hz}, n = 1, 2, \ldots$$

The summation of the two components is sampled at 4096 Hz. Fig.6 shows the synthesis and processing of the signal. The sinusoidal wave shown in graph (a) is used to emulate a signal from a turbine flowmeter. The other component shown in graph (b) consists of a series of sharp pulses occurring at a different frequency to emulate pulses possibly caused by the additional displacement due to a fault in the races of the bearings which support the rotor. The combination of the two components is plotted in graph (c) in which the existence of the pulses is barely detected by eye because the magnitude of the pulses is only one percent of the amplitude of the sinusoidal wave. However, after the processing with the algorithm for extracting the periodic component, the dominating periodicity is extracted and the residual is enhanced dramatically as seen in graph (d).

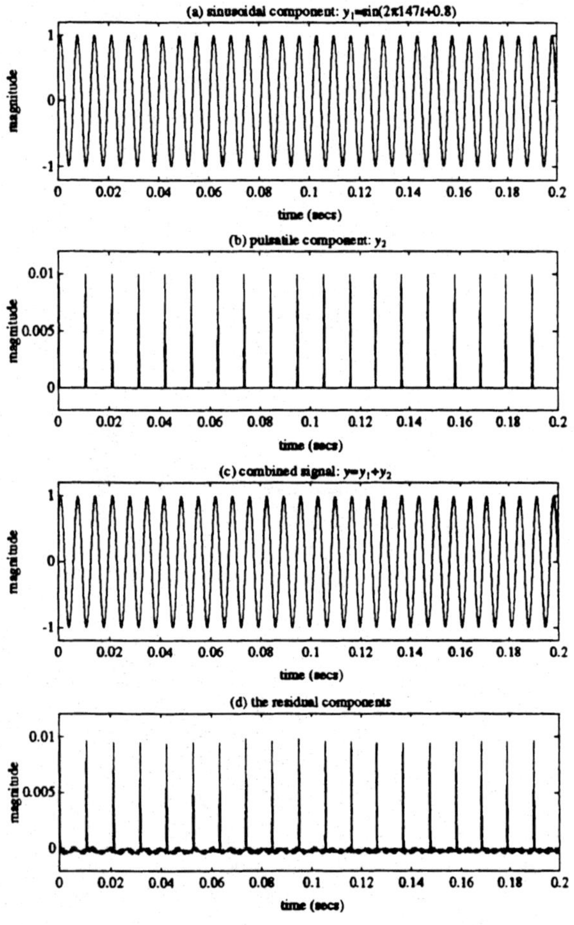

Fig. 6. Application to a synthesized signal; (a) component y_1; (b) component y_2; (c) combination y; (d) residual after the periodicity is removed

3.2. *Application to measured signals*

The algorithm was used to process the signal from a turbine flowmeter in the laboratory flow rig. The model used in the experiments was a KDG Mobrey product MX1/40/500 with a 38 mm diameter with two journal bearings supporting a rotor with six blades. The measurements were taken under two different conditions, namely constant and pulsatile flow. Pulsation in the flow was produced by a Ling 407 electromagnetic vibrator driving a brass bellows.

The signal generated by the flowmeter is shown in Fig.7(a) and seems very regular due to the good quality of its rotor. But pulse period analysis showed that the cycles of the signal are not precisely equal. Therefore, the procedure for unifying the periods was used before extracting the periodicity. In Fig.7(b), sharp peaks at regular intervals in the residual signal suggest that the shaft was not spinning smoothly in the journal bearings, causing extra motion either tangentially or radially or both. The effectiveness of the algorithm can be evidenced by the signal taken under the pulsatile flow condition. Fig.7(c) shows the frequency spectrum of the signal with pulsatile flow. It is evident that the spectrum consists of a dominant peak at 0.1 times the sampling frequency f_s, the second harmonic at 0.2 f_s and two sidebands around the dominant peak. The sidebands are caused by the flow pulsation which causes both frequency and amplitude modulation of the signal[5]. Fig.7(d) shows the residual of the signal after unification of the periods and removal of the periodic components. It is apparent that the sharp peaks are still clearly seen. Fig.7(e) is the spectrum of the residual signal under a pulsatile flow condition. It is apparent that the dominant peak and the second harmonic have been removed. Noticing the scale of graph (e), it is seen that the magnitude of the sidebands has been reduced by 85%. The remaining parts of the sidebands represent the amplitude modulation because the amplitude of the signal was not modified by period unification.

The algorithm was also used to analyse the flow noise caused by a partially closed ball valve. The flow measurements were obtained using a DRUCK

PTX510-20 0-200 kPa pressure transmitter. The waveforms and spectra of the *a.c.* components of the signals under reference and abnormal conditions are shown in Fig.8(a), (b), (g) and (h). It is apparent that the waveform has been significantly changed by the interference of the valve, but the amount of noise caused by the valve is masked by the dominant periodic component which is caused

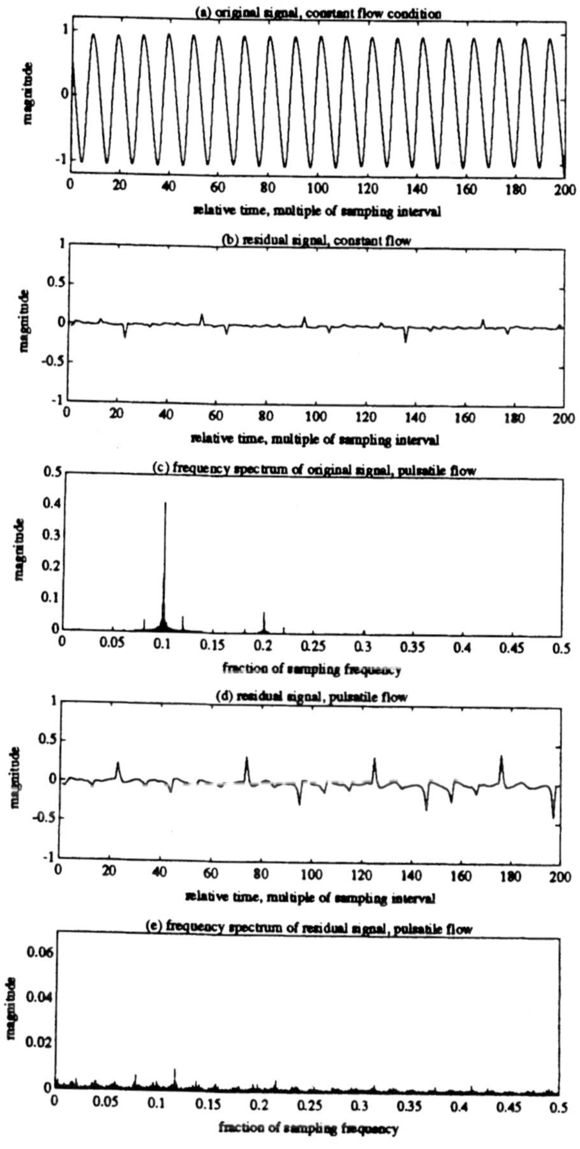

Fig. 7. Turbine flowmeter signals

by the blades of the pump impeller. The spectra show that the periodic components consist of a series of harmonics which would be difficult to remove. When applying the algorithm introduced in this paper, together with the processing for unifying the periods of the signal, the periodic components were satisfactorily extracted as shown in Fig.8(c) and (i), for which the corresponding spectra are shown in Fig.8(i) and (j). The waveforms and spectra of the residuals are plotted in Fig.8(e), (f), (k) and (l). Although the difference between graphs (e) and (k) is not apparent, graphs (f) and (l) shown in the frequency domain, indicate that the partially closed valve evidently increased the stochastic noise, in addition to distorting the flow pressure waveform.

4. CONCLUSION

An algorithm for extracting periodic components has been developed which overcomes the restriction that the period has to be an integral multiple of sampling intervals. The procedure for unifying the periods allows frequency modulation components to be abstracted. The paper describes how the algorithm has been used to identify interesting features in a signal which is dominated by a periodic component. The ability to do this has proved of value for assessing faults in flow instrumentation and process plant.

ACKNOWLEDGEMENT

The authors wish to thank Professor E. H. Higham for his valuable guidance and thank KDG and Druck for providing some of the equipment used in the experiments.

REFERENCES

[1] Kendal, M. and Ord, K.J., "Time Series", 3rd Edition, Edward Arnold, 1990.

[2] Lynn, P. A., "Digital Signal Processing and Noise", Macmillan, 1992.

[3] Zhu, H. and Atherton, D.P., "A Specially Designed Algorithm for Separating the AM and FM Components in Oscillatory Signals from Flowmeters", Proceedings of CIE International Conference on Signal Processing, Beijing, China, 26 October 1993.

[4] Zhu, H., "Signal Analysis and Change Detection of Flow Measurement for Condition Monitoring and Fault Diagnosis", D. Phil. Thesis, University of Sussex, UK, 1993.

[5] Amadi-Echendu, J. E. and H. Zhu, "On monitoring the condition of turbine flowmeters", IEEE Transactions on Instrumentation and Measurement. Vol.41, No.1, February 1992.

FUZZY LOGIC: SIGNAL CONDITIONING FOR SO_2 SENSOR

K.SRINIVAS*[1], W.WOLF**, A.CZAJOR***

* Central Electronics Engineering Research Institute, Electronics Systems Division, Pilani - 333031, India.

** Institut für Mathematik und Datenverarbeitung (Prof. Dr.-Ing. U. Appel), Universität der Bundeswehr München, D-85577 Neubiberg, Germany.

*** Institut für Meß- und Automatisierungstechnik (Prof. Dr.-Ing. H.-R. Tränkler), Universität der Bundeswehr München, D-85577 Neubiberg, Germany.

Abstract: In the last decade researchers are actively working on development of gas sensors as their applications have increased from industrial automation to household applications. The measurement of SO_2 is important for industries as well as for pollution control. The development of low cost sensors is facing problems with reliability. The main problem is that they are cross-sensitive and have nonlinear characteristics, which requires compensation as well as linearisation by signal conditioning. For this purpose complex mathematical models of SO_2 sensors are designed which are still under test. Alternatively this work pursues this problem by using fuzzy logic; as a first step, various effects are individually considered, and the performance of the fuzzy models are compared to the available mathematical models.

Key Words: Fuzzy systems, gas sensors, signal processing, modelling, cross sensitivity, sulphur-dioxide, humidity, temperature.

1. INTRODUCTION

The industrial societies being faced with the economic realities of productivity, quality, reliability of the products and the stringent vigilance on pollution control are placing major demands on existing manufacturing technologies to improve production methods. For this the basic requirement is the availability of low cost and precise measurement systems, particularly in the area of gas measurement. Until now mostly for measurement of gases optical methods were used, majority of them are off-line. These methods provide highly precise measurements associated with high costs of the system. On the other hand, we need low cost and precise sensors for operating in real time and for commercial applications. One solution for this is using semiconductor type gas sensors or measuring the changes in sensitive layer surfaces to the gas. The problem with these sensors is that they are not only sensitive to one gas but also to some other gases present in the surrounding atmosphere, i.e. cross sensitivity. The success of fuzzy logic in solving complex and vaguely defined problems triggered this work, to explore the application of fuzzy logic in gas sensors signal conditioning.

One of the methods used in signal conditioning is to correct the cross sensitivity by means of a mathematical correction algorithm. The problem with the gas sensors is that the influencing parameters are manyfold and also the effects are nonlinear, hence handling them mathematically is very difficult.

1.1. *Fuzzy Sets*

In the last decade fuzzy logic applications have rapidly increased from consumer electronic goods to industrial automation. Fuzzy logic applications

[1]Presently working at Institut für Meß- und Automatisierungstechnik (Univ. Prof. Dr.-Ing. H.-R. Tränkler), Universität der Bundeswehr München, D-85577 Neubiberg on DAAD fellowship.

can be divided into three major areas: fuzzy logic for control, pattern recognition and expert systems. It has proven its advantages for control and pattern recognition tasks which cannot be easily described by mathematical modelling. Many researchers and authors discussed fuzzy sets and fuzzy logic in detail, (e.g. Zimmermann,1991; Dubious *et al.*, 1980), so here only the required definitions of fuzzy sets for the problem are briefly described according to Zadeh (1987).

Consider temperature, and define terms **high** if it is above 50 °C and **low** if it is below 50 °C, as shown in the Fig.1(a) then 49.5 °C belongs to **low** accordingly. In practice we do not distinguish sharply between the meaning of the words **low** and **high**, because we see no justification for it.

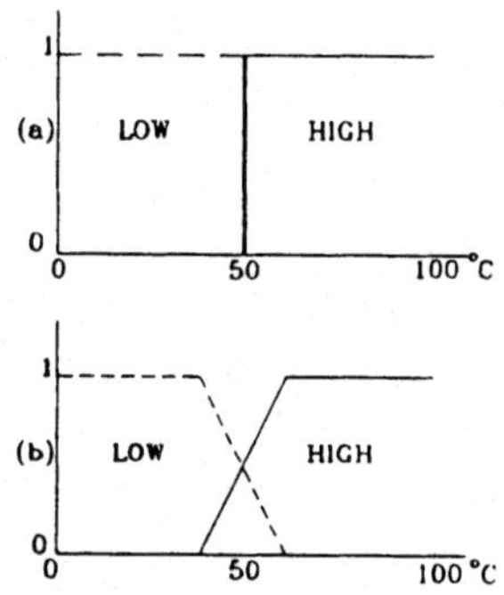

Fig. 1: Graphical representation of terms High and Low in a) Boolean logic and b) fuzzy logic

A **fuzzy set** is a generalised set that allows for various degrees of different set memberships rather than requiring **yes** or **no** memberships, i.e. values within a certain range can belong to two classes as shown in the example of Fig.1(b).

Classical set is normally defined as a collection of elements or objects $x \in X$. Each single element can either belong to or do not belong to a set A, $A \subseteq X$ defined as follows:

$$X = \{x_1, x_2, x_3, x_4, x_5, x_6\}; \qquad A = \{x_1, x_3, x_5\}$$

If we represent A by accompanying element x_i of X with its **characteristic function** values:

$$A = \{x_i, \mu(x_i)\},$$

then $\mu(x_i)$ takes values of 0 or 1, i.e. $\mu(x_i) = \{0,1\}$, in the classical set theory, but values in the interval [0,1] with fuzzy sets.

The **union** of two fuzzy sets A and B with respective membership functions μ_A and μ_B, a fuzzy subset C, is the smallest fuzzy set containing both A and B, whose membership function is related to those of A and B by:

$$\mu_C(x) = \text{Max}\,[\mu_A(x), \mu_B(x)], \quad x \in X$$

The **intersection** of A and B is the largest fuzzy set which is contained in both A and B. The intersection of two fuzzy sets A and B with respective membership functions μ_A and μ_B is a fuzzy set C, whose membership function is related to those of A and B by

$$\mu_C(x) = \text{Min}[\mu_A(x), \mu_B(x)], \quad x \in X$$

The **complement** of a fuzzy set is denoted by $\neg A$ and is defined by

$$\mu_A' = (1 - \mu_A)$$

1.2. *Advantages of fuzzy logic*

Fuzzy logic is a powerful tool in formulating heuristics or values and decisions in the form of algorithms. The common spoken language can be represented with nearly precise mathematical models using fuzzy logic, hence the system is more descriptive, and is easy to understand. Conventional and fuzzy design methods for a control application is shown in Fig.2.

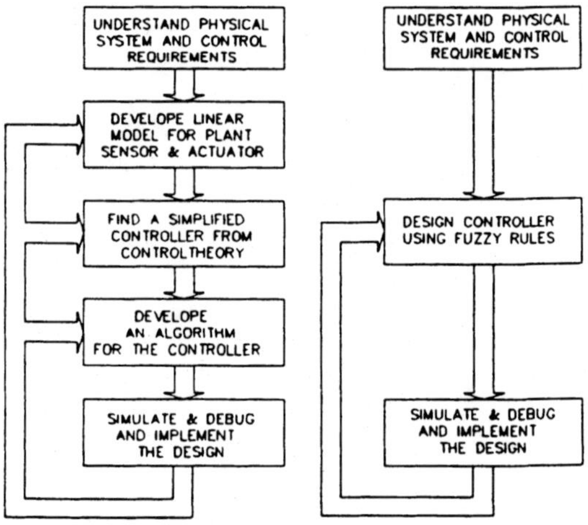

Fig. 2: Conventional (left) and Fuzzy (right) design methods

The two methods look similar but with fuzzy the design cycle is simpler. In fuzzy logic the non-linearity is built in rules, membership functions and inference processes, which results in simpler implementation. Rules are not like a look-up table because the fuzzy arithmetic interpolates the shape of non-linear functions.

2. SENSOR SIGNAL CONDITIONING

A sensor is an electronic device that transforms a physical phenomenon into an electrical signal. The physical phenomenon to be measured is not isolated from the environment, hence a sensor signal conditioning unit must be able to isolate the signals from interfering quantities. For conventional methods, tremendous efforts and investments are made to weed out all interferences except the desired sensing parameter (Tränkler, 1992). A simple block diagram of a sensor transducer is shown in the Fig.3.

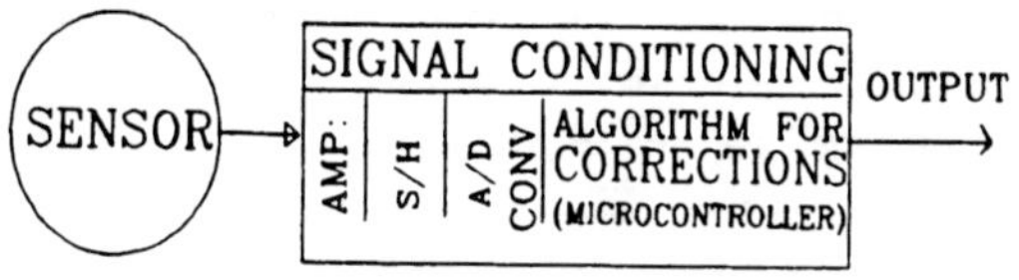

Fig. 3: Schematic diagram of the sensor signal conditioning

The general task in the sensor signal conditioning is to convert the sensor signal (related to physical phenomena) to a stable output, which can be used for further evaluation. This process may use any one or all of the following steps: linearisation, correction of cross sensitivities, hysteresis error correction, filtering of interfering quantities, dynamic error correction, creep correction, drift and offset corrections.

2.1. *Cross Sensitivity*

Almost all gas sensors are sensitive not only to the measuring quantity, but also to some influences like ambient temperature and humidity.

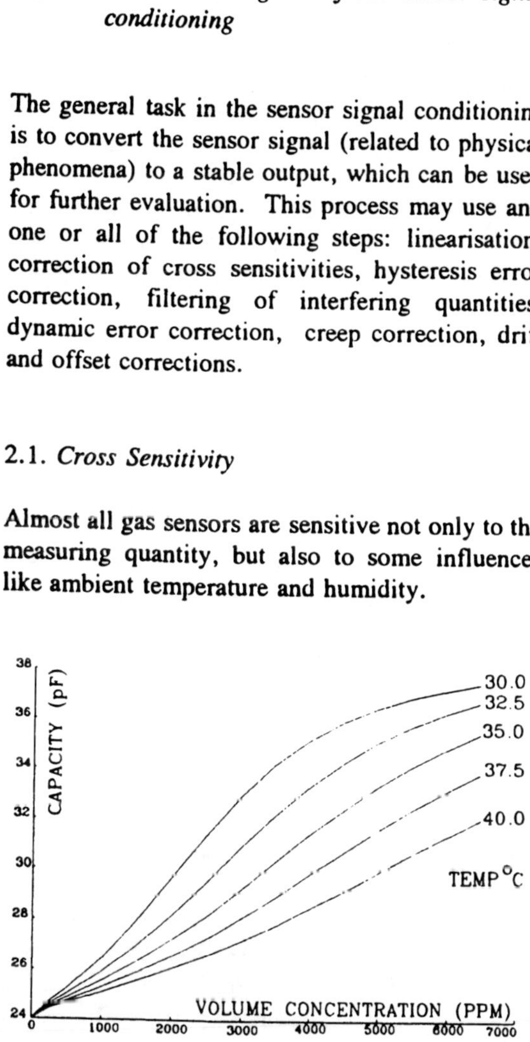

Fig. 4: SO_2 sensor characteristics showing the non-linearity and influence of temperature

Temperature effect: Consider the SO_2 sensor; its working principle is based on measurement of capacitance of the dielectric layer which is sensitive to the SO_2 gas (i.e. the changes in gas concentration alters the dielectric properties of the sensitive layer). The characteristics of the sensor are shown in Fig.4; it can be seen that the output capacitance of the sensor changes with the concentration of the gas and also prominently changes with the temperature of the gas. This is due to the adsorption and desorption of the gas molecules in the sensitive layer which is dependent on temperature of the sensitive layer.

Hysteresis: Various sensor principles are characterised by non-unique, spread response curves, depending on the current direction of change in the input signal.

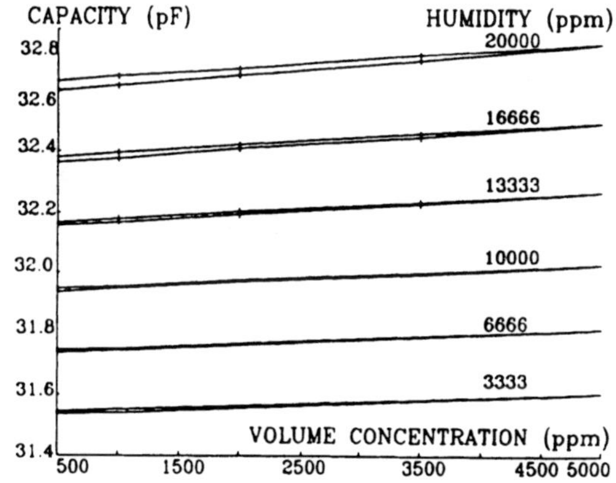

Fig. 5: The effect of humidity on SO_2 sensor and its hysteresis characteristics

Typical representation of such an example is shown in Fig.5. The characteristics of the gas sensors are influenced by humidity and showing hysteresis. The hysteresis effect is clearly indicated in the top curve of the Fig.5 for humidity range of 0 to 20000 ppm in total gas. Here the measurements were taken from 0 - 20000 ppm in steps of 3333 ppm and back to 0. For correcting the above effects, first step is to analyze the sensor behaviour and its dependency on the measuring and the interfering quantities, then the sensor characteristics are described by a physical model which corrects the influences of interfering quantities.

For hysteresis correction the history of the measuring quantity taken into account (i.e. the sequence of local extreme of the measuring quantity in time must be involved in the calculation of the corrected value). In most instances it is very difficult to find such physical model functions because the measuring effects of sensors - if they are known at all- are very complex and, therefore, not easy to express by mathematical equations.

But even if no physical model is known, or if such an existing basic model becomes unusable due to parasitic effects (e.g. manufacturing tolerances and environmental interferences), the sensor characteristics can stlll be described by a heuristic mathematical model which is adapted to the actual sensor behaviour.

3. MODELLING OF SO_2 SENSOR

Keeping in view of the problems, the modelling of sensor is necessary to correct the nonlinearity, temperature and humidity dependency for getting the correct output. But the correction of all parameters by mathematical modelling is very complex and difficult to implement, till now only individual parameter correcting models are available. The successful applications of fuzzy logic in complex problems which cannot be represented by mathematics, triggered us to use it for developing a fuzzy model which represents all parameters. As a first step, individual effects, e.g. temperature effect, nonlinearity and humidity are considered for development of fuzzy models.

The characteristics of the SO_2 sensor (as shown in Fig.4. and Fig.5.) working on interdigital capacitor (IDC) principle with NNDE/PTMS sensitive layer as dielectric, is non-linear and strongly dependent on temperature and humidity (see reference Lin *et al.* (1991)). Two different fuzzy models of sensor characteristics were developed, one for correcting the temperature and non-linearity effects and the other for correcting the humidity influence.

A **fuzzy model** is defined by a set of linguistic relations which together form an algorithm for determining the output of the model describing the characteristic function of the sensor. Fuzzy logic based system block diagram is shown in Fig. 6.

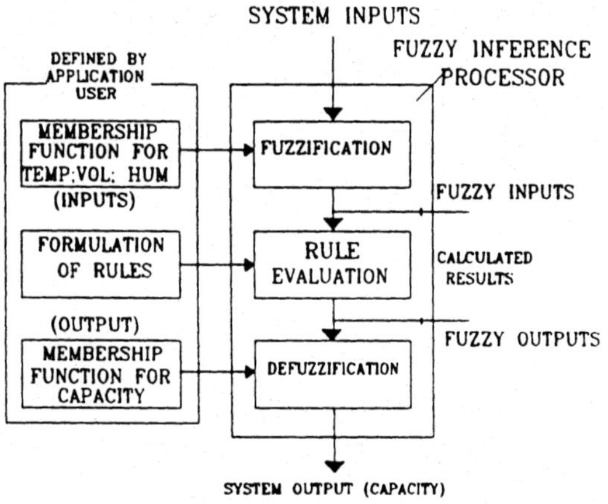

Fig. 6: Block schematic of the fuzzy logic based system design

The operation of **fuzzification** has, in general, the effect of transforming a nonfuzzy set into a fuzzy set or increasing the fuzziness of a set, or converting technical quantities (base variables) into verbal meanings (linguistic variables).

This connection between a technical specification and the linguistic expression for an item is the first requirement. The next requirement is the means of combining these linguistic variables. In **fuzzy logic** similar to the Boolean logic operators AND or OR are now available. The degree of truth of two premises connected by an or can for example be established with the maximum degree of truth between the two premises.

The production rules are basic elements in a fuzzy information process. It consists of preconditioned **IF** part and a concluding **THEN** part as shown in the example given below. The calculating method with which rules are processed is called **fuzzy inference** and consists of two steps: (i) The first step of fuzzy inference **aggregation** determines the degree to which the complete IF-part of the rule is fulfilled. (ii) the second calculation step of each production rule **composition** uses the validity of the precondition to determine the validity of the conclusion. **Fuzzy Associative Maps** (FAM) inference is used in which each rule is assigned a Degree of Support representing the individual importance of the rule. Thus the rules themselves can be "fuzzy" meaning, with a validity between 0 and 1.

example: **IF temperature = low AND capacity = low THEN Volume =low**

The result produced from the evaluation of fuzzy rules is, of course, fuzzy. Membership functions are used to retranslate the fuzzy output into a crisp value. This retranslation is known as **defuzzification** and can be performed using several methods: Centre of Area (COA), Centre of Maximum (CO-M), Mean of Maximum (MOM). For details of the definitions of membership functions and formulation of rules see Srinivas *et al.* (1993).

4. RESULTS AND CONCLUSIONS

The fuzzy models were developed using the experimental calibration data taken at humidity 0 ppm for temperature and non-linearity correction and for humidity dependence at 30°C. The results of the models were analyzed in two aspects. One is in respect to the behaviour of model with the definition of membership functions and use of different aggregation and composition operators in inference mechanism and the other is comparing it with mathematical models.

4.1. *Effect of membership functions and Operators*

The results show that, if the number of terms defined for linguistic variables is increased, the transfer characteristics are more closer to the original transfer characteristics, but the computation time is increased and the required memory used is increased, since the number of rules is increased.

Table-1: The results of the model with the different number of terms used for the linguistic variables

No.of Terms	Characteristics of the curve	Memory used	Computation Time
3	Not good	Less	Fast
5	Average	Medium	Medium
7	Good	High	Slow

For this application the best results were obtained with the operators Min-Max with 0 value for aggregation and Gamma for composition. In this case also as the gamma value increases the outputs are smoothed, but the computation time is increased. Some of the observed results are tabulated in Table 1 and Table 2.

Table-2: The results of the model characteristics of the sensor with different values given to the operators

Operator used	Characteristics of the curve	Computation time
Min-Max gamma=0	Good	Low
Min-Max gamma=.3	Very good	Medium
Min-Max gamma=1	Good	High

4.2. *Comparison with mathematical models*

The mathematical model developed for correcting the effects of nonlinearity and temperature dependence by Horn (1994) was derived from the physical knowledge using the theory of adsorption at the sensitive layer. The fuzzy model is derived from the characteristic curves of the sensor. Both models were simulated with the predefined inputs. The results of both models are good and the obtained characteristics of the sensor with the models are shown in Fig.7 together with the experimental data. The error was calculated for both models, and the maximum error obtained is less than 4% in both cases except for the mathematical model at the initial conditions. The calculated average error for the mathematical model is 0.013531 and for the fuzzy model is 0.005385. The error calculated for one temperature of 35°C is shown in Fig. 8.

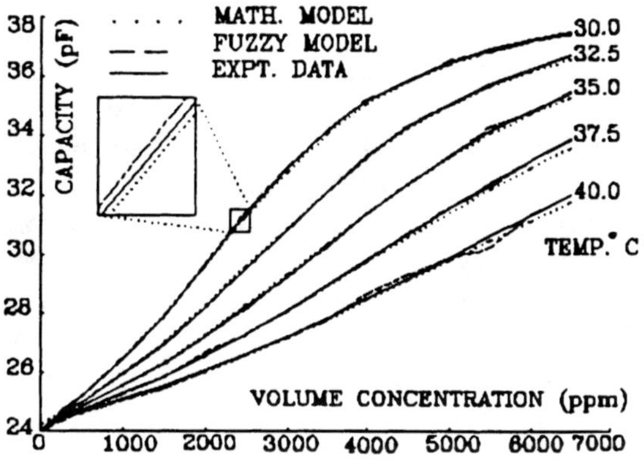

Fig. 7: The results of the mathematical and fuzzy models describing the effect of temperature and non-linearity of the SO_2 sensor, and also the experimental curve

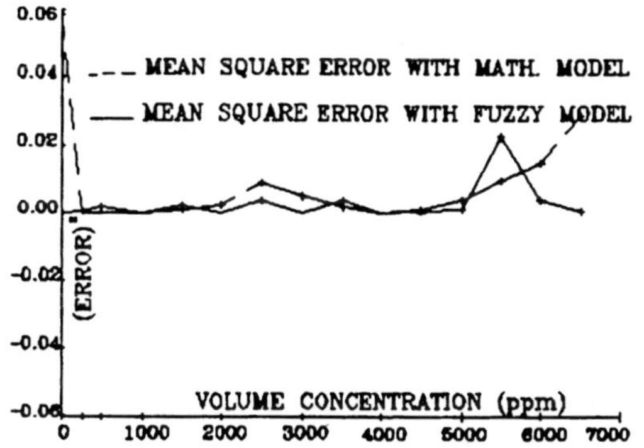

Fig. 8: Error calculated for the sensor characteristics with the mathematical model and fuzzy model

Another mathematical model for correcting the humidity effect was developed by Pichlmaier (1994) based on adsorption at the surfaces.

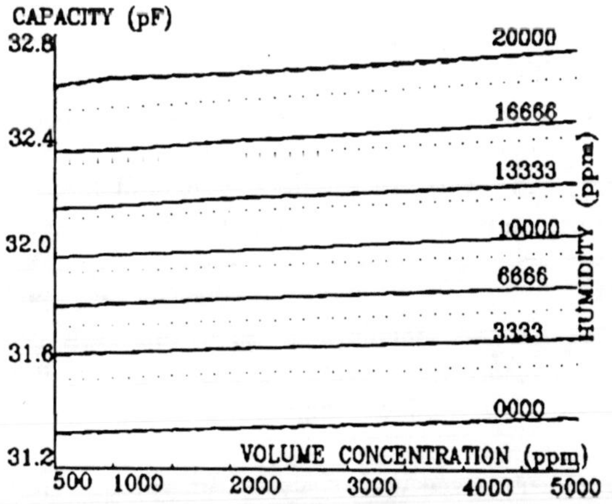

Fig. 9: The results of the mathematical and fuzzy models for correcting the humidity effect and the experimental data.

The results of the mathematical and fuzzy models

were shown in Fig.9 together with the experimental data. The mathematical model was developed by simplifying the nonlinear characteristics and other effects, hence we can see the error with mathematical model is larger and also there is a shift in the model output in comparison to the original data. Results obtained with fuzzy model are better. The average errors were calculated in this case too and the errors for one humidity range were shown for both models in Fig.10 The calculated average error for mathematical model is 0.005157 and for fuzzy model is 0.000013.

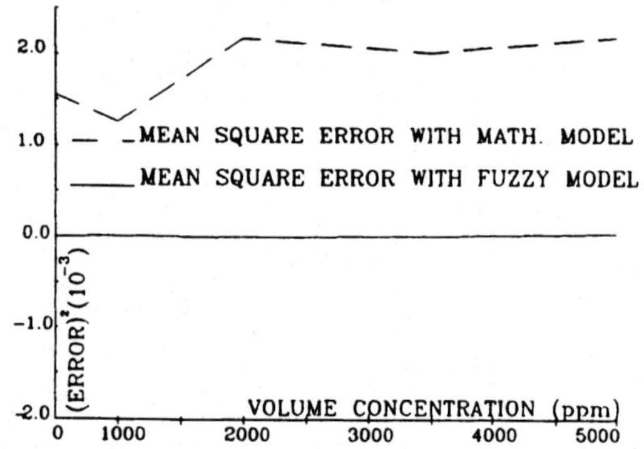

Fig. 11: Error calculated with the mathematical model and fuzzy model for humidity correction

The software used for the development of these models is Fuzzy TECH from Inform GmbH, Aachen. The software supports NeuroFuzzy modelling, and it requires the basic membership functions and rule blocks and definition of operators like aggregation and composition as well as the inputs and outputs. But we have not got good results for the above examples with neuro fuzzy learning methods.

If we see the working principles of gas sensors, the sensor is influenced by many interfering quantities, which rises difficulties in making reliable measurement systems. Another solution for the above is using sensor arrays with different sensitive layers for different gases and infer the information using fuzzy logic. This can be used to estimate the component gases in a mixture (Agapito *et al. (1993)*). Considering the above facts and needs the development of fuzzy inference mechanism to abstract the data from multiple sensor data is under progress.

Another perspective of using fuzzy logic in the area of development of precise sensors by combining fuzzy inference and imprecise low cost sensors, so that the system costs will be reduced and precise sensor systems can be used in the industry. Finally we may not require precise sensors if we are using fuzzy controllers in process industries, since the controllers can take fuzzy inputs instead of crisp signals, which we may call as fuzzy sensors.

5. REFERENCES

Abdelrahaman, M. (1990). Fuzzy Sensors for Fuzzy Logic. *Control Engineering*, Dec., 50-51.

Agapito, J.A., L.de. Agapito, M. Schneider, R. Garcia and T.de. Pedro (1993). Fuzzy logic applied to gas sensors. *Sensors and Actuators B*,15 105-109.

Constantin von Altrock. (1991). Fuzzy Logic für die Sensorik. *Sensor Report*, 4, 39-42.

Dubois, D. and H. Prade (1980). Fuzzy sets and systems: Theory and Applications. New York: Academic press.

Fuzzy Tech 2.2 Reference manual, INFORM GmbH Aachen.

Horn, M. (1994). Physical description for the sensor principle of SO_2. *To be published in the Proc. of the Fifth Int. Meeting on Chemical Sensors.*

Lin, J., S. Möller, and E. Obermeier. (1991). Thin film gas sensors with organically modified silicates for the measurement of SO_2, *Sensors and Actuators B*,5, 219-221.

Pichlmaier, J. (1994). Compensation of Humidity on the measurement of SO_2 by Signal Processing. *To be published in the proc. of the Eurosensors VIII.*

Srinivas, K., W. Wolf, and A. Czajor. (1993). Applications of fuzzy logic in chemical gas sensors. *Proc. of the DFMRS.* 77-91.

Tränkler, H.-R. (1992). Signal processing. In: Sensors: A comprehensive survey (Göpel, W., J. Hesse, and J.N. Zemel, Eds.), VCH, Weinheim.

Zadeh, L.A. (1987). Fuzzy Sets and Applications (Yager, R., S. Ovchinnikov, R. Tong and H. Nguyen, Eds.) Wiley, New York

Zimmermann, H.J. (1991). Fuzzy Set Theory and Its Applications. 2nd ed. Kluwer Academic Publishers, Dordrecht.

Acknowledgements:

The authors are very much thankful to M. Horn and J. Pichlmaier for providing the timely help and data about the models and calibration. The first author is very much indebted to Prof. Tränkler for providing him the opportunity to work at his institute and also thankful to the group members, for giving the help and time to discussions. He also expresses his gratitude to DAAD, Bonn for providing the fellowship to carry out this work.

ON THE USE OF VIRTUAL INSTRUMENTS IN REALIZING ADAPTIVE CONTROLLERS

A. SOUMELIDIS*, CS. BÁNYÁSZ*, L. KEVICZKY* and I. VAJK**

*Computer and Automation Research Institute, Hungarian Academy of Sciences, H-1111. Budapest, Kende 11-13.
**Department of Automation, Technical University of Budapest, H-1111. Budapest, Goldmann Gy. sq. 3.

Abstract. A 'Virtual Instrument'-based approach of realizing an adaptive PID controller is presented. The principle of the virtual instruments (VIs) - based upon the hardware and software basis provided by National Instruments - proved to be an effective and convenient tool to realize multiple parallel operating adaptive PID controllers. An effective, numerically robust adaptive PID algorithm has been realized, which observes the process, estimates the parameters of a second order model with dead-time, and redesigns the PID controller corresponding to the model. The VI approach of realization results in a set of building blocks which can easily be integrated to a control system corresponding to the user demands.

Key Words. Control systems; PID control; adaptive control; identification; process parameter estimation; programming environments; virtual instruments.

1. INTRODUCTION

The idea of 'Virtual Instrument' - a means of constructing programs looking like conventional instruments - is an ideal approach for building systems for measurement, data acquisition and control. It offers a great deal of components of instruments to acquire data from plug-in boards and programmable instruments, to analyze data, to control the process and to present the results through graphical interfaces. One of the most advanced representation of this idea is the National Instruments' LabVIEW®. A new family of adaptive controllers has been developed based upon this system. This paper presents in detail one element of them, an adaptive PID controller, with the speciality that the delay time of the system is estimated with a new algorithm in the adaptation step. In the first part of the paper a brief introduction to the principle of 'Virtual Instruments' is given, then the realization aspects of digital controllers are discussed. In the second part the adaptive algorithm is presented, and finally a simple simulation example is given.

2. THE PRINCIPLE OF VIRTUAL INSTRUMENTS

An adaptive controller family has been realized by applying the principle of 'Virtual Instrument' introduced by National Instruments in its advanced programming environment LabVIEW®. Virtual Instrument (VI) essentially means a software which looks like a real instrument (e.g., a measuring instrument such as a digital voltmeter or an oscilloscope). The control panel of the instrument is realized on the computer screen by applying the interactive graphical representation of control and indicator elements common in real instruments, e.g., potmeters, switches, buttons, lights, pointers, etc. The user can operate on these elements by using any pointing device (e.g. mouse). The software environment offers methods how to connect control panel elements with the different operations of the instrument software.

In LabVIEW® the principle of Virtual Instruments means not only the method how to construct control panels of 'high fidelity', but offers a new way of building complicated software systems by applying reusable building blocks. The method is well fitted to the block-diagram view of system structure natural to most engineers, and does not use conventional programming techniques, instead offers interactive graphical tools for constructing hierarchical block-diagrams. LabVIEW® contains an extensive set of elementary functions, for example mathematical, logical, string processing, etc. functions, commonly used constants, as well as programming structures for realizing conditional branches and cycles, also offers extensive libraries covering several fields, e.g., elementary mathematical functions, functions for matrix computation, signal processing, or statistics all of them realized as Virtual Instruments (National, 1993a-b-c). Hardware elements produced by

National Instruments (e.g., analog input and output cards, digital input and output cards, IEEE488 interface cards, etc.) can be handled by Virtual Instruments - the Data Acquisition VIs - which can easily be inserted in complicated schemes to realize connections with the real world (National, 1993d, 1994).

3. REALIZING CONTROLLERS AS VI

Direct computer control of real world processes can be implemented by applying the following steps (considering only the simplest case):

- analog to digital conversion of the measured process variable,
- digital control algorithm,
- digital to analog conversion to generate control signal.

Digital control algorithm has parameters which can be set initially and changed during the operation. For example a simple PID algorithm must have the following properties:

- switching between automatic and manual mode of operation,
- setting Setvalue for automatic and Manual Setvalue for manual mode,
- indicating the value of the Process Variable and the Manipulated Variable,
- setting the parameters of the controller.

The adaptive controllers can of course have additional parameters and operating modes to set. The conventional programs for direct computer control apply computer programming techniques to configure the system and set parameters of the controller. The first form of interactivity has been realized by using character oriented screens. Setting and modifying parameters have been done by entering numerical data, similarly the program can display also numerical data to inform the operator. Operators who have been accustomed in handling buttons, switches, potmeters, and observing pointers and lights are confused when they must work with numerical values. Numerical values are not so expressive than blinking red lights or a pointer entering in the red zone of the scale. Decision making upon numerical values is more difficult and time consuming process. Interactive graphical user interfaces (the operating system of the Apple's Macintosh was the pioneer among them) have given the possibility to realize easy-to handle user-friendly interfaces which facilitate fast decision making and interaction.

The LabVIEW® realization of controllers offer the facilities described above, as an example let's see the front panel of an Adaptive PID controller on Fig. 1 The layout of the controller panel is similar to that of compact industrial PID controllers. The user can easily set operating modes, change the setvalues, as well as supervise the operation. Setting and changing the control parameters as well as the identification and adaptation of the parameters is not the task of an operator, and it cannot be done frequently, hence numerical inputs are far enough to place on a separate so-called Service Panel - see on Fig. 2 - which can be recalled optionally.

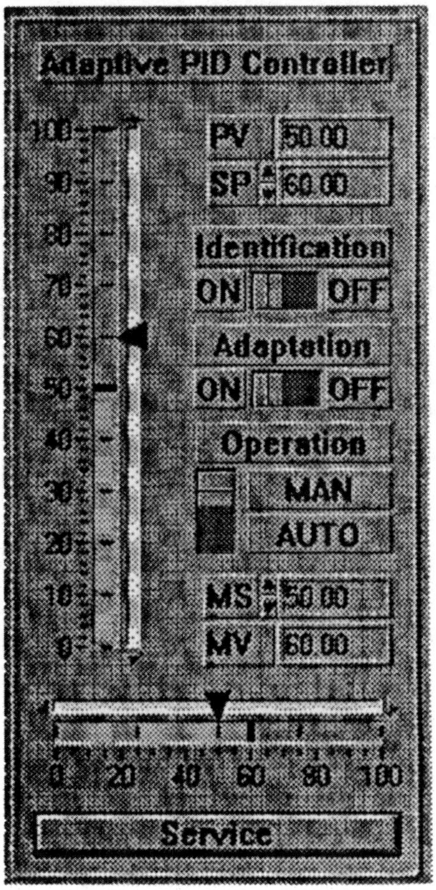

Fig.1. Control Panel for adaptive PID

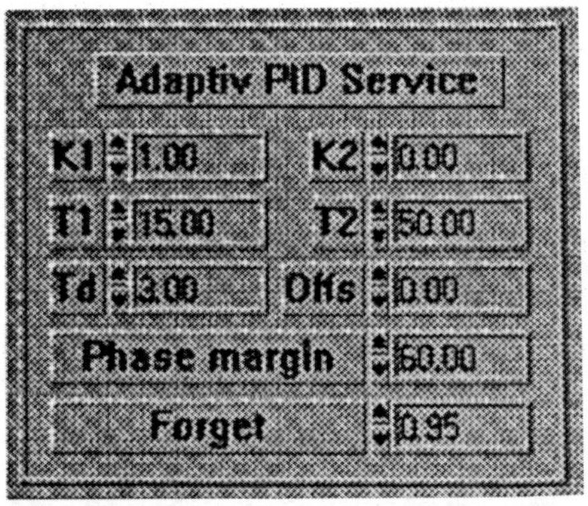

Fig. 2. Service Panel for adaptive PID

The Adaptive PID Virtual Instrument can be treated as a building block with three inputs and one output: the Process Variable (PV) input and the Manipulated Variable as output (MV) and as input (MV') - it is necessary for avoiding windup effect and realizing jumpless switching, and one additional input necessary for discretization, Sample Time (see Fig. 3).

Fig. 3. The adaptive PID building block

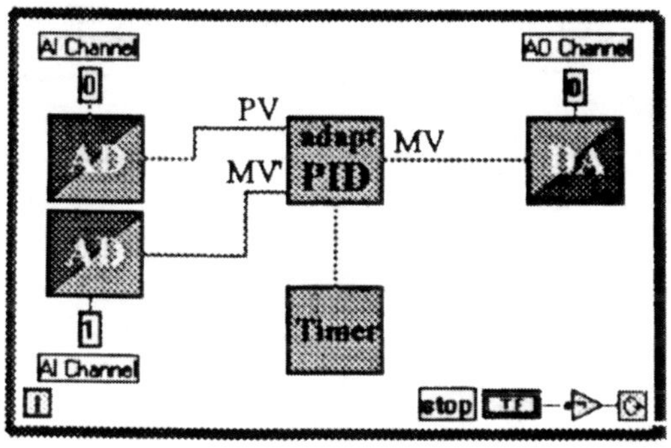

Fig. 4. Realization of one control loop

A control application can be composed of a controller VI and analog input and output VI-s by providing serial periodic execution with a given frequency. A simplified scheme can be seen on the Fig. 4. The controller and analog I/O VI-s are placed in an infinite While cycle which contains a timing element. Multiple independent controller loops can be placed in the same application which operate on different channels of the analog input and output card. For example if a multifunction I/O card is used (MIO-16 series of the National Instruments) containing 16 analog input and two analog output channels complemented with an analog output card (AO-6 at National Instruments) containing 6 analog output channels an eight-loop controller application can be realized. The number of the control loops of course depends also upon the relation between sample frequencies applied on several processes and the run-time of the controller algorithm.

4. THE ADAPTIVE PID ALGORITHM

It is assumed that the controlled process can be approximated by a second order, stable, linear lumped parameter system with additive offset and noise:

$$y(t) = w(t) * u_m(t) + offset + n(t) \tag{1}$$

where

$$L(w(t)) = W(s) = \frac{K_0 + sK_1}{1 + sT_1 + s^2T_2} e^{-sT_d} \tag{2}$$

The following notation is used:

L	-	Laplace transformation,
s	-	Laplace operator,
*	-	convolution integral,
y(t)	-	measured process output,
w(t)	-	impulse response of the process,
$u_m(t)$	-	measured process input,
offset	-	offset,
n(t)	-	noise,
K_0, K_1 T_1, T_2	-	parameters of the second order part of the model,
T_d	-	delay time of the process.

The drift is slowly varying. The spectrum of the noise is wider than that of the noiseless process output. The input signal, the output of the controller u(t) may be limited by the technology and the effective input can be measured. The parameter estimation and the control of the process are based on sampled data. The sampling time for control is fixed. It is chosen relatively small. Its value is determined by the load of the computers.

The process is assumed to be controlled by the PID controller of an often used structure. Its parameters are updated at each identification step. The transfer function of the applied continuous PID controller is:

$$Y_c(s) = \frac{P(1 + sT_i + sT_iT_D)}{sT_i(1 + s\alpha T_D)} \tag{3}$$

where

P	- proportional gain of the controller,
T_i	- integral time constant,
T_D	- derivative time constant,
α	- constant for not ideal realization of the derivative action (default α=0.2).

For the identification of the continuous linearized process we determine the best fitting discrete time model from the input and output measurements. The structure of the model used for identification is:

$$y_t = \frac{B(q^{-1})}{A(q^{-1})} u_t^m + offset + n_t \tag{4}$$

where

$$\frac{B(q^{-1})}{A(q^{-1})} = \frac{b_1 q^{-1} + \ldots + b_m q^{-m}}{1 + a_1 q^{-1} + \ldots + a_n q^{-n}} \tag{5}$$

In several applications m=10 and n=2 were selected. From the estimated parameters a_i, b_i the continuous description can be determined. The continuous description is used for controller design, and can be displayed for the operator to verify the correct operation of the estimation process.

The sampling time for identification is different to that of the control. It is the integer multiple of the sampling time for control (h_c). It is chosen using the apriori information of the process parameters. (The apriori information can be overwritten by the aposteriori information. Thus the sampling time for identification can be automatically redesigned. The sampling time used for identification is not less than:

- h_c (It is larger than or equal to the control sampling time.),
- $2T_d/m$ (It depends on the delay time.),
- $0.1/\omega^*$ (It depends on the time constants of the process.)

where ω^* is the crossover frequency of $\sqrt{2} * W'(j\omega) / K_0$. W'(s) is the delay free part of the transfer function W(s). On the basis of the third condition if the process is a first order system, the sampling time is not less than the one tenth of the time constant to reduce the numerical problem. For the generation of the parameter estimation database the primary variables are averaged according to multiplicity between the sampling time for identification and for control.

Thus the control system uses three different descriptions:

- continuous time model - continuous time process model for communication between the operator and the virtual instrument and for the regulator design;
- discrete time model based on identification sampling time - the discrete time process model obtained by the identification;
- discrete time model based on control sampling time - this description is used to generate the manipulated variable.

These descriptions are realized in the algorithm steps discussed in the followings.

4.1. Identification of a continuous process

The identification of the continuous process is realized in two steps:

- Identification of the discrete system for sampled data, then
- Calculation of an equivalent continuous-time description by a discrete-continuous transformation.

Identification of the discrete system. It is worthwhile to choose such a method that guaranties an unbiased estimation with relative small variance and at the same time the calculation complexity is relatively small. The calculation time of identification can be reduced by decreasing the number of the estimated parameters. So it is worth using an identification method that estimates only the dynamics of the system and does not estimate the parameters of the noise. For the estimation the instrumental variable method was chosen.

On the basis of (4) the following model is used:

$$y_t = -a_1 y_{t-1} - a_2 y_{t-2} + b_1 u^m_{t-1} + b_2 u^m_{t-2} + \ldots + b_m u^m_{t-m} + offset' + n'_t \qquad (6)$$

where

$$offset' = (1 + a_1 + a_2) offset$$

and

$$n'_t = n_t + a_1 n_{t-1} + a_2 n_{t-2}$$

Equation (6) can be written in vector form:

$$y_t = \varphi_t^T \theta + n'_t \qquad (7)$$

Here

$$\theta = [a_1, a_2, b_1, b_2, \ldots, b_m, offset']^T$$

and

$$\varphi_t = [-y_{t-1}, -y_{t-2}, u^m_{t-1}, u^m_{t-2}, \ldots, u^m_{t-m}, 1]^T$$

As Söderström *et al.* (1987) showed, a suitable instrument selection is

$$\zeta_t = K(q)[-y_{t-1-n}, -y_{t-2-n}, u^m_{t-1-n}, u^m_{t-2-n}, \ldots, u^m_{t-m-n}, 1]^T \qquad (8)$$

in closed loop if the noise can be approximated by a moving average process of order n. Since it was assumed that the spectrum of the noise n_t is wide, so the noise n'_t is approximately a moving average process of order 2 in this application. The estimated parameters can be used for filtering the variables to improve the estimation. In the adaptive controller the recursive version of the instrumental variable method is applied. To make the algorithm numerically less sensitive the UDL factorization is used. The algorithm of the UDL factorization can be found in Keviczky *et al* (1985).

Due to the poor excitation in the closed loop, little information enters in certain directions of the parameter space. The uniform forgetting eventually causes covariance blow up which can lead to unstable controller adaptation. The directional forgetting (Kulhavy, 1987) prevents the blow up effect. The directional forgetting combined with the

UDL factorization was used for the identification. In the parameter estimation different forgetting factors were applied for the offset and the other parameters.

Discrete-continuous transformation. The exact transformation between the different process descriptions is difficult to realize. It requires iteration since the delay time of the continuous process is not exact multiple of the sampling time used for identification. From the overestimated model (6) we want to get an approximation of a second order system with delay. The used approximation is based on the model matching at zero frequency principle. The principle of equivalence requires the matching of the zero, first and second order derivatives of the original and the modified transfer functions. Schematically we can write

$$\frac{B(z^{-1})}{A(z^{-1})} \approx \frac{b_1' z^{-1} + b_2' z^{-2}}{1 + a_1 z^{-1} + a_2 z^{-2}} z^{-d} \tag{9}$$

where the delay time d is not integer, it is a real value, i.e., we approximate the process by a second order discrete model with non integer discrete delay. For the low frequency approximation it is enough to investigate the numerators of the polynomials.

$$\begin{aligned} B(z^{-1})\Big|_{\omega=0} &= \sum_{i=1}^{m} b_i = c_0 = b_1' + b_2' \\ \frac{d\,B(z^{-1})}{d\,\omega}\Big|_{\omega=0} &= \sum_{i=1}^{m} b_i\, i = c_1 = (d+1)b_1' + (d+2)b_2' \\ \frac{d^2\,B(z^{-1})}{d\,\omega^2}\Big|_{\omega=0} &= \sum_{i=1}^{m} b_i\, i^2 = c_2 = (d+1)^2 b_1' + (d+2)^2 b_2' \end{aligned} \tag{10}$$

In the above three equations we have three unknown parameters. In (10) c_0, c_1, c_2 and auxiliary variables were introduced for simpler calculation. From (10) the parameters b_1', b_2' and d can be determined without iteration

$$d^2 + d\left(3 - 2\frac{c_1}{c_0}\right) + \left(2 - 3\frac{c_1}{c_0} + \frac{c_2}{c_0}\right) = 0 \tag{11}$$

$$b_2' = c_1 - (d+1)c_0 \qquad b_1' = c_0 - b_2'$$

The positive root of the second order equation is used as the equivalent discrete delay time. From the parameters of the discrete system the continuous description can be easily obtained (Bányász et al, 1993).

4.2. Controller design

The design of the controller is based on the estimation of the continuous process. The process is assumed to be controlled by an often used PID controller (3). The discrete form of PID is

$$Y_c(z) = \frac{b_0^c + b_1^c z^{-1} + b_2^c z^{-2}}{1 + a_1^c z^{-1} + a_2^c z^{-2}} \tag{12}$$

The approximated relationships (not using exponential functions) between the parameters of the continuous and discrete controllers are:

$$\begin{aligned} a_2^c &= \frac{T_D \alpha / h_c}{1 + T_D \alpha / h_c} \qquad a_1^c = -(1 + a_2^c) \\ b_2^c &= P(1 - a_2^c)\frac{T_D}{h_c} \\ b_1^c &= -P(1 - a_2^c)\left(1 + 2\frac{T_D}{h_c}\right) \\ b_0^c &= P(1 - a_2^c)\left(1 + \frac{h_c}{T_i} + \frac{T_D}{h_c}\right) \end{aligned} \tag{13}$$

To eliminate the windup effect and to use a jumpless switching the controller output is calculated by

$$u_t = b_0^c e_t + b_1^c e_{t-1} + b_2^c e_{t-2} - a_1^c u_{t-1}^m - a_2^c u_{t-2}^m \tag{14}$$

The design of the controller is based on the closed loop behavior of the system

$$G(s) = \frac{P(1 + sT_i + s^2 T_i T_D)}{sT_i(1 + s\alpha T_D)} \frac{K_0 + sK_1}{1 + sT_1 + s^2 T_2} e^{-s(T_d + h_c)} \tag{15}$$

The delay time of the loop is increased by the sampling time. The integral and derivative parameters are chosen such a way that the poles of the system are cancelled. By the selection of the controller gain a given phase margin (Φ) is guarantied:

$$\begin{aligned} &\alpha = fixed \qquad (default\ 0.2) \\ &T_i = T_1 \qquad\qquad T_D = T_2 / T_1 \\ &-\pi / 2 + \Phi = arctg(\omega_c K_1 / K_0) - \\ &\qquad arctg(\omega_c \alpha T_D) - (T_d + h_c)\omega_c \\ &P = \omega_c T_i / K_0 \end{aligned} \tag{16}$$

In order to eliminate the arctg calculation, the arctg(x) function is approximated by

$$arctg(x) \approx \begin{cases} x & \\ \frac{\pi}{4} - \frac{1}{2} + \frac{x}{2} & \\ \frac{\pi}{4} + \frac{1}{2} - \frac{1}{2x} & \\ \frac{\pi}{2} - \frac{1}{x} & \end{cases} \quad if \quad \begin{aligned} &0 \le x \le \frac{\pi}{2} - 1 \\ &\frac{\pi}{2} - 1 \le x \le 1 \\ &1 \ge \frac{1}{x} \ge \frac{\pi}{2} - 1 \\ &\frac{\pi}{2} - 1 \ge \frac{1}{x} \ge 0 \end{aligned} \tag{17}$$

during the solution of (16). (Generally an iterative method can be used for the solution.)

5. A SIMULATION EXAMPLE

A simple simulation example is presented now. A simulation environment has been developed with testing purposes. Any type of controller can be inserted in a closed loop with a simulated process which can be built up of first and second order, as well as delay time elements. In our illustrative example a first order element has been used with delay. Setvalue was driven with a square pulse. The adaptive controller was operated to reach optimal performance in the closed loop, then one of the parameters (namely the delay time) of the process was changed, which resulted in periodicity and considerable overshoot in the transients. On Fig. 5 the adaptation process can be seen after the change on parameters has occurred. It can be observed that four or five transients were enough for adaptation process to eliminate the changes and to reach optimal performance.

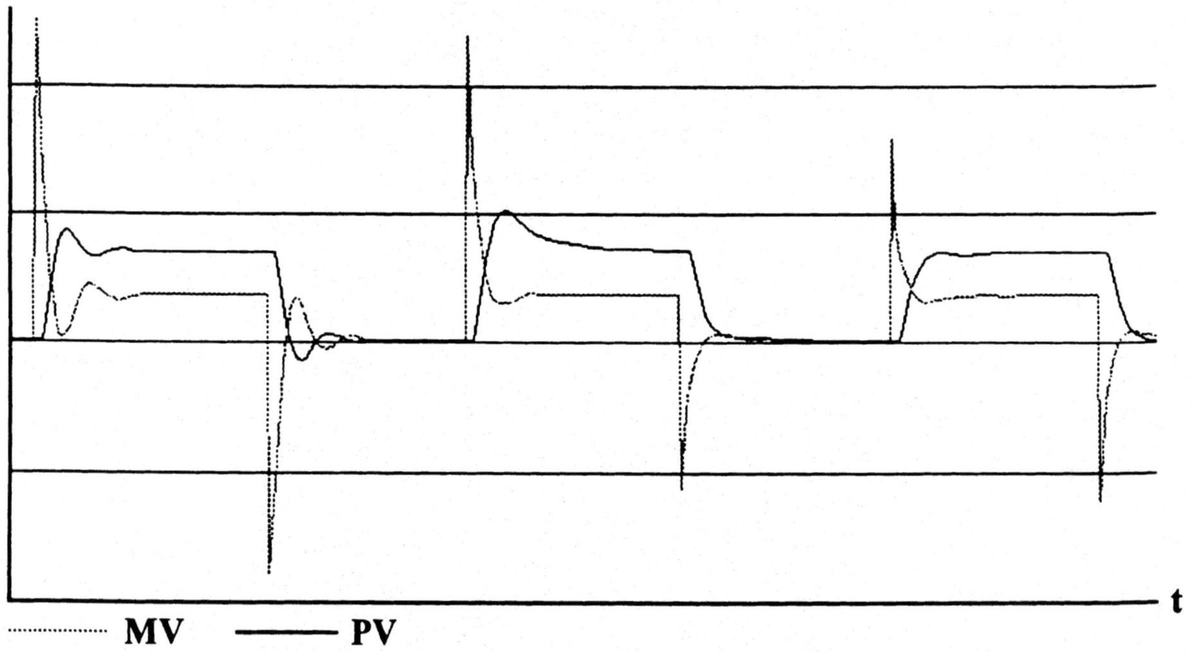

Fig. 5. The adaptation process: an illustrative example

6. CONCLUSIONS

A virtual instrument approach in realizing digital controllers results in a low cost, easy to use, user-friendly solution for process control applications. An adaptive PID controller - with a new algorithm in adaptation against delay time changes - has been presented as an element of a Controller VI set, ready to use in control applications. The Controller VIs are available on Macintosh and Microsoft Windows 3.1 version of LabVIEW®.

7. REFERENCES

Bányász, Cs. and L. Keviczky (1993). Design of adaptive PID regulators based on recursive estimation of the process parameters. *Journal of Process Control*, Vol. 3, 1, pp. 53-59.

Bányász, Cs., I. Vajk and L. Keviczky (1994). Handling delay-time in low cost controllers. presented in *SICICA'94*, June 8-10. 1994. Budapest.

Keviczky, L., I. Vajk, J. Hetthessy (1985). INTELLICON: An industrial multiloop adaptive regulator. *Proc. IFAC Identification and System Parameter Estimation 1985*, York, pp. 1267-1272.

Kulhavy, R. (1987). Restricted exponential forgetting in real-time identification. *Automatica*, 23, pp. 589-600.

Ljung, L. (1987). System identification, theory for the user. *Prentice Hall, Inc.*

National Instruments (1993a). LabVIEW® User Manual. *National Instruments*, Austin, TX.

National Instruments (1993b). LabVIEW® Function Reference Manual. *National Instruments*, Austin, TX.

National Instruments (1993c). LabVIEW® Analysis VI Reference Manual. *National Instruments*, Austin, TX.

National Instruments (1993d). LabVIEW® Data Acquisition VI Reference Manual. *National Instruments*, Austin, TX.

National Instruments (1994). IEEE-488 and VXIbus Control, Data Acquisition, and Analysis Catalog, *National Instruments*, Austin, TX.

Söderström, T., P. Stoica and E.Trulsson (1987). Instrumental variable methods for closed loop system. *Proc. 10th IFAC Congress*, Munich., Vol. X., pp. 363-368.

A SYNTACTIC AND CONTEXTUAL EDGE DETECTOR

C. NERRIEC*, I. ALVAREZ**, A. DIEZ**, J.A. CANCELAS**, G. OJEA** AND R. GONZALEZ**

*TIMEAT, Rennes, France
**University of Oviedo, ISA, Gijón, Asturias, Spain

Abstract. This paper present an image segmentation method which may be applied to various tasks as natural segmentation of monochromatic, colored and even three dimensional images. This algorithm is based on syntactic recognition. The particular novelty of this algorithm lies in the use of an efficient grammar to find edge models with an unknown a priori context. In order to improve and eventually to correct the detection, a set of rules is applied on the edges image. Experimental results are shown.

Key Words. edge detection, segmentation, image processing, formal languages, automata theory.

1. INTRODUCTION

A fundamental stage in most computer vision systems is the generation of the description of an image, which is more useful than the set of pixels. The objective of segmentation is the extraction of visual indexed in an image, that permits to split the image in subsets of points called regions. This is not generally considered as a goal by itself, but is an efficient method to transform data allowing it's analysis by a higher level process.

A great variety of "image segmentation" techniques permits to achieve this goal. In general, they are based on the search of local discontinuities or on the detection of parts of the image which present homogeneous characteristics. Those techniques generate different types of algorithms and this fact leads to obtain different results

The edge detectors can be associated in two different families. The first one considers an edge as a set of edge points, while the second one considers edges as lines consisting on an ordered set of points. Edge detectors can be classified according to the extension of the neighborhood analyzed when the image is processed. The importance of the context is to be emphasized during the analysis phase of the image for edge extraction.

In spite of the important number of works related to image segmentation, few algorithms work with the approach of an indeterminate context. In fact, most of them are conceived to solve specific problems, and therefore only analyzes a limited context.

The objective of the present work is to define a contextual approach for image segmentation. The underlying idea in this approach is to describe an edge without being limited a priori by a fixed window of interest in the image. In other words, the brightness step between the background and an object in the scene can be displayed in the image by an unknown number of pixels.

In section 2 a formal definition of image segmentation and a brief memory on grammar and its associated language are present. Section 3 contains the definition of edge modeling and the extracting method. In section 4 rewriting rules methodology is presented to improve edges detection. Section 5 is dedicated to present a measure of the method quality.

2. EDGE MODELING

2.1. Segmentation definition

If the intuitive notion of edge is commonly accepted by everybody, a more formal and constructive definition of this concept is more delicate. The goal of this section is then to give a formal definition of segmentation and to solve the intrinsic problems of this definition.

There is not an absolute segmentation. The quality of a segmentation results is generally judged bye human semantic criteria difficult to implement. In addition, this quality is usually dependent on ulterior treatments. Anyway, the goal is generally to describe the image in regions or region edges: projections of objects or parts of them present in the scene.

Formal definition. Every pixel in the image belongs to only one region, or its frontier between two regions. The segmentation algorithm will then treat every point in the image.

The regions are convex. This implies they are taken in care of a neighborhood (context).

Segmentation is defined by several criteria to be respected. Certification of this criteria gives a necessary and sufficient condition to obtain a segmentation. This segmentation is not unique.

2.2. Grammatical Inference. (Fu, 1974; Fu, 1975a; Fu, 1975b).

V_T is a finite set of symbols used to generate the strings of interest. V^*_T consist of all strings of interest, including the null one λ. L, $L \subseteq V^*_T$ is a language. If L is finite, the language is finite, otherwise is infinite.

Definition 1. A grammar G is a 4-tuple $G = (V_N, V_T, R, \sigma)$ where:

- V_N is a finite set of non terminal symbols.
- V_T is a finite set of terminal symbols.
  - $V_N \cup V_T = V$
  - $V_N \cap V_T = \varnothing$
- $\sigma \in V_N$ is the start symbol.
- R is a finite set of rewriting rules of production.

A string $x \in V_T^*$ is generated by a grammar G, $\sigma \xrightarrow{n} x$, where n is the number of rewriting rules from R to apply to obtain the string x starting with σ. The set of sentences defined by the grammar G is denoted as:

$$L(G) = \left\{ x \middle| x \in V_T^*,\ \sigma \xrightarrow{n} x \right\} \qquad (1)$$

and is called a phase structure language.

Definition 2. Let $G = (V_N, V_T, R, \sigma)$ be a grammar. If for every production $\alpha \to \beta$, $\{\alpha \to \beta\} \subset R$ $A = (S, V_T, \delta, s_o, F)$, $\alpha \in V^*$ and $\beta \in V^*$ the grammar G is a type 2 grammar.

If every production in the set R, of the form $A \to B$ or $A \to a$, $(A,B) \in V_N^2$ and $a \in V_T$ then the grammar G is a type 3 grammar.

The relationship between a language and a grammar is an injection. A deterministic finite automation $A = (S, V_T, \delta, s_o, F)$ can be obtained from a type 3 grammar, where:

- S is a finite set of states.
- V_T is the alphabet of the grammar G.
- δ is a mapping from SxV_T of S.
- s_o, $s_o \in S$ is the initial state.

8F is the set of final states ($F \subset S$).

A sentence $x = a_1 a_2 \ldots a_n$ is accepted when:

$$\delta(a_n, \delta(a_{n-1}, \ldots, \delta(a_1, s_o))) \in F \qquad (2)$$

3. EDGES EXTRACTION

Many studies have been done in the domain of image segmentation (Haddon & Boycd, 1990a; Haddon & Boyce 1990b; Deriche, 1990; Nevatia 1980; Nerriec 1990) to evaluate an observed scene. The objective of segmentation is then to determine the projection of the different objects composing the scene.

The quality of the segmentation is maximum if every region corresponds to a single object and vice verse.

Although the property of homogeneity of an object is normally known, it is difficult to characterize the homogeneity of a surface where luminosity is degraded, or when the object has a complex texture.

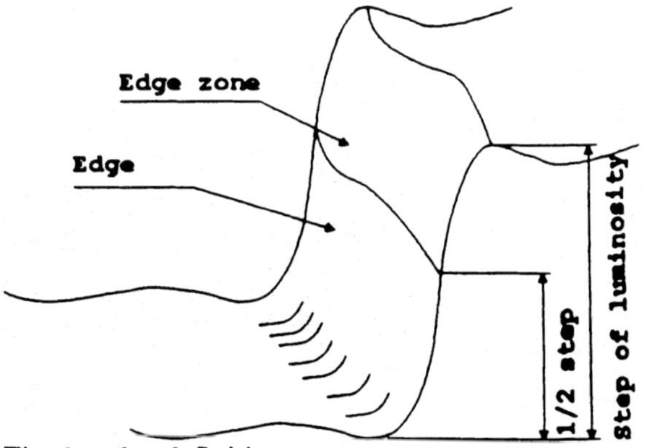

Fig. 1: edge definition.

The presented algorithm is based on the homogeneity discontinuities between different projections of objects of the scene in an image plane. Edges models are then defined as an increase or decrease of luminosity.

Luminosity discontinuities detection is formalized by a type 2 grammar. The use of such a grammar allows to take care of an unknown a priori context for the two main directions of the image (rows and columns) for the discontinuity to be detected. Therefore, a representation by a finite states automata of the grammar for model extraction is possible.

The model extraction does not permit to extract immediately the edge of an object. The presented modelization gives better information: it defines a zone where the edges are placed, but does not find the pixel itself. A decision criteria must determine the representative point of the edge inside the detected edge zone.

To be able to maintain the advantage of integer arithmetic developed for the edge zones extraction, the following criteria has been adopted to extract the edge line.

Edge definition. The edge representing one edge zone will be the curve that joins all the pixels whose luminosity is closer to half the step of luminosity in the direction of analysis.

To be noticed that every pixel in the image belongs to an edge zone. It become then necessary to maintain only the significant edge zones applying a selection criteria.

Selection criteria. Only those edge zones where the medium luminosity step, in a length ε of the zone, is greater than a given contrast step, are conserved.

fig. 2: unidirectional analysis of LENA image after applying selection criteria.

Figure 2 represent an unidirectional analysis of an image after application of these criteria on the edge zones.

Edges intersections are normally related to noise phenomena.

4. REWRITING RULES

Three kind of rewriting rules can be defined. Type 1 rules corrects detection errors due to segmentation automata. These rules use a reduced context. Type 2 rules allows the solution of those cases when there exits an incoherence between different directions of analysis. Type 3 rules improve the final aspect of the detection.

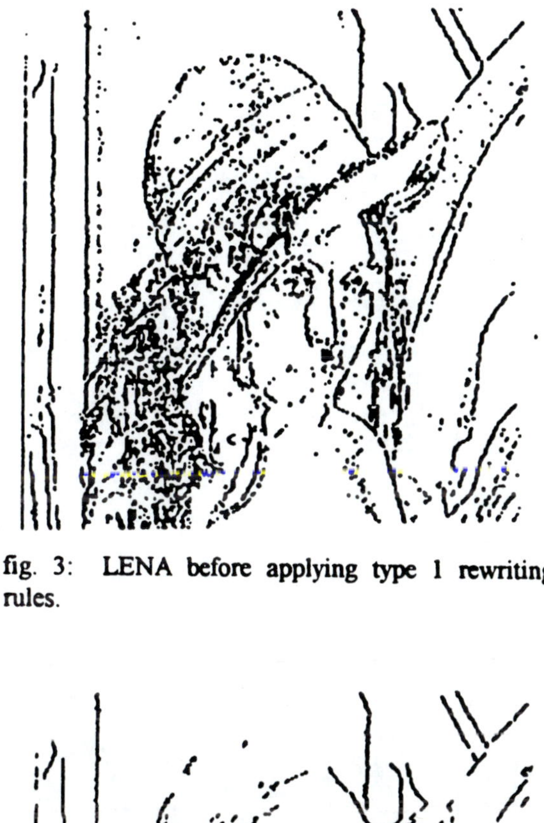

fig. 3: LENA before applying type 1 rewriting rules.

fig. 4: LENA after applying type 1 rewriting rules.

Type 1 rules are independent among them but the application order of those rules has influence on the final result. They have been developed to eliminate non realistic configurations, mainly in the intersection of two edges. For example, convergence and divergence of edges, or isolated edges areas.

Type 2 rules remove double representation of edge zones detected by several directions of analysis.

Type 3 rules has been developed to improve the quality of detection, i.e. allow edges closing and removal of small length edges.

fig. 5: LENA before applying all rewriting rules.

fig. 6: LENA after applying all rewriting rules.

5. METHOD QUALITY

Peli and Malah (1982) had proposed several qualitative measures to study the performance of a given detector. Tests should be run in the image shown in fig. 7.

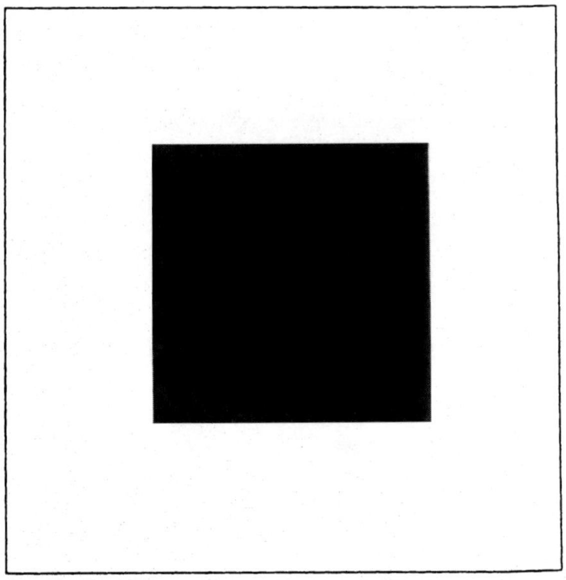

fig. 7: Test image.

Applying the detector explained above, achieved results are:

1) Every edge pixel was detected
2) Every pixel detected as an edge pixel was, in fact, an edge pixel.
3) Signal to noise ratio is zero.
4) The mean width of edges is one pixel.
5) Pratt's criterion is one.
6) Quadratic deviation is zero.
7) Mean absolute value of the deviation is zero.

All those measures attain their best possible value. The achieved segmented image is shown in fig. 8.

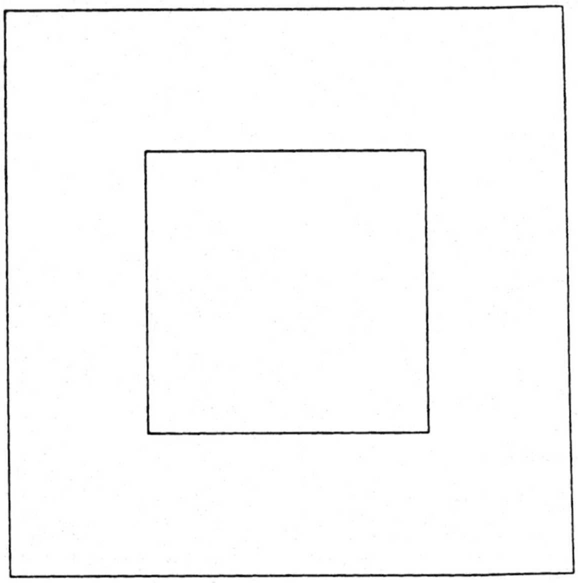

fig. 8: Detected edges in the test image.

For an image slightly contaminated with Gaussian noise, detector performance is excellent, but when standard deviation is greater than ten the quality of this method suffers a little decrement.

6. CONCLUSION

In this paper an image segmentation method and its application on various problems has been presented. The originality of this approach resides on the following facts.

- A separation of the model and the algorithm allows to generalize the detection.

- An edge detection in an environment unknown a priori.

- A better performance of detection by application of rewriting rules that increase edges length and correct false detection.

- An updated data structure gives this method a feasible complexity in its algorithms. In fact, edge zones extraction is done by an integer arithmetic.

- A regular hierarchization of results allows also to simplify error detection.

7. REFERENCES

Deriche, R. (1990). Fast algorithms for low level vision. IEEE T-PAMI, **12**, 78-87.

Fu, K. S. (1974). *Syntactic Pattern Recognition*. Prentice-Hall, London.

Fu, K. S. (1975a). Grammatical inference: introduction and survey (part I). IEEE T-SMC, **5**, 95-111.

Fu, K. S. (1975b). Grammatical inference: introduction and survey (part II). IEEE T-SMC, **5**, 409-423.

Haddon, J. F. and Boyce, J. F. (1990a). Unification of segmentation and edge detection. IEEE Proceedings, **137**, 129-135.

Haddon,J. F. and Boyce, J. F. (1990b). Image segmentation by unifying region and boundary information. IEEE T-PAMI, **12**, 929-948.

Nevatia R. and Babu R. (1980) Linear feature extraction and description. Computer Graphics and Image Processing.

Nerriec, C. (1990). Clasificador estadístico para la extracción de los contornos. Universidad de Oviedo. Gijón.

Peli T. and Malah, D. (1982). A study of edge detection algorithms. Computer Graphics and Image Processing, **20**, 1-21.

NAVIGATION SYSTEM FOR A MOBILE ROBOT BY MEANS OF AN INFRARED SYSTEM

R. GARCIA, M. MAZO, F. ESPINOSA, P. REVENGA, J. GARCÍA

Departamento de Electrónica. Universidad de Alcala de Henares. Alcala de Henares. Madrid. España
Tfno: +34-1-8854810 Fax: +34-1-8854814 E-mail: eldep@alcala.es.

Abstract. This paper describes how an infrared system could improve a mobile robot following a path using an odometric system. The infrared system takes out the accumulative errors given by the odometrics, and the tracking errors are minimized using a Kalman filter .

Key words. Mobile Robots, Infrared sensors, Kalman Filters, Guidance systems, Navigation.

1. INTRODUCTION

Integrating odometer, infrared waves and telemeter in the guidance of a mobile robot is a control problem, when minimizing the position errors. An algorithm for the following of paths allows moving the vehicle along routes set by the user, before hand.

The error increases with time as the noise coming from an odometric system shows, when estimating position. This noise is a Wiener process. A Kalman filter is used to eliminate this measurement noise, as much as possible.

The reliability of the position estimation increases in relation to the odometric system, but it also decreases with time. Feeding back the position coordinates obtained through the output of the Kalman filter, reduces the error in relation to the real path. This error increases with time as already explained. The problem is solved using an external sensor with an infrared system and a telemeter.

2. THE SENSOR SYSTEM

The mobile robot consists of a platform with two freewheels, and two fixed motive ones [Mazo M, 1992]. It could be considered as a black box of which the inputs are the desired values of the angular velocities of the wheels. The outputs are the real angular velocities which don't differ much from the former values, due to the existence of the internal velocity control.

The infrared and telemetric system basically consist of a few beacons, an infrared receiver and a laser telemeter.

It is necessary to take into account that the position determination requires a distance and angle measurement from the robot to the beacons.

3. CONTROL OF THE DRIVING WHEELS ANGULAR VELOCITIES.

The control of the wheel angular velocities, comes from the control with a PID of the robot mechanical model, so Plant 1 as it is shown in Fig. 1.

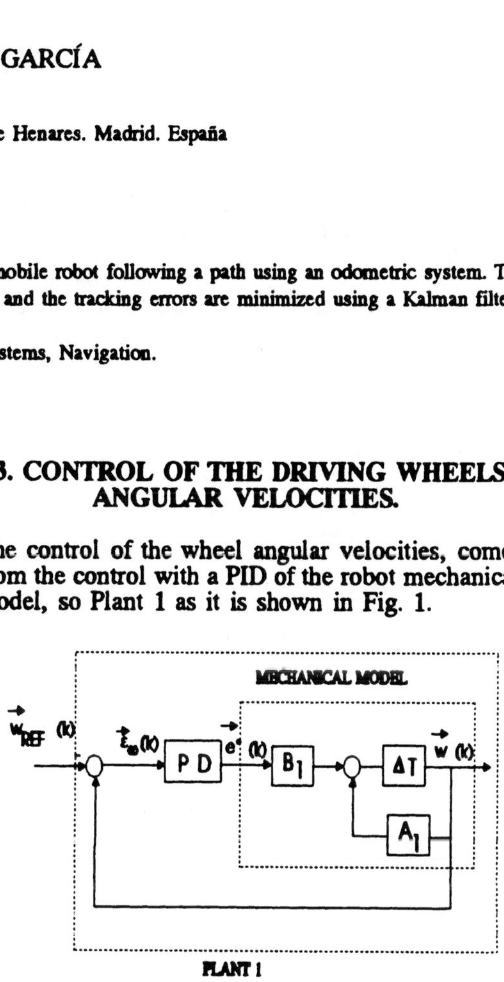

Figure 1. Control of the angular velocities of the wheels.

The mechanical model is:

$$\vec{w}(k+1)=A_1\vec{w}(k)+B_1\vec{e}^*(k) \qquad (1)$$

which expressed in matrix form:

$$\begin{bmatrix} w_R(k+1) \\ w_L(k+1) \end{bmatrix} = \begin{bmatrix} a & 0 \\ 0 & a \end{bmatrix} \begin{bmatrix} w_R(k) \\ w_L(k) \end{bmatrix} + \begin{bmatrix} b & 0 \\ 0 & b \end{bmatrix} \begin{bmatrix} e_R(k) \\ e_L(k) \end{bmatrix} \qquad (2)$$

$$a = e^{-\frac{T_1}{T_m}} \quad ; \qquad b = nk_0(1-e^{-\frac{T_1}{t_m}})$$

The control is carried out by means of a PD and the control signal is:

$$\vec{\varepsilon}^*(k)=k_{pw}\vec{\varepsilon}_w(k)+\frac{k_{dw}}{T}\vec{\varepsilon}_w(k)-\frac{k_{dw}}{T}\vec{\varepsilon}_w(k-1)= \qquad (3)$$
$$=k_{1w}\vec{\varepsilon}_w(k)-k_{2w}\vec{\varepsilon}_w(k-1)$$

and the equation in closed loop is:

$$\vec{w}(k+1)=A_{1w}\vec{w}(k)+A_{2w}\vec{w}(k-1)+B_{1w}\vec{W}_{REF}(k) \qquad (4)$$

The robot is reduced to a "black box" of which inputs and outputs are the angular velocities of the driving wheels connected by means of the previous vector equation [Yoshimasa, 1987].

4. CONTROL OF ROBOT THE GENERALIZED VELOCITIES.

The current aim is controlling the longitudinal and angular velocities of the robot. So, it is necessary to calculate the direct model, which obtains the mentioned velocities from the angular velocities of the driving wheels. The model is the one shown of Fig. 2.

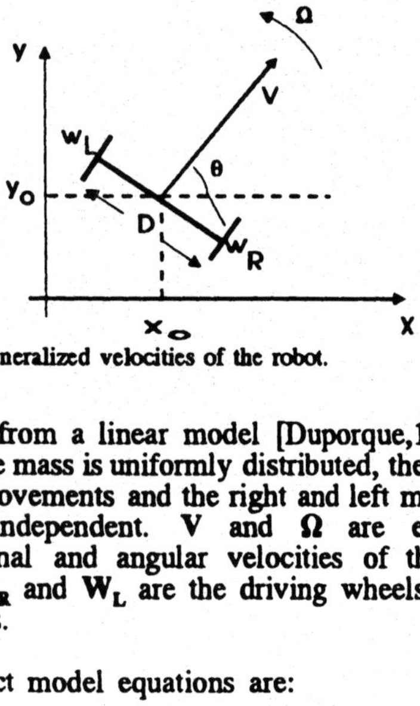

Fig. 2 Generalized velocities of the robot.

Starting from a linear model [Duporque,1990], in which the mass is uniformly distributed, there are no lateral movements and the right and left motors are totally independent. **V** and **Ω** are each the longitudinal and angular velocities of the robot, while $\mathbf{W_R}$ and $\mathbf{W_L}$ are the driving wheels angular velocities.

The direct model equations are:

$$V=RW_T=\frac{R}{2}(w_L+w_R) \qquad (5)$$

$$\Omega=\frac{2R}{D}w_G=\frac{2R}{D}\frac{w_L-w_R}{2}=\frac{R}{D}(w_L-w_R) \qquad (6)$$

The control loop of the velocities **V** and **Ω** is indicated in Fig. 3. Introducing the inverted model, in which equations are inverse to those of the direct model, allows maintaining the symmetry of the equations and also the independent control of the movement and rotation velocities.
The combined equations formed by the inverted model (MI), the plant and the direct model (MD), are the same as those for plant 1 but changing the input and output variables. The only problem is now to control a plant, through a control PD, of which vector equation is:

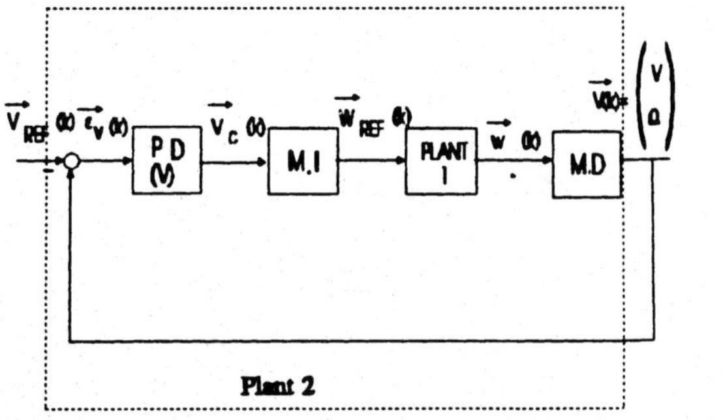

Fig. 3.- Control of the generalized velocities.

$$\vec{V}(k+1)=A_{1w}\vec{V}(k)+A_{2w}\vec{V}(k-1)+B_{1w}\vec{V}_c(k) \qquad (7)$$

The whole subsystem would be equivalent to a plant-2, of which the inputs and outputs are indicated in Fig. 3. The vector equation is:

$$\vec{V}(k+1)=A_{1v}\vec{V}(k)+A_{2v}\vec{V}(k-1)+B_{1v}\vec{V}_{REF}(k) \qquad (8)$$

These equations would permit the user to drive the robot by means of its movement and rotation velocities, which is a more logical way than doing it with the angular velocities of the wheels.

5. ODOMETRIC CONTROL OF THE POSITION

Knowing the longitudinal and angular velocities of the robot, that is the output of plant 2, the odometer and the transformation matrix $\mathbf{T_R}^{-1}$ transform these signals into increments in the cartesian coordinates Δx and Δy. Adding them to the former position gives the cartesian coordinates of the current position [Watanabe, 1989]. These coordinates are compared with the reference ones, obtaining the error signal.

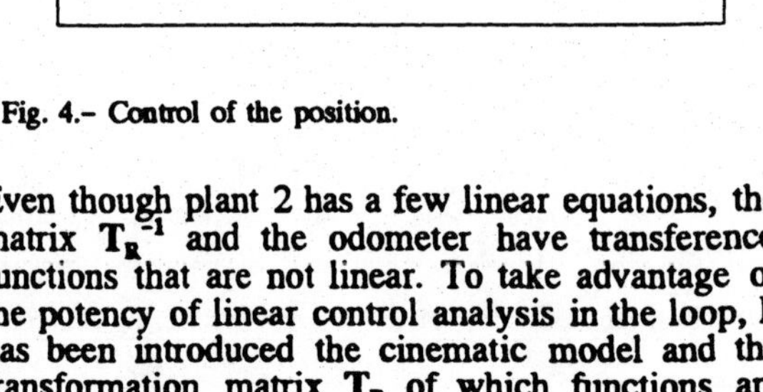

Fig. 4.- Control of the position.

Even though plant 2 has a few linear equations, the matrix $\mathbf{T_R}^{-1}$ and the odometer have transference functions that are not linear. To take advantage of the potency of linear control analysis in the loop, It has been introduced the cinematic model and the transformation matrix $\mathbf{T_R}$ of which functions are inverted to the former ones, in a way that the whole block marked with a pointed line has a linear transference function.

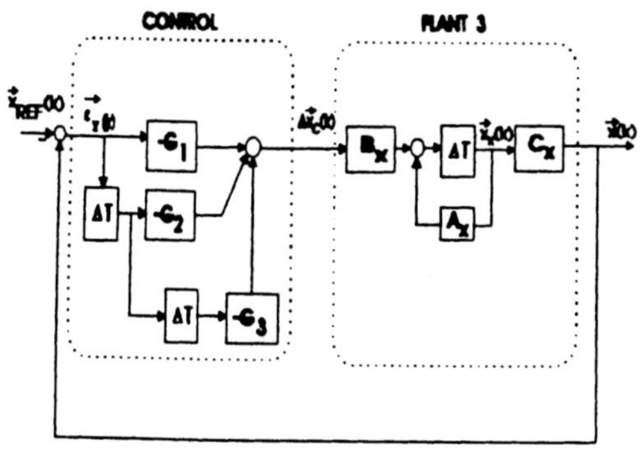

Fig. 5.- Control signals of the position.

The control diagram is reduced to the one shown in Fig. 5. That can be expressed in state space , with a state vector of six components and the matrixes $\mathbf{A_x}$ of 6x6, $\mathbf{B_x}$ of 6x2 and $\mathbf{C_x}$ of 2x6, with constant and known coefficients. The system is controllable and observable.
The state and output vectors would be:

$$\vec{x}_x(k)=\begin{bmatrix} x(k) \\ y(k) \\ x(k-1) \\ y(k-1) \\ x(k-2) \\ y(k-2) \end{bmatrix} \qquad \vec{x}(k)=\begin{bmatrix} x(k) \\ y(k) \end{bmatrix} \tag{9}$$

6. TRACKING OF THE PATHS

The tracking algorithm that has been used consists of following a clothoidal path.
This strategy allows the movement to be very accelerated when the robot is off-path, and practically uniformed when it is on the path. At the end of the straight section the movement can be slow down as required just by reducing the approximation distance.

7. THE KALMAN FILTER

The equations of plant 3 (Fig. 4), in state space, are:

$$\vec{x}_x(k+1)=A_x\vec{x}_x(k)+B_x\Delta\vec{x}_c(k) \tag{10}$$

$$\vec{x}(k)=C_x\vec{x}_x(k) \tag{11}$$

and considering the internal noise that can be generated in the system and the error produced in the measurement, the equations turn out to be:

$$\begin{gathered} \vec{x}_x(k+1)=A_x\vec{x}_x(k)+B_x\Delta\vec{x}_c(k)+\vec{v}(k) \\ \vec{z}(k)=\vec{x}(k)+\vec{\mu}(k)=C_x\vec{x}_x(k)+\vec{\mu}(k) \end{gathered} \tag{12}$$

The measures given by the odometer are contaminated by noise (which is a random process), It could be tried to eliminate it as much as possible through a Kalman filter as shown in Fig. 6.

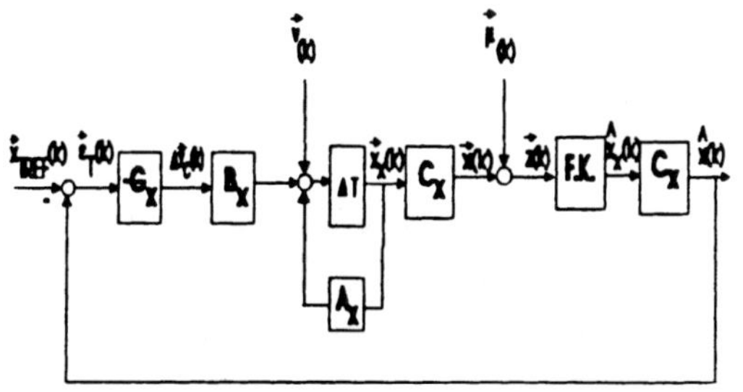

Fig. 6.- Improvement of the position through a Kalman filter.

The application of the Kalman filter requires that the measure and system noises to be white, with known matrixes of covariance, which imply obtaining a few equations that model these two noises.

The previous analysis of measure and system noises, allows the implementation of the Kalman filter [Maravall,1990].

The system is defined by six state variables, which is perfectly logical since in order to understand the system it is required to know the position, velocity and acceleration in relation to the **x** and **y** axis. Out of the six state variables chosen, only the two first, –the coordinates at the instant k–, are affected by the measurement noise in that moment, since the coordinates in the instants **k–1** and **k–2** are not affected by introduced noise measuring in the instant **k**. This fact makes it possible to apply a Kalman filter of reduced size. It not possible to apply the algorithm of the Kalman filter since one of the conditions is that the measure noise ought to be a white noise and the measure noise results a Wiener process.

Introducing the noise models of the system(the status vector will have stochastic components) and transforming the output equation. The equations could be expressed in such a way that the resulting noise will be white. So a Kalman filter could be used [Chui,1990].

The way of comparing the precision of the measure between the different sensors, is by means of its covariance matrices that physically represent the uncertainty with which the sensors obtain the position measures.

The geometric position of the points where the covariance matrix are the same that the squared error, has the equation:

$$\begin{bmatrix} x(k)-\hat{x}(k) \\ y(k)-\hat{y}(k) \end{bmatrix}^T \begin{bmatrix} p_{11}(k) & p_{12}(k) \\ p_{21}(k) & p_{22}(k) \end{bmatrix}^{-1} \begin{bmatrix} x(k)-\hat{x}(k) \\ y(k)-\hat{y}(k) \end{bmatrix}=1 \tag{13}$$

This equation can be expressed, with the help of the parameter **t**, as:

$$\begin{bmatrix} x-\hat{x} \\ y-\hat{y} \\ t \end{bmatrix}^T \begin{bmatrix} \begin{vmatrix} p_{11}(k) & p_{12}(k) \\ p_{21}(k) & p_{22}(k) \end{vmatrix}^{-1} & \begin{matrix} 0 \\ 0 \end{matrix} \\ \begin{matrix} 0 & 0 \end{matrix} & -1 \end{bmatrix} \begin{bmatrix} x-\hat{x} \\ y-\hat{y} \\ t \end{bmatrix}=0 \tag{14}$$

which can be written:

$$[x \cdot y \;\; t]\, \boldsymbol{D} \begin{bmatrix} x \\ y \\ t \end{bmatrix} = 0 \tag{15}$$

this last expression is the general equation of a cone.

Studying the coefficients of the matrix **D** indicates that it is a real ellipse, which expressed in its reduced form is:

$$\lambda_1 x^2 + \lambda_2 y^2 + \frac{D_{33}}{|D|} = 0 \tag{16}$$

or else:

$$\frac{x^2}{a^2} + \frac{y^2}{b^2} = 1 \tag{17}$$

allows to calculate the length of the semi-axis:

$$a = \sqrt{-\frac{D_{33}}{\lambda_1 |D|}} \qquad b = \sqrt{-\frac{D_{33}}{\lambda_2 |D|}} \tag{18}$$

the values of λ_1 and λ_2 are the eigenvalues of the matrix **P(k)**. These error distribution ellipses are therefore being used, in order to compare the different covariance matrixes.

8. CORRECTION WITH INFRARED WAVES

The global scheme, in which it is indicated how the estimation of position has been corrected with the external infrared sensor, is the one presented in Fig.7.

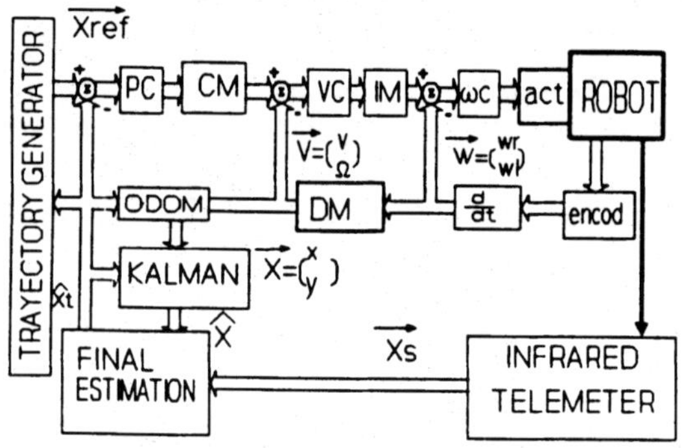

Fig.7 - Global Control System

It can be seen in this block diagram, the control loop of the robot wheels angular velocities, called **plant 1**, the second control loop of the generalized velocities, the third control loop of the position coordinates, as well as the path generator and the Kalman filter. The infrared sensor and the telemeter operate in parallel.

The positions obtained by the Kalman filter and by the infrared sensor, both with their corresponding covariance matrices get to the final position arrive estimator.

The optimum total estimation is the one that complies with its covariance matrix is the "parallel" of the Kalman filter variance matrixes and the sensor ones. The final estimation value of this covariance matrix is $\mathbf{p}_t(\mathbf{k})$:

$$\hat{p}_T(k) = Q_T(k) P^{-1}(k) \hat{p}(k) + Q_T(k) Q_s^{-1}(k) \bar{p}_s(k) \tag{19}$$

it can easily proved that the uncertainty associated to the measure $\mathbf{p}_t(\mathbf{k})$ is:

$$Q_T^{-1}(k) = P^{-1}(k) + Q_s^{-1}(k) \tag{20}$$

These final estimation coordinates are those that are feed back, but keep in mind that, in the same way as the coordinates of the Kalman filter are obtained in each sampling period (250ms), the coordinates of the sensor are obtained in a variable time period that may be longer than 5s.

The covariance matrix value of the sensor depends on the way it is used. Two strategies have been followed and in Fig. 8 the receiver is situated in the coordinates origin and the marker on board the robot.

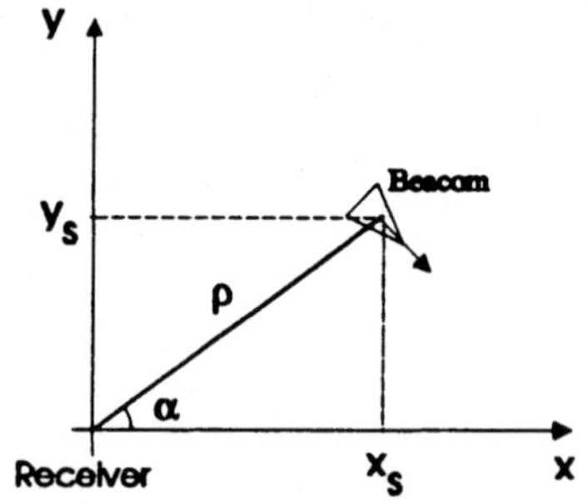

Fig. 8.- Infrared receiver and telemeter positioned externally.

The value of the variances in the measures ϱ and α has been calculated by repeating a fixed measure 2000 times. Taking the main value as a reliable measure, an error file will be obtained from which the variance is.

$$\sigma_\rho^2 = 1{,}5.10^{-5} mm^2$$
$$\sigma_\alpha^2 = 2{,}1.10^{-4} mrad^2 \tag{21}$$

Therefore:

$$x_s = \rho \cos\, \alpha$$
$$y_s = \rho\, sen\, \alpha \tag{22}$$

the variables α and ρ are affected by noise:

$$\Delta x_s = \Delta\rho \cos\alpha - \rho\, sen\alpha\, \Delta\alpha$$
$$\Delta y_s = \Delta\rho\, sen\alpha + \rho\cos\alpha\, \Delta\alpha \tag{23}$$

or else:

$$\Delta \bar{p}_s(k) = \begin{bmatrix} \cos\alpha & -\rho\, sen\alpha \\ sen\alpha & \rho\cos\alpha \end{bmatrix} \begin{bmatrix} \Delta\rho \\ \Delta\alpha \end{bmatrix} \tag{24}$$

The covariance matrix of the sensor would be:

$$Q_s(k) = E\{\Delta \bar{p}_s(k) \Delta \bar{p}_s^T(k)\} =$$

$$= \begin{bmatrix} \sigma_\rho^2 \cos^2\alpha + \rho^2 \sigma_\alpha^2 sen^2\alpha & \sigma_\rho^2 sen\alpha\cos\alpha - \rho^2 \sigma_\alpha^2 sen\alpha\cos\alpha \\ \sigma_\rho^2 sen\alpha\cos\alpha - \rho^2 \sigma_\alpha^2 sen\alpha\cos\alpha & \sigma_\rho^2 sen^2\alpha + \rho^2 \sigma_\alpha^2 \cos^2\alpha \end{bmatrix} \tag{25}$$

Fig. 9 shows the infrared receiver on board the robot and two beacons externally positioned.

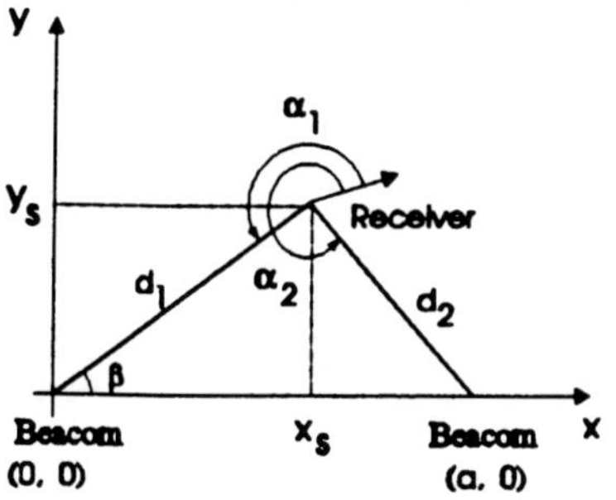

Fig. 9.- Receiver on board the robot.

The covariance matrix will be:

$$Q_s(k)=\begin{bmatrix} \sigma_\rho^2\left(1-\frac{d_2^2 sen^2\alpha}{a^2}\right)+\sigma_\rho^2\frac{(d_1 d_2 sen^2\alpha)^2}{a^2(a^2-d_2^2 sen^2\alpha)}+2\sigma_\alpha^2\frac{(d_1 d_2 sen\alpha\cos\alpha)^2}{a^2(a^2-d_2^2 sen^2\alpha)} & \frac{d_2}{a^2}\sigma_\rho^2 sen\alpha\sqrt{a^2-d_2^2 sen^2\alpha}-\frac{d_1^2 d_2 sen^3\alpha}{a^2\sqrt{a^2-d_2^2 sen^2\alpha}}\sigma_\rho^2-2\sigma_\alpha^2\frac{d_1^2 d_2^2 sen^2\alpha\cos\alpha}{a^2\sqrt{a^2-d_2^2 sen^2\alpha}} \\ \frac{d_2}{a^2}\sigma_\rho^2 sen\alpha\sqrt{a^2-d_2^2 sen^2\alpha}-\frac{d_1^2 d_2 sen^3\alpha}{a^2\sqrt{a^2-d_2^2 sen^2\alpha}}\sigma_\rho^2-2\sigma_\alpha^2\frac{d_1^2 d_2^2 sen^2\alpha\cos\alpha}{a^2\sqrt{a^2-d_2^2 sen^2\alpha}} & \frac{d_2^2 sen^2\alpha}{a^2}\sigma_\rho^2+\frac{d_1^2 sen^2\alpha}{a^2}\sigma_\rho^2+\frac{d_1^2 d_2^2\cos^2\alpha}{a^2}2\sigma_\alpha^2 \end{bmatrix} \quad (26)$$

In both cases these matrices do not increase with time. Initially the measure obtained by the Kalman filter is more reliable, but as time increases, the uncertainty of this measure becomes bigger whereas the one obtained by the infrared sensor will have a constant covariance matrix.

When the measure has been obtained with the sensor, its matrix of covariance is compared to the one of the Kalman filter by means of the uncertainty ellipses semi-axes. In case this is considerably smaller, it indicates that the Kalman filter measurement is more accurate than that one of the sensor. The final estimation, being the parallel, will practically be as the measurement of the Kalman filter. When the sensor covariance matrix is a lot smaller than the one of the Kalman filter, the final estimation will be that one of the sensor. In case that the sensor covariance matrix is comparable to the one of the Kalman filter, the final estimation will be $p_T(k)$, as in the combination previously indicated, so that its covariance matrix would be the "parallel" of the other two.

9. RESULTS

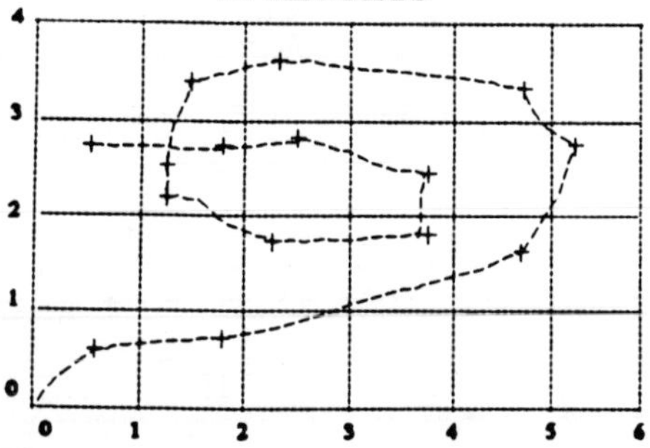

Fig. 10- Results obtained.

Fig. 10 show real paths followed by the robot guided by odometer with **n**=1m and E_μ=90cm. Note the practically uniform separation between the different sampling instants.

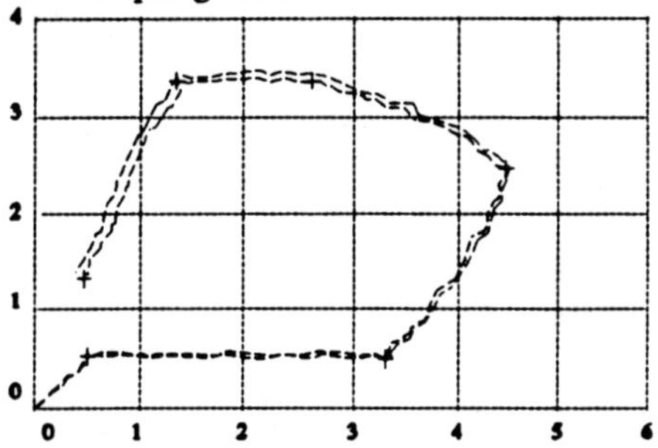

Fig. 11.- Measures with the odometer and the Kalman filter.

Fig. 11 indicates the obtained positions with the odometer and with the Kalman filter. It can be seen how the odometer measurements and the Kalman filter are practically the same, during short sections.

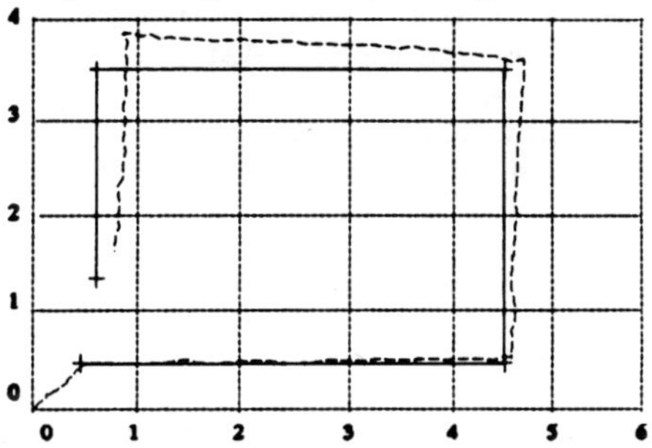

Fig. 12.- Diferences between the odometer and the Kalman filter (increased).

Changing the covariance matrix **R(k)**, in Fig. 12, the difference between the odometer and the position estimated by the Kalman filter is increased and it can be seen how the deviation of the odometer in relation to the path increases with time and thus is bigger than the error of position obtained by the Kalman filter.

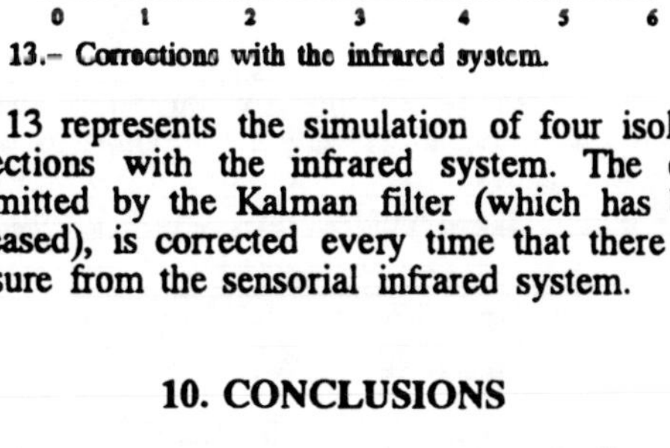

Fig. 13.- Corrections with the infrared system.

Fig. 13 represents the simulation of four isolated corrections with the infrared system. The error committed by the Kalman filter (which has been increased), is corrected every time that there is a measure from the sensorial infrared system.

10. CONCLUSIONS

The total block diagram of the robot guidance is shown in Fig. 7.

The four control loops with different sampling periods for each one are also shown. In the three first ones the sampling period is constant, while the last one uses a variable sampling period that depends on the instant that the infrared system has got its measure available.

The precision of the measure with the infrared system is at its maximum when the robot is standing still. That could be avoided if the speed of obtaining the measures would be very high, near milliseconds. The only way of obtaining these velocities is through two receivers set as path markers, but the high cost of those receivers has made this solution impossible. The substitution of these receivers for video cameras sensitive to infrared waves could be an interesting solution.

The whole study has been carried out with linear models so that their calculate power could be used, but It could be possible that working with a non-linear model would have given better results, but naturally the calculation difficulties would have been bigger.

The different loops control is not the only possible way and could be improved, but the fundamental part of the study has been focused on the odometric system noise. In order to eliminate the noise it has been tried the Kalman filter but it has also been shown that filter covariance matrix increases slowly with time without a permanent rule.

The odometric system has proved to be a very high precision, since the robot can be working during a few minutes without an excessive error.

Depending on the sampling instant, it can not be guaranteed that the system produces a smaller measurement error. It has been chosen to make a final estimation which is a combination of the positions obtained by the Kalman filter and the infrared system. This final estimation is made in such a way that its covariance matrix is the "parallel" of the two systems covariance matrixes, so that the two measurement are used in an optimal way. In any case of both slipping or noises of the system will disguise the measure of the Kalman filter, therefore the measure of the sensorial infrared system is more reliable than what theory sets up.

The fact that the robot knows exactly its own position is very important for achieving more complex functions as for example, the detection of obstacles and the way of avoiding them, environments with crossings or with other robots, making up paths to reach predetermined destinations, searching for paths when the robot is guided by vision and has lost its way, incorporation of manipulating arms over the mobile vehicle etc.

The horizon of possibilities is so wide at the moment that every step at any stage points towards the whole final aim: making up machines with more and more similar behaviour to that one of the human beings'.

11. REFERENCES

Chui, C.K. and Chen G. (1990). *Kalman filtering with real time applications.* Springer Verlag.New York.

Duporque V. (1990). Techniques classiques de commande des robots mobiles. *Seminaire Les Robots Mobiles.* La Défense. París.

Maravall D. (1990). *Control, Estimación y Filtrado de Sistemas.* Colegio Oficial de Ingenieros de Telecomunicación. Madrid.

Mazo M., García R, Revenga P., Ureña J and Santiso E. (1992). Mobile Robot with multisensor integration. In: *Symposium on Intelligent Components and instruments for control applications.* Malaga. p401–406.

Watanabe Y. and Yuta S. (1989). Estimation of position and its uncertainty in the dead reckoning system for the wheeled mobile robots. *20th ISIR.* Tokyo.p205–210.

Yoshimasa G. and Stentz A. (1987) Mobile Robot Navigation: the CMU system. *IEEE expert.* nº 4. p320–324.

NEURAL NETWORK LOCAL NAVIGATION OF MOBILE ROBOTS IN A MOVING OBSTACLES ENVIRONMENT

J. GOMEZ-ORTEGA, E. F. CAMACHO and J. QUERO

Dpto. Ing. de Sistemas y Automática, Univ. de Sevilla, Avd. Reina Mercedes s/n, Spain.
Fax: +34-5-4556849, E-mail:juango@esi.us.es

Abstract. This paper presents a local navigation method based on generalized predictive control. A modified cost function to avoid moving and static obstacles is presented. An Extended Kalman Filter is proposed to predict the motions of the obstacles. A Neural Network implementation of this method is analysed. Simulation results are shown.

Key Words— Mobile robots; guidance system; obstacle avoidance; neural Network; predictive real-time control; extended Kalman filter.

1. INTRODUCTION

One of the most important issues in the design and development of intelligent mobile robots is the navigation problem. This consists of the ability of a vehicle to plan and execute collision-free motions within its environment.

This problem can be divided into two hierarchical levels. The higher level, called global navigation or path planning, is concerned with the generation of a trajectory (space-time) from an initial configuration to a goal configuration, avoiding the known static and mobile obstacles in the environment. At this level, only those obstacles whose situation and motion are previously known are taken into account. Although some works presented in the literature consider the kinematic and dynamic models of the vehicle (Shiller and Gwo 1991), most of them only consider the geometric approach to the problem. Their computation time is not acceptable for real-time control of mobile robots. Some well known solutions are proposed by: Fujimura and Samet (1989), based on including time as one of the dimensions of the model world. This allow them to regard the moving obstacles as being stationary in the extended world; Kant and Zucker (1986) proposed a solution based on the decomposition of the trajectory planning problem (TPP) into two subproblems: the path planning problem (PPP), which is concerned with planning the path to avoid stationary obstacles, and the velocity planning problem (VPP), which is concerned with planning the velocities along the path to avoid moving obstacles. Although this reduces the complexity of the global problem, this solution does not change the predefined path and it cannot avoid moving obstacles with colinear trajectories to the robot's. Erdmann and Lozano-Perez (1987) proposed a solution based on a planner for moving objects that constructs a configuration space each time an object in the scene changes its velocity.

The lower level, called local navigation or guidance, is concerned with driving the vehicle through the trajectory generated by the global planner, now avoiding the unexpected obstacles (static or moving), and compensating the uncertainty of the configurations and motions data used by the global planner, using the real-time enviroment information provided by the sensor system. Usually, at this level, the vehicle kinematics and/or dynamics , and kinematics constraints (such as maximum velocity or acceleration) are considered by the system. Kant and Zucker (1988) have proposed a modification to their algorithm, adding a low control module, which compensates the uncertainty in the velocities of the moving obstacles considered by the planner. The solution proposed by Kathib (1989) is based on the *artificial potential field* (APF) approach, which is one of the most popular methods of real-time static obstacle avoidance. Another approach based on APF has been proposed by Borenstein and Koren (1989). They use the concept of the *certainty grid* to obtain the repulsive forces from the data for the sensors. Griswold and Eem (1990) ,based on Kant and Zuckers approach, propose a solution for un expected moving objects. The uncertainty in velocity and direction of moving obstacles are considered simply as "noise". This noise suggests that speed and direction angles of moving obstacles, relative to the robot, should be considered random variables, with a predetermined distribution, at a fixed time. Wang and Tsai (1991) use a modified least-mean-squared-error classification algorithm (used in pattern recognition) to compute a local collision-free navigation path among moving obstacles with no *a priori* position information, in an indoor corridor environment. The trajectories of moving obstacles are predicted by a real-time LMSE estimation algorithm, and the speed value of the robot is determined by the manoeuvering board technique used for nautical navigation. Papageorgiou and Steinkogler (1993) have proposed an optimal control approach to the problem of moving vehicles in changing environments. This approach is the most similar to the proposed one, although they give a numerical solution to the minimization problem while here, a neural network (NN) solution is presented. Papageorgiou and Steinkogler give times of less than a second to solve the optimal problem in some examples, using a very simple kinematic model (heading is not considered). A more complex model, which takes into account the heading of the robot and the velocities of both driving wheels, is considered here. With this model, the numerical solution requires too much computation for real-time.

2. GENERALIZED PREDICTIVE CONTROL

The *Generalized Predictive Control (GPC)* proposed by Clarke *et al.* (1987) is an optimal control technique, that has inspired much research work in the recent years. The objetive of the GPC is to drive future system outputs (in our case the robot position and orientation) close to their desirable values in some sense, bearing in mind the control activity required to do so. This is done using a *receding horizon* approach for which, at each sample instant, using a prediction model to generate a set of predicted outputs, some appropriate quadratic function J of the future errors and controls is minimized, assuming that after some *control horizon* H further increments in control are zero. Only the first control is applied, resulting in a control law that belongs to the class known as *Open-Loop-Feedback-Optimal control.* The cost function J can be of the form:

$$J(H,\Delta V)=\sum_{i=1}^{H}[X(t+i)-X_d(t+i)]^2$$

$$+\sum_{i=1}^{H}\lambda[\Delta V(t+i-1)]^2$$

where X is the vector of predicted outputs, X_d is the vector of desired values for X, V is the control variables vector, and λ is a weighting factor. Some research has been done aiming to apply this technique to the path tracking problem (Ollero and Amidi 1991).

This paper proposes a modification of the cost function J to include a term that penalizes the proximity to any obstacle (static or moving) in any of the H next sample instants. This leads to an obstacle avoidance with low control cost. The idea is that if it is noticed that a collision may be produced in the future, it will begin to avoid this situation with smooth control actions. The problem can be defined as follows: given a trajectory (space-time), drive the robot to follow it using the on-line sensor data, avoiding the unexpected obstacles found in the environment, and compensating the uncertainties in the data (positions of the obstacles) used by the global planner. The new cost function J is (see Fig. 1):

$$J(H,\Delta V)=\sum_{i=1}^{H}[X(t+i)-X_d(t+i)]^2$$

$$+\sum_{i=1}^{H}(\lambda_1([\Delta V_r(t+i-1)]^2+[\Delta V_l(t+i-1)]^2)$$

$$+\lambda_2[V_r(t+i-1)]-V_l(t+i-1)]^2)$$

$$+\sum_{j=1}^{NMO}(\sum_{i=1}^{H}\frac{\xi_1}{P_1(t+i)[dist(X(t+i),XMO^j(t+i))]^2})$$

$$+\sum_{j=1}^{NSO}(\sum_{i=1}^{H}\frac{\xi_2}{P_2(t+i)[dist(X(t+i),XSO^j(t+i))]^2})$$

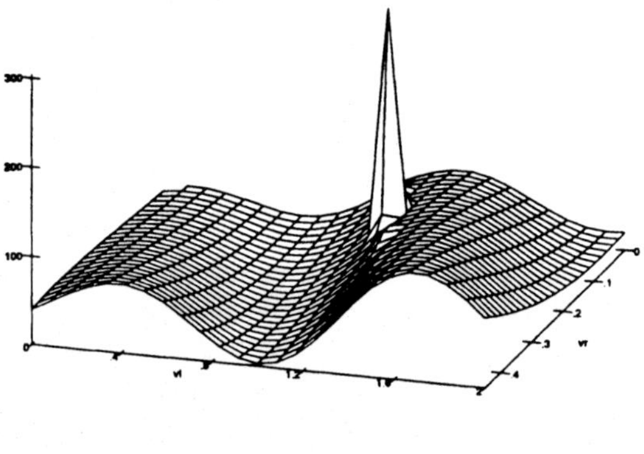

FIG. 1. Objetive Function $J(H,\Delta V)$ for H=1 and one static obstacle

where $X(i)=\{x(i),y(i),\theta(i)\}$ is the position and orientation vector of the robot in the sample instant i, $X_d(i)=\{x_d(i),y_d(i),\theta_d(i)\}$ is the desired position and orientation vector for the control horizon, and $XMO^j(i)=\{xmo^j(i),ymo^j(i)\}$ and $XSO^j(i)=\{xso^j(i),yso^j(i)\}$ are the positions of the moving and static obstacles in the sample instant i. V_r and V_l are the right and left velocities of the two driving wheels, which are the control variables. NMO and NSO are the number of moving and static obstacles respectivly, and λ_1, λ_2, ξ_1 and ξ_2 are weighting factors. P_1 and P_2 are the covariance matrices of the predictions of the future positions of the obstacles, and *dist* is the euclidean distance between the robot and the obstacles.

For this formulation a model is needed to predict the future positions and orientations of the robot and a model to predict the positions of the moving obstacles. The following kinematic model (which corresponds to a differential-drive vehicle) is used for the first issue:

$$\theta(k+1) = \theta(k) + AT$$

$$x(k+1) = x(k) + \frac{V}{A}(\sin(\theta(k) + AT) - \sin(\theta(k)))$$

$$y(k+1) = y(k) - \frac{V}{A}(\cos(\theta(k) + AT) - \cos(\theta(k)))$$

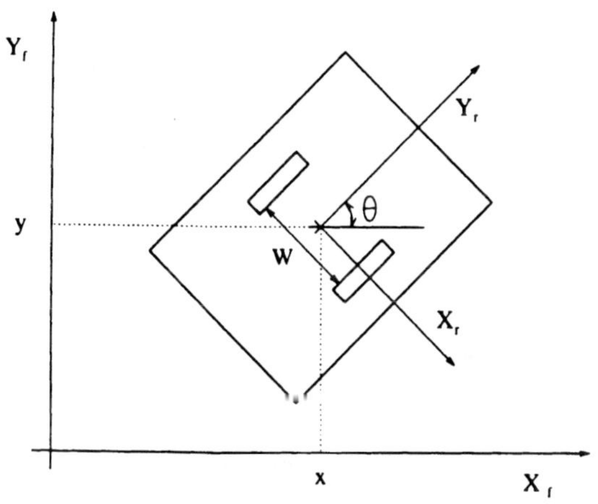

FIG. 2. Reference Frame

where x, y, θ are the position and orientation of the robot in a fixed reference frame, $A = \frac{V_r - V_l}{W}$ and $V = \frac{V_r + V_l}{2}$. T is the sample time and W is the distance between wheels (see Fig. 2).

3. MOBILE OBSTACLE MOTIONS PREDICTION

To predict the positions of the moving obstacles in the future, when the velocity of the obstacle is assumed to be constant between sampling intervals, the following non linear model can be considered:

$$x_o(k+1) = x_o(k) + V(k+1)cos(\theta_o(k+1))$$

$$y_o(k+1) = y_o(k) + V(k+1)sin(\theta_o(k+1))$$

$$V(k) = ((x_o(k) - x_o(k-1))^2 +$$

$$(y_o(k) - y_o(k-1)^2)^{1/2}$$

$$\theta_o(k) = arctan(\frac{y_o(k) - y_o(k-1)}{x_o(k) - x_o(k-1)})$$

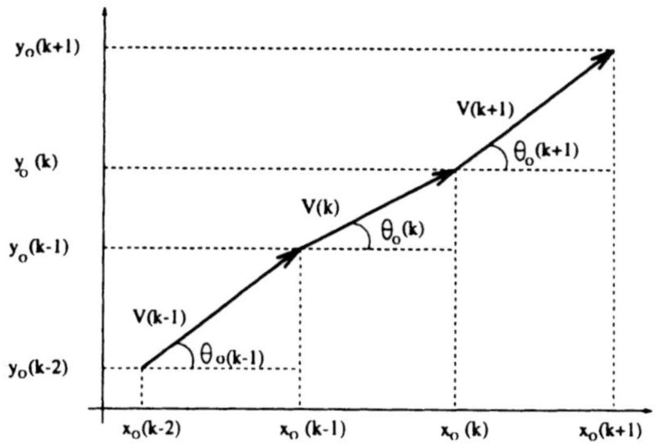

FIG. 3. Obstacle motion parameters

where x_o, y_o are the position of the mobile obstacle, V is the lineal velocity between two positions and θ_o is the angle of the velocity vector in respect to the horizontal x axis (see Fig. 3).

Now, the following hypothesis are made:

$$V(k+1) = V(k)$$

$$\Delta\theta_o(k+1) = \Delta\theta_o(k) = \theta_o(k) - \theta_o(k-1)$$

The *Extended Kalman Filter* approach is applied to this model to predict the future positions of the mobile obstacle and their covariance matrices $P_1(k+j|k)$. The result of the propagation cycle for H periods of prediction are the following equations:

$$\hat{x_o}(k+j|k) = \hat{x_o}(k|k) + V(k)T\sum_{i=0}^{j-1} cos(\mu(i))$$

$$\hat{y_o}(k+j|k) = \hat{y_o}(k|k) + V(k)T\sum_{i=0}^{j-1} \sin(\mu(i))$$

$$\mu(i) = 2arctan(\frac{\hat{y_o}(k+i|k) - \hat{y_o}(k+i-1|k)}{\hat{x_o}(k+i|k) - \hat{x_o}(k+i-1|k)})$$

$$-\arctan(\frac{\hat{y_o}(k+i-1|k) - \hat{y_o}(k+i-2|k)}{\hat{x_o}(k+i-1|k) - \hat{x_o}(k+i-2|k)})$$

With the actualization cycle only $\hat{x_o}(k+1|k+1), \hat{y_o}(k+1|k+1)$ and the actualized covariance matrix $P_1(k+1|k+1)$ are calculated, because only measures of the $k+1$ instant are available; but these reduce the covariance matrix for the next predictions.

To take into account the prediction uncertainties, the distances between the robot and the obstacles are penalized with the covariance matrices obtained from the Kalman Filter equations. In the computation of the robot-obstacles distances, it will be assumed that the obstacles have a circular form.

4. THE NEURAL NETWORK APPROACH

The minimization of the cost function J cannot be obtained in real time with numerical methods. So, a Neural Network solution is proposed, which guarantees real time for the robot control. The Neural Network approach for robot guidance has been proposed by other researches (Pomerlau 1990), (Meng and Picton 1992).
The architecture of the NN controller consist of a single hidden layer backpropagation network.

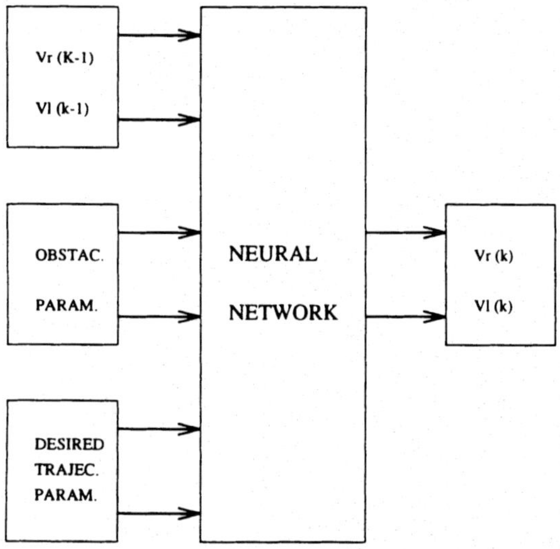

FIG. 4. Neural Network Scheme

The input layer consists of three modules (see fig. 4). The first one includes two network inputs associated with the control values in the last sample instant. The second module corresponds to the obstacle parameters: two inputs for each static obstacle (distance and orientation), and five inputs for each moving obstacle $(x_o(k), y_o(k), V(k+1), \theta_o(k+1), \Delta\theta_o(k+1))$. The third module includes ten network inputs which correspond to the parametrization of the desired trajectory in the next H sample instants. These parameters consist of the location and orientation of the first point of the local trajectory, the curvature of the next H points and an average desired velocity. The output layer consists of two nodes which correspond to the control command: the left and right wheels velocities.
The training module consists of a backpropagation scheme, shown in fig. 5, where the training patterns are solved by a GPC module which uses a numerical method to generate the outputs. The training patterns are selected properly to represent all the possible situations of driving among obstacles. Also, a local reference system is used to reduce the parameters range of variation.
The NN approach will be as follows:

- At time k, obtain the obstacles position parameters from the sensor system. Predict the future positions of the moving obstacles.

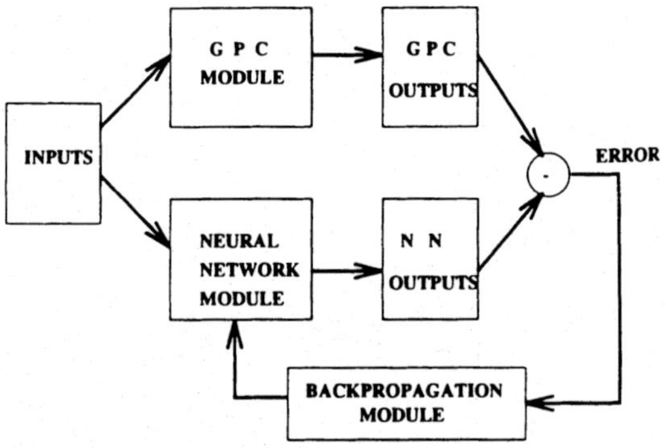

FIG. 5. Neural Network training Scheme

- Using the NN, predict the future H positions of the robot as if no obstacles where encountered.

- Detect if a collision may be produced in the next H sample instant.

- if not, apply the first NN control outputs to the vehicle.

- if a collision may be produced, then use the predicted positions and orientations of the robot as the new desired trajectory, and compute the NN again, now considering the obstacles.

The use of the predicted positions of the robot as a new desired trajectory is justified by the reduction of the number of training patterns necessary to obtain a good performance because a symetry analysis has can be made. In this analysis it has been supposed that the robot is always on the desired path.
Finally, it is important to notice that this method solves, at the same time, the problems of finding avoidance local trajectories and the control one. This guarantees that the trajectory followed by the robot is continuous in curvature, avoiding discontinuities in the wheels velocities, which produces errors between the predicted and the real position of the vehicle.

5. RESULTS

The proposed control structure has been tested by simulation with a model of the *Labmate* mobile robot (T. R. C. 1989). The NN used consisted of seventeen input neurons corresponding to the past control actions, past mobile obstacles positions and future refernce trajectory. The hidden layer was composed of 35 neurons and the output layer consisted of two neurons corresponding to the left and right wheel velocities for the mobile robot.
The NN was trained in a supervised manner as described previously. The control horizon choosen for the GPC was made equal to six. And the

weighting factors were given the following values: $\xi_1 = 3, \xi_2 = 3, \lambda_1 = 63, \lambda_2 = 10$. Where λ_1 and λ_2 correspond to the weights of the module of the velocity and to the angular velocity respectively. The high value of λ_1 is to make sure that the GPC will not choose the *easy* solution of stopping the robot and wait for the mobile object to pass. The relatively high value of λ_2 ensures a smooth trajectory. Fig. 6 shows the evolution of the error function $E(i) = \sum_{i=1}^{N}(O_d(i) - O(i))^2$ of the training phase, where N is the number of training patterns, and $O_d(i)$ and $O(i)$ are the desired and network output for the iteration i.

The simulated results obtained for three situations, different from the training cases, can be seen in fig. 7. The trajectories shown on the left hand side of fig. 7 correspond to the solutions obtained when applying the GPC controller while the trajectories shown on the right hand side correspond to the solutions obtained with the NN. Fig. 7.1 corresponds to an obstacle which is coming towards the mobile robot, while figures 7.2 and 7.3 correspond to obstacle trajectories crossing the mobile robot trajectory. The trajectories shown in fig. 7.3 correspond to a case where the initial position of the mobile robot is not sitting on the desired path. In all cases, the GPC and NN make the mobile robot take the necessary evasive control actions. As expected, the NN reproduces the behaviour of the GPC control quite well and takes only small fraction of the computation time required for solving the GPC which has to be solved using a numerical optimization algorithm (powell method has been used here).

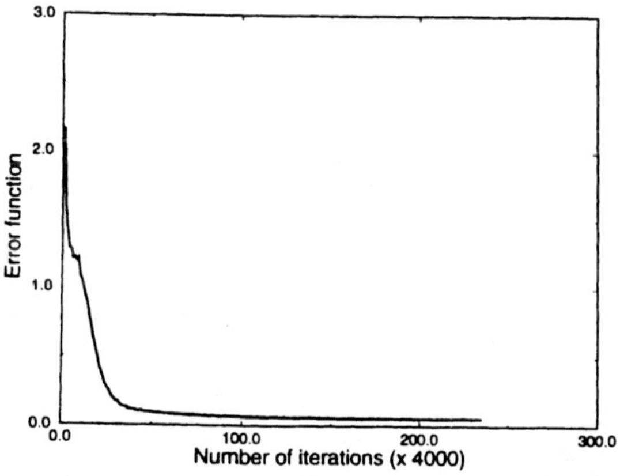

FIG. 6. Evolution of the error function

6. CONCLUSIONS

A local navigation method based on the GPC control algorithm has been proposed. A new cost function which penalizes the inverse of the distance between the robot and the moving and static obstacles has been presented. A prediction method for the future positions of the mobile obstacles has been studied. Finally a Neural Network implementation of the GPC method has been proposed to obtain real-time performance. Simulation results have been presented.

7. ACKNOWLEDGEMENT

The authors would like to acknowledge the *CICYT* for funding this work under grants TAP93-0408 and TAP93-0581.

REFERENCES

Borenstein, J. and Y. Koren (1989). *Real-Time Obstacle Avoidance for Fast Mobile Robots.* IEEE trans. on System, Man, and Cybernetics, vol. 19, n0 5, pp 1179-1187.

T. R. C. (1989). *Labmate reference manual.* (Transitions Research Corporation).

Clarke, D. W., C. Mohtadi and P. S. Tuffs (1987). *Generalized Predictive Control - Part I. The Basic Algorithm.* Automatica, vol 23, no 2, pp 137-148.

Erdmann, M. and T. Lozano-Perez (1987). *On multiple moving objects.* Algorithmica, vol. 2, no. 4 pp 477-521.

Fujimura, K. and H. Samet (1989). *A Hierarchical Strategy for Path Planning Among Moving Obstacles..* IEEE trans. on robotics and automation, vol. 5, no. 1, pp 61-69.

Griswold, N. C. and J. Eem (1990). *Control for mobile Robots in the Presence of Moving Objects.* IEEE trans. on Robotics and Automation, vol 6, no 2, pp 263-268.

Kant, K. and S. W. Zucker (1986). *Toward Efficient Trajectory Planning: The Path-Velocity Decomposition..* The Int. journal of Robotics Research, vol. 5, no. 3, pp 72-89.

Kant, K. and S. W. Zucker (1988). *Planning Collision Free Trajectories in Time-varying environments: A Two-Level hierarchy..* Proc. IEEE Int. Conf. on Robotics and Automation, pp 1644-1649.

Kathib, O. (1989). *Real-time Obstacle Avoidance for Manipulators and Mobile Robots.* The Int. Journal of Robotics Research, vol.5, no. 1, pp 90-98.

Meng, H. and P. D. Picton (1992). *Obstacle Avoidance Using a Neural Network Controller and Visual Feedback.* Proc. SICICA 92, pp 563-568.

Ollero, A. and O. Amidi (1991). *Predictive Path Tracking of Mobile Robots. Aplications to the CMU Navlab..* Proc. IEEE Fifth Int. Conf. on Advanced Robotics (Pisa), vol 2, pp 1081-1086.

Papageorgiou, M. and A. Steinkogler (1993). *Real-Time Optimal Control of Moving Vehicles in Changing Environments.* Proc. of the 12th IFAC World Congress (Sydney), vol 3, pp 107-112.

Pomerlau, D. A. (1990). *Neural Network Based Autonomous Navigation, (Vision and Navigation. The Carnegie Mellon Navlab).* C.E. Thorpe Editor., Kluwer Academic Publishers, pp 83-93.

Sharma, R. (1992). *Locally Efficient Path Planning in an Uncertain, Dynamic Environment Using a Probabilistic Model.* IEEE trans. on Roboics and Automation, vol 8, no 1, pp 105-110.

Shiller, Z. and Y. Gwo (1991). *Dynamic Motion Planning of Autonomous Vehicles.* Trans. IEEE on Robotics and Automation, vol 7, no 2, pp 241-249.

Wang, L. and W. Tsai (1991). *Collision Avoidance by a Modified Least-Mean-Square-Error Scheme for Indoor Autonomous Land Vehicle Navigation.* Int. Journal of Robotics Systems, vol 8, no 5, pp 677-698.

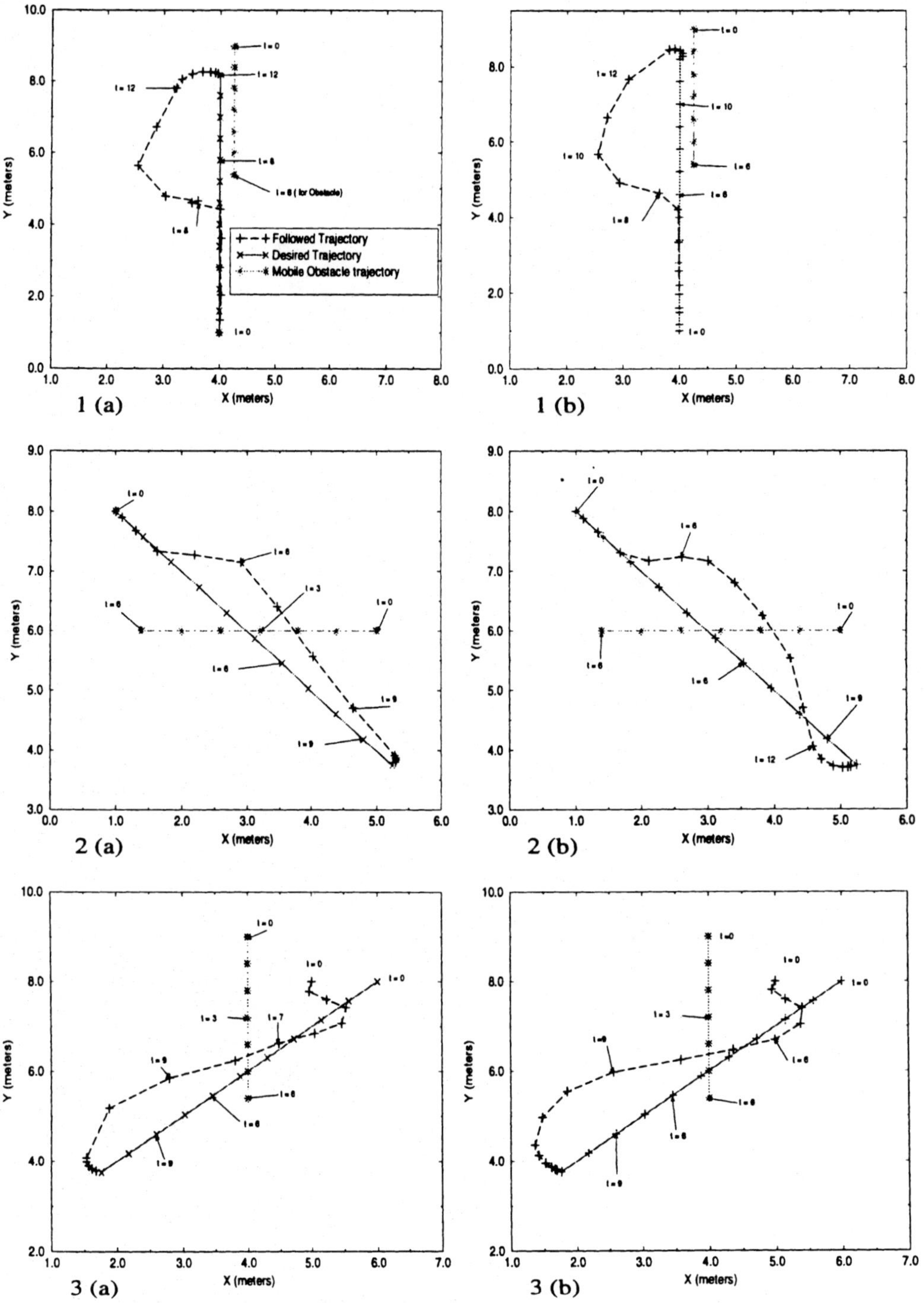

FIG. 7. Results.The left figures correspond to the numerical solved GPC. The right figures correspond to the Neural Network solution.

A COMBINED PATH GENERATION AND TRACKING CONTROLLER FOR MOBILE ROBOTS[1]

J. L. MARTÍNEZ* and A. OLLERO**

*Universidad de Málaga. Departamento de Ingeniería de Sistemas y Automática. Plaza El Ejido, 29013 Málaga (Spain). Fax:+34-5-2131413. Email: jlmartinez@ctima.uma.es
**Universidad de Sevilla. Departamento de Ingeniería de Sistemas y Automática. Avenida Reina Mercedes, 41012 Sevilla (Spain). Fax:+34-5-4556849. Email: aollero@etsii.us.es

Abstract. A new method to automatically generate paths for wheeled mobile robots is proposed. The paper also includes the application to RAM-1, a mobile robot testbed. The control system of an autonomous vehicle involves several functions such as perception, path planning, path tracking and low level control. In this paper an intelligent component of such a control system is presented. This component incorporates path generation, to link the passing points or route points defined by a path planner, and path tracking to compute the control commands so that the vehicle can follow the generated path.

Key Words. Mobile robots, path generation, path tracking, fuzzy control, real time vehicle control system.

1. INTRODUCTION

The ability of a real vehicle to track a path depends on the characteristics of the path. Thus, continuity in position, orientation and curvature are basic constraints to be satisfied by the path generator. However, smooth variation in curvature is still needed because of the relation with the characteristics of the required steering motion.

Numerous path generation methods have been presented in the literature. In the simplest methods a path is composed of a sequence of straight lines (Moravec, 1983). This kind of path is only continuous in position. The continuity in orientation can be achieved by combining straight lines with circular arcs (Hongo, 1985).

Subsequently, other methods have used quintic cartesian polynomials (Shin, 1990), polar polynomial curves (Nelson, 1989), Bezier's polynomials (Rombaut *et al.*, 1991), B-splines (Muñoz *et al.*, 1993a) and β-splines curves (Muñoz *et al.*, 1993b). These paths are continuous in position, orientation and curvature, but it is not always easy to ensure a smooth curvature transition.

More sophisticated techniques use clothoid curves segments (Kanayama and Miyake, 1985), (Shin *et al*, 1993) and cubic spirals (Kanayama and Hartman, 1989). Smooth variation in curvature can be obtained directly. The drawback is that it is not an easy matter to generate these curves because there are no closed form expressions available and they must be evaluated using numerical methods. Thus, computer requirements may preclude its application in real time.

On the other hand, path tracking consists on the computation of steering commands in order to follow the generated path. One of the most simple path tracking methods is the so called "pure pursuit" (Amidi, 1990).

Differences between path tracking and path generation are not clear. In some path tracking techniques, like the quintic polynomial fit (Shin, 1990) or the generalized predictive path tracking method (Ollero and Amidi, 1991), approach trajectories are built in order to follow a given path.

The new intelligent path generation and tracking method proposed in this paper is based on the recursive application of the pure pursuit technique to generate a path, followed by the application of the same technique to compute the control command to follow the path.

The method has been applied successfully in the RAM-1 control architecture. RAM-1 (see Fig. 1) is a mobile robot, designed and built for research and applications in indoor and outdoor industrial environments (Ollero *et al.*, 1993). In this paper, the figures showing the application of the proposed method were obtained from experiments with RAM-1.

This paper is organized as follows. In the next section the proposed method is described. Sections 3, 4, 5 reports the generation of approach, local and global

[1]This research has been partially supported by the CICYT projects TAP93-0581, and TAP93-0608.

paths respectively, by using the new method. The results of the experiments with the real vehicle are also presented in the paper.

Fig. 1. The autonomous mobile robot RAM-1

2. THE PROPOSED METHOD

When a vehicle is following a path, its control system knows its actual posture by means of the odometric system. Mobile robot posture is composed of position, orientation, curvature and speed. So, the vehicle control system can be commanded to store the postures that it reaches at regular distances. The list of recorded postures forms a path. This path has the following properties:

- It is continuous in position, orientation, curvature, and speed. This is because the vehicle cannot execute discontinuous motions.
- The path has a smooth transition in curvature and speed, that the mobile robot is physically capable to execute.
- In the new path the postures are evenly separated.

Thus, if there is some path to follow and an initial posture for the vehicle, then it is possible to simulate the behaviour of the robot, when following the path with a path tracking method, in order to generate a new path. The pure pursuit technique has been chosen for this purpose because it is very easy to compute and performs well enough.

In the pure pursuit path tracking technique, the path is tracked by repeatedly fitting circular arcs to different goal points on the path to be followed as the vehicle moves forward (see Fig. 2).

In the basic pure pursuit method the vehicle dynamics is not considered. However, when it is appropiately tuned, using fuzzy logic, this technique has demonstrated high perfomance (Ollero *et al.*, 1994).

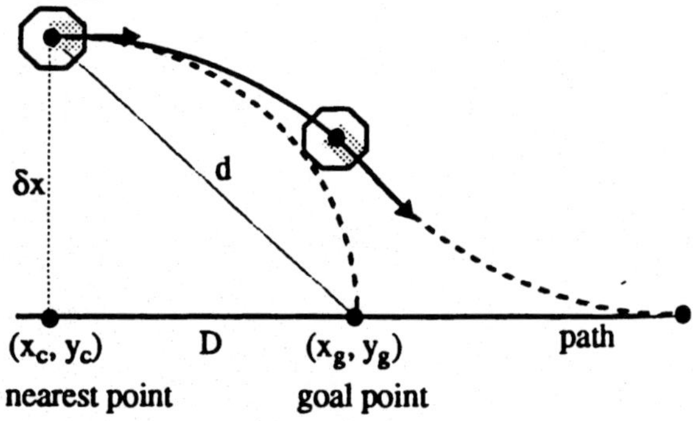

Fig. 2. Path tracking with pure pursuit

Let $(x_0, y_0, \phi_0, \gamma_0)$ be the initial posture of the mobile robot, and $\{(x_i, y_i, \phi_i, \gamma_i), i>0\}$ be the list of postures that define a path to follow. Since pure pursuit does not directly use the orientation and curvature of the path, it is not mandatory for the path to be followed to have these components defined.

Then, the algorithm to compute a new path $\hat{L} = \{(\hat{x}_i, \hat{y}_i, \hat{\phi}_i, \hat{\gamma}_i), i \geq 0\}$ is:

1. Initialize the new generated path:

   1.1 Let the first point of the path be equal to the initial posture:

$$\hat{x}_0 = x_0 \quad \hat{y}_0 = y_0 \quad \hat{\phi}_0 = \phi_0 \quad \hat{\gamma}_0 = \gamma_0 \tag{1}$$

   1.2 Let the number of postures recorded of the path n be equal to 1.

2. Generate the n-th posture of the path. Repeat until the last generated posture (the simulated actual position of the vehicle) reaches the path to follow:

   2.1 Choose a goal point (x_g, y_g) in the path to follow somewhere ahead of the vehicle:

      2.1.1 Look for the point in the path closer to the vehicle (x_c, y_c).

      2.1.2 Compute a certain lookahead distance D.

      2.1.3 Let the goal point (x_g, y_g) be the point over the path to follow, that is separated from the nearest point (x_c, y_c) the lookahead distance D (see Fig. 2).

   2.2 Compute the desired curvature of the vehicle γ_d, by using the pure pursuit control law:

$$\gamma_d = -2\frac{\delta x}{d^2} \tag{2}$$

where δx is the x displacement of the goal point in vehicle coordinates and d is the distance between the current position of the mobile robot and the goal point (see Fig. 2):

$$\delta x = (x_g - x_{n-1})\cos\phi_{n-1} + (y_g - y_{n-1})\sin\phi_{n-1}$$

$$d = \sqrt{(x_g - x_{n-1})^2 + (y_g - y_{n-1})^2} \qquad (3)$$

2.3 Simulate the behaviour of the robot when trying to reach the desired curvature, and it has travelled the distance Δs between two consecutive postures at a certain constant speed v:

2.3.1 Compute the curvature of the new posture. If it is considered that the response of the steering system to reach γ_d from γ_{n-1} is not perfect, and it can be approximated by a first order model, then:

$$\gamma_n = \gamma_{n-1} e^{-\frac{\Delta t}{\tau}} + \gamma_d \left(1 - e^{-\frac{\Delta t}{\tau}}\right) \qquad (4)$$

where τ is the steering time constant, and Δt is the time that is needed to travel from one posture to the following ($\Delta t = \Delta s/v$).

2.3.2 Compute the new position and orientation of the vehicle:

$$x_n = x_{n-1} - \Delta s \ \sin\phi_{n-1}$$

$$y_n = y_{n-1} + \Delta s \ \cos\phi_{n-1} \qquad (5)$$

$$\phi_n = \phi_{n-1} + \Delta s \ \gamma_{n-1}$$

These equations represent a good approximation to the vehicle motion when Δs is small enough (Amidi, 1990).

2.4 Increment the number of generated postures n.

3. APPROACH-PATH GENERATION

Approach paths are used by a mobile robot when the vehicle has a large orientation error or it is far from the path to follow.

In these situations, the path to be generated must link the actual posture of the vehicle with a posture in the path to follow, which is called the meeting posture (see Fig. 3).

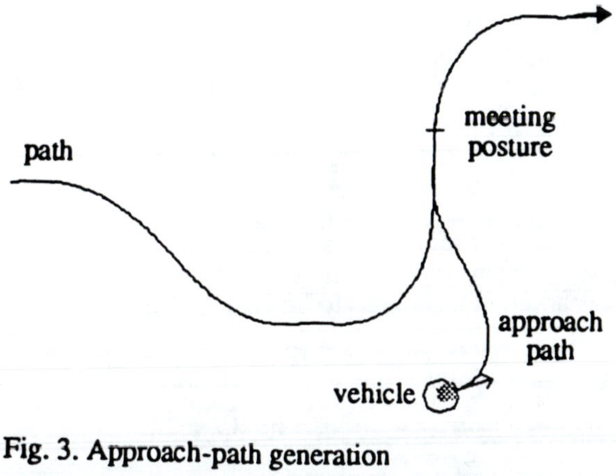

Fig. 3. Approach-path generation

The application of the proposed algorithm to generate an approach-path is straightforward, and can be done in real time by the vehicle controller. The curvature of the generated path in Fig. 3 is shown in Fig 4. The vehicle begins with zero curvature, and ends with the curvature of the path on the meeting point.

The location of the meeting point depends on the value of the chosen lookahead distance. If D is large then the approach-path would be smoother but it would also be longer. Shorter paths can be generated if the lookahead distance is reduced.

In order to reduce the steering wheel variation, approach-paths are used with the predictive path tracking method (Ollero and Amidi, 1990). In this situation only a few points (the maximum costing horizon) of an approach-path from the actual position of the vehicle has to be generated.

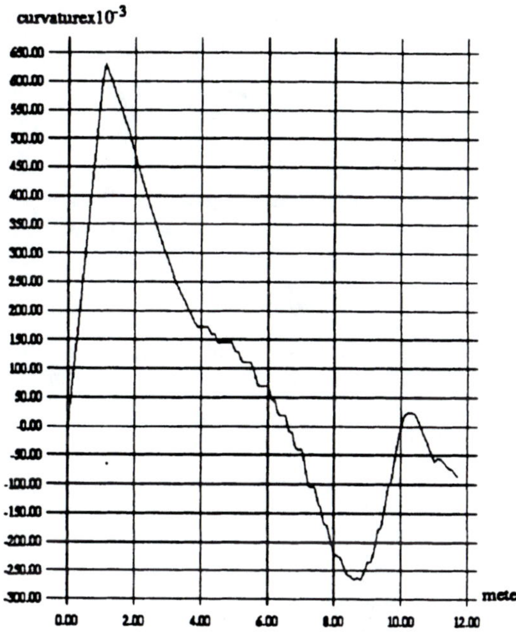

Fig. 4 Curvature of the approach-path

4. LOCAL PATH GENERATION

The generation of a local path is required when the mobile robot is following a path and an obstacle is detected on the path. Then the perception system has to specify a point (x_a, y_a) to avoid the obstacle (Brumitt *et al.*, 1993).

Thus, the objective is to generate a local path, that originates from the actual position of the vehicle in the path, pass over the avoidance point and finish in the path to follow behind the obstacle (see Fig. 5).

To make the local path pass far enough from the obstacle, the orientation at the avoidance point ϕ_a is taken to be equal to the orientation of the nearest posture in the path.

The local path is composed of two sections. The first is generated from the avoidance point, but in the opposite direction, trying to reach the path in front of the obstacle. The second section also starts at the avoidance point, but looking forward to the path behind the obstacle.

The initial curvature in the avoidance point is the largest curvature generated by pure pursuit when initially generating both sections.

The second section of the path has the right direction,

but for the first section, it is necessary to change it's orientation and curvature, which is done simply by adding 180º to the orientation, and changing the sign of the curvature:

$$\hat{\phi}_i \leftarrow \hat{\phi}_i + \pi$$
$$\hat{\gamma}_i \leftarrow -\hat{\gamma}_i \tag{6}$$

The local path can be obtained by the real time vehicle controller, without the need to stop the mobile robot.

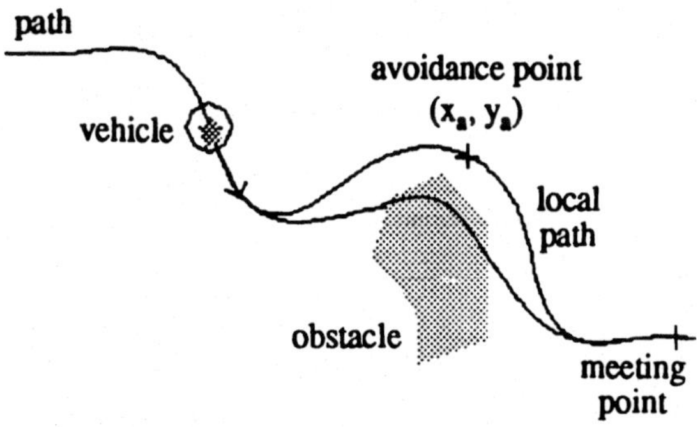

Fig 5. Local path generation

Figure 6 shows the curvature of the generated local path. The transition between the local path and the path to follow is continuous in position, orientation and curvature.

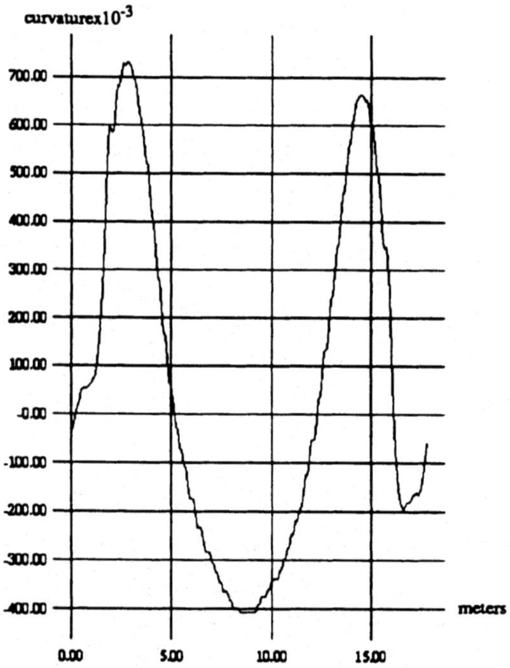

Fig 6. Curvature of the local path

The method uses a fuzzy supervisory rule-based technique to automatically adapt the lookahead distance D, which is the critical parameter of the pure pursuit strategy (Ollero *et al.*, 1994). The heuristic rules to apply are the following:

- Begin with the maximum possible lookahead for the initial conditions (which means the minimum curvature).
- Decrease D as the vehicle approaches the meeting point in the path to follow.
- Change the lookahead distance, according with the variation of errors in vehicle´s position, orientation or curvature with respect to the goal point. Increase D if the errors grow, and decrese otherwise. The error in position is the most significant, while the error in curvature is the least.

5. GLOBAL PATH GENERATION

Typically, path planning is divided into two problems. Firstly, it is necessary to use the information about the world to obtain a sequence of objective points, that the vehicle must follow in order to progress toward a goal, without collisions. Secondly, for the path tracking control of the vehicle, these objective points are too separate, so it is necessary to generate a more detailed path.

Let $\{(P_1, P_2, \ldots, P_m), P_i=(x_i, y_i)\}$ be a sequence of m objective points. The global path generation problem consists of the development of an easily trackable fine grained path, that passes over the objective points in the specified order.

In order to apply the proposed method, a predetermined path to follow and a completely defined initial posture are needed. To meet the first condition, it is necessary to generate a temporary path, which is formed by joining consecutive objective points with segments of straight lines. Let S_i be the segment of straight line which begins in P_i and ends in P_{i+1}.

To meet the second condition, it is necessary to define an orientation ϕ_i and curvature γ_i for each of the objective points P_i.

Let θ_i be the orientation of the straight line that goes through the objective points P_i and P_{i+1}. Let d_i be the distance between P_i and P_{i+1}:

$$\theta_i = \operatorname{atan}\left(\frac{x_i - x_{i+1}}{y_{i+1} - y_i}\right)$$
$$d_i = \sqrt{(x_{i+1} - x_i)^2 + (y_{i+1} - y_i)^2} \tag{7}$$

The orientation at the initial and final postures are the same as the line between the two first objective points, and the line between the two last way points respectively:

$$\phi_1 = \theta_1$$
$$\phi_m = \theta_{m-1} \tag{8}$$

The curvature at the initial and end objective points are chosen in both cases to be zero ($\gamma_1=\gamma_m=0$).

For an intermediate objective point the orientation is calculated with the following expression:

$$\phi_i = \frac{\theta_{i-1}\ d_i + \theta_i\ d_{i-1}}{d_i + d_{i-1}} \tag{9}$$

So that, the variation of orientation is proportional to the available distance:

$$\frac{\theta_i - \phi_i}{d_i} = \frac{\phi_i - \theta_{i-1}}{d_{i-1}} = \frac{\theta_i - \theta_{i-1}}{d_i + d_{i-1}} \tag{10}$$

The global path is composed of sections that are built simulating the behaviour of the vehicle when it begins at an intermediate objective point looking for the previous or next way point.

The curvature of an intermediate objective point is calculated as the biggest curvature generated by the pure pursuit technique when applying the proposed method in both directions. This value avoids oscillations in curvature, otherwise present in the generated path near the intermediate points.

For each intermediate way point P_i ($1<i<m$) two sections are generated $T_{i,1}, T_{i,2}$. Both sections begin in the objective posture P_i, but the first $T_{i,1}$ with the opposite direction, trying to reach $T_{i-1,2}$ or the S_1 segment if i=2. The second section $T_{i,2}$ is generated trying to reach the segment S_i (see Fig. 7).

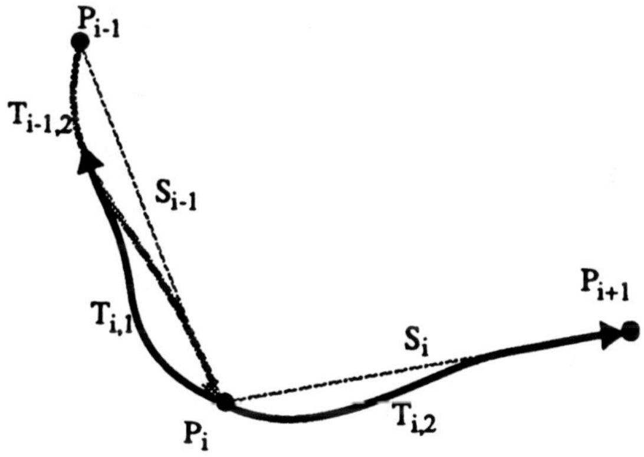

Fig. 7. Piece-wise path generation

If $T_{i,1}$ follows S_{i-1} instead of $T_{i-1,2}$ then it would be necessary to generate $T_{i-1,2}$ and $T_{i,1}$ to meet in the middle of S_{i-1}, which would imply a bigger curvature in the path, which in turn means to increase the control activity when tracking the path.

Global path generation can be done by using the same heuristic rules applied to choose the lookahead distance for local path generation.

A global path generated from a sequence of seven points is shown in Fig. 8. The curvature of the path can be observed in Fig. 9.

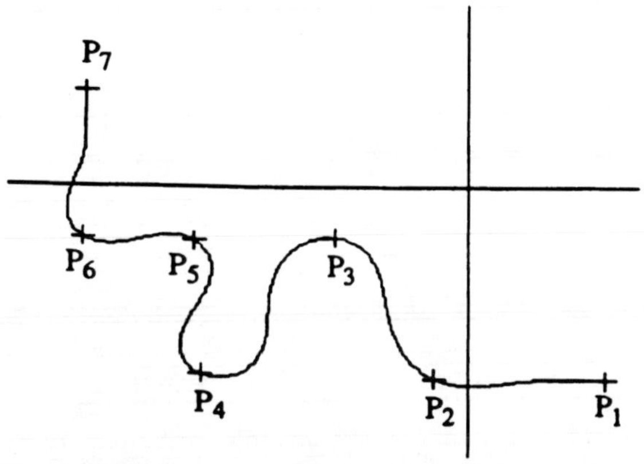

Fig. 8. Global path generation

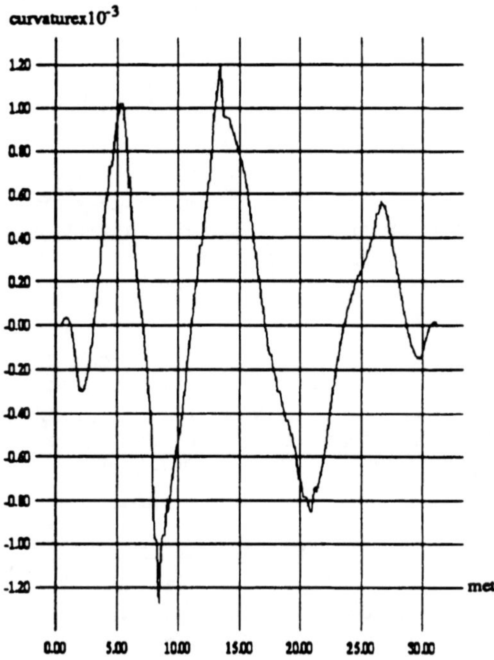

Fig. 9 Curvature of the global path

6. CONCLUSIONS

The method proposed in this paper is able to cope with the kinematic and dynamic characteristics of the steering system. Furthermore, the control system can compensate for perturbations due to vehicle-ground interaction, when tracking a path. Fig. 10 shows the tracking of a global path at a speed of 1.2 m/s executed by RAM-1.

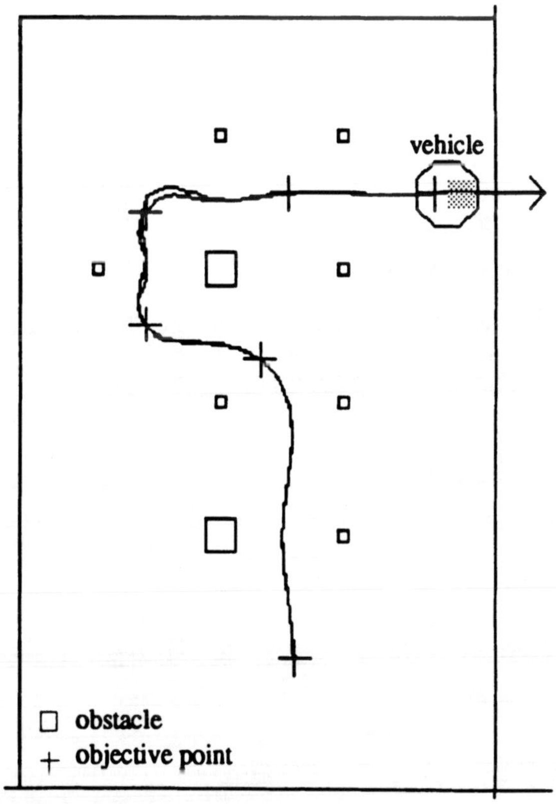

Fig. 10. Global path tracking by RAM-1

The proposed intelligent method can be efficiently implemented, making it suitable for real time control. The approach-path and local-path generation has been applied successfully in the real time controller of RAM-1. The global path is generated before the vehicle moves.

This method has also been used to build appropiate paths to be tracked, when the points are given by a visual tracker, which follows external features such as lines or walls. This is because the list of points given by this kind of tracker usually has gaps or is overlapped.

As an additional advantage, the generated paths always sweep outside the acute angles formed by the straight line connection of the objective points. This property is useful to keep the vehicle far away from obstacles that are usually on the inside of angles.

7. REFERENCES

Amidi O. (1990). Integrated Mobile Robot Control. *Technical Report CMU-RI-TR-90-17*. Carnegie Mellon University. Robotics Institute.

Brumitt B., R. C. Coulter, A. Kelly, and A. Stenz (1993). A System for Autonomous Cross-Country Navigation. *Intelligent Components and Instruments for Control. Proc. SICICA'92 IFAC Symposium. A. Ollero and E. F. Camacho eds.* Pergamon Press.

Hongo T., H. Arakawa, G. Sugimoto, K. Tange, and Y. Yamamoto (1985). An Automatic Guidance System of a Selfcontrolled Vehicle. The Command System and Control Algorithm. *Proc. IECON*, pp. 535-540.

Kanayama Y. and N. Miyake (1985). Trajectory Generation for Mobile Robots. *Proc. of the Third International Symposium on Robotics Research*, Gouvieux, France, pp. 333-340.

Kanayama Y. and B. Hartman (1989). Smooth Local Path Planning for Autonomous Vehicles. *Proc. of the IEEE International Conference on Robotics and Automation*, pp. 1265-1270.

Moravec H. P. (1983). The Stanford Cart and the CMU Rover. *Autonomous Robot Vehicles. I. J. Cox and G. T. Wilfong eds.* Springer-Verlag.

Muñoz V., J. L. Martínez, and A. Ollero (1993a). New Continuous Curvature Local Path Generators for Mobile Robots. *Intelligent Components and Instruments for Control. Proc. SICICA'92 IFAC Symposium. A. Ollero and E. F. Camacho eds*, pp. 455-460. Pergamon Press.

Muñoz V., A. Ollero, and M. Paz (1993b). A New Integrated Path Planning and Generation Real Time System for Mobile Robots. *Proc. First IFAC International Workshop on Intelligent Autonomous Vehicles*, Southampton, U. K., pp. 452-457.

Nelson W. (1989). Continuous Curvature Paths for Autonomous Vehicles. *Proc. IEEE Int. Conference on Robotics and Automation*, pp. 1260-1264.

Ollero A. and O. Amidi (1991). Predictive Path Tracking of Mobile Robots. Applications to the CMU Navlab. *Proc. Fifth International Conference on Advanced Robotics*, Pisa, Italy, Vol. II, pp. 1081-1086.

Ollero A., A. Simón, F. García, and V. Torres (1993). Integrated Mechanical Design of a New Mobile Robot. *Intelligent Components and Instruments for Control. Proc. SICICA'92 IFAC Symposium. A. Ollero and E. F. Camacho eds.*, pp. 461-466. Pergamon Press.

Ollero A., A. García-Cerezo, and J. L. Martínez (1994). Fuzzy Supervisory Path Tracking of Mobile Robots. *Control Engineering Practice*, num. 2, vol. 2. Pergamon Press.

Rombaut M., A. Segovia, D. Meziel, and A. Preciado (1991). Displacement of a Mobile Robot in a Known Environment. *IMACS-MTCS Symposium*, Villaneuve d'Ascq, France.

Shin D. H. (1990). High Perfomance Tracking of Explicit Paths by Roadworthy Mobile Robots. *PhD Thesis*. Carnegie Mellon University.

Shin D. H., S. Singh, and W. Whittaker (1993). Path Generation for a Robot Vehicle using Composite Clothoid Segments. *Intelligent Components and Instruments for Control. Proc. SICICA'92 IFAC Symposium. A. Ollero and E. F. Camacho eds.* Pergamon Press.

MOTION ESTIMATION FOR A MOBILE ROBOT USING VISUAL AND RANGE DATA

R. LHERBIER, X. TU and B. DUBUISSON

Université de Technologie de Compiègne, URA CNRS - Heudiasyc, BP 649, 60206 Compiègne Cedex, FRANCE

Abstract. We study the problem of recovering the motion of a mobile robot in an unknown environment. The method used merges 2 kinds of 3D data, telemetric range data and 3D planes observed in a scene by stereovision system and checked by rangefinder. To recover the motion parameters, we use an Iterated Extended Kalman Filter. In order to reduce computation time, we focus one's attention on periodic use of vision. The algorithm is divided into three parts: local map creation, motion recovering and relocalization. This technique has been tested on synthetic noisy data and shows that the method is sufficient for simple and complex planar motions.

Key words. Robots; Motion Estimation; Multisensor Fusion; Extended Kalman Filter

1. INTRODUCTION

Motion analysis is a very important research field in robotics, and precise results of analysis lead to direct interpretation of 3D scenes. Mobile robot navigation, construction of world model, scene segmentation and object tracking are some examples of work in which knowlegde of motion is important.

The determination of the three-dimensional structure and the motion of objects is often derived from image data. These results usually can be got by two approaches. The first one is called *optical flow* (Adiv, 1985) and is often based on monocular vision (2D data). The second one is called *feature tracking* (Ayache, 1991; Zhang, 1992; Young, 1990). Then, the motion is computed with 3D data coming from, for example, a stereoscopic system. Both approaches present advantages and disadvantages. But they have a common default: the computation time consumming. In fact, in real applications, not to reduce the robot capacity (for example, limitation of speed), we have to use fast algorithms. In the case of a feature tracking approach (which seems to us more interesting), features extraction and features matching are time handicap. For this reason, stereovision data will be used only time to time when necessary. The aim of this work is to merge information provided by the trinocular stereoscopic vision system with the one got from other sensors. Some work combining 3D vision and information coming from other sensors have been developed for recovering the motion with such a mixed sensor system. For example, Castagliola (1991) and Kriegman (1987) use 3D vision and ultrasonic sensors and Chenavier (1992) uses the vision and odometry in a known environment. Grandjean (1989) and Moutarlier (1989) use 3D data extracted from trinocular vision system and laser range sensor for 3D modelling of scenes.

A fusion method of informations coming from a fast sensor (telemeter) and from a stereoscopic sensor is used to recover the robot motion. The robot navigates in an unknown environment containing only rigid objects. Only one hypothesis about the motion is made: 3D object motions are rigid. A local environment map is built up using 3D planes obtained with 3D segments issued from the vision. In order to compute the robot motion, we use this map and range measures.

The following section presents features description and extraction of 3D planes. Their fusion with range data through an Iterated Extended Kalman Filter (IEKF) to recover the motion are explained in Section 3. The model of motion is also presented in this section. Results are included in Section 4 and are followed by conclusions in Section 5.

2. GENERAL DESCRIPTION

2- 1 *Brief description of sensors*

The perception system consists of a trinocular vision system and an infrared rangefinder. After extraction of contours and their approximation by 2D segments in images, a stereovision algorithm calculates the parameters of the 3D line segments and their covariance matrices from three images. Our laboratory is equiped with a hardware system, called ASTRID™ and developed by ITMI Corp., which gives three-dimensionnal-segments at the rate of

1 Hz. The infrared telemeter, called ORS-1 and developped by ESP technologies Inc, delivers a succession of measures in a polar coordinate system (ρ,θ). It rotates from 0 to 360 degrees around a vertical axis and therefore, forms a virtual plan (called laser plane). A covariance matrix is associated with each range measure.

2-2 *Statistical tools for fusion*

Every noisy feature is considered as a random vector denoted by **x**. Then, **x** is characterized by $(\hat{x}, \Lambda_x)$ where $\hat{x}$ is an estimation obtained from the observations and Λ_x is its covariance matrix with $\Lambda_x = E[(\hat{x}-x)(\hat{x}-x)^T]$.. This approach was also chosen in Ayache (1989) and in Durrant-White (1988).

Due to the non linear relationship between noisy features, we apply the Extended Kalman Filter to estimate the different parameters. A good study of this problem is given in Ayache (1991).

In order to verify a non linear relationship $\mathbf{f}(x_1,x_2) = 0$ between two noisy features $(\hat{x}_1, \Lambda_{x1})$ and $(\hat{x}_2, \Lambda_{x2})$, we use a non linear hypothesis test based on the Mahalanobis distance which follows a χ^2 probability law with n degrees of freedom, n being equal to the dimension of **f** (Ayache, 1989).

2-3 *3D line segment representation*

A line segment in 3D space can be represented by different ways. It is usually represented by its endpoints or by its direction vector and its midpoint. Thoses parameters are not reliable due to the approximations during preprocessing and partial occlusions. In order to limit these problems, we have chosen a non minimal representation. Such that, a 3D line segment is represented by

- an infinite supporting line, L
- its midpoint, M
- its length, l

Each parameter is used at different stage of the algorithm.

An infinite line is represented by using the Plücker coordinate representation (Fig. 1).

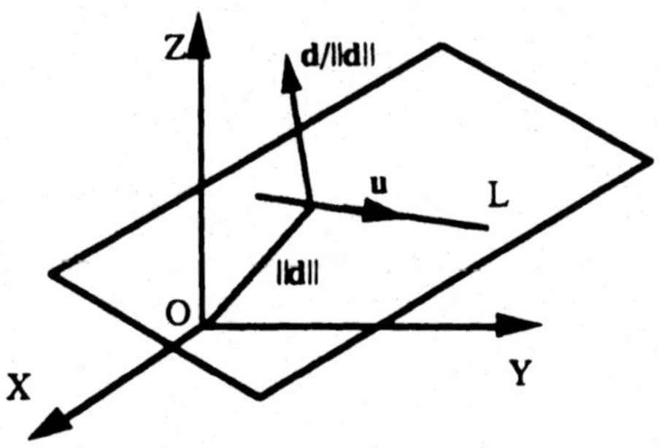

Fig.1. A plücker coordinates representation

It uses 6 parameters which are obtained from two vectors (**u**, **d**): **u** is the unit direction vector and norm of **d** is the distance from the line to the origin of the coordinates system, direction of **d** is parallel to the normal of the plane passing through the origin and the line. This is also a non minimal representation.

A 3D line segment is then described by 10 parameters (6 parameters for L, 3 for M and 1 for length). A covariance matrix is associated with this representation.

2-4 *Local map creation from planes*

From the segments in the scene, and in order to create a local map of the environment, we are interested by the surfaces intersecting the laser plane.To obtain plane equations, we only consider vertical line segments and we use an hypothesize-and-verify method. Each couple of nearby 3D vertical line segments is associated with a plane. The scene is supposed to be composed of succesive planes in a clockwise sense around the vertical axis passing through the viewpoint.

A 3D plan can be decribed as follow (Fig. 2)

$$\mathbf{OM}.\mathbf{v} = d \qquad (1)$$

where

M is a point of the plane.
v is the unit normal vector of the plane.
d is the distance from the origin to the plane.
. is the scalar product.

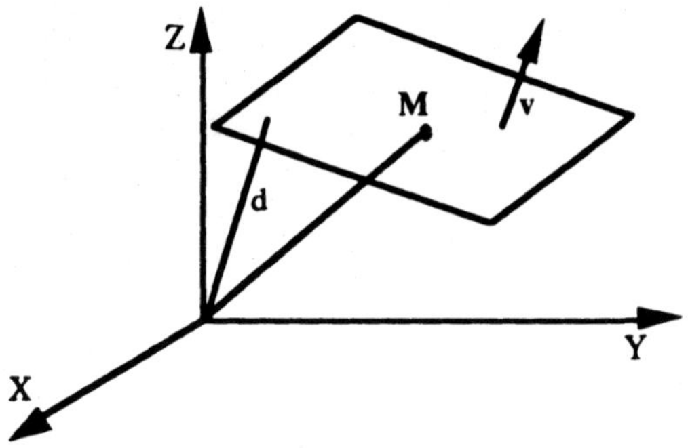

Fig.2. The plane representation

v,d and their covariance matrices Λ_v and Λ_d are computed from a Kalman filter with measures issued from supporting line of vertical line segments. By this method, equations of all potential planes are calculated. The verification is done using infrared telemetric data.

Limit points, given by the intersection point between supporting line segments and laser plane, are included for each plane (**v**,d) to obtain a surface. We compute easily the angle θ used by the telemeter in order to reach this potential plane from the viewpoint (θ is calculated to shoot in the middle of both segments). Then, an hypothesis test is done comparing the distance of the midpoint of the potential plane and the real distance ρ given by the telemetric measure. So, the existence of the plane is checked.

All the checked planes (v,d) will be useful to create a local map of the environment. From this local map and using, at different times, telemetric measures we will recover robot motion. The use of this map requires a relocalization at different positions during the motion of the robot.

3. DETERMINING 3D MOTION OF RIGID OBJECTS

3-1 Representation of 3D motion

We are only interested in the case of rigid motion. It is well known that a 3-D motion can be decomposed into a translation and a pure rotation around an axis passing through the origine of the coordinate system.

In general, the three-dimensionnal motion of an object is descibed by the equation

$$\mathbf{P}' = \mathbf{R}\mathbf{P} + \mathbf{t} \quad (2)$$

In this equation, **P** and **P'** are the position vectors of an object point at time instants i and i', respectively, **R** is a 3X3 matrix called the rotation matrix, **t** is the translation vector.

So the motion can be uniquely determined by 12 parameters. This rigid motion can be represented in an optimal manner with a 6-dimensionnal vector $\mathbf{m} = \left[\mathbf{t}^T \ \mathbf{r}^T\right]^T$ (see Zhang (1992) for more details) where **t** is the translation vector and **r** represents the rotation, direction of **r** is the rotation axis, norm is equal to the rotation angle.

3.2 Plane movement

Let P(v,d) be a plane described in the coordinates system Oxyz

$$\mathbf{OM}.\mathbf{v} = \mathrm{d}$$

The same plane is written as follow in an another coordinates system O'x'y'z'

$$\mathbf{O'M'}.\mathbf{v'} = \mathrm{d'}$$

This change of coordinates system can be depicted by (**R**,**t**) and we obtain, as explained in Ayache (1991)

$$\mathbf{O'M'} = \mathbf{ROM} + \mathbf{t}$$
$$\mathbf{v}' = \mathbf{R}\mathbf{v}$$

As M' lies on the plane

$$(\mathbf{ROM} + \mathbf{t}).(\mathbf{R}\mathbf{v}) = \mathrm{d'} \quad (3)$$

From these equations and since the rotation preserves scalar product

$$\mathbf{v}' = \mathbf{R}\mathbf{v} \quad (4)$$
$$\mathrm{d}' = \mathrm{d} + (\mathbf{R}\mathbf{v}). \quad (5)$$

3-2 Iterated Extended Kalman Filter

Using (4), (5) and only telemetric measures, the motion parameters can be calculated.

The parameters of a plane P in the scene at time i (after motion) are given by $\mathbf{y}_P^i = \left[\mathbf{v}_P^{iT} \ d_P^i\right]^T$. Parameters of the same plane at time i-1 (before motion) are given by $\mathbf{y}_P^{i-1} = \left[\mathbf{v}_P^{i-1T} \ d_P^{i-1}\right]^T$.

Using the usual filtering terminology, the 6-dimensional vector $\mathbf{x}^i = \left[\mathbf{t}^T \ \mathbf{r}^T\right]^T$ is called the state vector and represents the motion to be estimated and the 8-dimensional vector $\mathbf{z}_P^i = \left[\mathbf{y}_P^{i-1T} \ \mathbf{y}_P^{iT}\right]^T$, is called the measurement vector.

The system measurement is assumed to be disturbed by an additive white noise η, so the real observed measurement $\mathbf{z}_P^{i'}$ is expressed as

$$\mathbf{z}_P^{i'} = \mathbf{z}_P^i + \eta^i ,$$

where

$$E\left[\eta^i\right] = 0 .$$
$$E\left[\eta^i \eta^{iT}\right] = \Lambda_\eta , \ \Lambda_\eta \text{ is a diagonal matrix}$$

At time i and for each plane P, (4) and (5) can be expressed as

$$\mathbf{f}(\mathbf{z}_P^i, \mathbf{x}^i) = \begin{bmatrix} \mathbf{v}_P^i - \mathbf{R}^i \mathbf{v}_P^{i-1} \\ d_P^i - d_P^{i-1} - (\mathbf{R}^i \mathbf{v}_P^{i-1}).\mathbf{t}^i \end{bmatrix} = 0 \quad (6)$$

Then, let P, the number of planes in the scene, and M, the number of range measures on each plane, we have to use a set of equations $\mathbf{f}({}_m\mathbf{z}_P^i, {}_m\mathbf{x}^i) = 0$ for p = 1,..,P and m=1,..,M , in order to recover the motion parameters.

Since the measurement function $\mathbf{f}(\mathbf{z}_P^i, \mathbf{x}^i)$ is nonlinear, we must linearize the system in order to apply a linearized Kalman filter. To reduce the effect of nonlinearities, we apply the EKF iteratively. Therefore, we must calculate the derivates of **f** with respect to **z** and **x**

$$\frac{\partial \mathbf{f}}{\partial \mathbf{z}} = \begin{bmatrix} -\mathbf{R}^i & 0_3^1 & \mathbf{I}_3 & 0_3^1 \\ -\mathbf{t}^{iT}\mathbf{R}^i & -1 & 0_1^3 & 1 \end{bmatrix}$$

$$\frac{\partial \mathbf{f}}{\partial \mathbf{x}} = \begin{bmatrix} 0_3^3 & -\delta(\mathbf{R}^i \mathbf{v}^{i-1})/\delta \mathbf{r} \\ -(\mathbf{R}^i \mathbf{v}^{i-1})^T & -\mathbf{t}^{iT}\delta(\mathbf{R}^i \mathbf{v}^{i-1})/\delta \mathbf{r} \end{bmatrix}$$

where $\mathbf{I}_3$ is the 3x3 matrix identity, 0_n^m is the nxm null matrix ,$\delta(\mathbf{R}\mathbf{v})/\delta\,\mathbf{r}$ is the derivate of **Rv** with respect to **r** and is given by

$$\delta \mathbf{R}\mathbf{v}/\delta\, \mathbf{r} = (\cos(\theta) - f(\theta))/\theta^2(\mathbf{r}\wedge\mathbf{v})\mathbf{r}^T + (\sin(\theta) - 2\theta g(\theta))/\theta^3(\mathbf{r}\wedge(\mathbf{r}\wedge\mathbf{v}))\mathbf{r}^T - f(\theta)\tilde{\mathbf{v}} + g(\theta)(-(\widetilde{\mathbf{r}\wedge\mathbf{v}}) + (\mathbf{r}.\mathbf{v})\mathbf{I}_3 - \mathbf{v}\mathbf{r}^T)$$

where θ is the rotation angle, $f(\theta) = \sin(\theta)/\theta$, $g(\theta) = (1-\cos(\theta))/\theta^2$, $\tilde{\mathbf{v}}$ is the antisymmetric matrix associated with $\mathbf{v}$ and $\wedge$ denotes the cross product.
The rotation matrix $\mathbf{R}$ is computed with the coordinates of the vector $\mathbf{r}$. The details are given in Zhang (1992).

The expectation and the covariance matrix of the new measurement noise ξ^i may be written

$$E[\xi^i] = 0\,,$$
$$\mathbf{W}^i = E[\xi^i\xi^{iT}] = \frac{\partial \mathbf{f}}{\partial \mathbf{z}}\Lambda^i\frac{\partial \mathbf{f}}{\partial \mathbf{z}}^T$$

Let us assume that set of planes P^{i-1} and P^i, before and after the motion are independent, then

$$\Lambda^i = \begin{bmatrix} \Lambda_{P^{i-1}} & 0 \\ 0 & \Lambda_{P^i} \end{bmatrix}$$

In order to simplify the kalman filter equations, we break down the problem. At first the optimal rotation is estimated and after we use this rotation to determine the translation. So each estimation becomes more simple and the solution at the optimization problem is, about numerical aspect, very reliable. The disadvantage is that the possible errors associated to the rotation computation have repercussions on the translation estimation.

3-3. v and d computation from range measures

Let P(v, d) be a plane. A surface Sp is obtained using P and the two segments with which we have built up this plane. Furthermore, at each instant i, the estimated value of the motion is known

$$\hat{\mathbf{x}}^{i-1} = \left[\hat{\mathbf{r}}^{i-1T} \quad \hat{\mathbf{t}}^{i-1T}\right]^T$$

First, the position of the surface is predicted at time i from the estimated motion parameters and the position of the same surface at time i-1. But it is important to know the covariance matrix of each intersection point between laser plane and a boundary segment. Both matrices are the same as the ones at time i-1 increased by combining uncertainty on motion and uncertainty on position (Fig. 3). The complete computation details can be found in Chesseman (1986).
The parameters (v and d) of the plane can be recovered with Kalman filter fed to several measures in each plane (typically between 5 and 10).
Since the number of the features (3-D planes) are small, typically between 5 to 20 for each scene, the matching step is considered as unnecessary. In fact this matching step is included in an hypothesis test where the prediction of plane position is the hypothesis and where the verification is given by range measures used to find the plane parameters

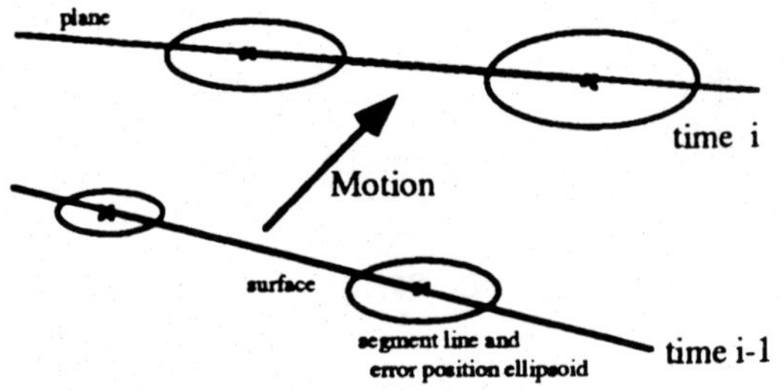

Fig.3. Plane position uncertainty evolution (representation in laser plane)

3-4 Motion computation

From local map, and after a robot movement, we use successive telemetric measures (the different angles of sight of the rangefinder are calculated to reach the planes of the secne) to estimate, via EKF, the motion. Since the robot moves, we take a new serie of telemetric measures to obtain a motion estimator better than previous.
The use frequency of these measures is selected to be a constant. After a number of motion estimation, we must relocalize the robot in order to limit the estimation errors. This relocalization is done with the vision system. The relocalization frequency is computed from position errors.

3-4-1 Appearance and disappearance of features. With the telemetric measures, the disappearance of planes is not a real problem. In fact, the rangefinder takes the measures in all directions. So, the planes can always be reached. If new features appear during the motion estimation, they are not used, and they will serve in the estimation procedure only after relocalization.
Since the camera are used, appearance and disappearance problems are important due to the visual field limitation. Thus, some planes can appear or disapear in the field of view. These problems are solved using relocalization. In fact, the appeared features during the estimation are considered as new features and the disappeared features are considered as being lost, but not definitely, in order to limit the extraction problems. So, only the features present in the scene at successive relocalization time are matched.

3-4-2 Relocalization by vision. First, the estimated planes are matched with the real observed planes. This matching is realized by a classical hypothesize-and-verify technique. A Mahalanobis distance is used to check the real matches. For this reason, the observed planes (by vision) which appear and the estimated 3D planes (by telemeter) which disapear have not to be matched.

After matching, the position error, Δ_{pos}, of the robot is computed, observing different couples of matched planes. So, we need only to match few planes. The error will be used during the calculation of vision setting to relocate the robot.
After relocalization, the new local map is extracted from vision. The estimated values and covariance matrices of the motion are the same as the ones before the relocalization and are used as initial value in the Kalman filtering to calculate the new motion estimation.

3-4-3 Vision setting. The technique, which allows us to set the vision is as following (Fig. 4)

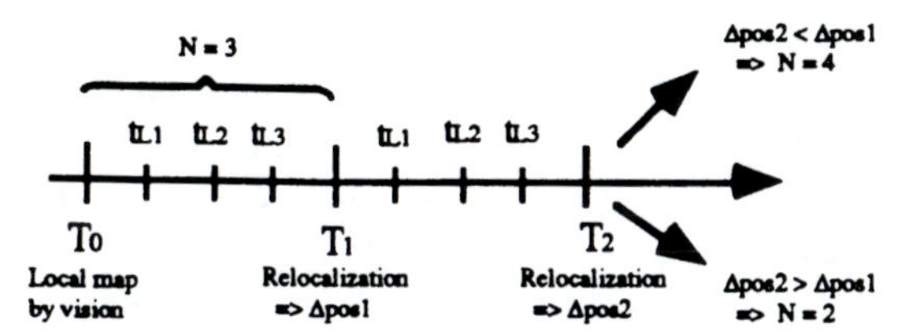

Fig.4. Technique of vision setting

- At First, we fix the number of samples N. So, after N motion estimations, we relocalize the robot to obtain the position error, Δ_{pos1}.
- After this relocalization, the motion is estimated with the same number N of samples. Then, Δ_{pos2} is calculated.
- Observing Δ_{pos1} and Δ_{pos2} (logically, $\Delta_{pos2}<\Delta_{pos1}$), we increase or decrease the period between 2 relocalizations (N is modified).
- With this new N, the same technique is used to increase (or decrease) the period of vision setting in order to relocate the robot.

4. EXPERIMENTAL RESULTS

For the reason of simplicity and in order to avoid undesirable effects of real frames, we test the algorithm using synthetic 3D frames. These data come from a software developed in our laboratory which allows to realize the simulation of various robot sensors (2D image data, 3D reconstruction, laser range finder, ultrasonic sensors). The robot is navigating in a polyhedric indoor environment. These synthetic data simulate the sensor performance (without noise). In order to obtain performance as realistic as possible, we only add noise in the measures. Results will be given using these synthetic scenes.
A simple motion (translation) and a complex motion (rotation plus translation) are presented. In two cases, the robot moves in 2D plane Oxz. We only present the convergence of filter between two relocalizations.

4-1 Translation

An estimation of the translation is computed by the method. There are 4 planes in the scene. The estimation is updated for each measure. Five measures are done on each plane. Figure 5 shows the result for a translation $\mathbf{t} = [22\ 0\ 22]^T$ (in centimetre). An initial guess of the state vector (representing **t**) is required as input to the filter. In our implementation, the state vector **x** is simply initialized to zero (*i.e.* $\hat{\mathbf{x}}_0 = 0$) but with a big covariance matrix (*i.e.* $\Lambda_{\hat{x}_0}[i][i] = 10^8$ and $\Lambda_{\hat{x}_0}[i][j] = 0$ for $i \neq j$).

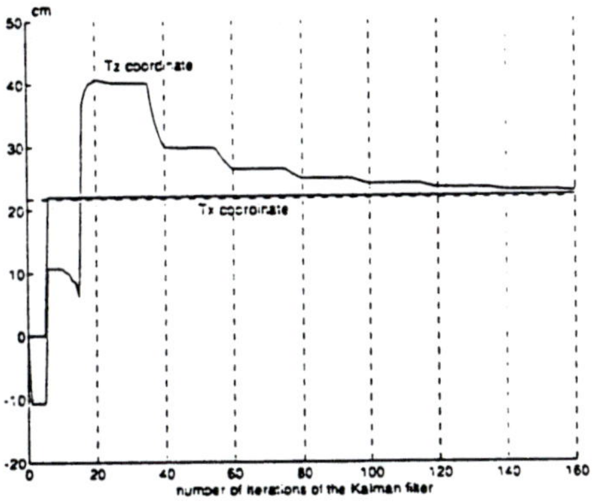

Fig.5. Convergence for a translation **t** = [22 0 22]

From Fig.5, we can see that the convergence is obtained after 7 telemetric measure series. The convergence for the z coordinate is more difficult due to noise resulting from the 3D reconstruction.

4-2 Compound motion

Now, an estimation of a compound motion is computed. There are 9 planes in the scene. The filter is iterated for each planes (due to the difficult to obtain **v**). The normal vector **v** of each plane is obtained from Kalman filtering with 5 measures on each plane. Figures 6 and 7 show the result for a motion $\mathbf{m} = [\mathbf{t}^T\ \mathbf{r}^T]^T = [-21\ 0\ -21\ 0\ 0.26\ 0]^t$ (where **t** is expressed in centimetre and **r** in radian). The initial state vector $\hat{\mathbf{x}}_0$ is initialized to zero with a covariance matrix as follows: $\Lambda_{\hat{x}_0}[i][i] = 10^8$ for i = 1, 2, 3, $\Lambda_{\hat{x}_0}[i][i] = 2$ for i = 4, 5, 6 and $\Lambda_{\hat{x}_0}[i][j] = 0$ for $i \neq j$. For this estimation the motion is decomposed in a rotation plus a translation.
Figure 6 represents the convergence for the optimal rotation and Figure 7 gives the results for the translation.
Figure 8 represents the position error of the robot. This is the difference between the real positions calculated from the real motion and from the estimated motion.

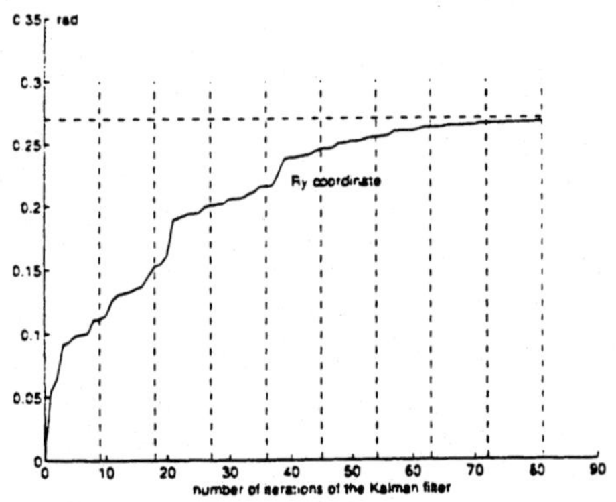

Fig.6. Convergence for the rotation of **m**

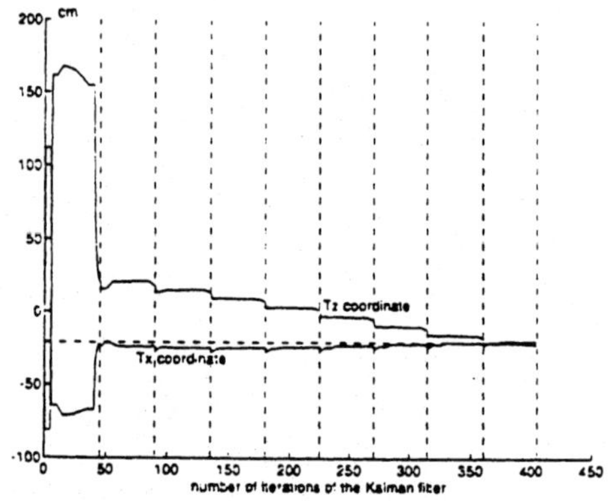

Fig.7. Convergence for the translation of **m**

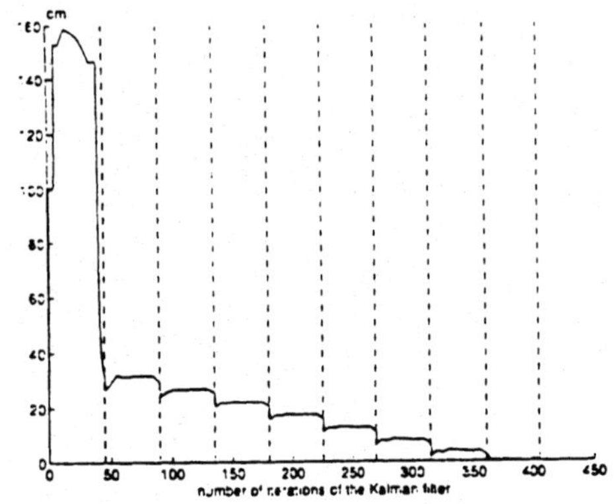

Fig.8. Position errors

5. CONCLUSION

In this paper, we have presented a method for recovering robot motion. We use the complementarity of 2 sensors: stereovision system and rangefinder. The results show the performance and robutness of the method for simple motion and complex motion. We also plan to apply the method to real measures while the robot navigates in a real indoor environment. In this case, the real time objective is more important. So, we must improve the sensors cooperation and study another techniques for vision setting.

6. REFERENCES

Adiv, G. (1985). Inherent ambiguities in recovering 3D motion and structure from a noisy flow field. *IEEE Int. Conf. on Computer Vision and Pattern Recognition, San Fransisco (USA)*, pp. 70-77.

Ayache, N. and O.Faugeras (1989). Maintening representations of the environment of a mobile robot. *IEEE Trans. on Robotics and Automation*, Vol. 5, no. 6, pp. 804-819.

Ayache, N. (1991). *Artificial Vision for Mobile Robots: Stereo Vision and Multisensory Perception.* MIT Press, Cambridge.

Castagliola, P. (1991). *Un systeme Incremental pour la Fusion Multi-Capteurs pour un Robot Mobile*. PhD Thesis, Compiegne University, URA CNRS.

Chenavier, F. and J. Crowley (1992). Position estimation for a mobile robot using vision and odometry. *IEEE Conf. on Robotics and Automation, Nice (France)*, pp. 2588-2593.

Chesseman, P. and R.C. Smith (1986). On the representation and estimation of spatial uncertainty. *The International Journal of Robotics Research*, Vol. 5, no. 4, pp. 56-68.

Durrant-Whyte, H.F. (1988). Uncertainty Geometry in Robotics. *IEEE Journal of Robotics and Automation*, Vol. 4, no. 1, pp. 23-31.

Grandjean, P. and A. Robert de Saint Vincent (1989). 3D modelling of indoor scenes by fusion of noisy range and stereo data. *IEEE Conf. on Robotics and Automation, Scottdale (USA)*, pp. 681-687.

Kriegman, D., E. Triendl and T. Bindford (1987). A mobile robot: sensing, planning and locomotion. *IEEE Conf. on Robotics and Automation, Raleigh (USA)*, pp. 402-408.

Moutarlier, P. and R. Chatila (1989). Stochastic multisensory data fusion for mobile robot location and 3D modeling *Proc. International Symposium on Robotics Research, Tokyo*.

Young, G. and R. Chellappa (1990). 3D motion estimation using a sequence of noisy stereo images: models, estimation and uniqueness results. *IEEE Trans. Pattern Analysis and Machine Intelligent*, Vol. 12, no. 8, pp. 735-759.

Zhang, Z. and O. Faugeras (1992). *3D Dynamic Scene Analysis*. Springer-Verlag, Berlin.

APPLICATION OF NEURAL NETWORKS TO IMAGE-BASED CONTROL OF ROBOT ARMS

CHRISTOPHE VENAILLE*, GORDON WELLS** and CARME TORRAS**

*THOMSON-CSF, Laboratoires Electroniques de Rennes, Avenue de Belle Fontaine, 35510 Cesson-Sévigé, FRANCE

**Instituto de Cibernética, Universidad Politécnica de Cataluña, Consejo Superior de Investigaciones Científicas, Diagonal 647, Barcelona 08028, SPAIN

Abstract. This paper describes experiments with the use of neural networks for image-based positioning of industrial robots. Firstly, an improved neural "visual-servoing" method based on geometric image characteristics is developed and analyzed. Subsequently, a more flexible and robust scheme is shown, based on global descriptors of the image.

Key Words. Neural Nets, Industrial Robots, Control System Design, Image Processing, Nonlinear Systems

1. INTRODUCTION

One of the main goals in present-day robotics research is to provide robots with a greater degree of autonomy through sensorimotor control. Visual control, in particular, has recently become a major focus of research, largely due to the advent of computers with sufficient speed to allow including image processing in the control loop.

Many authors have tackled the problem of visual positioning of robot manipulators, for applications ranging from inspection, grasping and assembly of parts to docking and navigation of autonomous guided vehicles. Most of the existing works have tended to rely on simple geometrical features extracted from images, such as points, lines or circles, together with projection transformations to analytically derive the mapping from 2D image space to the robot cartesian coordinates (Abidi, 1990; Mandel, 1987; Kabuka, 1987; Chaumette, 1991; Espiau, 1992).

Although existing approaches based on analytical methods have produced useful results, they all make similar assumptions about the images observed as well as the robot-camera system used, and consequently they share similar limitations. Firstly, all depend on simple geometric features which are assumed to be always visible and extractable in the camera image. In addition, the physical relationship between all object features must be known in order to derive the projection transformations. Furthermore, since more than one cue is typically used, matching must be performed to find the correspondence between each feature in the observed image and a feature in the reference image or physical scene. Finally, in order to derive the complete transformation from image features to robot coordinates, the precise relationship between the camera coordinates and those of the end effector must be known, requiring calibration of the camera as well as knowledge of its intrinsic parameters. Obviously, most of these assumptions place strong restrictions on the observed scene, and the operating conditions and robustness to noise of the resulting system.

Because of their ability to learn complex nonlinear mappings, neural networks are particularly suited for directly mapping sensory information to control actions. The mapping between 2D image feature deviations and robot position or joint angles is a highly nonlinear relationship which depends on the type and relative positions of the object features, the camera-robot relationship, the intrinsic camera parameters, and the robot kinematics. A neural network may be used to learn the entire transformation implicitly based on training examples, thus avoiding explicit computation of all intermediate transformations. A first work in this area was that of Hashimoto (1992) who used backpropagation networks to learn the mapping between the image deviations of 4 projected points of a viewed object with respect to a "reference" or desired image, and the corresponding joint angles of a robot with a camera mounted on its end effector. Once trained, using a series of images taken at randomly selected locations near the reference position, the neural network produced an estimate of the joint angles required to return the robot to the reference position from any other initial location within a limited range. Hashimoto's work showed the capability of a neural network to perform visual positioning with a degree of accuracy similar to that obtained using analytical techniques.

However, many possibilities by which neural networks can be exploited to produce more general-purpose visual positioning systems remain open for research. For example, training neural networks to map from more global image descriptors to robot commands could make visual servoing methods much less sensitive to the presence and extractability of a few simple geometric features in the observed scene. Similarly, the need for explicit feature matching might be avoided, as well as increasing robustness to obstructed features, noisy images, and changing reference positions.

The aim of this work was to develop more flexible and robust visual positioning methods based on neural networks. In all the experiments performed, the overall objective was to train a neural network to represent the mapping between the variations, with respect to a desired "reference" image, of several prespecified image features or descriptors as seen from a camera mounted on the robot end effector, and the 3D cartesian position and orientation of the end effector relative to the reference pose. Unlike in earlier neural approaches (Hashimoto, 1992), the robot's inverse kinematics were not included in the mapping. In this way aspects related to the visual mapping problem could be studied independently from the robot kinematics, which are usually well known anyway, and the additional benefit was achieved that the learned transformation is completely independent from the chosen reference position, and may therefore be used to correctly position the robot relative to the object regardless of their initial locations. Section 2 describes experiments performed using 4-point images as in previous works. Section 3 describes the use of closed contours encoded with Fourier descriptors as inputs to the neural network.

2. FOUR-POINT IMAGE EXPERIMENTS

2.1. *Image Generation*

For initial experiments, a simulated reference image containing four points was used. These points can be imagined as four distinct features extracted from the image of a real object, and were chosen based on the assumption that four points suffice for determining the position and orientation of an arbitrary 3D object (see for example Horaud, 1989; Abidi, 1990). The four feature points were arranged to form a square of width 100 mm in the simulated "scene", as shown in Figure 1. The reference position chosen was with the camera's viewing axis perpendicular to the center of the square formed by the four points, at a distance of 750 mm.

A software simulator was developed to model the projected images that would be seen from a CCD camera mounted on a robot arm. This "moveable-camera simulator" allowed the rapid generation of realistic training sets for varying object and camera characteristics, and the subsequent testing of neural networks trained on these data by simulating the movements the robot would make in 3D space to approach the pre-defined reference position and displaying the resulting images. The model used for the simulated camera was the well-known "pinhole" projection model, as illustrated in Figure 1. The model equations may be found in Abidi (1990).

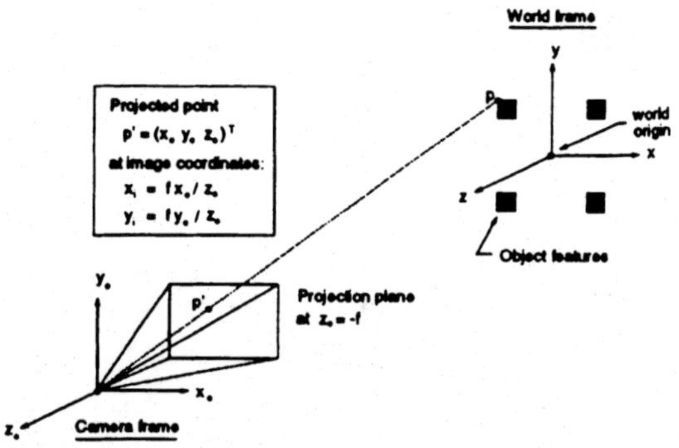

Fig. 1. The pinhole projection camera model used to generate simulated 4-point and closed-contour images.

2.2. *Neural Network Training*

Training examples were generated by choosing 1000 random camera poses within the ranges -50 to 50 mm translation and -3 to 3 degrees rotation for all three coordinate axes (which ensure that the four points are always visible in the resulting images) and generating the corresponding images of the four projected feature points using the simulated camera program. The x,y-coordinates in the reference image of the four points were subtracted from those in the observed image to produce *8 feature "offsets"*, which were used as the inputs to the neural network. The desired outputs of the neural network were the *6 cartesian coordinates* (position and orientation) of the simulated camera, with respect to the reference frame, from which the observed image was viewed. An additional test set of 250 randomly generated images was used for training validation.

These data were used to train a 3-layer backpropagation network with 8 inputs, 30 hidden units, and 6 outputs. The network was used with a *tanh* activation function in the hidden layer, a linear output-layer transfer function, and the delta learning rule. Training was continued until no further decrease in the RMS output error for the test set could be achieved, which occurred after approximately 200,000 training iterations. The final RMS error was 0.18 .

2.3. *Visual Positioning Tests*

Once trained, the network was used in an iterative visual positioning scheme to test its ability to guide the camera back to the reference position from any given initial position and orientation. This process is illustrated in Fig. 2. The camera is first placed at a randomly selected initial position in the vicinity of the reference position. The feature offsets are computed for the four points in the resulting image and input to the neural network. The network outputs a vector of six coordinates representing its estimate of the current camera position and orientation. The camera is then "moved" toward the reference position in the negative direction of these coordinates. The new image and feature offsets are computed at this new camera position. The average relative change in the image offsets is then calculated as the absolute value of the difference between the averages of the 8 offsets in the current and previous iterations. If this value is below a prespecified convergence threshold, the process is stopped. Otherwise, approach movements are continued until this condition is met.

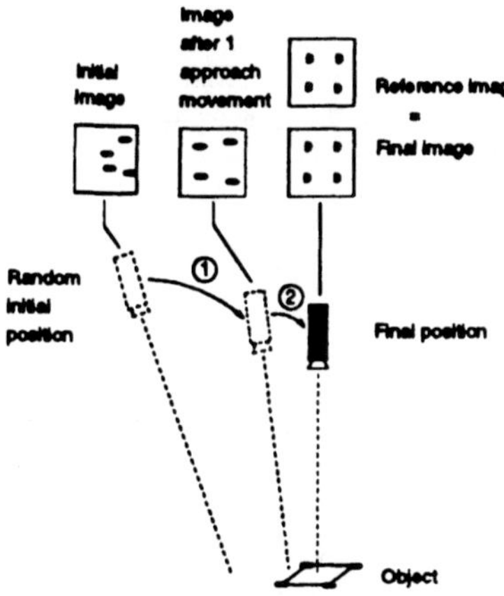

Fig. 2. Visual positioning of a robot-mounted camera using iterative approach movements.

In initial tests, it was observed that the method often converged to positions in which the observed image very closely resembled the reference image (2D error < 0.5 pixels), but which were somewhat displaced from the reference position (3D translation error $\leq$ 12 mm, rotation error $\leq$ 1.5 degrees). Upon closer inspection, it was discovered that many images resulting from translations on either the x or y axis were almost identical to those resulting from rotations around the opposite axis. These "multiple solutions" in the data accounted for the relatively poor training results, and for the significant 3D positioning errors obtained in the tests. The "uniqueness" of the 4-point images in the data set was increased by simply moving one of the four points perpendicularly back away from the plane of the other three by a distance of 100 mm, giving more relative depth to the four features of the simulated "object". Using the image data base computed with this new feature arrangement, the network training time was reduced to only 80,000 iterations, with a final RMS output error of 0.03 . This improved feature arrangement was maintained to produce the test results shown here.

Generalization test results are shown in Table 1. Column 1 shows results for initial positions within the same range of displacement from the reference position as used for training. Columns 2 and 3 show generalization results for 2 and 3 times this range. Values are averages for 100 random initial positions, and standard deviations are shown in parentheses. The convergence threshold used was 0.001 pixel. The final average 2D positioning error in the image was less than half a pixel, and the final average 3D positioning error was less than 2 mm translation and 0.2 degrees rotation (averages for the 3 coordinate axes), after an average of 8 movement iterations. For initial positions within a range of displacement from the reference position 2 and 3 times larger than that used to generate the training images (columns 2 and 3), the network generalized quite well and was able to achieve the same final positioning accuracy in just a few more approach movements.

Table 1 Generalization test results for 4-point images.

Generalization ratio	1	2	3
2D error (pixels)	0.30	0.30	0.32
	(0.07)	(0.07)	(0.08)
3D T error (mm)	1.51	1.50	1.58
	(0.55)	(0.59)	(0.63)
3D R error (deg)	0.11	0.10	0.10
	(0.05)	(0.05)	(0.05)
Movements	8	7	10
	(6)	(5)	(6)

Table 2 shows the effect of varying the convergence threshold (relative average 2D error for 2 consecutive movements) on the number of required iterations. Values are averages for 100 random initial positions within the same range as used for training. When the convergence threshold was relaxed, convergence was achieved in successively fewer iterations, as expected, but the final positioning accuracy remained practically the same, presumably because the 2D resolution used was limited to 1 pixel.

Table 2 Convergence-threshold test results for 4-point images.

Convergence threshold (pixels)	0.01	0.1	1.0
2D error (pixels)	0.30	0.30	0.29
	(0.08)	(0.09)	(0.14)
3D T error (mm)	1.46	1.63	1.67
	(0.59)	(0.63)	(0.64)
3D R error (deg)	0.11	0.11	0.11
	(0.05)	(0.05)	(0.04)
Movements	8	7	2
	(5)	(5)	(0.4)

3. FOURIER DESCRIPTOR EXPERIMENTS

3.1. *Image Generation*

In this set of experiments, the geometrical image features were substituted with global image descriptors as inputs for the neural network. Images were restricted to simple objects, highly contrasted with respect to the background, for which it was possible to extract an external silhouette. Fourier descriptors were used to encode the shape of the resulting closed contour.

For initial tests, a simulated scene of a flat object was used, which consisted of a set of 28 3D points in a plane, joined by a smooth curve so as to form an arbitrary closed contour representing the extracted silhouette of the object. Projected binary images of this shape were then generated using the same simulated camera program as before.

In tests using a real robot/camera system in a real

environment, various simple objects were used, such as flat metal parts, the cylinder head of a car engine, and a small water valve, as shown in Fig. 4. The objects were placed on a white translucent table illuminated from below to enhance contrast.

A 6-axis manipulator (GT Productique) with a camera mounted on its end effector was used to acquire images in the real environment (Fig. 3). A control program, implemented on a SUN workstation, digitized images at a rate of 40 ms per image and sent positioning commands to the robot in cartesian coordinates. Inverse kinematic calculations were performed by a low-level controller running on a PC linked to the SUN workstation by a serial port.

3.2. *Image Processing*

Before computing the Fourier descriptors of the real images, several preprocessing operations were performed on each image to extract the external contour of the object. The object was first segmented from the background to reduce background noise. A binary image with an adaptive threshold was obtained, and connected components were then computed. The biggest object (excluding the background) was identified, and its external contour was extracted.

For both the simulated and real images, the shape of the silhouette was encoded by sampling N equidistant points along the contour and calculating their Fourier descriptors. Each 2D point U_m may be considered a 2D vector in complex space ($U_m = x_m + jy_m$). The Fourier descriptors are defined as:

$$D_n = \frac{1}{N}\sum_{m=0}^{N-1} U_m e^{-j2\pi\frac{nm}{N}} \quad . \tag{1}$$

This defines a list of N descriptors which encode the frequency content of the shape.

It may be observed that these coefficients are not invariant to basic geometrical transformations such as translation, rotation and scaling, nor to the initial point chosen on the contour at which to begin the calculation. In pattern recognition applications, it is usually desired to normalize the Fourier coefficients so as to make them invariant to geometric transformations. For visual servoing, however, variance to translation, rotation and scaling must be preserved, and, in order to avoid matching between the observed and reference images, invariance to the starting point on the contour is desired.

Invariance to the starting point is achieved by shifting the starting point by k points, so as to obtain a new set of points U'_m defined by:

$$U'_m = U_{(m+k) mod N} \tag{2}$$

The Fourier descriptors become:

$$D'_n = D_n e^{2j\pi\frac{kn}{N}} \tag{3}$$

To obtain the same coefficients as before, each descriptor D'_n must be multiplied by $e^{-j\pi\Phi'_1}$, where Φ'_1 is the phase of coefficient D'_1 .

All coefficients D'_i are now transformed into the new normalized coefficients C_i:

$$C_n = D'_n e^{-j\pi\Phi'_1} \tag{4}$$

For the purposes of the tests performed here, only the first 32 (low-frequency) coefficients were used. This was because the neural network required a fixed number of inputs, and the remaining coefficients mostly correspond to high frequencies, which encode very small details in the shape, in addition to noise.

3.3. *Neural Network Training*

As in the 4-point experiments, training sets of both simulated and real images were generated for the neural network by choosing 1000 random initial poses in the vicinity of the reference position, in this case within the ranges of -50 to 50 mm translation and -5 to 5 degrees rotation for the three coordinate axes. The reference position was chosen arbitrarily so as to have the camera located approximately perpendicular to the object's surface at its center, at a distance of 750 mm.

For the observed image at each pose, the closed contour of the object was extracted as explained above, and a set of 512 equidistant points were sampled from the contour to compute the Fourier descriptors. The network inputs were simply the differences between each of the corresponding normalized Fourier coefficients (64 terms for 32 complex coefficients) for the reference and observed images. As with the 4-point images, the desired outputs were the 6 cartesian coordinates of the camera. A set of 500 additional randomly generated images were used as a test set for training validation.

These data were used to train a 3-layer backpropagation network as before, but with 64 inputs, 30 hidden nodes, and 6 outputs. Training for the simulated images was continued for 500,000 iterations, when the final RMS error for the test set was 0.10 .

3.4. *Visual Positioning Tests*

The trained neural networks were tested using the same iterative visual positioning scheme as in the 4-point experiments.

For the network trained using simulated images, 500

random initial positions were tested, both for the same and for twice the range of displacement from the reference position as used for training. These same tests were repeated for two different convergence thresholds. Results are shown in Table 3. As before, a good 2D accuracy was achieved: less than 0.2 pixel. However, the resulting 3D errors were somewhat high: from 4 to 29 mm translation and between 1.83 and 5.60 degrees rotation. Values varied widely from one test to another, however, as the large standard deviations indicate. This difficulty in 3D positioning could be explained by the planarity of the simulated contours used, which might give rise to the same "non-uniqueness" problem as described for the 4-point images. This hypothesis was supported by the improved 3D accuracy achieved in the tests with real 3D (non-planar) objects. The results for generalization and relaxing the convergence threshold showed the same effect on the number of movement iterations as in the 4-point experiments, yet the final 3D errors were approximately doubled.

Table 3 Test results for simulated Fourier-descriptor images.

Generalization ratio	1		2	
Convergence threshold	0.0001	0.001	0.0001	0.001
2D error	0.11 (0.02)	0.11 (0.02)	0.13 (0.03)	0.13 (0.04)
3D T error in X	14 (10)	14 (10)	29 (19)	28 (19)
3D T error in Y	15 (10)	15 (11)	29 (19)	29 (20)
3D T error in Z	4 (3)	4 (3)	9 (8)	8 (7)
3D R error in X	2.84 (1.31)	2.90 (1.38)	5.58 (2.50)	5.60 (2.60)
3D R error in Y	1.90 (1.22)	1.95 (1.31)	5.58 (2.50)	3.63 (2.43)
3D R error in Z	1.83 (1.25)	1.87 (1.25)	3.71 (2.42)	3.66 (2.43)
Movements	4 (1.1)	3 (0.5)	5 (1.0)	4 (0.3)

An example of visual positioning with respect to a real object (a water valve) is illustrated in the photographs in Figs. 3 through 10. The accurate 2D convergence of the observed images onto the reference image is easily seen. Compilation of results using real images is still underway, but have so far shown somewhat improved performance in terms of 3D positioning accuracy than for the simulated contours.

4. CONCLUSIONS

The main aim of this work was to explore the possibilities of using neural networks to perform visual positioning of a robot manipulator with a greater degree of flexibility, efficiency and robustness than has so far been achieved in previous works. First, backpropagation networks were trained to position a robot based on the observed deviations of 4-point images, like those used in existing works based on analytical and neural methods. This allowed preliminary experimentation and comparison with previous results, especially that of Hashimoto (1992), to which a variation was applied to make the trained network independent of the original reference image used for training. Subsequently, the use of *global image descriptors* as network inputs was explored. This represents a novel and more advantageous use of the full potential of neural networks for learning complex nonlinear mappings, which in this case could potentially extend the applicability of visual servoing to a much wider range of image types and operating conditions than are possible when easily extractable geometrical cues are required. Experiments were first performed using Fourier descriptors, to encode closed contours extracted from the observed images, as network inputs, both in simulation and then on a real robot. These tests demonstrated the feasibility of using non-geometric features, as well as the possibility of eliminating the matching step. Experiments with geometric moments are currently underway, which offer the potential of extending the method to images in which a predominant closed contour is not present.

Acknowledgements

Research in neural networks and their applications to process control at the Instituto de Cibernética is partially supported by the research grant CICYT-TIC91-0423 of the Spanish Science and Technology Council. THOMSON-CSF at Rennes and the Instituto de Cibernética are currently developing neural applications for robot control under ESPRIT project No. 6715 "Robot Control based on Neural Network Systems", partially supported by the European Community.

5. REFERENCES

Abidi, M. and R. Gonzalez (1990). "The use of multisensor data for robotic applications", *IEEE Trans. on Robotics and Automation*, Vol. 6, No. 2.

Chaumette, F., P. Rives, and B. Espiau (1991). "Positioning of a Robot with respect to an Object, Tracking it and Estimating its Velocity by Visual Servoing", *Proceedings of the IEEE International Conference on Robotics and Automation*, Sacramento, April, 1991.

Espiau, B., F. Chaumette, and P. Rives (1992). "A New Approach to Visual Servoing in Robotics", *IEEE Transactions on Robotics and Automation*, June, 1992.

Hashimoto, H., T. Kubota, M. Kudou, and F. Harashima, "Self-Organizing Visual Servo System Based on Neural Networks", *IEEE Control Systems*, April, 1992.

Horaud, R., B. Conio, O. Lebouleux (1989) "An Analytic Solution For the Perspective 4-Point Problem", *Computer Vision, Graphics and Image Processing*, **44**, 1989, pp. 33-44.

Kabuka, M. and A. Arenas (1987) "Position verification of a movile robot using standard pattern", *IEEE Journal of Robotics and Automation*, Vol. RA-13, No. 6, Dec., 1987.

Mandel, K. and N. Duffie (1987). "On-line compensation of mobile robot docking errors", *IEEE Journal of Robotics and Automation*, Vol. RA-3, No. 6, Dec., 1987.

Fig. 3. Real robot/camera system used for visual positioning with respect to a water valve.

Fig. 4. Reference image of water valve.

Fig. 5. Observed image at random initial position.

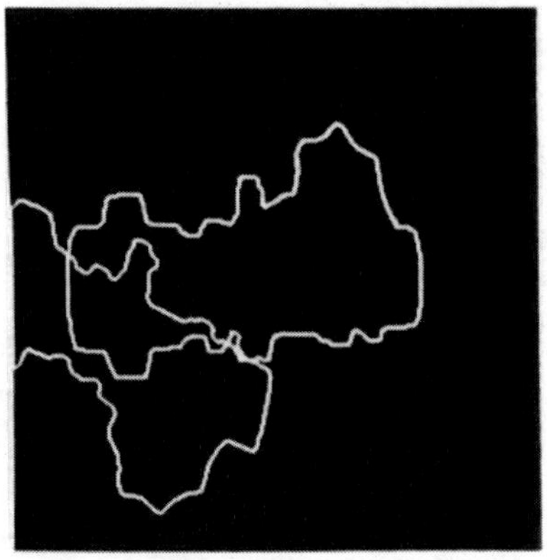

Fig. 6. Reference and observed contours at initial position.

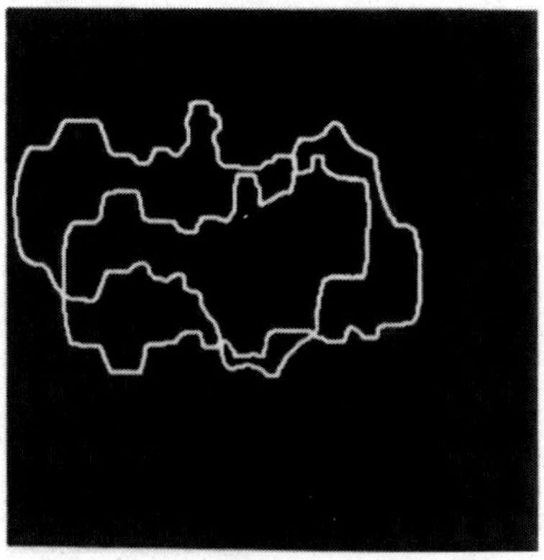

Fig. 7. Reference and observed contours after 1 movement.

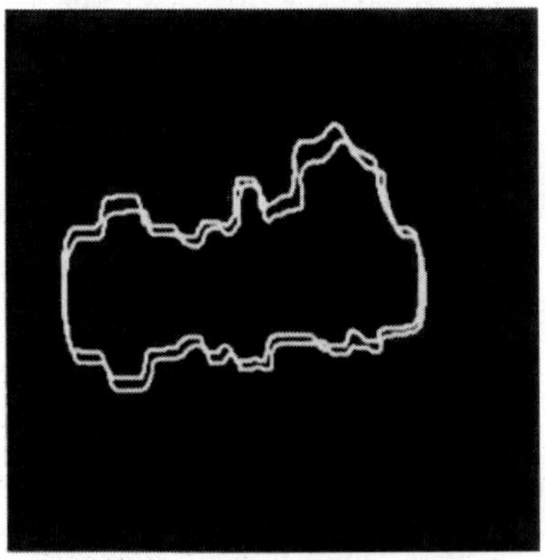

Fig. 8. Reference and observed contours after 5 movements.

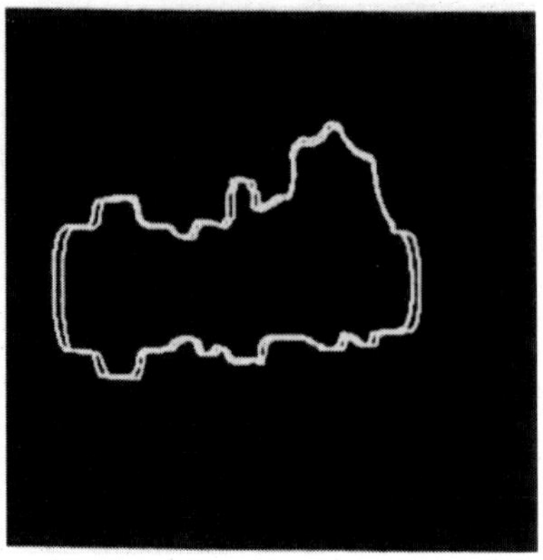

Fig. 9. Reference and observed contours after 7 movements.

Fig. 10. Final image after convergence.

HANDLING DELAY-TIME IN LOW COST CONTROLLERS

CS. BÁNYÁSZ*, I. VAJK** and L. KEVICZKY*

*Computer and Automation Institute, Hung. Academy of Sciences, H-1111 Budapest, Kende u 13-17
**Department of Automation, Technical Univ. of Budapest, H-1111 Budapest, Goldmann Gy. sq. 3

Abstract. Recent investigations proved that the structures and parameter domains of discrete-time equivalent models of continuous-time linear processes play important role in advanced modeling and control methodologies. Besides the well-known transformation formulas a new technique is introduced how to represent even fractional time delays. The proper relationships are discussed for **δ**, modified bilinear and step response equivalent transformations and further generalizations are shown. The introduced modeling technique can be applied in system identification and/or dynamic system simulation.

Key Words. Delay-time estimation; continuous-discrete model transformation.

1. INTRODUCTION

Sampled data control systems have a very important role in advanced modeling, simulation and control methods, mostly because a discrete-time technique can easily be realized by digital computers or microprocessors. The discrete-time approach is very popular between scientists, too, because it is very easy to get computer simulation results using the generally recursive formulas. However, it is easy to forget that in the practice the discrete control algorithms work together with continuous processes and a discrete-time model can be considered equivalent to the continuous one only at the sampling instants. The parameters of a discrete model, therefore the parameters of an optimal regulator and of course, the intersampling properties of the transient processes strongly depend on the selection of the sampling interval h, and the applied method of discrete-continuous transformation or the resulting model structure. It is much easier to get nice digital simulation results using a carefully selected discrete model, than to investigate, what is the chance to obtain that model by using a reasonable h at a given continuous process. The appropriate choice of h and the applied model structure is especially very important in the design of adaptive and optimal control systems, where sometimes a bad selection may eliminate the improvement caused by the optimality.

We can talk about the equivalence of a discrete $G(z^{-1})$ and a continuous transfer function $H(s)$ only requesting the equivalence of the output responses for given input excitation or assuming a given order hold serially connected to the continuous process or applying a given order approximation for the intersampling input signal, which assumptions are equivalent (Keviczky, 1979).

The most commonly used step response equivalent (**SRE**) transformation is given by

$$G_{\mathrm{SRE}}\left(z^{-1}\right)=\frac{1-z^{-1}}{z^{-1}}\mathcal{Z}\left[\mathcal{L}^{-1}\left\{\frac{1}{s}H(s)\right\}\right]^{*} \tag{1}$$

which means the application of a zero-order hold (or step-wise approximation of the input signal at the discretization of the state equations).

Another widely used transformation method is the bilinear (**BL**) transformation (Sinha, 1972) which is based on the first term of the Laurent series approximation of the well known

$$s=\frac{1}{h}\ln z \tag{2}$$

relationship. This approximation gives the simple

$$\frac{1}{s}\cong\frac{h}{2}\frac{1+z^{-1}}{1-z^{-1}} \quad \text{or} \quad s\cong\frac{2}{h}\frac{1-z^{-1}}{1+z^{-1}} \tag{3}$$

transformation formulae corresponding to a trapezoidal integration rule approximating the continuous time integrators. The continuous-discrete transformation is given by

$$G_{\mathrm{BL}}\left(z^{-1}\right)=Y(s)\Big|_{s=\frac{2}{h}\frac{1-z^{-1}}{1+z^{-1}}} \tag{4}$$

The bilinear transformation distorts the frequency scale. This effect can be eliminated by determining the roots of the numerator and denominator of the continuous time system obtained by the substitution (3), thereupon by rescaling the imaginary part μ with the complex roots according to the relation

$$\omega = \frac{2}{h}\mathrm{tg}\left(\frac{h\mu}{2}\right) \tag{5}$$

and by multiplying the new roots with each other (Latzel and Wiegond, 1971).

Another problem with the bilinear transformation is that it always gives an all pass delay free transfer (distortion) on high frequencies which is difficult to interpret. These problems can be considerably decreased by applying a rectangular approximation

$$\frac{1}{s} \cong h\frac{z^{-1}}{1-z^{-1}} \tag{6}$$

instead of the trapezoidal one at the first integrator in the scheme of the continuous time system (Vajk, 1985). The continuous-discrete transformation can finally be performed by

$$G_{\mathrm{MBL}}\left(z^{-1}\right) = \frac{2z^{-1}}{1+z^{-1}} Y(s)\Big|_{s=\frac{2}{h}\frac{1-z^{-1}}{1+z^{-1}}} \tag{7}$$

what we will call as a modified bilinear (**MBL**) transformation (Vajk, 1985).

Another approach to compute the continuous-discrete transformation is based on the so-called **δ**-transformation (Goodwin *et al.*, 1986). In spite of some recent publications, the origin of this transformation can be led back to early publications of this field (see e.g. Farmanfarma, 1957). Avoiding a detailed analysis, this approach is used here only for another equivalent form. The **δ**-transformation is a simple substitution of the Laplace variable s by δ, i.e.:

$$s \leftarrow \delta = \frac{z-1}{h} \tag{8}$$

where z is the forward $\left(zx(k) = x(k+1)\right)$, z^{-1} is the backward $\left(z^{-1}x(k) = x(k-1)\right)$ shift operator. It can be well seen that δ is a forward difference operator corresponding to the well-known first order Euler-approximation. Applying (6) the continuous-discrete transformation is obtained by

$$G_\delta\left(z^{-1}\right) = Y(s)\Big|_{s=\frac{z-1}{h}} \tag{9}$$

2. MODELING DIGITAL REGULATORS

Most of the industrial digital regulators (including compact controllers, D/A converter based real-time interfaces, etc.) applies a zero order hold sequentially connected after the regulator, or their operational conditions correspond to this situation. These circumstances involve that a **SRE** transformation of the continuous regulator transfer function exactly describes its dynamic time-domain operation.

Consider the most general form of a continuous time **PID** regulator

$$Y_{\mathrm{r}}(s) = K\left(1 + \frac{1}{sT_{\mathrm{i}}} + \frac{sT_{\mathrm{d}}}{1+sT}\right) \tag{10}$$

where the differential action is not ideal, represented by a time constant T. Therefore this form is sometimes denoted by **PIDT**, too. The **SRE** discrete transfer function given by (1) is

$$G_{\mathrm{r}}\left(z^{-1}\right) = \frac{p_0 + p_1 z^{-1} + p_2 z^{-2}}{\left(1-z^{-1}\right)\left(1+az^{-1}\right)} = \frac{P\left(z^{-1}\right)}{\left(1-z^{-1}\right)\left(1+az^{-1}\right)} \tag{11}$$

which has four parameters

$$p_0 = K\left[1 + \frac{T_{\mathrm{d}}(1+a)}{h}\right] \tag{12}$$

$$p_1 = K\left[\frac{h}{T_{\mathrm{i}}} - (1-a) - 2\frac{T_{\mathrm{d}}(1+a)}{h}\right] \tag{13}$$

$$p_2 = K\left[a\left(\frac{h}{T_{\mathrm{i}}} - 1\right) + \frac{T_{\mathrm{d}}(1+a)}{h}\right] \tag{14}$$

$$a = -e^{-h/T} \tag{15}$$

unambiguously determined by the parameters of the continuous form. (The same technique can be applied for more sophisticated regulator transfer functions.)

3. MODELING CONTINUOUS LINEAR PROCESSES

In this section we summarize the **SRE**, **MBL** and **δ**-transformation (discrete→continuous) forms of the simplest first and second order linear continuous-time transfer functions and they will be given in a form of

$$Y(s) = \frac{\beta(s)}{\alpha(s)}e^{-s\tau} \rightarrow G\left(z^{-1}\right) = \frac{B\left(z^{-1}\right)}{A\left(z^{-1}\right)}z^{-d} \tag{16}$$

Here τ is the time delay of the continuous process and d is that of the discrete one. First the delay free (τ=0 or d=1) then the integer and fractional delay cases are discussed.

First order process

$$\frac{\beta_0}{1+\alpha_1 s} \rightarrow \frac{b_0 z^{-1}}{1+a_1 z^{-1}} \tag{17}$$

<u>**SRE transformation**</u>

$$b_0 = \beta_0\left(1-e^{-x}\right) \quad a_1 = -e^{-x} \quad x = h/T \tag{18}$$

MBL transformation

$$b_0 = \frac{2\beta_0 h}{h+2\alpha_1} \qquad a_1 = \frac{h-2\alpha_1}{h+2\alpha_1} \tag{19}$$

δ-transformation

$$b_0 = \frac{\beta_0 h}{\alpha_1} \qquad a_1 = \frac{h}{\alpha_1} - 1 \tag{20}$$

Second order process

$$\frac{\beta_0 + \beta_1 s}{1+\alpha_1 s + \alpha_2 s^2} \rightarrow \frac{(b_0 + b_1 z^{-1})z^{-1}}{1+a_1 z^{-1} + a_2 z^{-2}} \tag{21}$$

MBL transformation

$$b_0 = \frac{2(\beta_0 h + 2\beta_1)h}{h^2+2\alpha_1 h + 4\alpha_2} \qquad b_1 = \frac{2(\beta_0 h - 2\beta_1)h}{h^2+2\alpha_1 h + 4\alpha_2} \tag{22}$$

$$a_1 = \frac{2(h^2 - 4\alpha_2)}{h^2+2\alpha_1 h + 4\alpha_2} \qquad a_2 = \frac{h^2 - 2\alpha_1 h + 4\alpha_2}{h^2+2\alpha_1 h + 4\alpha_2} \tag{23}$$

δ-transformation

$$b_0 = \frac{\beta_1 h}{\alpha_2} \qquad b_1 = \frac{(\beta_0 h - \beta_1)h}{\alpha_2} \tag{24}$$

$$a_1 = \frac{\alpha_1 h - 2\alpha_2}{\alpha_2} \qquad a_2 = \frac{h^2 - \alpha_1 h + \alpha_2}{\alpha_2} \tag{25}$$

Time delay processes

Remembering the meaning of z^{-1} the multiplying backward shift operator z^{-d} $\left(z^{-d}x(k) = x(k-d)\right)$ corresponds to a dead time lag having a time delay $(\tau_0 = dh)$ integer multiple of h. This is very rare in the practice and we mostly have to deal with non integer or so-called fractional delay when computing the continuous discrete transformation. The usual theory based on the z or δ-transformation can not handle non integer powers, therefore we propose another approach when the structure of the linear discrete model is changed instead. This change means the increase of the order of the numerator by one and calculating the parameters of the equivalent model by a *model matching* at zero frequency.

In a general case the time delay t can be written in the form

$$\tau = \tau_0 + \Delta\tau = (d' + \gamma)h \qquad 0 \le \gamma \le 1 \tag{26}$$

where d' is integer. The discrete transfer function $G(z^{-1})$ for $\gamma = 0$ has the form

$$G(z^{-1}) = \frac{B(z^{-1})}{A(z^{-1})} z^{-d} \tag{27}$$

where z^{-d} means a backward shift operator (delay) of d and

$$d = d' + 1 \tag{28}$$

Here d' corresponds to the integer delay and the additional unity delay (+1) comes from the structure of the numerators of the delay free $G(z^{-1})$. (Note that for δ-transformation $d = d' + n$ should be applied instead of (28), where n is the order of the system, because of the special structure of the numereator of the discrete transfer function.) For $\gamma \ne 0$ the numerator of (27) will change as

$$G(z^{-1}) = \frac{B'(z^{-1})}{A(z^{-1})} z^{-d} \tag{29}$$

according to the "*model matching* at zero frequency principle" proposed. This principle of equivalence requires the matching of the zero, first and second order derivatives of the original and the modified numerator polynomials, i.e.,

$$B\left(z = e^{j\omega h}\right)\Big|_{\omega=0} = B'\left(z = e^{j\omega h}\right)\Big|_{\omega=0} \tag{30}$$

$$\left.\frac{dB(j\omega)}{d\omega}\right|_{\omega=0} = \left.\frac{dB'(j\omega)}{d\omega}\right|_{\omega=0} \tag{31}$$

$$\vdots$$

The suggested method provides the approximate equivalence of the two sides of

$$b_0 z^{-\gamma} \Leftrightarrow b_0' + b_1' z^{-1} \tag{32}$$

$$(b_0 + b_1 z^{-1})z^{-\gamma} \Leftrightarrow b_0' + b_1' z^{-1} + b_2' z^{-2} \tag{33}$$

at low frequencies for the first and second order cases, respectively. The matching requirements result in the following system equation

$$\begin{aligned} b_0' + b_1' + b_2' &= P = b_0 + b_1 \\ b_1' + 2b_2' &= Q = \gamma b_0 + (\gamma+1)b_1 \\ b_1' + 4b_2' &= R = \gamma^2 b_0 + (\gamma+1)^2 b_1 \end{aligned} \tag{34}$$

for second order case, which has the solution of

$$b_0' = \frac{2P + R - 3Q}{2} \qquad b_1' = 2Q - R \qquad b_2' = \frac{R-Q}{2} \tag{35}$$

For first order case the equations are

$$b_0' + b_1' = P = b_0 \qquad b_1' = Q = \gamma b_0 \tag{36}$$

and the solution is

$$b_1' = Q \qquad b_0' = P - b_1' \tag{37}$$

4. IDENTIFICATION OF CONTINUOUS LINEAR PROCESSES

In general one can not expect that the continuous transfer functions of linear dynamic processes under investigation are available. One of the possible techniques to obtain these forms is the system identification and/or process parameter estimation methodology. In this area we determine the best fitting discrete time model of the process first mostly from input/output measurements, then calculate an equivalent continuous-time model by an appropriate discrete-continuous transformation. This second step means the inverse process of the previously discussed transformation algorithms.

The different identification techniques generally estimate the parameters of $\hat{A}(z^{-1})$, $\hat{B}(z^{-1})$, $\hat{C}(z^{-1})$, $\hat{D}(z^{-1})$ and $\hat{d}$ in a linear **ARMAX** model

$$y(k) = \frac{B(z^{-1})}{A(z^{-1})} z^{-d} u(k) + \frac{C(z^{-1})}{D(z^{-1})} e(k) \tag{38}$$

by one of the widely applied methods as: **LS** (Least-Squares), **ML** (Maximum Likelihood), **IV** (Instrumental Variable), etc.

In this section we summarize the **SRE**, **MBL** and **δ**-transformation (discrete$\rightarrow$continuous) forms of the simplest first and second order linear discrete-time transfer functions. First the delay free then the integer and finally the fractional delay cases are discussed.

First order process

$$\frac{b_0 z^{-1}}{1 + a_1 z^{-1}} \rightarrow \frac{\beta_0}{1 + \alpha_1 s} \tag{39}$$

<u>**SRE** transformation</u>

$$\beta_0 = \frac{b_0}{1 + a_1} \qquad \alpha_1 = \frac{-h}{\ln(-a_1)} \tag{40}$$

<u>**MBL** transformation</u>

$$\beta_0 = \frac{b_0}{1 + a_1} \qquad \alpha_1 = \frac{h}{2} \frac{1 - a_1}{1 + a_1} \tag{41}$$

<u>**δ**-transformation</u>

$$\beta_0 = \frac{b_0}{1 + a_1} \qquad \alpha_1 = \frac{h}{1 + a_1} \tag{42}$$

Second order process

$$\frac{(b_0 + b_1 z^{-1}) z^{-1}}{1 + a_1 z^{-1} + a_2 z^{-2}} \rightarrow \frac{\beta_0 + \beta_1 s}{1 + \alpha_1 s + \alpha_2 s^2} \tag{43}$$

<u>**MBL** transformation</u>

$$\beta_0 = \frac{b_0 + b_1}{1 + a_1 + a_2} \qquad \beta_1 = \frac{(b_0 - b_1) h}{2(1 + a_1 + a_2)} \tag{44}$$

$$\alpha_1 = \frac{(1 - a_2) h}{1 + a_1 + a_2} \qquad \alpha_2 = \frac{(1 - a_1 + a_2) h^2}{4(1 + a_1 + a_2)} \tag{45}$$

<u>**δ**-transformation</u>

$$\beta_0 = \frac{b_0 + b_1}{1 + a_1 + a_2} \qquad \beta_1 = \frac{b_0 h}{1 + a_1 + a_2} \tag{46}$$

$$\alpha_1 = \frac{(2 + a_1) h}{1 + a_1 + a_2} \qquad \alpha_2 = \frac{h^2}{1 + a_1 + a_2} \tag{47}$$

Time delay processes

It is important to know that we always must count the possibility of a non integer time delay at process identification. Therefore one should always have to identify a higher order numerator at the discrete transfer function. After having obtained the *over-estimated* $\hat{b}_0' + \hat{b}_1' z^{-1}$ parameters for a first and $\hat{b}_0' + \hat{b}_1' z^{-1} + \hat{b}_2' z^{-2}$ for a second order process model the numerator parameters of the delay free transfer function can easily be obtained by the equations (30) and (31) again.

For a first order process we have to solve the following system equations

$$\begin{aligned} \hat{b}_0' + \hat{b}_1' &= P = \hat{b}_0 \\ \hat{b}_1' &= Q = \hat{\gamma} \hat{b}_0 \end{aligned} \tag{48}$$

to obtain $\hat{b}_0$ and $\hat{\gamma}$. The solution is

$$\hat{\gamma} = \frac{\hat{b}_1'}{\hat{b}_0' + \hat{b}_1'} \qquad \text{and} \tag{49}$$

$$\hat{b}_0 = \hat{b}_0' + \hat{b}_1' \tag{50}$$

If $\hat{\gamma} < 0$ or $\hat{\gamma} > 1$ then the transformation does not exist.

For a second order case we have to solve the nonlinear system equations

$$\begin{aligned} \hat{b}_0' + \hat{b}_1' + \hat{b}_2' &= P = \hat{b}_0 + \hat{b}_1 \\ \hat{b}_1' + 2\hat{b}_2' &= Q = \hat{\gamma} \hat{b}_0 + (\hat{\gamma} + 1) \hat{b}_1 \\ \hat{b}_1' + 4\hat{b}_2' &= R = \hat{\gamma}^2 \hat{b}_0 + (\hat{\gamma} + 1)^2 \hat{b}_1 \end{aligned} \tag{51}$$

to obtain $\hat{b}_0, \hat{b}_1$ and $\hat{\gamma}$. First the second order equation

$$\hat{\gamma}^2 + \left(1 - 2\frac{Q}{P}\right) \hat{\gamma} + \frac{R - Q}{P} = 0 \tag{52}$$

must be solved for $\hat{\gamma}$ and the two roots must be analyzed. If $0 \le \hat{\gamma}_1 \le 1$ and $0 \le \hat{\gamma}_1 \le \hat{\gamma}_2 \le 1$ then the greater $\hat{\gamma}_i$ should be $\hat{\gamma}$. If only one root fulfills that $0 \le \gamma_i \le 1$ then this $\hat{\gamma}_i$ is $\hat{\gamma}_i$. Finally, if none of the $\hat{\gamma}_i$'s falls into the allowed interval, then the transformation does not exist. Having known $\hat{\gamma}$ we calculate

$$\begin{aligned} \hat{b}_0 &= (1+\hat{\gamma})P - Q \\ \hat{b}_1 &= Q - \hat{\gamma}P \end{aligned} \tag{53}$$

Finally the estimated continuous time delay can be obtained by

$$\hat{\tau} = \left(\hat{d} - 1 + \hat{\gamma}\right)h \quad \text{for both cases.} \tag{54}$$

Note that the above method is easy to extend for higher order systems, when a higher order polynomial equation must be solved and its roots are more difficult to analyze.

5. SIMULATION EXAMPLE

The above transformation relationships can be used for adaptive tuning of (e.g., **PID**) regulator parameters in case of unknown time delay.

In our example the **PID** regulator is applied in a closed-loop system given in Fig. 1. This figure shows the continuous process given by the transfer function $Y(s) = \beta(s)e^{-s\tau}/\alpha(s)$, the zero order holding (Z.O.H.) element and the discrete transfer function $G_r(z^{-1}) = P(z^{-1})/(1-z^{-1})$ of the regulator (this form corresponds to (11) with T=0, i.e., a=0).

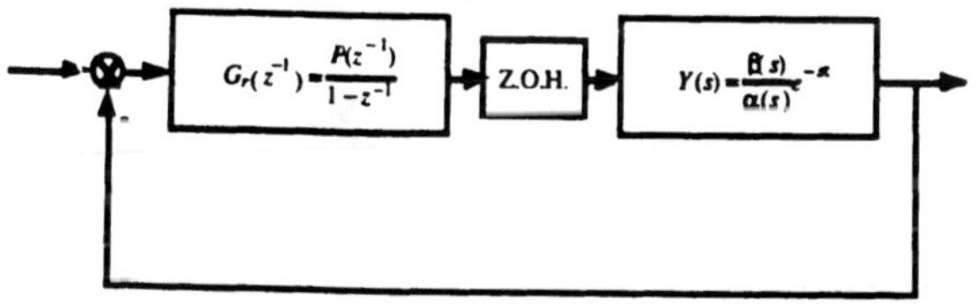

Fig. 1. The block scheme of the closed loop system

In the industrial practice it is a good approximation to use a second order time delay lag as continuous process transfer function in most of the cases, i.e.,

$$Y(s) = \frac{K_P}{1 + 2\xi T s + T^2 s^2} e^{-s\tau} \tag{55}$$

The δ-transformation of this process is given by

$$G(z^{-1}) = \frac{b_0 z^{-2}}{1 + a_1 z^{-1} + a_2 z^{-2}} z^{-d'} = \frac{b_0}{1 + a_1 z^{-1} + a_2 z^{-2}} z^{-d} \tag{56}$$

according to the previously discussed relationships (see Fig. 2).

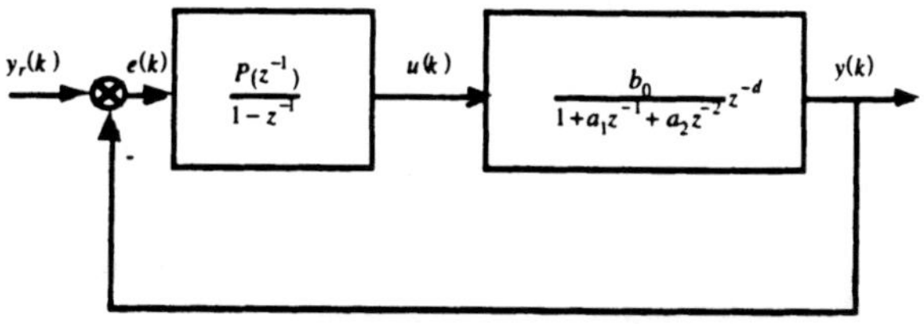

Fig. 2. A second order plant by δ-transformation

The direct estimation of the discrete time (integer) delay makes an **LS** (Least-Squares) estimation strongly nonlinear that is very sensitive for the initial conditions and for the selected recursive technique. (Practically this is the case at a continuous time estimation, however, a rational fraction approximation (e.g., a low order Pade) can improve the linearity, while the higher order derivatives make the method noise sensitive.)

The most widely suggested method estimates

$$\hat{B}_M(z^{-1}) = \hat{b}_1 z^{-1} + \dots + \hat{b}_M z^{-M} \tag{57}$$

instead of $\hat{B}(z^{-1})z^{-\hat{d}}$ with an assumed delay $\hat{d}$. Here M is the possible maximum value of $\hat{d}$. The best approximating $\hat{B}(z^{-1})$ and $\hat{d}$ can be found by probing different values of $\hat{d}$ in a second estimation phase.

The special structure of $G(z^{-1})$ in (56) allows to introduce a very simple new method to estimate $\hat{b}_0$ and $\hat{d}$. This method also estimates the parameters of $\hat{B}_M(z^{-1})$. For this reason the well-known **RLS** (**R**ecursive **L**east-**S**quares) equations can be used with the parameter vector

$$\hat{q} = \left[\hat{b}_1, \dots, \hat{b}_M, \hat{a}_1, \hat{a}_2\right]^{\mathrm{T}} \tag{58}$$

and observation vector

$$f(k-1) = \left[u(k-1), \dots, u(k-M), -y(k-1), -y(k-2)\right]^{\mathrm{T}} \tag{59}$$

Having obtained $\hat{B}_M(z^{-1})$, the equivalent

$$\hat{B}(z^{-1}) = \hat{b}_0 z^{-\hat{d}} \tag{60}$$

numerator (parameter and delay) can be determined by the previously discussed *model matching* at zero frequency. For the determination of the two unknown parameters: $\hat{b}_0$ and $\hat{d}$, the zero and first order derivatives of $\hat{B}_M(j\omega)$ and $\hat{B}(j\omega)$ will be required to be equal at ω=0. This means the solution of (53) and (54), which gives

$$\hat{\bar{d}} = \hat{d} + \gamma = \frac{1}{\hat{b}_0} \sum_{i=1}^{M} i\hat{b}_i = \frac{\sum_{i=1}^{M} i\hat{b}_i}{\sum_{i=1}^{M} \hat{b}_i} \tag{61}$$

Note that $\hat{\bar{d}}$ is not yet an integer number here, therefore the exact calculation of $\hat{d}$ would require $\hat{\gamma}$, which involves the usage of

$$\hat{B}'(z^{-1}) = (\hat{b}_0 + \hat{b}_1) z^{-\hat{d}} \tag{62}$$

instead of (60). However, in the adaptive practice the best integer approximation of (61) is generally used, i.e.,

$$\hat{d} = \mathbf{INT}\left(\hat{\bar{d}} + 0.5\right) \tag{63}$$

In the suggested adaptive **PID** algorithm special design formulas (Keviczky and Bányász, 1992) are used with $\hat{b}_0$ and $\hat{d}$ besides the estimated $\hat{a}_1, \hat{a}_2$ for the determination of the optimal regulator parameters.

Example: Consider a second order process with discrete transfer function given by

$$G(z^{-1}) = \frac{0.1}{1 - 1.3z^{-1} + 0.4z^{-2}} z^{-2} \tag{64}$$

where the true time delay d=2 changes to d=3 at k=45 and is unknown for the adaptive regulator. (Here $\boldsymbol{R}_0 = 100\boldsymbol{I}$, w=0.95 and $\hat{\boldsymbol{q}}_0 = \boldsymbol{0}$ was used at the simulation (Keviczky and Bányász, 1992).) The square wave reference signal excitation had a time period 10. The maximal order M=4 was used for $\hat{B}_M(z^{-1})$.

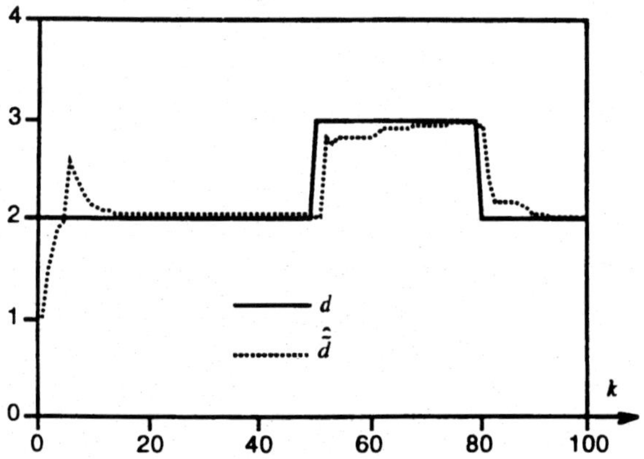

Fig. 3. Adaptation of a changing time delay d

Figure 3 shows the true d and estimated non integer delay $\hat{\bar{d}}$. Figure 4 shows the estimated integer delay $\hat{d}$ too. The latter figure shows the speed of adaptation, which is quite fast.

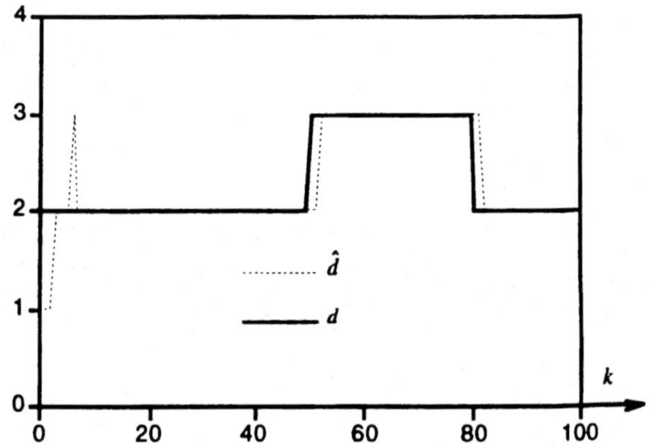

Fig. 4. Estimation of the discrete time delay $\hat{d}$

6. CONCLUSIONS

Our paper presents a survey of the simple methods for continuous-discrete and discrete-continuous transformation of process models used for modeling, identification and control. The transformation formulas for simple first and second order processes are also given using **SRE**, **MBL** and **δ**-transformation.

The main contribution of the paper is to introduce a new method how to take the time delay into consideration even if it is not the integer multiple of the sampling interval. This new approach augments the order of the numerator of the discrete transfer function by one and provides the equivalence of the zero and first order derivatives of the frequency characteristics at zero frequency.
A simple simulation example is also given how these transformation formulas and the new *model matching principle* at zero frequency can be used for adaptive regulator algorithms.

7. REFERENCES

Farmanfarma, G. (1957). Analysis of linear sampled-data systems with finite pulse width: open loop. *Trans. of AIEE*, Vol. 75, 808-819.

Goodwin, G. C., R. Lozano Leal, D. Q. Mayne and R. H. Middleton (1986). Rapprochement between continuous and discrete model reference adaptive control. *Automatica*, Vol. 22, 199-207.

Keviczky, L. (1979). *On the transfer functions of sampled continuous systems*. Report, Center for Control Sciences, University of Minnesota.

Keviczky, L. and Cs. Bányász (1992). An adaptive PID regulator based on time delay estimation. *31st IEEE Conference on Decision and Control*, Tucson, USA, 3243-48.

Latzel, W. and R. Wiegond (1971). DIGSI - A simulation language for dynamic processes and DDC. *IFAC Symposium*, *DISCOP*, Győr (Hungary).

Sinha, N. K. (1972). Estimation of transfer function of continuous system from sampled data. *Proc. IEEE*, Vol. 119.

Vajk, I. (1985). *Implementation problems of adaptive regulators*. Report, Computer and Automation Research Institute, Hungarian Academy of Sciences.

FLEXIBLE LOW-COST MULTIFUNCTION INSTRUMENT

L. TRYBUS, Z. ŚWIDER and M. ŚNIEŻEK

Technical University, Department of Electrical Engineering, 35-959 Rzeszów, Poland

Abstract. A multifunction controller has been developed for the automation of small and medium-scale plants. The main features are: substantial number of inputs and outputs, autonomous configuration by using the faceplate, flexibility of applications due to replaceable regulatory and logic function blocks, peer-to-peer communication. The latter allows one to control larger, highly interactive plants by means of a network of small instruments. Over 200 controllers have been installed in industry so far, ranging from steam turbine control to milk pasteurization.

Key Words. Microcomputer-based control; control equipment; process control; distributed control; communication control applications

1. INTRODUCTION

Digital control devices of the early 90s can be classified as follows: (1) PID regulators, (2) smart transmitters, servos, recorders - having a PID algorithm, (3) small programmable logic controllers - PLCs, (4) DIN-seized multifunction instruments, (5) big multifunction controllers and PLCs (crate housed). The multifunction instruments, this paper is concerned with, may operate as PID regulators, logic controllers, or both, although not with such short scan as PLCs. The software comprises about 100 function blocks which are connected into suitable strategy during configuration (see e.g. Lucas, 1986). Such instruments are manufactured by Siemens, Micon, Toshiba, Yamatake-Honeywell, Bailey, and a few other leading companies.

From the user viewpoint the following features of the multifunction instruments are most important:

- *Number of inputs and outputs*, as it determines the number of PID loops and logic control circuits available, so the cost of automation.
- *Autonomous configuration*, by using the faceplate only, without any additional device, such as hand-held configurer, lap-top PC, etc.
- *Flexibility of applications*, which is considered total, if all resources of the instrument can be used either for regulation or for logic control.
- *Peer-to-peer communication*, independent of *master-slave*, for automation of larger plants by means of a network of connected instruments.

SIPART DR24 Multifunction Unit from Siemens, which has up to 8 AI, 4 AO, 14 BI, and 16 BO (AI - analog input, BO - binary output, etc.), is the leader as far as the number of IOs is concerned. DR24, Moore's MYCRO 352 and 383, Yokogawa's YS100, Hartmann-Braun's Digitric P, and Fuji's FC-PNA can be configured autonomously. Micon's S-32, Toshiba's EC300, Fischer-Porter's MC5000, and Bailey's Command Series provide total flexibility, as each of their software blocks may be assigned either a regulatory or a logic function. S32, MYCRO 352, MC5000, Command Series, and T640 from Eurotherm (formerly Turnbull) are equipped with peer-to-peer communication (called also *horizontal*

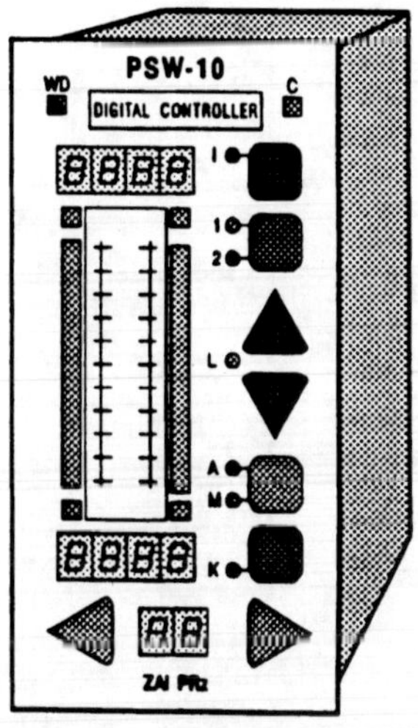

Fig. 1. PSW-10 multifunction controller

or *lateral*). These excellent instruments determine world standard in automation technology. Note, however, that none of them incorporates all the features specified above, at least at present.

A PSW multifunction instrument developed a few years ago by the authors, under the assumptions of flexibility and low cost, is described in this paper (PSW - the Polish abbreviation for *programmable multifunction controller*). Low-cost automation is a reality in Eastern Europe, cut off until the late 80s from the VLSI components and modern technology. The PSW faceplate (Fig.1) is similar to Siemens SIPART instruments. The controller is manufactured by ZPDA ZAP Ostrów Wlkp. and ELMONT Biała Podl. The earlier model with Z80 microprocessor was denoted PSW-8, the new one with Z84 microcontroller is denoted PSW-10. Several Polish universities use it in student labs (textbook - Trybus, 1992).

2. HARDWARE

Inputs, outputs. The PSW consists of a microcomputer board, 2 process interface boards, a faceplate, and a power pack. The interface boards are: *analog* with 8 AI + 4 AO (PSW-8 had 6 AI + 2 AO), *binary* with 20 BI + 16 BO. This is even somewhat more than in Siemens DR24. The two boards are *interchangeable*, so the following versions of PSW are available:

- Standard: 8 AI + 4 AO, 20 BI + 16 BO
- Binary: 40 BI + 32 BO
- Analog: 16 AI + 8 AO.

The interchangeability increases flexibility. Usually, a standard PSW is configured for 3 or 4 PID loops and 10 to 15 logic circuits. A binary version is used as a PLC; the analog as a data-acquisition station or process calculator.

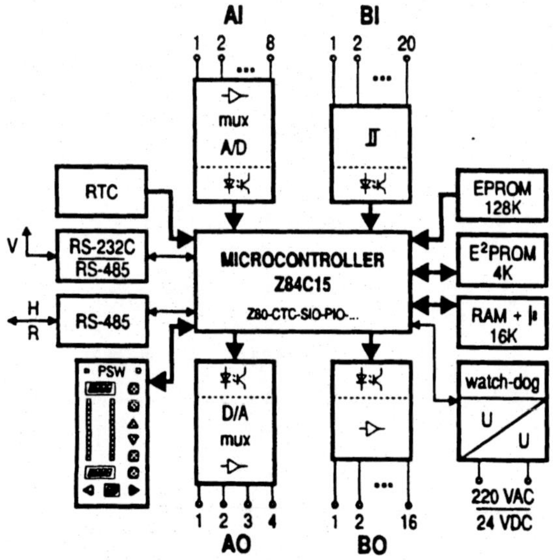

Fig. 2. Hardware structure of the PSW-10

Faceplate. It consists of: (1) 3 digital displays - 2 four-digit, 1 two-digit, (2) 2 LED bars (linear displays), (3) 6 individual LEDs, (4) 1 watch-dog LED - WD, (5) 8 keys (pushbuttons) to which none, one, or two LEDs are assigned (Fig.1). LED bars can be used as *bargraphs* or *point displays*, or as *sets of separate LEDs* (every second or a pair of three). In the latter usage the LEDs indicate logic states. The faceplate elements can be freely used in the configured strategy. It is another factor of flexibility.

Microcomputer. Hardware structure of the PSW-10 is shown in Fig.2. The Z84C15 microcontroller incorporates Z80 CPU (18.4/2 MHz), CTC - counter/timer, SIO - serial IO, PIO - parallel IO, and 2 other units (Zilog, 1990; CTC, SIO, PIO were separate chips in PSW-8). The EPROM 128K, E^2PROM 4K, and CMOS-RAM 16K (with battery back-up) are banked. The microcomputer is equipped with the real-time clock RTC for time-of-day dependent strategies. Other important chips are the following: faceplate drivers (2xMAX), watch-dog/voltage controller (MAX), decoder (GAL), RS-485/RS-232C and RS-485 line drivers at the SIO.

Analog, binary boards. 12-bit A/D and D/A converters with serial data transfer are used (MAX). Analog outputs are monitored by the inputs (a 16-channel multiplexer is used, DG). Binary output drivers (TDA) signal overload or short-circuit. A watch-dog output is provided to configure redundant systems. Both boards are opto-isolated from the microcomputer.

3. FUNCTION BLOCKS

Functional diagram of the PSW-10 is shown in Fig.3 (standard version). As for every multifunction controller, it consists of input, algorithmic, and output blocks. The IO blocks drive the hardware components (Fig.2, see also Sec.5). The algorithmic blocks include:

- *80 simple blocks*, to each of which any of *48 simple functions* can be assigned. The functions are arithmetic and logic, switches, comparators, etc. (Table 1). They are software equivalents of simple conventional control elements.
- *40 complex blocks*, each of which can realize any of *35 complex functions*, provided that enough memory is available. PID algorithm, servo positioner, PROM logic table are examples of these functions (Table 2). They are equivalents of more complicated control devices.

There are also:

- *Universal parameters*: linear, decadic, and binary - 16 of each; linear are adjusted linearly, decadic by decades (e.g. time constants of periods).

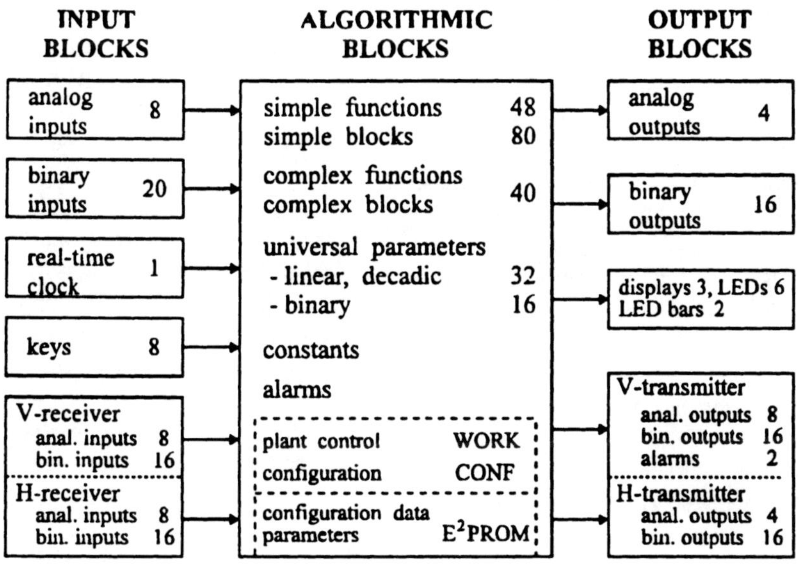

Fig. 3. Functional diagram

Table 1. Simple functions (48; software v.3.0)

Arithmetic functions (14)	***Logic functions (7)***
ABS absolute value	AND product
ADD addition	ANDN product with neg. inp
AMP differential amplifier	EXOR exclusive-or
DIV division	NAND product negation
ENT integer part (entier)	NOR sum negation
EXP exponential	NOT negation
LINE linear function	OR sum
LOGD logarithm base 10	
LOGN natural logarithm	***Flip-flops, pulsers (12)***
MUL multiplication	COUN counter
POWR power (10^x)	DELS delay by 1 scan
RAND random generator	DFF D flip-flop
SQR square root	GENR square wave gener.
SUB subtraction	JKFF JK flip-flop
	PDUR pulse duration
Switches, selectors (7)	PULS pulser
AMEM analog memory	RSFF RS flip-flop
ASWI analog switch	SHOT 1 scan pulse
BMEM binary memory	TFF T flip-flop
BSWI binary switch	TIMR timer
COMP comparator	TOTI totalization/pulse
MASE maximum selector	
MISE minimum selector	***Miscellaneous (4)***
	BACK signal returned
Filters, signal analyzers (4)	DEBA dead-band
FILT first-order filter	HOUR time conversion
DIFR differentiation	LIMT limiter
MAT max. over time	
MIT min. over time	

Table 2. Complex functions (35)

Regulation (6)	***Logic control (10)***
DEL delay	BCNT binary counter
IA integrator	CODR coder
PID controller	DCD decoder
POS servo positioner	LCAL logic calculator
SERV servo simulator	LCAS logic calculator
SP set-point driver	NUMB binary-to-real
	PROM logic table
Approximation (5)	3GAT 3× AND, OR, ...
FUNA parabolic approxim.	3NSD 3× NOT, SHOT, ...
FUNL linear approxim.	3TIM 3× TIMR, PULS,...
3ANS 3× ABS, SQR, ...	
3LIN 3× LINE	***Sequencer (3)***
3MTH 3× ADD, SUB, ...	FUNB binary function
	SEQ sequencer clock
Multiplexers, comparators	SEQC sequencer controller
AMUX analog multiplexer	
BMUX binary multiplexer	***Additional parameters (2)***
3AMT 3× AMEM, ...	PD decadic (4)
3CDS 3× comparators	PL linear
3CHL 3× comparators	
	Miscellaneous (2)
Filters (2)	BALF binary alarms
ADFI adaptive filter	FLOW calculator
3FIL 3× FILT (lag)	

- *Constants* - analog and binary: -1.0, -0.5, ..., 1.25, LOW, HIGH.
- *Alarms:* power-on, configuration (off-line), IO failures, battery down, communication failures.

During configuration, the IO blocks, algorithmic blocks with assigned functions, universal parameters, constants, and alarms are connected together into suitable strategy. The total number of functions, 48+35=83, exceeds the average for multifunction instruments (about 50), being close to Bailey's Command Series. Graphical symbols of functions and IO notation resemble Siemens controllers. Examples are shown in Fig.4. A block number is placed in the upper left corner (b__ - simple, c__ - complex); a position in the right (p__ - execution order, see the next section).

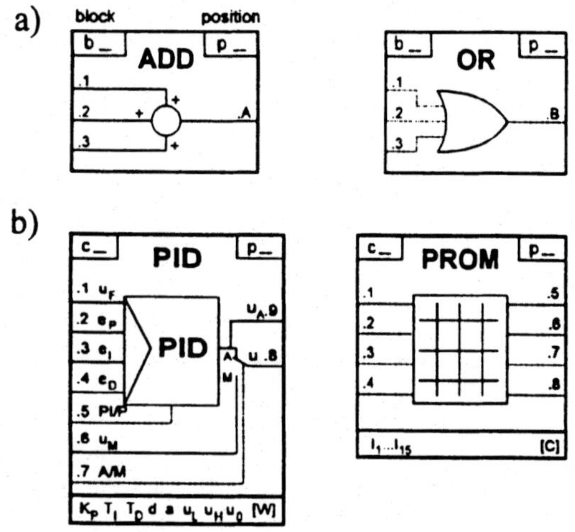

Fig. 4. Sample functions: a) simple, b) complex

Memory for complex blocks. E^2PROM+RAM memory spaces occupied by the simple blocks are the same. The complex blocks, however, occupy spaces relevant to their functions. This feature is called *dynamic assignment of memory* (*static* is the case for simple blocks). For instance, the BMUX function, binary multiplexer, or IA, integrator, require small space; PID algorithm or PROM table - medium; DEL, delay - large. Hence one may assign BMUX or OR to all 40 blocks, PID and PROM to about 20, and DEL to just 4. Similar estimates hold for 4 other instruments with dynamic assignment of memory (Micon's S-32, Toshiba's EC300, Fischer-Porter's MC5000, Bailey's C.S.).

Flexibility. Since each algorithmic block may realize any of the corresponding functions (provided that there is enough memory for a complex function), the whole E^2PROM and RAM memories can be, in extreme cases, assigned either to regulation or to logic control. This guarantees total flexibility of applications, as in case of the 4 instruments mentioned above. In some applications the PSWs were able to handle 6 PID loops or about 30 logic circuits.

4. CONFIGURATION

The PSW can be configured autonomously from the faceplate, as the instruments from Siemens, Moore, Yokogawa, Hartmann-Braun, and Fuji. The configuration consists of the phases listed below (also Fig.5). It is performed off-line, in conversational manner, proceeding through a series of questions and answers (menu). Configuration data and parameters are stored in E^2PROM. The chip is replaceable and can be installed in another controller, if desired. The phases are as follows:

- NAME - strategy name (number)
- DEFN - assignment of functions to blocks
- CONN - input-output connections
- POSN - positions of the blocks
- ONPA - on-line parameters (initial values)
- OFPA - off-line parameters
- OPTN - optional configurations
- RUN - start of WORK.

The positioning POSN, not present in some controllers (Siemens has introduced it), allows one to modify the block execution order without reconfiguring the entire strategy. It helps especially in case of logic algorithms. The OPTN phase consists of *preset*, *load-from-PC*, and *EPROM-predefined-strategy* options. Calculation scan depends

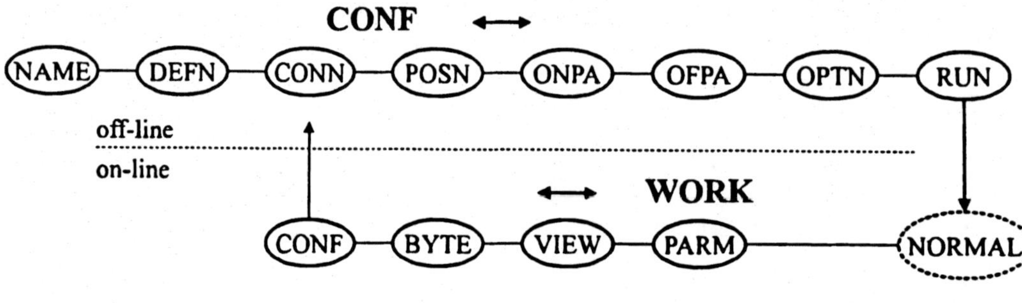

Fig. 5. Configuration phases and WORK modes

primarily on the number of floating-point operations. Typically, the scan is about 0.2 s for regulation and 50 ms for logic control.

Configuration analysis. During RUN (Fig.5) the PSW analyzes whether: (1) all inputs are connected, (2) at least one output of each block is connected to an input, (3) all blocks are positioned. If not connected input has been found, the controller indicates an error and returns to the CONN phase. The second and third case causes warnings, but if acknowledged, the controller assumes WORK anyway.

WORK (on-line). After RUN the controller begins NORMAL operation, i.e. it controls the plant, drives the faceplate, and communicates. One may switch to PARM, VIEW, and BYTE modes (all on-line), or return to configuration (if allowed). In PARM the on-line parameters may be changed (initial values have been set in ONPA). VIEW displays inputs of the blocks, what helps while testing. E^2PROM bytes may be modified in BYTE.

A PC can also be used for configuration by means of a simulator package. The faceplate is displayed on the monitor, numeric keys of the keyboard represent the PSW pushbuttons. Configuration data can be transmitted if the controller is in the *load-from-PC* mode (OPTN). A graphical configuration based on OrCAD is currently being developed.[1]

5. COMMUNICATIONS

The PSW instruments may operate in an RSA distributed control system (Polish abbreviation for DCS). An 486/SVGA PC expanded with a 8xRS-232C/485 card is used as an operator station. The SCADA-type software is called ORSA ("display" of RSA). The PSW has 2 communication channels (Figs.1,2) - vertical (*master/slave*) and horizontal (*peer-to-peer*). The second may be alternatively used for redundancy (*master/slave*). An RSA topology with the three communications is shown in Fig.6 (three are also 3 other instruments belonging to the RSA: a general purpose PID regulator, data-acquisition station, dedicated two-channel regulator for buildings).

Vertical communication. The protocol is quite simple. The computer question consists of 5 bytes

SOH	CNO	CMD	KEY	CRC

where: SOH - Start of Header, CNO - code and controller number, CMD - requested data identifier,

[1] The authors appreciate the help of Prof. W.Halang from FernUniversität, Hagen.

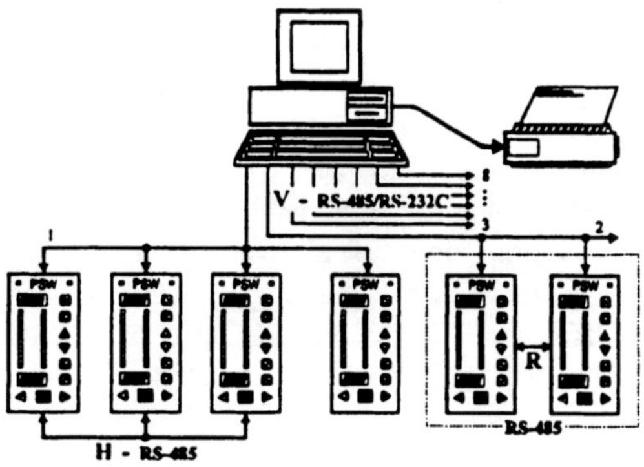

Fig. 6. RSA distributed system with vertical, horizontal and redundant communications

KEY - positions of the keys (emulation), CRC - check-sum (CRC-8). Successive bits of the CMD identify: 1 - transmitter inputs, 2 - faceplate readings, 3 - receiver inputs, 4 - plant IOs, 5 - configuration data (128B packs). The KEY byte and the controller's answer to CMD=2 allow the PC to emulate the faceplate. The controller answers with the frame

SOH	CNO	DATA	CRC

where the DATA correspond to the CMD. Note that due to the bit-wise CMD there is little provision for transmitting individual variables or parameters. Whereas it may sometimes be restrictive, the important advantage is that to develop a communication driver, *the controller memory map is not required.* Thus the driver may be very simple. The individual variables and parameters are accessed by emulating the faceplate.

Horizontal communication. The PSWs exchange data over the H-bus (Fig.6). Here there is only one type of frame, i.e.

SOH	No	N	TOKEN	CODE	DATA	CRC

where the DATA depend on the CODE. No is the sender's number (1,...,8), N - total number of bytes, TOKEN - number of the instrument which should transmit next (*token-passing* technique is used). The controllers select appropriate variables from the data appearing on the bus. Details are given in Trybus *et al.* (1994). Recall that the horizontal communication is necessary for the automation of larger, highly interactive plants. The network of small instruments may have computing power similar to big multifunction controllers or PLCs. Besides, the dependability of such network is greater, since failure of one instrument does not stop the others.

Communication of redundant pair. 2 PSWs are able to provide uninterrupted automatic control of critical processes. When the main instrument fails, the back-up assumes operation. Inputs and outputs of the pair are connected in a tandem configuration. The redundancy implemented by using a communication channel does not allow even for momentary interruption of the output signals (Bailey's Batch and Sequence C.S. instruments have such feature). In case of the PSW pair the 5-byte frames are exchanged every 50 ms

SOH	DATA	CRC

The DATA consist of binary outputs (16) and some logic variables (8). If the main controller has not sent a frame, the back-up rises up its outputs immediately and assumes transmitting as the master. It is also possible to configure the redundant pair by using the watch-dog signals. However, brief interruptions of the outputs appear after the main fails.

6. APPLICATIONS

Over 200 PSW instruments have been installed in industry so far. Applications include a steam turbine, boilers, water degassing and deironing, sewage plant, water treatment, municipal heating systems, various chemical installations, milk pasteurization, cheese manufacturing, some mechanical equipment and machines. The PSWs have turned out particularly convenient for small installations, up to about 100 signals, where a system of 3 or 4 instruments communicating horizontally is sufficient. The users appreciate that the same equipment performs regulation, logic control, and plant monitoring, what reduces the cost of training and maintenance. The ease with which the strategies can be modified has been acknowledged.

Sulphur oiling. The first bigger application of the RSA distributed system took place at the sulphur oiling facility of SIARKOPOL Grzybów in south-eastern Poland. Oiled sulphur powder is a component of radial tires and quality rubber products. The process is relatively difficult to control, since sulphur powder, if handled without care, very quickly clogs the pipes and equipment. The system consists of 6 PSWs, 1 standard and 5 binary (Sec.2), and an 386/SVGA PC with 8×RS-232C/422 card and a printer. 2 PSWs communicate with automatic scales. Total number of inputs is 168, outputs - 105. The controllers have been configured mostly for logic and sequential operations. 4 analog and 43 binary signals are transmitted over the horizontal bus. As a whole, these signals provide coordination of the controllers in this sequential, interactive process.

Future enhancements of the PSW will include self-tuning, PROFIBUS communication (DIN 19245), and modifications of functions to conform with the new IEC 1131-3 standard. A prototype with relay self-tuning (Hägglund and Åström, 1991) is being tested.

7. REFERENCES

Hägglund, T., and K.J. Åström (1991). Industrial adaptive controllers based on frequency response techniques. *Automatica*, **27**, 599-610.

Linzenkirchner, E. (1980). Ein adaptives Filter für gestörte Messignale. *Regelungstechnische Praxis*, **22**, 392-395.

Lukas, M.P. (1986). *Distributed Control Systems.* Van Nostrand Reinhold, New York.

Trybus, L., M. Śnieżek, and Z. Świder (1994). Horizontal communication of PSW-8 multifunction controller. *Control Engineering Practice* (revised).

Zilog, Inc. (1990). Z84C13/C15 CMOS Intelligent Peripheral Controller. Campbell, USA.

COMPUTER AIDED TRACTOR/IMPLEMENT COUPLING

M. GRAEF and G. JAHNS

Institute of Biosystems Engineering, Federal Agricultural Research Centre, 38116 Braunschweig, Germany

Abstract: Coupling tractors and implements cause severe accidents in agriculture. Automatic couplers are on the market, but they have a limited catching area. Because of restricted visibility, the operator often fails to meet this area. Therefore a system is proposed which is able to measure the distance and direction continuously, to calculate the required steering angle and to display it to the operator.

Key words: Tractor, Implement, Coupling, Occupational Safety, Computer Simulation, Neural Network, Fuzzy Control, Ultrasonic Measuring System, On–Board Computer, Monitoring

1. INTRODUCTION

Objectives are to improve occupational safety and driver comfort by technical means during coupling and uncoupling of agricultural tractors and implements.

The use of tractors and implements in agriculture is characterized by frequently changed implements, a task with the permanent risk of severe accidents [1,2]. To reduce this risk automatic couplers have been introduced. For proper operation the tractor has to approach the implement closely enough to bring the automatic couplers into action.

If the operator fails to drive the tractor into the catching area where the automatic couplers come into action, then he or another person often steps into the space between tractor and implement to adjust the implement position and to connect the coupling elements manually.

In the following, technical means are proposed to enable the operator to safely back up his tractor into the catching area of the couplers, even if the coupling area is not visible from his position. By these means working conditions are also improved.

2. SOLUTION

To avoid the presence of the operator or any other person between tractor and implement during coupling; to avoid mounting or dismounting the tractor, and to avoid the operator twisting while driving backwards it is necessary:

1. to use automatic couplers for tractors and implements
2. to provide the right information to the operator, so that he can back the tractor up precisely into the catching area of the couplers even if he is not able to observe the implement.

To achieve this, a system continuously has

- to measure distance and direction between tractor and implement;
- to calculate the steering angle required to drive the tractor into the catching area and
- to display the required steering angle to the operator.

The main parts of the project discussed here are shown in Figure 1. Distance and angle between tractor and implement can either be measured (left top) or generated by a model (left bottom). Three different ways to predict the required steering angle for successful coupling are indicated (middle), while at the right a user interface for manual control (top) or an automatic steering control is depicted.

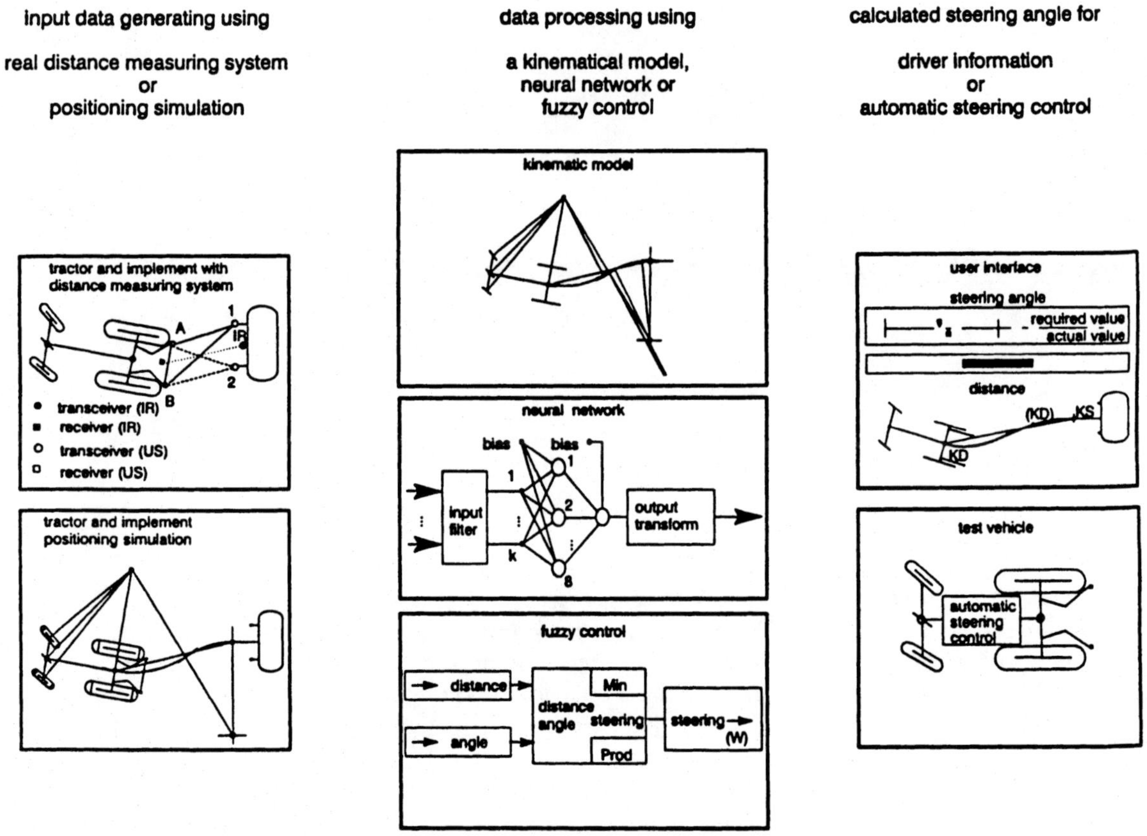

Fig. 1. Data generating, processing and use for tractor and implement coupling

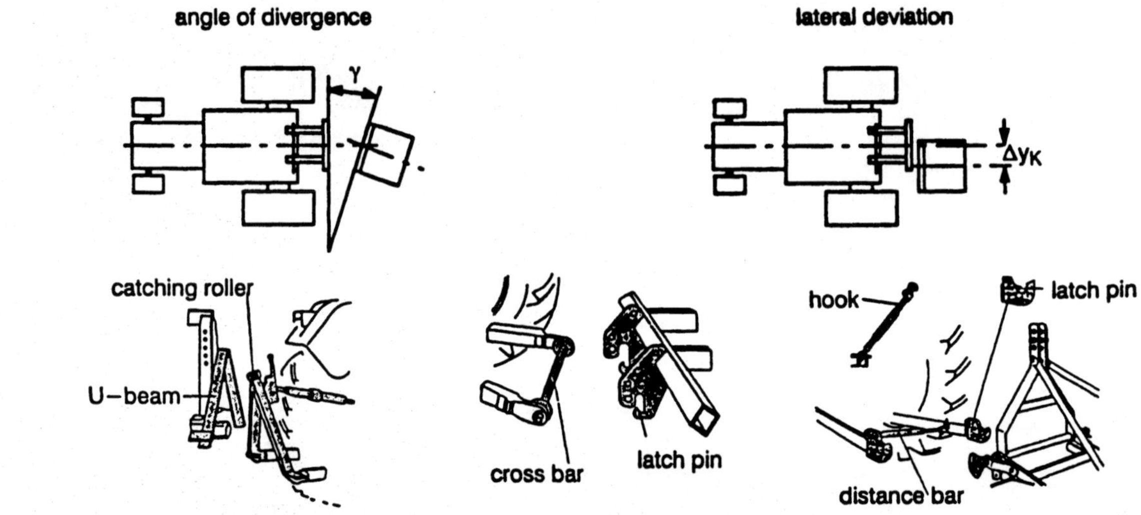

	A–frame coupler	two–point coupler type	link coupler
γ	12 °	4 °	2,5 °
Δy_K	160 mm	50 mm	30 mm

Fig. 2. Automatic couplers and threshhold values for angle of divergence (γ) and lateral deviation (Δy_K) limiting the catching area, beyond which the automatic couplers come into action

2.1. Automatic couplers

Automatic couplers are already [3] on the market, especially for threepoint hitches. They are standardized or expected to be standardized in the near future [4]. Automatic couplers have a catching area. This is the area within which the automatic coupling mechanism comes into action. This area is defined by the design of the couplers. The Swedish Institute of Agricultural Engineering in Uppsala has investigated tractor/implement hitchings and the effect of hitch design on the threshold values for the angle of divergence and the lateral deviation between tractor and implement. Because only beyond these values do the automatic couplers come into action [5, 6].

From left to right Figure 2 shows the A–frame coupler, the two–point coupler and the three–point free–link hitch or link coupler. The latter is the most commonly used automatic coupler. This coupling system is standardized and can be used for nearly all purposes. It neither limits the necessary space nor shifts the center of gravity of the implement.

The way a tractor approaches an implement depends on its constructive maneuverability and the field of vision of the operator. The maneuverability is set by dimensions of the tractor like (Figure 3) wheel base l, wheel track s and radius of turn r_{min}. The distance of the instantaneous center of rotation P from the rear axle center with respect to the wheel base is the determining parameter of tractor maneuverability.

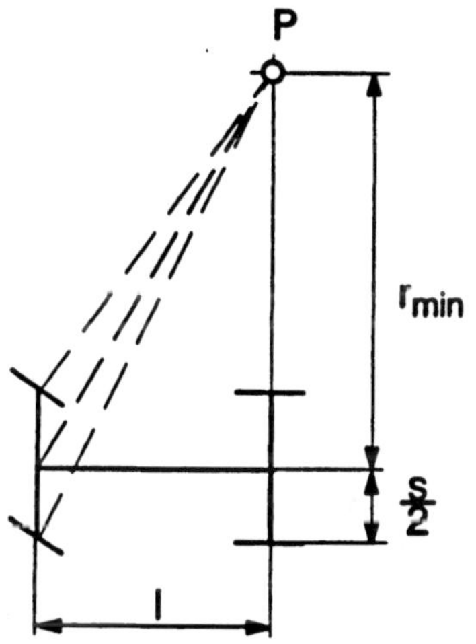

Fig. 3. Tractor model

The second condition affecting the coupling maneuver is the operator's field of vision. It is well understood and verified by investigations that the visibility severely affects the steering accuracy [7]. Adequate visibility is hindered by the increasing size of modern tractors, the operators' cab and its position.

2.2. Providing the right information

In order to provide the required information to the operator, it is necessary to measure angle and distance between tractor and implement, process these data and display the results in an ergonomic way.

2.2.1. Measurement of angle and distance between tractor and implement

For experimental purposes a laboratory test system has been set up. This system consists of two ultrasonic transmitters fixed to the implement and two receivers mounted on the tractor. From both transmitters an ultrasonic signal is transmitted alternatively. The beginning of each transmission is relayed by an infrared transmitter from the implement to the tractor, where an infrared receiver is mounted. Four distances between these two transmitters and two receivers are measured.

They are used to calculate the relative distance of tractor and implement. This system has been proved sufficiently for laboratory tests, but because active elements are used on the implement it is obviously not appropriate enough for agricultural practice.

Therefore together with the Institut für zerstörungsfreie Prüfverfahren (IfzP) of the Fraunhofer Gesellschaft (FhG) in Saarbrücken a system using only active elements on the tractor is under development. This system consists of an array of ultrasonic transmitters and one receiver mounted on the tractor. All elements on the implements are simple passive reflectors. The phased array principle well known for radar application is used to focus the ultrasonic beam and scan it across the area behind the tractor to detect distance and direction of the implement [8]. The system has no moving mechanical parts. For practical use a measuring range of about 5m and an angle of about 30° seems to be sufficient.

2.2.2. Data processing to predict the steering angle and distance.

Assuming the tractor has been driven into the measuring range of the system, then distance and direction between tractor and implement will continuously be measured. From this, the desired steering angle to reach the catching area is calculated and displayed to the operator. Three different ways to predict the desired steering angle have been investigated: a kinematic model, a neural network and a fuzzy–controller.

To test and compare these alternatives, a simple tractor model according to Figure 3 was used. This is justified because driving backwards to couple an implement is only done at low speeds and angle and distance is measured frequently. Therefore it is not necessary to take slippage or other peculiarities into account.

Kinematic model. The kinematic model is based upon the driving behaviour and the steering behaviour of tractors when backing up. The developed computer program calculates the feasible driving curve according to the kinematic characteristics of the tractor steering system. For the calculation it is necessary to take into account

- the discrete control speed of the steering angle
- a parallel positioning of the tractor coupling elements relative to those of the implement at the end of the driving curve
- a geometrical compensation of different distances from the starting point to the coupling position.

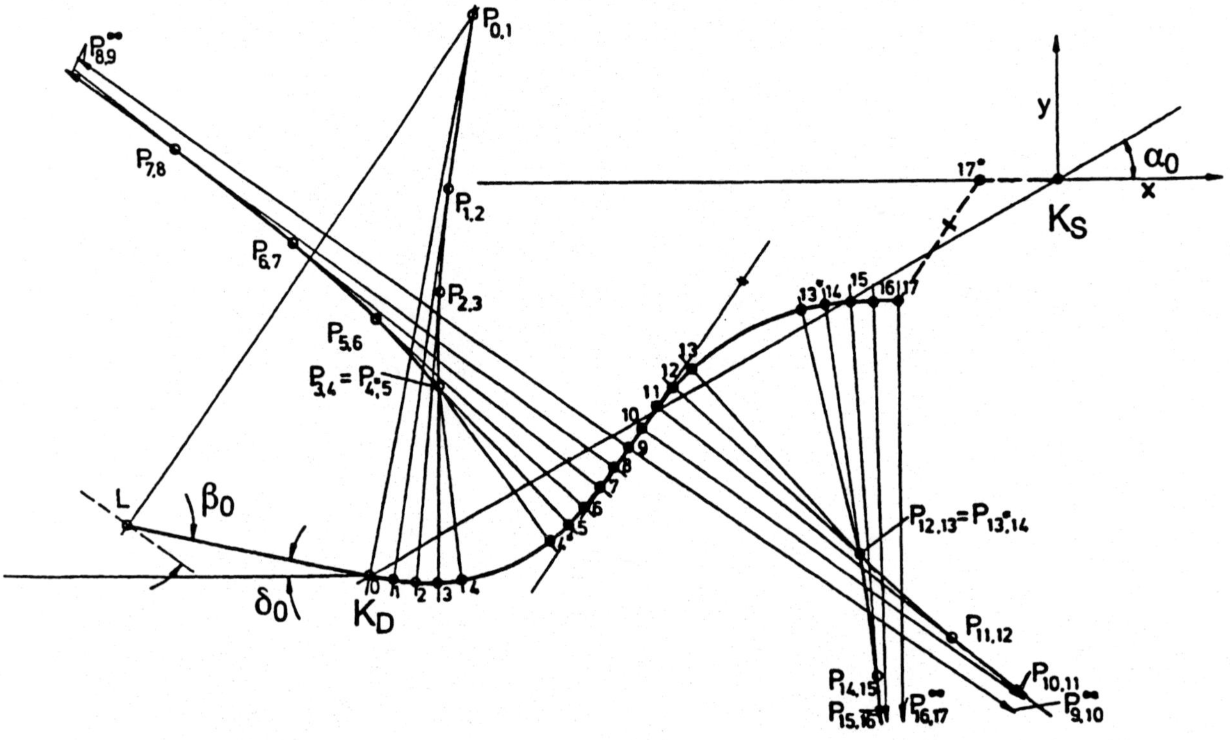

Fig. 4. Kinematic design of driving curve segments, centrodes and line segments for compensation of distance to the coupling point (point of inflection W = 9)

Figure 4 in particular shows the principle design of the driving curve with its centrodes and straight line segments for geometrical compensation of different distances from the starting point to the coupling position. The initial position is characterized by the tractor/implement distance $K_D K_S$; by the appertaining angle α_0 between $K_D K_S$ and the longitudinal axis of the implement; by the angle δ_0 between the x–axis and the longitudinal tractor axis. The latter is always the tangent to the driving curve. The driving curve is divided into circular arches, transition curves and straight lines to fulfill the above mentioned requirements. To get a smooth curve, transition conditions are part of the computer program. As mentioned, distance $K_S K_D$ and angle measurement as well as calculation of the required steering angle are performed frequently while backing up.

The kinematic solution requires expert knowledge in kinematics to design the model. It offers precise solutions and additionally can predict if it is possible to reach the coupling point from the starting point.

Neural network. A neural network is a model free estimator. It learns by examples. Therefore the data used to train the network have to be chosen carefully. To avoid expensive practical measurements to create these data the kinematic model according to Figure 3 has been used. A feed–forward network was used with four inputs for the four distances measured, eight hidden neurons and one output neuron for the desired steering angle. Different backpropagating learning algorithm were applied. The trained neural network was able to predict the required steering angle to guide the tractor into the desired catching area of the automatic couplers. The achieved accuracy was within the demanded limits (Figure 2).

Fuzzy–control. A fuzzy–controller may be an appropriate solution to avoid extensive efforts to model a system, or in cases where it is not possible to model a system. In order to establish a fuzzy–controller it is suffizient to articulate common or expert knowledge of a subject area. This holds true for driving a vehicle, because everybody able to drive a car or tractor will be able to describe the rules applied in common language. The fuzzy–controller design has two linguistic variables, the tractor/implement distance $K_D K_S$ and the appertaining angle α_0 (compare Figure 4). The output is the required steering angle.
Minimal effort was needed to set up a rapid prototype by using a fuzzy development system which generated the C–Code of the controller. This prototype did go beyond all threshold values (Figure 2) for different starting positions. To achieve this a fine tuning was necessary.

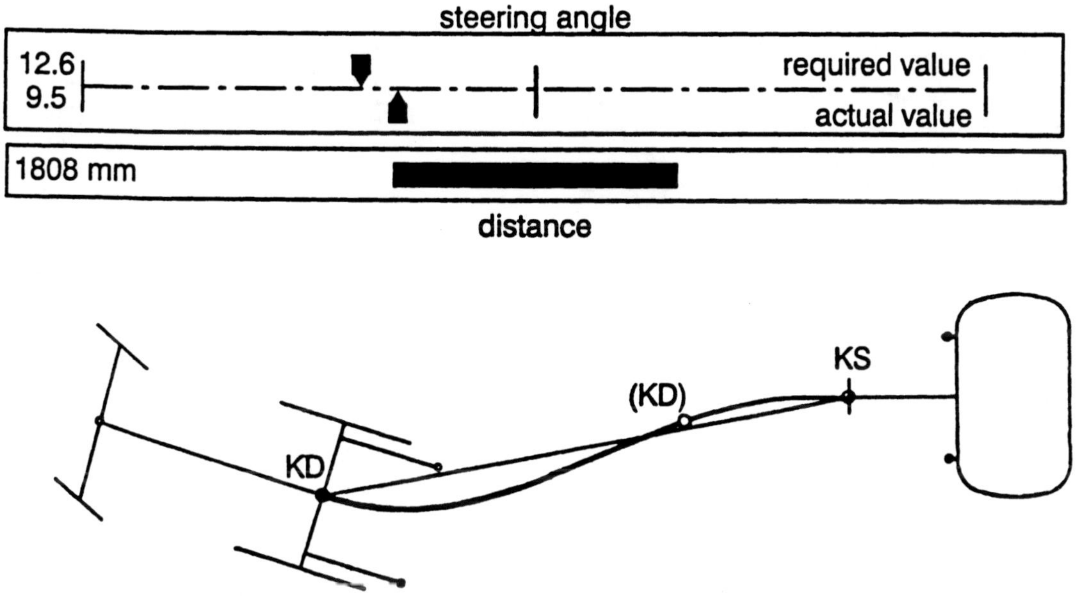

Fig. 5. Operator information displayed on a monitor

2.2.3. Operator information

Universal driver information systems have been discussed and investigated for along time [9, 10]. Now the first are on the market including on–board computers and central graphic displays. Therefore, the following concentrates only on the manner in which the necessary information should/ could be displayed to the operator. The required steering angle is displayed on the top and the actual steering angle immediately below in Figure 5. Both are displayed in an analog manner, because this makes it easier to detect deviations and changes. In addition the actual values may be given (far left). The operator has to align both marks by acting on the steering wheel. Just below the steering angles, the distance between tractor and implement is displayed using a bar, and additionally, the actual values are indicated (left). For clarity a tractor symbol at the starting position, the path, and the target, the implement, are shown.

Together with the system, the operator performs a control loop. It is mainly a question of expense to replace the operator by a controller.

3. Final remarks

The discussed computer aided system for tractor/implement coupling has been proposed to improve occupational safety and operator comfort. It tackles the main problem during coupling: the restricted visibility. To avoid excessive costs the system is based on devices already on the market or expected in the near future, like automatic couplers and on–board computers with universal graphic displays. Only the required measuring system will cause additional hardware costs. The system under development should only have passive elements like reflectors on the implements, and the parts on the tractor should have no moving mechanical elements. Different principles to calculate the required steering angle from the measured distances and directions between tractor and implement have been investigated. The application of classic kinematic modelling to this problem is possible and has proved to be superior to neural network and fuzzy control application.

4. References

[1] Heidt, H. and G. Groh (1984). Schwerpunkte landwirtschaftlicher Arbeitsunfälle und Möglichkeiten zu ihrer Verhütung. *Landtechnik*, **39**, No. 1, pp. 35–40.

[2 Hammer, W. and G. Thaer (1990). Unfälle bei Bestellarbeiten. *Landtechnik*, **45**, No. 1, pp. 41–44.

[3] Geisthoff, H. (1974). Ein automatisches Kuppelverfahren für landwirtschaftliche Geräte. *Grundlagen der Landtechnik*, **24**, No. 3, pp. 87–89.

[4] ISO/DIS 11001 (1991). Agricultural wheeled tractors and implements – Three-point hitch couplers. *International Organization for Standardization.*

[5] Aas, M., J. Bergström and O. Norén (1985). Lyftsnabbkopplingar för traktor och redskap. *Jordbrugstekniska Institutet, Meddelande*, No. 406, Uppsala.

[6] Carlson, B. and S. Persson (1982). Analysis of tractor–implement hitching and effect of hitch design: a proposal. *JTI–Rapport*, No. 41, Uppsala.

[7] Bottoms, D.J. (1983). The interaction of driving speed, steering difficulty and lateral tolerance with particular reference to agriculture. *Ergonomics*, **26**, No. 2, pp. 123–239

[8] Gebhardt, W. et al. (1979). Ultraschallfeldsteuerung, Fehlerklassierung und Fehlerrekonstruktion mittels "Phased Arrays". *Materialprüfung*, **21**, No. 12, pp. 437–443

[9] Jahns, G. and H. Speckmann (1985). Ein Bordcomputerkonzept für Schlepper und angekoppelte Geräte zur Optimierung landwirtschaftlicher Produktionsprozesse. *Grundlagen der Landtechnik*, **35**, No. 2, pp. 47–54

[10] Jahns, G. and H. Speckmann (1992). An open Agricultural BUS Standard. *ASAE–Paper*, No. 92 3006

STRATEGIES OF MODEL REFERENCE ADAPTIVE CONTROL WITH ACTIVE LEARNING PROPERTIES

N. M. FILATOV and H. UNBEHAUEN

Automatic Control Laboratory, Faculty of Electrical Engineering, Ruhr University, D-44780 Bochum, Germany

Abstract. The problems of synthesis and analysis of model reference adaptive control for systems with active learning are considered. New active adaptive control algorithms for systems with implicit and explicit reference model are designed. The interconnections between direct and indirect active adaptive control systems are discussed and it is shown that the suggested method is uniformal for both types of systems. The convergence properties of the designed algorithms are analysed. Using the martingale convergence theory the solutions for global stability of the algorithms are presented. A simulated example is used to demonstrate the potential and superiority of the algorithms.

Key Words. Control system synthesis; convergence analysis; dual control; model reference adaptive control; nonminimum-phase systems; simulation; stochastic systems.

1. INTRODUCTION

The problems of quality improvement of adaptive control and the design of acceptable algorithms for practical applications are the most important ones at the modern stage of development of adaptive controllers (Åström, 1987; Unbehauen, 1993). Using the active adaptive approach, which is based on the principle of dual control, allows to improve the control quality and to decrease the adaptation time (Åström, 1987; Chan and Zarrop, 1985; Zhivoglyadov and Filatov, 1990; Zhivoglyadov *et al.*, 1993a,b). The implementation of active adaptive control (AAC) based on a multicriterial approach provides algorithms which are simple and which guarantee improved control quality (Zhivoglyadov and Filatov, 1990; Zhivoglyadov *et al.*, 1993a,b). The improvement of the control quality by AAC is attained as a result of interactions of caution and active learning (probing) properties (Bar-Shalom, 1981). The peculiarity of the multicriterial approach to AAC results as a compromise solution from the optimisation of two criteria which correspond to the losses due to both control and estimation, respectively (Zhivoglyadov, 1985).

It should be mentioned that the active adaptive control approach is used up to now only for systems with indirect adaptation. In most classical model reference adaptive control systems (MRAS), however, the certainty equivalence (CE) assumption is applied and the current accuracy of the parameter estimates (and the accuracy of regulator tuning) are not taken into consideration. The CE assumption leads often to insufficient control quality in the adaptation time. Adaptive control with implicit reference model (or, in other words, direct adaptive pole placement control) does not use the error between system output and model output and it is not clear which performance index should be used for the control optimisation. These problems must be solved to extend the active adaptive approach for MRAS.

The statement and solution of the new synthesis problem of AAC for systems with implicit reference model is presented in Sec. 2. A new criterion for control optimisation in systems with implicit reference model will be discussed. The problem of AAC for systems with explicit reference model is considered in Sec. 3. It is well-known that indirect adaptive systems can be obtained from direct ones as a special case (Egardt, 1980; Schmid, 1979). The problems of interconnections between indirect and direct active adaptive control (DAAC) are considered in Sec. 4. In addition the peculiarities for the application of both approaches for nonminimum phase plants are discussed. The first solution of the problem of global stability of AAC was obtained by Radenković (1988) for the generalised dual minimum variance controller. In Sec. 5 the solution of the global stability problem for the new DAAC is presented. Sec. 6 discusses the comparative performances of the DAAC and the standard approach based on the CE assumption using computer simulations of a stochastic system. Conclusions are given in Sec. 7.

2. SYSTEMS WITH IMPLICIT REFERENCE MODEL

Consider the following discrete-time feedback controller for a SISO system

$$U(z) = \frac{S(z^{-1})}{R(z^{-1})}E(z) = \frac{1+s_1 z^{-1}+\ldots+s_{n_S} z^{-n_S}}{r_0 + r_1 z^{-1}+\ldots+r_{n_R} z^{-n_R}}E(z),\ r_0 \neq 0, \quad (1)$$

where $E(z) = W(z) - Y(z)$, $W(z)$ is the z-transform of the reference signal, $Y(z)$ and $U(z)$ are the corresponding z-transforms of the output and input signals of the plant. It is assumed that using the controller according to eq. (1) the closed loop system equation with additional noise disturbance in the plant will be of the type

$$A(z^{-1})Y(z) = B(z^{-1})W(z) + \frac{C(z^{-1})}{S(z^{-1})}\Xi(z), \qquad (2)$$

where the polynomials $A(z^{-1})$ and $B(z^{-1})$ represent the desired closed loop tracking behaviour (reference model) of the system, $\Xi(z)$ is the z-transform of a white gaussian random sequence $\xi(k)$ with zero mean and covariance σ_ξ^2. The polynomials are specified by

$$\begin{aligned} B(z^{-1}) &= b_1 z^{-1} + \ldots + b_{n_B} z^{-n_B}, \\ A(z^{-1}) &= 1 + a_1 z^{-1} + \ldots + a_{n_A} z^{-n_A} \\ C(z^{-1}) &= 1 + c_1 z^{-1} + \ldots + c_{n_C} z^{-n_C}. \end{aligned} \qquad (3)$$

For the direct adaptive control the controller parameters which are unknown owing to the unknown plant model have to be estimated in real-time. The model equation (2), after transformation as

$$W(z) = \frac{A(z^{-1})}{B(z^{-1})}Y(z) - \frac{C(z^{-1})}{S(z^{-1})}\Xi(z),$$

is substituted in eq. (1) and after some algebraic manipulations follows the equation for estimation

$$\begin{aligned} &S(z^{-1})\left[A(z^{-1}) - B(z^{-1})\right]Y(z) \\ &= R(z^{-1})B(z^{-1})U(z) + C(z^{-1})\Xi(z). \end{aligned} \qquad (4)$$

After introducing the filtered values of the input and output signals

$$\begin{aligned} \overline{Y}(z) &= \left[A(z^{-1}) - B(z^{-1})\right]Y(z), \\ \overline{U}(z) &= zB(z^{-1})U(z), \end{aligned} \qquad (5)$$

eq. (4) can be written as

$$S(z^{-1})\overline{Y}(z) = R(z^{-1})\overline{U}(z) + C(z^{-1})\Xi(z). \qquad (6)$$

In order to obtain the vector form of eq. (6), the following vectors are introduced

$$p = \left[r_0 \ldots r_{n_R} \vdots -s_1 \ldots -s_{n_S}\right]^T = \left[r_0 \vdots p_0^T\right]^T \qquad (7)$$

$$\begin{aligned} m(k) &= \left[\overline{u}(k) \ldots \overline{u}(k-n_R) \vdots \overline{y}(k) \ldots \overline{y}(k-n_S+1)\right]^T \\ &= \left[\overline{u}(k) \vdots m_0^T(k)\right]^T, \end{aligned} \qquad (8)$$

where the signals $\overline{u}(k)$ and $\overline{y}(k)$ are the representation of $\overline{U}(z)$ and $\overline{Y}(z)$ in discrete time domain correspondingly, k is the discrete time. In the case of time-varying parameters of the plant the controller parameter vector (7) should be time-varying too. Consider the case of stochastic parameter drift:

$$p(k+1) = p(k) + \varepsilon(k), \qquad (9)$$

where $\varepsilon(k)$ is a drift noise vector with zero mean and covariance matrix $Q_\varepsilon(k)$. In the case when $C(z^{-1}) = 1$ eq. (6) can be represented after inverse transformation to discrete-time domain using eqs. (7) and (8) as follows

$$\overline{y}(k+1) = p^T(k)m(k) + \xi(k), \qquad (10)$$

For parameter estimation of the model according to eqs. (9) and (10) the following Kalman filter approach is applied:

$$\hat{p}(k+1) = \hat{p}(k) + q(k+1)\left[\overline{y}(k+1) - \hat{p}^T(k)m(k)\right] \qquad (11)$$

$$q(k+1) = P(k)m(k)\left[m^T(k)P(k)m(k) + \sigma_\xi^2\right]^{-1} \qquad (12)$$

$$P(k+1) = P(k) - q(k+1)m^T(k)P(k) + Q_\varepsilon(k) \qquad (13)$$

where

$$\begin{aligned} \hat{p}(k) &= \mathrm{E}\{p(k)|\Im_k\}, \\ P(k) &= \mathrm{E}\{(p-\hat{p}(k))(p-\hat{p}(k))^T|\Im_k\}, \end{aligned} \qquad (14)$$

$$\Im_k = \{y(0) \ldots y(k) \vdots u(0) \ldots u(k-1)\}, \; \Im_0 = \{y(0)\}. \qquad (15)$$

$\Im_k$ is the set of input and output values at time k. It is assumed here that the initial values $\hat{p}(0)$ and $P(0)$ for eqs. (11)-(13) are given. From the eqs. (11)-(13) the algorithm with forgetting factor as well as the constant trace algorithm can be obtained as special cases by appropriately choosing the covariance matrix $Q_\varepsilon(k)$ of the drift noise in eq. (13).

It is clear, that the control quality would be improved if the system output would be the same as the response of the input signal of the unknown controller with not adjusted parameters. The filtered nominal output of the system is defined as response of the input signal of the unknown controller according to eq. (1) with not adjusted parameters and regression vector $m(k-1)$ when no disturbances act onto the system

$$\begin{aligned} \overline{y}_n(k) = {} & r_0\overline{u}_n(k-1) + \ldots + r_{n_R}\overline{u}(k-n_R-1) \\ & - s_1\overline{y}(k-1) - \ldots - s_{n_S}\overline{y}(k-n_S) \end{aligned} \qquad (16)$$

where

$$u_n(k) = -\frac{1}{r_0}\left[p_0^T\tilde{m}_0(k) - e(k)\right] \qquad (17)$$

$$\tilde{m}_0(k) = \left[u(k-1) \ldots u(k-n_R) \vdots e(k-1) \ldots e(k-n_S)\right]^T. \qquad (18)$$

Eq. (17) has been obtained from eq. (1) after inverse transformation to discrete time domain. In accordance with the above mentioned multicriterial approach the following two criteria are introduced to derive the control law:

$$J_k = \mathrm{E}\{(y_n(k+1) - y(k+1))^2|\Im_k\}, \qquad (19)$$

$$J_k^a = -\mathrm{E}\{(\overline{y}(k+1) - \hat{p}^T(k)m(k))^2|\Im_k\}. \qquad (20)$$

The second criterion according to eq. (20) is used for acceleration of the parameter estimation process (Chan and Zarrop, 1985; Zhivoglyadov, 1993a) by increasing the innovation value in eq. (11). The AAC algorithm will be obtained after solving the bicriterial optimisation problem of eqs. (19) and (20) for the system described by eqs. (1), (2), (10) and

(17) using the constraints for eq. (20) around the optimal solution for eq. (19) as follows:

$$u(k)=\underbrace{\arg\min}_{u(k)\in\Omega_k} J_k^a \tag{21}$$

$$\Omega_k=[u_c(k)-\theta(k);u_c(k)+\theta(k)],\ \theta(k)=\eta\, f_k\{P(k)\},\ \eta\geq 0 \tag{22}$$

$$u_c(k)=\underbrace{\arg\min}_{u(k)} J_k, \tag{23}$$

where f_k is a scalar function of a matrix (for example f_k may be chosen as $f_k\{P(k)\}=\mathrm{tr}\{P(k)\}$) and $u_c(k)$ is the cautious control. The parameter $\theta(k)$ defines the value of the constraint area Ω_k for optimisation in eq. (21) and the value of the exiting learning signal at the same time. This value depends on the parameter η and trace of the covariance matrix $P(k)$ as in eq. (22). The covariance matrix can be described by the following structural elements

$$P(k)=\begin{bmatrix} P_{r_0}(k) & p^T_{r_0p_0}(k) \\ p_{r_0p_0}(k) & P_{p_0}(k)\end{bmatrix}, \tag{24}$$

where $p_{r_0p_0}(k)$ is a vector of the dimension n_R+n_S, $P_{r_0}(k)$ is a scalar and $P_{p_0}(k)$ is a matrix of the dimension $(n_R+n_S+1)\times(n_R+n_S+1)$. After substituting eqs. (10) and (16) by eq. (19) and taking into account eq. (5) one has

$$\begin{aligned} J_k&=\mathrm{E}\{(y_m(k+1)-y(k+1))^2|\Im_k\}\\ &=\mathrm{E}\{(r_0b_1u_m(k)-r_0b_1u(k))^2|\Im_k\}.\end{aligned} \tag{25}$$

After substituting eq. (17) by eq. (25) and minimisation, the following cautious control is obtained:

$$u_c(k)=\frac{\hat{r}_0(k)e(k)-\left[\hat{r}_0(k)\hat{p}_0^T(k)+p^T_{r_0p_0}(k)\right]\bar{m}_0(k)}{\hat{r}_0^2(k)+P_{r_0}(k)}, \tag{26}$$

where $\hat{p}(k)=\left[\hat{r}_0(k)\vdots\hat{p}_0^T(k)\right]^T$. After substituting eq. (10) by eq. (20) and minimisation of eqs. (21) and (22) the DAAC algorithm follows:

$$\begin{aligned} u(k)&=u_c(k)\\ &+\theta(k)\mathrm{sign}\{b_1P_{r_0}(k)\bar{u}_c(k)+b_1p^T_{r_0p_0}(k)m_0(k)\}.\end{aligned} \tag{27}$$

It is necessary to point out that the DAAC given by eq. (26) and (27) depends not only on the parameter estimates but also on their covariance. Thus, the regulator provides the necessary caution and active learning (probing) properties which are very important for the quality of adaptive control, and the structure of the cautious controller according to eq. (26) is the same as in eq. (1).

3. SYSTEM WITH EXPLICIT REFERENCE MODEL

Consider the system with the control law according to eq. (1) and the following explicit reference model

$$A(z^{-1})Y_m(z)=B(z^{-1})W(z), \tag{28}$$

where $Y_m(z)$ is the output of the reference model $y_m(k)$ represented in z-domain. A criterion for control optimisation of the following type is defined:

$$J_k=\mathrm{E}\{(y_m(k+1)-y(k+1))^2|\Im_k\}. \tag{29}$$

The bicriterial optimisation problem described by eqs. (20) and (29) with constraints as in eqs. (21)-(23) has to be solved to derive the AAC algorithm with explicit reference model as finally described by eq. (33). After inverse transformation of eqs. (5) and (6) they can be written in the following form for $C(z^{-1})=1$:

$$\begin{aligned}\bar{u}(k)&=b_1u(k)+\hat{u}(k),\\ \hat{u}(k)&=(B(z^{-1})-b_1z^{-1})u(k-1),\end{aligned} \tag{30}$$

$$\begin{aligned}\bar{y}(k)&=y(k+1)+\hat{y}(k),\\ \hat{y}(k)&=(A(z^{-1})-B(z^{-1})-1)y(k+1),\end{aligned} \tag{31}$$

$$\begin{aligned}y(k+1)&=a_1r_0u(k)+p_0^Tm_0(k)\\ &+r_0\hat{u}(k)+\hat{y}(k)+\xi(k).\end{aligned} \tag{32}$$

After solving the optimisation problem according to eqs. (20)-(23) and (29) the following control law will be obtained:

$$\begin{aligned}u_c(k)=&\left(\hat{r}_0(k)y_m(k)-\left[\hat{r}_0(k)\hat{p}_0^T(k)+p^T_{r_0p_0}(k)\right]m_0(k)\right.\\ &-(\hat{r}_0^2(k)+P_{r_0}(k))\hat{u}(k)\\ &\left.+\hat{r}_0(k)\hat{y}(k)\right)\left((\hat{r}_0^2(k)+P_{r_0}(k))b_1\right)^{-1}.\end{aligned} \tag{33}$$

$$\begin{aligned}u(k)&=u_c(k)\\ &+\theta(k)\mathrm{sign}\{b_1P_{r_0}(k)\bar{u}_c(k)+b_1p^T_{r_0p_0}(k)m_0(k)\}.\end{aligned} \tag{34}$$

Thus, the cautious control, as derived in eq. (33) for systems with explicit reference model, has another structure than the desired controller (according to eq. (1)). Only under the condition of exact estimates and in the case of noise absence the controller described by eqs. (33) and (34) leads to the structure of eq. (1).

4. NONMINIMUM PHASE SYSTEMS AND INTERCONNECTIONS BETWEEN ADAPTIVE CONTROL SYSTEMS STRUCTURES

In the system described by the eqs. (1) and (2) the undisturbed plant model has the following form

$$A^0(z^{-1})Y(z)=B^0(z^{-1})U(z). \tag{35}$$

The stability of the feedback system depends on the location of the zeros of the polynomial

$$P(z^{-1})=R(z^{-1})A^0(z^{-1})+S(z^{-1})B^0(z^{-1}). \tag{36}$$

All zeros of $P(z^{-1})$ must be situated outside the unit disk in the z-domain. It is essential to point out, that the desired dynamics of eq. (2) can be supplied for the nonminimum phase plant if $B(z^{-1})$ has the form

$$B(z^{-1})=S(z^{-1})B^0(z^{-1}). \tag{37}$$

In order to obtain the indirect AAC from DAAC according to eqs. (26) and (27) or (33) and (34), one should consider a unit delay reference model $A(z^{-1})=1,\ B(z^{-1})=z^{-1}$. In this case the controller described by eq. (1) takes the following form for a minimum phase plant:

$$U(z) = \frac{z^{-1}A^0(z^{-1})}{B^0(z^{-1})(1-z^{-1})}E(z). \tag{38}$$

Further, for eq. (16) follows

$$y_n(k) = w(k-1), \tag{39}$$

and eq. (22) changes to

$$J_k = \mathrm{E}\{(w(k) - y(k+1))^2 | \Im_k\}. \tag{40}$$

The model according to eq. (6) takes the form of the plant model.

After solving the bicriterial optimisation problem, eqs. (21)-(23), (20) and (40), for the plant model with $C(z^{-1}) = 1$, the controller given by eqs. (26) and (27) takes the form of the known indirect AAC (Zhivoglyadov and Filatov, 1990; Zhivoglyadov *et al.*, 1993a,b)

Hence, the AAC can be obtained from the direct one by means of a unit delay reference model, and passive adaptive control systems can be derived from AAC systems after applying the CE assumption. In addition, for passive adaptive control system it is well known that indirect adaptive systems can be obtained from direct ones as a special case (Egardt, 1980; Schmid, 1979). Therefore, all above mentioned systems can be obtained from the DAAC structure by means of the CE assumption or by applying a unit delay reference model.

5. CONVERGENCE PROPERTIES OF DAAC

Consider the convergence properties of the DAAC according to eqs. (26) and (27). Let the learning signal value in eq. (22) have the property

$$\sum_{k=1}^{\infty} \theta^2(k) < \infty, \tag{41}$$

and introduce the following assumptions:

(A1) There exist a large $K > 0$ and small $\rho > 0$ so that for any matrix $P \geq 0$ and $k > 1$ the following inequality holds

$$K + \frac{m^T(k)Pm(k)}{\rho} \geq \hat{m}^T(k)P\hat{m}(k), \tag{42}$$

where

$$m(k) = \tilde{m}(k) + \hat{m}(k). \tag{43}$$

(A2) $1 - m^T(k)P(k+1)m(k)\sigma_\xi^{-2} > 0$,

(A3) $|w(k)| < \infty$.

(A4) Polynomials, $A(z^{-1})$, $C(z^{-1})$, $A(z^{-1})R(z^{-1})$, $A(z^{-1})S(z^{-1})$, and $R(z^{-1})A^0(z^{-1}) + S(z^{-1})B^0(z^{-1})$ have zeros strictly outside the unit disk of the z-domain.

Consider the plant with fixed parameters ($Q_\varepsilon(k) = 0$). Than from eqs. (12) and (13) follows

$$\sigma_\xi^2 P^{-1}(k+1) = \sigma_\xi^2 P^{-1}(k) + m(k)m^T(k). \tag{44}$$

Eqs. (26) and (27) can be represented in the form

$$\hat{p}^T(k)\tilde{m}(k) = e(k) - \alpha(k), \tag{45}$$

where

$$\alpha(k) = \frac{1}{\hat{r}_0(k)} p_1^T(k)\tilde{m}(k) + \frac{(\hat{r}_0^2(k) + P_{r_0}(k))}{\hat{r}_0(k)}\tilde{u}(k), \tag{46}$$

$$\tilde{u}(k) = \theta(k)\mathrm{sign}\{b_1 P_{r_0}(k)\overline{u}^c(k) + b_1 p_{r_0 p_0}^T(k)m_0(k)\}, \tag{47}$$

$$p_1^T(k) = [P_{r_0}(k) \vdots p_{r_0 p_0}^T(k)]. \tag{48}$$

Furthermore, the following notations:

$$\begin{aligned}\tilde{p}(k+1) &= \hat{p}(k+1) - p \\ &= \tilde{p}(k) + \sigma_\xi^2 P(k+1)m(k)\chi(k),\end{aligned} \tag{49}$$

$$\chi(k) = \overline{y}(k+1) - \hat{p}^T(k)m(k), \tag{50}$$

$$V(k) = \frac{\tilde{p}^T(k)P^{-1}(k)\tilde{p}^T(k)}{h(k)}, \tag{51}$$

$$h(k) = (\ln v(k))^{1+\delta_1},\ 0 < \delta_1 < \frac{1}{2}, \tag{52}$$

$$v(k) = v(k-1) + m^T(k)m(k),\ v(0) > 1$$

$$\eta(k) = \tilde{p}^T(k)m(k), \tag{53}$$

where eq. (46) has been obtained from eqs. (11)-(13). The model according to eq. (6) than takes the form

$$\overline{y}(k) - C(z^{-1})\xi(k) = p^T m(k). \tag{54}$$

Theorem 1. Let the assumptions (A1)-(A4) hold. Then for the algorithm according to eqs. (11), (12), (26), (27), (44) and (54) with probability 1

$$0 < \lim_{N\to\infty} \frac{1}{N}\sum_{k=1}^{N} m^T(k)m(k) < \infty, \tag{55}$$

$$\lim_{N\to\infty} \frac{1}{N}\sum_{k=1}^{N} y_d^2(k) = \overline{c}\sigma_\xi^2, \tag{56}$$

where

$$y_d(k) = y_n(k) - y(k),\quad \overline{c} = 1 + c_1^2 + \ldots + c_{n_c}^2. \tag{57}$$

Proof. From eqs. (43), (49) and (51) follows that

$$\begin{aligned}V(k+1) = V(k) &+ \frac{\eta^2(k)\sigma_\xi^{-2}}{h(k)} + \frac{2\eta(k)\chi(k)\sigma_\xi^{-2}}{h(k)} \\ &+ \frac{m^T(k)P(k)m^T(k)\chi^2(k)\sigma_\xi^{-4}}{h(k)},\quad \text{(a.s.)}\end{aligned} \tag{58}$$

and eqs. (46), (49) and (50) lead to

$$\chi(k) = -\eta(k) + C(z^{-1})\xi(k), \tag{59}$$

After substituting $\chi(k)$ in eq. (58) by eq. (59) and applying the conditional expectation operator the following inequality is obtained:

$$\begin{aligned}\mathrm{E}_\xi\{V(k+1)|\Im_k\} &\leq V(k) \\ &+ (1 - m^T(k)P(k)m^T(k)\sigma_\xi^{-2})\frac{-\eta^2(k)\sigma_\xi^{-2}}{h(k)}\end{aligned} \tag{60}$$

$$+\frac{m^T(k)P(k)m^T(k)\sigma_\xi^{-2}}{h(k)}\bar{c}, \quad \text{(a.s.)}$$

where $E_\xi\{.\}$ is the expectation operator with respect to the noise $\{\xi(k)\}$. Now, after introducing

$$\Omega(z)=z\overline{Y}_n(z)-z\overline{Y}(z)+C(z^{-1})\Xi(z), \tag{61}$$

from eqs. (43), (45), (53) and (54) follows

$$\frac{\omega(k)}{b_1}=\eta(k)+\tilde{p}^T(k)\hat{m}(k)+\alpha(k), \tag{62}$$

where $\Omega(z)$ is z-transform of $\omega(k)$. From this equation using the assumption (A1) the following inequality is obtained:

$$-\eta^2(k)\le-\bar{\rho}\frac{\omega^2(k)}{b_1^2}+K_1\alpha^2(k)+K_2, \tag{63}$$

where $\bar{\rho}>0$ is a small value, $0<K_1,K_1<\infty$. Introducing eq. (63) into eq. (60) leads to

$$\begin{aligned}E_\xi\{V(k+1)|\Im_k\}&\le V(k)-\frac{\bar{\rho}\rho_1\omega^2(k)}{b_1^2h(k)}\\&-(1-m^T(k)P(k+1)m(k)\sigma_\xi^{-2}-\rho_1)\frac{\bar{\rho}\omega^2(k)\sigma_\xi^{-2}}{b_1^2h(k)}\\&+(1-m^T(k)P(k+1)m(k)\sigma_\xi^{-2})\frac{K_1\alpha^2(k)\sigma_\xi^{-2}}{h(k)}\\&+\frac{m^T(k)P(k+1)m^T(k)\sigma_\xi^{-2}}{h(k)}\bar{c}\\&+(1-m^T(k)P(k+1)m(k)\sigma_\xi^{-2})\frac{K_2\sigma_\xi^{-2}}{h(k)}, \quad \text{(a.s.)}\end{aligned} \tag{64}$$

where $\rho_1>0$ is again a small value. The introduction of

$$\begin{aligned}g(k)=\sum_{i=1}^{k}\omega(i)(1-m^T(i)P(i+1)m(i)\sigma_\xi^{-2}-\rho_1)\omega(i)\\+K_3\ge 0,\end{aligned} \tag{65}$$

where $0\le K_3<\infty$, leads for positive semidefiniteness of $g(k)$ to the assumption (A2). So the assumption (A2) is equivalent to the strict passivity of the operator $(1-m^T(i)P(i+1)m(i)\sigma_\xi^{-2})$. From eqs. (64) and (65) thus follows

$$\begin{aligned}E_\xi\{V(k+1)|\Im_k\}+\frac{\bar{\rho}g(k)}{b_1^2h(k)}&\le V(k)+\frac{\bar{\rho}g(k-1)}{b_1^2h(k-1)}-\frac{\bar{\rho}\rho_1\omega^2(k)}{b_1^2h(k)}\\&+(1-m^T(k)P(k+1)m(k)\sigma_\xi^{-2})\frac{K_1\alpha^2(k)\sigma_\xi^{-2}}{h(k)}\\&+\frac{m^T(k)P(k+1)m^T(k)\sigma_\xi^{-2}}{h(k)}\bar{c}\\&+(1-m^T(k)P(k+1)m(k)\sigma_\xi^{-2})\frac{K_2\sigma_\xi^{-2}}{h(k)} \quad \text{(a.s.)}.\end{aligned} \tag{66}$$

Taking into account eqs. (41), (46), (47) and Lemma A.1 of Radenković (1988) one has

$$\sum_{k=1}^{\infty}(1-m^T(k)P(k+1)m(k)\sigma_\xi^{-2})\frac{K_1\alpha^2(k)\sigma_\xi^{-2}}{h(k)}<\infty. \tag{67}$$

Lemma 3 of Chen and Guo (1986) provides

$$\frac{m^T(k)P(k+1)m^T(k)\sigma_\xi^{-2}}{h(k)}\bar{c}<\infty. \tag{68}$$

Using Lemma 3.1 of Radenković (1988) leads to $\lim_{k\to\infty}h(k)=\infty$, and taking into account eq. (68) it is easy to see that

$$(1-m^T(k)P(k+1)m(k)\sigma_\xi^{-2})\frac{K_2\sigma_\xi^{-2}}{h(k)}<\infty. \tag{69}$$

On the basis of the martingale convergence theorem follows from eqs. (66)-(69)

$$\sum_{k=1}^{\infty}\frac{\omega^2(k)}{h(k)}<\infty. \quad \text{(a.s.)} \tag{70}$$

From eqs. (1), (2) and (61), it is easy to see that

$$\begin{aligned}ASY(z)&=-(A-B)S\Omega(z)+BSW(z)\\&\quad+(A-B)SC\Xi(z),\end{aligned} \tag{71}$$

$$ARU(z)=(A-B)S[\Omega(z)+W(z)-C\Xi(z)], \tag{72}$$

where the arguments of the polynoms are omitted for notation simplicity. The boundedness of the input and output variables in the mean square sense and statement that

$$\lim_{N\to\infty}\frac{1}{(\ln N)^{1+\delta_1}}\sum_{k=1}^{N}\omega^2(k)=0 \quad \text{(a.s.)} \tag{73}$$

follow from (A1) and eqs. (70)-(72) as in Radenković (1988). The asymptotic optimality of the algorithm, eq. (56), as well as inequality (55) follow from eq. (73), as in Caines and Lafortune (1984). Thus the proof is completed.

It is necessary to point out that inequality (42) of assumption (A1) for Theorem 1 holds always for systems with indirect adaptation. As seen in the above-mentioned theorem, assumption (A1) together with assumption (A2) impose the conditions to the evaluation of the sequence $m^T(k)P(k+1)m^T(k)$ and to the interconnections between the regression vectors $m(k)$ and $\hat{m}(k)$=$m(k)$-$\tilde{m}(k)$. Furthermore, for global stability in the conditions of the Theorem 1, there are no special direct requirements for the system noise dynamics $C(z^{-1})$ as compared with various indirect adaptive controllers based on recursive parameter estimation (Gawthrop, 1980; Goodwin *et al.*, 1981; Radenković, 1988).

6. SIMULATION EXAMPLE AND COMPARISONS

Consider the following stable plant

$$\frac{B^0(z^{-1})}{A^0(z^{-1})}=\frac{0.1387\,z^{-1}+0.0889\,z^{-2}}{1-1.036z^{-1}+0.2636\,z^{-2}}.$$

Let the reference model according to eq. (2) and performance indices for evaluation and comparison be

$$\frac{B(z^{-1})}{A(z^{-1})}=\frac{0.3z^{-1}}{1-0.2867z^{-1}}, J_1=\sum_{k=1}^{100}(y-K_w y_m)^2, J_2=\sum_{k=1}^{100}u^2.$$

The reference input is given by $w(k)=1$ for $0<k\le 20$,, $40<k\le 60$, $80<k\le 100$ and $w(k)=-1$ for $20<k\le 40$, $60<k\le 80$. The initial values of estimated parameters are $\hat{s}_1(0)=0.1$, $\hat{r}_0(0)=\hat{r}_1(0)=0.1$ and the initial values $m(0)$ are

equal to zero. The adaptive control algorithm according to eqs. (26) and (27) ($P_0 = I_3$, $\theta(k) = \eta \mathrm{tr}\{P(k)\}$, $\eta = 0.1$, $\sigma_\xi^2 = 0.0001$, $Q_s(k) \equiv 0$,) is applied, and the results are shown in Figs. 1 and 2. For DAAC $J_1 = 0.4114$, $J_2 = 125.9$ (Fig. 1) and for the control algorithm based on the CE assumption (when $P(k) \equiv 0$ in eqs. (26) and (27)) $J_1 = 4.825$ and $J_2 = 224.7$ (Fig. 2). It should be noted, that the reference model has the gain factor K_w=0.4206. This simulation example indicates that the DAAC provides better control quality at the beginning of the adaptation due to the caution and active learning properties. The direct adaptive control, however, based on the CE assumption (standard approach) gives large overshoot at the beginning of the adaptation.

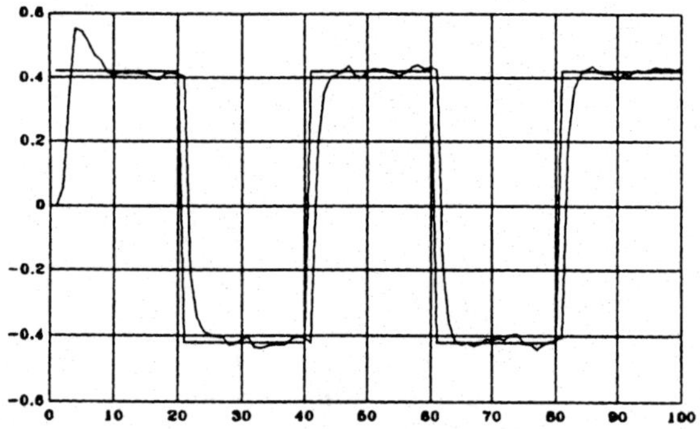

Fig. 1. System output and setpoint for DAAC

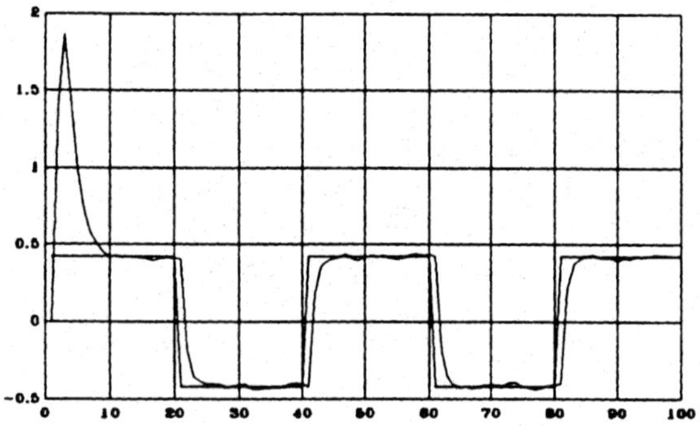

Fig. 2. System output and setpoint for control based on CE assumption

7. CONCLUSIONS

In this paper the formulation and solution of the synthesis problems of AAC systems with implicit and explicit reference model are presented. The synthesis is based on the development of the multicriterial approach and on the introduction of a new criterion for control optimisation in systems with implicit reference model. The interconnections between direct and indirect, passive and active adaptive control have been shown. It has been demonstrated that the derived AAC system with reference model includes the generalisation of known indirect AAC systems and passive adaptive control systems with direct and indirect adaptation. The solution for global stability of the algorithms has been presented. The potential of the proposed method has been demonstrated by a simulation example.

8. ACKNOWLEDGEMENT

N. M. Filatov thanks the Alexander von Humboldt Foundation for financial support.

9. REFERENCES

Åström, K.J. (1987). Adaptive feedback control. *Proc. of the IEEE*, **75**, 2, 185-217.

Bar-Shalom Y. (1981). Stochastic dynamic programming: caution and probing, *IEEE Trans. Autom. Control*, **26**, 1184-1194.

Caines, P.E., and S. Lafortune (1984). Adaptive control with recursive identification for stochastic linear systems. *IEEE Trans. Autom. Control*, **29**, 312-321.

Chan, S.S., and M.B. Zarrop (1985). A suboptimal dual controller for stochastic systems with unknown parameters. *Int. J. Control*, **41**, 507-524.

Chen, H.F. and L. Guo. (1986). Convergence rate of least-squares identification and adaptive control for stochastic systems. *Int. J. Control*, **44**, 1459-1476.

Egardt, B. (1980). Unification of some discrete time adaptive control schemes. *IEEE Trans. Auto. Control*, **25**, 693-697.

Gawthrop, P.J. (1980). On the stability convergence of a self-tuning controller. *Int. J. Control*, **31**, 973-998.

Goodwin, G.C., P.J. Ramadge, and P.E. Caines (1981). Discrete time stochastic adaptive control. *SIAM J. Control and Optim.*, **19**, 829-853.

Radenković, M.S. (1988). Convergence of the generalised dual control algorithm. *Int. J. Control*, **47**, 1419-1441.

Schmid, Chr. (1979). Ein Beitrag zur Realisierung Adaptiver Regelungssysteme mit dem Prozessrechner. Doctoral Thesis, Ruhr-Universität Bochum, F.R.G.

Unbehauen H. (1993). Some critical remarks to acceptance of adaptive control in industrial practice. *Proc. of 1993 IEEE International Conference on Systems, Man and Cybernetics*, Le Touquet - France, **3**, 447-451.

Zhivoglyadov, V.P. (1985). Active adaptation and identification in dual control systems. *Identification and System Parameter estimation. Proc. 7th IFAC/IFORS Symp.*, Glasgow, 593-597.

Zhivoglyadov, V.P. and N.M. Filatov (1990). Synthesis and comparison of actively and passively adaptive control algorithms. *Proc. IFAC Workshop on Evaluation of Adaptive Control Strategies in Industrial Application*, Tbilisi, 9-14.

Zhivoglyadov, V.P., G.P. Rao and N.M. Filatov (1993). Application of δ-operator models to active adaptive control of continuous-time plants. *Control - Theory and Advanced Technology*, **9**, 1, 127-137.

Zhivoglyadov, V.P., G.P. Rao and N.M. Filatov (1993). Active adaptive control algorithm with BPF. *Prep. 12th IFAC World Congress* , Sydney, Australia, **10**, 401-404.

ADAPTATION AND SELF-LEARNING OF NONLINEAR MULTIVARIABLE SYSTEMS

I.V.MIROSHNIK*, V.O.NIKIFOROV*

**Institute of Fine Mechanics and Optics, Department of Automatic and Remote Control, Sablinskaya 14, Saint-Petersburg, 197101, Russia*

Abstract. It is proposed a self-learning control strategy for the spatial motion along an arbitrary smooth unknown curve. On the basis of geometrical approach a unified structure of the control system is derived. An algorithm of system adaptation and an approximation mechanism are chosen to support the training motions along the curve and to estimate undetermined functional components of the control law.

Key Words. Nonlinear systems, spatial motion, adaptive control, self-learning

1. INTRODUCTION

The dynamic plant spatial motion along prescribed curves, surfaces and other multidimensional manifolds is often used to ensure desired properties of the control system such as optimality, disturbances decoupling, local linearity and coordination of the plant outputs. Such a behavior is reached by methods of the modern nonlinear control theory intensively developed in the last years (Isidory, 1989). The system design principles proposed allow one to obtain nonlinear control law which ensures system stabilization about a given manifold and an acceptable mode of local motion (Miroshnik, 1990; Miroshnik and Nikiforov, 1991a; Byrnes and Isidory, 1991). The law introduces feedback loops on the deviations from the manifold (errors) and nonlinear functional terms specially chosen for every analytically described manifold.

The probable situation when a description of the manifold is uncertain or *a priori* unknown makes the conventional control problem undetermined and demands involving more sophisticated control principles. The most vivid example of undetermined task for mechanical systems is presented by that of path tracking along physically detectable space curves (Khosla *et al.* , 1985; Lizunov *et al.* , 1987; Miroshnik and Nikiforov, 1994). The equations of the path being unknown and only current measurements of the deviations being available, the above mentioned functional terms of the control law cannot be calculated directly. Unlike the traditional problem of the parameter uncertainty resolved by adaptation techniques (Miroshnik, 1990; Miroshnik and Nikiforov, 1991a), we are faced here with a functionally undetermined problem that has to be solved by means of self-learning.

This paper concerns the undetermined problem of path tracking for a limited class of controlled plants. The self-learning strategy provides on-line adjustment of the nonlinear control law during training motions along the curve and fine path tracking after the adjustment. The consideration of a simplified case will allow us to avoid the known analytical difficulties of the general synthesis procedure and emphasize the main features of that kind of problems.

The approach is successfully extended to the computer-aided design of nonlinear systems when simulation training experiments and self-learning can be used to exclude the cumbersome procedures of analytical design and finally to obtain relatively simple unified nonlinear controllers.

2. THE PROBLEM STATEMENT

Let us consider a dynamic model which is the most typical of mechanical plants:

$$\dot{x}_1 = x_2 \tag{1}$$

$$\dot{x}_2 = f(x_1, x_2) + G(x_1, x_2)u \tag{2}$$

where x_1 and x_2 are m-dimensional columns of the state variables, u is an m-dimensional control vector, $f(x_1, x_2)$ $G(x_1, x_2)$ are the vector/matrix functions and $G(x_1, x_2)$ is invertible.

The desired path of the spatial motion is formed by segments of smooth regular curves specified by equations of the form:

$$\varphi(x_1) = 0 \tag{3}$$

where $\varphi(\cdot)$ is an *(m-1)*-dimensional vector function such that $rank(\partial\varphi/\partial x_1) = m-1$. Equation (3) establishes desired relationships between the plant variables and specifies the typical problem of the plant outputs coordination (Miroshnik, 1990). It requires special output variables to be introduced as those of the ***longitudinal motion***

$$s = \psi(x_1) \tag{4}$$

and the ***transversal motion*** $\{\varepsilon_j\} = \varepsilon$

$$\varepsilon = \varphi(x_1) \tag{5}$$

The functions $\varphi(\cdot)$ and $\psi(\cdot)$ are assumed to be chosen so that on curve (3) the matrix

$$M = M(s) = \begin{vmatrix} \partial\psi/\partial x_1 \\ \partial\varphi/\partial x_1 \end{vmatrix} \tag{6}$$

is orthogonal, i.e. $M^{-1} = M^T$. The rows of the matrix M specify the moving basis that on curve (3) can be considered as a solution of the Frene-like equation:

$$\dot{M} = \dot{s}\, \Xi(\xi)\, M \tag{7}$$

where Ξ is a skew-symmetrical tridiagonal matrix that can be presented in the form

$$\Xi = \sum_{i=1}^{m-1} E_i^{i+1}\xi_i \tag{8}$$

$E_j^i = \{e_\nu^\rho\}$ is a basis skew-symmetrical matrix with coefficients

$$\begin{aligned} e_\nu^\rho &= (-1)^{\rho+\nu+1}, \; if \;\; \nu = i, \rho = j, \\ e_\nu^\rho &= (-1)^{\rho+\nu}, \;\;\; if \;\; \nu = j, \rho = i, \\ e_\nu^\rho &= 0, \qquad\qquad otherwise \end{aligned}$$

$\xi = \{\xi_j\}$ $(j = \overline{1, m-1})$ is a vector of geometrical invariants.

The variables $\varepsilon_j(t)$ being orthogonal deviations from the curve must be reduced by appropriate control techniques. The variable $s(t)$ is a local coordinate chosen as the length of the path. The desired mode of the motion along the curve is given by the demands $s^*(t)$ or $\dot{s}^*(t)$. Let us emphasize the more stringent requirements for the dynamic properties and accuracy of the transversal motion in comparison with the quality of the longitudinal control.

The undetermined task of path tracking is connected with controlled motions of the sensor-equipped plant along a physically detectable path when the analytical description is not given and therefore the control law contains unknown variables $\xi_j(s)$ being estimated during the motion. The problem could be solved by a conventional adaptive technique if

$$\dot{\xi}_j = (\partial\xi_j/\partial s)\cdot\dot{s} \cong 0 \tag{9}$$

However the assumption (9) is correct only for low-speed motion where $\dot{s} \cong 0$ or motion along straight lines and circle-like curves where $\partial\xi/\partial s = 0$.

In the more general case we deal with not parametric but functional uncertainty of the control law that puts forward the problem of the system learning. The self-learning strategy is to provide the restoration of the undetermined functions in the course of controllable training motions along the prescribed path, with the on-line measurements of the curve deviations ε_j being the main information for the control and self-learning. The problem to design the acceptable control laws and algorithms of function estimation is stated as the main goal of this paper.

3. CONCEPTION OF THE SPATIAL MOTION CONTROL

Let us now briefly describe the general procedure of the conventional space motion control law design (Miroshnik, 1990). First of all we have to define an extended vector of the longitudinal motion $[\sigma_1, \sigma_2]^T = [s, \dot{s}]^T$ with additional coordinate

$$\sigma_2 = (\partial\psi/\partial x_1) \cdot x_2 \qquad (10)$$

and an extended vector of the transversal motion $[e_1^T, e_2^T]^T = [\varepsilon^T, \dot{\varepsilon}^T]^T$ with additional coordinates

$$e_2 = (\partial\varphi/\partial x_1) \cdot x_2 \qquad (11)$$

Expressions (4) (5) (10) and (11) introduce the nonlinear transformation of the state variables used for construction of a specific model of the spatial motion. Such a model obtained by differentiating (10) and (11) with respect to time and substituting (7) has the form

$$\begin{vmatrix} \dot{\sigma}_1 \\ \dot{e}_1 \end{vmatrix} = \begin{vmatrix} \sigma_2 \\ e_2 \end{vmatrix} \qquad (12)$$

$$\begin{vmatrix} \dot{\sigma}_2 \\ \dot{e}_2 \end{vmatrix} = \sigma_2 \cdot \Xi(\xi) \cdot \begin{vmatrix} \sigma_2 \\ e_2 \end{vmatrix} + M \cdot G(x_1, x_2) u \qquad (13)$$

The general control law is chosen as

$$u = G^{-1} \cdot \{\tilde{u} - f(x_1, x_2)\} \qquad (14)$$

where $\tilde{u}$ is a spatial motion control vector. The change of the last vector coordinates is defined as

$$\tilde{u} = M^T \cdot \begin{vmatrix} u_s \\ u_e \end{vmatrix} \qquad (15)$$

where u_s is a signal of ***longitudinal motion control*** and u_e is a vector of ***transversal motion control.*** Equation (13) takes the form

$$\begin{vmatrix} \dot{\sigma}_2 \\ \dot{e}_2 \end{vmatrix} = \sigma_2 \cdot \Xi(\xi) \cdot \begin{vmatrix} \sigma_2 \\ e_2 \end{vmatrix} + \begin{vmatrix} u_s \\ u_e \end{vmatrix} \qquad (16)$$

where matrix $\Xi(\xi)$ can be divided into the blocks

$$\Xi = \begin{vmatrix} 0 & h^T \\ -h & \Xi' \end{vmatrix}$$

and $h = h(\xi_1) = [\xi_1\ 0 \ldots 0]^T$, $\Xi' = \Xi'(\xi')$, $\xi' = \{\xi_j\}$, $(j = \overline{2, m-1})$. Equations (12) and (16) specify the interacting models of the longitudinal and transversal motion which can be written separately:

$$\dot{\sigma}_1 = \sigma_2, \quad \dot{\sigma}_2 = h^T(\xi_1) e_2 \sigma_2 + u_s \qquad (17)$$

$$\dot{e}_1 = e_2, \quad \dot{e}_2 = \sigma_2 \Xi'(\xi') e_2 - h(\xi_1)\sigma_2^2 + u_e \qquad (18)$$

The transversal controller is derived from the requirement to stabilize the model (18), so that

$$u_e = \sigma_2^2 \cdot h(\xi_1) - v \qquad (19)$$

where

$$v = K_1 \cdot e_1 + K_2 \cdot e_2 \qquad (20)$$

is an output error vector, K_1 and K_2 are feedback matrices. Then the closed loop model is written as

$$\dot{e}_1 = e_2, \quad \dot{e}_2 = -K_1 e_1 + (-K_2 + \sigma_2 \Xi') e_2. \qquad (21)$$

Its stabilization is insured by an appropriate choice of the matrices K_1 and K_2.

The stabilization of the curve motion being reached, the model (17) can be considered as a simplest SISO plant. The solution to the appropriate control problem does not involve any peculiarity and will be omitted.

It is worth noting that the description of the general controller (14) is determined by the structure and dynamic properties of the plant only, with the transformation (15) and the transversal controller (19) depending only on a prescribed curve.

4. SELF-LEARNING CONTROL

The solution of the self-learning control problem is based on the following general thesis. The undetermined matrix $M(s)$ is considered as a solution of equation (7), the vector $\xi(s)$ being the vector of unknown parameters of this equation. The main difficulty is connected with probable fast variations of the parameters ξ_j and can be avoided by deceleration of the motion during the self-learning stage.

Let us introduce an auxiliary model

$$\dot{\widehat{M}} = \left(\sigma_2 \cdot \Xi(\widehat{\xi}) + \Delta\right) \cdot \widehat{M} \tag{22}$$

where $\widehat{M}$ is an estimate of the unknown matrix M with initial value $\widehat{M}_0$, $\widehat{\xi}$ is an estimate of the vector function $\xi(s)$, Δ is a skew-symmetrical matrix of the auxiliary feedbacks. This model is involved into the control system that causes the need for the additional feedback.

In this case the transformation of the control variables is executed as

$$\widetilde{u} = \widehat{M}^T \cdot \begin{vmatrix} u_s \\ u_e \end{vmatrix} \tag{23}$$

and the transversal controller is described by the expression

$$u_e = \sigma_2^2 \cdot h(\widehat{\xi}_1) - v \tag{24}$$

where $h(\widehat{\xi}_1) = [\widehat{\xi}_1\ 0\ 0 \ldots 0]^T$ and v is the output error vector calculated by formula (20). In order to construct the auxiliary error model let us define the error matrix of the form:

$$\begin{aligned} D &= M \cdot \left(M - \widehat{M}\right)^T \\ &= I - M\widehat{M}^T \end{aligned} \tag{25}$$

and the parameter error vector $\widetilde{\xi} = \{\widetilde{\xi}_j\} = \xi - \widehat{\xi} \quad (j = \overline{1, m-1})$. The difference $M - \widehat{M}$ and error $\widetilde{\xi}$ being sufficiently small, the matrix D is a skew symmetrical such that

$$D = D(d) = \sum_{i=1}^{m-1} \sum_{j=i+1}^{m} E_i^j d_i^j$$

where the elements d_i^j of the matrix D can be united into the $m(m-1)/2$-dimensional vector $d = \{d_i^j\}$. Then differentiating (25) and taking equations (7) and (22) into account we obtain the following auxiliary error model

$$\begin{aligned} \dot{D} = \sigma_2 \left(\Xi(\xi) \cdot D - D \cdot \Xi(\xi)\right) - \\ -\sigma_2 \cdot \Xi(\widetilde{\xi}) + \Delta \end{aligned} \tag{26}$$

On the other hand, for the small deviations e_1 and e_2 the model of the transversal motion takes the form

$$\begin{aligned} \dot{e}_1 &= e_2, \quad \dot{e}_2 = -K_1 e_1 - \left(K_2 - \sigma_2 \Xi'\right) e_2 \\ &- \sigma_2^2 h(\widetilde{\xi}_1) + u_s d' \end{aligned} \tag{27}$$

where $d' = [d_1^2\ d_1^3 \ldots d_1^m]^T$.

In such a manner the extended error model of the undermined control problem includes the models (27) and (26) where variables e_1, e_2 and d can be considered as the states, the signal v is an output error, terms $\sigma_2^2 h(\widetilde{\xi}_1)$ and $\Xi(\widetilde{\xi})$ are disturbances. The coefficients δ_i^j of the skew-symmetrical matrix

$$\Delta = \Delta(\delta) = \sum_{i=1}^{m-1} \sum_{j=i+1}^{m} E_i^j \delta_i^j$$

can be united into the $(m-1)m/2$-dimensional auxiliary feedback vector

$\delta = \{\delta_i^j\}$. To stabilize the steady-state solution of the extended error model we construct this vector as

$$\delta = -u_s \cdot (K \cdot v) \qquad (28)$$

where K is the matrix of the auxiliary feedback coefficients. Stabilization can be achieved by means of the hyperstability theory.

For the disturbing terms of the considered model to be rejected and its zero steady-state solution to be ensured, adaptive techniques involved in a general self-learning strategy are used.

Let us introduce the initial estimate $\xi^o(s)$ of the unknown function $\xi(s)$. Such a pre-estimate is supposed to be known, with the following expression being correct

$$\frac{d}{dt}\left(\xi - \xi^0\right) = \frac{\partial}{\partial s}\left(\xi(s) - \xi^0(s)\right) \cdot \sigma_2 \cong 0$$

This restriction specifies the requirement for the plant slow motion if the estimation errors are considerable and the possibility of the plant to go back to the normal speed after closer estimating $\xi(s)$.

The on-line estimate $\widehat{\xi}$ is formed as

$$\widehat{\xi} = \xi^o(s) + \xi_{AD}^1 \qquad (29)$$

where ξ_{AD}^1 is a vector of adjustable parameters generated by an adaptation algorithm. This algorithm is derived to minimize the following criterion

$$J_1 = \frac{1}{2} \cdot \widetilde{\xi}^T \cdot Q \cdot \widetilde{\xi}$$

where $Q = WW^T$, $W = \partial e_1 / \partial \widetilde{\xi}$ is a matrix of sensitivity functions. Elements of W can be generated on-line by a sensitivity model derived from equations (26) and (27). Then the gradient technique of the criterion minimization leads to the adaptation law:

$$\dot{\xi}_{AD}^1 = K_A \cdot W \cdot \widehat{e}_1 \qquad (30)$$

where $K_A = K_A^T > 0$, $\widehat{e}_1 = W^T \widetilde{\xi}$ is the modified error calculated indirectly in accordance with a method described by Miroshnik and Nikiforov (1991b). The use of the algorithm (30) allows one to accelerate the rate of the parameter convergence by virtue of an appropriate choice of matrix K_A and therefore to insure at least the rough equality $\widehat{\xi} \cong \xi$ even if ξ is slowly varying with time.

Thus the learning trajectory motion is supported by adjustable dynamic model (22) stabilized by feedback (28) and simultaneously tuned by the adaptation algorithm (30).

Throughout the learning stage the current estimate $\widehat{\xi}$ is also used for the restoration of unknown function $\xi(s)$. It is supported by an approximation procedure realizing on-line identification of the unknown vector $\theta = \{\theta_j\}$ of the approximation polynomial $\xi(s) = Z^T(s) \cdot \theta$, where $Z(s)$ is a matrix of suitable functions on s. For this purpose the following criterion is used:

$$J_2 = \frac{1}{2} \int_0^t \widetilde{\theta}^T(t) \cdot P(\tau) \cdot \widetilde{\theta}(t)\, d\tau \qquad (31)$$

where $\widetilde{\theta}(t) = \theta - \widehat{\theta}(t)$, $P(\tau) = Z(\tau) \cdot Z^T(\tau)$, $\widehat{\theta}(t)$ is a vector of adjustable parameters. Then

$$\xi_{AP} = Z^T(s) \cdot \widehat{\theta}$$

is the estimate of the approximation polynomial. To realize the identification procedure one must use $\widehat{\xi}_{AD}^1$ instead of the unknown function ξ and therefore the difference $\widehat{\xi}_{AD}^1 - \xi_{AP}$ is the error of approximation.

The minimization of (31) means that the parameter $\widehat{\theta}(t)$ chosen at time t minimizes the integral square of the Euclidian norm of the error $\widehat{\xi}_{AD}^1 - \xi_{AP}$ on all past data (Narendra and Annaswamy, 1989). This property makes the criterion especially useful for deriving the approximation algorithm. Such a physically realizable algorithm can be presented in the form:

$$\dot{\widehat{\theta}} = K_{AP}\{\int_0^t Z(\tau)\left(\widehat{\xi}_{AD}^1 - \xi_{AP}\right) d\tau +$$

$$+\int_0^t P(\tau)\widehat{\theta}(\tau)d\tau - \int_0^t P(\tau)d\tau\widehat{\theta}(t)\}(32)$$

where $K_{AP} = K_{AP}^T > 0$ is a time varying matrix specially chosen to provide the boundedness of the right side of (32).

It is highly plausible that after the first training motion along the explored segment of the unknown curve the accurate estimate ξ_{AD}^1 is not achieved and thereby the rough approximation ξ_{AP} is obtained only. For the high accuracy to be reached, the multi-step self-learning strategy is applied so that

$$\widehat{\xi}^k = \xi_{AP}^{k-1} + \xi_{AD}^k$$

where k is a number of the learning motion along the explored curve segment. In a few steps the approximation procedure provides $\xi_{AP} = \widehat{\xi}$. In such a manner sequential adaptive approximation of the unknown functions $\xi_s(t)$ are realized.

Intensive digital simulation of the proposed control strategy has been performed. The self-learning motions along *a priori* unknown paths formed by analytical plane curves (straight line, circle, ellipse, exponential curve and so on) have been investigated. Simulation results have demonstrated stable motion with acceptable accuracy and fast parameter convergence of the adaptation and approximation algorithms during already the first training motion.

5. CONCLUSION

The discussed strategy of self-learning control provides careful training motions of the plant along an *a priori* unknown curve and on-line adjustment of the control law in the course of self-learning. After self-learning stage the precise curve tracking is insured for an arbitrary mode of the longitudinal motion. For these aims an auxiliary dynamic model describing the behavior of the moving basis of the curve as well as algorithms of adaptation and on-line approximation have been included into the control system. The problems of stability of the extended system and parameter convergence briefly outlined in this paper introduce the main difficulties of the undetermined control tasks and should be treated in more detail.

5.REFERENCES

Byrnes, C.I. and A. Isidory (1991). Asymptotic stabilization of minimum phase nonlinear systems. *IEEE Trans. on Autom. Control*, **36**, 1112-1137.

Isidory A. (1989). *Nonlinear Control Systems.* Springer Verlag, New York.

Khosla, P.K., C.P. Neuman and F.B. Prinz (1985). An algorithm for seam tracking application. *Int. J. Robotics Res.*, **4**, 27-41.

Lizunov, A.B., A.M. Formal'skiy and A.Yu. Sneider (1987). Motion control of a manipulator having a proximity sensor along the object shape. *Automation and Remote Control*, **5**, 48-57 (in Russian).

Miroshnik, I.V. (1990). *Coordinating Control of Multivariable Systems.* Energoatomizdat, Leningrad (in Russian).

Miroshnik, I.V. and V.O. Nikiforov (1991a). Adaptive control of nonlinear plant spatial motion. *Automation and Remote Control.* **9**, 78-87 (in Russian).

Miroshnik, I.V. and V.O. Nikiforov (1991b). Acceleration of a gradient algorithm convergence. *Izvestiya Vuzov of the USSR -Priborostroenie.* **8**, 26-34 (in Russian).

Miroshnik, I.V. and V.O. Nikiforov (1994) Adaptive control and selflearning of robot trajectory motion. *The 4th IFAC Symposia on Robot Control, Capri, Italy* (submitted).

Narendra, K.S. and A.M. Annaswamy (1989). *Stable Adaptive Systems.* Prentice Hall, New York.

START-UP OF PROCESS PLANTS BY CONSERVATIVE LEARNING BASED QUALITATIVE REASONING

Zs. Csáki, K.M. Hangos

Systems and Control Laboratory Computer and Automation Institute, Hungarian Academy of Sciences
1518 Budapest, P.O.Box 63., HUNGARY, email: h681csa@ella.hu, h1245han@ella.hu

Abstract. A model conservative learning method is proposed in this paper which is able to transfer knowledge between different qualitative differential equation models of the same szstem. This is an essentialstep of generating operating procedures when multiple models with overlapping validity regions must be used to describe the controlled szstem over a wide operating range. The method is based on our previous results on introducing learning into Extended Qualitative Simulation by Bayesian identification and output prediction.

Key Words. Qualitative Reasoning; learning; process control; Bayesian identification; start-up procedures.

1. INTRODUCTION

According to the international experience one of the main problems of generating start-up operating procedures is to satisfy global safety critical requirements in the presence of uncertain or incomplete knowledge about the process plant. A possible solution of this problem is to integrate one of the incomplete knowledge representation methods with the traditional techniques for generating operating procedures (Fusilio, Powers, 1988; Laksmanan, Stephanopoulos, 1988).
Qualitative Simulation (QSIM (Kuipers, 1989; Weld 1990) has recently become one of the most popular ways for representing incomplete knowledge. Its aim is to distinguish the significant differences in possible system behaviours. Besides of representing parameter uncertainties, partially known dependencies can also be described by qualitative functions which are sets of functions constrained by some apriori knowledge. The advantage of this representation is that the model equations conserve the structure of the engineering knowledge about the system in mathematical terms by directing attention to the unknown parts.
At the same time difficult theoretical problems have arisen in the practical applications (Csaki, Hangos, 1992; Hangos, Csaki, Jorgensen 1992; Kuipers, 1989) which initiated efforts for improving the original methods in different ways (Csaki, Hangos, 1992; Kay, Kuipers, 1993; Kuipers, 1991; Weld, de Kleer, 1990). Some of these problems are as follows. From the knowledge representation side different model uncertainties can be present (e.g. undiscovered important landmarks with undiscovered proper corresponding values, too wide envelope functions, etc.), which are difficult to identify. Furthermore the amount of incomplete knowledge is proven to be increasing during the reasoning process. From the algorithmic side the main problem is the existence of spurious behaviours. From the viewpont of control applications a need to extend the constraints of the Qualitative Simulation with actuators (e.g. input variables) and to relate qualitative time to the real one have been arisen too.
In order to handle control related applications the EQSIM (Extended Qualitative Simulation (Csaki, Hangos, 1992)) method has been developed and applied to the problem of generating start-up procedures. Moreover Bayesian identification and output prediction (Peterka, Bayesian, 1981) has been used to introduce learning into Qualitative Simulation and to improve its reasoning by better prediction. The aim of this paper is to propose a conservative learning method which is able to transfer knowledge between different models of the same system. This is an essential step of generating operating procedures, when multiple models with overlapping validity regions must be used to describe the controlled system over a wide operating range.

2. BAYESIAN IDENTIFICATION AND PREDICTION BASED LEARNING IN QUALITATIVE SIMULATION

In order to introduce learning into Qualitative Simulation and to improve its reasoning by better prediction, Bayesian identification and output prediction is used. The aim is to eliminate branching from the Qualitative Simulation by evaluating a one-step-ahead output prediction so that the simulation follows the most expected branch according to the measured data up to a given time point. Since there is a quite natural relationship between intervals (of QSIM and EQSIM) and distributions with finite support, the main idea of this method is to transform our qualitative information into probabilistic terms. Then the prediction is generated by Bayesian output prediction rather than by P- and I-transition tables of QSIM (Kuipers, 1989), and the prediction is followed by an identification (parameter estimation or learning) step. Finally the knowledge is transformed back to qualitative terms.

Throughout in this section a unique model for describing the physical system is assumed and qualitative functions are assumed to be described by linear envelope functions. A qualitative value of a variable X at a given time t is characterized by a pair

Qval(X,t) = <Qmag(X,t), Qdir(X,t)>

where $\mathrm{Qmag}(X,t) \in Z^X$ the magnitude of the variable with the landmark set $L^X = \{\ell_i | \ell_i < \ell_{i+1},\ i = 1,2,\ldots,n-1\}$, Z^X is the set of possible connected intervals (ℓ_i, ℓ_j) and $\mathrm{Qdir}(X,t) \in \{\mathrm{inc},\ \mathrm{dec},\ \mathrm{std}\}$ the direction of the variable.

2.1. Definition: The *probabilistic representation interpretation* of the qualitative variable X assuming uniform distribution on the given (ℓ_i, ℓ_j) interval at time $t \in \mathbf{T}$ is defined as:

$$p(X_t) = \begin{cases} \chi\left(\ell_i \le \mathrm{Qmag}(X,t) \le \ell_j\right)\dfrac{1}{\ell_j - \ell_i} & ,\text{if } j = i+1 \\ \chi\left(\mathrm{Qmag}(X,t) = \ell_i\right) & ,\text{if } i = j \end{cases} \qquad (2.1.)$$

where χ (.) is the indicator function of a real interval.

2.2. Definition. Let **P** be the set of probability density functions over Z^X where the elements of **P** are uniform distributions over finite intervals (elements of Z^X). Then, φ_X: $Z^X \to \mathbf{P}$ is a *formal transformation function from qualitative representation to uniform distribution type probabilistic representation* for the qualitative variable X at time $t \in \mathbf{T}$:

$$\varphi_X\left(\left(\ell_i, \ell_j\right)\right) = \begin{cases} \chi\left(\ell_i \le \mathrm{Qmag}(X,t) \le \ell_j\right)\dfrac{1}{\ell_j - \ell_i} & ,\text{ if } j = i+1 \\ \chi\left(\mathrm{Qmag}(X,t) = \ell_j\right) & ,\text{ if } i = j \end{cases} \qquad (2.2.)$$

From a mathematical point of view a bijective function is needed for the qualitative-uniform distribution type probabilistic representation transformations, and φ_X is clearly bijective. But in order to compute the qualitative direction of X from a given distribution, we have to remember the previous qualitative value of X.

2.3. Definition: ψ_X: $Z^X \times Z^X \to \{\mathrm{inc},\ \mathrm{std},\ \mathrm{dec}\}$ is a *change detection function* of the variable X at time t where the first parameter is the qualitative magnitude of X at time t, the second parameter is the qualitative magnitude of X at time t-1:

$$\mathrm{Qmag}\ (X,\ t) = (\ell_i, \ell_j),$$

$$\mathrm{Qmag}\ (X,\ t\text{-}1) = (\ell_p, \ell_q)$$

$$\psi_X\left(Qmag(X,t), Qmag(X,t-1)\right) = \begin{Bmatrix} inc, & if \quad \ell_q \le \ell_i \\ dec, & if \quad \ell_p \ge \ell_j \\ std, & otherwise \end{Bmatrix} \qquad (2.3.)$$

Problem statement:

Let us investigate the control synthesis problem stated as follows:

$$\underline{\dot{x}}(t) \in \underline{F}(\underline{x}(t), \underline{U}(t, \underline{x}(t)), t, \underline{\theta}, \underline{\gamma}) \qquad (2.4.)$$

$$(t \in [a,b])$$

$$\underline{y}(t) \in \underline{G}(\underline{x}(t), \underline{\theta}, \underline{\gamma}),$$

$$\underline{x}(t_0) = \xi\ (t_0 \in [a,b])\ ,\ \underline{x}(t_u) \in M,\ \underline{U}(t,\underline{x}) \subset P(t).$$

where

$\underline{x} \in \mathbf{R}^n$ - the state vector, real vector valued function,

$\underline{y} \in \mathbf{R}^o$ - the output vector, real vector valued function,

intv W^i - is the set of i dimensional vectors $\underline{w} = [w_1, \ldots, w_i]^T$ where each element w_j is an interval $(w_{j\ell}, w_{jn}), w_{j\ell}, w_{jn} \in W$,

$\underline{U} \subset$ intv $\mathbf{R}^m$ - the set of possible inputs, continuous set-valued function,

P: (a,b) $\to$ intv $\mathbf{R}^m$ - the constraint function for the possible inputs,continuous set-valued function,

$\underline{\theta}\in\Theta\subseteq$ intv $\mathbf{R}^s$ - the parameter interval vector,

$\underline{\gamma}\in\Phi\subseteq$ intv C^r - the vector of unknown functions where for each element of $\underline{\gamma}\in\Phi$, γ_i (i=1,2,...,r) is a convex compact set of continuous, monotonous, real functions,

$\underline{\xi}\in\mathbf{R}^n$, - the initial state,

$M\subset\mathbf{R}^n$ - the set of acceptable goal states,

$\underline{F}$, $\underline{G}$ - the generalized state transition and output functions, differential inclusions: $\underline{F}(\underline{x}, \underline{U}, t, \underline{\theta}, \underline{\gamma}) = \{\underline{f}(\underline{x}, \underline{u}, \underline{t}, \underline{\vartheta}, \underline{\eta})_{\underline{u}\in\underline{U}, \underline{\vartheta}\in\underline{\theta}}, \underline{\eta}\in\underline{\gamma}\}$ and f is a real vector valued state transition function, and

$$\underline{G}(\underline{x},\underline{\theta},\underline{\gamma},)\{\underline{g}(\underline{x},\underline{\vartheta},\underline{\eta},)_{\underline{\vartheta}\in\underline{\theta},\underline{\eta}\in\underline{\gamma}}\}$$

The vector of unknown functions ($\underline{\gamma}$) is described by qualitative functions (i.e. $M^+()$ and $M^-()$) represented by envelope functions where the envelopes are given by exact numerical functions. For simplicity, it is assumed that the envelope functions are linear and

$$M^+(0)=0 \text{ and } M^-(0)=0 \qquad (2.5.)$$

$$a * X \le M^+(X) \le b * X, \quad 0 \le a \le b \le MAX\ \infty \qquad (2.6.)$$

The aim is to predict the system behaviour so that the difference of the gradients of the corresponding envelope functions (i.e. |a-b| in (2.6.)) has to be narrowed by using measured data. These gradients will form the unknown parameter vector $\underline{\theta}$. So the control synthesis problem (2.4.) has been reduced to problem (2.8.) below. If it is assumed that there are m different qualitative functions, then $|\underline{\theta}|$ = m, so that:

$$\underline{\theta} = [\,(a_{1l},a_{1u}),\ (b_{1l},b_{1u}) \ldots ,(a_{ml},a_{mu})\ (b_{ml},b_{mu})]^T \qquad (2.7.)$$

Applying the notation of envelope functions to (2.4.), it can be reformalized as follows:

$$\underline{f}_l(\underline{x}(t),\underline{u}(t),t,\underline{\theta}_l) \le \dot{\underline{x}}(t) \le \underline{f}_u(\underline{x}(t),\underline{u}(t),t,\underline{\theta}_u)$$

$$\forall \underline{u}(t) \in \underline{U}(t,\underline{x}(t)) \qquad (2.8.)$$

$$\underline{g}_l(\underline{x}(t),\underline{\theta}_l) \le \underline{y}(t) \le \underline{g}_u(\underline{x}(t),\underline{\theta}_u)$$

$\underline{x}(t_0)=\underline{\xi}$ $(t_0\in[a,b])$, $\underline{x}(t_u)\in M$, $\underline{U}(t,\underline{x}) \subset P(t)$, $(t \in [a, b])$

where

$\underline{\theta}_l$ - real vector of the gradients which forms the lower bound for $\dot{\underline{x}}(t)$,

$\underline{\theta}_u$ - real vector of the gradients which forms the lower bound for $\dot{\underline{x}}(t)$.

This substitution generates a relation among the gradients of the envelope functions: two envelope functions are in relation if they are on the same side of (2.8.). This is an equivalence relation and generates a classification among the gradients of the envelope functions where the corresponding envelopes (2.6.) are in different classes. It is important to note that $\underline{\theta}_l$ and $\underline{\theta}_u$ are not necessarily equal to the vector of gradients of the lower and upper envelope functions, when an $M^+()$ qualitative function has a minus sign in (2.8.)

The uniform distribution will be only assumed for the ($\underline{\theta}_l$, $\underline{\theta}_u$) pairs, i.e. the unknown function is uniformly distributed within the upper and lower envelope functions. Then applying the propositions for sum and product of uniformly distributed functions, the output distribution can be computed, what will not necessarily have uniform distribution. But the result distribution will be estimated by uniform distribution so that the endpoints of the non-zero values of the achieved probability function can be computed applying interval algebra, and the estimated probability density function will be computed on the basis of these supports.

2.1. Algorithm for determining the Bayesian system model

1. The inputs of the algorithm

The qualitative model equations given in the form of (2.8.)

2. The algorithm

Step 1. Transform the qualitative differential equations (2.8.) into qualitative difference equations (2.9.) by transforming the equations to discrete time, i.e.

$$\underline{f}_l\left(\underline{x}_{(\tau-1)},u_{(\tau)}\underline{\theta}_{(l)}\right) \le \underline{x}_{(\tau)} - \underline{x}_{(\tau-1)} \le \le \underline{f}_u\left(\underline{x}_{(\tau-1)},\underline{u}_{(\tau)},\tau,\underline{\theta}_u\right) \qquad (2.9.)$$

Step 2. Transform the qualitative model (2.9.) to Bayesian parametric system model (2.11.) applying the uncertainty interpretation transformation functions (2.2.) and (2.1.) where z is the number of qualitative difference equations:

$$\underline{f}_l(\underline{x}_{(\tau-1)},\underline{u}_{(\tau)},\tau,\underline{\theta}_l) \le \underline{x}_{(\tau)} - \underline{x}_{(\tau-1)} \le \le \underline{f}_u(\underline{x}_{(\tau-1)},\underline{u}_{(\tau)},\tau,\underline{\theta}_u) \qquad (2.10.)$$

$$p\left(\underline{y}_{(\tau)}\middle|\underline{u}_{(\tau)},D_{(t1)}^{(\tau-1)},\underline{\theta}\right) = \qquad (2.11.)$$

$$= \prod_{i=1}^{z} \chi\left\{\left(\underline{f}_{i,l}\left(\underline{x}_{(\tau-1)},\underline{u}_{(\tau)},\tau,\underline{\theta}_l\right) \le \underline{x}_{(\tau)}\underline{x}_{(\tau-1)} \le \le \underline{f}_{i,u}\left(\underline{x}_{(\tau-1)},\underline{u}_{(\tau)},\tau,\underline{\theta}_u\right)\right)\right\}\cdot$$

$$\cdot \frac{1}{\underline{f}_{i,u}\left(\underline{x}_{(\tau-1)},\underline{u}_{(\tau)},\tau,\underline{\theta}_u\right) - \underline{f}_{i,l}\left(\underline{x}_{(\tau-1)},\underline{u}_{(\tau)},\tau,\underline{\theta}_l\right)}$$

It is important to substitute $\underline{x}(\tau)$ and $\underline{x}(\tau-1)$ by $\underline{y}(\tau)$ and $\underline{y}(\tau-1)$ applying the equation $\underline{y}(\tau) = g(\underline{x}(\tau), \underline{\theta})$ in order to employ the reachable measurements (see the example in Section 4).

3. The output of the algorithm:
The Bayesian system model (2.11.).

Since there are apriori lower and upper limits for each unknown variable (2.6.), divided into two groups ($\underline{\theta}_l$, $\underline{\theta}_u$), the possible behaviour within these ranges is under investigation. Moreover, the Bayesian system model also contains constraints for $\underline{\theta}$, i.e. through the measurements. This can be illustrated by two m-dimensional coordinate systems (since $|\underline{\theta}|=2m$) where each measurement means a constraint, an (m-1) dimensional plane (because of the linearity) in the m-dimensional space. The planes of the apriori constraints with the plains of the constraint equations define an m-dimensional polygon in each coordinate system (see Fig. 2.1.).

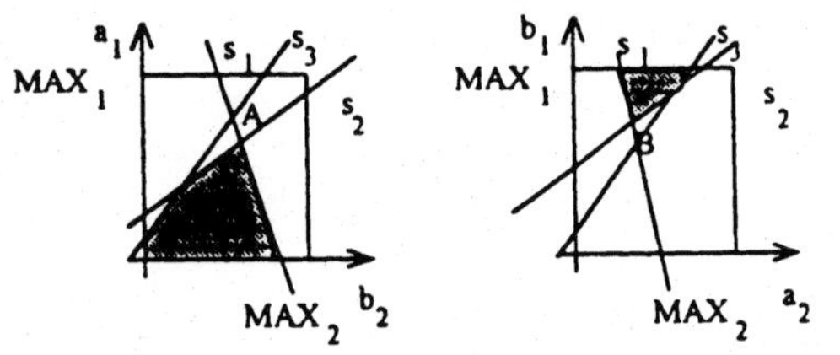

where s_1, s_2, s_3 are the constraint generated by the measurements and $\underline{\theta}_l = [b_2,a_1]^T$ and $\underline{\theta}_u = [a_2,b_1]^T$

Fig. 2.1. The parameter space with measurements

2.1. Proposition. The polygon above is convex.
Proof: If the polygon were non-convex, it would mean that there was a constraint (a plane) of which appropriate values could lie on each side. But at every time t a new constraint plain is introduced which leaves only one part of the intersected original polygon.•

2.2. Learning algorithm for Qualitative Simulation (LQSIM): using Bayesian identification and prediction.

1. The inputs of LQSIM:
(a) The qualitative system model extended by acuator and action description.
(b) An initial time point symbol, t_0, and qualitative values for each qualitative variables at t_0.
(c) A goal state description.
(d) The Bayesian system model (2.11.) as computed from (2.8.) according to the algorithm 2.1.
(e) The apriori estimation for a qualitative functions (2.6.) as defined in the Bayesian framework.

$$p_0(\underline{\theta}) = \chi(0 \le a_1 \le b_1 \le MAX_1 < \infty \,\&\, \ldots \,\&\, 0 \le a_m \le b_m \le MAX_m < \infty) \qquad (2.12.)$$

2. The algorithm
LQSIM repeatedly compares the real system state to the goal state and selects the variables contradicting with the goal state requirements, and then, tries to fulfill these missing conditions by actions, as in the case of EQSIM. But the prediction of the next system state is different. In the forward chaining step the Bayesian one-step ahead prediction will be used instead of the transition tables and filters.
Compare the initial state (which is the set of qualitative values at the initial time point) with the goal state. Repeat the following steps until the goal state is reached or it is possible to find any action for fulfilling the missing condition.

Step 0. Initial step. Determine the real system state in on-line case from measurements or in off-line case from an input file or applying the result of the previous reasoning step. Allow the user to change the model equation set of the simulation which reflects the real system state.

Step 1. Select a variable from the goal state description whose requirement is not fulfilled. Determine the direction of the required change.

Step 2. Backward chaining step. Search for an actuator or an input variable which has the desired effect for the unsatisfied variables (in the goal state). This backward chaining has been based on the connection of the state / internal / input variables described by the qualitative model equations.

Step 3. In case of more than one promising actuator/input variable, show the possibilities to the user to select from among them.

Step 4. Perform the action associated with the selected actuator and direction.

Step 5. Learning steps (Step 5.-7.) - evaluation of a maximum likelihood point estimate for the unknown parameters constrained by the characteristic functions in (2.11.). In order to find the distribution corresponding to the Bayesian system model with the highest probability, restrict the area, described in the characteristic function, within the parameter space by the measurements first. That is, take the actual measurement and apply it as a constraint on the parameter space restricted by the previous measurements.

Step 6. Apply a vertex enumeration algorithm like [Chen, Hansen, 1991] to find the point in the generated polygon for the maximization of the probability over the set of restricted possible parameter values (2.13.).

Step 7. Then, the output prediction can be performed.

Step 8. Transform back to qualitative description applying φ_x^{-1} and ψ_x as in (2.3.) and interval or sign algebra to determine the new system state and determine the value for all the other variables by applying the interval or sign algebra.

3. The outputs of LQSIM:

(a) At each distinguished time point a set of advice towards the goal state is supplied.

(b) A behaviour tree which branches according to the different pieces of advice (and not according to the possible system behaviour).

3. CONSERVATIVE LEARNING FOR MULTIPLE MODEL BASED QUALITATIVE REASONING

This section is devoted to the generalization of the previous one when different, even structurally different models should be used to describe the behaviour of the system over the investigated operating region. Here Learning algorithm 2.2. can be analogously applied but on the basis of the compound system model. On the other hand, the main problem here is how the learnt knowledge can improve the quality of the new model.

On the basis of the investigations of the previous section, let us investigate the control synthesis problem (2.8.). For investigation of the *space of the qualitative system model* the following definitions will be introduced.

3.0. Definition. Let S denote the set of possible system state values.

Remark Note that in the case of qualitative models if the number of landmarks belonging to a variable x(t) is l_x, then the number of its possible qualitative states is $(2l_x-1)^3$. So

$$|S| = \prod_{i=1}^{|\underline{x}|}\left(2l_{x_i}-1\right)^3$$

3.1. Definition. For simplicity let **M** denote the qualitative system model as it is restricted in the previous chapter (see (2.8.))

3.2. Definition. A given qualitative system model (**M**) is also characterized by a *state validity region* (in notation: $V_{\underline{x}}$), which is a subset of S. This connection is denoted by $V_{\underline{x}}|\mathbf{M}$, and the model **M** can be applied in case of $\underline{x}' \in V_{\underline{x}}|\mathbf{M}$ only.

3.3. Definition. For parametrized models an *extended validity region* $V|\mathbf{M}$, can be defined analogously where $V|\mathbf{M} = (V_{\underline{x}}|\mathbf{M}) \times (V_{\underline{\vartheta}}|\mathbf{M})$, where $\underline{\vartheta}$ is real vector valued parameter.

Usually the given system model is not suitable for a significant shift in operation conditions. In this case a new model is needed, which contains the new system state in its validity region. There exist two cases for two models with non-empty common state validity regions.

3.4. Definition. Let us assume that two models are given in the following forms: $\mathbf{M}1$ with extended validity region $V1=V|\mathbf{M}1$, and $\mathbf{M}2$ with $V2=V|\mathbf{M}2$. In principle, there are two cases at model change:

i) $V1 \cap V2 \neq 0$ and $V_{\underline{x},1} \cap V_{\underline{x},2} \neq 0$ (where $V_{\underline{x},i}=V_{\underline{x}}|\mathbf{M}i$, i=1,2). Here the models have *overlapping* state validity regions, and within the common region the models have to give the same response for the same input in qualitative sense.

ii) Its special case when $V_{\underline{x},1} \cap V_{\underline{x},2}=\{\underline{x}^0\}$. Here the models have *non-intersecting* state validity regions, but $V_{\underline{x},1}, V_{\underline{x},2}$ are connected subsets of S, through the connecting state $\underline{x}^0$.

In both cases it can be assumed that the *model change is smooth enough*, i.e. the following two conditions hold.

i) The two given models $\mathbf{M}1$ and $\mathbf{M}2$ have non-empty common state validity region: $V_{\underline{x},1} \cap V_{\underline{x},2} \neq 0$.

ii) Furthermore, $\underline{\theta}_1'$, $\underline{\theta}_2'$, $\underline{u}_1'(t)$, $\underline{u}_2'(t)$, $\underline{x}_1'(t)$, $\underline{x}_2'(t)$ can be selected so that both models conserve their structure, and

$$\dot{\underline{x}}(t) = \underline{f}_1(\underline{x}_1'(t), \underline{u}_1'(t), t, \underline{\theta}_1') =$$
$$= \underline{f}_2(\underline{x}_2'(t), \underline{u}_2'(t), t, \underline{\theta}_2').$$

Note that these requirements do not mean that the models are equivalent.

One of the main problems of learning at model change is that usual learning methods assume a unique model. When a significant shift in operation conditions occurs, the old system model is generally no longer valid. A new system model valid in the new system state is selected, but the relevant part of the learnt knowledge in the previous model usually cannot be used in the new one. In other words, learning usually starts from a fixed general initial state at any model change. To avoid this artificial complete forgetting, the following key problems have to be solved:

i) How can the past data serve for teaching the new model as it starts to be valid?

ii) How can the learnt old model serve for teaching the new one?

3.1. Applying Past Data for Teaching the New Model

For solving the knowledge transfer problem between models, the Bayesian approach will be applied, since the problem reduces here to the definition and computation of the p.d.f. $p(\underline{\theta}|D)$ where D is a subset of the past history ($D^{(t)} \supseteq D$). From our viewpoint, the important question is which subset of $D^{(t)}$ can be applied for teaching the new model, which data-pairs $D(\tau) = (\underline{u}(\tau), \underline{y}(\tau))$ from the past history is valid even in the neighborhood of the new model. After determining this D subset, the new qualitative system model can be taught on this subset by the Bayesian learning method. For this purpose, the new qualitative system model has to be transformed to a probabilistic system model applying the knowledge transformation method defined in section 2.

In the case of models with overlapping state validity regions the solution is quite simple. First of all, a subset of past data ($D^{(t)} \supset D$) belonging to the validity region of the new model can be used as common knowledge source. Here the aprior knowledge of $\underline{\theta}_2$ in **M**2 can be computed as it is described in (3.2.) where $D(\tau) \in D$ means that the production is performed for data-pairs (measured at time τ) belonging to the above defined set D:

$$p(\underline{\theta}_2|D) = \frac{\prod_{D(\tau)\in D} p\left(\underline{y}_{2,(\tau)} \middle| \underline{u}_{2,(\tau)}, D^{(\tau-1)}\right) p\left(\underline{\theta}_2 \middle| D^0\right)}{\int \prod_{D(\tau)\in D} p\left(\underline{y}_{2,(\tau)} \middle| \underline{u}_{2,(\tau)}, D^{(\tau-1)}\right) p\left(\underline{\theta}_2 \middle| D^0\right) d\underline{\theta}_2} \qquad (3.2)$$

3.2. Applying the Learnt Previous Model for Teaching the New One

Teaching the new model before its application means embedding different types of apriori knowledge (past history, partial system knowledge, etc.) to the system model. This embedding has been solved in the framework of Bayesian identification and prediction (Kárny, Halousková, 1992). The idea of the solution is to generate *fictitious data* from all kind of prior knowledge. In case of applying a previous model for teaching, the method can be generalized so that it uses the old model as a simulator to generate fictitious data for the new model. Then, this data can be used for teaching purposes as it is described in the previous section. The most important open question here is to select the input data for the simulation. It has to be chosen from the common state validity region of **M**1 and **M**2 (i.e. $\underline{x} = \underline{x}^0$ in case of non-intersecting validity regions) and with parameters from the common parameter region $V_{\underline{\theta},1} \cap V_{\underline{\theta},2}$. This generated fictitious data will form the set D. So, (3.2.) can be applied for learning in the overlapping and non-intersecting cases as well. In both cases overlapping regions exist in the space of extended system states, i.e. in the space of the system states and/or in the space of the parameters and we have qualitative models, so we are able to generate data by a model, **M1**, to another one, **M2**, so that the common (intersecting) system state value(s) and points from the common parameter space are selected. This method has the advantage of teaching the new system model through the apriori given corresponding value pairs if the uncertainties in the new model are described by envelope functions.

Note that the proposed method assumes that the model **M**1 is able to predict or simulate $\underline{x}(t)$ and $\underline{y}(t)$ given a sequence of inputs $\underline{u}(t)$ t= t0, t0+1, ... In the case of Qualitative Simulation it is automatically fulfilled, because the model equations are qualitative differential equations or algebraic constraints.

4. ACKNOWLEDGEMENT

This research was supported by the Hungarian National Science Foundation through grant 2577 which is gratefully acknowledged.

5. REFERENCES

Csáki, Zs. Identification as Learning in Reasoning about Incomplete Knowledge of Physical Systems. *Proceedings of the European Conference on Artificial Intelligence*, 714-718, 1992.

Csáki, Zs., Hangos, K.M. Lessons learned from a prototype operator advisor system for real-time process control. *Workshop Notes of the Workshop on Advances in Real-time Expert Systems Technologies at ECAI'92*, 23-31, 1992

Fusilio, R., Powers, H. Operating Procedure Synthesis Using Local Models and Distributed Goals. *Comput. chem. Engng*, **12**, 1023-1034, 1988.

Hangos, K.M., Csáki, Zs., Jørgensen, S.B. Qualitative Model-based Intelligent Control of a Distillation Column. *Engineering Application of Artificial Intelligence*, **5**, 431-440, 1992.

Kárny, M. and Halousková, A. User Supplied Information in the Design of Linear Quadratic Gaussian Self-Tuning Controllers, *Proc. ACASP'92*, 1992

Kay, H., Kuipers, B.J. Numerical Behavioural Envelopes for Qualitative Simulation. *Proc. AAAI-93*, in press, 1993.

Kuipers, B. Qualitative Reasoning: Modeling and Simulation with Incomplete Knowledge. *Automatica*, **25**, 571-585, 1989

Kuipers, B. Higher order derivative constraints in qualitative simulation. *Artificial Intelligence*, **51**, 343-379, 1991

Laksmanan, R., Stephanopoulos. Synthesis of Operating Procedures for Complete Chemical Plants. *Comput. chem. Engng*, **12**, 985-1021, 1988.

Peterka, V. Bayesian Approach to System Identification. In *Trends and Progress in System Identification (P. Eykhoff ed.), Pergamon Press*, 239-304, 1981

Weld, D.S. and de Kleer, J. (eds.). Readings in Qualitative Reasoning about Physical Systems. *The Morgan Kaufmann*, 1-720, 1990

KNOWLEDGE-BASED SYSTEM IDENTIFICATION - COMPONENTS OF THE EXPERTISE

F. LOUAGE*, T. P. DOBROWIECKI** and B. PATAKI**

*Free University of Brussels, Dienst ELEC, Pleinlaan 2, B-1050 Brussels, Belgium
**Technical University of Budapest, Dept of Measurement & Instr. Eng., Müegyetem rkp. 9, H-1521 Budapest, Hungary

Abstract. The system identification problem is analysed from the perspective of the required (expert) knowledge. Recent years witnessed efforts to introduce artificial intelligence techniques to this field. Considering the problem on the whole, however, the developed expert systems covered important but limited issues, with similarly limited expert knowledge. The software system supporting various identification activities can aim at different levels of automation, related to the increased volume and complexity of knowledge. System identification problem turns out upon the analysis to be a knowledge-intensive, hard AI problem, where many essential developments are still to come. Accordingly an expert system conforming with a number of the introduced ideas is presented.

Key Words. Knowledge-engineering; expert systems; artificial intelligence; modelling; system identification

1. INTRODUCTION

System identification isn't just yet another application field for the artificial intelligence (AI) tools. Its importance lies in the fact that identifying systems from experimental data, i.e. performing an inductive modelling of the environment, constitutes an elementary and ever-present step of the scientific cognitive process. On the other hand, the richness of the system identification theories and the increasing complexity of the interdisciplinary applications call for more and more expertise regarding the choice, usage and verification of the modelling approaches. The AI tools in this field appear thus both necessary and opportunistic. Their usefulness lies in their ability to describe and handle that part of knowledge, which is difficult or even impossible to describe efficiently by an analytical approach. The border between the analytical description and the AI tools is fluent. Once a particular bit of knowledge is clarified enough to reveal its algorithmic structure, it becomes an element of a well-established analytical theory. Many topics, however, essential to the system identification practice, are still resistant to attempts of a meaningful algorithmisation and this fact gives merit and future to the extensive research in the knowledge-based systems

This work is supported by the Belgian government as a part of the Belgian programme on Interuniversity Poles of attraction (IUAP20) initiated, by the Belgian State, Prime Minister's Office, Science Policy Programming.

The role a computerised SI tool plays during the system identification session and consequently the extent and the nature of tasks performed by it depends upon the volume and nature of the embedded knowledge. Last years have seen considerable efforts in designing SI expert systems. Numerous prototypes appeared, together with surveys, which witnesses already a certain maturity of this particular field (Meier zu Farwig and Unbehauen, 1991). Considering the identification problem on the whole, issues covered by those systems, though important, were limited, and similarly only limited expert knowledge found its way into them. Naturally the most difficult problems will appear if one considers a full automation of the system identification session. Such systems will be demanded when and where the expert resources are really scarce. This situation becomes increasingly realistic in interdisciplinary fields and certain areas of scientific research.

2. MODELLING AND MEASUREMENT

The primary characteristics of the modelling (and thus of system identification) process is its intrinsic relation with the measurement and the well-known iterative structure. Data missing from the preliminary model are seeked through experiments, i.e. by purposeful planning, execution and evaluation of the measurements (Eykhoff, 1974; Ljung, 1987). Iterations are the consequence of problems revealed during the final model validation. An inadequate model indicates usually that the data processing should be done better, that perhaps new measurements should be obtained, or in the extreme cases that the whole experimental approach is wrong and has to be re-evaluated. Iterative experiments are also intrinsic to the more advanced, complex identification problems. A limited initial (a priori) knowledge usually does not permit to plan for the optimal experiments, which would serve results of final quality (accuracy) satisfying the user. The empirical information and hence the model of the identified system is built up rather gradually (through a series of experiments), until enough knowledge is collected to consider optimal measurements of the final (parametric) model of the subject (Schoukens and Pintelon, 1991). System identification activities are therefore intrinsically related to the planning of physical experiments, handling and interpreting signals, finally to the processing of signal information into the framework of the particular model class. Components of these skills, summing up to the working knowledge of an expert will be analysed in more detail in the following.

3. COMPONENTS OF THE EXPERTISE

When a problem domain is considered as a potential AI application, firstly so called knowledge-level analysis should be performed. Knowledge level is devoted to the characterisation of the knowledge contained in the problem without saying how it should be represented and subsequently implemented (Steels, 1990). There are many ways to represent knowledge but even the best implementation won't help if the system is handicapped by a badly chosen representation scheme. Being familiar with the peculiarities of the knowledge and knowing which concepts and relations are essential creates necessary criteria for the selection of the appropriate AI tools.

What skills are needed to run system identification tasks ? An expert will usually draw his knowledge from:

Signal & system theory. Abstract representations of the concept of signal and how these descriptions are related to the system models.

Signal processing algorithms. Algorithmic realisation of various signal transformations.

System identification theory. Hierarchy of models and identification strategies, also optimisation procedures and model validation criteria

Instrumentation. Functional and qualitative knowledge of instruments, knowledge of how abstract functionality is distorted in particular instruments, what are the available instrument pools and how to diagnose instruments or to substitute one instrument for another.

Numerical (software) packages. Knowledge of how different implementations of the particular signal processing algorithms are related, what are the essential differences, side effects and free parameters influencing the quality of the computed results.

Qualitative understanding of signals. Ability of how to "read" signals, i.e. how to interpret their visual appearance regarding the functionality of instruments and how to relate these observations to the signal and system theory.

Measurement "common sense" knowledge. Coping with the physical dimensions of the instruments, evaluating noises, grounding problems, loads and other environmental problems (e.g. EMC).

General planning skills. Planning of experiments, balancing solutions regarding hardware and software, balancing different goals like real-time aspect, duration of the experiment, costs, accuracy, etc.

General modelling skills. Knowledge of how different system identification problems are related, i.e. how to reason by analogy and how to relate problems born in physically different problem-domains.

History of successes and failures. Experience about various border cases, history of past sessions in case of particularly interesting systems or experimental conditions.

The main problem in automating system identification tasks is that the related knowledge and control expose a mixture of abstraction levels and heterogeneous approaches, escaping any kind of traditional representation. Even the briefest analysis reveals that the mathematical (analytical) knowledge is used concurrently with shallow heuristics, numerically intensive computations interweave with the subsequent symbolical interpretation. Well-structured knowledge is mixed with logic-like assertions, "sure" knowledge appears together with uncertain knowledge of a very difficult (frequently subjective) nature. There is a knowledge body that is strictly local concerning the problem (as testing an instrument for example), but part of the reasoning applies to the whole experimental environment and the history of the experiments (e.g. is the noise level in the measurements adequate). An additional difficulty is that almost every concept should be represented in a number of ways and at different levels of abstraction (multi-level modelling), considering that different tasks require a different "look" at the objects participating in the system identification session. The deepest level of modelling is concerned usually with the physical laws, at the "surface" models describe only qualitative and heuristic relations. Particularly interesting and prone in consequences is that running the experiment involves not only strict "expert" knowledge, but also a volume of "common sense" knowledge (Rich and Knight, 1991), which is burdensome to represent and handle.

Similarly confused and difficult is the control of the problem solving. Strictly algorithmic steps and well-ordered sequences of activities interweave with almost pure "heuristic search". Every activity draws from some shallow heuristic knowledge, the general scheme of system identification, however, is almost purely deterministic and iterative.

The experimental part of the system identification session exhibits a rich body of typical and specialised tasks, covering

- decision making, e.g. choosing the topology of the measurement set-up, selecting appropriate instruments, making consistent decisions regarding excitations, setting of instruments and processing algorithms, etc.;
- design, e.g. making the measurement set-ups, building the control of the instruments;
- interpretation, e.g. "reading" signal shapes, relating observed signal features to the knowledge about systems, "reading" cost functions and modelling tests;
- planning, e.g. the sequence of experiments;
- diagnosis, e.g. testing instruments, leads, expected shapes of the measured signals, noise levels, etc.

This last category of tasks is of special importance, since the quality of an experiment (and thus that of the final model) depends upon an enormous number of factors, with the influence of the low quality decisions frequently hidden and indirect.

4. INTELLIGENT SI ENVIRONMENT

Expert systems belonging to the "first wave" of intelligent identification tools automated expert activities only within a very limited scope of an off-line processing of the measurement data. Those systems worked with pre-measured signals and had no means to reconsider the modelling process (i.e. to plan better experiment, to get better data, etc.) once the processing and the model were ready (Bergeon *et al.*, 1991; Gentile *et al.*, 1990; Haest *et al.*, 1990, 1991; Larsson and Persson, 1991; Nagy, 1990). The built-in expertise was limited generally to the evaluation of the estimated model, regarding only possible modifications to the structure.

Systems taking responsibility for the acquisition of data appeared later (Niederlinski *et al.*, 1991). Despite the presence of the measurement hardware, however, the expert usage of the instruments was only fractional, and far from the full coverage of activities observed in human expert behaviour. Furthermore essential decisions regarding the organisation of the experiment (in this case the choice of excitation) remained the prerogative of the user.

System identification expert systems equipped with a flexible measurement hardware and assuming responsibility for the whole range of the experiment related activities are still to come. The main challenge facing them is not the conduct of a specific measurement and the subsequent system identification, but the planning for the whole structure of experiments, with many of the efforts directed to establish those experimental conditions, within which a considered model class will remain valid.

To address fully the above mentioned issues a new kind of integrated programming environment was developed, equipped with means to maintain a full control over the measurement environment, and thus able to automate the full cycle of the system identification problem. It has been organised around three separate components of knowledge, namely:

(1) library of the measurement technical methods and numerical packages;
(2) object-oriented signal & system representation tools, handy to build models of the physical experimental environment;
(3) heuristic knowledge underlying the decision making in the system identification problems.

The environment has been specified following the generic task analysis of the system identification problem (Chandrasekaran *et al.*, 1992; Steels, 1991; Yen and Lee, 1993) (Fig. 1). The computational model of object-oriented representation corresponds well with the physical nature of systems and signals (e.g. Fig. 2). Heuristic knowledge processing has been distributed along this object hierarchy in embedded rule-based knowledge sources. Special system objects couple it to the physical instruments (through GPIB or similar busses), and to numerical packages.

With the help of this programming environment and the frequency-domain system identification toolbox (Kollár, 1991), a prototype of a frequency-domain identification expert system (FRIES) has been developed. FRIES resides on a Macintosh computer, which also serves as a controller for the instrument hardware. Basic configuration covers a Wavetek arbitrary wave form signal generator, programmable filters and Nicolet data acquisition unit (Fig. 3). The user is asked for some basic and non-measurable information (e.g. is his system SISO or MIMO). After this the identification session progresses in a fully automated way (the user can volunteer more information, but the system is able to recover it from additional measurements). The expert system

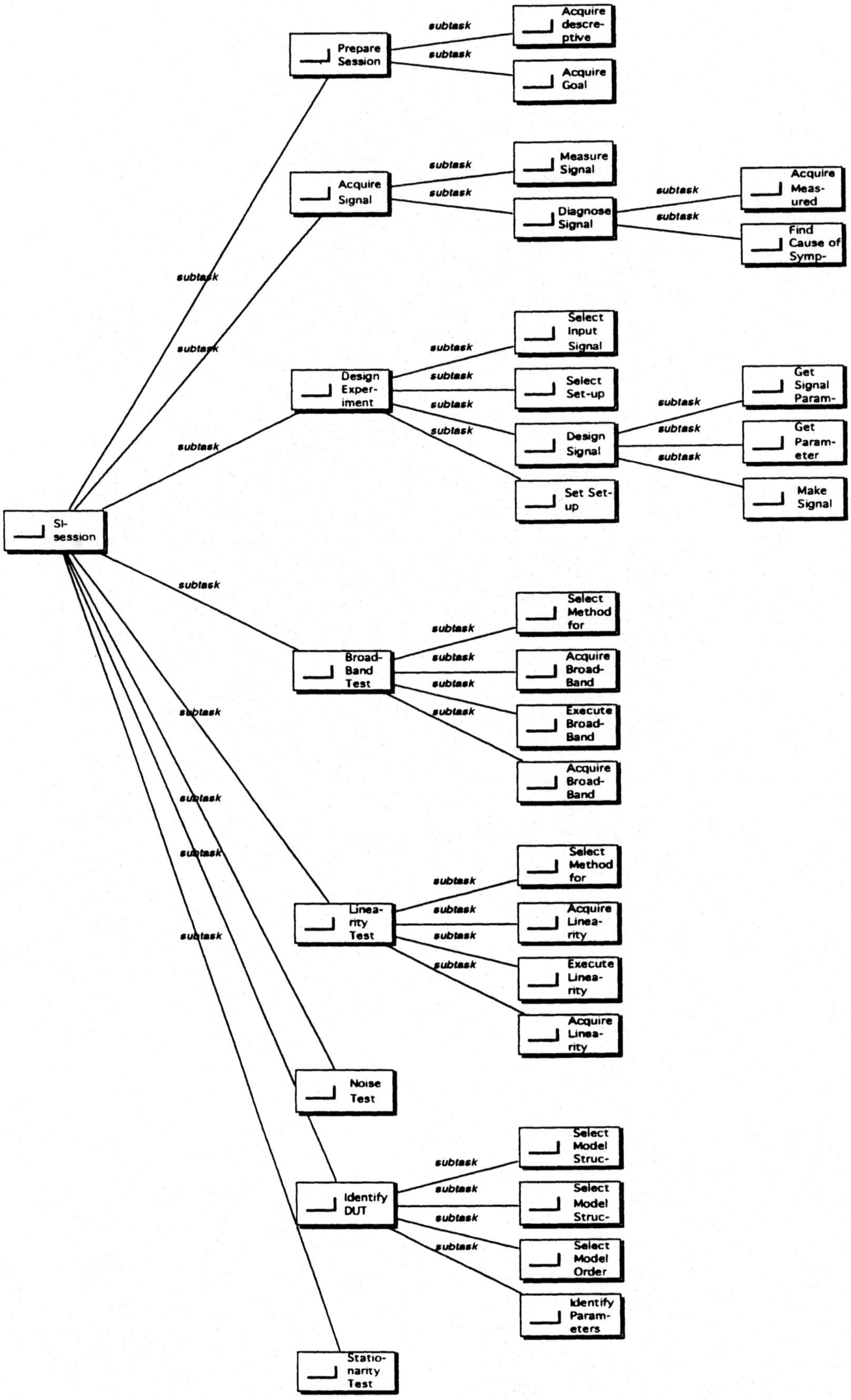

Fig. 1. Task structure of the identification session (prepared with the support of the KREST model-oriented task analysis workbench (KREST, 1993).

evaluates which basic measurements should be done (noise conditions, linearity limits for the model, etc.) and subsequently organises control over instruments and evaluates the measurement data. Upon the incoming information the identification session is expanded with new, more detailed measurements (frequency bands, broad-band view of the system, etc.), until the system can guess a good starting structure. At that point the usual parameter estimation - structure modification iteration begins. For physical systems with simple structure the developed expert system performs at expert level. For more problematic cases the knowledge bases of the expert system are actually extended.

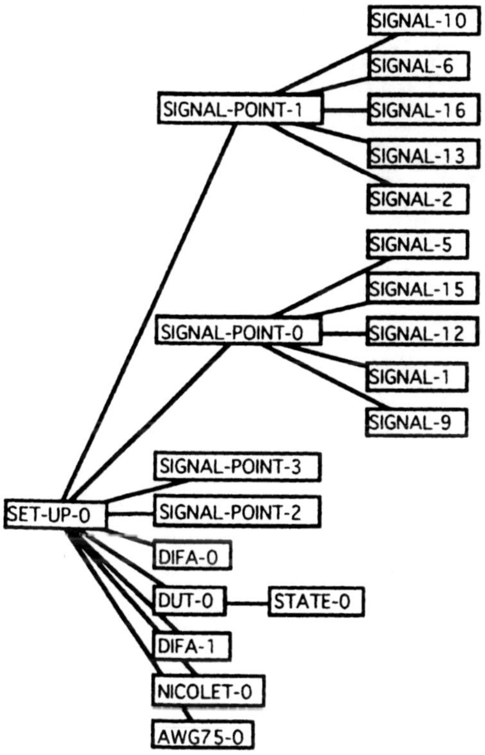

Fig. 2. Internal representation of a set-up

5. CONCLUSIONS

The single foremost conclusion is that AI tools are a natural way to tackle the system identification problems. The uncertainty inherent to the model validating criteria leaves margin where the best structure can be found but with a heuristic search. Although the future research can clarify the issue of choosing the initial structure better, automation of the experiments opens new prospects for AI tools.

To equip computer system with the measurement hardware is however not enough. The issue of the measurement drastically changes the conceptual background of the expert knowledge. The emphasis shifts from the expertise related to identification to the maintenance of experiments, reasoning about signals and measurement instrumentation (sensors, peripherals). In this perspective system identification expert systems, intelligent measurement system, industrial monitoring systems, etc. can be considered as special cases of a more general class of the intelligent modelling system.

From the perspective of the expert system development the consequences are severe. A relatively homogenous and limited body of knowledge (space of structures) gives place to the aggregation of elaborate skills, which diversity brings system identification problem to the very first line of the expert system research. Certain aspects of the discussed problem domain, e.g. construction and diagnosis of the technical artefacts (experimental set-ups), representation of the common sense measurement knowledge, multi-level model-based reasoning, belong to current basic research in knowledge-based systems. In consequence, development of the intelligent system identification tools, pressed by the demands of automation, will necessarily draw from the most advanced of the AI technologies, and thus bears considerable importance for the applied AI research as well.

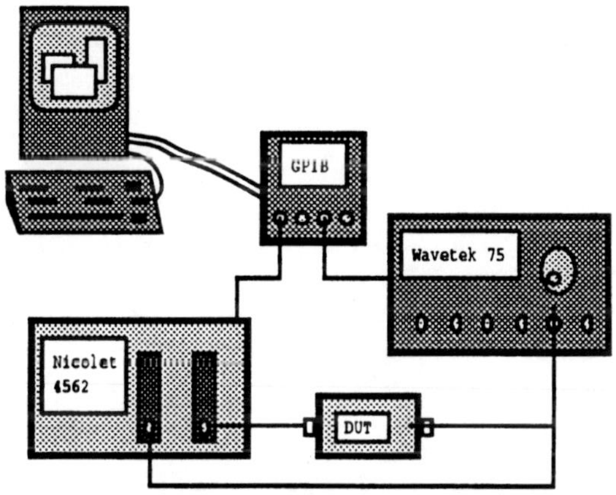

Fig. 3. Experiment setting with FRIES

6. REFERENCES

Bergeon, B. *et al.* (1991). Generic expert system for identification of industrial processes, *9th IFAC/IFORS Symposium*, 8-12 July, 1991, Budapest

Chandrasekaran, B., T.R. Johnson and J.W. Smith (1992). Task-Structure Analysis for Knowledge Modelling, *Communications of the ACM*, **35**, No 9, September, 124-137

Eykhoff , P. (1974). *System identification, Parameter and State Estimation*, Wiley & Sons

Gentile, S., A.Y. Barraud, and K. Szafnicki (1990). SEXI: An Expert Identification Package, *Automatica*, **26**, No 4, 803-809

Haest, M. *et al.* (1990). ESPION: an Expert System for System Identification, *Automatica*, **26**, No. 1, 85-95

Haest, M. *et al.* (1991). A branch-and-bound approach to the identification problem, *9th IFAC/IFORS Symposium on Identification and System Parameter Estimation*, Budapest, 8-12 July, 924-929

Kollár, I. (1991). Frequency domain system identification toolbox, *IFAC/IFORS Symposium on Identification and System Parameter Estimation*, 8-12 July, Budapest, 1591-1595

KREST (1993). KREST Manual 2.5, August 25, 1993, AI Lab, Free University of Brussels.

Larsson, J.E. and P.Persson (1991). An Expert System Interface for an Identification Program, *Automatica*, **27**, No 6, 919-930

Ljung, L. (1987). *System identification. Theory for the user*, Prentice Hall,

Meier zu Farwig, H. and H. Unbehauen (1991). Knowledge-based system identification, *IFAC/IFORS Symposium on Identification and System Parameter Estimation*, 8-12 July, Budapest, 20-28

Nagy, P. (1990). *The ID TOOL EXPERT Documentation*, Report LiTH-ISY-I-1104, Dept of Electrical Eng., Linköping University, 1990-07-30

Niederlinski, A. *et al.* (1991). EFPI - An integrated intelligent environment for system and signal identification, *IFAC/IFORS Symposium on Identification and System Parameter Estimation*, 8-12 July, Budapest, 567-572

Rich, E. and K. Knight (1991). *Artificial intelligence*, Second Edition, McGraw-Hill Book Comp.

Schoukens, J. and R. Pintelon (1991). *Identification of linear systems. A practical guideline to accurate modelling*, Pergamon Press

Steels, L. (1990). The Components of Expertise, *AI Magazine*, Summer

Steels, L. (1991). *Model-oriented task analysis*, March 1991, AI Lab, Free University of Brussels, Construct Project, Workpackage 3

Yen, J. and J. Lee (1993). A Task-Based Methodology for Specifying Expert Systems, *IEEE Expert*, February , 8-16

A FUZZY MONITORING AND DIAGNOSIS PROCESS TO DETECT EVOLUTIONS OF A CAR DRIVER'S BEHAVIOR

M.-A. PELTIER * and B. DUBUISSON **

*ECP/EPAP, Grande Voie des Vignes, 92295 Chatenay Malabry Cedex, France

**URA CNRS 817, Université de Technologie de Compiègne, BP 649, 60206 Compiègne Cedex France

Abstract. This paper describes a monitoring and diagnosis process for the detection of evolution of a system whose characteristics vary with time. This is an adaptive fuzzy pattern recognition algorithm that achieves progressive and on-line learning of the system states. A neural net based architecture associated with a real-time algorithm for the detection of abrupt changes provides a diagnosis of evolution that may also be predictive. The algorithm was applied to the monitoring of a car driver's behavior. Results on real data are reported.

Key words. Fuzzy systems; pattern recognition; neural nets; diagnosis; monitoring; learning systems; time-varying systems; human factors; automobiles

1. INTRODUCTION

Driving a car is a complex task that requires the driver to maintain a high level of performance for a few minutes to several hours. Either a lack of attention or a performance decrement may have severe consequences, even if it lasts a very short period of time. In such a case, the driver may not be able to react on time to any unpredicted event to which an immediate and precise response is needed.

Indeed, it would be very interesting to monitor the driver in order to prevent from undesirable states before it is too late. In other words, the evolution of the driver's behavior is more important to detect than his behavior itself. This can be achieved even if the reason of the evolution is not known.

A car driver monitoring process must also be non-constraining. This implies the driver's behavior to be observed only through his activity on the man-vehicle interface. This approach is reasonable under the hypothesis that any change in the driver status influences his driving activity. It also means that initial and final states cannot be identified as physiological states such as hypovigilance or drowsiness.

Finally, the driver's activity is influenced by a large number of factors that depend on the driver himself (driving experience, physiological status, ...) as well as his environment (traffic density, rain, night, ...). Thus, it should not be easy to derive general rules describing a relationship between physiological states and their consequences on the driving activity. Furthermore, some of the driver's states cannot be observed in real driving conditions for security reasons. To go ahead with this idea, let us say that the only observable state is the "normal" one and it is supposed to be driver specific. Since no prior information is available, the diagnosis system must be able to learn the driver specific characteristics "on the road" and adapt its decision rule accordingly.

This paper describes a complete monitoring and diagnosis procedure that could apply to any process or system for which no prior knowledge is available. This is an adaptive fuzzy pattern recognition process that focuses on the detection of evolution and lies on the following idea:

Suppose that the system is being observed from time $t = 0$. Since there is no prior knowledge available, the first computed patterns are seen as a reference behavior. Then, any pattern falling far from this reference class in the feature space is interpreted as a sign of evolution. It may announce the oncoming of a new state that has to be learnt by the diagnosis process, in order to be recognized later. The implementation of this idea is based on a model of human states as fuzzy classes and involves three main tasks to be achieved, namely the recognition of known states, the detection and learning of emerging states, and the on-line detection of evolution.

Diagnosis and fuzzy pattern recognition are defined in section 2. The interest of fuzzy sets for the diagnosis of time evolving processes is highlighted. Section 3 is dedicated to the description of each step of the diagnosis process. The application of the process to the detection of evolution of a car driver's behavior is given in section 4. Experimental results on real data are reported.

2. FUZZY PATTERN RECOGNITION AND DIAGNOSIS

Let S be a technological system or process, aim of which is to perform a task T. S has at least two operating modes: a normal one and an abnormal one. The normal mode is the desired one, i.e. the operating mode under which the task is fully and correctly performed. The abnormal mode is usually composed of various states, depending on the fault that occured on S: the process may be completely stopped, or the task partially or badly performed. S is observed through a set of sensors that provide relevant information about its status and the way T is performed.

Given these data, the aim of a diagnosis procedure is to determine the operating mode of the system being monitored, in order to detect and identify the fault and possibly make a decision. This involves having prior information on the system, either in the form of a model or an expert and statistical knowledge. In that latter case, this knowledge is generally incomplete so the diagnosis process must be able to learn emerging new states, and adapt its decision rule accordingly.

The aim of pattern recognition is to develop classification algorithms, in order to associate a new observation x with only one among M possible classes (Fukunaga, 1972). An observed state at time t is represented by a d-component vector in the feature space E_d. The problem of pattern recognition is to find a decision rule, i.e. a relationship between the feature space E_d and a finite decision space of known size $\Omega = \{\omega_i\}_{i=1,\ldots,M}$, where ω_i stands for class i.

Thus, making a diagnosis of a system S can be identified as a pattern recognition problem in which the classes are the operating modes of S, and the pattern vector is composed of relevant information provided by the sensors (Dubuisson, 1990a). Each component then corresponds to a parameter which is directly measured on a signal or extracted from it. The main difference between pattern recognition and diagnosis lies in the fact that the former deals with a finite set of classes, when the latter must adapt its decision rule to the emergence of new ones. To do so, a diagnosis process must be able to decide that a given pattern x does not belong to one of the M existing classes, but represents a state which is still unknown. Assuming boundaries of all the classes to be well-known, a distance reject option may be implemented, as illustrated on Fig.1 (Dubuisson, 1990b).

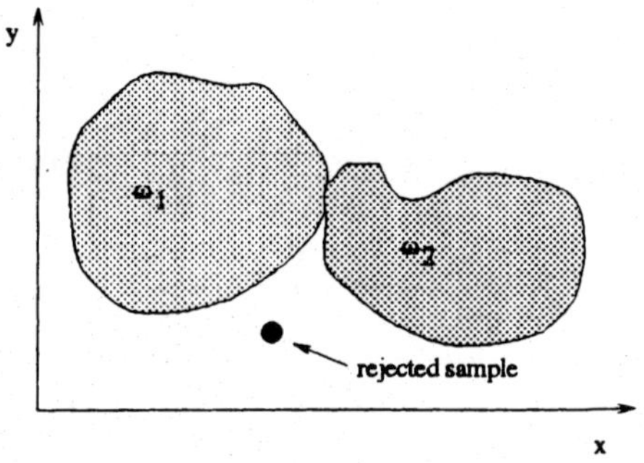

Fig. 1. Example of a distance rejection

Now the problem of the detection of evolution can be re-stated in the context of pattern recognition for diagnosis. Let the driver be the system being monitored, X is a pattern vector composed of parameters that reflect the driver activity on the man-vehicle interface. This vector is sampled at regular time intervals Δt, providing a time sequence of patterns $\{x_{j\Delta t}\}_{j=0,1,\ldots} = \{x_j\}_{j=0,1,\ldots}$ that describes the driver status with time. Detecting an evolution then consists in detecting tracks of the time sequence in the feature space, either between two known classes (case 1), or from a known class to an unknown one (case 2), as illustrated on Fig 2. Hence, following an evolution clearly involves being able to locate patterns with respect to the center of gravity and boundaries of the classes.

Fig. 2. Evolutions in the feature space

This location problem cannot be solved in the context of classical pattern recognition, where classes are described by a characteristic function whose value is 1 for any pattern that belongs to the class, and 0 otherwise. Such an exclusive classification procedure does not provide information about the degree of compatibility of a pattern x to the class ω. Indeed, a pattern located in the center is clearly more representative of the class than a pattern which is close to the boundaries. The fuzzy set theory (Zadeh, 1965) has been introduced in pattern recognition to overcome this limitation (Bezdek, 1981). A fuzzy class is described with a continuous function μ taking value in [0, 1]. This membership function is a generalization of the characteristic function. It takes the value 1 if x is a prototype of the class, 0 if no similarity can be found between x and the prototypes, and an intermediate real value otherwise, depending on a measure of similarity between x and the prototypes. Thus, the membership function value can also be interpreted as a proximity measure of x to the prototypes, allowing to follow the evolution of patterns between or within classes in the feature space. This interpretation is the basis of the diagnosis process described in the next section.

3. DETECTING AN EVOLUTION

Each state of the driver is modeled by a fuzzy class. As mentioned in the introduction section, the diagnosis process involves three main steps to be implemented: recognition of known states, detection and learning of new classes, and on-line detection of evolution.

3. 1. *Recognition of Known States*

Since there is no prior information available on the system states, the shape of the membership function associated with each class is unknown and has to be approximated. A method based on a particular architecture of neural networks has been proposed to compute the values of the membership functions (Peltier and Dubuisson, 1993a). As many networks are implemented as there are classes, each of them being associated with one class. The so-called Gaussian Potential Function Networks (GPFN) (Lee and Kil, 1991) are trained upon a learning set composed of:

- a core made of at least n_{min} prototypes for which the desired output value is 1,
- a non-limited number of patterns that are close enough to the prototypes to take part of the learning set, but which cannot be considered as prototypes. The desired output for them lies in [0.8, 1[, depending on a similarity index to the prototypes (Peltier and Dubuisson, 1993b).

Once a network is trained, any input pattern x gives an output value in [0, 1]. It is 1 if x falls within the learning set, and decreases non-linearly as its distance to the prototypes increases.

Assuming that a reference class has been built that describes the initial state of the driver, the GPFN provide a way of computing the membership function of a pattern x to this first class. The problem now is to detect evolutions out from this reference and identify the oncoming new state(s). In other words, the problem is to build the learning sets that may give birth to new classes.

3. 2. *Detecting and Learning New Classes*

x is rejected in distance from the reference class if it is too far from its prototypes, i.e. if its membership function value to this class is under a given threshold. Generalizing this condition to more than one class leads to the following rejection rule:

$$Reject\ x\ if\ max_{i=1,\dots,M}\mu_i(x) < T_r \qquad (1)$$

where M is the number of known classes and $\mu_i(x)$ is the membership value of pattern x to the class ω_i. The value of T_r is independant from the class.

New learning sets have to be created as soon as an evolution is detected, enabling successive evolutions occuring on a small time interval to be taken into account. The idea is to build them by on-line aggregation of rejected patterns. Its implementation requires the definition of two "on-line" decision rules. The first one is an aggregation criterion. It consists in deciding if the state representation of a pattern x_j rejected at time $j\Delta t$ is the same as that of previously rejected samples. The second one is a stopping rule, it is used to initiate the creation of a new class and the training step of a new network. This rule involves the definition of a minimum size for the learning sets, which corresponds to the shortest possible duration of the system states.

A class is assumed to represent a human state. So its learning set should be composed of patterns having the same meaning in terms of human behavior, even if this behavior cannot be identified. This suggests to build our criterion on a distance measure between representatives of the same state. Furthermore, a given state is supposed to exist for a minimum amount of time (a few seconds to a few minutes) If this state is not stable, we are expecting an evolution. So any observable new behavior is supposed to be described by patterns that are also close in time. Space and time are combined into a single criterion through the concept of *k-time Nearest-Neighbors* (Peltier and Dubuisson, 1993b).

The aggregation process does not stop as soon as the minimum size has been reached. It adapts to the space and time structure of the data, allowing the process to go on as long as the observed state remains stable.

The creation of the reference class follows the same aggregation process as the creation of the others. This process is initiated upon the first computed sample. Then, since no class has been learnt yet, the rejection rule is defined with respect to the groups of previously aggregated patterns.

The neural-based architecture associated with the aggregation process allows the detection and learning of new states as the driver status changes. Thus, assuming that M classes $\{\omega_i\}_{i=1,M}$ have been detected, the output $\{\mu_i(j\Delta t)\}_{j>=0}$ of each network $i = 1, M$ describes the behavior of the system with respect to the class ω_i as time goes on. $\{\mu_i(j)\}_{j>=0}$ is called the *membership signal* to the class ω_i. As illustrated on Fig. 3, it is a one-dimensional time signal composed of stable phases interrupted by abrupt changes. Oscillations of the signal around its mean value reflect internal moves of the system within the class. A decreasing value indicates that the system is leaving state i, when an increase reflects a movement towards the corresponding class. An evolution from state i to state j introduces a decrease of μ_i, possibly associated with an increase in μ_j if ω_j is a known class. Following the evolution of the system then consists in detecting and interpreting abrupt changes of the M signals $\{\mu_i\}_{i=1,M}$.

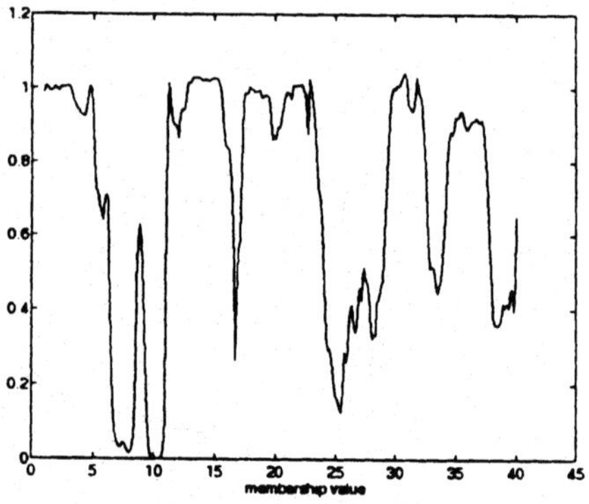

Fig. 3. Example of a membership signal

3. 3. *On-line Detection of Evolution*

The CUSUM algorithm is a signal processing method that detects abrupt changes of a one-dimensional time signal by detection of a change in its statistic properties. Let $\{x_j = e_j + n_j\}_{j\geq 0}$ be a discrete step edge $\{e_j\}$ with additive white gaussian noise $\{n_j\}$ with probability distribution $\mathcal{N}(0, \sigma^2)$. $\{x_j\}$ is a stochastic signal, it follows a gaussian distribution $\mathcal{N}(\theta, \sigma^2)$ where $\theta = \theta_0$ before the step edge and $\theta = \theta_1$ after it. Thus, detecting the abrupt change is equivalent to performing a two-hypothesis test:

$$\begin{cases} H_0: & \theta = \theta_0 \\ \text{vs} & \\ H_1: & \theta = \theta_1 \end{cases} \tag{2}$$

A decision function $g(j\Delta t) = g_j$ is computed at each time instant $j\Delta t, j \geq 0$:

$$\begin{cases} g_0 = 0 \\ g_j = sup\ (0, g_{j-1} + x_j - \theta_0 - \frac{\nu}{2}) \end{cases} \tag{3}$$

Its value remains 0 as long as $\theta = \theta_0$ and increases as soon as H_1 is valid, as shown on Fig. 4. A threshold $h > 0$ on the value of g defines an alarm time $t_a = min_{j>0}\{j : g_j \geq h\}$ that warns of a detection. This algorithm assumes θ_0, θ_1 and σ to be known or easy to estimate.

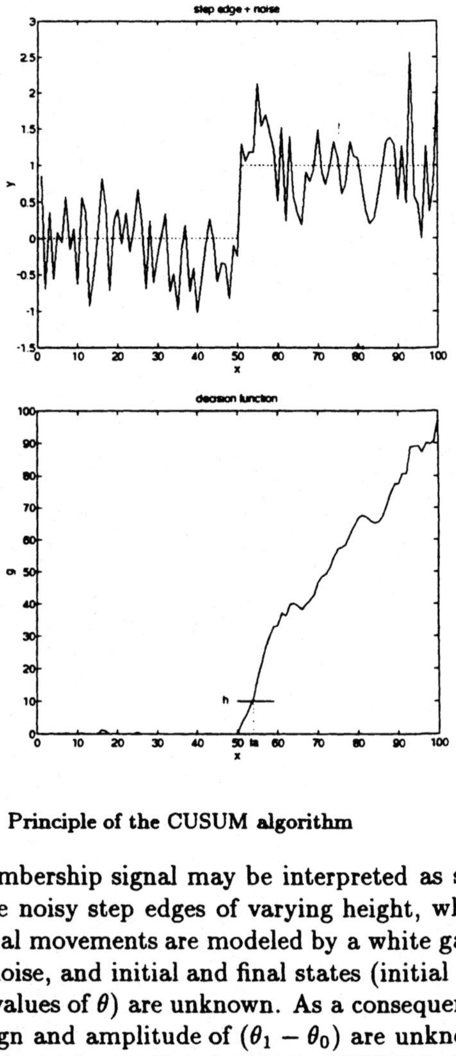

Fig. 4. Principle of the CUSUM algorithm

A membership signal may be interpreted as successive noisy step edges of varying height, where internal movements are modeled by a white gaussian noise, and initial and final states (initial and final values of θ) are unknown. As a consequence, the sign and amplitude of $(\theta_1 - \theta_0)$ are unknown and cannot be predicted. So the CUSUM algorithm may be adapted in the following way:

- estimate σ a priori as the mean amplitude of internal movements within classes;
- set $\nu = |\theta_1 - \theta_0|$ as the minimum step height that you wish to detect;
- use two decision functions g_j^+ and g_j^- dedicated to the detection of rising and falling edges respectively:

$$g_j^+ = sup(0, g_{j-1}^+ + x_j - \theta_0 - \frac{\nu}{2}) \quad (4)$$

$$g_j^- = sup(0, g_{j-1}^- - x_j + \theta_0 - \frac{\nu}{2}) \quad (5)$$

Both functions are computed simultaneously. A step edge is detected as soon as:

$$max(g_j^+, g_j^-) > h \quad (6)$$

The sign of the variation of the function is immediately deduced.

- the initial value θ_0 of θ before a step edge is the final value θ_1 of θ after the previous one. Thus, only an estimate of θ_0 is needed, it is computed after each detected step edge as the mean value of the membership signal measured just after the detection of the edge.

The modified CUSUM algorithm is given below. Let $\{\mu_j\}_{j\geq 0}$ be a membership signal:

1. initialize $\theta_0 = \mu_0$, $k = 1$, $S_0 = \mu_0$,
2. for $j \geq 0$,
   - update the estimate of θ_0: $S_0 = S_0 + \mu_j, k = k + 1, \theta_0 = S_0/k$.
   - compute g_j^+ et g_j^- according to Eq. (4) et (5).
   - if $max(g_j^+, g_j^-) > h$: an edge is detected. Its sign is given by the decision functions.
   - re-initialize θ_0: $k = 0$, $S_0 = 0$.

ν is estimated so as to tune the time instant t_a of the detection. The smaller the value of ν, the earlier the detection of the step edge. Since step edges should not be confused with noise, this value should be chosen as a trade off between early detection of evolutions and false alarms. Figure 5 illustrates the effect of the value of ν on t_a, indicated with "*" on the plots.

The algorithm may thus be predictive, allowing a change in the driver status to be detected as soon as it induces a variation of a membership signal. But this prediction capability is clearly limited to a few time intervals, since an evolution of the driver's behavior may not be detected before it has influenced the driving activity. Anyway, a smaller value of ν leads to an earlier detection of evolutions and even to multiple detections of steps of large amplitude. This is an interesting property, for it reinforces the diagnosis of evolution and increases the robustness of the method.

A complete algorithm is now available. It was applied on real data, as shown in the next session.

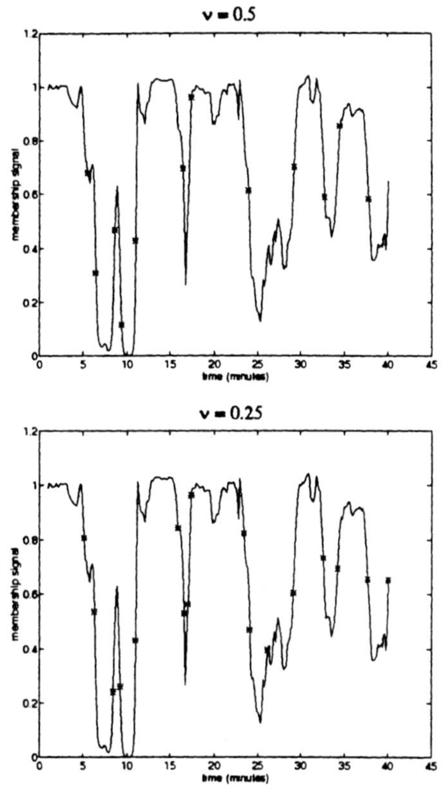

Fig. 5. Influence of ν on t_a

4. APPLICATION

The method was applied to detect evolutions of a car driver's behavior. Experiments were conducted on real road with various drivers. A car was equipped with a set of sensors connected to a data acquisition system, providing information about the driver's behavior on the man-vehicle interface. People were asked to drive for several hours on a motorway, so as to minimize the influence of the environment on the driver's behavior and make sure that changes in his status will be experienced by the subject. The data were digitalized and recorded on an on-board computer for later processing.

An example of a result obtained with one of the drivers during the first 40 minutes of the experiment is presented below. The behavior is described through a pattern vector of dimension 7 updated every 8 seconds. The resulting sequence is presented as inputs to the algorithm. Figure 6 shows the evolution of the network outputs with time. Three classes were detected and learnt. The date of creation of the associated networks is pointed out on the plots. Note that patterns that are presented before the creation of a network should not have a membership value to the associated class. The signals drawn on the plots have been re-calculated a posteriori at the end of the experimental session.

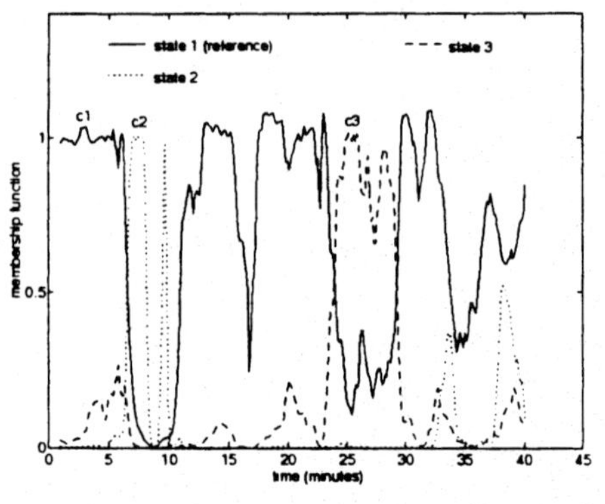

Fig. 6. Example of result with three detected classes

This plot first shows that the algorithm is able to detect and learn the driver's states as his status changes. The existence of a reference state is confirmed, it is observed at the beginning of the monitoring session and the driver returns to it as soon as it has left the other classes. This phenomenon can also be observed later on in the experiment. Note that membership signals are quite symmetrical, providing a way of following the driver's status with respect to the known classes. A move towards an unknown class leads to its detection and learning.

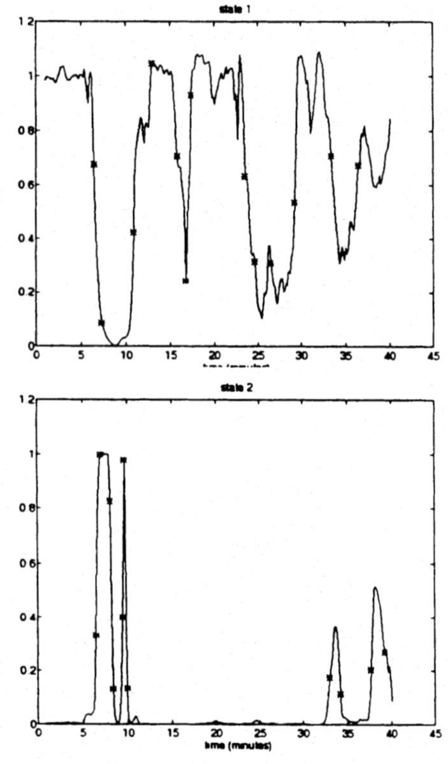

Fig. 7. Step edge detection on two simultaneous membership signals

An on-line detection of the evolutions is obtained with the filtering algorithm described in section 3.3. Figure 7 shows the time instants "*" at which edges are detected for the first two membership signals filtered simultaneously. All the relevant steps have been detected. Furthermore, a simultaneous interpretation on both signals allows to identify the nature and direction of the evolutions. Thus, an interpretation procedure is now needed, in order to automatically define the nature of the evolution of the system (evolution between two known classes, identification of initial and final states, detection of a new class, stabilization, ...) and possibly make a decision.

5. CONCLUSION

This paper describes a complete monitoring and diagnosis process that achieves a progressive non-supervised learning of the system characteristics. This process was applied to the detection of evolution of a car driver's behavior. It also applies to any system or process for which no prior knowledge is available.

6. REFERENCES

Bezdek, J.C. (1981). *Pattern recognition with fuzzy objective function algorithms.* Plenum Press, New York.

Dubuisson, B. (1990a) *Diagnostic et reconnaissance des formes*, Hermès.

Dubuisson, B. (1990b). Decision with reject option. In: *Proceedings Eusipco.* Barcelone.

Fukunaga, K. (1972). *Introduction to statistical pattern recognition.* Academic Press, New York.

Lee, S., and R.M. Kil. (1991). A gaussian potential function network with hierarchically self-organizing learning. *Neural Networks.* Vol. 4, pp. 207-224.

Peltier, M.A., and B. Dubuisson. (1993a). A human state detection system based on a fuzzy approach. *Proceedings Tooldiag'93, International Conference on Fault Diagnosis.* Vol.2, pp 645-652.

Peltier, M.A., and B. Dubuisson. (1993b). A fuzzy clustering algorithm based on the k-nearest neighbors'rule for the detection of evolution. *Proceedings IEEE-SMC'93.*

Zadeh, L.A. (1965). Fuzzy Sets. *Information and Control*, vol. 8, pp. 338-353.

A SIMPLE IDENTIFICATION METHOD FOR THE ORDER OF THE STREJC MODEL AND ITS APPLICATION TO AUTOTUNING

S. SKOCZOWSKI, A. OSADOWSKI

Technical University of Szczecin, Institute of Control Engineering
Sikorskiego 37, 70-313 Szczecin, Poland

Abstract. An approximate method of parameter identification for the Strejc linear model is presented. The order and consequently the remaining parameters of the model are determined from the process step response initial stage approximated by a parabola. The method applicability to the autotuning of PID controllers is demonstrated by means of simulation techniques. Some test results of identification of selected electroheated control plants are presented.

Key Words. Identification; modelling, PID control; self-tuning regulators

1. INTRODUCTION

It is supposed that the real control plant dynamics may be represented approximately, after linearization, by a stable linear model given by the following transfer function

$$K(s) = \frac{b_m s^m + \ldots + b_1 s + b_0}{s^n + a_{n-1}s^{n-1} + \ldots + a_1 s + a_0} e^{-s\tau}, n > m$$

where τ is the known time delay. (1)

The so-called mean time constant (Xu,1989) is associated with the model (1), and is given by

$$\bar{T} = \frac{n}{a_{n-1}} \tag{2}$$

with

$$a_{n-1} = -\sum_{j=1}^{n} p_j \tag{3}$$

where $p_j (j = 1,2,\ldots n)$ are poles of the transfer function (1).

Of special interest are aperiodic processes which are devoid of the differential component. They represent a wide class of real control plants described by a multi-time-lag model (Skoczowski,1984;Nishikava *et al.*,1984)

$$K_w(s) = \frac{b_0 e^{-s\tau}}{\prod_{i=1}^{n}(s + \frac{1}{T_i})} = \frac{k e^{-s\tau}}{\prod_{i=1}^{n}(1 + sT_i)} \tag{4}$$

for which the mean time constant, according to (1), is determined by

$$T_{min} < \bar{T} = \frac{n}{\sum_{i=1}^{n} \frac{1}{T_i}} < T_{max} \tag{5}$$

where

$$T_{max} = \max\left[-\frac{1}{\delta_1}, \ldots, -\frac{1}{\delta_n}\right]$$

$$T_{min} = \min\left[-\frac{1}{\delta_1}, \ldots, -\frac{1}{\delta_n}\right] \tag{6}$$

$$\delta_j = Re[p_j]$$

The smaller is the dispersion of time constants, the more the model (4) approaches the time-lag bearing Strejc model

$$K_s(s) = \frac{b_{0_M} e^{-s\tau}}{(s + \frac{1}{T_M})^{n_M}} = \frac{k_M e^{-s\tau}}{(1 + sT_M)^{n_M}} \tag{7}$$

Despite its simplicity (only 4 parameters: gain K_M, time constant T_M, order n_M, and, if necessary, time delay τ) the model (7) takes account remarkably well of the multitude of time constants, unlike the simple and widely used Küpfmüller model being a special case of (7) when n=1. For the Strejc model (7) we have

$$a_{n-1} = \sum_{i=1}^{n} \frac{1}{T_i} = \frac{n}{\bar{T}} = \frac{n}{T_M} \tag{8}$$

The simplicity of both models is essential for a simplified identification for practical purposes, as, for autotuning and self-adaptivity of industrial controllers. Some techniques for parameter estimating of both models from the complete step response h(t) are described in the literature of the subject. For the Strejc model, particularly, tables are given (Strejc,1959), from which the model parametres can be determined, provided the complete step response of the plant to be modelled is known.

It might be well to point out that the approximate method for estimating of the model order, hence, of the rest of the Strejc model parameters, described in (Skoczowski,1977) and then in (Skoczowski, 1983; Skoczowski 1984):

$$n_M \approx 1 + 2\pi\left(\frac{t_p}{T_a}\right)^2, \quad \bar{T} \approx \frac{t_p}{n-1}, k = h(\infty) \tag{9}$$

where

t_p is the time within the space of which the step response reaches its inflexion point and T_a is the inflexional subtangent to h(t) (see Fig. 4), is much more practical in use than that of (Strejc, 1959).

In (Xu, 1989) a method for estimating of $\bar{T}$ and $n_M = n - m$ for the transfer function (1), is described.

In this paper a new practical method for order evaluation of the models (4) and (7) based on discrete samples of the initial portion of the plant step response is presented (Skoczowski, 1992a).

In the case that the process under study is represented by the transfer function (1) the proposed method allows us to evaluate the order $n_M = n - m$. A quick information about the order value may be very helpful. Knowing the process order one can estimate the remaining process (7) parameters, which may be utilized for autotuning purposes.

A limitation of the method is that it is a deterministic one, therefore it may be susceptable to random disturbances caused, for instance, by a limited resolution of A/D converters. However, this limitation may be greatly reduced, as will be shown later.

2. EVALUATING THE ORDER AND THE REMAINING PARAMETERS OF THE MODEL (7).

The aim of the study is to evaluate the reduced order $n_M = n - m$ of the model during the start-up phase of the plant unit step response. Due to the long-division method applied to polynomials in the transfer function (1), the unit step response of the system can be written as follows:

$$h(t) = b_m \frac{(t-\tau)^{n-m}}{(n-m)!} + \left(b_{m-1} - b_m a_{n-1}\right)\frac{(t-\tau)^{n-m+1}}{(n-m+1)!} + \left(b_{m-2} - b_{m-1}a_{n-1} + b_m\left(a_{n-1}^2 - a_{n-2}\right)\right)\frac{(t-\tau)^{n-m+2}}{(n-m+2)!}\cdots \tag{10}$$

with h(t)=0 for $0 \le t \le \tau_M$.

For the simplified model (7) we get:

$$h(t) = b_{0_M} \frac{(t-\tau)^{n_M}}{n_M!} - b_{0_M} a_{n_M-1} \frac{(t-\tau)^{n_M+1}}{(n_M+1)!} + b_{0_M}\left(a_{n_M-1}^2 - a_{n_M-2}\right)\frac{(t-\tau)^{n_M+2}}{(n_M+2)} + \cdots \tag{11}$$

It can be seen that most of the information about the model order is carried by the first terms of (11). This implies that the initial slope of the step response curve can be approximated by a reduced order n_M parabola (see Fig.1).

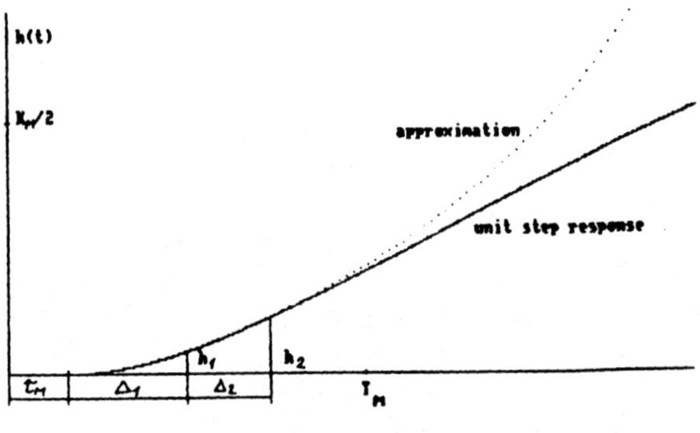

Fig. 1. Parabolic approximation by three terms of (11) of the plant step response for the model with n_M= 3

In the particular case of the Strejc model (7) we have

$$a_{n_{M-1}} = \frac{n_M}{T_M}$$

$$a_{n_{M-2}} = \frac{n_M(n_M - 1)}{2T_M^2} \tag{12}$$

Taking into account the first three terms of the expansion (11) and combining them with (12) one can calculate approximate values of the two first discrete samples of the step response h(t) for $t = \Delta$ and $t = 2\Delta$, where Δ is the sampling period. After simple manipulations, the samples quotient h_1/h_2 becomes equal to

$$\frac{h_2}{h_1} \approx 2^{n_M} \varphi\left(n_M, \frac{\Delta}{T_M}\right) \tag{13}$$

where

$$\varphi\left(n_M, \frac{\Delta}{T_M}\right) \approx \frac{1 - \frac{2n_M}{n_M+1}\left(\frac{\Delta}{T_M}\right) + \frac{2n_M}{n_M+2}\left(\frac{\Delta}{T_M}\right)^2}{1 - \frac{n_M}{n+1}\left(\frac{\Delta}{T_M}\right) + \frac{n_M}{2(n_M+2)}\left(\frac{\Delta}{T_M}\right)^2} \tag{14}$$

Taking into account the fact that values of $\varphi(\cdot)$ approximately equal 1 for small Δ/T_M and n_M, as well as that in practice we assume the order n_M to be integer, the relation (6) provides a simple expression determining the order of the model (7)

$$n_M \approx \frac{\ln \frac{h_2}{h_1}}{\ln 2} \tag{15}$$

Table 1. Values of φ (Eq.14) as a function of Δ/T for various model orders n_M.

$\Delta/\overline{T}$	n_M	φ
0,1	1	0,952714536
	2	0,936776491
	3	0,928879310
0,2	1	0,911764706
	2	0,882129278
	3	0,86774920
0,3	1	0,878612717
	2	0,838905775
	3	0,820448878
0,4	1	0,854838710
	2	0,810344828
	3	0,791443850
0,5	1	0,842105263
	2	0,800000000
	3	0,785714286

The character of the function $\varphi(\cdot)$ implies, as a rule, rounding the order up to the nearest integer. This is exhibited in Table 1 (Skoczowski, 1992a).

If we know the order n_M we can easily estimate the remaining parameters of the model (7) at the very beginning of the unit step response.

3. SIMULATION OF THE AUTOTUNING PROCEDURE OF A PID CONTROLLER

The main difficulties in practical use of the method are caused by the existence of disturbances, lack of knowledge of τ, limited resolution of A/D converters, non-linearity of the plant, dispersion of time constants etc. The deterministic identification of the model order n_M so effected may be of practical use, even though the measured values of h_1 and h_2 are subject to disturbances, because the result of Eq.(5) requires to be rounded up to the nearest integer.

In order to decrease the noise susceptability one can raise the h_1 and h_2 , as shown in (Osadowski, 1993), or use the averaged values of h_1 and h_2 stemming from a series of measurements (Skoczowski, 1992b).

$$T_M \approx \frac{\Delta n_M}{n_M+1} \cdot \frac{1}{1 - 2^{n_M} \frac{h_1}{h_2}} \tag{16}$$

$$K_M \approx \frac{h_1^2}{h_2}\left(\frac{2}{\Delta}\right)^{n_M} n_M!\, T_M^{n_M} \tag{17}$$

Since these estmates are susceptible to disturbances, an optimization procedure has been adopted in (Osadowski, 1993), where the values obtained from (16) and (17) are regarded as starting ones only.
The starting values T_M and K_M determined by (16) and (17) can be derived after averaging a series of calculations obtained from a series of measurements during which the sampling rate should be increased at the identification stage, as was proposed in (Skoczowski, 1992b).
Figure 2 depicts the effects of series averaging and optimization in the presence of disturbances. The disturbances are visible in the initial phase of the step response, shown on an exaggerated scale, during which the identification occurs (Osadowski,1993).

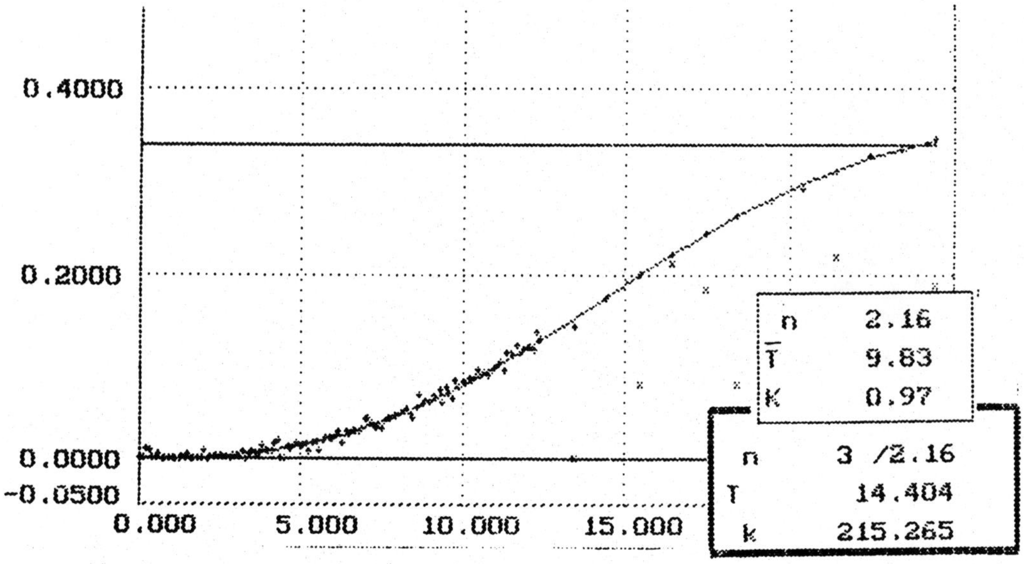

Fig. 2. Simulation of the identification in the presence of random disturbances for a plant with n=3, T_1=12, T_2=10, T_3=8, k=1

By minimizing the square deviation for a pair of samples or for a series of pairs of samples a much better accuracy of identification of T_M and K_M has been achieved. This allows the PID controller to be adequately tuned. An example of simulation of a self tuning controller based on this identification method in the presence of the mentioned factors is shown.

Typical work of such a controller is illustrated in Fig.3, where *a* is the controlled variable and *b* is the controlling action for a plant with n_M=3, T_M=10, K_M=1, hysteresis and random disturbances.

The first stage, when the controlling action is constant, is that of identificatcion and self tuning and the second stage is that of control.

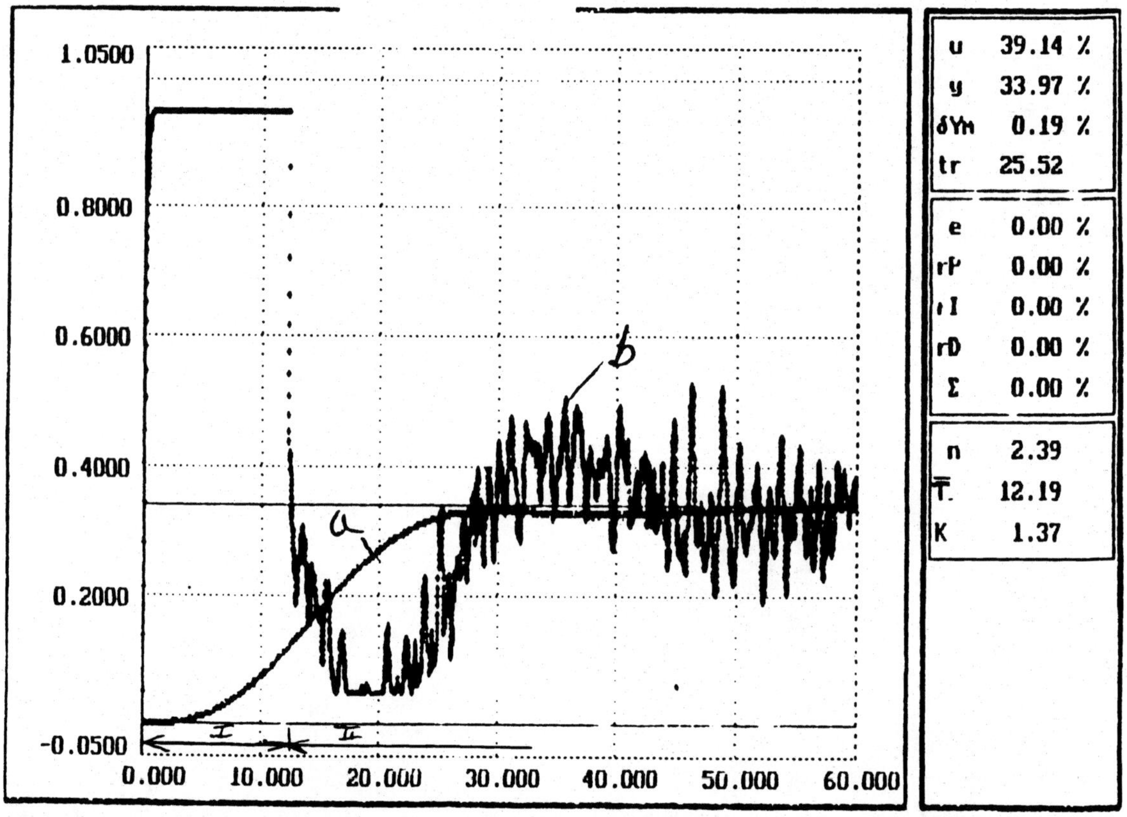

Fig. 3. The controlled variable and the controlling action at the self tuning and regulation stage in the presence of random disturbances, obtained as results of control (Osadowski, 1993).

4. MODIFICATION OF THE DESCRIBED METHOD AND EXPERIMENTAL TESTS.

Another technique to determine n_M and K_M utilizes the results obtained in (Skoczowski, 1977). The principle consists in memorizing of increments $\Delta h/\Delta=\alpha$ until the inflexion point of the step response h(t) is found (see Fig.4). In what follows the values of t_p, $\alpha_{max}= K_M/T_a$ (see Fig.4) and n_M (providing $n_M>1$) are essential to the approximate determining of T_M and K_M (on the strength of Eq.(9), given the model order n_M):

$$T_M \approx \frac{t_p}{n_M - 1} \tag{18}$$

$$K_M \approx \alpha_{max} t_p \sqrt{\frac{2\pi}{n_M - 1}} \tag{19}$$

This technique has not been tested in (Osadowski, 1993) and is mentioned here in order that one could possibly utilize it. A limitation of the method is that it keep us waiting till the instant t_p when the inflexion of the step response h(t) occurs.

The model (7) is most adequate for control plants with selfregulation, as in this case only $b_0(m=0)$ occurs in the trasfer function numerator, and then $n_M=n$. Such a plant has been simulated in (Osadowski, 1993). Tests were conducted using the PC-card PCL 718 enabling the step response samples to be gathered (along with a 12-bit A/D converter) and laboratory electroheated plants, as well (Śmigielski, 1993).

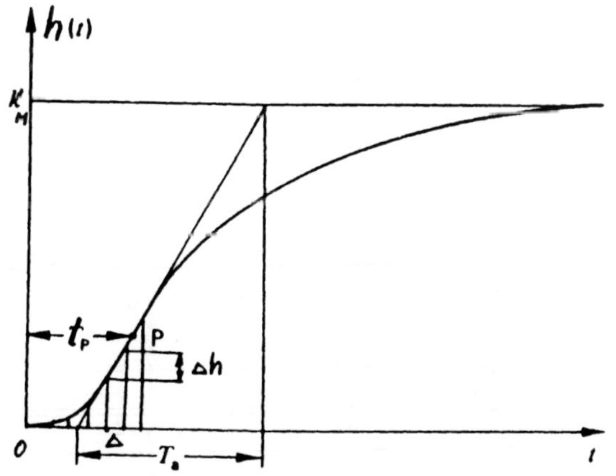

Fig.4. Sampling of the initial stage of the plant step response to determine α_{max}

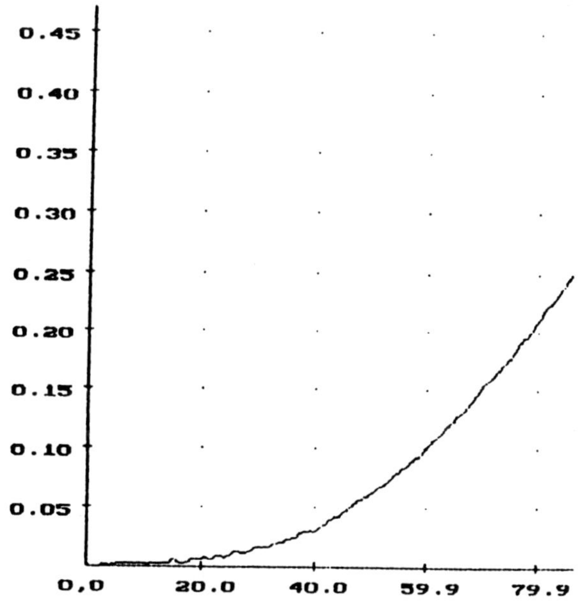

Fig. 5a The initial portion of the step response measured for the plant A.

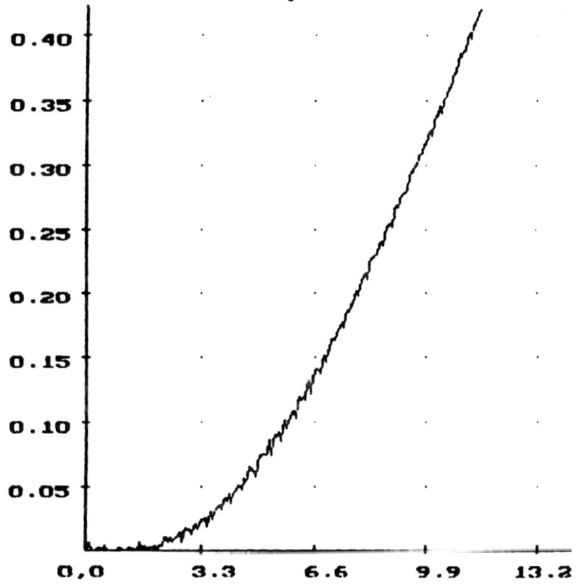

Fig. 6a. The initial portion of the step response measured for the plant B.

For example, in Fig. 5a the initial stage of a plant step response is depicted. The complete step response for the same plant is shown in Fig. 5b for comparison. The real step response is labelled 1, the step response for the model (7) with parameters obtained after optimization is denoted 2, and 3 is used for the step response of the model (7) with parameters estimated after averaging the series.

Similarly, in Fig. 6a and 6b the results for the plant B are displayed. Figures 7 and 8 applies to the same plants, with the difference that the inflexion point method is adopted. The curve 1 is the real step response, and the curve 2 is the step response for the model being identified.

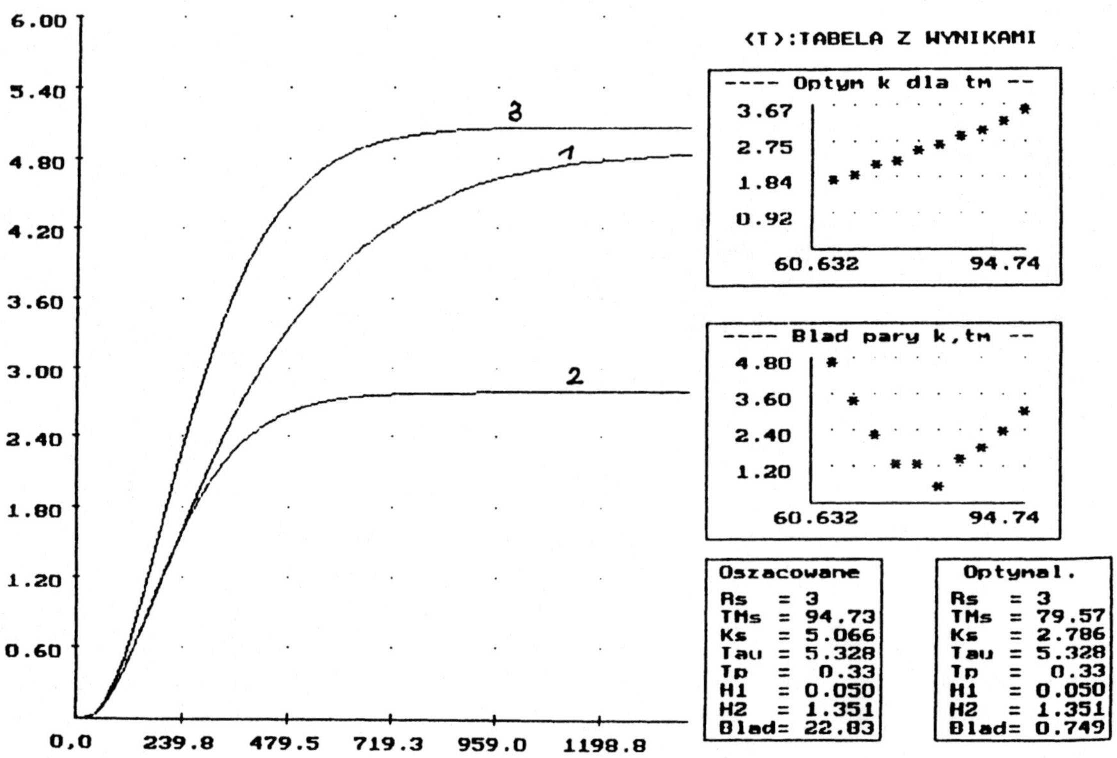

Fig. 5b. Results of the identification of the plant A carried out using the described method.

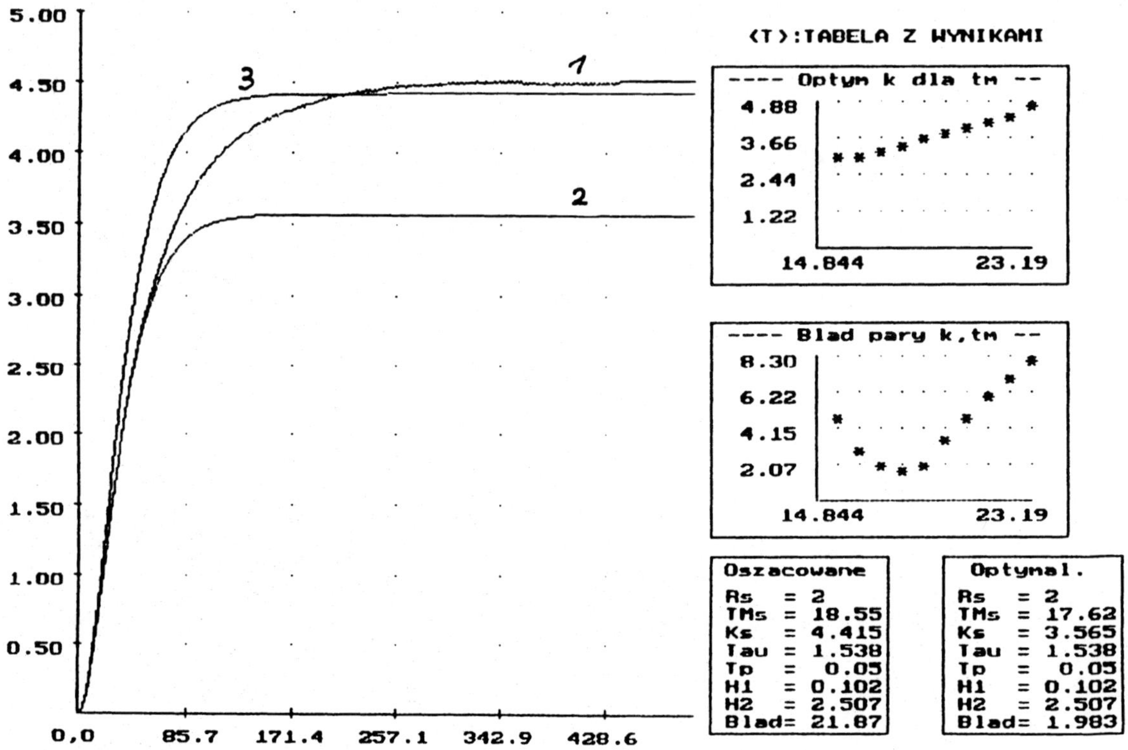

Fig. 6b. Results of the identification of the plant B carried out using the described method.

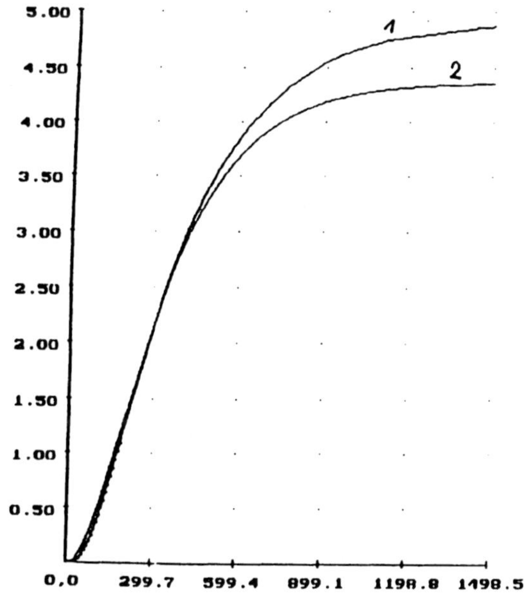

Fig.7. Results of the identification of the plant A carried out using the modified method.

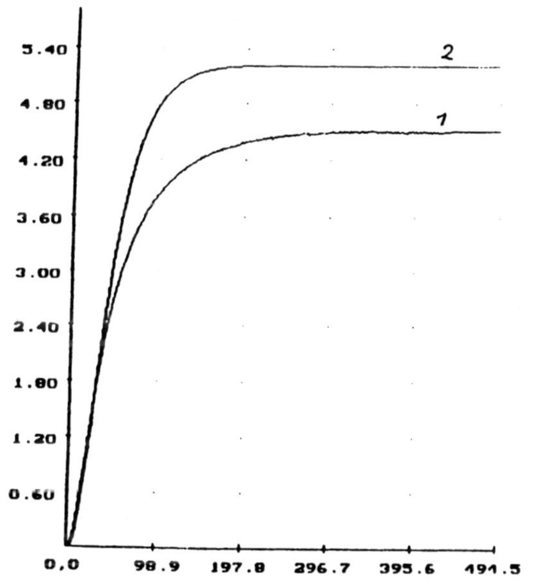

Fig. 8. Results of the identification of the plant B carried out using the modified method.

5. CONCLUSIONS.

As stated above, the described method is not free from some limitations. Among them one should number the fact that the method is in principle a deterministic one and that it is based on the assumption of steady state being in existence for initial conditions of the identification stage. Even with such limitations the proposed method may find practical use, particularly when disturbances and measurement noises are relatively small. The most obvious advantage of the method is that it permits easy and quick evaluation of the model order. Having an arbitrary taken process model and knowing its order the remaining model parameters can be determined simultaneously. This information may be utilized for autotuning of the controller.

Such a control may yield satisfactory results, even though the identification performed is not quite fine, provided the PID algorithm, noted for that it does not require the exact knowledge of the process model, is applied.

6. REFERENCES

Nishikava, Y., N. Saunomiya, T. Ohta and H. Tanaka (1984). A method for autotuning of PID control parameters. *Automatica (IFAC)*, **20**, 321-332

Osadowski, A. (1993). *A selftuning controller adopting a plant identification technique based on first samples of the step response.* (In Polish). M. Sc. thesis. Institute of Control Engineering, Technical University of Szczecin, Szczecin, Poland.

Skoczowski, S. (1977). *Two-position tempera-ture control.* (In Polish). WNT, Warszawa.

Skoczowski, S. (1983). Einige Bemerkungen zur Approximierung von Regelstrecken mit Ausgleich. *Regelungstechnik*, **31**, 231-234.

Skoczowski, S. (1984). Einfache Modelle der Temperaturregelstrecken und ihr Brauchbarkeitsmass. In: *Proc. 10th UIE Congress, Stockholm, Sweden.*

Skoczowski, S. (1992a). A simple method for identification of the Strejc model based on first discrete samples of step responses. (In Polish). *Kwartalnik Elektroniki i Telekomunikacji*, **38**, 293-303.

Skoczowski, S. (1992b). A simple method for identification of the Strejc model. (In Polish). In: *Proc. 1st Symposium on Simulation, Measurement and Diagnostics in Electroheating* (Technical University of Białystok, Białystok, Poland), 143-152.

Strejc, V. (1959). Näherungsverfahren für aperiodische Übergangscharakteristiken. *Regelungstechnik*, 7, 124-128.

Śmigielski, S. (1993). *Experimental study of a simplified identification method for the Strejc model using selected electroheated control plants.* (In Polish). M. Sc. thesis. Institute of Control Enginering, Technical University of Szczecin, Szczecin, Poland.

Xu, N.-S. (1989). Approximate evaluation of the sampled model parameters for a linear continuous-time system with relatively small sampling period. *Ins. J. System Sci.*, **20**, 811-838.

NONLINEAR OBSERVERS FOR DISTILLATION COLUMNS

A. RIX[*] , C. SKRCZYPCZEK[*] and H. GELBE[*]

[*]*Institute of Process and Plant Design, Technical University of Berlin, Federal Republic of Germany*

Abstract. A nonlinear observer, capable of accurate on-line-calculation of the product and feed composition of a binary distillation column is presented. The observer is based on a nonlinear reduced-order process model. In simulation studies the observer shows better performance, better robustness and easier adaption to real plant dynamics than an observer of conventional design. Experiments prove the high accuracy and flexibilty of the new observer under real-life conditions. The observer was used to test advanced feedforward/feedback control systems for distillation columns in experiments.

Key Words. Nonlinear observer; distillation control; model oder reduction; on-line simulation; feedforward control

1. INTRODUCTION

Distillation is the most important separation process in chemical industry. To reduce the energy requirements of the process, all product concentrations must be controlled (Luyben, 1975). In practice, this will seldom be possible, because composition measurements are normally not available.

Therefore, on most distillation columns, temperatures of selected trays are controlled as a simple alternative. This may lead to unsatisfactory control performance, especially when disturbances in the feed composition occur.

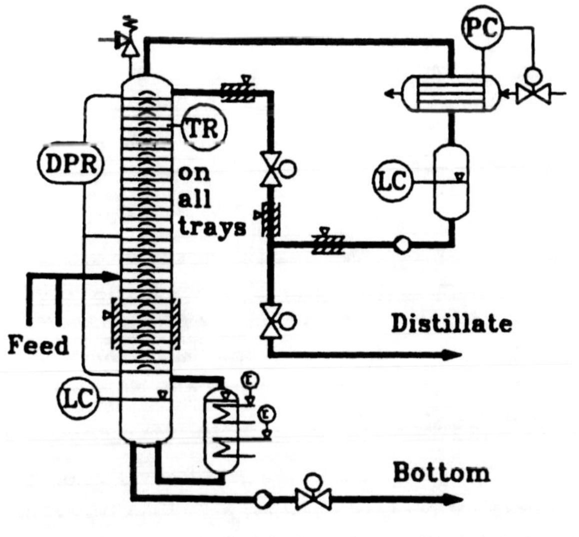

Fig. 1. Distillation Pilot Plant

To avoid this problem, an observer can be employed to estimate product compositions from easily measured variables. An improved design strategy for a nonlinear observer is presented and applied to a binary distillation pilot plant. Accurate estimates of the concentrations of the product streams and the feed composition are provided, which can be used as input to feedforward/feedback composition controllers.

2. EXPERIMENTAL PLANT

Experiments were carried out on a pilot-plant-scale distillation column (Fig. 1, Table 1).

Table 1 Column Specifications

mixture	methanol-water
product purities	x_D=1-x_B=.995 mol/mol
feed composition	$x_F = 0.18 \ldots 0.32$mol/mol
feed flow	F = 27 l/h
number of trays	$n_{trays} = 28$
feedtray	$n_F = 10$
pressure	athmospheric, 1 bar
column diameter	100 mm

The column is equipped with two separate feed lines to allow defined disturbances in feed flowrate, temperature and composition. All measurements are digitalized on a process control system Teleperm M coupled to a VAXCluster.

3. PLANT MODEL

The plant model used is based on the following assumptions:

- Dynamic component balances
- Constant liquid holdup
- Negligible vapour holdup
- Static enthalpy balance
- Boiling temperature on all trays
- Ideal gas phase
- Linear pressure profile
- Constant Murphree tray efficiency.

The model is implemented within a SpeedUp environment.

The vapour-liquid-equilibrium is described by the Wilson and Antoine equations for activity coefficients or saturation pressures.

To save computation time, the order of the column model is reduced following the idea of compartmental analysis. Nonlinear simulations show that full and reduced order models match excellently if static trays are described by full static models instead of the compartment separation functions proposed by Benallou *et al.* (1986). Only the number of dynamic equations is reduced; no algebraic equations are eliminated. This approach has the advantage, that temperature and composition can be calculated on all trays. The reduced model contains full information about the state of the column. Reducing the model order from 30 to 7 saves about 50% computation time.

4. OBSERVER THEORY

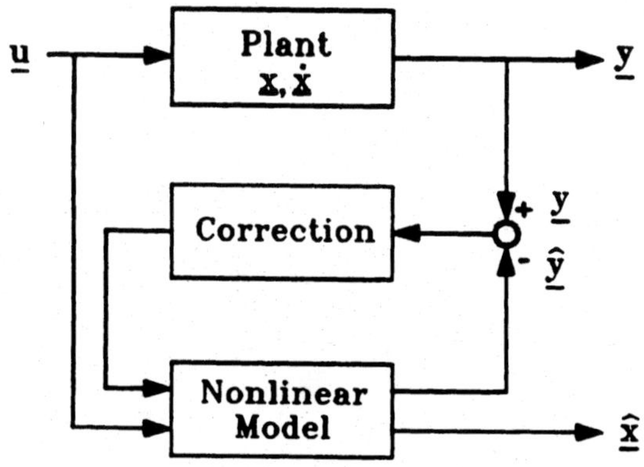

Fig. 2. Observer block diagramm

To estimate product compositions, the reduced-order model is calculated on-line (see Fig. 2). The measured plant outputs - the tray temperatures $\underline{T}$ - are compared to the calculated tray temperatures $\hat{\underline{T}}$ of the model. The estimation error $\underline{e}_{obs} = \underline{T} - \hat{\underline{T}}$ can be used to ensure the convergence of the model states to those of the plant via the correction function $\underline{K}$. This yields the dynamic observer equation for tray j:

$$Holdup_j \frac{d\,\hat{x}_j}{d\,t} = \hat{L}_{j+1}\,\hat{x}_{j+1} - \hat{L}_j\,\hat{x}_j + \hat{V}_{j-1}\,\hat{y}_{j-1} - \hat{V}_j\,\hat{y}_j + K_j\,(T_j - \hat{T}_j) \tag{1}$$

$\underline{K}(\hat{\underline{x}}, \underline{u}, t)$ generally being a nonlinear time-variant function of the operation point and the state. Methods to calculate $\underline{K}$ are given by Zeitz (1977), but are not suitable for high-order models. In this work it is assumed, that K is a constant to be determined from experiments.

Component balances on static trays are calculated without correction.

The tray temperature $\hat{T}$ is calculated from the bubble point equation

$$\sum y_i = \sum \frac{\gamma_i(\hat{x}_i, \hat{T})\, p_{oi}(\hat{T})}{p}\, x_i \tag{2}$$

5. APPLICATION TO BINARY DISTILLATION

To solve the observer model for the complete column, the feed concentration x_F is needed as input. If x_F cannot be measured, it has to be estimated additionally.

Lang and Gilles (1990) proposed a linear relationship of the form

$$\frac{d\,\hat{x}_{F,1}}{d\,t} = C_a\,(T_a - \hat{T}_a) + C_b\,(T_b - \hat{T}_b) \tag{3}$$

where the index a stands for a tray somewhere above and b for a tray below the feed tray. C_a and C_b are further constants to be determined experimentally.

A simpler approach by Riehl *et al.* (1989) avoids this problem by observing the rectifying section of the column only. Thus, the feed concentration is not needed at all and more easily measured variables can be used as an input to the observer. In this work, Riehl's approach is applied to both ends of the column

As is shown in Fig. 3, the rectifying section is observed from tray 17 upwards. The internal flows are calculated from the distillate and reflux flowrates. The composition $\hat{y}_{16}$ and specific enthalpy $\hat{h}^V_{16}$ of the vapour entering tray 17 is determined by flash-calculation on tray 16. On this tray temperature and pressure must be measured.

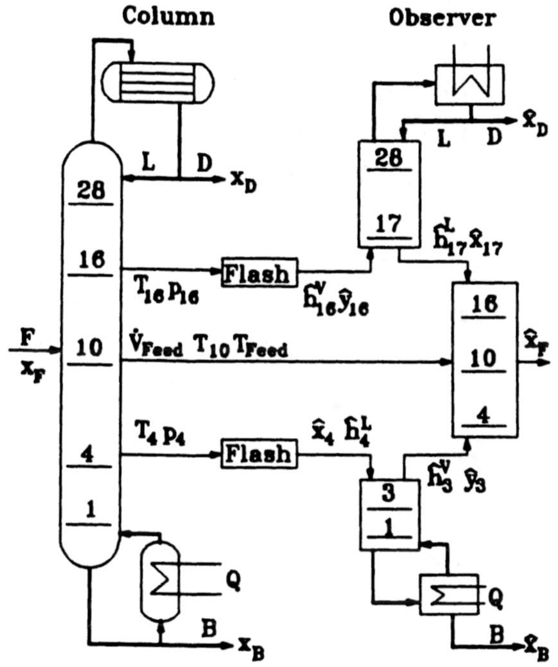

Fig. 3. Observer structure

The bottoms product composition is estimated by an analogous observer calculating the column from tray 3 down. The concentration $\hat{x}_4$ and the specific enthalpy $\hat{h}_4^L$ of the liquid entering tray 3 is calculated from measured temperature and pressure on tray 4.

Estimating the product compositions with two independent observers yields several serious advantages compared to conventional observer design:

- The feed composition is not needed to estimate product compositions.
- More measured variables (eg. all flows and two tray temperatures) can be used as inputs.
- The use of tray temperatures T_4 and T_{16} as inputs automatically leads to convergence of plant and model states, because they contain information about the current dynamic state of the column.
- The importance of the correction function $\underline{K}$ to ensure the convergence of the observer is thereby reduced.
- The determination of tray efficiencies for the rectifying and the stripping section to match plant measurements is simplified.

A third observer for the estimation of the feed composition can be added.

Inputs to the observer model for the feed section are the estimated values of flowrate, composition and enthalpy of the two streams leaving the product observers, as well as feed flowrate, feed temperature and the temperature of the feedtray.

No correction function (Eqn.3) is needed because setting of the temperature of the feedtray as input allows direct calculation of the feed composition.

Note, that no variables from the observer for feed composition are transferred to the observers for product compositions. Thus, the estimation of product compositions cannot be affected by possible deviations in estimation of feed composition.

6. SIMULATION RESULTS

Nonlinear simulation was used to compare the performance of the new and the conventional observer design strategies. In simulation all model parameters are known. This allows to examine the robustness of both designs in the presence of modelling and measurement errors inevitable in practical application.

In general the new observer shows higher accuracy and better robustness than a conventional observer for the whole column. As already mentioned, the subdivision of the observer model into three parts allows the use of plant outputs as observer inputs.

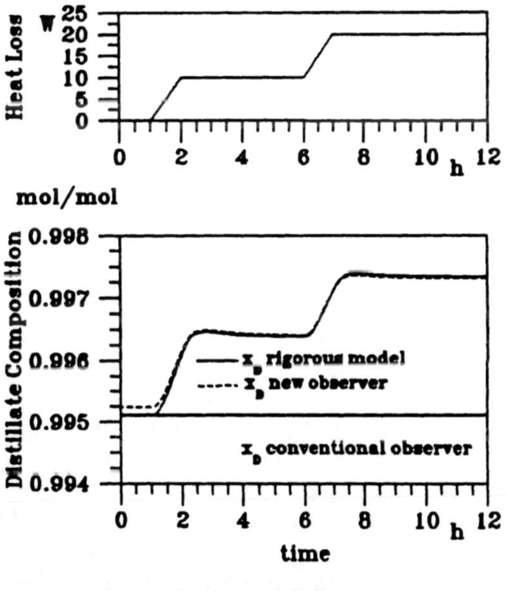

Fig. 4. Effect of Heat Losses in the Column on Distillate Quality

To point out this advantage, both the conventional and the new observer are computed as pure on-line simulators without any corrections ($K_j = 0.0$).

An important example for a disturbance, that cannot be measured, is heat loss on the column trays, that might be caused by rain. Heat losses

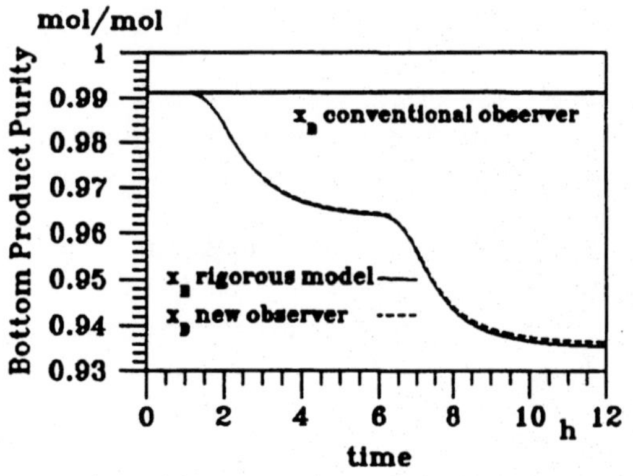

Fig. 5. Effect of Heat Losses in the Column on Bottom Product Quality

lead to additional condensation on the trays, thus increasing the internal reflux ratio as well as the distillate purity, whereas the bottom product purity is deteriorated (if L and V as manipulated variables are kept constant).

As the thick lines in Fig. 4 and Fig. 5 show, the conventional observer can only react to changes in the feed or in the manipulated variables L and V. Consequently, the effects of heat loss are not detected.

The new observer on the other hand uses the temperatures of the trays 16 and 4 as inputs. As condensation on the trays changes the temperature profile of the column, the disturbance is detected by the new observer. The estimated compositions, given by broken lines in Figs 4 and Fig. 5 hardly deviate from the rigorous model's results.

7. EXPERIMENTAL RESULTS

The new observer was validated in open-loop experiments.

Distillate and bottom product flowrates are used to control the levels of the reflux drum and the reboiler. A time step of 48sec is sufficient to solve the observer model on a VAXstation 3100. All measurements are stored on harddisk. Thus, observer parameters can be fitted by trial and error by rerunning the observers with stored experimental data.

Measured concentrations are depicted as diamonds in the following figures. Distillate compositions are obtained by Karl-Fischer-titration and given by filled diamonds. The concentration of methanol in the bottom product is measured by gas chromatography (unfilled diamonds), while feed composition is determined via density measurements (half-filled diamonds).

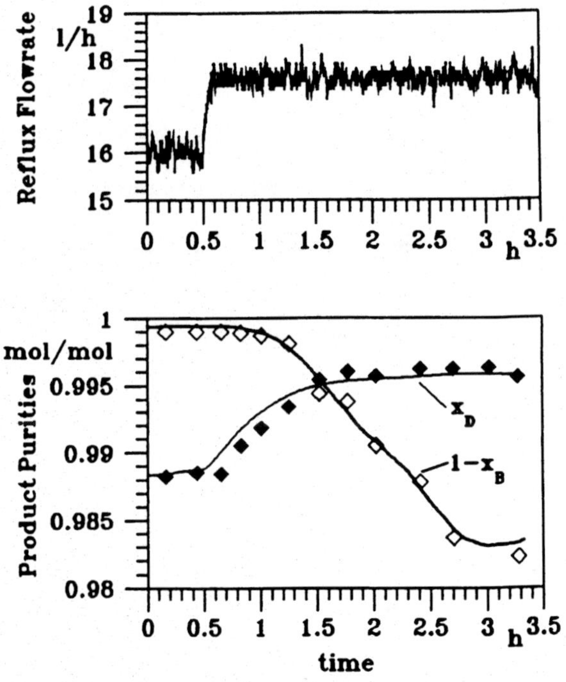

Fig. 6. Reflux disturbance

Fig. 6 shows the response of the column to a 10% increase of the volumetric reflux flowrate from 16 to 17.6 liters per hour, while reboiler duty is kept constant. Due to the neglected level dynamics, the estimation of the distillate composition is about 5min too fast.

Although no hydrodynamics are considered, the time delay of about 15min, the disturbance takes to affect the bottom product, is accurately observed. The observer will only react, when the temperature of the 4th tray is affected, thus including correct information about the time delay.

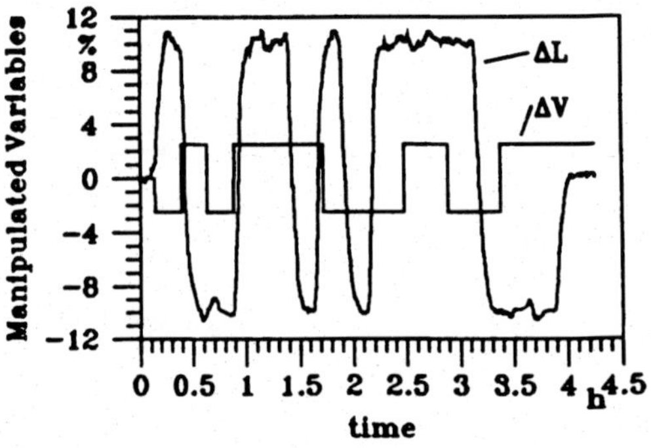

Fig. 7. PRBS-Signal: Manipulated Variables

Sensible values for the correction constants K_j are determined by disturbing the manipulated variables reflux flowrate and reboiler duty by PRBS-signals of amplitudes of 10% and 2.5% respectively (Fig. 7).

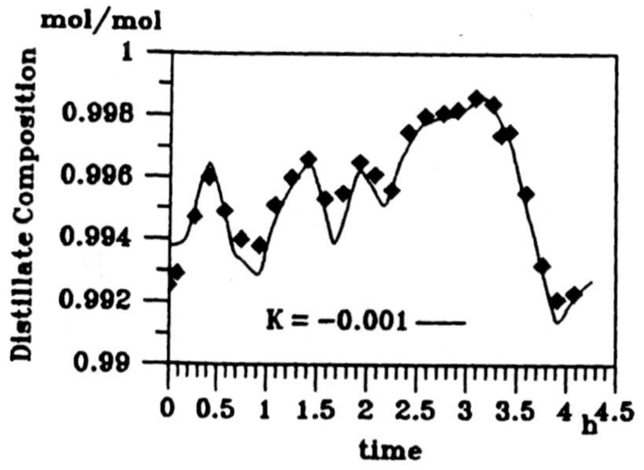

Fig. 8. PRBS-Signal: Distillate concentration

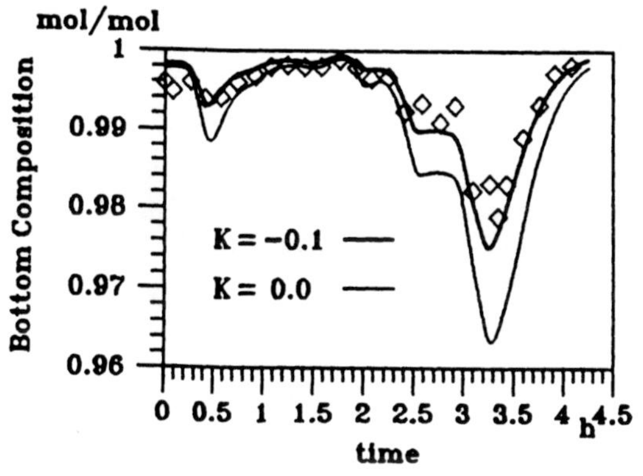

Fig. 9. PRBS-Signal: Bottoms product concentration

The response of the bottoms product composition is shown in Fig. 9. For a Murhree tray efficiency of $\eta = 0.6$ and $K_2 = 0.0 kmol/(sK)$ - the case of pure on-line simulation without correction - the estimated concentration reacts too fast. Setting $K_2 = -0.1 kmol/(sK)$ yields very good dynamic agreement of measured and observed concentrations, simultaneously decreasing the differences between measured and estimated tray temperatures. Higher values of K_2 do not improve the observer dynamics any further, but increase computation time.

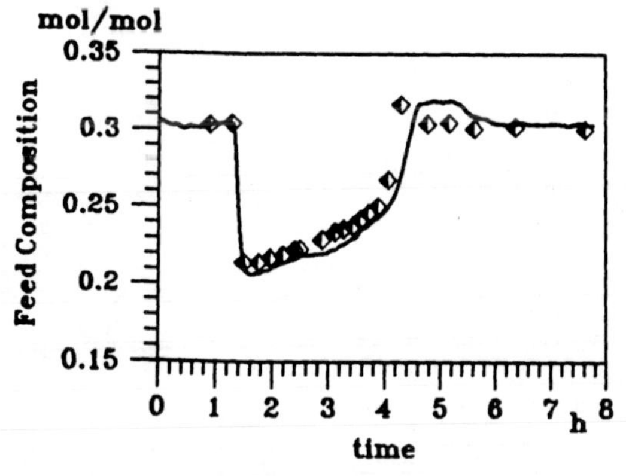

Fig. 10. Feed composition disturbance

The distillate composition can be estimated very well with $K_{18} = K_{26} = 0.001 kmol/(sK)$ and $\eta = 0.9$, as shown in Fig. 8.

Similarly good results are obtained for the feed composition. Figs 10 to 12 give results for an experiment with severe disturbances in feed composition.

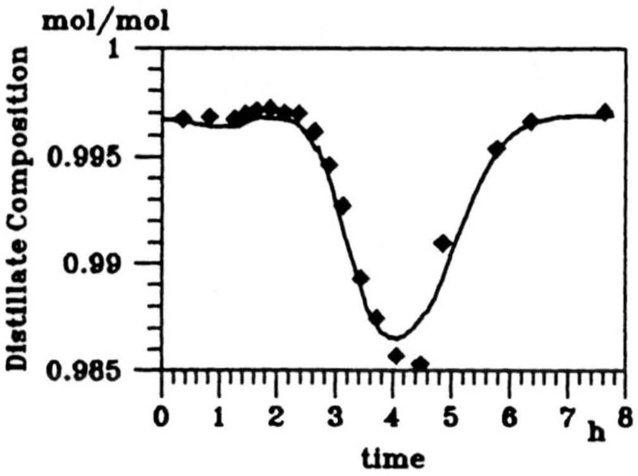

Fig. 11. Distillate concentration

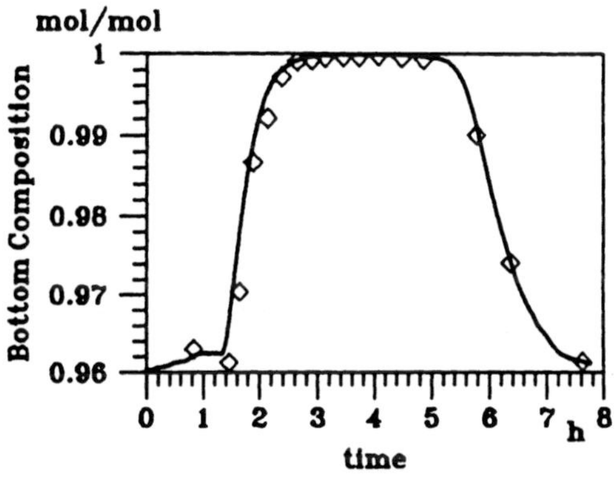

Fig. 12. Bottoms product concentration

At time t=1.3h (see Fig. 10), feed concentration is changed from 30.3 to 21.2 mol% methanol by switching from feed tank 1 to tank 2. The methanol concentration in feed tank 2 is gradually increased by feeding the distillate (methanol) back into the tank, while bottom product (water) is fed into a product tank. At time t=4.5h tank 2 is almost empty, (therefore x_F rises exponentially) and the plant is fed again from tank 1, which has not changed composition. The step change of feed composition is observed within the first time step.

Both product compositions are estimated with high accuracy on a wide operation range (see Figs. 11 and 12).

8. CONCLUSION

An improved observer design strategy for distillation columns is introduced and applied to a binary distillation pilot plant.

The proposed observer shows high accuracy, robustness and excellent dynamic behaviour over a wide range of operation points and under realistic conditions. Not only product and feed compositions, but the whole concentration profile of the column is estimated.

By a rather small time step of 48sec, almost continuous estimates for the three interesting concentrations are available as inputs to concentration controllers and for feedforward control models.

The observer was used successfully for experiments on multivariable feedforward/feedback control strategies (Broll and Gelbe, 1992), (Broll, 1993).

An important advantage of the proposed observer design strategy is its applicability to one point control. If only one product purity is to be controlled, it is not necessary to observe the whole column.

9. SYMBOLS USED

B bottom flowrate
D distillate flowrate
F feed flowrate
K observer correction function
L reflux flowrate
p pressure
T temperature
V vapour flowrate
$\dot{V}$ volumetric flowrate
x liquid mole fraction
y vapour mole fraction

Subscripts

i component number
j tray number
L liquid phase
V vapour phase
$\wedge$ estimated value
oi pure component

10. REFERENCES

Benallou, A., D.E. Seborg and D.A. Mellichamp (1986). Dynamic compartmental models for separation processes. *AIChE Journal* **7**, 1067–1077.

Broll, J. (1993). Steuerung von Destillationskolonnen mit Hilfe nichtlinearer Modelle. Diss.. TU Berlin.

Broll, J. and H. Gelbe (1992). Controlling binary high purity distillation columns. In: *Preprints Distillation and Absorption, Birmingham '92* (ICHEME, Ed.). pp. B1–B10.

Lang, L. and E.D. Gilles (1990). Nonlinear observers for distillation columns. *Computers Chem. Engng.* **11**, 1297–1301.

Luyben, W.L. (1975). Steady-state energy conservation aspects of distillation column control system design. *Ind. Eng. Chem. Fundam.* **14**, 321.

Riehl, F.F., P. Perne, S. Paune and F. Steffens (1989). Model-based control systems for distillation columns. In: *Preprints DYCORD+'89* (IFAC, Ed.). pp. 32–35.

Zeitz, M. (1977). Nichtlineare Beobachter für chemische Reaktoren. In: *VDI-Fortschr.-Berichte*. Vol. 27. VDI-Verlag. Düsseldorf.

A NEW VARIABLE STEP SIZE ADAPTATION ALGORITHM FOR NEURAL NETWORKS

ANNAMÁRIA R. VÁRKONYI-KÓCZY

Technical University of Budapest, Department of Measurement and Instrument Engineering, H-1521 Budapest, Hungary

Abstract. A number of recent publications on learning systems show that adaptation rate issues are still very important research topics since practical problems require always higher and higher learning rates. It has been shown that variable step size methods can provide better convergence speed than fixed ones. This paper concentrates on variable step size methods related mainly to the different versions of the least mean square (LMS) algorithm including the backpropagation method for training feedforward neural networks, and introduces a new approach to the design of the adaptation mechanism.

Key Words. Adaptive systems; neural nets; recursive estimation; iterative methods; convergence analysis

1. INTRODUCTION

A number of recent publications on learning systems show that adaptation rate issues are still very important research topics since practical problems require always higher and higher learning rates. It has been shown that variable step size methods can provide better convergence speed than fixed ones. This paper concentrates on variable step size methods related mainly to the different versions of the least mean square (LMS) algorithm (Widrow *et al.* (1985)) including the backpropagation method for training feedforward neural nets, and introduces a new approach to the design of the adaptation mechanism.

The backpropagation (BP) method (see e.g. Werbos (1990)) is a learning procedure for multilayered, feedforward neural nets to map a set of inputs to a set of outputs. Learning is carried out by iteratively adjusting the coupling coefficients in the network so as to minimize the differences between the the actual output and the target vectors. During the learning procedure the input vector is propagated forward to determine the output which is compared with the target vector resulting in an error signal. The error is backpropagated through the network in order to adjust the coupling coefficients. The whole procedure is repeated until the networks responds for each input vector with an output vector that is sufficiently close to the desired one. Unfortunately the BP method requires a selection of appropriate parameters by the user, that is mainly executed with a trial-and-error process. Since one of the advantages of neural nets is the ease with which they can handle novel or poorly understood problems, it is essential to consider automated and robust learning methods with good performance. Almost every method (see e.g. Battiti (1992)) suffers from the problem of the proper step size. In this paper this aspect is investigated within the framework of LMS-related algorithms.

Section 2 provides a review of the variable step size algorithms covering not only the conventional LMS-related methods, but also the typical step size considerations for infinite impulse response (IIR) adaptive filters. Section 3 investigates the momentum technique widely used with the backpropagation algorithms, while Section 4 is devoted to alternative optimization techniques.

Section 5 presents those considerations which result in a family of new algorithms with variable step size. The key idea of this approach is the direct, nonlinear relation of the learning rate to the actual value of the gradient. Such an approach seems to be able to handle the convergence speed versus misadjustment (Widrow *et al.* (1985)) problem.

2. VARIABLE STEP SIZE ALGORITHMS

The LMS-related variable step size methods attempt to improve the convergence of the LMS algorithm while preserving the steady-state performance, and try to be more effective at tracking in nonstationary environments. For simplicity, in this paper, the different approaches to this step size problem are treated as alternatives for the adaptive linear combiner (see e.g. Widrow *et al.* (1985)), however their purpose may considerably differ depending on the actual LMS-like method applied. The adaptive linear combiner with N inputs and one output produces a sequence

$$y_k = W_k^T X_k \qquad (1)$$

where W_k and X_k denote the column vectors of the tap weights, and the N input signal samples at the kth iteration. The output error at time k is given by

$$e_k = d_k - W_k^T X_k \qquad (2)$$

where d_k stands for the kth sample of the desired signal.

The LMS algorithm updates the taps of the combiner according to

$$W_{k+1} = W_k + 2\mu e_k X_k \qquad (3)$$

where μ is the step size parameter. The step size must be selected to balance the contradictory requirements of fast convergence and small steady-state error.

Formally the variable step size algorithms use a coefficient update of the form

$$W_{k+1} = W_k + 2M_k e_k X_k \quad (4)$$

where typically

$$M_k = \text{diag}\,(\mu_{0,k}, \mu_{1,k}, \ldots, \mu_{N-1,k}) \quad (5)$$

providing individual step sizes for the coefficients at time k. An interesting study concerning both theoretical and practical aspects is given in Evans (1993). Here first the case of increasing the step size by multiplying the previous step size by a factor $\alpha>1$ and that of decreasing by dividing the previous step size by α is considered:

$$\mu_{i,k} = \alpha\mu_{i,k-1}, \text{ or } \mu_{i,k} = \mu_{i,k-1}/\alpha \quad (6)$$

where for easy implementation α is usually taken to be 2. The second method adapts the step size by arithmetic steps ($\alpha'>0$):

$$\mu_{i,k} = \mu_{i,k-1} + \alpha', \text{ or } \mu_{i,k} = \mu_{i,k-1} - \alpha'. \quad (7)$$

The step size for the ith coefficient is decreased if a specified number of consecutive sign changes occur in the correction term of (3), and it is increased if a specified number of consecutive updates are of the same sign. The step size is not changed if the thresholds are not exceeded.

Another interesting variable step size algorithm is the α-LMS rule (see e.g. Widrow et al. (1990)). In this case a "common" step is applied, but it is normalized with respect to the input signal energy:

$$\mu = \alpha/|X_k|^2 \quad (8)$$

Stability is ensured if $0<\alpha<2$, but for practical applications $0.1<\alpha<1$ can be suggested. The LMS algorithm of the form (3) can have the alternative name μ-LMS. The α-LMS algorithm seems to better than the μ-LMS when the eigenvalues of the input autocorrelation matrix R are highly disparate, however the μ-LMS has the advantage that it will always converge in the mean to the minimum mean squared error solution, while α-LMS may converge to a somewhat biased solution.

The learning rate problem is also very important in the case of adaptive IIR filtering. The simplest adaptive filters apply the structure coming from the difference equation form, which can be considered as two linear combiners, the first attached to a single delay line having the input samples at its entrance, while the second one weights the delayed samples of the output, and thus generates feedback. Usually for this latter combiner individual adaptation rates are applied (e.g. Widrow *et al.* (1985)). The first possible suggestion is the extension of the normalizing concept of the α-LMS rule. Mikhael *et al.* (1986) proposes further refinements. The so-called individual adaptation algorithm utilizes as step size

$$\mu_{n,k} = 0.5\alpha|x_{k-n}|/(\sum_{i=0}^{N}|x_{k-i}|^3),\ n=0,1,\ldots,N \quad (9)$$

for the "input" combiner, while

$$\rho_{m,k} = 0.5(1-\alpha)|y_{k-m}|/(\sum_{i=1}^{M}|y_{k-i}|^3),\ m=0,1,\ldots,M \quad (10)$$

for the "output" combiner ($0<\alpha<1$). These individual step sizes, however, do not reduce misadjustment, thus can be recommended only where the noise level is relatively low.

Mohammad (1993) introduces a gradient-based variable step size algorithm. For simplicity gradient estimation of the LMS algorithm is applied:

$$\mu_{n,k} = \mu\,(1 - \exp(-\gamma|e_k x_{n-k}|)),\ n=0,1,\ldots,N \quad (11)$$

$$\rho_{m,k} = \rho_m(1 - \exp(-\gamma|e_k y_{m-k}|)),\ m=1,2,\ldots,M \quad (12)$$

where μ is the "common" step size for the delayed input samples if the error is large, while ρ_m stands as individual step size for the delayed output samples if the error is large. The constant coefficient $\gamma>0$ is the so called damping factor. The experimental verification shows that in the case of adaptive IIR filtering this gradient-based approach is very efficient if the input signal of the filter is not stationary, or contains strong impulsive noise.

At the price of higher complexity the LMS/Newton algorithms provide better performance. Diniz *et al.* (1992) offers an efficient approach for the computation of the optimum convergence factor. In the case of adaptive finite response filters (FIR) a dramatic reduction in convergence time could be achieved. A similar approach is proposed in Du *et al.* (1992), where block-smoothed gradient is applied as direction for descent, and an optimum block-adaptive learning rate is introduced for each block to minimize the approximate output mean-square error of the block. This ideas are elaborated to the backpropagation algorithm without momentum term (see Section 3). With further computational burden Kollias *et al.* (1989), Azimi-Sadjadi *et al.* (1992), and Bello (1992) suggest enhanced training algorithms for feedforward networks. Similarly the training algorithms based on the (extended) Kalman filter approach (Scalero *et al.* (1992), Singhal *et al.* (1989), and Palmieri *et al.* (1990)) seem to be quite efficient. All these contributions provide in a sense "optimum" step sizes.

Recently the applicability of classical numerical optimization methods has been widely investigated (see e.g. Johansson *et al.* (1992)). It turned out, that in particular, the conjugate gradient method ia easily adapted to the backpropagation learning problem. Conjugate gradient methods compute search direction as a linear combination of the current gradient vector and the previous search direction. Starting with a gradient vector the new search direction is

$$d^{(k+1)} = -\text{grad} + \beta^{(k)}d^{(k)}, \quad (13)$$

where $\beta^{(k)}$ is the scalar parameter chosen to assure that the sequence of vectors $d^{(k)}$ satisfy a mutual conjugacy condition. The major advantages of the conjugate gradient methods are their speed and simplicity. They are much faster than steepest descent, and do not suffer from the inefficiencies and possible instabilities that arise from using a fixed step size, as in the conventional backpropagation method. However, they are slower than the second order Newton and quasi-Newton minimization methods. The details of these methods, an their applicability in learning can be found e.g. in the studies of Johansson *et al.* (1992), Makram-Ebeid *et al.* (1990), and Moller (1993).

3. THE MOMENTUM TECHNIQUE

The application of the momentum technique is a further attempt to increase the robustness of the BP algorithm (see e.g. Widrow *et al.* (1990)). The momentum technique introduces in the weight change vector an additional term proportional to the corresponding weight change from the previous iteration. This means that (3) is replaced by

$$W_{k+1} = W_k + 2\mu(1-\eta)e_k X_k + \eta(W_k - W_{k-1}) \quad (14)$$

where the momentum constant $0 \le \eta < 1$ is in practice usually set to around 0.8 and 0.9. The momentum technique low-pass filters the weight updates and thereby tends to resist drastic weight changes caused by gradient noise or high spatial frequencies in the mean squared error surface. However, the momentum technique doesn't solve the step size problem, the step size μ remains a small constant value, thus the momentum technique used alone is of little importance if the gradient estimates are noisy. Combined strategies, where both time-varying step size and the momentum technique are present provide better results (e.g. Widrow *et al.* (1990)).

4. ALTERNATIVE OPTIMIZATION TECHNIQUES

There are several other approaches to improve the performance of the BP algorithms. Ooyen (1992) suggests the modification of the error function applied. It is typical with the sigmoid-like activation units, that if the output "comes" from their saturation range, the derivative of the output with respect to the weights will be extremely small, and completely unable to produce an error signal strong enough to significantly adjust the coupling strengths. Ooyen (1992) gives proposes such an error function definition, which results in a more sensitive output layer. His development is devoted to $1/(1+\exp())$ like activation characteristics: here below the case of th() like activation units will be presented. The new error definition for th() like activation units has the following form:

$$e_k = \frac{1}{2}(1+d_k)\ln(1+y_k) + \frac{1}{2}(1-d_k)\ln(1-y_k). \qquad (15)$$

With this error definition the parameter update equation of the output layer coefficients coincides with (3). This modification improves final convergence (Ooyen (1992)).

Another type of optimization can be that of the network architecture, which is extremely important to reduce the computational load. Two different approaches have been proposed in literature to achieve this aim. In the pruning approach, a network is trained and then analyzed by unit activity or a relevance measure over weights. Nonrelevant weights or units are subsequently removed by an iteration process which interleaves removing of weights/units with learning to recover the network damage. The other approach tries to force out pointless connections from the final network by introducing a tendency to decay to zero for each connection during learning. The tendency to decay to zero is obtained by adding a term to the error function that penalizes the growth of the weights. Sperduti *et al.* (1993) proposes to extend the BP framework by adding a gradient descent on the steepness parameters of the sigmoids. There is an additional learning procedure for the steepness parameters as well. The slope of the sigmoid of an activation unit can be considered as an index of the sensitivity of the unit to the inputs, and units that are not sensitive can be removed from the network because they are useless. In Sperduti *et al.* (1993) it is shown, that the gradient descent over the steepness parameters corresponds to the use of a network with fixed steepness parameters where in the parameter adaptation a quadratic term of the gradient is also present. This latter has some relation to the proposed step size algorithm in the subsequent Section.

5. THE NEW VARIABLE STEP SIZE ALGORITHM

As it is indicated in the above Sections, there are several approaches to the calculation of the search direction and to that of the learning rate. In the case of the BP algorithms, which often behave very badly on large-scale problems, and whose success depends on the user dependent learning rate and possibly momentum constant, it is important to find proper, less user dependent methods to provide fast convergence, and robustness. If we follow the basic LMS approach of (3), the application of the variable step size is the only way to influence the overall performance of our learning algorithm. Fast convergence, and small steady-state error, however, are contradictory requirements. If we accept as search direction approximate derivatives, then we can relate (to certain extent) step size and distance from the optimum. If we are far from the optimum the derivatives are larger, larger steps can be applied, however, low level derivatives may indicate that we are near to the optimum, and only very small steps are acceptable. As it is shown in Section 2, Mohammad (1993) suggested for adaptive IIR filtering a gradient dependent step size technique (see (11), (12)). Here a more sensitive step size function will be presented:

$$\mu_k = \nu\,|\mathrm{grad}_k|^{\beta(1-\mathrm{th}(|\mathrm{grad}_k|))} \qquad (16)$$

where ν, and β are "constants of the user", however control an automatism in the step size generation. Obviously the th() function in the exponent can be replaced with other functions of similar behaviour, like e.g. the exponential in (11) and (12). For small gradients (15) gives

$$\mu_k \approx \nu\,|\mathrm{grad}_k|^{\beta}, \qquad (17)$$

while for larger gradients

$$\mu_k \approx \nu \qquad (18)$$

that can be a limiting value for μ. As an illustrative example the XOR problem has been investigated. On Fig.1 the performance of the conventional BP algorithm, while on Fig.2 the same with Mohammad's variable step size technique is presented with different user dependent parameter settings. Fig.3 shows the behaviour of the XOR problem with the new algorithm, and finally Fig.4 gives the comparison.

6. CONCLUSIONS

In this paper the step size problem of the LMS-like learning algorithms are investigated. It turned out that gradient based variable step size techniques can improve the overall performance, even if they need user defined "external" parameters. There are, however, several open questions concerning efficient learning in large-scale networks, and stimulating further research.

7. REFERENCES

Azimi-Sadjadi, M.R., R.-J. Liou (1992). Fast Learning Process of Multilayer Neural Networks Using Recursive Least Squares Method. *IEEE Trans. on Signal Processing*, Vol. 40, No.2, pp. 446-450.

Battiti, R. (1992). First- and Second-Order Methods for Learning: Between Steepest Descent and Newton's Method. *Neural Computation*, 4, pp. 141-166.

Bello, M.G. (1992). Enhanced Training Algorithms, and Integrated Training/Architecture Selection for Multilayer Perceptron Networks. *IEEE Trans. on Neural Networks*, Vol.3, No.6, pp. 864-874.

Diniz, P.S., L.W.P. Biscainho (1992). Optimal Variable Step Size for the LMS/Newton Algorithm with Application to Subband Adaptive Filtering. *IEEE Trans. on Signal Processing*, Vol.40, No. 11, pp. 2825-2829.

Du, L.-M., Z.-Q. Hou, Q.-H. Li (1992). Optimum Block-Adaptive Learning Algorithm for Error Back-Propagation Networks. *IEEE Trans. on Signal Processing*, Vol. 40, No. 12, pp. 3032-3042.

Evans, J.B., P. Xue, B. Liu (1993). Analysis and Implementation of Variable Step Size Adaptive Algorithms. *IEEE Trans. on Signal Processing*, Vol.41, No. 8, pp. 2517-2535.

Johansson, E.M., F.U. Dowla, D.M. Goodman (1992). Back-propagation Learning for Multilayer Feed-Forward Neural Networks Using the Conjugate Gradient Method. *International Journal of Neural Systems*, Vol. 2, No. 4, pp. 291-301.

Kollias, S., D. Anastassiou (1989). An Adaptive Least Squares Algorithm for the Efficient Training of Artificial Neural Networks. *IEEE Trans. on Circuits and Systems*, Vol. 36, No.8, pp. 1092-1101.

Mikhael, W.B. et al.(1986). Adaptive Filters with Individual Adaptation of Parameters. *IEEE Trans. on Circuits and Systems*, Vol. CAS-33, pp. 677-686.

Mohammad, Th. N. (1993). Improved Adaptive Signal Processing Algorithms. *PhD. Thesis*. Hungarian Academy of Sciences, Budapest, Hungary.

Moller, M.F. (1993). A Scaled Conjugate Gradient Algorithm for Fast Supervised Learning. *Neural Networks*, Vol. 6, pp. 523-533.

Ooyen, A., B. Nienhuis (1992). Improving the Convergence of the Back-Propagation Algorithm. *Neural Networks*, Vol. 5, pp. 465-471.

Palmieri, F., S.A.Shah (1990). Fast Training of Multi-layer Perceptrons using Multilinear Parametrization. *Proceedings of the IJCNN-90*, Washington, pp. 696-699.

Scalero, R.S., N. Tepedelenlioglu (1992). A Fast New Algorithm for Training Feedforward Neural Networks. *IEEE Trans. on Signal Processing*, Vol.40, No.1, pp. 202-210.

Singhal, S., L. Wu (1989). Training Feed-Forward Networks with Extended Kalman Filter. *Proceedings of 1989 Int. Conf. on ASSP*, Glasgow, Scotland, pp. 1187-1190.

Sperduti, A., A. Starita (1993). Speed Up Learning and Network Optimization With Extended Back Propagation. *Neural Networks*, Vol. 6, pp. 365-383.

Werbos, P.J. (1990). Backpropagation Through Time: What it Does and How to Do it. *Proc. of the IEEE*, Vol. 78, No. 10, pp. 1550-1560.

Widrow,B., S.D.Stearns (1985). *Adaptive Signal Processing*. Prentice-Hall, Englewood Cliffs, NJ. 1985.

Widrow, B. (1990). 30 Years of Adaptive Neural Networks: Perceptron, Madaline, and Backpropagation. *Proc. of the IEEE*, Vol. 78, No.9, pp. 1415-1442.

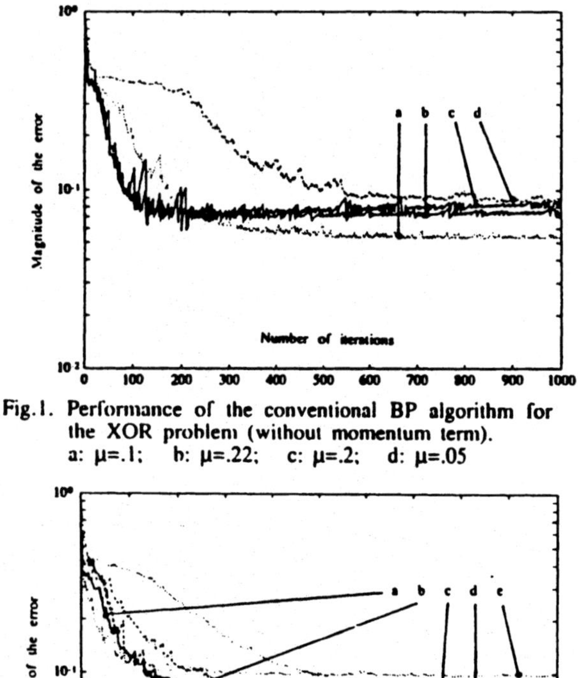

Fig.1. Performance of the conventional BP algorithm for the XOR problem (without momentum term).
a: μ=.1; b: μ=.22; c: μ=.2; d: μ=.05

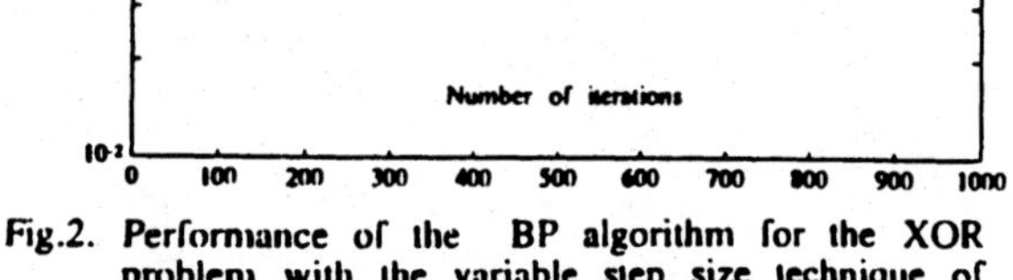

Fig.2. Performance of the BP algorithm for the XOR problem with the variable step size technique of Mohammad. a: μ=.22, γ=5; b: μ=.2, γ=10: c: μ=.1, γ=10; d: μ=.05, γ=10; e: μ=.22, γ=10

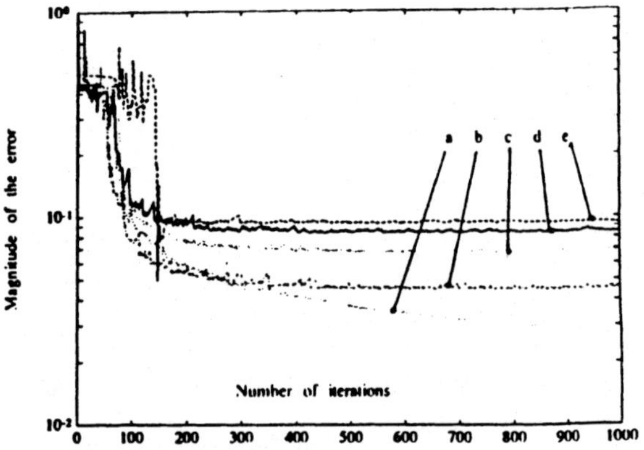

Fig.3. Performance of the BP algorithm for the XOR problem with the new variable step size method suggested in this paper. a: ν=.6, β=.5; b: ν=.6, β=.4; c: ν=.6, β=.6; d: ν=.5, β=.5; e: ν=.7, β=.5

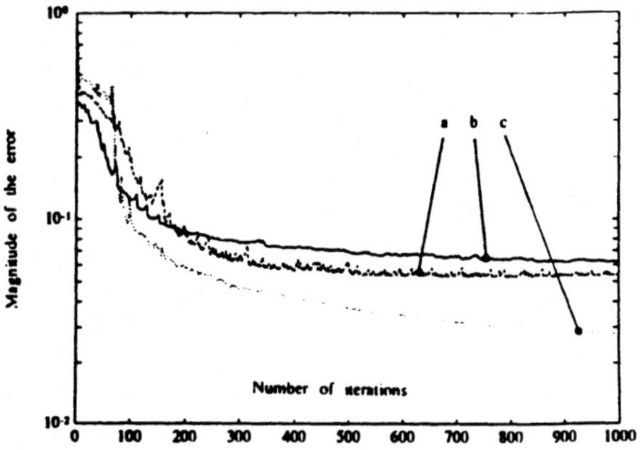

Fig.4. Comparison of the different step size methods
a: BP μ=.1
b: BP with the Mohammad method μ=.22, γ=5
c. BP with the new method ν=.6, β=.5

ESTIMATION OF RESONANT TRANSFER FUNCTIONS IN THE JOINTS OF AN INDUSTRIAL ROBOT

G. FERRETTI, G. MAGNANI and P. ROCCO

Dipartimento di Elettronica e Informazione - Politecnico di Milano
Piazza Leonardo da Vinci, 32 - 20133 Milano, Italy

Abstract. Torsional flexibility in the joints is responsible for the typical high frequency modes detectable in the motion of industrial robots. In this work a systematic procedure for the estimation of the elastic dynamics is presented, and some aspects relevant to a correct design of the identification experiment are pointed out. It is shown that advanced software tools (mechanical CAD, symbolic computation) can be useful for this purpose. An experimental setup consisting of a real industrial robot (COMAU SMART3S) interfaced to a PC is also described and some experimental results are presented.

Key Words. Control technology; identification; industrial robots; microprocessor control; modeling

1. INTRODUCTION

In the design of the position control for industrial robots it is customary to refer to a rigid model of the mechanical structure: the links of the robot are assumed to be rigid bodies and each motor rigidly connected to the corresponding link. While the assumption of link rigidity can be accepted in the frequency ranges (<30 Hz) usually required to the position control, the hypothesis of joint rigidity turns to be inadequate to give an explanation of vibratory modes usually detected in the motion of industrial robots. It is also important to take into account the elastic behavior of the joints when attempting to increase the performances of the position control, by speeding up the system response.

Physical sources of joint elasticity can be found in the deflection of drive shafts (Sweet and Good, 1985), (Anderson, 1985) and in the elasticity affecting the gears, in particular Harmonic Drives (Asada and Youcef-Toumi, 1987) widely used in industrial robots. It is then customaty to lump the elasticity of the robot in the joints and to write a model of an industrial robot where each joint is represented by means of a torsional spring between a link and the corresponding motor (Spong, 1987).

Although the importance of joint flexibility has been widely recognised in the literature, and a great deal of work is being done to find theoretical solutions to the control problem of robots affected by elasticity, very few attempts (Henrichfreise, 1988) have been reported to find procedures suitable for the identification on industrial (not research) robots of the physical parameters, such as joint stiffness and damping, that describe the model of the flexible joint. This is mainly due to the difficulty in carrying on experiments on robots equipped with standard industrial controllers: not only it is difficult or even impossible to add excitation (pseudo-random) signal to either the position set-points or to the driving signals, but also the frequencies at which the position measurements can be recorded are usually too small for the identification of the mechanical resonances. Moreover, there are some aspects (static friction, coupling among links, gravity effects), affecting the reliability of the results of the identification experiment, that are not clearly taken into account in the literature.

One main contribution of this work concerns the demonstration of the feasibility of identification experiments on an industrial robot (COMAU SMART 3S). The robot is endowed with an open version (Dogliani *et al.*, 1993) of its commercial controller: this version offers facilities to interface the VME-based controller of the robot to a 486 Personal Computer and lets the PC access some internal basic variables of the controller and modify other variables (set-points) at high frequency (up to 1 kHz). It is also possible to make experiments on a single joint while maintaining the other ones mechanically braked: this is particularly useful in experiments concerning local properties of the joints, such as the flexibility.

Based on this experimental setup, a procedure to design identification experiments has been developed that carefully takes into account some aspects peculiar to industrial robots; then, experiments have been performed on the robot to identify the transfer function of each joint.

The paper first recalls the model of the robot affected by joint elasticity (Section 2), then Section 3 presents some criteria for a correct design of the identification experiment; the experimental setup is described in Section 4, while Secion 5 is devoted to the presentation of the experimental results. Some concluding remarks are finally given in Section 6.

2. DYNAMIC MODEL OF A ROBOT WITH ELASTIC JOINTS

The following dynamic model of a robot affected by joint elasticity, valid under several commonly adopted assumptions (Spong, 1987), is considered:

$$\mathbf{M}(\mathbf{q})\ddot{\mathbf{q}} + \mathbf{C}(\mathbf{q},\dot{\mathbf{q}}) + \mathbf{g}(\mathbf{q}) = \mathbf{P}^{-1}\boldsymbol{\tau}_t \tag{1}$$

$$\boldsymbol{\tau}_t = \mathbf{K}(\mathbf{q}_m - \mathbf{P}^{-1}\mathbf{q}) + \mathbf{D}(\dot{\mathbf{q}}_m - \mathbf{P}^{-1}\dot{\mathbf{q}}) \tag{2}$$

$$\mathbf{J}_m\ddot{\mathbf{q}}_m + \mathbf{D}_m\dot{\mathbf{q}}_m + \boldsymbol{\tau}_t = \boldsymbol{\tau}_m \tag{3}$$

where $\mathbf{q}$ is the links generalized coordinates vector, $\mathbf{q}_m$ is the motors displacements vector referred to the motor shafts, $\mathbf{M}(\mathbf{q})$ is the robot inertia matrix, $\mathbf{C}(\mathbf{q},\dot{\mathbf{q}})$ accounts for Coriolis and centrifugal terms, while $\mathbf{g}(\mathbf{q})$ accounts for gravitational terms, $\mathbf{J}_m = diag\{J_{mi}\}$, $\mathbf{D}_m = diag\{D_{mi}\}$, $\mathbf{K} = diag\{K_i\}$, $\mathbf{D} = diag\{D_i\}$, $\mathbf{P} = diag\{\rho_i\}$, being J_{mi}, D_{mi}, K_i, D_i, ρ_i, the motor inertia, the motor viscous coefficient, the joint stiffness and damping, and the gear ratio ($\rho_i < 1$) of the i-th joint, respectively, $\boldsymbol{\tau}_m$ is the motor torques vector and $\boldsymbol{\tau}_t$ is the vector of the torques transmitted by each flexible element.

The linearized model, valid in a neighbourhood of a given configuration, is also introduced since it turns to be useful for the purpose of the identification. The equilibrium point corresponding to a given fixed value of the motor torques ($\boldsymbol{\tau}_m = \bar{\boldsymbol{\tau}}_m$) is defined by the following relations (being $\mathbf{q} = \bar{\mathbf{q}}$, $\mathbf{q}_m = \bar{\mathbf{q}}_m$, $\dot{\mathbf{q}} = 0$, $\dot{\mathbf{q}}_m = 0$):

$$\mathbf{g}(\bar{\mathbf{q}}) = \mathbf{P}^{-1}\bar{\boldsymbol{\tau}}_m \ , \tag{4}$$

$$\bar{\mathbf{q}}_m = \mathbf{P}^{-1}\bar{\mathbf{q}} + \mathbf{K}^{-1}\bar{\boldsymbol{\tau}}_m \ , \tag{5}$$

and the linearization of the model (1-3) yields (denoting with δx the variation of the quantity x from the equilibrium value):

$$\mathbf{M}(\bar{\mathbf{q}})\,\delta\ddot{\mathbf{q}} + \boldsymbol{\Gamma}(\bar{\mathbf{q}})\,\delta\mathbf{q} = \mathbf{P}^{-1}\,\delta\boldsymbol{\tau}_t \tag{6}$$

$$\delta\boldsymbol{\tau}_t = \mathbf{K}\,(\delta\mathbf{q}_m - \mathbf{P}^{-1}\,\delta\mathbf{q}) + \mathbf{D}\,(\delta\dot{\mathbf{q}}_m - \mathbf{P}^{-1}\,\delta\dot{\mathbf{q}}) \tag{7}$$

$$\mathbf{J}_m\,\delta\ddot{\mathbf{q}}_m + \mathbf{D}_m\,\delta\dot{\mathbf{q}}_m + \delta\boldsymbol{\tau}_t = \delta\boldsymbol{\tau}_m \tag{8}$$

where $\boldsymbol{\Gamma}(\mathbf{q}) = \partial\mathbf{g}/\partial\mathbf{q}$.

Finally, by Laplace-transforming eqs. (6-8) and eliminating variables $\delta\mathbf{q}_m$ and $\delta\boldsymbol{\tau}_t$, one obtains:

$$\mathbf{A}_{\bar{\mathbf{q}}}(s)\,\delta\mathbf{q} = \mathbf{B}(s)\,\delta\boldsymbol{\tau}_m \ , \tag{9}$$

where

$$\mathbf{A}_{\bar{\mathbf{q}}}(s) = s^2\mathbf{M}(\bar{\mathbf{q}}) + \boldsymbol{\Gamma}(\bar{\mathbf{q}}) + \mathbf{H}(s) \ , \tag{10}$$

$$\mathbf{B}(s) = \mathbf{P}^{-1}\,(\mathbf{K} + s\mathbf{D})\,[\,s^2\mathbf{J}_m + s(\mathbf{D}_m + \mathbf{D}) + \mathbf{K}\,]^{-1} \tag{11}$$

$$\mathbf{H}(s) = \mathbf{P}^{-2}\,(\mathbf{K} + s\mathbf{D})\,[\,s^2\mathbf{J}_m + s\mathbf{D}_m\,]\,[\,s^2\mathbf{J}_m + s(\mathbf{D}_m + \mathbf{D}) + \mathbf{K}\,]^{-1}. \tag{12}$$

3. DESIGN OF THE IDENTIFICATION EXPERIMENT

The local nature of the joint elasticity (note that in (2) the torque delivered by each joint depends only on the position of the motor and of the link corresponding to the same joint) suggests a SISO approach to the identification: in other words, the goal is to identify the transfer function $G_i(s)$ from the motor torque $\delta\tau_{mi}$ of each joint to the corresponding motor velocity $\delta\,\dot{q}_{mi}$:

$$G_i(s) = \frac{\delta\dot{q}_{mi}}{\delta\tau_{mi}} \tag{13}$$

Note that $G_i(s)$ is the transfer function of the system controlled by each regulator of the decentralized control architecture usually adopted in the industrial robots. While the velocity data $\dot{q}_{mi}$ can be easily computed by simple differentiation of the position measures, the motor torque τ_{mi}, which is proportional to the motor current for brushless motors, is actually imposed by means of a large bandwidth analog current loop. Neglecting the dynamics of the current loop (much faster than the dynamics related to mechanical phenomena), for each joint the transfer function $\hat{G}_i(s)$ from the current set point δI_{di} to the motor velocity $\delta\dot{q}_{mi}$ has to be identified:

$$\hat{G}_i(s) = \frac{\delta\dot{q}_{mi}}{\delta I_{di}} \tag{13}$$

and $\hat{G}_i(s) = k_{ti}\,G_i(s)$, being k_{ti} the gain between motor current and torque.

However, in order to successfully apply this straightforward identification technique, special care must be taken in designing experiments, in particular to minimize the influence due to the dynamic interaction among joints and to the static friction.

3.1 *Choice of the configuration*

Although the system made up by the motors and the transmissions is actually decoupled, this is not true for the mechanical structure of the robot. Note, in fact, that the symmetric matrices $\mathbf{M}(\bar{\mathbf{q}})$ and $\Gamma(\bar{\mathbf{q}})$, and thus the transfer function matrix $\mathbf{A}_{\bar{\mathbf{q}}}(s)$ too, are not diagonal. By virtue of (9) this means that even perturbing the motor torque of a single joint, the motion of the other links cannot be excluded, so that the transfer function to be identified would be affected by the dynamics of all the d.o.f. of the robot, invalidating the parametric identification task. Moreover, the dynamic interaction cannot be removed by simply blocking (with mechanical brakes) all the motors but the one under identification, since the links would be anyway free to move, due to the elastic element between the motor and the link.

To tackle this problem, it is therefore necessary to accurately choose the configuration $\bar{\mathbf{q}}$ so as to eliminate, or at least to minimize, the dynamic coupling among the links. For this, observe that both if perturbing only one (the i-th) motor torque ($\delta\tau_{mj} = 0,\ \forall\, j \neq i$) and if mechanically braking all the motors but the one (the i-th) subject to the identification (this condition is equivalent to $J_{mj} \to \infty\ \forall\, j \neq i$), eq. (9) should be rewritten as follows:

$$\mathbf{A}_{\bar{\mathbf{q}}^i}(s)\ \delta\mathbf{q} = \mathbf{E}^i\ B_{ii}(s)\ \delta\tau_{mi} \tag{14}$$

where $\mathbf{E}^i$ is a n-dim. column vector made up by all zeros but the i-th entry whose value is 1, while $\mathbf{A}_{\bar{\mathbf{q}}^i}(s)$ is the result of the evaluation of matrix $\mathbf{A}_{\bar{\mathbf{q}}}(s)$ when $\bar{\mathbf{q}} = \bar{\mathbf{q}}^i$, being $\bar{\mathbf{q}}^i$ the configuration to be adopted in the identification of joint i. The decoupling conditions

$$\delta q_j = 0 \quad , \quad \forall\, j \neq i$$

can be expressed as

$$\delta\mathbf{q} = \mathbf{E}^i\ \delta q_i \tag{15}$$

Substitution of (15) in (14) yields the conditions:

$$[\mathbf{A}_{\bar{\mathbf{q}}^i}(s)]_{ij} = 0 \quad , \forall\, j \neq i$$

which translates into (remember that $\mathbf{H}(s)$ in (10) is diagonal):

$$\mathbf{M}(\bar{\mathbf{q}}^i)_{ij} = 0,\ \ \Gamma(\bar{\mathbf{q}}^i)_{ij} = 0\ ,\ \forall\, j \neq i\ . \tag{16}$$

Satisfaction of conditions (16) is not a trivial task due to the complexity of the terms in the two matrices (Fu *et al.*, 1987). In a first approximation, variations of the gravitational terms have been neglected ($\Gamma(\bar{\mathbf{q}}^i)_{ij} = 0$) since small displacements from the equilibrium condition are considered. On the contrary, off-diagonal terms of the inertia matrix can become of the same order of magnitude of the diagonal terms (Fu *et al.*, 1987; Khosla and Kanade, 1988), so that it is necessary to minimize their effect on the dynamic coupling among the links. For this, the following procedure has been followed:

1) Computation of the inertial parameters (mass, center of mass, inertia moments) by means of a CAD package (COSMOS/M™) endowed with solid modeling tools;

2) Symbolic manipulation, based on Mathematica™ (Wolfram, 1988), of the dynamic equations of the robot in order to compute the terms of the inertia matrix of the robot;

3) For each link i, search of the configuration $\bar{\mathbf{q}}^i$ in which the off-diagonal terms of the inertia matrix become negligible: this step can be performed by trial and is inevitably approximated;

4) Assuming $\mathbf{M}(\mathbf{q}) \approx \mathbf{M}(\mathbf{P}\mathbf{q}_m)$, the motors of the robot must be positioned in the configuration $\bar{\mathbf{q}}_m{}^i = \mathbf{P}^{-1}\,\bar{\mathbf{q}}^i$.

Thus it can be seen that a careful approach to the experiment design requires some conceptual effort and use of advanced software tools.

3.2 *Trajectory excitation*

Although the open version (see Section 4) of the robot controller allows to make open-loop experiments (i.e. experiments with the position control disconnected), for the joints affected by gravity this kind of experiments is unfeasible, since an uncontrolled motion of the robot would be very difficult to handle, with serious risks of damaging the mechanical devices in case of collisions with the environment. Therefore all the experiments have been made with the position control closed by means of standard PID algorithms.

The choice of the excitation signal has been guided by considerations related to the static friction that affects transmission of industrial manipulation devices. To avoid effects of the static friction on the experimental results, the trajectories (position set-points) have been excited by superimposing on a ramp trend of position (constant velocity) (see Fig.

1) a random signal, without determining any inversion of the motion. The excitation signal is obtained by integrating a white noise, to avoid a direct differentiation of the random signal by the derivative action of the PID, which could yield sudden changes in the control signal.

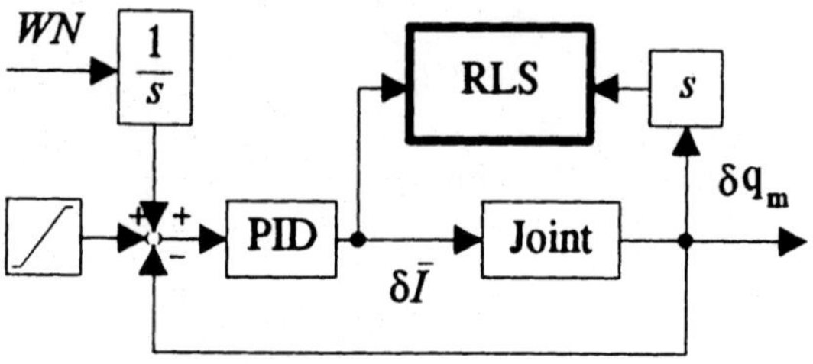

Fig. 1. Identification scheme

Note that the input signal used in the identification is anyway the current set-point.

3.3 *Dependence of the identified parameters on the configuration*

The identified model for the transfer function $\hat{G}_i(s)$ takes the following form:

$$\hat{G}_i(s) = \frac{\mu_i}{1 + T_i s} \frac{s^2/\omega_{nzi}^2 + 2\zeta_{zi} s/\omega_{nzi} + 1}{s^2/\omega_{npi}^2 + 2\zeta_{pi} s/\omega_{npi} + 1} \quad (17)$$

Therefore the system exhibits a first order dynamics, usually at low frequency, due to inertial effects, and a second order dynamics, at higher frequency, characterized by a couple of poles and a couple of zeros complex and conjugate. At low frequencies ($s \to 0$) and as the frequency of the resonance increases (ω_{np}, $\omega_{nz} \to \infty$) the model tends to the rigid one. However, all the parameters in (17) and, in particular, the natural frequencies of both the poles and the zeros, still depend on the robot configuration $\bar{\mathbf{q}}^i$: as the configuration changes, the inertial load on the axis $(\mathbf{M}(\bar{\mathbf{q}}^i)_{ii})$ changes too, together with the dynamics associated to the elastic phenomenon. It follows that, to correctly take into account the effects of the elasticity when determining a fixed parameters regulator, one should obtain, by means of many experiments, the whole range of the resonance frequencies, or at least find the configuration where the resonance frequency is lowest, which is the most critical value.

On the other hand, model (17) is consistent with the model obtained from first principles: in fact, manipulating the equations of the robot affected by elasticity in the joints (Section 2), it is possible to obtain the following expression for the transfer function under identification:

$$\hat{G}_i(s) = k_{ti} \frac{\rho_i^2 J_{li} s^2 + D_i s + K_i}{\rho_i^2 J_{li} J_{mi} s^3 + (J_{ti} D_i + \rho_i^2 J_{li} D_{mi}) s^2 + (J_{ti} K_i + D_{mi} D_i) s + D_{mi} K_i}$$

where $J_{li} = J_{li}(\bar{\mathbf{q}}^i) = \mathbf{M}(\bar{\mathbf{q}}^i)_{ii}$ and $J_{ti} = J_{ti}(\bar{\mathbf{q}}^i)$, being $J_{ti}(\bar{\mathbf{q}}^i) = J_{mi} + \rho_i^2 J_{li}(\bar{\mathbf{q}}^i)$.

A comparison between the two models yields a system of nonlinear equations whose numerical solutions are the estimates of the physical parameters of the model. Once these parameters are computed, a model suitable for the simulation of the non linear robotic system will be available. Moreover, it will be possible to compute the minimum value of the resonance frequency, thanks to the knowledge of the dependance of J_{li} on the configuration $\bar{\mathbf{q}}$, as given by the symbolic model of the robot.

4. EXPERIMENTAL SETUP

The experiments have been made on the industrial robot SMART3S, manufactured by COMAU, endowed with an open version of its controller C3G 9000.

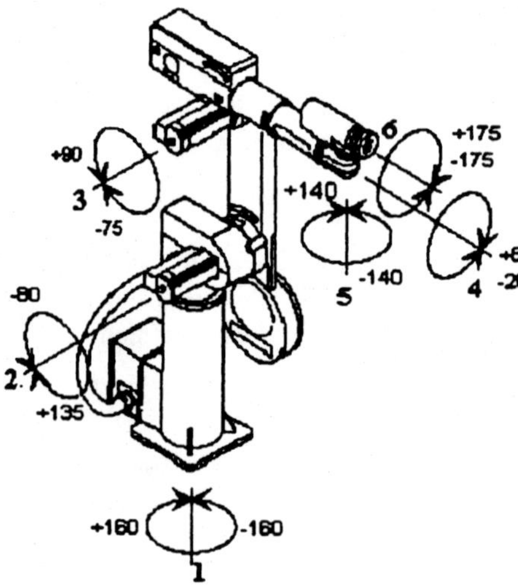

Fig. 2. Robot COMAU SMART3S

The robot SMART 3S (Fig. 2) is a 6 d.o.f., 6 Kg payload manipulator, mainly used for arc welding operations. Among its characteristics it is worth recalling the motion transmission by means of high ratio reductors and the use of brushless motors with absolute position measurement devices (resolvers, resolution 5.49 10^{-3} *deg*).

The controller C3G 9000 is based on a Motorola VME bus and consists of a board (RobotCPU), with a Motorola 68020 CPU, devoted to man-machine interface, and another board (ServoCPU), with a Motorola 68020 CPU and a Texas TMS 320C25 DSP. The latter board performs trajectory generation, kinematic inversion and axis control.

In the open version, the controller is interfaced to a 486–50Mhz PC by means of a bus-to-bus connection made up by two boards manufactured by BIT3. There are basically two ways for the PC to interact with the robot controller: one has a time basis of 10 *ms* and maintains the axes position control of the industrial controller (letting the PC modify or substitute the set-points), while the other one allows the PC to dialogue with the C3G9000 every *ms*, but bypasses the industrial position control.

The latter tool was used for the identification experiments, and the algorithms of the PID position controllers were rewritten in the PC. Every *ms* the PC receives an interrupt from the C3G, reads the motor position in a memory shared with the Motorola ServoCPU, computes a new current set-point (on the basis of a precomputed position set-point and of a suitable PID control law) and writes the current set-point into the shared memory. Both the motor position and the current set-point are stored every *ms*. These data are then analyzed off-line with common software packages for data analysis. In particular, Matlab™ Identification Toolbox (Ljung, 1986) was used in this work. Data analysis consists in numerical differentiation of position data to obtain velocity data, data detrend, choice of model structure, RLS identification of the parameters and model validation.

5. EXPERIMENTAL RESULTS

Following the procedure described in Section 3, the configurations suitable for each identification experiment have been determined. The result of this study is summarized in Table 1, where the same coordinate system as in Fig. 2 has been adopted.

	$\bar{q}^1$	$\bar{q}^2$	$\bar{q}^3$	$\bar{q}^4$	$\bar{q}^5$	$\bar{q}^6$
$\bar{q}^i_1$	0	0	0	0	0	0
$\bar{q}^i_2$	90	-50	-50	0	0	0
$\bar{q}^i_3$	90	-75	-75	90	90	90
$\bar{q}^i_4$	0	0	0	0	0	0
$\bar{q}^i_5$	0	0	0	0	0	0
$\bar{q}^i_6$	0	0	0	0	0	0

Table 1. Experimental configurations

Table 2 reports the identified values (in SI units) for the gain μ_i, for the time constants T_i, for the resonance frequencies (ω_{pi}, ω_{zi}), and the dampings (ζ_{pi}, ζ_{zi}) of both the poles and the zeros complex and conjugate (in the configurations reported in Table 1). All the values are averaged on many experiments. Note that the values reported for the sixth axis are poorly reliable, since it was not possible to make experiments with a payload suitable for the identification of the resonance.

	axis 1	axis 2	axis 3	axis 4	axis 5	axis 6
μ_i	8.1247	7.6000	36.012	124.86	113.41	52.639
T_i	0.0513	0.0294	0.1016	0.0707	0.1114	0.0690
ω_{pi}	129.14	127.64	151.79	541.27	694.54	2184.4
ω_{zi}	58.845	85.124	118.51	481.86	650.74	1321.9
ζ_{pi}	0.2482	0.2228	0.2175	0.1828	0.0591	0.0819
ζ_{zi}	0.7160	0.3287	0.3421	0.2332	0.0478	0.0893

Table 2. Identified values

Figs. 3 to 8 report, for each axis, the Bode diagram of the identified frequency response, from which the importance of the resonant dynamics due to the elastic effects should be apparent.

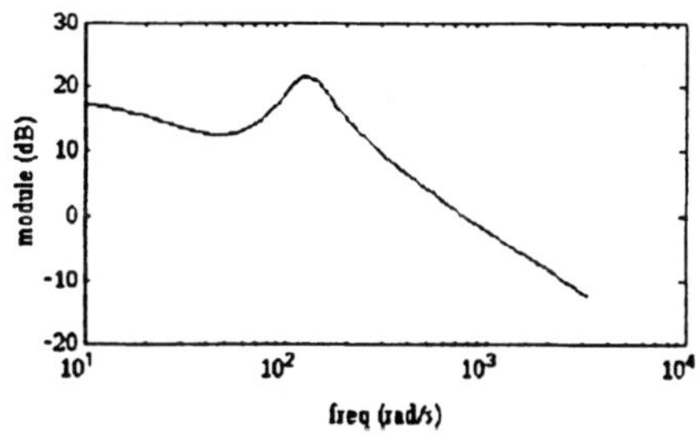

Fig. 3. Identified frequency response for axis 1

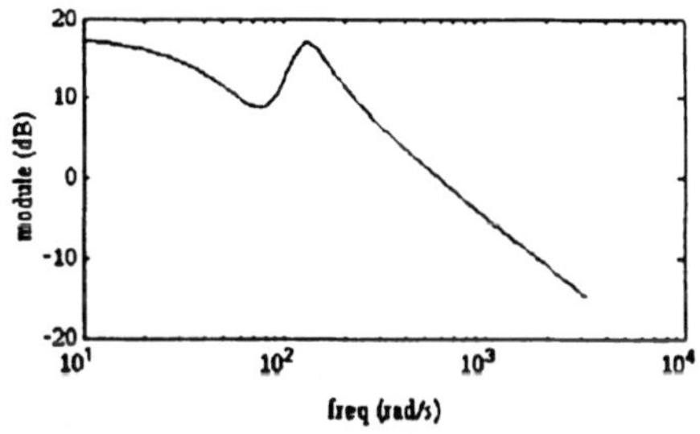

Fig. 4. Identified frequency response for axis 2

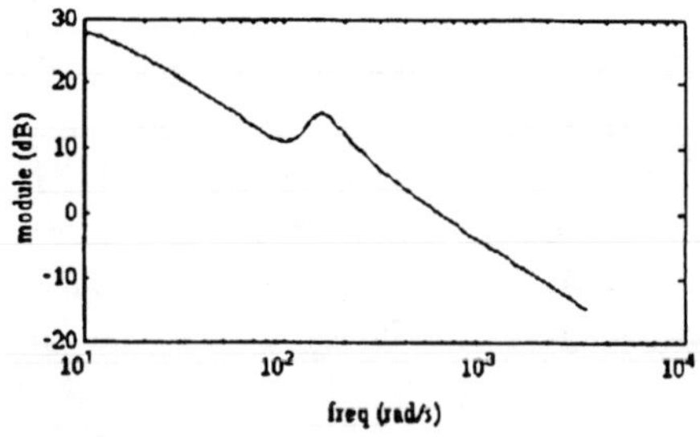

Fig. 5. Identified frequency response for axis 3

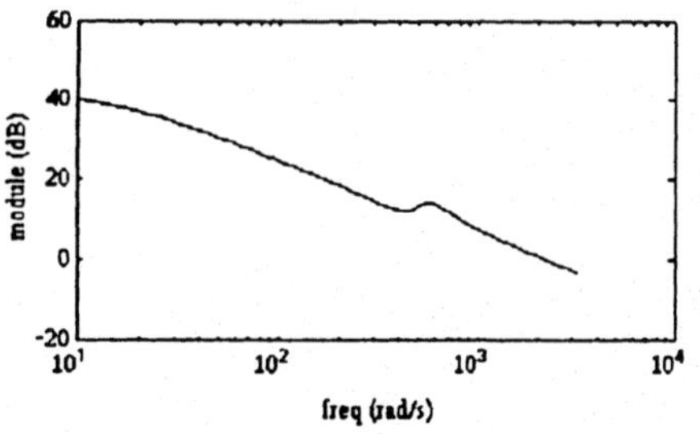

Fig. 6. Identified frequency response for axis 4

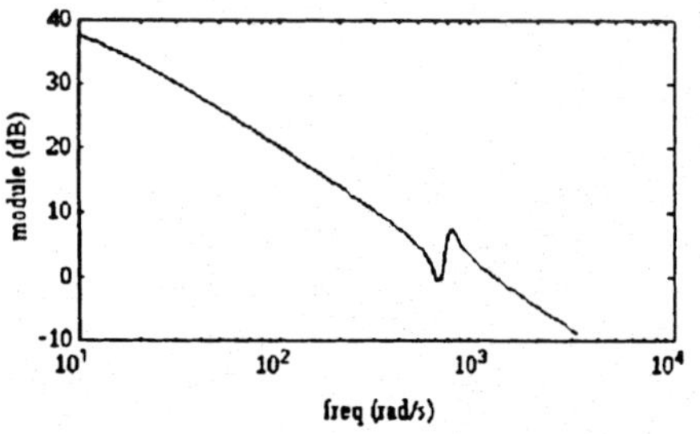

Fig. 7. Identified frequency response for axis 5

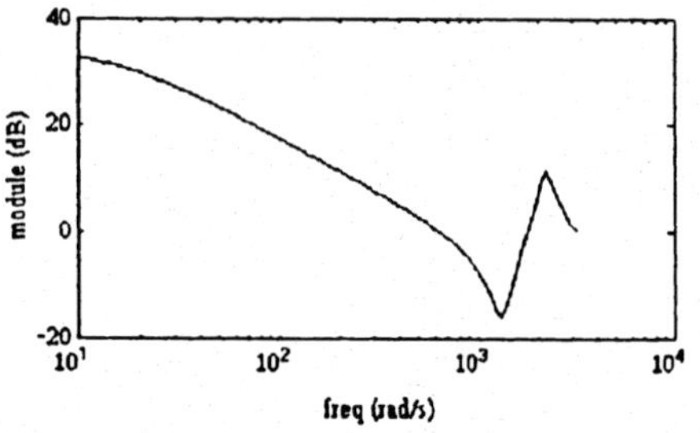

Fig. 8. Identified frequency response for axis 6

6. CONCLUSIONS

The identification of the resonant transfer functions in the joints of industrial robots has been discussed in this paper. Mechanical coupling among the links and static friction in the transmissions involve an accurate design of the identification experiment, particularly with regard to the choice of the robot configuration and of the excitation signal. A special experimental setup is also needed, that gives the possibility to input excitation signals and to record output variables at suitably high frequency (e.g. 1 kHz). The open version of the COMAU controller serves this purpose, allowing the interface of the industrial controller to the most widespread hardware platform, the PC.

An accurate validation of the experimental results is being carried out: it consists of comparison of the obtained physical parameters with the values got through a completely different approach, based on the interaction of the manipulator with the environment, and described in (Ferretti *et al.*, 1994).

7. ACKNOWLEDGEMENTS

This work was done with support of "Progetto Finalizzato Robotica" of CNR (Italian National Research Board) under contracts 91.01933.67 and 91.01995.67.

8. REFERENCES

Anderson, G.L. (1985). Stability of a manipulator with resilient joints. *Journ. Sound Vibration*, **101**, 463-480.

Asada, H., and K. Youcef-Toumi (1987). *Direct Drive Robots Theory and Practice.* MIT Press, Cambridge, MA.

Dogliani, F., G. Magnani and L. Sciavicco (1993). An open industrial robot controller. *IEEE Robotics & Automation Newsletter*, August issue.

Ferretti, G., C. Maffezzoni, G. Magnani and P. Rocco (1994). Joint stiffness estimation based on force sensor measurements in industrial manipulators. *ASME J. Dyn Sys., Meas., and Contr.*, in press.

Fu, K.S., R.C. Gonzalez and C.S.G. Lee (1987). *Robotics: Control, Sensing, Vision and Intelligence*. McGraw-Hill.

Henrichfreise, H. (1988). The control of an elastic manipulation device using DSP. *Proceedings of the American Control Conference*, Vol. 2, 1029-1035, Atlanta, Georgia.

Khosla, P.K. and T. Kanade (1988). Experimental evaluation of nonlinear feedback and feedforward control schemes for manipulators. *Int. J. Robotics Res.*, 7, 18-28.

Ljung, L. (1986). *System Identification Toolbox: User's Guide*. The Mathworks.

Spong, M. W. (1987). Modelling and control of elastic joint robots. *ASME J. Dyn Sys., Meas., and Contr.*, **109**, 310-319.

Sweet, L.W.and M.C. Good (1985). Re-definition of the robot control problem. *IEEE Control System Magazine*, **5**, 18-25.

Wolfram, S. (1988). *Mathematica™: A System for Doing Mathematics by Computer.* Addison Wesley.

SMART ASYNCHRONOUS DRIVES : ADVANCED CONTROL AND FAULT DETECTION AND IDENTIFICATION

M. BAYART*, J.P. CASSAR*, M. STAROSWIECKI*, B. SEMAIL**, C. ROMBAUT**

* *LAIL URA CNRS 1440D - UFR d'IEEA - Université des Sciences et Technologies de Lille, 59655 Villeneuve d'Ascq - France*
** *LEEP - École Centrale de Lille - 59651 Villeneuve d'Ascq - France*

Abstract. The proposed paper presents the application of a model based FDI method to the supervision of an asynchronous drive. The obtained results allow to define local essential instrumentation for its advanced control and supervision.

Keywords. Vector control of asynchronous machines, Smart actuators, Fault Detection and Identification, Structure design of FDI procedures, Instrumentation Design

1. INTRODUCTION

One of the principal characteristic features of smart instruments - sensors and actuators - is that they integrate communication and computation capabilities. Communication capabilities are linked with the development of industrial communication networks of the fieldbus type (Leterrier and Thomesse, 1990). Computation capabilities allow, on one hand, the improvement of the field instrument primary functions, and on the other one, the distribution of a number of functions which were in the past either implemented on centralized architectures, or not implemented at all (Staroswiecki, 1992)

In the case of smart actuators, the primary functions are concerned with power transformation (from the electrical field into the mechanical one for electrical drives, for example), and the improvement of their characteristics is measured by shorter response times, better position or speed precision and stability, better perturbances decoupling, better power efficiency ... The integration of microcontrollers and advanced control algorithms allows to obtain such results (Staroswiecki and Bayart, 1994).

However, actuators are in general integrated within a global system with higher level production goals. The correct achievement of these goals is obtained as long as the actuator is able to process correctly the operating requests which are communicated to it by the higher levels. In that sense, any actuator fault or malfunction may have heavy consequences on the security of the men who operate the process, or on the security and productivity of the process itself. Fault Detection and Isolation (FDI) techniques are developed in order to avoid as far as possible such situations (Patton and *al.*, 1989).

The proposed paper presents the design procedure and the results obtained for the advanced control and the supervision of asynchronous drives. In the first part of the paper, we present a FDI method based on the determination of Analytical Redundancy Relations through structural analysis, then we show, in Section two, how it can be applied to the induction motor. The exploitation of the results, given Section three, allows to define the local instrumentation which has to be implemented within the asynchronous motor in order to improve its self-checking capacities at low extra costs.

2. STRUCTURAL ANALYSIS

Model based FDI classicaly makes use of state-space or transfer function representations of the system to be supervized. They easily apply to the case of linear systems (parity equations (Patton and Chen, 1991), observers (Wunnenberg and Frank, 1988; Staroswiecki and *al.*, 1991; Cocquempot, 1993) and are characterized by the fact that they consider the definition of the measurement system as an input of the FDI problem. However, the design of FDI procedures to be integrated in smart actuators must also include the design of the minimal cost instrumentation which will allow to obtain the desired FDI specifications. Structural models are well suited for the simultaneous study of the system instrumentation and of the FDI capabilities it allows (Staroswiecki and Declerck, 1989). In this paper, we are interested in model based FDI methods.

The method we develop generates Analytical Redundancy Relations (ARR) on the basis of a model which describes the knowledge we have about the system. The generated ARR residuals have to reflect the failure effects with a maximal sensitivity on one hand, and on the other one, they have to present a minimal sensitivity to the initial state, perturbations and modelling errors which are not known. Moreover, the set of the residuals has to inform on the failure localization.

2.1. Residuals Generation

The structural analysis is based on a functional model of the process which provides the various relations which govern it, and which are called process-functions. These relations link known variables such as the controls applied to the process or measured data from sensors, as well as unknown variables.

For example, let us consider the following system :

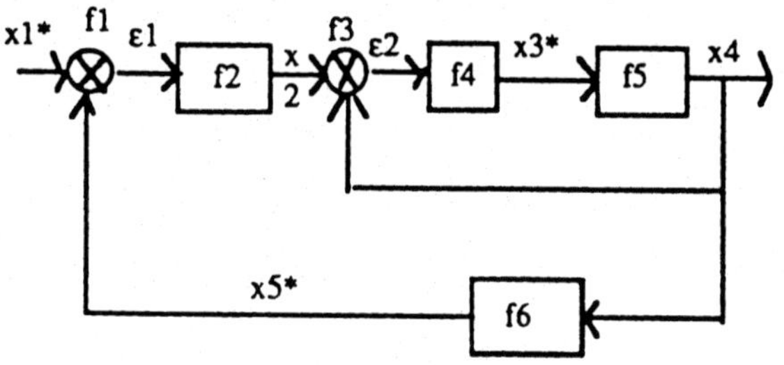

Fig. 1. Example of a functional scheme

The different process-functions are : f1, f2, f3, f4, f5, f6. We suppose that the only known data are the measured variables x1, x3 and x5 (they are marked by *) . The other variables which appear on the figure are unknown.

From this information, we can build the structural table defined by the following binary relation :

$$S : \{F\} \times \{Z\} \rightarrow \{0, 1\}$$

$$(fi,\ zj) \rightarrow S(fi,\ zj\} = 1 \text{ iff } zj \text{ appears in the relation fi}$$

$$S(fi, zj\} = 0 \text{ otherwise}$$

with {F} the set of the process-functions and {Z} the set of the variables.

The advantages of the structural model is that no hypothesis is required about the relations nature (linear, non linear, qualitative, ...). Moreover, the structural representation we use doesn't differentiate the variables from their derivatives or integrals. The structural representation of the system may be enlarged adding the measurement-functions which provide the known variables. The aim of this enlarged representation is to make easy the study of several types of sensors implementation. Applied to the given example, we obtain the table on fig. 2.

In this table, the ¿ indicates a matching on the graph which is associated to the structure. In fact, the existence of ARR has been shown to be the consequence of the existence of complete matchings on over-determined subsystem of the overall system (Decleck, 1991).Thus, the structural table allows to determine the Analytical Redundancy Relations of the process. A RRA is a function which links only known variables and whose veracity can consequently be checked in real-time. An example of the determination process of an ARR from the given example is given figure 3.

Variables		known			unknown						
		mx1	mx3	mx5	x1	x3	x5	x2	x4	ε1	ε2
Process functions	f1				1	0	1	0	0	1	0
	f2				0	0	0	1	0	¿	0
	f3				0	0	0	¿	1	0	1
	f4				0	1	0	0	0	0	¿
	f5				0	1	0	0	¿	0	0
	f6				0	0	1	0	1	0	0
Measurement functions	fmx1	1			¿						
	fmx3		1			¿					
	fmx5			1			¿				

Fig. 2. Structural table of the example

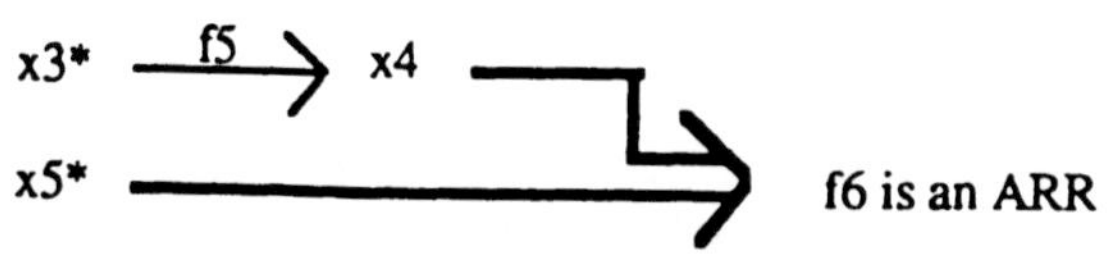

Fig. 3. ARR obtained from the example

2.2. Fault detection and isolation

The fault detection is based on the idea that since the ARR describe the normal operation of the process, they should not be verified in the case when some fault appears (at least the part of them which is sensitive to that fault). In fact the various ARR are of the form $R_i(Z_i^*, \varepsilon_i) = 0$, where Z_i^* represents the subset of those known variables which appear in the computing process of the ARR R_i, and ε_i are the noises, the modelling and data processing errors, the unknown inputs, etc. Because of the presence of these terms, the ARR are never strictly verified, and some decision procedures have to be implemented in order to decide wether or not the fault detection decision should be taken. These decision procedures are based on thresholds which depend on some measure of the noise probabilities, the models errors, etc. Different works are devoted to the proper determination of these thresholds which must not be too small, because it would increase the false alarm rate, and not too great because it would provide bad detection performances (Clark, 1989, Cassar and *al.*, 1992a).

The isolation problem is solved in an elegant way by the use of structured ARR (Gertler and Singer, 1990) : one creates ARR whose sets Z_i^* are different from each other, so that the isolation of a known variable fault is performed on the basis of its signature recognition. The signatures table gives the links between the ARR and the known variables. More precisely, it informs on the presence (given by 1) or absence (given by 0) of the variable in the ARR. Following the same exemple, the signature table is given by the figure 4.

	fmx1	fmx3	fmx5
ARR 1	0	1	1
ARR 2	1	1	1
ARR 3	1	0	1

Fig. 4. Signatures table of the sensors faults

A failure on sensor fmx1 is thus detected when the two ARR 2 and 3 are not verified, while ARR 1 remains true. A generalization of this technique can be proposed considering as known variables not only the sensors outputs but also some known system parameters : in fact, if some ARR depends on system parameters, any change in their nominal value will cause the value of the ARR to deviate from zero.

Our laboratory has developed a software tool called D^3 ("Détection et Diagnostic des Défaillances" = "Fault Detection and Diagnosis") which provides the calculation sequences of the set of all the possible ARR which characterize a system described by its structural table. The application of the presented method to an asynchronous drive will now be developped. It requires first a modelling step.

3. MODELLING OF AN ASYNCHRONOUS DRIVE AND ITS CONTROL

Park transformation is used for system modelling. The result is a set of non-linear differential equations taking into account the control as well as the electrical and the mechanical phenomena. Numerous approaches exist for the vector control of induction machines (Leonard, 1985). In the reported study, our interest was focused on the closed loop control using speed, intensity and flux loops. The control variables are the input voltages. The proposed control uses classical PI controllers associated to an on line estimation of some characteristic variables.

The figures 5 and 6 give the functional scheme of the motor and the control.

From this modelling, we obtain a structural table which indicates the relations between the variables and the functions. For a given configuration of known and unknown variables, the D^3 software provides the set of the ARR which characterize the motor supervision possibilities, so that they can be evaluated from the point of view of the detectability and isolability of the various faults which could occur. It becomes thus possible to compare several instrumentation schemes from the point of view of the FDI possibilities they allow.

4. OBTAINED RESULTS

The FDI specifications we have used were the following :
- detect, and isolate any modification in the values K_1, K_2, K_3 of the three control gains
- detect any modification of the friction coefficient f
- detect and isolate any modification of the statoric and rotoric resistances rs and rr, which may vary with temperature.
- detect and isolate any malfunction of the implemented sensors.

The vectorial control algorithm needs one sensor for the measure of the speed and two sensors for the measure of two among the three intensities. The first study leads to determine the FDI performances which can be obtained using this minimal instrumentation.

Since the control instrumentation does not allow to follow the specifications, we have designed different FDI procedures using extra sensors, namely : a third intensity sensor, and voltage sensors. Flux and/or temperature sensors have not been considered, because of their high cost which would increase too much the cost of the smart actuator.

The table given on figure 7 sums up the obtained results. Each column is concerned with one run of the D^3 software, under a different instrumentation scheme. The first group of lines is connected with the parameters to be supervized, while the second group is concerned with the implemented sensors. The meaning of the figures in the table is the following :
- X indicates an absent sensor,
- 1 indicates a parameter or a sensor which can be monitored, ie. whose fault can be detected and isolated
- 0 indicates a parameter or a sensor whose fault cannot even be detected
- Decimal numbers indicate a parameter or a sensor whose fault can be detected but only partly isolated. In fact, this case arises when the same signature is obtained for different faults, making the detection possible but leaving some indetermination on the isolation process. The given number indicates then the isolation percentage (1/ Number of variables having the same signature).

For the six first cases, the load torque is considered to be unknown, while for the tests from 7 to 12 it is supposed to be known as a function of the speed.

The table shows that it is in general impossible to dissociate the parameter rr and the speed sensor. In the cases 5, 6, 8, 9, 11 and 12 their common signature differs from the other ones, which allows a 50% isolation. A fault on one or the other can be detected, but more precise determination cannot be done.

The test 1 concerns the minimal instrumentation which is necessary for the vectorial control. In this case, only the voltage inverter can be monitored.

In the test 2, an intensity sensor has been added. This allows not only the monitoring of the voltage inverter, but also of the three intensity sensors.

The implementation of a voltage sensor (test 3) allows the monitoring of rs, but insures only partly this sensor supervision.

In the test 4, no satisfying result is obtained through the introduction of two voltage and two intensity sensors.

The test 5 gives good results. Three intensity and two voltage sensors allow to monitor the voltage inverter faults as well as their own ones. The speed sensor and the rotoric resistance rr are detectable but not isolable.

Adding a voltage sensor in the test 6 does not seem at first to improve the previous results. However, we remark that this extra sensor can be monitored, and the computation of the distances between the different signatures shows that they are increased, making the FDI system more robust with respect to the decision procedures errors (Cassar and *al.* 1992b).

The supposed knowledge of the load torque, all along the next cases, makes it possible to monitor the friction coefficient f.

In the test 7, the intensity sensors are not monitored.

The test 8 (measure of the third current), allows to monitor the intensity sensors and the other parameters except the rotoric resistance and the speed.

The other tests do not provide better results.

5. CONCLUSION

The reported results show the FDI possibilities of an asynchronous machine and its control system as a function of its instrumentation. It is worth to notice that the FDI specifications are not limited to detect, and isolate any modification in the values K_1, K_2, K_3 of the three control gains, of the friction coefficient and of the statoric and rotoric resistances, but include, for each of the introduced sensors, the detection and the isolation of its own faults. If it appears, for example, that the introduction on an extra sensor only allows its own supervision, then it is clear that it presents no interest for the smart actuator!

Among the twelve instrumentation configurations which have been studied, we find that the most interesting one relatively to the FDI specifications includes three intensity and two voltage sensors. In this case, all the possible faults can be detected and isolated except the rotoric resistance and the speed sensor, whose faults can be detected but not isolated from each other. Since the control instrumentation already needs two of the three intensity sensors, one sees that significant improvements of the actuator functionnalities - namely the intoduction of the self-checking capabilities which are allowed by the local implementation of suitably designed FDI procedures - can be obtained at relatively low extra costs.

The advanced vectorial control and the local FDI possibilities can easily be implemented on a specific Digital Signal Processor ; they provide the asynchronous drive with a great part of the functions of an intelligent actuator.

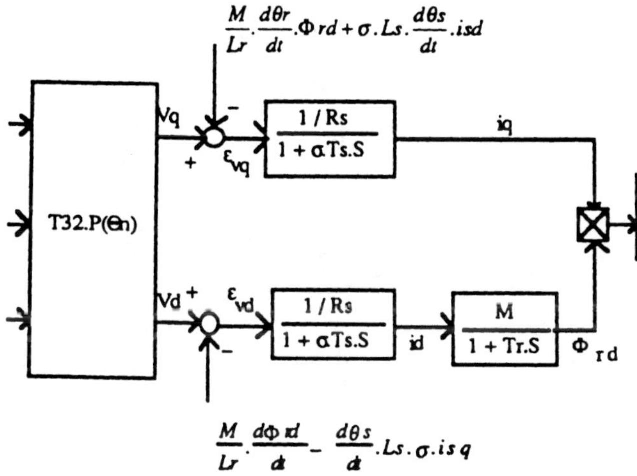

Fig. 5. Modelling of the motor

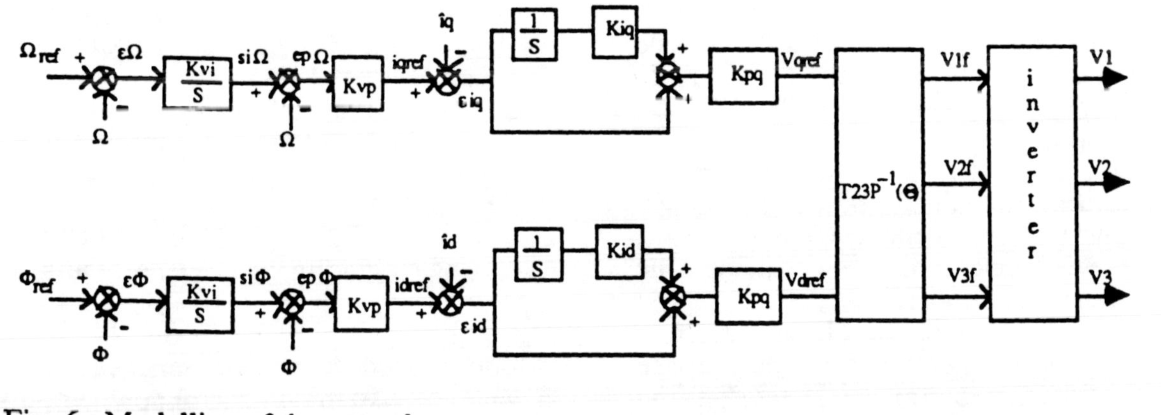

Fig. 6 : Modelling of the control

	1	2	3	4	5*	6	7	8*	9	10	11	12
k1	1	1	1	1	1	1	1	1	1	1	1	1
k2	1	1	1	1	1	1	1	1	1	1	1	1
k3	1	1	1	1	1	1	1	1	1	1	1	1
f	0	0	0	0	0	0	1	1	1	1	1	1
rs	0.2	0.33	1	1	1	1	1	1	1	1	1	1
rr	0.2	0.33	0.33	0.25	0.5	0.5	0.25	0.5	0.5	0.25	0.5	0.5
Ω	0.2	0.33	0.33	0.25	0.5	0.5	0.25	0.5	0.5	0.25	0.5	0.5
i1	0.2	1	1	0.25	1	1	0.25	1	1	0.25	1	1
i2	0.2	1	1	0.25	1	1	0.25	1	1	0.25	1	1
i3	X	1	1	X	1	1	X	1	1	X	1	1
v1	X	X	0.33	1	1	1	X	X	1	1	1	1
v2	X	X	X	1	1	1	X	X	X	1	1	1
v3	X	X	X	X	X	1	X	X	X	X	X	1

Fig. 7. Monitoring possibilities according to the instrumentation

6. REFERENCES

Cassar J.P., Staroswiecki M., Ferhati, R., (1992a), Multivalued Logic Voting Scheme for Residual Evaluation in Failur Detection and Isolation Systems. IFAC/IFIP/IMACS Symp. on Artificial Intelligence in Real Time Control, Delf, The Netherlands, June.

Cassar, J.Ph., M. Staroswiecki, P. Declerck (1992b), "Structural Decomposition of Large Scale Systems for the Design of Failure Detection and Isolation Procedures" System Science XI, Wroclaw, Poland, September.

Clark, R.N., (1989), State Estimation Schemes for Instrument Fault Detection, Chapter 2 in : Patton R.J. and al., Fault Diagnosis in Dynamic Systems, Theory and Application, Prentice Hall.

Cocquempot, V. (1993), "Surveillance de processus industriels complexes, Génération et optimisation des relations de redondance analytiques", Thèse de Doctorat de l'Université de Lille 1, France, February.

Declerck, P., (1991), Analyse structurale et fonctionnelle des grands systèmes. Application à une centrale PWR 900 MW., Thèse de Doctorat de l'Université de Lille 1, France, December.

Gertler J. J., Singer D. (1990), A New Structural Framework for Parity Space Equation Based Failure Detection and Isolation. Automatica, vol 26, 381-388.

Leonard, W. (1985), Control of Electrical Drives, Springer Verlag.

Leterrier, P. and J. P. Thomesse (1990), Le réseau FIP : concepts, services et protocoles. Capteurs ISIL, Liège, Belgium, February.

Patton, R. J., P. M. Frank, R. N. Clark, (1989), Fault Diagnosis in Dynamical Systems, Theory and Application. Prentice Hall.

Patton, R.J., J. Chen, (1991), A Review of Parity Space Approaches to Fault Diagnostic. Proceedings of the SAFE PROCESS'91 IFAC/IMACS Symposium, Baden Baden, Germany, September 10-13,, vol 1, pp 239-255.

Staroswiecki, M., P. Declerck, (1989), Analytical redundancy in non linear interconnected systems by means of structural analysis, IFAC AIPAC, Nancy 3-5 July.

Staroswiecki, M., V. Cocquempot, J. Ph. Cassar, (1991) Optimal Design of FDI Systems via Parity Space and Observer Based Approaches. Proceedings of IECON'91, Kobe, Japan, October 28-November 1,pp. 143-148

Staroswiecki, M. (1992), Les systèmes et le concept d'instrumentation intelligente. Ingénieurs et Scientifiques de France, 2 - 92, pp. 35 - 42.

Staroswiecki, M., M. Bayart, (1994), Actionneurs intelligents, Hermès, Paris, to appear.

Wünnenberg, J., P. M. Frank, (1988), Model Based Residual Generation for Dynamic Systems with Unknown Inputs observers, Proceedings of 12th IMACS World Congress on Scientific Computation, Paris, Vol 2, pp. 435-437.

A DRIVEN LASER-BEAM FOR AN ACTIVE SENSOR ON ROBOT CONTROL

G. QUICK and P. C. MÜLLER

University of Wuppertal, Safety Control Engineering, Gaußstraße 20, D-42097 Wuppertal, FRG

Abstract. A high accuracy measuring system for robots with revolute joints is developed using a CCD line sensor and a laser-beam driven by an extra drive. An implementation of this system is based on standard electronic components. The principle is explained on a system with one degree of freedom. Substitution of standard position sensor on robots by this measuring system in combination with an extended joint control makes high absolute positional accuracy as well as vibration control possible.

Key Words. Closed-loop systems; Direct digital control; Incremental rotary encoder; Laser-beam; Line sensor; Mechatronic; Position measurement; Robots; Stepping motors; Vibration measurement

1. INTRODUCTION

With the manufacturing by CAM systems often off-line programmed robots are used. For these robots, there is a necessity for high accuracy in absolute positioning of the robot links. However, most robots have an insufficient accuracy in absolute positioning. The real characteristics of these manipulators are given by friction, backlash, elasticity in the gears, and elasticity in the links with the result of some error in positioning. Relating to the end effector of the robot, dynamic errors during the movement based on vibrations are existent supplementary to these errors. Some special kind of controller design such as observer technique (Müller and Ackermann, 1986; Henrichfreise, 1987) and preview control can improve parts of the system characteristics, particularly dynamic effects. But however, sensors are used supplementary for an exact finding of the disturbances. In these cases there are many problems for example with the placement of strain gauges. The avoidance of disturbances is another possibility for higher quality of robots . Backlash of gears can be prevented by application of direct drives (Bühler, 1989) but with the consequence of an extra torque ripple. Errors in gears can be measured and compensated by robot control (Pritschow *et al.*, 1992) but changes during the process based on wear make necessary measurements for several times. All sensors and special implementations are normally working the way, that specific errors can be compensated. Sensing errors originated by joint or beam is distinguished especially. Different sensor systems are required in order to determine initiated errors in all.

Another possibility for the detection of position and orientation of the tool center point (TCP) is the insertion of an independent measuring system (ASL CTD; Weule and Reichling, 1987). This kind of sensor system determines all errors initiated by the robot. Tracking the TCP with lasers by geodesic and triangulation principles makes realizable real time operation for on linc control. Nevertheless these systems are suitable for quality tests of robots, but the implementation on robots in production is to costly. The necessary direct view from TCP to the lasers outside of the robot is the reason for a non practical use. An attempt to an integrated measuring system for servo control with an inclinometer was done by Saito and Shinohara (1983). In this case each joint drive is separately equipped by a measuring system. Friction in the inclinometer and a frequency response about 35 Hz are problems with the system. Moreover the system can measure only position error on the same degree of freedom than the joint. An optical measuring system for a robot at all was presented by Lewis *et al.* (1993). One laser-beam generated at the bottom of a robot with revolute joints goes through the inside of the beams and joints. Driven prisms in the joints feed forward the laser-beam up to the TCP. Supplementary beam splitting mirrors in combination with position detecting cameras are mounted in every joint. This makes possible relating to two degrees of freedom the measurement of the position and the orientation of the joints. Another method with measurement on three degrees of freedom using a laser-beam as well was presented by Quick and Müller (1993). The sensor system makes measurable small changes in position and orientation from one position to a second position. For determination the relative transformation from one joint to the next joint along the kinematics the direction of the sensor system can change by a drive. So this technique has to be implemented for each joint separately and position and orientation of the TCP can be calculated. An extension for measurement of all 6 degrees of freedom are given by Quick and Müller (1994). With the following view to the drive of the measuring system the sensing part is reduced to the degree of freedom corresponding to the measuring drive respectively corresponding to the joint.

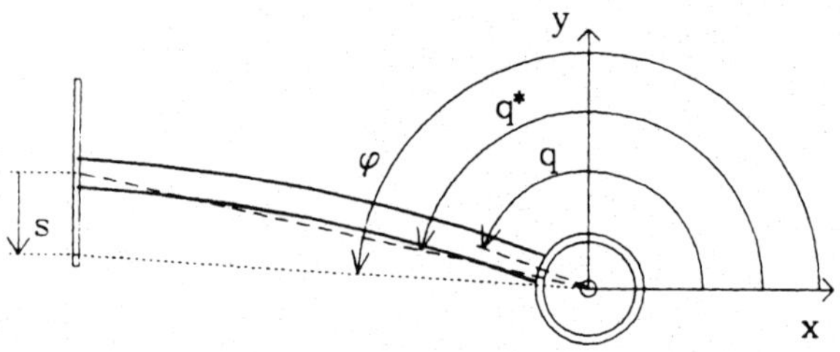

Fig. 1. Sectional drawing

Figure 1 represents the one-link system with a revolute joint taken into consideration. The real position q^* of the tip is initiated by the joint angle q. This angle results from the motor angle p, gear ratio, and errors of the gear. The aim of controlling the robot joint and beam by a structure given in Fig. 2 for example demonstrates the function of the measuring system again. Instead of usually controlled angles p and q, there is a direct control of the tip which is reflected by q^*.

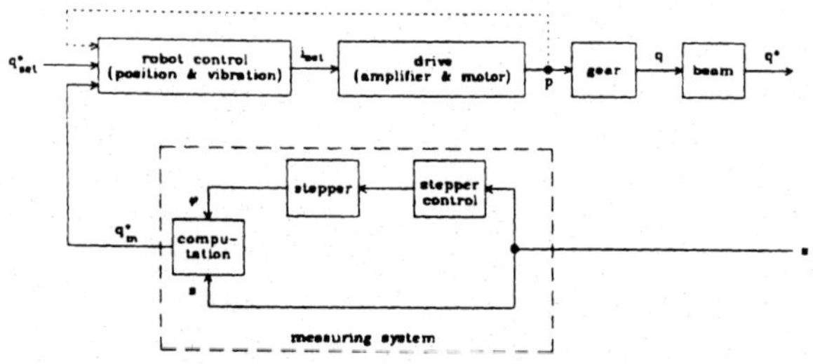

Fig. 2. Robot control

2. STRUCTURE AND MODELING

After explaining the idea of the full system, now, some meaningful details to the implementation and a high performance are given. The principle set up for the one-link system is sketched in Fig. 3. The driven laser-beam has to be mounted at the joint and a position sensitive detector (PSD) is connected to the tip of this beam and the base of the next joint, respectively. The separate drive for the laser-beam is an important characteristic. The axis position of this drive is on the same line as the axis of the joint drive. But however, there is no direct connection between the laser and the joint. Therefore reaction forces of the driven joint cannot influence the accuracy of the measuring system. Together with the knowledge of the actual angle φ of the laser-beam drive, the measurement of the lighted position s on the PSD makes it possible to calculate q^* representing the exact position of the tip of the beam.

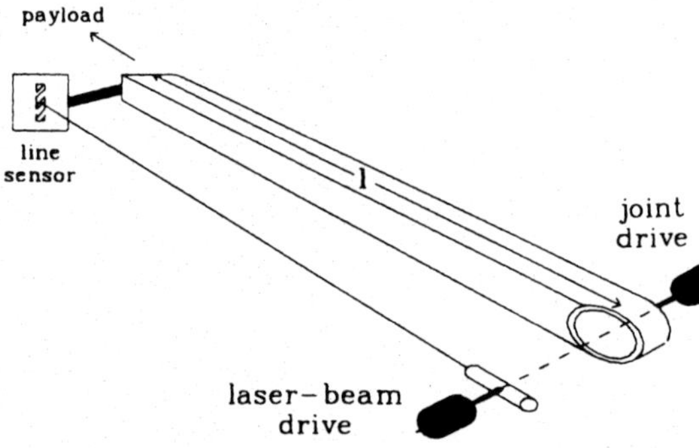

Fig. 3. One-link system

The actual laser-beam drive consists of a 5-phase hybrid stepping-motor. Because of a constant load this brushless motor is characterized by the possibility of positioning with an open-loop control. Moreover the costly and complicated control for some other kinds of brushless motors is dropped out because the necessary low power motor would be uneconomic anyway. The motor rating depends on the features of the joint drive. The stepping motor must be able to generate an angle speed $d\varphi/dt$ and an angle acceleration $d^2\varphi/dt^2$ similar to the dynamics of the joint drive. For these derivations

$$d\varphi/dt \geq dq/dt, \tag{1}$$

$$d^2\varphi/dt^2 \geq d^2q/dt^2 \tag{2}$$

have to be fulfilled. The constant low load

$$J = J_l + J_e + J_m \tag{3}$$

is given by the laser J_l, the incremental shaft encoder J_e to detect the angle φ, and the motor J_m.

Mounting the laser to the motor shaft at an equilibrium position has the effect that there is necessary no additional torque as for acceleration and bearing friction. So with the specific self-holding torque T_D of hybrid stepping motors the angle φ is in steady-state without power supply too. The shaft encoder precisely determines the angle φ so depending on the stepping motor the disturbing torque ripple cannot affect the accuracy. Also loosing steps would not affect the precision. Selecting a 5-phase motor makes possible a step angle θ, e. g. 10mrad, by full-step mode. Taking a 4-phase full-step mode gives results in a practicability of an electric damping by the fifth phase. The equations of motion are represented by

$$J\frac{d^2\varphi}{dt^2} + T_f(\frac{d\varphi}{dt}) + T(\varphi) = 0, \quad (4)$$

$$\begin{aligned} T(\varphi) &= k_\tau \sum_{j=1}^{5} i_j \sin(A\varphi - \frac{j-1}{5}\pi) \\ &+ \frac{i_j}{|i_j|} T_D \sin(10A\varphi), \end{aligned} \quad (5)$$

$$u_j = Ri_j + L\frac{di_j}{dt} + k_\tau \frac{d\varphi}{dt}\sin(A\varphi - \frac{j-1}{5}\pi), \quad j = 1, ..., 5, \quad (6)$$

where T_f represents the friction inside the stepping motor, k_τ is the torque-current constant, and A is the number of torque-angles curve per revolution of motor. On the other hand the laser-beam detector must be considered. With regard to an exclusive digital signal processing, the PSD is realized by a charged coupled device (CCD) line sensor. Thus local discretization is already realized. The sensitive area has a length of

$$s_{max} > l\cos\theta, \quad (7)$$

$$s_{max} = -s_{max}. \quad (8)$$

For the discretization of the detected light level, the CCD sensor signal is connected to a A/D-converter. Because it is of interest to know, whether some CCD pixels are illuminated or not, for this special case the converter can be realized by a suitable comparator. The operation mode of this kind of line sensors, i. e. taking the sensor information by a clock driver, makes it easy to determine the position where the laser beam strikes applying some clock synchronized counters. The counters and the complete synchronization as well are programmed on a logic cell array (LCA). An interpolation module has to be mentioned for a high resolution of φ measurement. This 2^{10}-interpolation together with the required counter is implemented on an application specific integrated

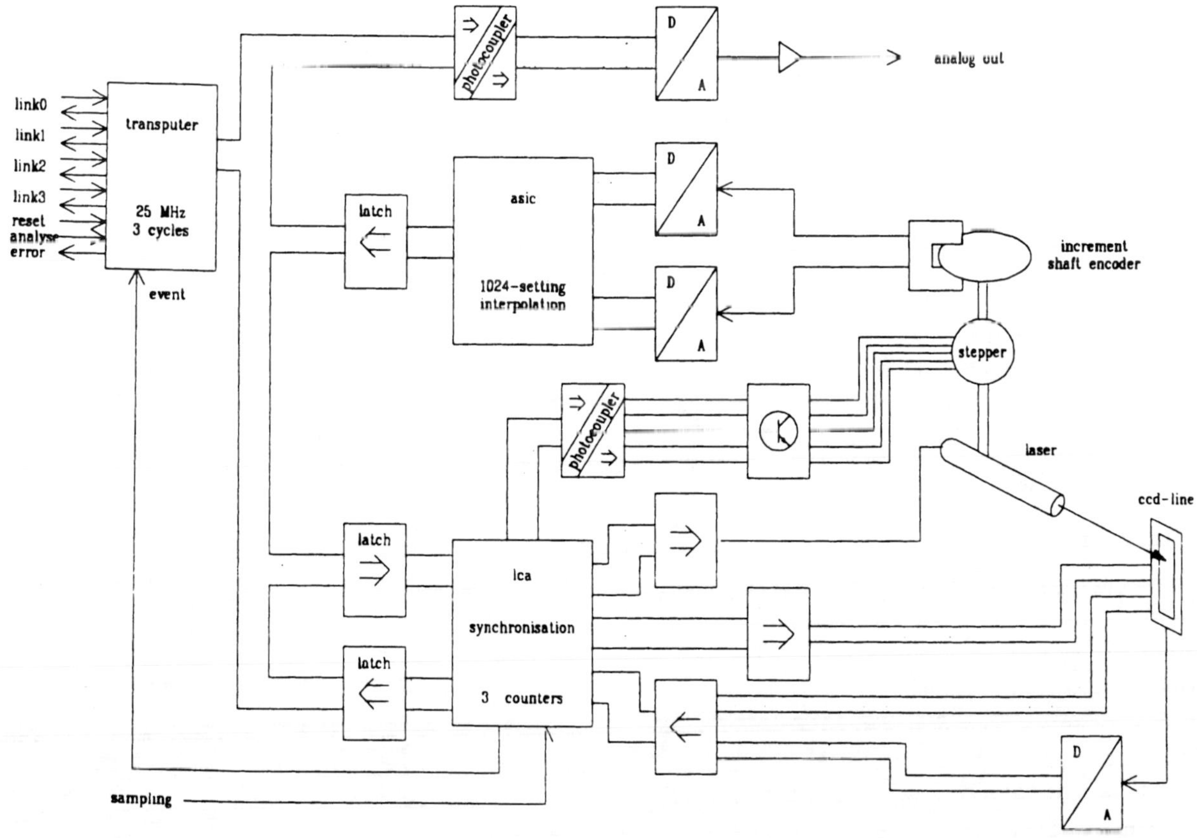

Fig. 4. Block diagram

circuit (ASIC). Both modules latch the angle φ during the laser-on phase after $t_{light}/2$. Further details on the connection of components and signals are demonstrated in the application for the measuring system and for the robot joint given in Fig. 4. The description of the timing diagram (Fig. 5) must be done with respect to joint drive

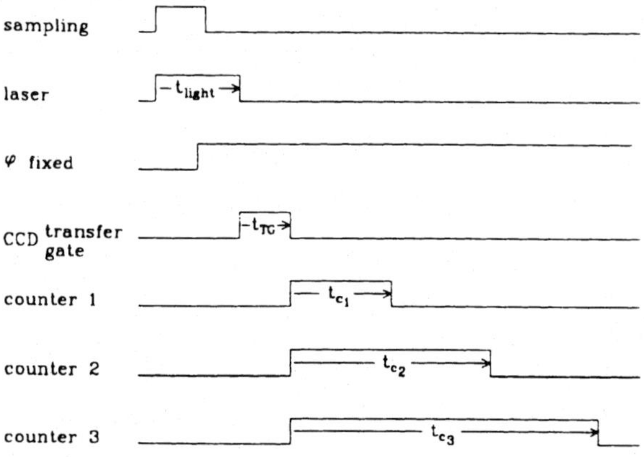

Fig. 5. Timing diagram

control. Due to the measured value s for joint drive control as well as for laser-beam drive control, the timing has to be synchronized by the sampling control of the digital joint drive control. So the rising edge of *sampling* synchronizes the timing of the measuring system given by f_{cl}. Synchronously with *sampling* the measuring cycle starts by switching on the laser-beam for a period of t_{light}. After exposing the CCD sensor the information is prepared during processing t_{TG} by the signal *CCD transfer gate*. After the transfer all counters start counting the signal f_{cl} also synchronizing the pixel transfer to the comparator. The counter reading is given by $c_i = t_{c_i} \cdot f_{cl}$. So *counter*1 determines the first lighted pixel and *counter*2 counts up to the first not illuminated pixel beside the lighted area. *Counter*3 counts during all transmitted pixel information and has a supervising function. If there are malfunctions like $c_1 > c_3$ or $c_2 > c_3$ an error is initiated. With the number of measuring pixels z_m, dummy pixels z_d, and the resolution of the sensor the value s is calculated by

$$s = (f_{cl}\,(t_{c_2} - t_{c_1}) - \frac{z_m + z_d}{2})\, d_{pixel}. \qquad (9)$$

For small angles $q - q^*$ without change of length l the real position of the tip

$$q^* = \varphi + \arccos(\frac{s}{l}) \qquad (10)$$

can be computed. The total time in all for one measuring cycle results in

$$t_{cycle} = t_{light} + t_{TG} + \frac{z_m + z_d}{f_{cl}}. \qquad (11)$$

It is also remarkable that all parts of the measuring system are directly coupled and there are no gears and no couplings which cause new errors. Thus a mechatronic system founded on the laser-beam measuring system and the laser-beam drive is the description of an active sensor for another mechatronic system given by the robot. The overall system consists of coupled systems, because the laser-beam follows the movement of the tip and the control for the joint drive is based on the values of the laser beam measurement.

3. CONTROL

In this case control is related to the stepping motor without the consideration of the joint drive control. At first it is important that the stepping motor is not to be controlled for point-to-point moves. Therefore a perturbation controller is rather necessary. For the discussion of state space control algorithms a description by direct-quadrature (DQ) would be very useful. Bodson *et al.* (1993) applied this technique to a high-performance control for a stepping motor. However this high quality control by nonlinear feedback needs measurement of the position and the current of the stepping motor. The nonlinear feedback has also the function of an electronic commutator. Both supplying by selecting signals and using an speed observer needs computer power supplementary. For disregarding feedback the stepping motor can work in a micro step mode. Additional calculations would be necessary too. Moreover micro-step mode seems to work bad without position feedback also. Static friction must be mentioned as founding and that is why using this technique would be useful on a pilot control. With the aim of an economical implementation the open loop controller for full step mode should be performed for this application. This guarantees a movement of the shaft because of the full step mode torque. In this case digital control for the stepping motor or rather the steps have to be synchronized to the sampling frequency f_{sample} of the joint drive control. Consequently the time between two single steps t_{step} is

$$t_{step} = \frac{n}{f_{sample}}, \qquad n \in N, \qquad (12)$$

and additionally for a very high starting upper cut-off frequency

$$f_{startmax}(J) > f_{sample},$$

$$t_{step} = \frac{1}{m\, f_{sample}}, \qquad m \in N, \qquad (13)$$

is also possible. Selecting the relation f_{sample} in contrast with resonant vibration of the mechanical system can guarantee a stabilized operation in every operation condition. On the other hand

f_{sample} is limited because of on the measuring system by

$$f_{sample} < \frac{1}{t_{cycle}}. \quad (14)$$

By reason of the used CCD line sensor the bandwidth of the measuring system is limited. An application to a continuous analog PSD appears to raise up the bandwidth which in this case is given by an A/D-converter. But on the other hand such sensors are slow acting having more non-linear characteristics.
The avoidance of all non-linearities can be effected with a differential opto-transistor. The stepping motor must exactly orientate the laser-beam to the optotransistor by the control. The angle q^* is calculated by the measurement φ only. Because of the inevitable error which could not be determined, the line sensor is essential for the measurement technique. Only the uncertainty of place initiated by exposure time t_{light} effectuates an extra error.

4. EXAMPLE

The measuring system example is characterized by the stepping motor, the incremental shaft encoder, the laser, and the CCD line sensor. The most important parameters are given to show the performance.

$$\begin{aligned}
A &= 1000 \\
T_{max} &= 140 \cdot 10^{-3} Nm \\
J_m &= 3.5 \cdot 10^{-6} kgm^2 \\
J_e &= 17 \cdot 10^{-6} kgm^2 \\
Encoder\ resolution &= 5000\ lines \\
J_l &= 6.25 \cdot 10^{-6} kgm^2 \\
t_{light} &= 5 \cdot 10^{-6} s \\
z_m &= 5000 \\
z_d &= 34 \\
d_{pixel} &= 7 \cdot 10^{-6} m \\
t_{TG} &= 10 \cdot 10^{-6} s
\end{aligned}$$

Application to an 1 m-beam makes realizable a resolution about 7 μm generated by the line sensor. Without consideration of blurring by exposure the accuracy is approximately 0.5 mm. The cycle time results in $t_{cycle} = 0.17\ ms$ so that a sampling frequency about 5900 Hz is permissible. Together with the sampling theorem the robot beam is admissible to have modes over 2500 Hz.

5. CONCLUSION

The presented measurement arrangement describes a sensor system for advanced robot control. In contrast to standard position sensors an extension with a stepping motor, a line sensor, and a laser makes possible a high accuracy position measurement of the robot tip. Instead of spending more costly joint drives it is more economic to implement such sensor components. The future goal consists of reducing the necessary length of the line sensor. A shorten line sensor cuts down t_{cycle} with the consequence of increasing f_{sample}. This is important to attained beams for vibration control with higher modes. This success can be received by feedback of state values of the joint drive control as well as by progressive stepping motor control with current and position feedback. Nevertheless using the available computing power for the joint-drive control is of higher priority.

6. REFERENCES

ASL CTD. Lasertrack system. *Technical documentation*, Bradville, England.

Bodson, Marc, John N. Chiasson, Robert T. Novotnak, and Ronald B. Rekowski (1993). High-Performance Nonlinear Feedback Control of a Permanent Magnet Stepper Motor. *Transactions on Control System Technology*, Vol. 1, No. 1, pp. 5 - 14.

Bühler, C. (1989). Regelung direkt angetriebener Roboterachsen am Beispiel eines sechsachsigen Industrieroboters. *Ph. D. Dissertation*, University of Dortmund, Germany.

Henrichfreise, H. (1987). Aktive Schwingungsdämpfung an einem elastischen Knickarmroboter. *Ph. D. Dissertation*, University of Paderborn, Germany.

Lewis, J., R. Gill, and A. S. White (1993). Development of an Optical System for Robot Control. *Proceedings of the 1993 IEEE/RSJ International Conference on Intelligent Robots and Systems*, Yokohama, Japan, July 26 - 30, pp. 723 - 726.

Müller, P. C. and J. Ackermann (1986). *VDI-Berichte*, Nr. 598, pp. 321 - 333.

Pritschow, G., H. Klingel, M. Bauder, and A. Horn (1992). Erhöhung der Bahngenauigkeit von Industrierobotern. *Robotersysteme* **8**, pp. 162 - 170.

Quick, G., and P. C. Müller (1993). An Optical Measurement System for High Accuracy Positioning of Robots. *Proceedings American Control Conference*, San Francisco, California, June 2 - 4, 1993, pp. 758 - 759.

Quick, G., and P. C. Müller (1994). Position Measurement on Robot Links with Prismatic and Revolute Joints. Accepted for presentation at IMEKO'94, Torino, Italy.

Saito, Yukio, and Takayoshi Shinohara (1983). An Angle Correction System for a Multi-Joint Robot. *Proceedings 4th International Conference on Assembly Automation*, Tokyo, Japan, pp. 39 - 48.

Weule, H., and B. Reichling (1987). Optisches Meßsystem zur Genauigkeitsprüfung von Industrierobotern. *Robotersysteme* **3**, pp. 189 - 198.

IMPROVEMENT OF FLEXIBILITY AND RELIABILITY OF AUTOMOBILE ACTUATORS BY MODEL-BASED ALGORITHMS

THOMAS PFEUFER

Technical University of Darmstadt, Institute of Automatic Control, Laboratory of Control Engineering and Process Automation, Landgraf-Georg-Str. 4, D-64283 Darmstadt
Phone: +49 6151 167405, Fax: +49 6151 293604, e-mail: pfeu@irt2.rt.e-technik.th-darmstadt.de

Abstract. Automobile actuators have to work very reliable. In extension to redundancy schemes the use of model-based fault detection methods is advantageous for the early detection of incipient faults in the process or in the signals. Two different approaches, the parity-space method and the parameter estimation approach, are selected for the investigations for the example of an automobile actuator driven by a d.c. motor. After a brief description of the theoretical basis the experimental application follows to several faults.

Key Words. Electric actuators; failure detection; parameter estimation; parity-space; redundancy

1. INTRODUCTION

Modern automobiles with a better efficiency, reduced exhaust emissions, improved safety or simply more comfort are often requiring an increasing number of actuators and sensors. This is necessary for the detailed control and supervision of each subprocess. Due to the hard ambient conditions such as a wide temperature range, vibrations and disturbances in signals and power supply, it is not always possible to fulfil the demands for fast and precise positioning with conventional designed actuating systems. Sometimes another problem is the decreasing level of the overall safety and reliability of the automobile which is caused by the increasing number of components. Therefore in the past ten years research efforts have been made to integrate microelectronics within the actuator and to give the actuator more intelligent functions (Isermann and Raab, 1992; Kiencke *et al*, 1992).

In general the design of a highly reliable actuator should include redundancy schemes for the most important subsystems, which cannot be made as reliable as necessary. This step has to take into account faults in sensors, actuators and also in the digital signal processing unit.

In extension to the established schemes, model-based techniques can be used to detect incipient process or signal changes (caused by faults) in an early stage during operation (Willsky, 1976; Isermann, 1984, 1993; Patton *et al.*, 1989; Frank, 1990; Gertler, 1991; Patton and Chen, 1991; etc.). The detection and isolation of the fault can then lead to appropriate actions like maintenance or replacement of components.

This contribution deals with two different methods for model-based fault detection, which are outlined in section 2. After the description of the theoretical basis the application of these methods to a d.c. motor driven automobile actuator follows. Possibilities and problems are discussed for the example of several selected experiments. Finally a short comparison between these two methods is made.

2. METHODS FOR FAULT DETECTION

This section describes briefly the theoretical basis of the investigated methods, the continuous time parameter estimation approach and the discrete time parity-space equations.

2.1. The Parameter Estimation Approach

The parameter estimation approach is based on the idea, that many process faults cause changes in process coefficients p like resistances, inductances, damping and friction coefficients (Isermann and Freyermuth, 1991). For the detection of changes in these coefficients one has to establish a theoretical dynamic process model which, for a linear process with lumped parameters, usually consists of a system of ordinary differential equations like

$$y(t) + a_1\dot{y}(t) + \dots + a_n y^{(n)}(t) = b_0u(t) + b_1\dot{u}(t) + \dots + b_m u^{(m)}(t) + c_0 \quad (1)$$

The process model parameters

$$\underline{\Theta}^T = [a_1 \dots a_n \; ; \; b_0 \dots b_m \; ; \; c_0] = f(\underline{p}) \quad (2)$$

depend on the physical process coefficients $\underline{p}$. Changes in these coefficients can be calculated by the estimation of the model parameters Θ_i based on measured signals y(t) and u(t) and subsequently by the use of the inverse relationship of eq. (2)

$$\underline{p} = f^{-1}(\underline{\Theta}) \quad (3)$$

A survey of methods for the estimation of the continuous time model parameters Θ_i was given by Young (1981). Several applications have shown, that the following procedure is suitable (Isermann and Freyermuth, 1991).

With the assumption that theoretical modelling yields a differential equation like eq. (1) which is linear in the parameters, the least-squares method may be applied. Thus one obtains the general LS-structure for the equation error e(t)

$$y(t) = \underline{\psi}^T(t)\underline{\Theta} + e(t) \quad (4)$$

with the signal vector

$$\underline{\psi}^T(t) = [-\dot{y}(t) \; -\ddot{y}(t) \dots -y^{(n)}(t) \; \vdots \; u(t) \; \dot{u}(t) \dots u^{(m)}(t) \; ; \; 1] \quad (5)$$

and the parameter vector $\underline{\Theta}$ (see eq. (2)). The input and output signals are sampled at discrete times $t=kT_0$. Contemplation of N+1 measurements lead from

$$y(k) = \underline{\psi}^T(k)\underline{\Theta} + e(k) \quad (6)$$

to the vector equation

$$\underline{y} = \boldsymbol{\Psi}\underline{\Theta} + \underline{e} \quad (7)$$

Finally minimization of the equation error yields the well-known nonrecursive LS-estimation algorithm

$$\hat{\underline{\Theta}} = [\boldsymbol{\Psi}^T\boldsymbol{\Psi}]^{-1}\boldsymbol{\Psi}^T\underline{y} \quad (8)$$

Recursive formulations of the LS-estimation algorithm are known as well. The discrete square root filtering approach in information form (DSFI) or UD-factorization for example are more suitable for implementation, because of the lower computational effort and the better numerical properties.

One main condition for the applicability of the continuous time parameter estimation approach is the knowledge of all input and output signals as well as their derivatives. Due to the fact that not all derivatives can be measured in practice, these values have to be calculated each sampling step with a suitable method e.g. a state variable filter (SVF) (Young, 1981).

$$F(s) = \frac{f_0}{f_0 + f_1 s + \dots + f_{\alpha-1}s^{\alpha-1} + s^{\alpha}} \; ; \; \alpha > n \quad (9)$$

The states of the filter correspond to the derivatives of the filtered signals. The parameters f_i can be chosen from a butterworth filter design where the cut-off frequencies have to be fitted to the largest process eigen-frequencies of interest (Peter and Isermann, 1989). For real-time applications its more appropriate to z-transform eq. (9) and to calculate the filtered signals and their derivatives in the discrete time domain. In order to avoid additional phase lags for the parameter estimation, input and output signals have to be filtered with the same state variable filter.

2.2. *The Parity-Space Approach*

The parity-space approach utilizes a kind of temporal analytical redundancy, i.e. the relationship among histories of sensor outputs and actuator inputs (Chow and Willsky, 1984; Patton and Chen, 1991; Gertler, 1991). Starting with the well known linear mathematical model in state space representation

$$\underset{(n,1)}{\underline{x}(k+1)} = \underset{(n,n)}{\boldsymbol{A}} \; \underset{(n,1)}{\underline{x}(k)} + \underset{(n,q)}{\boldsymbol{B}} \; \underset{(q,1)}{\underline{u}(k)} \quad (10)$$

$$\underset{(m,1)}{\underline{y}(k)} = \underset{(m,n)}{\boldsymbol{C}} \; \underset{(n,1)}{\underline{x}(k)} + \underset{(m,q)}{\boldsymbol{D}} \; \underset{(q,1)}{\underline{u}(k)} \quad (11)$$

a set of redundancy relations can be generated by observing the inputs and outputs over a window of p+1 sampling steps and evaluating eq. (10),(11) from time k-p to time k.

$$\begin{bmatrix} \underline{y}(k-p) \\ \underline{y}(k-p+1) \\ \vdots \\ \underline{y}(k) \end{bmatrix} = \begin{bmatrix} \boldsymbol{C} \\ \boldsymbol{CA} \\ \vdots \\ \boldsymbol{CA}^p \end{bmatrix} \underline{x}(k-p) + \begin{bmatrix} \boldsymbol{D} & 0 & \cdots & 0 \\ \boldsymbol{CB} & \boldsymbol{D} & \cdots & 0 \\ \vdots & \ddots & \ddots & \vdots \\ \boldsymbol{CA}^{p-1}\boldsymbol{B} & \boldsymbol{CA}^{p-2}\boldsymbol{B} & \cdots & \boldsymbol{D} \end{bmatrix} \begin{bmatrix} \underline{u}(k-p) \\ \underline{u}(k-p+1) \\ \vdots \\ \underline{u}(k) \end{bmatrix} \quad (12)$$

$$\underline{Y}(k) = T\underline{x}(k-p) + Q\underline{U}(k) \qquad (13)$$

In order to get rid of the unknown state vector $\underline{x}(k-p)$, eq. (13) is multiplied with a row vector $\underline{\omega}^T$, which has to fulfil the following constraint

$$\underline{\omega}^T T = \underline{0} \qquad (14)$$

This yields a so called parity equation

$$r(k) = \underline{\omega}^T \underline{Y}(k) - \underline{\omega}^T Q \underline{U}(k) \qquad (15)$$

The residual r(k) is zero in the ideal case (no noise and model uncertainty) and it becomes nonzero if a fault has occurred.

The design of a parity equation is primarily determined by eq. (14), which holds a set of n homogeneous equations for the calculation of (p+1)m elements in the vector $\underline{\omega}^T$. Therefore some degrees of freedom are left which can be used for the design of a set of linearly independent parity equations.

$$\underline{r}(k) = \mathbf{\Omega}\underline{Y}(k) - \mathbf{\Omega}Q\underline{U}(k) \qquad (16)$$

From the theorem of Cayley-Hamilton follows, that for a great enough p there is allways a vector $\underline{\omega}^T$ and a matrix Ω respectively, which fullfils eq. (14). The order p of the parity equation (15) is directly related to the length p+1 of the data window. An optimal p can be chosen separately for each output y_i according to the rank of the corresponding observability matrix. One obtains the generalized parity-space (Chow and Willsky, 1984), which includes a maximum of redundancy information and a maximum amount of independent parity equations.

The isolation of a component fault is made easy by giving the so called parity vector $\underline{r}(k)$ special features. One approach is based on the design of a structured parity space, where the number of parity relations is equal to the total amount of measurable inputs and outputs and each relation is isolated from an other component. Thus a fault always violates all relations with exception of the isolated one (Gertler and Singer, 1990).

3. EXPERIMENTAL RESULTS

The practical investigations were conducted for the example of a throttle valve actuator driven by a d.c. motor, which is used in a car to control the air mass flow through the intake manifold into the cylinders. This automotive actuator is embeded in various control systems such as traction control, velocity control and gear control. It has to work highly reliable, because a fault can cause an unintentional stop or run away of the engine.

3.1. *Theoretical modelling*

Fig. 1 shows the mechanical function of the actuator. A permanently excited d.c. motor is working with a gear against the torque of a spring for the angular deflection of the throttle valve.

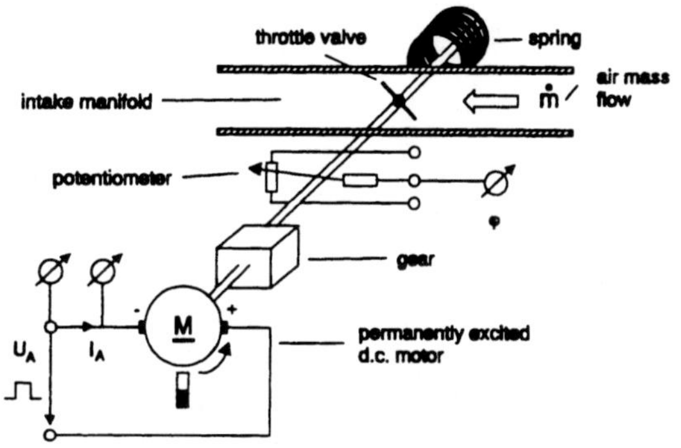

Fig. 1. Function scheme of the investigated d.c. motor throttle valve actuator

The motor is driven by a puls width modulated (PWM) armature voltage U_A which is measured as well as the resulting armature current I_A. The angular deflection φ is measured in the range [0..105°] by a potentiometer.

Theoretical modelling and measurements have shown, that the model structure illustrated in Fig. 2 is a sufficient basis for fault detection. In order to get easier relations, the gear was modelled as a proportional factor with the reduction ratio ν.

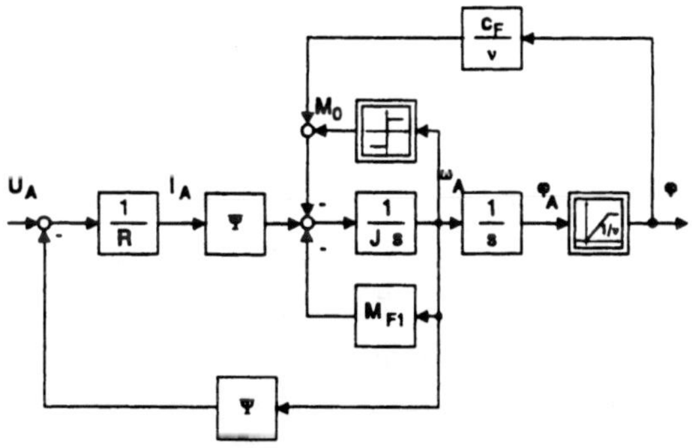

Fig. 2. Mathematical model of the d.c. motor throttle valve actuator

The inductance was also neglected because the electrical time constant is much smaller as the mechanical one. Other parameters are the armature resistance R, the magnetic flux linkage Ψ, the inertia constant J, the viscous friction coefficient M_{F1} and the spring constant C_F. Dry friction parts, which depend on the sign of the angular velocity are included in the signal M_0 as well as the spring pretension and external load torques.

3.2. *Fault detection with parameter estimation*

The mathematical process model consists of an electrical and a mechanical subsystem. As the inputs and outputs of the subsystems are all measurable, two estimation equations can be obtained.

$$\begin{aligned} U_A &= R \cdot I_A + \Psi \cdot \nu \cdot \dot{\varphi} + c_{0e} \\ &= [I_A \;\; \nu\dot{\varphi} \;\; 1]\,[\Theta_{e1} \;\; \Theta_{e2} \;\; \Theta_{e3}]^T \end{aligned} \tag{17}$$

$$\begin{aligned} \ddot{\varphi} &= \frac{1}{J\nu}(\Psi \cdot I_A - M_{F1} \cdot \nu \cdot \dot{\varphi} - \frac{C_F}{\nu} \cdot \varphi) + c_{0m\pm} \\ &= [-\varphi/\nu^2 \;\; -\dot{\varphi} \;\; I_A/\nu \;\; 1]\,[\Theta_{m1} \;\; \Theta_{m2} \;\; \Theta_{m3} \;\; \Theta_{m4}]^T \end{aligned} \tag{18}$$

All measured signals are filtered with a state variable filter for the calculation of the derivatives. Then least squares parameter estimation with the UD-factorization approach yields parameter estimates in each sampling step (T_0=10 ms). The calculation of the physical process coefficients $\underline{p}$ becomes easy from (17), (18)

$$\begin{gathered} R=\Theta_{e1} \qquad \Psi=\Theta_{e2} \qquad c_{oe}=\Theta_{e3} \\ J=\frac{\Theta_{e2}}{\Theta_{m3}} \quad C_F=\Theta_{m1}J \quad M_{F1}=\Theta_{m2}J \quad c_{0m\pm}=\Theta_{m4} \end{gathered} \tag{19}$$

The d.c. value c_{0e} is assumed to be zero in the fault free case, otherwise it includes additive faults in the electrical subsystem. Additive faults, coulomb friction, spring pretension and external load torques are included in the d.c. value $c_{0m\pm}$ of the mechanical subsystem. The nonlinear friction characteristic leads to different estimation values dependent of the direction of motion, which is denoted by index ±.

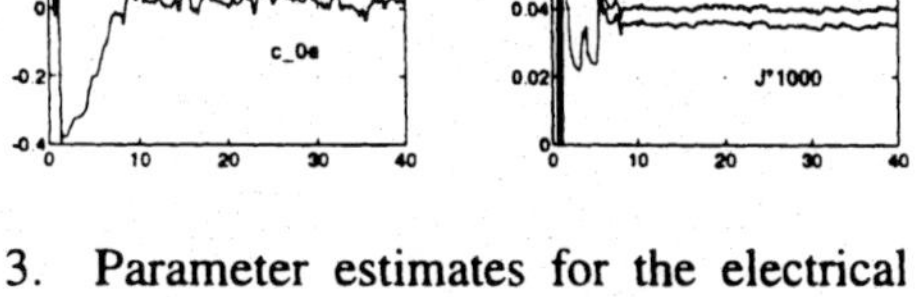

Fig. 3. Parameter estimates for the electrical and the mechanical subsystem in the fault free case

For fault detection at first it is necessary to calculate the nominal parameter values by a pre-identification. Fig. 3 shows a good convergence for the estimated process coefficients, while the process was excited in the open loop with a PRBS-signal. A further reduced variance of the coefficients can be obtained by filtering or calculation of mean-time values.

In the sequel a set of experiments is outlined, where the process was excited in the open loop with a PRBS-signal. The faults were introduced in all cases at time point T_F=15 s, which is marked in the plots by a dash-dotted line.

a) Parameter fault in the armature resistance R (+20%)

The armature resistance R was increased by 0.2 Ω with an additional resistor. According to the expectations, only the estimated coefficient R increases (Fig. 4). The other estimated process coefficients (Ψ, c_{0e}) are transiently disrupted until the estimation has converged.

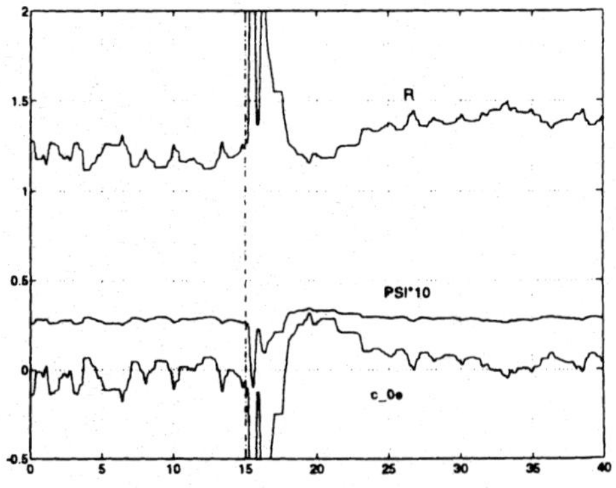

Fig. 4. Parameter fault in the armature resistance R (+20%)

b) Additive sensor fault on the process input U_A (+10%)

A sensor fault was introduced by the addition of a constant offset of 0.4 V to the measured process input. Constant signal faults are reproduced in the estimation equation (17) by the d.c. value c_{0e}. Thus the obtained time history of c_{0e} (Fig. 5) corresponds to the expectations.

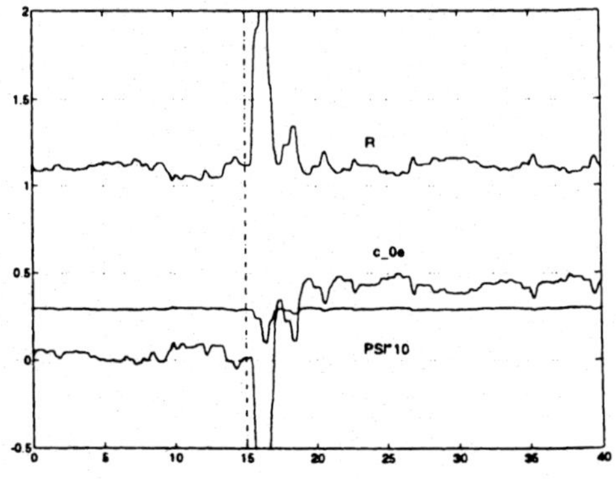

Fig. 5. Additive sensor fault on the process input U_A (+10%)

c) Multiplicative sensor fault on the process output φ (+10%)

In order to simulate a multiplicative fault, the measured process output signal φ was multiplied by the constant factor 1.1. A change of the measurement gain leads to a change of all estimated parameters which are related to the faulty signal or any derivative of them. From eq. (17),(18) follows, that constant changes in the coefficients Ψ, J, M_{F1} and C_F can be expected. This is shown also in Fig. 6.

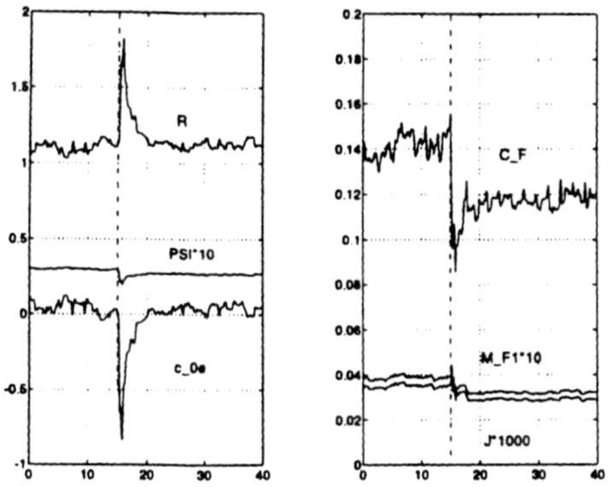

Fig. 6. Multiplicative sensor fault on the process output φ (+10%)

3.3. *Fault detection with parity-equations*

The basis for the design of a parity-space is the following 2nd order state space model of the investigated actuator. Here the d.c. value M_0 of Fig. 2 is assumed to be an input.

$$\begin{bmatrix} \dot{\omega}_A \\ \dot{\varphi}_A \end{bmatrix} = \begin{bmatrix} -\frac{\Psi^2}{JR}-\frac{M_{FI}}{J} & -\frac{C_F}{\nu^2 J} \\ 1 & 0 \end{bmatrix} \begin{bmatrix} \omega_A \\ \varphi_A \end{bmatrix} + \begin{bmatrix} \frac{\Psi}{JR} & -\frac{1}{J} \\ 0 & 0 \end{bmatrix} \begin{bmatrix} U_A \\ M_0 \end{bmatrix} \quad (20)$$

$$\begin{bmatrix} \varphi \\ I_A \end{bmatrix} = \begin{bmatrix} 0 & \frac{1}{\nu} \\ -\frac{\Psi}{R} & 0 \end{bmatrix} \begin{bmatrix} \omega_A \\ \varphi_A \end{bmatrix} + \begin{bmatrix} 0 & 0 \\ \frac{1}{R} & 0 \end{bmatrix} \begin{bmatrix} U_A \\ M_0 \end{bmatrix} \quad (21)$$

Calculation of the rank of the observability matrix yields for both measurable outputs (φ, I_A) a rank p_i=2. Therefore a 2nd order parity-space was developed starting with the z-transformed state space model of (20),(21). The elements of the matrix Ω have been chosen in such a way that a parity-space with 4 structured residuals is obtained, where each residual is decoupled from an other input or output signal. Investigations by simulation have shown, that the residuals r_i react to sudden constant and continual additive faults as outlined in Table 1.

Table 1 Fault patterns of the parity-space residuals

	r_1 [φ]	r_2 [I_A]	r_3 [U_A]	r_4 [M_0]
$\Delta\varphi$	-	_⌐‾	_⌐‾	_⌐‾⌐_
ΔI_A	_⌐‾	-	_⌐‾	_⌐‾
ΔU_A	_⌐‾	_⌐‾	-	_⌐‾
ΔM_0	_⌐‾⌐_	_⌐‾	_⌐‾	-

It is conspicuous that the residuals r_1 and r_4 are only transiently deflected by a continual constant signal offset ΔM_0 and $\Delta\varphi$ respectively, while all other residuals are continually deflected .

In the following the experimental investigations on an input and an output fault are described. The residuals were calculated for square-wave process excitation in the closed loop with a sampling period T_0=1 ms. All residuals were normalized by division through their fault detection threshold. The fault occurrence time point at T_F = 1 sec is marked by a dash-dotted line. The term in the title of the residual plots denotes the decoupled signal.

a) Additive sensor fault of the process input U_A (+10%)

Corresponding to the previous experiment with the parameter estimation, a constant offset of 0.4 V has been added to the measured process input U_A. The residual r_4 reacts as expected (Fig. 7). It becomes continually deflected, while the residual r_1 is only transiently disrupted. In contrary to the simulation study r_2 do not react.

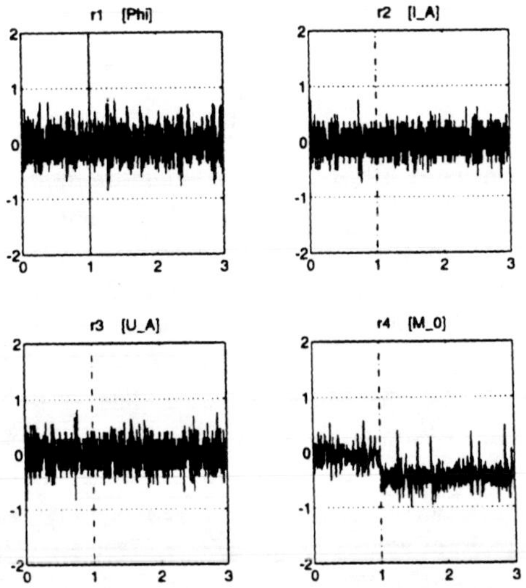

Fig. 7. Parity-space residuals for an actuator offset (U_A): +10%

b) Additive sensor fault of the process output I_A (+10%)

An additive sensor fault of the process output I_A leads to almost the same pattern of residuals as in the previous experiment. The residual r_1 is transiently disrupted without any detection delay. The residuals r_2 and r_3 remain unaffected, while r_4 is continually deflected (Fig. 8).

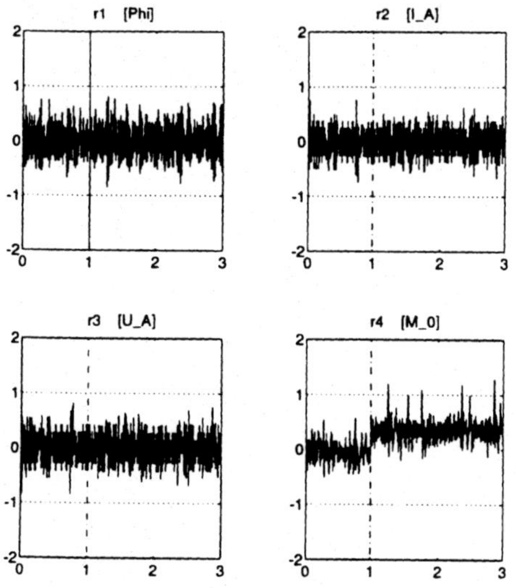

Fig. 8. Parity-space residuals for a measurement sensor offset (I_A): +10%

The results in Fig. 7 and Fig. 8 show that the residuals are very noisy in the fault free case. Therefore it is quite difficult to detect and isolate small offset faults by continually deflected residuals. Reasons for the difference of the achieved results in comparison to the simulation study are uncertainties in the estimated parameters and the severe friction characteristic of the actuator, which leads to a partially neglected nonlinearity in the process parameters.

4. CONCLUSION

Two principally different methods for model-based fault detection were investigated. The application to a d.c. motor driven automobile actuator with several selected faults has shown, that both methods are suitable for fault detection. It depends on the kind of the faults to take into account, which fault detection method is the more suitable one. While the parameter estimation approach is mainly suitable for the detection of multiplicative (parameter) faults, the parity space method is more appropriate for the detection of additive (signal) faults (see also Höfling and Pfeufer, 1994).

The parameter estimation approach makes the fault decision easier than the parity space approach and it seems to be the more robust method with respect to model uncertainty. But on the other hand parity equations achieve a much smaller delay of detection and they are more suitable for on-line and real-time application due to their rather simple algorithm.

5. REFERENCES

Chow, E. Y., and A.S. Willsky (1984). Analytical redundancy and the design of robust detection systems. *IEEE Transactions on Automatic Control*, **AC-29**, 7, pp. 603-614.

Frank, P.M. (1990). Fault diagnosis in dynamic systems using analytical and knowledge-based redundancy - A survey and some new results. *Automatica*, **26**, pp. 459-474.

Gertler, J. (1991). Analytical redundancy methods in failure detection and isolation. *IFAC-Symposium SAFEPROCESS'91*, Baden-Baden, Vol.I, pp. 9-21.

Gertler, J., and D. Singer (1990). A new structural framework for parity-equation based failure detection and isolation. *Automatica*, **26**, pp. 381-388.

Höfling, T., and T. Pfeufer (1994). Detection of additive and multiplicative faults - parity space vs. parameter estimation. *IFAC Symposium on Fault Detection, Supervision and Safety for Technical Processes - SAFEPROCESS '94*, Espoo, Finland.

Isermann R. (1984). Process fault detection based on modelling and estimation methods: A survey. *Automatica*, **20**, pp. 387-404.

Isermann R., and B. Freyermuth (1991). Process fault diagnosis based on process model knowledge. Transactions of the ASME, *Journal of Dynamics, Measurement and Control*, Vol. 113.

Isermann R., and U. Raab (1992). Intelligent actuators - ways to autonomous actuating systems-. *IFAC Symposium on Intelligent Components*, Malaga, Spain.

Isermann R. (1993). On the applicability of model based fault detection for technical processes. *12th IFAC-World-Congress*, Sydney.

Kiencke, U., Lacher, F., Schreiber, H., and M. Ulm (1992). Mikroelektronik im Kraftfahrzeug, neue Anwendungen und Trends. *Automatisierungstechnische Praxis*, **34**, 5, pp. 231-238.

Patton, R.J., Frank, P.M., Clark, R.N. (eds) (1989). *Fault diagnosis in dynamic systems - Theory and applications*. Prentice Hall (Control Engineering Series).

Patton, R.J., Chen, J. (1991). A review of parity space approaches to fault diagnosis. *IFAC-Symposium SAFEPROCESS'91*, Baden-Baden, Vol.I, pp. 239-255.

Peter, K.H., and R. Isermann (1989). Parameter-adaptive PID-control based on continuous-time process models. *IFAC Symposium on Adaptive Control and Signal Processing*, Glasgow.

Willsky A.S. (1976). A survey of design methods for failure detection in dynamic systems. *Automatica*, **12**, pp. 601-611.

Young P.C. (1981). Parameter estimation for continuous time models - a survey. *Automatica*, **17**, pp. 23-39.

VS - CASCADE CONTROL

A. BALESTRINO*, D.GALARDINI# and A. LANDI*

*DSEA - Dipartimento di Sistemi Elettrici e Automazione, Universitá di Pisa, Via Diotisalvi 2, Pisa, 56126, Italia.
#CESAME - Université Catholique de Louvain, Bât. Euler, Av.G.Lemaitre,4-6,B-1348 Louvain-la-Neuve, Belgique

Abstract. A variable structure cascade control is presented in order to avoid the increasing loss in the bandwidth of the outer loops, typical drawback of classical cascade control. In the choice of the coefficient of the VS-controller the compatibility of the command signals with the physics of the controlled plants is also guaranteed. As a test case a DC-motor speed and position control is considered.

Key Words. Cascade control; variable structure control; rule based controllers; process control

1. INTRODUCTION

In several applications cascade control (Buhler, 1979; Leonhard, 1985) is a well-established control technique. Designers agree that the most effective control method for electrical drives is a cascade scheme with an inner current control loop and an outer speed control loop, but cascade control is a powerful technique also for processes with a multi-loop hierarchical structure. Typically a series of nested loops is closed starting from the innermost to the outermost by imposing the largest bandwidth to the innermost loop and decreasing progressively the bandwidth going towards the outer loop. In a classical cascade control design an iterative procedure can be followed: at the beginning the innermost loop is designed going through the intermediate loops from the lower to the higher control levels. The main advantage of this design technique is that a single control loop is dealt with at each step: moreover any intermediate variable can be constrained in a natural way by limiting its reference signal. Unfortunately a multi-loop cascade scheme leads to an increasing loss in the bandwidth towards the outer loops. Nowadays the higher performance demand may require comparable bandwidths among the loops, invalidating the classical cascade control approach. A first solution was proposed by Leonhard (Fig.1): the inner loops should be activated by additional feed-forward signals derived from a reference generator. A quite natural development of this solution was the introduction of an additional adaptive control loop to a classical cascade scheme in order to counteract strong parametric variations or external disturbances (Courtiol and Landau, 1975; Lozano and Landau, 1981; Balestrino *et al.*, 1983; Stephan *et al.*, 1991).

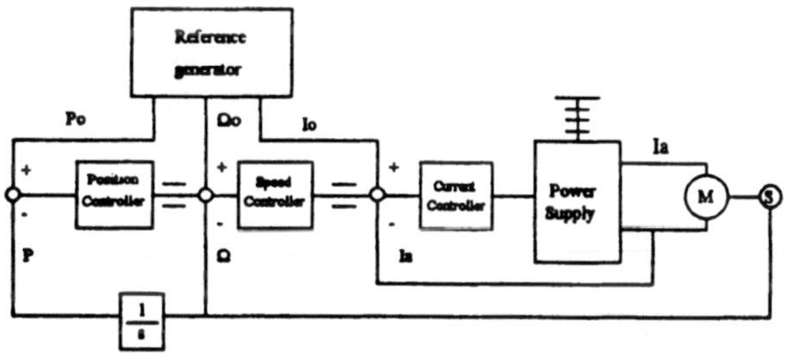

Fig.1. Cascade control structure of drive with feed-forward

In recent years the availability of microprocessors and of fast power switches at low cost allows to improve the technology of regulators and to consider nonlinear control techniques. A large amount of literature is concerned with sliding mode control (Utkin, 1974, 1993; Buhler, 1986) and a variety of VS controllers has been proposed.

In this paper the controllers of the outer loops are designed following the variable structure (VS) control approach (Balestrino and Landi, 1992), leading to a VS-cascade control. In fact the use of a VS-PI controller for the outer loops, in which the proportional and integral gains are functions of the error of the inner loops, leads to an easy tuning of the controller parameters for high order systems in such a way that a separation between the bandwidths is not required. A different problem, more important than the previous one, is given by the compatibility of the reference variation with the physics of the controlled plant. Such problem, mainly due to input saturation and to constrained variables, is more demanding in the case of large parameter variations and external disturbances. The VS-cascade control, with an additional prefiltering action, solves the problem in a simple and effective way. As a case study a DC-motor with an inner speed loop and an outer position loop is considered. Performance and robustness properties in the presence of a prefiltering action and/or of the VS-controller are analysed by means of several simulations.

2. VS-PI CONTROLLER

In this section a suitable modification to the standard proportional integral (PI) regulator by using variable structure laws is proposed. The structure of the proposed control laws is similar to the one (VS-PI) proposed by Balestrino *et al.* (1989, 1990, 1992, 1993) with proportional and integral actions. The overall logic of the VS-PI controller is briefly described by the following rules:

- for very large errors the integral control action vanishes and the proportional action approximates a relay control; the proportional gain must be chosen with upper bounds within the stability region and the input saturation gives a further bound on the maximum proportional action allowed;

- for small errors the proportional action is standard and the integral one becomes predominant: the zero steady-state error is guaranteed by the large integral action.

The general form of the VS-PI controller is:

$$C(s)= k_p(e)e + \frac{k_i(e)}{s} \tag{1}$$

where e is the error variable, i.e. the difference between the desired output r and the actual response y. Different coefficient variations as function of the error can be chosen: a particular simple choice has been proposed as follows

$$k_p(e) = \frac{K_p}{c_p}\,\frac{c_p}{1+c_p|e|} \tag{2}$$

$$k_i(e) = \frac{K_i}{c_i}\,\frac{c_i}{1+(c_ie)^2} \tag{3}$$

The smoothed transition from relay to proportional action due to (2) and (3) eliminates the typical drawback of relay control due to chattering with a very simple and easily implementable solution. Moreover during the transient, saturation is reached mainly owing to P action and windup and bumpless transfer problems (Hanus *et al.*, 1987; Campo and Morari, 1990) are effectively reduced. Values of the controller parameters are chosen in order to guarantee stability condition for the controlled process: once this condition is respected the tuning of the VS-PI is not critical at all. With reference to the simple double loop scheme of Figure 2, a VS-PI controller acts in the outer loops. The PI coefficients may be chosen following a nonlinear law as in (2), (3), but the error e must be now depending on the error of the nearest inner loop. In such a way a separation between the bandwidths in the different loops is no more requested.

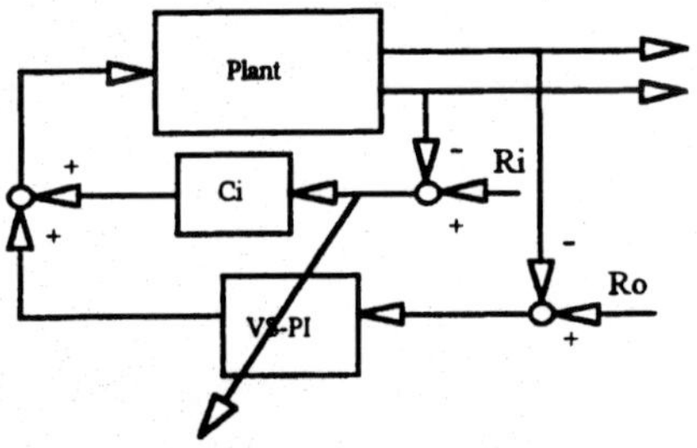

Fig.2. VS-cascade control structure: double loop scheme

2. PREFILTERING ACTION

In the control scheme illustrated in Fig. 3 a proportional controller with input saturation is considered. The external reference signal r is not fed directly to the input of the control system, but through a prefilter in order to tune the input variable r^*.

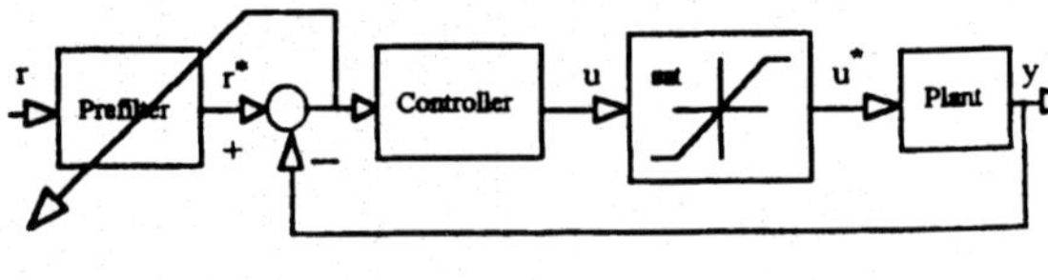

Fig.3. Prefiltering scheme

A prefiltering law may be chosen as:

$$r^* = r(k_1 + k_2 \exp(-e^2/2\sigma^2)) \tag{4}$$

or as

$$r^* = r(h_1 + \frac{h_2}{1+h_3|e|}) \tag{5}$$

The tuning of the parameters is not critical. For instance a practical choice of the parameter σ^2 is:

$$\sigma^2 = 0.01\,\frac{\text{saturation level}}{\text{controller proportional gain}} \tag{6}$$

Indeed the value of the parameter σ^2 can vary in a wide range without affecting the final responses.
As a case study consider the third order plant:

$$P(s)= \frac{k_m}{s[(sL_a+R_ak_m)(sJ_m+B_m)+k_mk_b]} \tag{7}$$

representing the transfer function of a DC-motor servo-drive, whose nominal parameters are shown in Table 1.

Table1 Nominal parameters of the tested process

Parameter	Value
Ra	*0.46 Ω*
km	*0.05 N m A*$^{-1}$
kb	*0.05 V rad*$^{-1}$*s*
Jm	*0.0025 kg m*
B	*0.0013 N m rad*$^{-1}$*s*
La	*0.0046 H*

As an example of prefiltering with the exponential law (4) the time responses in the case of a square wave reference are shown in Fig.4 and in Fig.5.

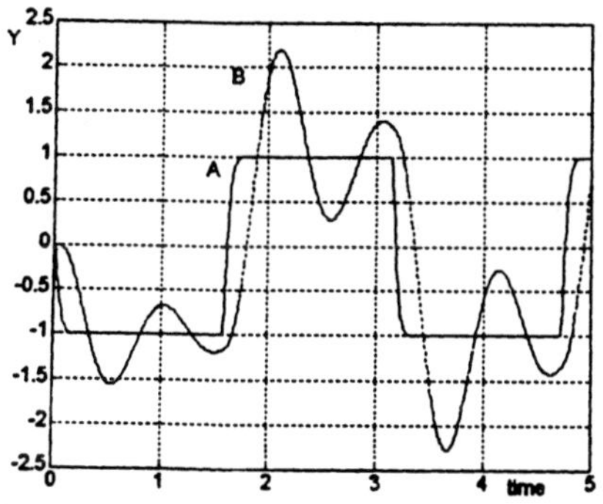

Fig.4. Time response (B) for square wave input (A)

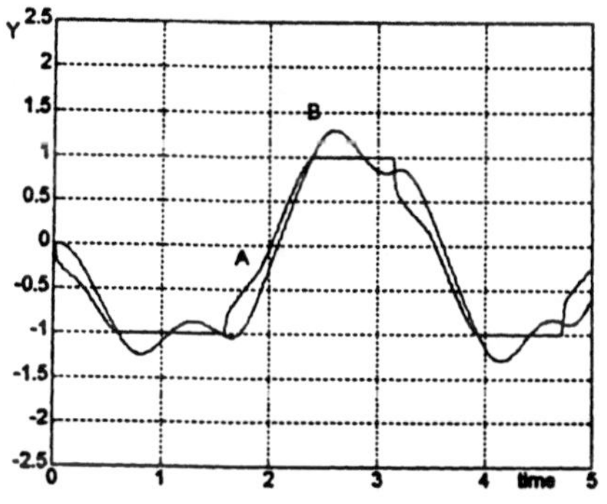

Fig.5. Time response (B) for prefiltered square wave input (A)

In Fig.4 a high overshoot of the time response with respect to the reference signal *r* indicates an inappropriate control action. If the proportional constant k_p (in the case study $k_p=1$) of the controller decreases, overshoot can be effectively reduced and the resulting response will be slower. In Fig.5 it can be noted that the system exhibits a considerable overshoot reduction with respect to the prefiltered reference signal *r**, even if k_p does not change. The prefiltering action reduces the sensitivity of the controlled plant with respect to parametric uncertainties in the process and to inappropriate tuning of the controller parameters. In this way the problem of the compatibility of the reference signals with the physics of the controlled plant in the presence of parametric variations, external disturbances and input saturations can be easily solved. Similar results can be obtained in the case of a prefiltering given by the expression (5). Several simulations have been performed in the case of square wave and of sinusoidal inputs, with different amplitudes and frequency, showing a remarkable improvement in the time responses with a prefiltering action.

3. VS-CASCADE CONTROLLER

Consider the double loop cascade control illustrated in Fig.6.

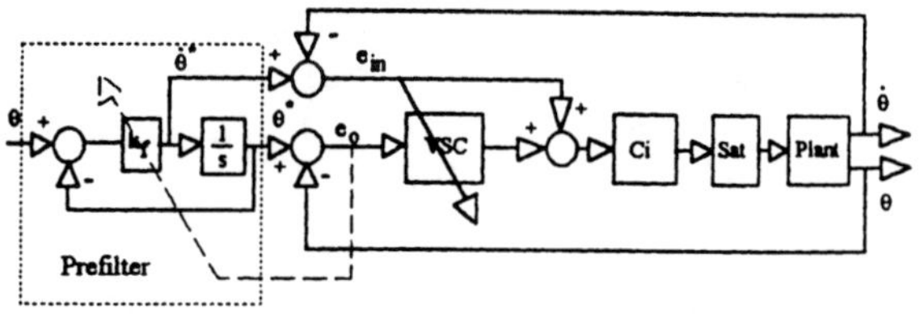

Fig.6. Double loop VS-cascade scheme with prefiltering

The controller of the outer loop may be designed following a modified version of the variable structure approach presented in section 2. Therefore the expression (1) becomes

$$C_o(s) = k_p(e_{in})e_o + \frac{k_i(e_o)}{s} \tag{8}$$

where e_{in} and e_o are the errors in the inner and in the outer loop respectively. The overall logic of the VS controller is now described by the following rules:

- for large errors in the innermost loop the outer controller approximates a small gain PI controller
- for small errors in the innermost loop the outer controller approximates a high gain PI controller.

The increasing loss in the bandwidth towards the outer loops is therefore avoided and a betterment in the counteraction of parametric variations (or external disturbances) is easily achieved. Using appropriate smooth functions for the integral and proportional coefficients, the typical problems of high gain systems (e.g. chattering and control discontinuity) are avoided. In the choice of the variable coefficients also the compatibility of the reference signals with the physics of the controlled plant must be considered. As in the previous section an exponential law for the proportional coefficient can be proposed

$$k_p = k_1 + k_2 \exp(-e_{in}^2/2\sigma^2) \tag{9}$$

For the integral term the proposed law is

$$k_i = \frac{K_i}{c_i}\frac{c_i}{1+(c_i e_{in})^2} \tag{10}$$

The coefficient σ^2 can be chosen following the same simple rule of (6). The prefiltering action is now less mandatory but may be still useful in the case of large reference variations. However the prefiltering action makes available a smoothed version of the derivative of the input reference.

As test case the process (7) is presented and several simulations has been performed considering a conventional cascade control, a VS-cascade control, with or without a prefiltering action. The innermost controller C_i is a standard PI regulator, while the outermost VS controller maintains only the proportional action.

In Fig.7 the position response for classical cascade control is shown. It must be remarked the performance deterioration due to excessive proportional action in the outermost loop. A further increasing of this action leads to instability.

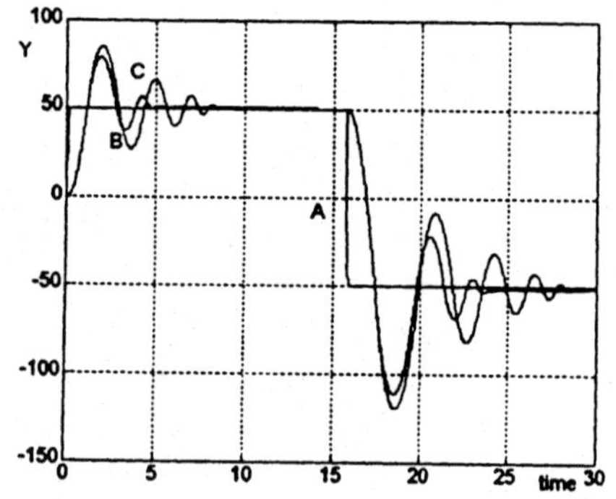

Fig.7. Time responses in the case of a classical cascade control. Case A: square wave set point; case B: time response if the outermost bandwidth is larger then the innermost one; case C: time response with equal bandwidths

In Fig.8 a comparison between classical cascade controller and VS-cascade controller with or without active prefiltering action is shown. The prefiltering action improves slightly the overshoot reduction.

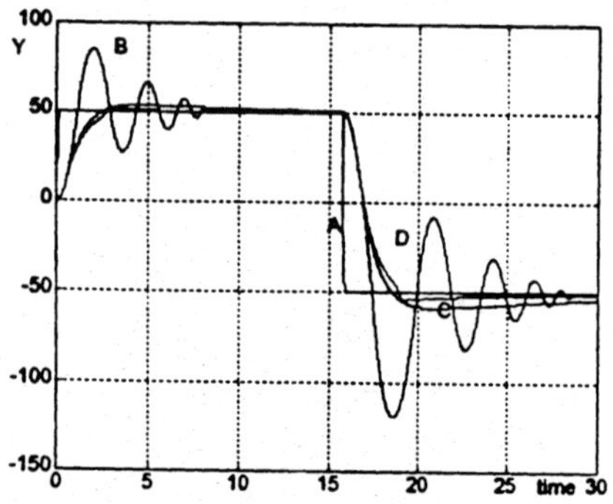

Fig.8. Time responses with a square wave set point (A). Classical cascade control with larger external bandwidth (B); VS-cascade control (C); VS-cascade control with active prefiltering action (D)

In Fig.9 a VS integral action is introduced in the innermost loop in order to reduce wind-up and bumpless transfer drawbacks. A smoother time response is shown and overshoot is totally avoided, at the expense of a small reduction of the rise time.

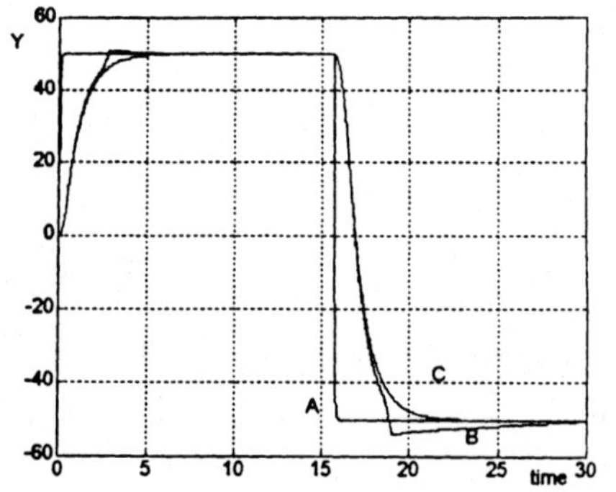

Fig.9. Time responses with a square wave set point (A). VS-cascade control (B); VS-cascade control with VS-I action (C)

In Fig.10a and in Fig.10b the two control actions are illustrated in the case of a classical cascade control and of a VS-cascade control with VS-I in the internal loop, respectively. In this last case the improvement in the control input is apparent and chattering is totally avoided. The saturation level is fixed at ± 20.

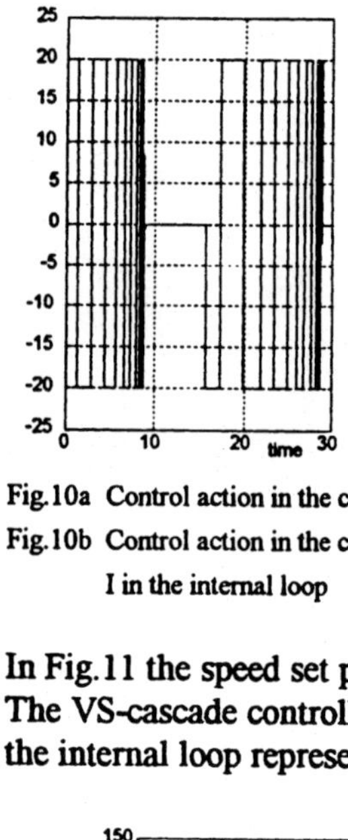

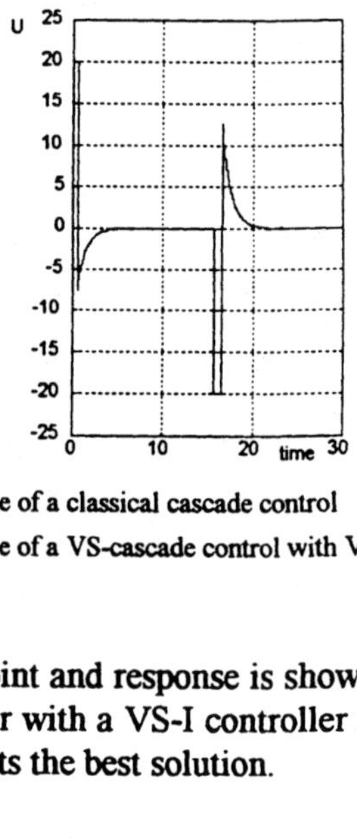

Fig.10a Control action in the case of a classical cascade control

Fig.10b Control action in the case of a VS-cascade control with VS-I in the internal loop

In Fig.11 the speed set point and response is shown. The VS-cascade controller with a VS-I controller in the internal loop represents the best solution.

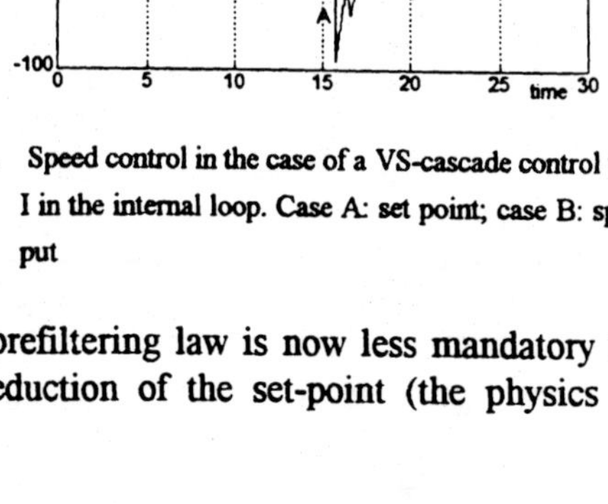

Fig.11. Speed control in the case of a VS-cascade control with VS-I in the internal loop. Case A: set point; case B: speed output

The prefiltering law is now less mandatory for the pre-reduction of the set-point (the physics of the plant is now respected through the combined action of the two VS-controllers), but it is remarkable that the prefiltering action gives an effective smoother speed reference signal.

A sudden large inertial variation during the simulations has been simulated in order to test the robustness of the proposed controller. The inertia value has been modified during the simulation and doubled in the interval from 10 to 23 sec. The VS-cascade control with VS-I action exhibits a very fast neutralisation of the disturbance, as shown in Fig.12.

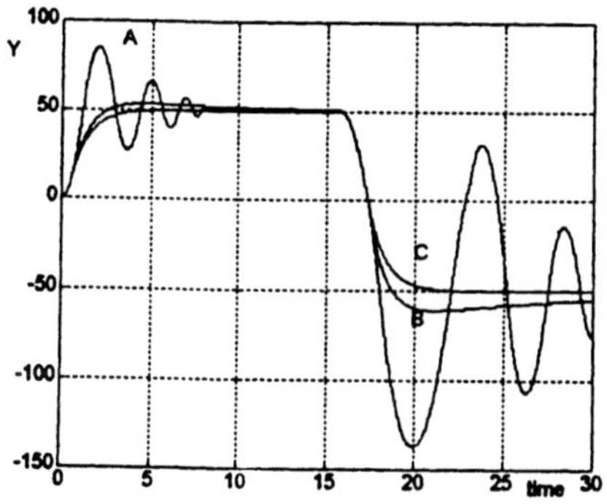

Fig.12. Time responses with a square wave set point and inertia variation. Cascade control (case A); VS-Cascade control (case B); VS-cascade control with VS-I action (case C). Inertia has been doubled (in t=10 sec) and returns to its nominal value (in t=23 sec)

All simulation results have been obtained by using the software package SIMULINK.

Several simulations have been performed in order to evaluate frequency responses of the system, typically by increasing the amplitude and/or the frequency of a sinusoidal reference signal. By using the proposed control the overall performance of the electrical drive is greatly improved over a classical design.

4. CONCLUSIONS

The proposed VS-cascade controller joins the advantages of the VS structure regulators with practical considerations about the compatibility of the command signals with the physics of the controlled plants. The controller parameters can be chosen by taking into account only few rough estimates of the saturation level. Such controllers can be easily implemented by using low cost microprocessors, so that their industrial use seems very attractive.

Acknowledgements

This work has been developed with the financial supports of MURST (Italian Ministry of University and Scientific Research).

6.REFERENCES

Balestrino, A., G. de Maria and L.Sciavicco (1983). Adaptive control design in servo-systems. *Third IFAC Symp. on Control in Power Electronics and Electrical Drives*, Lausanne, 125-131.

Balestrino, A., M.Innocenti and A.Landi (1989). Variable structure conventional controllers. *IFAC Low Cost Automation*, Milan, 187-191.

Balestrino, A. and A. Landi (1990). Intelligent variable structure control in electrical drives. *2nd IEEE Int. Workshop on Intelligent Motion Control*, Istanbul, **2**, 719-722.

Balestrino, A. and A. Landi (1992). Low cost robust nonlinear controllers. *Proc.of the IFAC Symp. on Intelligent Components and Instruments for Control Applications*, Malaga, pp.723-728.

Balestrino, A., A.Landi, G.Nardi and C.Scali (1993). Performance of variable structure PI controllers in the presence of uncertainty and saturation nonlinearities. *XII IFAC World Congress*, Sidney, **8**, 527-532.

Buhler, H., (1979). Electronique de reglage et de commande. Traité d'electricité, Vol.XVI, Georgi Ed., Lausanne.

Buhler, H., (1986). Reglage par mode de glissment. Presses Politechniques Romandes, Lausanne.

Campo, P.J. and M. Morari (1990). Robust control of processes subject to saturation nonlinearities. *Comp. &Chem Engng*, **14**, 343-358.

Courtiol, B. and I. Landau (1975). High speed adaptation system for controlled electrical drives. *Automatica*, **11**, 119-127.

Hanus, R., M. Kinnaert and J.L. Henrotte (1987). Conditioning technique, a general anti-windup and bumpless transfer method. *Automatica*, **6**, 729-739.

Leonhard, W., (1985). Control of electrical drives, Springer-Verlag, Berlin.

Lozano, R. and I. Landau (1981). Redesign of adaptive control schemes. *Int. J. Control*, **33**, 246-268.

MATLAB 4.0 with SIMULINK 1.2 (1993). The MathWorks, Inc. 24 Prime Park Way, Natick, MA 01760.

Stephan, R.M., V. Hahn, J. Dastych and H. Unbehauen (1991). Adaptive and robust cascade schemes for tyristor driven DC-motor speed control. *Automatica*, **27**, 449-461.

Utkin, V.I., (1974). Sliding mode and their application in variable structure systems. MIR Publ., Moscow.

Utkin, V.I., (1993). Sliding mode control design principles and applications to electric drives. *IEEE Trans. Ind. Electr.*, **40**, 23-36.

AN EDUCATIONAL PLATFORM FOR VARIABLE-STRUCTURE CONTROL OF SAFETY-RELATED SYSTEMS [1]

W. A. HALANG and M. WITTE

FernUniversität Hagen, Faculty of Electrical Engineering, D-58084 Hagen, Germany

Abstract. A model of robust variable-structure controllers and their use in distributed systems for optimisation of sub-processes is presented. The observer concept is introduced as a basis for variable-structure control. A platform to transfer the formal model into applications is described. With emphasis on student education, this platform provides an easy to use environment for control software design, correctness proof, and implementation as well as for its validation in a realistic testbed (conveyor belt model).

Key Words. Variable-structure control; safety-related systems; smart microcontroller; special-purpose architecture; education; software-implementation

1. INTRODUCTION

Current trends in data acquisition and control are characterised by a move towards intelligent digital sensors connected to distributed controllers using local area networks. In addition to overcoming the many problems associated with analogue transmission of low-level signals, this offers considerable potential for the design of advanced control systems involving interconnected local controllers, co-ordinated by global controllers. This has the practical advantage of allowing local controls to be designed and implemented independently to achieve local requirements such as accuracy, and then supplementing systems with higher-level controls to implement heat and mass balance, global optimisation etc. In addition, distributed systems possess a high degree of redundancy, ensuring that computer failures do not result in overall system failure as in centralised control systems. Furthermore, in safety-related situations such as hazardous areas in petroleum refineries, this offers scope for implementing robust control algorithms to overcome the disadvantages of conventional control, namely, that basic assumptions such as plant linearity, model structure and parameters, rarely hold for any length of time, and changes in operating level, load or disturbances invalidate the design, with consequent loss in accuracy and integrity, often resulting in instability. Modern multivariable design methods, which require estimation of all states to achieve the best control, are often highly sensitive to structural uncertainty. Since this affects other parts of large-scale control schemes through interacting paths, the global effect of a single controller deterioration may have far-reaching consequences for safety.

Implementation of an accurate and robust form of control for large-scale systems is required for any safety-related application avoiding the use of complex and unreliable multivariable, self-regulating, adaptive algorithms by the use of variable-structure (VS) controllers (Utkin, 1976).

The corresponding control algorithms have a non-linear form switching the controller gains in a discontinuous manner such that the system response trajectory is directed towards a predefined hyperplane in state space. When the hyperplane is reached, a sliding mode is achieved in which the trajectory switches about the hyperplane and slides down it to an equilibrium point, viz., the desired set-point. Once in sliding mode, systems react fast and possess a remarkable degree of insensitivity to structural uncertainty, plant disturbances,

[1] This work was supported under the Collaborative Research Grant CRG 900535 of the NATO Division of Scientific and Environmental Affairs.

or errors in state estimation. The effects of interactions between loops are then accommodated by interconnecting several controllers together to co-ordinate the controls subject to retaining integrity.

In this paper a concept for controlling safety-sensitive applications is worked out and prepared for teaching. According to a formal model, an environment is established allowing to transfer variable-structure control algorithms into an application of limited complexity. To meet the applications' safety requirements it is an integral part of the here presented training programme to subject any particular control and observer algorithm to a rigorous, formal correctness proof employing predicate calculus.

2. VARIABLE-STRUCTURE CONTROL ALGORITHM

Variable-structure control systems have been described in the literature especially by the Russian authors Emil'yanov and Utkin in the 1960ies and 1970ies. The state-variable equation of systems,

$$\dot{x} = Ax + Bu \quad (1)$$

is subjected to the control law $u = k^T x$, with $x \in \Re^n$ and $u \in \Re^m$, $m < n$. The coefficients of k depend discontinuously on the sign of $\sigma(x) \cdot x$, where $\sigma(x) = p \cdot x$ is a hyperplane of the state-space such that the trajectory x slides along σ. When this occurs the system is said to be in sliding mode. A stable sliding mode exhibits robustness in the presence of disturbances and parameter variations in the model as well as tighter control over the transient response leading to reductions in energy used.

The controller gain is given by

$$k = K + \Delta K \text{ with } K = A^T p / B^T p. \quad (2)$$

This control constrains the trajectory x to the hyperplane σ.

Recent work suggests that it may be substantially advantageous to decentralise process control systems using networking. Decentralisation can lead to cheaper, independent, and more powerful subsystems. At the same time it has become apparent that multivariable variable-structure controllers are increasingly expensive to implement on account of their complexity. Decentralised variable-structure controllers solve this problem by utilising local area networking on the following pattern. By splitting the system matrix, (1) can be written as

$$\dot{x} = Ax + Bu + Hx \quad (3)$$

and then decomposed into n subsystems

$$\dot{x}_i = A_i x_i + B_i u_i^l + B_i u_i^g + h_i x_i, i = 1, \ldots, n. \quad (4)$$

The sum $u = u_i^l + u_i^g$ is the overall system control and h_i represents interactions between subsystems. By constraining

$$u_i^g = -(B_i^T B_i)^{-1} B_i^T h_i x, \quad (5)$$

a VS controller can be designed for the local subsystem

$$\dot{x}_i = A_i x_i + B_i^T u_i^l. \quad (6)$$

If the system is decomposable, n scalar variable-structure controllers obviate the complexities of the multivariable case and lead to improved performance. Recent developments show that scalar VS controllers are more effective and versatile than multivariable ones, especially in the design of large systems decomposed into interconnected subsystems. The remainder of our discussion is therefore confined to scalar control.

For the above stated model (1) the sliding hyperplane $\sigma(x) = 0$ is feasible, i.e. it exists and is reachable, if

$$\sigma(x) \cdot \dot{\sigma}(x) < 0 \quad (7)$$

holds. With $\sigma(x) = p \cdot x$, this is the case if the feedback gains K_i satisfy the following inequalities:

$$\text{if } x_i p^T x < 0 \text{ then } K_i > (A^T p)_i / B^T p, \quad (8)$$

$$\text{if } x_i p^T x > 0 \text{ then } K_i < (A^T p)_i / B^T p. \quad (9)$$

Applications of VS control include servo regulations yielding fast, robust response with zero steady-state error, manoeverability control of spacecraft and VSTOL aircraft, the control of interconnected synchronous machines, energy management, and pollution control. Low order MIMO or SISO VS controllers have been applied in electrical power generation load balancing and frequency control as well as for protection control. After initially dominating in hydroelectric power stations and for the control of microscopic cell populations, the majority of VS utilisations can presently be found in the domain of mechanical systems, e.g. robots and manipulators implemented with microprocessors or the well-known overhead crane problem.

3. OBSERVER CONCEPT

Control and supervision of processes requires detailed information about the current states of the processes. Despite recent advances in sensor technology, important indicators of process states such as the chemical composition of a reaction mixture often cannot be measured directly and, hence, are not available for control purposes. Even if a variable can be measured on-line, the sensor dynamics may be too slow in comparison with the process dynamics resulting in poor control. Typical examples for that are temperature measurements. With the observer concept modern control theory provides a method for reconstructing unmeasurable variables. Missing information is obtained with the help of a mathematical process model simulated on-line and in parallel to the process. Measurable variables are used to update the model. As a result, the model follows the process as closely as possible. Thus, estimations of unknown quantities can be derived or sensor delays can be compensated. Furthermore, estimation provides more insight into the process and is a valuable tool for the early detection of faults and hazardous process states. These methods tend to be computationally demanding, so that their application could only begin with the advent of computers and modern process control systems.

3.1 Mathematical Process Model

The most important part of an estimator is a mathematical process model, with which the variations of process states are simulated on-line and in parallel to an actual process. The mathematical model also includes descriptions of the sensors. As input information, the model obtains values of all input variables and measurable disturbances and then predicts values for the internal input state x^* and output state y^*. Since there are errors in the initial input state x_0, the nonmeasurable disturbances of process w, and the measurements as well as there are modelling inaccuracies, the predicted output y^* and the measured value y of the output variable differ. The states of the model are corrected by a suitable feedback of this difference $y^* - y$, which is also called innovation. Thus, the estimated states closely follow the actual process states.

The accuracy of the estimation is determined by the accuracy of the model and the quality of the information obtained from the measurements. Both good measurements and accurate mathematical models are equally important preconditions for the use of observers. An observability analysis must be used to determine to which extent process states can be reconstructed from given measurements. The analysis may reveal that other quantities have to be measured or that the accuracy of measurements has to be increased.

Although still not the standard methodology, the new methods provided by modern control theory are increasingly used in difficult and complex processes for improving their operation, to meet tight product specifications, or to meet other requirements. One aim of applying advanced control methods is the improvement of product quality or to keep the product quality at a constant high standard. By increasing stability, however, these methods also make it possible to operate technical processes closer to the limits of stability or product specification and, hence, they provide more freedom in selecting and designing an apparatus.

3.2 Computing Tools for Process Control

Unfortunately, for a number of reasons, modern control methods are not widely used, yet. It is desirable to implement control algorithms as close to the process as possible. At low hierarchical levels, however, most of the presently available process control systems mainly provide discretised versions of classic PID controllers. These can be used to realise small fixed-parameter controllers, but it is not possible to implement complex control and observer algorithms.

Most modern controls involve mathematical formulae that can no longer be understood intuitively. Their design utilises computer-aided tools and is carried out in several stages. Apart from adaptive schemes, determining optimal controller parameters requires knowledge about the process, such as explicit state space models or frequency curves. So, the first stage is modelling the process, which can be done analytically or experimentally. Simulation runs are then carried out to study the general process behaviour and to verify a model by comparing simulated with experimental data. For a given model optimal control parameters are now calculated according to the specified optimisation criterion. Before implementation, simulation runs of process and controller show if the designed controller works satisfactorily or allow a choice between different controller types or structures. With regard to the implementation of an observer, for on-line simulation and estimation, the equations of a mathematical model have to be solved in real time, which often requires considerable computing power. Full process models can be too large to be solved on-line and have to be reduced. Even then, the required computing power imposes too heavy a burden on, or often exceeds the capacities of, contemporary process control systems.

4. TRAINING AND EDUCATION PLATFORM

Therefore, it appears appropriate to implement variable-structure controls and process observers on autonomous, dedicated microcontrollers, each assigned to exactly one control loop. To this end, a low-cost device had to be found, suitable both as an educational platform for the convenient implementation and validation of modern control techniques as well as for industrial use in large-scale control systems for safety-critical applications. The KAT-Ce (Himmerröder *et al*, 1986) microcomputer was selected, which comes as a kit to be self-assembled. This computer fits on a single Euro-sized board, employs an MC 68000 chip as processor and can be equipped with up to 128 kByte RAM and (EP)ROM each. The KAT-Ce contains an analogue-to-digital converter to perform measurements in the range from 0V to 2.55V. In order to perform control output, the computed incremental change in the control signal is fed into an integrated digital-to-analogue converter, whose output drives an actuator. Furthermore, the board comprises a timer, two separately programmable serial ports, and a 24 bits wide parallel, synchronous bus expansion interface. The latter allows for the attachment of user-specific devices and analogue peripherals of higher resolution.

A feature particularly important for training purposes is that the KAT-Ce can be run with different, specially designed operating systems and application software, that are currently available on the market. This software needs to be provided in ROM.

4.1 General Purpose Operation Mode

First of all, there is a small, but powerful monitor system interprated. As the KAT-Ce lacks its own peripheral devices, the monitoring kernel can address mass storage media, I/O-devices and interface ports of the host that it is attached to. This host is used both as a terminal for user interaction and for storing self-developed applications. The latter can be worked out with the supplied 68000 assembler or the built-in Pascal compiler. Sufficient for simple control algorithms and student education, Pascal is easy to learn and allows simple I/O-handling as well as comfortable data manipulations. More importantly, this programming language is widely spread on commonly used programming platforms, such as personal computers or the Macintosh line.

4.2 Real-Time Application Mode

Alternatively, the KAT-Ce comes with two EPROMs containing a complete PEARL (DIN 66 253, 1981, 1982 and 1989) programming system and the real-time operating system RTOS-UH. Conceived by control engineers, PEARL is easy to use and to read and exhibits the best expressive power among all available real-time languages (Halang and Stoyenko, 1990). This allows students to concentrate fully on the control aspects when they are to implement a VS control strategy and a process observer. The same holds with respect to programming and operating the KAT-Ce, which is carried out via one of the serial ports from standard personal computers or any other platform that can be driven as a terminal, on which the students can employ the tools they are already accustomed to.

The only complex operations occurring in the considered VS control and observer algorithms are floating point addition and multiplication. According to the speed requirements of the loops controlled, they are either implemented in software (of the RTOS-UH run-time system) or by an arithmetic co-processor chip, which can be attached to the expansion bus.

Using a conveyor belt model as a testbed, laboratory experiments are carried out, controlling the belt's speed with the methods and the microcontroller as outlined above.

5. AN EDUCATIONAL ENVIRONMENT

5.1 Implementation of Algorithm

In students laboratory classes, a formerly developed variable-structure control algorithm has to be adjusted and implemented using the PEARL (DIN 66 253) programming language supplied with the KAT-Ce. Typically, such a control program reads as follows:

```
MODULE CONTROLLER;

SYSTEM;
TERM: A1;

PROBLEM;
SPC TERM DATION INOUT ALPHIC CON-
TROL(ALL);
```

```
DCL N INV FIXED INIT(10);

DOT: PROCEDURE(XI () INV FLOAT IDENT,
YI () INV FLOAT IDENT)
RETURNS ( FLOAT );
DCL WERT FLOAT INIT(0.0);

FOR I FROM 1 TO N REPEAT
WERT:=WERT+ XI(I) * YI(I);
END;
RETURN(WERT);
END;

MDOT: PROCEDURE( Mat (,) INV FLOAT
IDENT,
Vec ( ) INV FLOAT IDENT,
Line INV FIXED ) RETURNS( FLOAT );
DCL WERT FLOAT INIT(0.0);

FOR I FROM 1 TO N REPEAT

WERT := WERT + Mat( Line, I ) * Vec( I );
END;
RETURN( WERT );
END;

DESIGNU : PROCEDURE( A ( , ) INV FLOAT
IDENT, b () INV FLOAT IDENT, c () INV
FLOAT IDENT, dk () INV FLOAT IDENT,
x () INV FLOAT IDENT ) RETURNS(FLOAT);
DCL bdotc FLOAT INIT(0.0);
DCL u FLOAT INIT(0.0);
DCL i FIXED INIT(1);
bdotc := DOT( b, c );
IF DOT(c,x) * bdotc < 0 THEN
WHILE i ≤ N REPEAT
IF( x(i) < 0.0 ) THEN
u := u + ( MDOT(A, c, i) / bdotc + dk(i)) *
x(i);
ELSE
u := u + ( MDOT(A, c, i) / bdotc - dk(i)) * x(i);
FIN;
i := i+1;
END;
ELSE
WHILE i ≤ N REPEAT
IF( x(i) < 0.0 ) THEN
u := u + ( MDOT(A, c, i) / bdotc - dk(i)) * x(i);
ELSE
u := u + ( MDOT(A, c, i) / bdotc + dk(i)) *
x(i);
FIN;
i := i+1;
END;
FIN;
RETURN( u );
END;

MODEND;
```

5.2 Formal Correctness Proof

For application in safety-related environments, a controller must be highly dependable. Therefore, its software may not contain any errors which, for software, are inherently design errors with a permanently latent presence. For these reasons formal correctness proofs of control programs are considered to be an integral part of any feasible implementation and, hence, emphasised in our environment. Predicate calculus (Deijkstra and Feijen, 1984) is used as formal correctness proof methodology.

The latter is based on Hoare-triples having the form

$$\{ P \} \; S \; \{ Q \}$$

in which S is a statement with predicate P as its pre-condition and predicate Q as its post-condition. For each statement in a given control program, such as the one listed in the previous section, an appropriate pre-condition and post-condition needs to be derived from the specification. Applying predicate calculus it then must be shown that the pre-condition P and the execution of the statement S imply the post-condition Q. Such proofs are generally rather lengthy and cannot be given here due to lack of space.

5.3 A Testbed for Validation

The laboratory experiment used as an educational testbed deals with keeping the speed of a conveyor belt at a constant level. Of course, any industrial use of conveyer belts demands individual consideration of the belts' speeds. The far reaching consequences of inadequate speed is easily imaginable. Many potential users of a conveyor will specify a speed that is unnecessarily high, as they do not appreciate that items will come off a conveyor belt at the same rate as they are put on. It is useless, sometimes even dangerous to operate too fast. The speed of a belt is dictated by the process and may require adjustments for changes in the process flow, e.g., different production rates or a wide scope of different loads, which can be encountered by applying the described model of VS control.

The schematic link between the personal computer being used as a terminal, the KAT-Ce, and the conveyor belt drive can be seen from Fig. 1.

The DC-shunt motor of the conveyor belt is supplied with a variable voltage of up to 180V, which requires an adjustment between the maximum input and output voltages of the KAT-Ce of about 2.5V towards 180V and vice versa. Furthermore, the required power for the actuation system exceeds the potential of the KAT-Ce's interface driver, so that a specially designed interface amplifier had to be developed. In order to cover the motor's operating voltage, the originally supplied eight bits wide digital-to-analogue (DAC) and analogue-to-digital (ADC) converter turned out to be insufficient. Therefore, both a higher resolution DAC and ADC, respectively, were added, which need to be addressed via bus expansion slot provided.

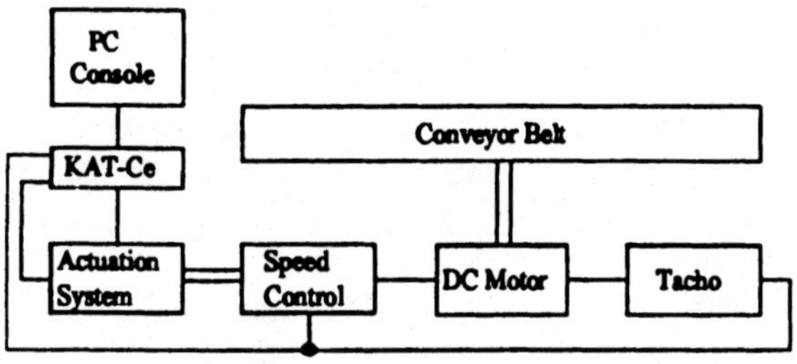

Fig. 1. Link between application console and conveyor belt

6. REFERENCES

Dijkstra, E.W., and W.H.J. Feijen (1984). *Een methode van programmeren*. Academic Service, The Hague.

DIN 66 253. *Programming Language PEARL*. Part 1 *Basic PEARL*, 1981, Part 2 *Full PEARL*, 1982, Part 3 *Multiprocessor-PEARL*, 1989. Beuth-Verlag, Berlin-Cologne.

Gerlach, J. and S. Frauenfelder (1992). *Formal correctness proof of a variable-structure control algorithm*. Internal report, Department of Computing Science, University of Groningen.

Halang, W.A. and A.D. Stoyenko (1990). *Comparitative Evaluation of High Level Real Time Programming Languages*. Real-Time Systems 2, pp. 365 – 382.

Himmeröder, H.-J., W. Mayer-Gürr and R.M. Toschke (1986). *c't-KAT-Ce — Ein 68000-Einplatinensystem*. c't 11, pp. 86 – 92 and c't 12, pp. 156 – 161. Verlag Heinz Heise, Hannover.

Utkin, I.V. (1976). *Control Systems of Variable Structure*. Wiley and Sons, Toronto.

ANALYSIS OF ACHIEVABLE PID PERFORMANCES VERSUS OPTIMAL LQR CONTROL

R. ARGELAGUET *#, M. PONS $, J. QUEVEDO *, J. AGUILAR MARTIN * +

** Dept. ESAII, Unive.Politècnica de Catalunya. ETSEIT, C/ Colom, 11, 08222 Terrassa, Spain*

+ LAAS-CNRS, 7, Av. Colonel Roche, 31077 Toulouse Cedex, France

$ Dept. Matemàtica Aplicada III, Universitat Politècnica de Catalunya EUPM, Avda. Bases de Manresa, 61-73, 08240 Manresa, Spain

 # Dept. Enginyeria Electrònica, Universitat Politècnica de Catalunya EUPM. Avda. Bases de Manresa, 61-73, 08240 Manresa, Spain

Abstract: Supervision of process control techniques can be based either on the observation of the real behavior or on the theoretical achievable performances of the closed loop system.
In this paper, the PID regulator is taken as an optimal LQR controller applied to an extended plant, in order to weight the integral of the output. By this statement of the problem, some properties of the closed loop system are studied as a function of the integral weight. As expected, the cost function of the PID increases with the weight, and we analyze its asymptotic behavior. The robustness loop margin decreases, and it is possible to determine which values of the parameters in the cost function define a PID with an acceptable margin. Moreover, the existence of an asymptotic PID is proved.

Keywords: Supervision, Three term-controller, LQR

1. INTRODUCTION

PID tuning methodology is oriented by practical desired performances (Ziegler Nichols, etc, ...). Nevertheless, none of those methods give a basis for a comparison neither they open a way to evaluate what are the achievable performances of the closed loop system.

On the contrary optimization methods (LQR, LQG, minimum time, ...), depend explicitly on a valuated criterion, therefore the performance can be located with respect to particular values.

When imbedding PID design problem in linear quadratic optimization theory, it appears that some robustness properties may diminish, so it is useful to be able to link robustness with the quadratic cost parameters. Conversely the components of the criterion may be analyzed in terms of the regulator design parameters.

2. PID AS AN OPTIMIZATION LQR PROBLEM

We consider a linear second order plant with input u(t), output y(t) and transfer function:

$$P_0(s)=\frac{K\left(1-\frac{T}{2}s\right)}{(1+\tau s)\left(1+\frac{T}{2}s\right)}=\frac{-b_1 s+b_0}{s^2+a_1 s+a_0} \qquad (1)$$

which approximates, at low frequencies, a first order plant with a pure delay T. The state space realization of

the plant is:

$$\dot{x}=Ax+Bu$$
$$y=Cx$$

where $x=(x_1\ \ x_2)^T$ is the vector

of state variables

$$A=\begin{bmatrix}0 & 1\\ -a_0 & -a_1\end{bmatrix}\quad B=\begin{bmatrix}-b_1\\ b_0+a_1 b_1\end{bmatrix}$$

$$C=[\ 1\quad 0\ ]$$

and the parameters b_1, b_0, a_1, a_0, are given by

$$b_0 = \frac{2K}{\tau T} \qquad a_0 = \frac{2}{\tau T}$$

$$b_1 = \frac{K}{\tau} \qquad a_1 = \frac{T + 2\tau}{\tau T}$$

If we assume that all states are measured and available for feedback, it is well known that there exist a unique full state gain matrix $F = (F_1\ F_2)$ such that the control law $u^* = -Fx$ minimizes the cost function

$$J_0(u) = \int_0^{\infty} (y^2 + \rho u^2)\, dt \qquad (2)$$

for any positive value of the parameter ρ.

For the plant we are considering (1), this control law can be expressed as

$$u_0^*(t) = -K_P\, y - K_D \frac{dy}{dt} \qquad (3)$$

that is to say, the linear quadratic optimal regulator is in this case a PD controller.

It is known that an integrator is needed so as to transform the system from "type 0" to "type 1", to achieve a zero steady-state error for a constant reference input.

This can be accomplished by augmenting the plant with the integral of the output and considering a new LQR problem, weighting the integral output in the cost function.

The augmented plant is:

$$A^* = \begin{bmatrix} A & 0 \\ C & 0 \end{bmatrix} \qquad B^* = \begin{bmatrix} B \\ 0 \end{bmatrix}$$

and the new cost function

$$J_\lambda(u) = \int_0^{\infty} (y^2 + \rho u^2 + \lambda z^2)\, dt$$

$$\text{where } z = \int_0^t y\, dt$$

is the new state variable, and $\lambda \geq 0$.
The control law

$$u_\lambda^* = -F_1 x_1 - F_2 x_2 - F_3 z$$

which minimizes J_λ can be expressed now by

$$u_\lambda^*(t) = -K_P\, y - K_D\, \dot{y} - K_I \int_0^t y\, dt$$

The controller F that we obtain depend obviously of the two parameters ρ, λ which appear in the expression of J_λ.

Fixed ρ, any value of $\lambda > 0$ defines a PID controller and, in order to know which is the best value of this parameter to obtain different design specifications, it is interesting to study the controller performance for different values of λ.

In order to determine the expression of the regulator parameters in terms of λ and ρ we have to solve the Riccati equation:

$$A^{*T} P + P A^* - \rho^{-1} P B^* B^{*T} P + Q = 0$$

where

$$Q = \begin{bmatrix} C^T C & 0 \\ 0 & \lambda \end{bmatrix}$$

In [1], the analytical solution of this equation, that is to say, the expression of the matrix P as a function of the parameters ρ, λ is obtained. From P we can explicitly obtain the parameters F_1, F_2, F_3 and therefore K_P, K_D, K_I of the controller in terms of ρ and λ.

3. ROBUSTNESS LOOP MARGIN

In order to measure the robustness of the system we will use the loop margin. The loop margin M is simply the minimum distance in the complex plane between the Nyquist plot of the loop gain and the critical point -1:

$$M = \min_{\omega} \left| 1 + P_0(j\omega) F(j\omega) \right|$$

The loop margin is connected to the sensitivity transfer function S, which relates the external input d to the output y

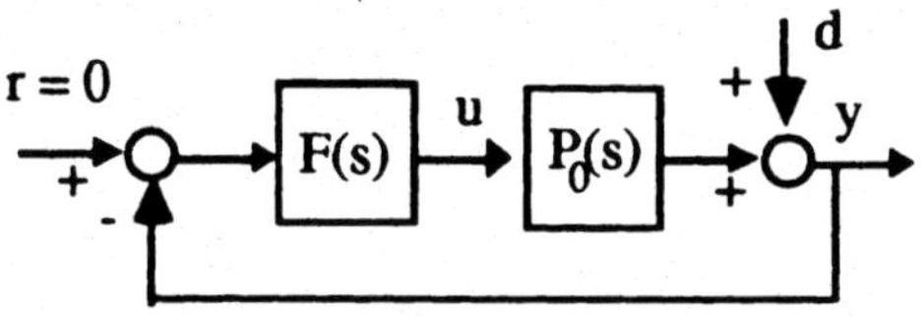

We can write $M = \frac{1}{\| S(j\omega) \|_\infty}$, where $\| \cdot \|_\infty$ indicates the infinity norm

It is not difficult to see that for the system composed of the plant (1) and the PID regulator defined from ρ, λ, the expression of the loop margin is:

$$M = \frac{1}{1 + b_1 F_2} \qquad (4)$$

If ρ is fixed and λ increases, we know that F_2 also increases and, therefore, M decreases as expected. In the limit, when $\lambda \to \infty$ we have $F2 \to \infty$ and $M \to 0$.

On the other hand, if λ is fixed and ρ increases, then F2 decreases, and when $\rho \to \infty$ we have $F2 \to 0$ and $M \to 1$.

In this way, we can compensate the loss of robustness caused by the introduction of the parameter λ by augmenting ρ.

Specifically, given any loop margin M1, it is necessary to take $\rho > \rho 1$ to obtain $M > M1$. Once ρ is selected, there exist λ_ρ such that $M > M1$ if and only if $\lambda < \lambda_\rho$

For example, to obtain a loop margin $M > 0.5$ we can take any

$$\rho > \frac{K^2}{\left(1+\frac{2\tau}{T}\right)\left(3+\frac{2\tau}{T}\right)}$$

and, once ρ is selected, choose any λ such that

$$0 \le \sqrt{\lambda} <$$

$$\frac{4\sqrt{\rho}}{KT}\left[1+\frac{2\tau}{T}-\frac{1}{2}\sqrt{\left(1+\frac{2\tau}{T}\right)\left(1+\frac{6\tau}{T}\right)+\frac{K^2}{\rho}}\right]$$

4. PID REGULATOR AND LQR COST

In this section we study the increase of the cost function when an integral weight $\lambda \neq 0$ is introduced. Specifically, for any value of the parameter ρ, we compare

$$J_0(u_0^*),\ J_0(u_\lambda^*) \text{ and } J_\lambda(u_\lambda^*)$$

It is clear that

$$J_0(u_0^*) \le J_0(u_\lambda^*) \le J_\lambda(u_\lambda^*)$$

All tree values are increasing functions of ρ, but their asymptotic behavior is very different.

While $J_0(u_0^*)$ tends to a finite limit as $\rho \to \infty$, $J_0(u_\lambda^*)$ and $J_\lambda(u_\lambda^*)$ tend to infinity. To be precise, it can be proved that

$$\lim_{\rho\to\infty} J_0(u_0^*)$$

$$= \frac{1}{2a_0a_1}\left((a_1x_1(0)+x_2(0))^2 + a_0x_1(0)^2\right)$$

that is to say

$J_0(u_0^*)$ tends asymptotically to a straight line which depends on the initial conditions of the state variables. On the other hand,

$$\lim_{\rho\to\infty} \frac{J_0(u_\lambda^*)}{\sqrt{\rho}} = \frac{\sqrt{\lambda}}{2a_0b_0}(a_1x_1(0)+x_2(0))^2 = \gamma$$

and it is not difficult to see that

$$\lim_{\rho\to\infty} \left(J_0(u_\lambda^*) - \gamma\sqrt{\rho}\right) \text{ is finite,}$$

what proves that $J_0(u_\lambda^*)$

tends asymptotically to a parabola.

Finally, $\lim_{\rho\to\infty} \frac{J_\lambda(u_\lambda^*)}{\sqrt{\rho}} = 2\gamma$ and $J_\lambda(u_\lambda^*)$

tends also to a parabola.

5. ASYMPTOTIC PID

For any positive parameters λ, ρ the closed loop transfer function for the corresponding PID controller is:

$$\bar{F}(\rho,\lambda) = \frac{P_0F}{1+P_0F} = \frac{F_2(b_0-b_1s)\left(s+\frac{b_1F_3}{b_0F_2}\right)}{s^2+\frac{b_1}{b_0}As+b_1F_3}$$

$$\text{where } A = \sqrt{a_0^2+\frac{b_0^2}{\rho}+2\frac{b_0^2}{b_1}F_3}\ .$$

It is not difficult to analyze the location of the poles as λ changes, for a fixed value of the parameter ρ.

. If $\lambda = \lambda_\rho$, where $\lambda_\rho = \frac{b_1^2\rho}{4b_0^4}\left(a_0^2+\frac{b_0^2}{\rho}\right)$,

There is a two order pole at

$$-s_\rho = \frac{b_1}{b_0}\sqrt{\frac{1}{2}\left(a_0^2+\frac{b_0^2}{\rho}\right)}$$

For $\lambda < \lambda_\rho$ there are two different negative real poles which tend respectively to $-2s_\rho$ and 0 when $\lambda \to 0$. In this way, it is clear that the closed loop transfer function for the PID obtained for $\lambda=0$ is

$$\bar{F}(\rho,0) = \frac{F_2(b_0-b_1s)}{s+2s_\rho}\ .$$

Finally for $\lambda > \lambda_\rho$ there are two complex conjugates poles which tend to infinity as λ increases. In this case, the transfer function tends to

$$\bar{F}(\rho,\infty) = \frac{(b_0 - b_1 s)(s + a_1)}{a_1 b_0}$$

when $\lambda \rightarrow \infty$.

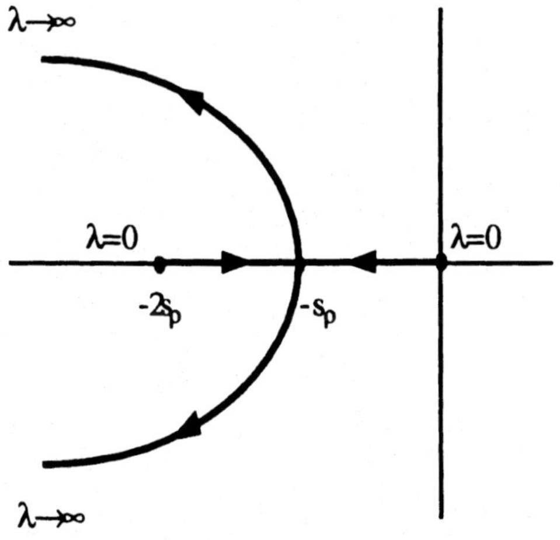

It can be proved that $\bar{F}(\rho,\infty)$ is the closed loop transfer function of a PID defined by the parameters: :

$$K_P^* = \lim_{\lambda \to \infty} K_P = \frac{b_0 + a_1 b_1}{b_1^2} = \frac{1}{K}\left(1 + 4\frac{\tau}{T}\right)$$

$$K_D^* = \lim_{\lambda \to \infty} K_D = \frac{1}{b_1} = \frac{\tau}{K}$$

$$K_I^* = \lim_{\lambda \to \infty} K_I = \frac{a_1 b_0}{b_1^2} = \frac{2}{KT}\left(1 + 2\frac{\tau}{T}\right)$$

It is interesting to notice that this PID limit is, in fact, independent of ρ. Thus , any PID obtained from any particular values of the parameters ρ, λ lies between the optimal LQR regulator obtained for $\lambda = 0$ (PD controller) and a PID limit obtained by increasing to infinity the integral weight.

5. ACKNOLEDGEMENT

This work has been supported by the Spanish Board for Research and Technology (CICYT) under contract TAP93-0596-C04, and the paper has been partially supported by the group for the Study and Research in Automatic Control (CERCA).

6. REFERENCES

Argelaguet, R., M. Pons, J. Quevedo, J. Aguilar Martin. Analysis of the PID controllers designed from optimal linear regulator theory. To appear.

Aström, K.J. (1991). Assessment of Performance of simple feedback loops. *Int. J. Adapt. Control & Sig. Proc.* No.5, pp. 3-19.

Athans, M. (1971). On the design of PID controllers using optimal linear regulator theory. *Automatica*, vol. 7, pp. 643-647. Pergamon Press.

Boyd, S., C. Barrat, S. Norman, (1990). Linear Controller Design: Limits of Performance Via Cònvex Optimization. Proc. IEEE, vol. 78, No.3.

Parker,K.T., B.A., M. Sc. (1972). Design of proportional integral derivative controllers by the use of optimal linear regulator theory. *Proc. IEEE*, vol.119, No.7.

THE SAMPLING PERIOD AS A CONTROL PARAMETER

I. LOPEZ, S. DORMIDO and F. MORILLA

UNED, Dpto de Informática y Automática, Avda Senda del Rey s/n, 28040 Madrid, Spain

Abstract. When a computer is used in the control of continuous processes, there arises an important parameter, sampling period T. In the analysis and design of digital control systems it is usual to transfer the results obtained for analog systems, doing T small. This procedure, valid for many processes, in some others can cause control deterioration and, besides, the advantages inherent to computer controlled systems can be lost. These are the purposes of the present paper: 1.-Give some rules for the selection of T as a control parameter. 2.-Compare the result obtained for the continuous case.

Key Words. PID control; Sampled data systems; Control system design; Digital control; Time-domain analysis.

1. INTRODUCTION

When a computer is used in the control of continuous processes, an important parameter is the sampling period T, because it has a great influence on the time response characteristics of the system (Åström and Wittenmark 1990).

The problem which arises is, once it is known the model of the process Gm(s), the time response characteristics and the controller Gc(z), to obtain the tuning formulas which allow to adjust the parameters of Gc(z) and the sampling period.

For the adjustment of the control parameters it can follow two procedures:

1.-To design the controller as if it were continuous and to adjust T, by means of simulation, in such a manner that the system response will be similar to the response obtained by continuous control.

2.-To consider the discrete behaviour of the system and to adjust T as a control parameter (Dormido *et al.* 1991).

The purpose of the present paper is twofold: 1.-To obtain the tuning formulas by using the tuning method by continuous and discrete pole placement (Åström and Hägglund 1988). 2.-To compare the results obtained, when you use both, in computer control.

The paper is organized as follows: The controllers and models to use are introduced in sections 2 and 3. The tuning method, formulas and criteria, are showed in sections 4, 5 and 6. Finally in sections 7 and 8 some examples and conclusions are given.

2. CONTROLLER

Due to its wide in industrial use, the controller will be of a PID type, in some of its versions P, PI, PD or PID (Shinskey 1989).

The basic algorithm of PID continuous control is

$$U(s) = Kc\left(1 + \frac{1}{sTi} + sTd\right)E(s) \quad (1)$$

and its discrete version can be expressed as

$$U(z) = Kc\left(1 + \frac{T}{Ti}\frac{z}{z-1} + \frac{Td}{T}\frac{z-1}{z}\right)E(z) \quad (2)$$

The continuous and discrete controllers can be given by the following expression

$$Gc(s) = S(s) / R(s) \quad (3)$$

$$Gc(z) = S(z) / R(z) \quad (4)$$

where S(s), R(s) and S(z), R(z) are given in tables 1 and 2 respectively.

Table 1 Continuous Controllers

Gc(s)	S(s)	R(s)
P	$S(s) = s_0$ $s_0 = Kc$	1
PI	$S(s) = s_0 s + s_1$ $s_0 = Kc \quad s_1 = \frac{Kc}{Ti}$	s
PD	$S(s) = s_0 s + s_1$ $s_0 = KcTd \quad s_1 = Kc$	1
PID	$S(s) = s_0 s^2 + s_1 s + s_2$ $s_0 = KcTd \; s_1 = Kc \; s_2 = \frac{Kc}{Ti}$	s

Table 2 Discrete Controllers

Gc(z)	S(z)	R(z)
P	$S(z) = s_0$ $s_0 = Kc$	1
PI	$S(s) = s_0 z + s_1$ $s_0 = \frac{Kc(Ti + T)}{Ti} \quad s_1 = -Kc$	z-1
PD	$S(s) = s_0 z + s_1$ $s_0 = \frac{Kc(T + Td)}{T} \quad s_1 = -\frac{KcTd}{T}$	z
PID	$S(s) = s_0 z^2 + s_1 z + s_2$ $s_0 = \frac{Kc(T\,Ti + Td\,Ti + T^2)}{Ti\,T}$ $s_1 = -\frac{Kc(T + 2Td)}{T} \quad s_2 = \frac{KcTd}{T}$	z^2-z

3. PROCESS MODELS

The processes will be continuous and described by means of

$$Gm(s) = B(s) / A(s) \tag{5}$$

If the behaviour of the D/A converter is that of a zero order holder, $Go(s)=(1-e^{-Ts})/s$, the discrete models will be obtained from (Phillips and Nagle 1984)

$$Gm(z) = (1 - z^{-1}) Z[Gm(s) / s] \tag{6}$$

resulting

$$Gm(z) = B(z) / A(z) \tag{7}$$

In the following the models of first and second order will be considered:

1.-Model 1

$$A(s) = s + a_1, \; B(s) = b_1 \tag{8}$$

$$A(z) = z + a_1', \; B(z) = b_1' \tag{9}$$

2.-Model 2

$$A(s) = s^2 + a_1 s + a_2, \; B(s) = b_1 \tag{10}$$

$$A(z) = z^2 + a_1' z + a_2', \; B(z) = b_1' z + b_2' \tag{11}$$

4. TUNING METHODS

Given Gm(s), Gc(z) and the time response characteristics, the parameters of Gc(z) and the sampling period T will be obtained by means of the following methods.

4.1. *Method by continuous pole placement*

Step 1: Calculate Gc(s) from Gc(z).

Step 2: From Gl(s) = Ga(s) / (1+Ga(s)) with Ga(s)=Gc(s)Gm(s), to deduce the system response Y(t). The Gl(s) parameters according to the characteristics specified can be obtained from y(t), and so the roots of the characteristic eq.

Step 3: Equalizing coefficients, to obtain from 1+Ga(s)=0 the relations between the roots obtained in step 2 and the parameters of Gm(s) and Gc(s).

Step 4: Calculate Gc(z) parameters and to adjust T in such a manner that the characteristics specified are verified.

4.2. *Method by discrete pole placement*

Step 1: Calculate Gm(z) from Gm(s) and (7).

Step 2: .-The roots location will be designed by the specified characteristics if these can be determined in the discrete domain, or, if the characteristics are referred to the system continuous behaviour from the roots location of a continuous system Gle(s), equivalent to Gl(z). In this case the sampling period T permit to specify one additional characteristic or the number of samples taken between different time instants.

Step 3: -Similarly as in the continuous case, to obtain the relations between roots location and the parameters of Gm(z) and Gc(z), and calculate Gc(z) parameters and T (Roffel *et al.* 1989).

5. TUNING FORMULAS

Stable systems with monotonous responses (if the order of the system is smaller that three) and/or underdamped response (if the order of the system is grater than one), are formed by controllers type PID and processes given by the models 1 and 2 given in section 3. The systems to be considered will be model 1 with P and PI controllers, and model 2 with P, PI, PD and PID controllers.

Table 3 Characteristic equations

SYST	CHARACTERISTIC EQ.
1 P	$s+a_1+s_0b_1=0$ $s+r_1=0$
1 PI	$s^2+(a_1+s_0b_1)s+s_1b_1=0$ $(s+r_1)(s+r_2)=0$ or $s^2+p_1s+p_2=0$
2 P	$s^2+a_1s+a_2+s_0b_1=0$ $(s+r_1)(s+r_2)=0$ or $s^2+p_1s+p_2=0$
2 PI	$s^3+a_1s^2+(a_2+s_0b_1)s+s_1b_1=0$ $(s+r_1)(s^2+p_1s+p_2)=0$
2 PD	$s^2+(a_1+s_0b_1)s+a_2+s_1b_1=0$ $s^2+p_1s+p_2=0$
2 PID	$s^3+(a_1+s_0b_1)s^2+(a_2+s_1b_1)s+b_1s_2=0$ $(s+r_1)(s^2+p_1s+p_2)=0$

5.1. *Continuous Control*

From (3) and (5) the characteristic eq. of the continuous system can be expressed

$$R(s)A(s)+S(s)B(s)=0 \tag{12}$$

R(s), S(s) are shown in table 1 and A(s), B(s) in (8) and (10). In table 3 the characteristic eq. for different systems are given, according to the parameters of Gm(s) and Gc(s). The root placements are obtained from the characteristics specified. As the system response will be stable and monotonous if the roots of the characteristic eq. are real, and underdamped if there are complex roots, it will be

$$r_i=1/\tau_i,\ p_1=2\delta\omega_n,\ p_2=\omega_n^2 \tag{13}$$

being τ_i the time constant, $0<\delta<1$ the damping ratio and ω_n the natural frequency.

By identifying coefficients in table 3 it is obtained table 4, where the s_i controller coefficients are related to the parameters of the models a_i, b_i and the roots placement r_i, p_i.

Table 4 Controller coefficients

SYST	S_0	S_1	S_2
1 P	$\frac{r_1-a_1}{b_1}$	-	-
1 PI	$\frac{r_1+r_2-a_1}{b_1}$ or $\frac{p_1-a_1}{b_1}$	$\frac{r_1r_2}{b_1}$ or $\frac{p_2}{b_1}$	-
2 P	$\frac{r_1r_2-a_2}{b_1}$ or $\frac{p_2-a_2}{b_1}$	-	-
2 PI	$\frac{r_1p_1+p_2-a_2}{b_1}$	$\frac{r_1p_2}{b_1}$	-
2 PD	$\frac{p_1-a_1}{b_1}$	$\frac{p_2-a_2}{b_1}$	-
2 PID	$\frac{r_1+p_1-a_1}{b_1}$	$\frac{p_2+r_1p_1-a_2}{b_1}$	$\frac{r_1p_2}{b_1}$

Table 5 Tuning formulas

SYST	SPECIFICATIONS	TUNING F.
1 P	r_1	$Kc=s_0$
1 PI	r_1, r_2 or p_1, p_2	$Kc=s_0$ $Ti=\frac{s_0}{s_1}$
2 P	one of r_1, r_2 or p_1, p_2 $a_1=r_1+r_2$ or $a_1=p_1$	$Kc=s_0$
2 PI	two of r_1, p_1, p_2 $a_1=r_1+p_1$	$Kc=s_0$ $Ti=\frac{s_0}{s_1}$
2 PD	p_1, p_2	$Kc=s_0$ $Td=\frac{s_0}{s_1}$
2 PID	r_1, p_1, p_2	$Kc=s_1$ $Ti=\frac{s_1}{s_2}$ $Td=\frac{s_0}{s_1}$

The parameters of Gc(s) are obtained from tables 1 and 4. In table 5 for different systems are showed the specifications to consider and the tuning formulas. The sampling period T will be obtained, by simulation, in such a manner that the system response will be simalar to the response obtained with a continuous controller.

5.2. *Discrete control*

From (4) and (7) the characteristic eq. of the discrete system is

$$R(z)A(z) + S(z)B(z) = 0 \qquad (14)$$

R(z), S(z) are shown in table 2 and A(z), B(z) in (9) and (11). Analogous by as in the continuous case it is deduced table 6 where the controller coefficients s_i are related to the parameters of the model a_i', b_i' and the roots placement r_i, p_i.

Table 6 Controller coefficients

SYST	S_0	S_1	S_2
1 P	$\frac{r_1 - a_1'}{b_1'}$	-	-
1 PI	$\frac{r_1 + r_2 + 1 - a_1'}{b_1'}$ or $\frac{p_1 + 1 - a_1'}{b_1'}$	$\frac{r_1 r_2 + a_1'}{b_1'}$ or $\frac{p_2 + a_1'}{b_1'}$	-
2 P	$\frac{r_1 r_2 - a_2'}{b_2'}$ or $\frac{p_2 - a_2'}{b_2'}$	-	-
2 PI	$\frac{p_1 + r_1 - a_1' + 1}{b_1'}$	$\frac{p_2 r_1 + a_2'}{b_2'}$	-
2 PD	$\frac{p_1 + r_1 - a_1'}{b_1'}$	$\frac{p_2 r_1}{b_2'}$	-
2 PID	$\frac{p_1 + r_1 + r_2 - a_1' + 1}{b_1'}$	$((r_1 + r_2)p_2 b_2' + r_1 r_2 (p_1 b_2' - p_2 b_1') + a_2 b_2') / b_2'^2$	$\frac{r_1 r_2 p_2}{b_2'}$

The parameters of Gc(z) are obtained from tables 2 and 6. In table 7 are showed for different systems the specifications to consider and the tuning formulas .

If the characteristics of the time response are given directly in the z-plane, they will be considered as r_i and/or p_i specifications. By contrary if the characteristics are referred to the continuous behaviour, the placement roots of the Gl(z) characteristic eq. will be obtained from the roots placement of the characteristic eq. of the continuous system Gle(s). So, when the Gl(z) characteristic eq. present real roots, $z = -r_i$, it is considered that the contribution to the time response system is equivalent to simple poles $s = -1/\tau_i$, in Gle(s). Therefore

$$r_i = -e^{-T/\tau_i} \qquad (15)$$

if Gl(z) characteristic eq. $z^2 + p_1 z + p_2 = 0$, presents complex roots it is considered that the contribution of these to the system response is equivalent to have complex poles $s = -2\delta\omega_n \pm \omega_n \sqrt{1-\delta^2}\, j$ in Gle(s), resulting

$$p_1 = -2e^{-\delta\omega_n T} \cos\left(\omega_n T \sqrt{1-\delta^2}\right),\ p_2 = e^{-2\delta\omega_n T} \qquad (16)$$

In discrete control the samplig period T is fixed by the specifications when the roots number of the characteristic eq. is greater than the number of the control parameters; this will happen for model 2. If the number of roots of the characteristic eq. is the same as the control parameters, T will not be fixed for the specifications; this will happen for model 1, where it will be considered as one more specification.

Table 7 Tuning Formulas

SYST	SP.	TUNING F.
1 P	r_1, T	$Kc = s_0$
1 PI	r_1, r_2, T or p_1, p_2, T	$Kc = -s_1 \quad Ti = \frac{s_1 T}{s_0 + s_1}$
2 P	r_1, r_2 or p_1, p_2	$Kc = s_0$ $T \rightarrow b_2'(r_1 + r_2 - a_1') - b_1'(r_1 r_2 - a_2') = 0$ $T \rightarrow b_2'(p_1 - a_1') - b_1'(p_2 - a_2') = 0$
2 PI	r_1, p_1, p_2	$Kc = -s_1 \quad Ti = \frac{s_1 T}{s_0 + s_1}$ $T \rightarrow p_2 + r_1 p_1 - a_2' + a_1' - s_0 b_2' - s_1 b_1' = 0$
2 PD	r_1, p_1, p_2	$Kc = s_0 + s_1 \quad Ti = \frac{s_1 T}{s_0 + s_1}$ $T \rightarrow p_2 + r_1 p_1 - a_2' - s_0 b_2' - s_1 b_1' = 0$
2 PID	r_1, r_2, p_1, p_2	$Kc = -(s_1 + 2s_2)$ $Ti = \frac{(s_1 + 2s_2)T}{s_0 + s_1 + s_2} \quad Td = \frac{s_2 T}{s_1 + 2s_2}$ $T \rightarrow a_2' - a_1' + b_1' s_1 + b_2' s_0 - p_2 - (r_1 + r_2)p_1 - r_1 r_2 = 0$

6. TUNING CRITERIA

The tuning criteria consist in placing the roots of the characteristic eq. in such a way that the specified characteristics are verified. It is supposed, that, for inputs of the step type, the system presents monotonous stable or underdamped responses. In this case the maximum number of independent characteristics that can be fixed it is equal to the number of the Gc(z) parameters. If it is used the formulas of table 7, or one more, if it is used T as an additional control parameter using formulas of table 12.

To obtain the Gc(z) parameters and T, proceed as it was showed in section 5, being the parameters of Gl(s) or Gle(s), δ, ω_n, τ_i and the following time response characteristics: rise time tr, the relation damped R, the oscillation period Ps and the maximum overshoot Mp (or others related to them). If it is used the table 12, it can be considered the number of samples between certain time instants t_i ($Nt_i=t_i/T$) as an additional characteristic. The determination of Y(t) is normally tedious, it can be fixed some characteristics by means of simulation or, in an approximate way, by means of relations obtained on simpler cases, or by using both procedures. For the systems considered, if the characteristics of time response are R, Ps, it can be obtained δ, ω_n as follows:

$$R = e^{-\frac{\delta\pi}{\sqrt{1-\delta^2}}}, \quad Ps = \frac{2\pi}{\omega_n\sqrt{1-\delta^2}} \qquad (17)$$

One limit case of the monotonous and underdamped responses is obtained for model 1 when the specification are T and all the roots of the discrete characteristic eq. are in the origin. The system response, to one step input, will be of the form showed in Fig. 1.a, for the P controller, or, of the form indicated in Fig. 1.b, for the PI controller. In both cases it can be used the time instant t in which it overtakes the steady state as characteristic the time response, to calculate T, obtaining T=t, for P control, and T=t/2, for PI control.

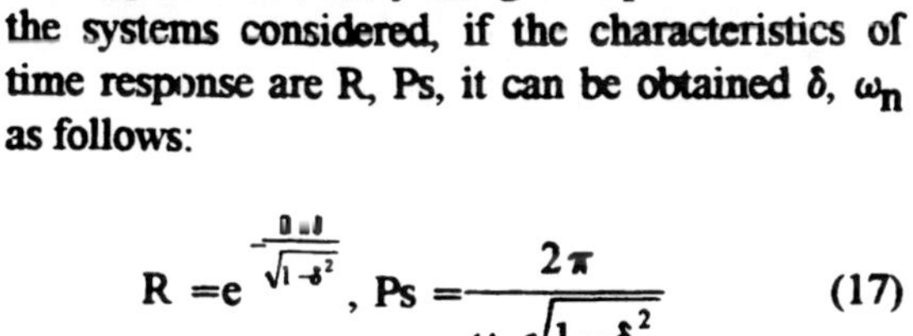

a) b)

Fig. 1. System Responses with P and PI control

7. EXAMPLE

Be Gm(s)=0.4/s(s+0.2) and the characteristics of the time response, for one unit step input, are an underdamped response, position error zero, R=1/4 and Ps according of the controller type. For the four controller type the system response are:

1.-For P controller, R=1/4 and table 7 results Ps=13.86 and Kc=0.54. If the sampling period T is reduced too enough in relation to Ps (e.g. T=0.1), the response of Fig. 2.a which verifies R=1/4 is obtained. If T is the same order as Ps, R=1/4 will not be verified as it is showed in Fig. 2.a, for T=1 and T=2. If table 12 is used, it is possible to consider two characteristics of the time responses R and Ps . The possible values of Ps can vary between that corresponding to the continuous case and the one obtained when the number of the samples until Ps is four (13.86<Ps<55.70). For R=1/4, Ps=14 and R=1/4, Ps=55 it is obtained the control parameters Kc=0.53, T=0.016 and Kc=0.06, T=12.92, respectively. The responses of the system are showed in Fig 2.b that verify the characteristics specified.

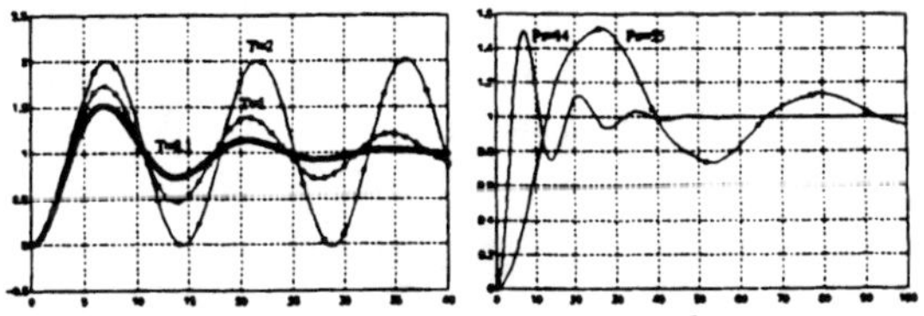

a) b)

Fig. 2. System Response with P control

2.-For PI controller and table 7, can be fixed two characteristics of the time responses. For R=1/4 and Ps=100 it is obtained the control parameters Ti=12.50, Kc=0.022 and τ_i=5.80. If the sampling period T is reduced too enough in relation to Ps, it is obtained the response showed in Fig. 3.a, which verifies the characteristics specified with Mp=63%. When T in crease the system becomes more unstable; in Fig. 3,a it is showed the response for T=15, that presents Mp>100%, R=1/2 and P=87. If used the table 12 the characteristics of the time responses similar to the continuous control with T bigger. For τ_1=15 (it will be always bigger than the continuous case), R=1/4, Ps=100, it is obtained T=16.38, Kc=0.021 and Ti=19.98, the response is showed in Fig. 3.a and it present Mp=85% and verifies the characteristics specified.

3.-For PD controller and table 7, with R=1/4, Ps=1 it is obtained Td=0.062, Kc=104, the response, for small T values in relation to Ps, is showed in Fig. 3.b, that verifies the specified characteristics with Mp=54%. When T increases the system becomes more unstable and it will not verify the specifications; in Fig. 3.b it is showed the response for T=0.1 which is almost unstable. If table 12 is used, similar characteristics are obtained to those of continuous control with T bigger. For τ_1=4 results T=0.20, Kc=3.47 and Td=3.93, the response is showed in Fig. 3.b with Mp=52%, which of verifies the characteristics specified.

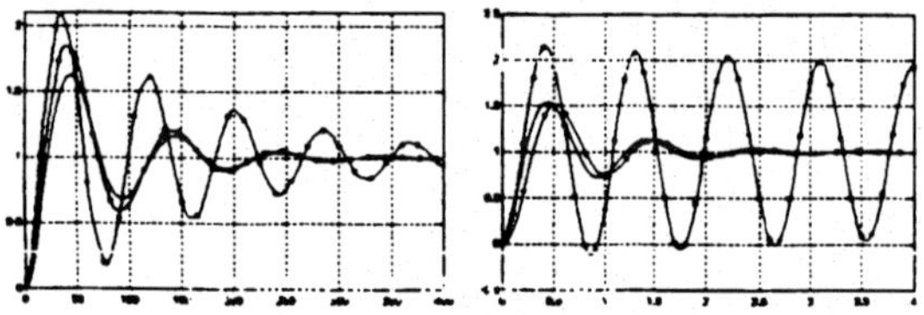

a) b)

Fig. 3. System Response with PI and PD control

4.-For PID controller and table 7, can be fixed three characteristic of the time response. If R=1/4, Ps=5, Mp=50%, results τ_1=1.25, Ti=1.59, Td=0.55, Kc=5.25, the response, T=0.01, is showed in Fig. 4, that verifies the specified characteristic. When T increase the system become more unstable; in Fig. 4 it is showed the response for T=0.1, that presents Mp>100%. If it is needed responses with bigger T, making use of table 12 for $\tau_1=\tau_2$=0.2 it is obtains T=0.1635, Kc=5.1690, Ti=0.6860 and Td=1.0757, the response is in Fig. 4, that verifies the specified characteristics.

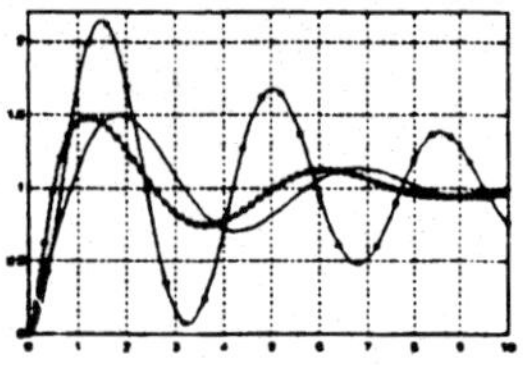

Fig. 4. System Response with PID control

8. CONCLUSIONS

1.-In the present paper is being showed how to apply the polo placement method, continuous and discrete, for the adjustment of the Gc(z) parameters and the sampling period T, in computer control for continuous processes. The tuning formulas for PID controllers and processes of first and second order have been obtained.

2.-The tuning formulas of continuous control are generally, more simple to apply than those of discrete control, but they lead to small T values that you have to adjust by means of simulation. Fast sampling can produce problems of computation time and it subjects the actuator to excessive non significatives changes. The greater complexity of the formulas of discrete control is not a disadvantage for the great computation power of present computers; on the other hand, it is possible to simplify these considering characteristics of the time response adequate.

3.-The tuning formulas of discrete control present the following advantages: 1) As the effect produced by T in the system response is included in them, they guarantee the attainment of the characteristics of time response without to having to resort to simulation. 2) The obtention of T is made as one more control parameter from the time response characteristics, and you can obtain a whole range of responses which vary from those similar to the ones of continuous control, when T tends to zero, to other types of responses. 3) The restrictions in the time response characteristics are less numerous than in the case of continuous control, because of the presence of T. 4) They permit to adjust the number of samples in relation to time response characteristics. 5) They permit to adjust T in a manner that is valid both for identification and control.

4.-When you modify T, the system changes its time response characteristics and in some cases if could become unstable. In general you can concluded that for constant values in the control parameters, when you increase the value of T the system will be faster with bigger Mp and bigger R, while if this are reduce produces the opposite effects.

9. REFERENCES

Åström, K.J. and T. Hägglund, (1988). Automatic Tuning of PID Controllers. ISA-Intrument Society of America.

Åström, K.J. and Wittenmark, B. (1990). Computer Controlled System, Theory and Design. Prentice-Hall International Editions. 2nd edition.

Dormido, S., López, I., Fernandez, J.L., Morilla F. and Aranda J. (1991). Design of Digital Control System using the Sampling Period as a Compensator. Procceding Vol. II of MELECON 91, IEEE, pp. 812-815.

Phillips, C.L. and Nagle, H.T. (1984). Digital Control Systems Analysis and Design. Englewood Cliffs, N.J.: Prentice-Hall.

Roffel, B., Vermeer, P.J. and Chin, P (1989). Simulation and Implementation of Self-Tuning Controllers. Prentice-Hall.

Shinskey, F.G. (1989). Process Control Systems: Aplication/ Desing/ Adjustment. McGraw-Hill Book Company. 3rd edition.

CONFIDENCE DEGREE ON THE ADVANCED STAND-ALONE CONTROLLERS

Joseba QUEVEDO and Albert GONZALEZ
Universitat Politècnica de Catalunya. Campus Terrassa
c/Colom,11. 08222 Terrassa
tel (34-3) 7398144
fax (34-3) 7398101
e-mail: joseba@esaii.upc.es

Abstract: This work presents the analysis of two advanced stand-alone controllers, that allow to tune automatically the control parameters attending some requirements of the controlled process response. The ES100 OMRON controller provides a fuzzy logic function (fine-tuning) to gradually inhibit the overshoot and the damping of the response and to improve the time response adjusting the control parameters. The 761 FOXBORO controller provides an adaptive control function, called EXACT, to adjust the control parameters from the information of the process in operating conditions.
This analysis has been carried out in a benchmark system composed by three real processes and different tests. The results obtained with these advanced controllers show a real improvement of the fuzzy logic and EXACT function but the dispersion of the results obtained with the different processes do not allow, by now, to preset very precise requirements of the process control performances.

Key words: performance assessment, adaptive control, fuzzy control, benchmark test of control systems, advanced control systems

1. INTRODUCTION

The incorporation of modern control methods in the new commercial controllers is already a reality. In fact, more than 50% of present controllers (Quevedo, 1990) integrate some kind of advanced control techniques (auto-tuning, adaptive, fuzzy, expert system,...).

The job of the process responsible, is not longer to manually tune the control parameters but to pre-fix one or some requirements of the controlled process, as the time response, overshoot, damping,...

Several recent contributions in the field of performance assessment and comparison of advanced control systems (Astrom, 1989; Hang, 1991; Larminat, 1991; Ninnes, 1991, Knapp,1991; Quevedo, 1993; Titli, 1993;...) confirm the importance of a rigorous study about the confidence of the advanced controllers and particularly in the accomplishment of the preset requirements. These contributions should give useful information to designers, about the improvements to be done in the new controllers and, also, to users, about the real performances of the commercial controllers.

The objective of this work is to do an experimental analysis of confidence degree of two controllers with capabilities to pre-fix some requirements in the controlled process. For this analysis we have selected the model 761 of FOXBORO and the model ES100 of OMRON which allow, with

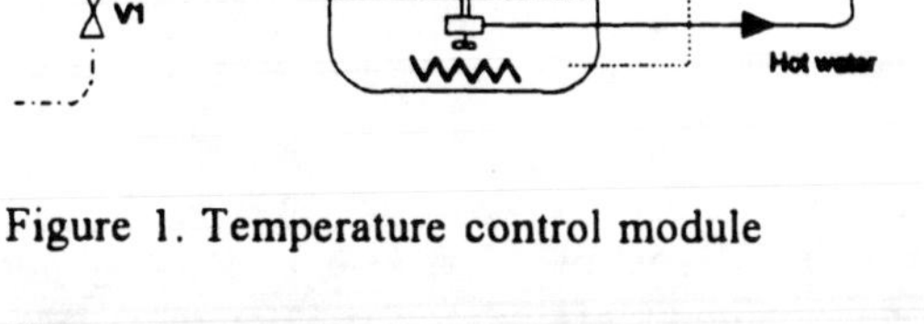

Figure 1. Temperature control module

different approaches, to define the desired transient response of the controlled process.
The two controllers have been applied to control three variables (level, temperature and flow) of a real system that will be presented in the next section. The section 3 will describe the ES100 controller from OMRON, the tests submitted and

the results obtained with its fuzzy interface module, that allows to fine tune the control parameters. The section 4 will describe the 761 controller from FOXBORO, the tests submitted and the results obtained with the EXACT (Expert Adaptive Controller Tuning) module. Finally, the section 5 will present the main conclusions of this analysis.

2. BENCHMARK SYSTEM

The benchmark system is a real laboratory process, that allows the control of temperature, level and flow of water. The temperature control module (figure 1) consists in two circuits of water and a heat exchanger. The primary is a closed circuit of hot water and the secondary is an open circuit of cold water. The controller manipulates a valve motor of the primary one in order to regulate the temperature of the secondary one at the output of a heat exchanger. There are four temperature sensors in different places of the primary and secondary circuits and another valve motor, in the secondary circuit, to modify the demand of water flow.

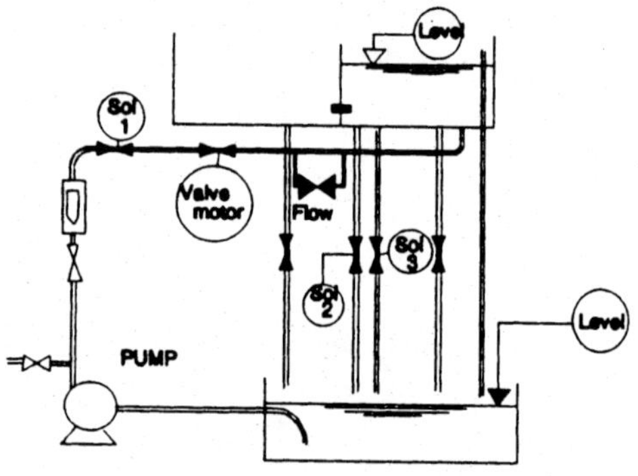

Figure 2. Level & flow control module

The module of level and flow control (figure 2) consists of two interconnected tanks and one pump that supplies the water from a general tank. The controller manipulates a valve motor to regulate either the tank level or the flow trough the supply pipe. There are some pipes with solenoid valves to provide water from the tanks to the general tank (closed circuit) and also different sensors of level and flow.

These modules can be combined in order to do a multivariable control. For example, using the secondary circuit of the temperature module to supply the water to the interconnected tanks of the level & flow module, the temperature and the level can be controlled simultaneously.

3. THE ES100 OMRON CONTROLLER

3.1. Outline of functions

The ES100 is a stand-alone digital PID controller with two functions to tune automatically the PID parameters. The first is an auto-tuning (AT) function that operates in closed loop generating a limit cycle. The optimum PID parameters are computed from the characteristics of the process variable.

A second function is called fine-tuning (FT). It allows to adjust the PID parameter changing the controlled process requirements through a friendly interface (figure 3). The user can set the following improvement required levels: hunting (damping) inhibit required level, overshoot inhibit required level and response improvement required level. After setting the levels of the requirements (0 to 5), FT function is executed and PID parameters are adjusted automatically and instantaneously. The FT uses fuzzy logic inference.

This controller can also work as a dual mode PID+Fuzzy control. The fuzzy control mode is only activated when there are large deviations of the response with respect to the set-point, due to external disturbances. In this work, the dual mode has not been studied.

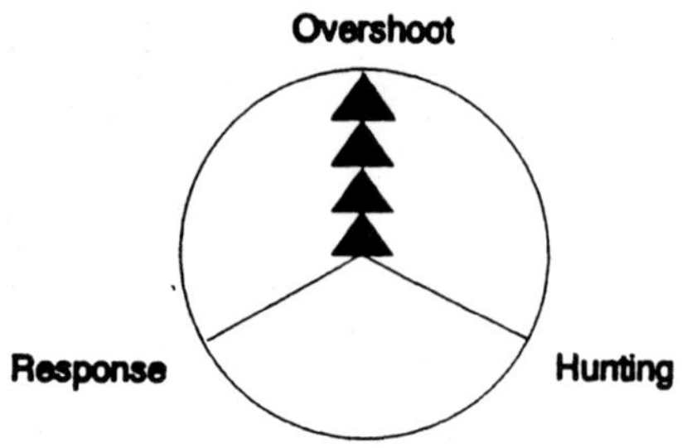

Figure 3. Overshoot fine tuning

3.2. Benchmark tests

The ES100 OMRON Controller has been submitted to control the level, temperature and flow of the real process in order to analyze the assessment performance of the fine-tuning function.

The tests for each process control consisted in the following sequence:

Process	Function	Improvement	Overshoot	Damping	$t_{10\text{-}90}$	t_s
Temperature	AT	--	15%	66%	29s	37s
Temperature	FT	overshoot	0%	0%	38s	58s
Temperature	FT	hunting	6%	0%	30s	32s
Temperature	FT	response	18%	53%	24s	∞
Level	AT	--	0%	0%	24s	25s
Level	FT	overshoot	0%	0%	50s	52s
Level	FT	hunting	0%	0%	37s	38s
Level	FT	response	0%	0%	22s	23s
Flow	AT	--	25%	100%	12s	∞
Flow	FT	overshoot	75%	100%	12s	∞
Flow	FT	hunting	0%	0%	14s	15s
Flow	FT	response	75%	100%	12s	∞

Table 1. Results of requirements with the ES100 OMRON

1) activation of auto-tuning function of the controller in closed loop with arbitrary initial parameters of PID
2) execution of a set of steps of the set-point (40-50-60%) and of the load (50-100%)
3)the analysis of the characteristics of the response (overshoot, damping and time response)
4) activation of fine-tuning function of the controller, requiring the maximum level (5) of the overshoot inhibit, and repeatition the phases 2) and 3) of the sequence.
5) activation of auto-tuning function of the controller in closed loop with arbitrary initial parameters of PID
6) activation of fine-tuning function of the controller, requiring the maximum level (5) of the hunting inhibit, and repeatition the phases 2) and 3) of the sequence.
7) activation of auto-tuning function of the controller in closed loop with arbitrary initial parameters of PID
8) activation of fine-tuning function of the controller, requiring the maximum level (5) of the response improvement, and repeatition the phases 2) and 3) of the sequence.

3.3. Results

The results (table 1) show firstly that the characteristics of the response are very different for the 3 processes studied, when only auto-tuning function is executed. While the overshoot and damping are null for level control, the overshoot is around 15% and the damping is more than 60% for temperature control and the response is a limit cycle for flow control .

The response time is analyzed using the rise time from the 10 to 90% of the response($t_{10\text{-}90}$) and by settling time (t_s)

The results of the fine-tuning function show a quite good correspondence between the requirement and the real improvement, except in the case of flow control where the requirements of overshoot inhibition or response improvement do not stabilize the controlled process.

The fine-tuning, with the time response requirement, always improves the rise time but does not always improve the settling time. For example in the temperature control the fine tuning of response time improves a little the $t_{10\text{-}90}$ but produces a limit cycle and then never achieves a steady state value ($t_s = \infty$)

The figure 4 shows the results obtained in temperature control after the auto-tuning function and the figure 5 shows the results after a fine-tuning requiring inhibit overshoot: the overshoot is improved but the response is slower.

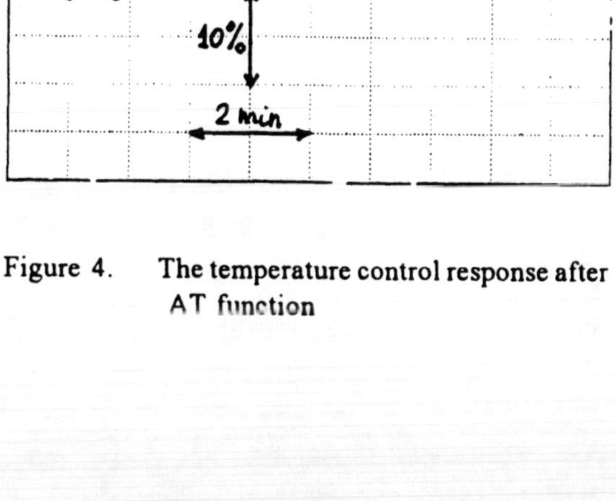

Figure 4. The temperature control response after AT function

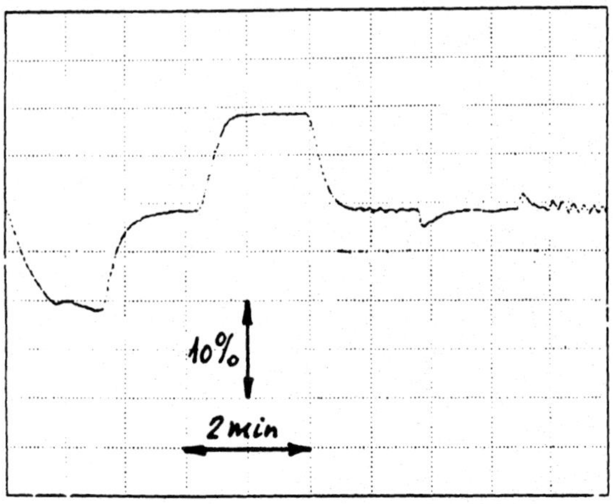

Figure 5. The temperature control response after FT with overshoot inhibit requirement

4. THE 761 FOXBORO CONTROLLER

4.1. Outline of functions

The model 761 is a stand-alone digital PID controller with auto-tuning and adaptive control in a module named EXACT. The auto-tuning function works in open loop applying a pulse of amplitude(BUMP) selected by the user. The controller computes the optimum PID parameters from the process response. Other information, as the noise band(NB) of the feedback signal and the maximum waiting time (WMAX) of the response are also computed in the auto-tuning.

The adaptive control function operates in closed loop when the deviation of the process variable to the set-point is bigger than twice the noise band. This function has some numerical parameters that the user should select. One is the requirement of the maximum overshoot allowed (OVR) and the other is the maximum damping allowed (DMP).

4.2. Benchmark tests

The 761 FOXBORO Controller has been submitted to control the level, temperature and flow of the real process in order to analyze the assessment performance of the EXACT module when different overshoot (OVR) and damping(DMP) are required.

The tests for each process control consisted in the following sequence:

1) activation of the auto-tuning function of the controller
2) execution of a set of steps in the set-point (40-50-60%) and in the load (50-100%)
3) analysis of the characteristics of the response (overshoot and damping of the response).
4) activation of the adaptive module of the controller, requiring a maximum overshoot of 50% (OVR=0.5) and a maximum damping of 50% (DMP=0.5), repetition 4 times the set of steps of the phase 2). Then the adaptive module is stopped and phases 2) and 3) with the last PID parameters are repeated.
5) activation of the auto-tuning function of the controller and repetition of phase 4) with OVR=0.1 and DMP=0.5
6) activation of the auto-tuning function of the controller and repetition of phase 4) with OVR=0.5 and DMP=0.1
7) activation of the auto-tuning function of the controller and repetition of phase 4) with OVR=0.1 and DMP=0.1

4.3. Results

The results (table 2) show, as the ones from the other controller do, that the performances of the response for the 3 processes studied are very different. The auto-tuningof this controller gives always a stable response for the three processes and the adaptive control improves generally all the performances of the response(overshoot, damping and time response), but if the requirements of the response are very tight (p.e. OVR=0.0 or/and DMP=0.1) the real performances don't accomplish the presets.

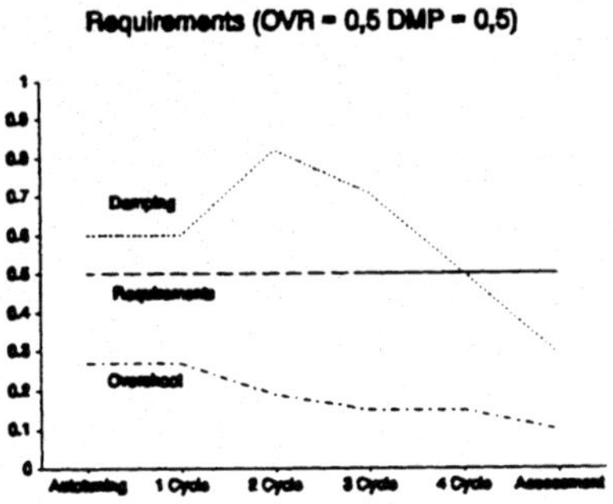

Figure 6. The evolution of the performances in adaptive control.

The experimental results don't give always the same results of overshoot and damping. For this reason, this table presents the mean values.

Finally, the improvement of the adaptive control (EXACT function) is obtained after a long period of changes in the set-point and in load. The figure 6 shows the evolution of the damping and the overshoot of the real response during the auto-tuning and in 5 cycles of the test 2).

Process	Function	Requirements OVR	DMP	Real parameters Overshoot	Damping	$t_{10/90}$
Temperature	AT	---	---	27%	60%	15s
Temperature	EXACT	0.5	0.5	10%	29%	14s
Temperature	EXACT	0.0	0.5	15%	61%	10s
Temperature	EXACT	0.5	0.1	6%	66%	16s
Temperature	EXACT	0.0	0.1	15%	50%	10s
Level	AT	---	---	21%	0%	85s
Level	EXACT	0.5	0.5	15%	0%	36s
Level	EXACT	0.0	0.5	7%	0%	22s
Level	EXACT	0.5	0.1	4%	0%	64s
Level	EXACT	0.0	0.1	24%	0%	73s
Flow	AT	---	---	21%	0%	20s
Flow	EXACT	0.5	0.5	9%	0%	8s
Flow	EXACT	0.0	0.5	10%	0%	8s
Flow	EXACT	0.5	0.1	5%	0%	10s
Flow	EXACT	0.0	0.1	0%	0%	15s

Table 2. Results of the 761 FOXBORO Controller

5. CONCLUSIONS

In this paper two advanced commercial controllers have been studied with different real processes. These controllers allow to define some requirements of the controlled process response.

The function Fine-Tuning of the ES100 OMRON controller and the function EXACT of the 761 FOXBORO controller improve the performance respect to their respective auto-tuning functions.

The results obtained in the different processes prove the big difficulty to accomplish with very concrete requirements. Only some qualitative requirements or some numerical band of desired performances can be defined a priori.

6. ACKNOWLEDGEMENT

This work has been supported by the Spanish Board for Research and Technology (CICYT) under contract TAP93-0596-C04, and the paper has been partially supported by the Group for the Study and Research in Automatic Control (CERCA).

7. REFERENCES

Astrom, K.J. (1989). Assessment of achievable performance of simple feedback loops. Dept. of Automatic Control. Lund Institute of Technology, ref. LUTFD2/(TFRT-7411), pages 1-18,1989.

Hang, C.C. and K.K. Sin. (1991). A comparative performance study of PID auto-tuners. *IEEE Control System Magazine*, n.8, pp. 41-47.

Knapp, T. and R.Isermann. (1991). Adaptive control of a benchmark example, *Proceedings of the 1st European Control Conference*, pp. 2358-2363, Grenoble.

Larminat, Ph de and O. Begovich. (1991). Linear adaptive control of a benchmark example, *Proceedings of the 1st European Control Conference*, pp. 2364-2367, Grenoble.

Ninnes, B., S.F.Graebe and G.C.Goodwi. (1991) *ECC adaptive control case study*, Proceedings of the 1st European Control Conference, Grenoble, pp. 2352-2357.

Quevedo, J., J. Armengol, T.Escobet and J.C.Hernandez. (1993) Evaluation and performances of auto-tuning and adaptive controllers, *Proceedings of the 2nd European Control Conference*, pp.1940-1944, Groningen.

Quevedo, J. Controladores digitales autónomos. (1990). *Automatica e Instrumentación*, n.206, pp 185-202.

Titli, A. (1993). About the design and the performance evaluation of fuzzy controllers, *Proceedings of European Workshop on Industrial Fuzzy Control & Applications*, Terrassa, 1993.

STRUCTURED NEURAL NETWORK FOR NONLINEAR DYNAMIC SYSTEMS MODELING

J. CODINA, R. VILLÀ and J. M. FUERTES

UPC-Facultat d'Informàtica de Barcelona, Department of Automatic Control and Computer Engineeering, Pau Gargallo 5, 08028 Barcelona. Catalonia

Abstract. The use of artificial neural networks (ANN) for nonlinear system modeling is a field where still there is much theoretical work to be done. A structured ANN which obtains neural models of nonlinear systems is presented. Those neural models are Fourier-series based. To check the goodness of the method, conventional difference equations are re-modeled via ANN and their respective input/outputs compared. Also their Fourier series expansion are compared. The Fourier coefficients being optimal for series truncation, this allows to estimate the goodness of the models obtained. Preliminary tests give encouraging results

Key Words. Neural nets; Nonlinear systems; Modeling; Control systems; State-space methods; Fourier analysis.

1. INTRODUCTION

Artificial neural networks (ANN) can be used, and are used to model the dynamics of a system or plant. (Yamada and Yabuta, 1993). Usually the ANN and the plant are feed with the same input signal (or input signals for multi-input systems) for training the neural network to model the plant behavior. The network output is then compared with the output from the plant and the error is used to update the weights of the synapses (see Fig. 1).

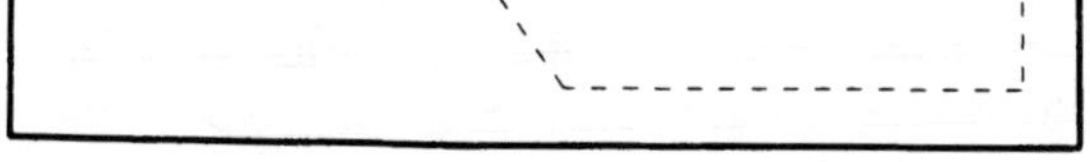

Fig. 1. ANN training structure for system modeling

A simple feed-forward neural network can be used, (Narendra and Parthasarathy 1990; Wu *et al.* 1992), if the ANN is feed with the current input, the previous inputs and the previous outputs. Using this configuration an input-output model of the plant is obtained, instead of a state space model. The advantage of this model is that a series-parallel model can be used while training the network. In a series-parallel model (Fig. 2) the previous outputs from the plant are used as inputs to the ANN instead of the outputs from the proper neural network. Once the model is obtained then the inputs can be taken from the previous outputs of the ANN obtaining and autonomous system.

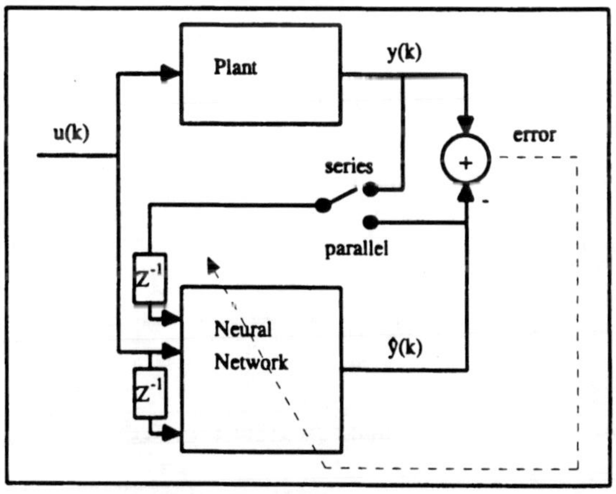

Fig. 2. Series-parallel training model

Using the series-parallel model configuration the obtained model has the following equation:

$$y(k) = F(u(k), u(k-1), u(k-2)\ldots, y(k-1), y(k-2)\ldots) \tag{1}$$

Different number of layers, activation functions and adaptation algorithms can be used. Qin *et al.* (1992) compares some configurations.

To obtain a space state model of a plant, a feed-forward ANN can also be used, assuming that the system state is accessible at any moment, (Nguyen and Widrow 1990; Anderson 1989). The resulting model is then a state space representation of the system dynamics:

$$[y(k), x(k+1)] = F(u(k), x(k)) \qquad (2)$$

In both cases the learning method is static, and usually the backpropagation algorithm is used. A dynamic discrete time ANN can also be used. This is the case of Jordan (1986) or Elman (1988) models where the outputs from the network or from the hidden layer respectively are feed back as new inputs to the network (Fig. 3).

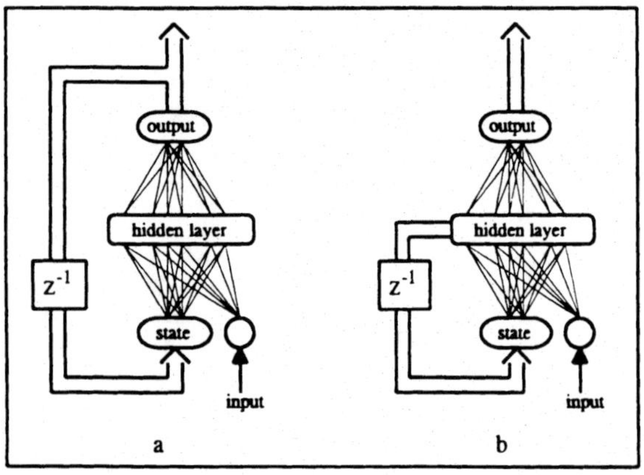

Fig. 3. a) Jordan network. b) Elaman network

The Jordan network is very similar to the serial-parallel model but it needs as many outputs as the system order. The Elman structure has some problems to learn systems where there is no direct connection between the input and the output, as happens when there are physical delays.

After the ANN training, we obtain a black box which acts as a system simulator. This neural network is often obtained as a first step in the design of neuronal controllers. There are difficulties in making a study of the properties of the plant from that model. The kind of activation functions, the number of layers, and the topology of the network are more a hinder than a help for this study.

2. STATE SPACE MODEL

The ability of extracting useful information about the real system from the simulated neural network model, is of primary interest. Although the existence of a neural network structure that allows the extraction of a set of equations is useful for the study of the system properties, it doesn't relate equations to physical phenomena.

A neural network with a structure inspired on the state space representation of a system could allow the application of modern control theory to SISO and MIMO systems. Beginning with linear discrete time systems we first obtained a neural network, (Codina *et al.* 1992) able to learn the matrices of the system state representation from the input-output properties (Fig 4). This model is a neural network with three layers, the input, the state and the output layers. The state and the output layers are connected to the input layer, composed by the actual inputs and state of the system.

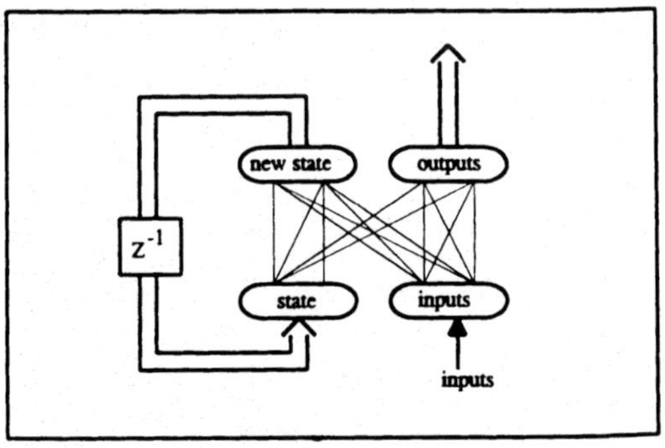

Fig. 4. Proposed linear structure.

To train the network, the backpropagation-through-time learning algorithm has been used (Werbos 1990), where the error is back-propagated from the output, through the present state, to the previous state. The weights updating is done in batch mode. That is, the weights are changed after a number N of input-output pairs have been processed. The number N is taken to be equal to the length of the input signal. With this methodology, and using linear neurons we can obtain from the synapses weights the matrices A, B, C and D of the state-space representation of the system:

$$\begin{aligned} x(k+1) &= Ax(k) + Bu(k) \\ y(k) &= Cx(k) + Du(k) \end{aligned} \qquad (3)$$

As the main interest of the application of neural networks in dynamic systems is their ability to learn and map nonlinear functions, we have expanded our previous model to deal with nonlinear systems. Expanding the linear model to nonlinear systems using sigmoids hinders further study of the obtained model. To solve this drawback we used a structured neural network based on Fourier series in order to approximate any nonlinear function in a bounded interval (Fig. 5).

A nonlinear discrete time system can be modeled by the following difference equations:

$$\begin{aligned} x(k+1) &= F(x(k), u(k)) \\ y(k) &= G(x(k), u(k)) \end{aligned} \qquad (4)$$

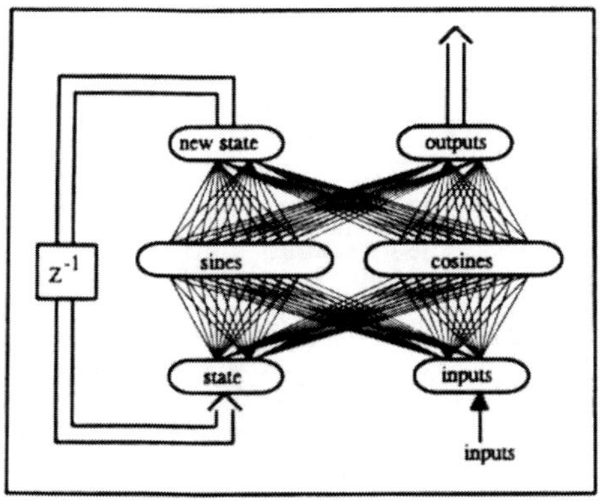

Fig. 5. Proposed nonlinear structure.

The Fourier series expansion allows us to express F and G as a weighted sum of sinusoidal functions, in a bounded interval:

$$\begin{aligned}\tilde{x}(k+1) = \sum_{n=0}^{N} \sum_{m=-M}^{M} & \tilde{A}_{\mathrm{F}n,m} \cos(nw_n\tilde{x}(k) + mw_m u(k)) \\ & + \tilde{B}_{\mathrm{F}n,m} \sin(nw_n\tilde{x}(k) + mw_m u(k))\end{aligned}$$

$$\begin{aligned}\tilde{y}(k) = \sum_{n=0}^{N} \sum_{m=-M}^{M} & \tilde{A}_{\mathrm{G}n,m} \cos(nw_n x(k) + mw_m u(k)) \\ & + \tilde{B}_{\mathrm{G}n,m} \sin(nw_n x(k) + mw_m u(k))\end{aligned} \quad (5)$$

Such series have the advantage that they form an orthogonal base of functions and, in particular, any coefficient can be added without changing the previous ones. So the network can be scalable in the sense of adding new sinusoidal elements, but beginning with the previous learned structure and weights. This model has similarities with a functional-link net (Pao 1990), but neurons with fixed weights are used in order to allow the application of the backpropagation-through-time algorithm. If the weights of the sine and cosine neurons are fixed in order to be $n\omega_0$ then this training method can be compared with the direct calculus of the Fourier coefficients (Codina *et al.* 1994).

In order to obtain the linear part of the system and to minimize the Gibbs phenomenon, we have included a linear part to the Fourier expansion, by using the c coefficients . This gives us a hybrid expansion:

$$\begin{aligned}\tilde{x}(k+1) = & c_1 x(k) + c_2 u(k) + \\ & \sum_{n=0}^{N} \sum_{m=-M}^{M} \tilde{A}_{\mathrm{F}n,m} \cos(nw_n\tilde{x}(k) + mw_m u(k)) \\ & + \tilde{B}_{\mathrm{F}n,m} \sin(nw_n\tilde{x}(k) + mw_m u(k))\end{aligned}$$

$$\begin{aligned}\tilde{y}(k) = & c_1 x(k) + c_2 u(k) + \\ & \sum_{n=0}^{N} \sum_{m=-M}^{M} \tilde{A}_{\mathrm{G}n,m} \cos(nw_n x(k) + mw_m u(k)) \\ & + \tilde{B}_{\mathrm{G}n,m} \sin(nw_n x(k) + mw_m u(k))\end{aligned} \quad (6)$$

Once the linear part of the model has been approximated, the network is expanded to include the nonlinear terms. This expansion will cause only a small change to the linear coefficients.

If the weights of the sine and cosine neurons are fixed then this training method can be compared with the direct calculus of the Fourier coefficients (Codina *et al.* 1994). In other words, the ANN is calculating the discrete Fourier transform of the nonlinear transfer function in a bounded interval. To improve the results, the number of points should be at least the number of Fourier coefficients we are calculating, to avoid multiple solutions, and should be within a regular sampling of the state space.

3. EXAMPLE

To test the ANN structure, a discrete-time nonlinear system has been simulated:

$$\begin{aligned}x(k+1) &= 2\left(\tfrac{x(k)+2}{4}\right)^2 + 0.5\frac{u^3(k)}{0.25+u^2(K)} + 1 \\ y(k) &= \tfrac{4}{\pi}\arctan(x(k)) + \tfrac{2}{1.61}\log(u(k)+3) + 1\end{aligned} \quad (7)$$

The learning procedure can not use a series-parallel model, where the previous outputs of the real system are used instead of those from the simulated system. This approach is not feasible using a state space structure, because neither the state of the real system nor the internal state representation taken by the ANN is known. Using the parallel model increases the number of training examples needed, and they are highly sensitive to initial conditions.

The methodology consists on the following steps.

1) A first approximation is obtained by a linear network, with a fixed training signal.

2) The model is then expanded using the Fourier terms, the network is trained to learn the coefficients for the same training signal. This step is repeated until the mean square error (MSE) decreases to a fixed order.

3) The NN is trained with different signals in order to make the system evolve through the whole working region of the state space.

4) If the number of Fourier coefficients is too small or the training signal in step 2 was not rich enough, then the third step results in an important differences of the MSE, between signals. If this happens then return to step 2 with a new training signal.

In our example we used as input a signal made up of 100 samples of a random step signal (Fig. 6), from which we obtain the desired output. As we use batching, the weighfs increments were calculated for each input-output pair but only updated at the end of the input signal processing.

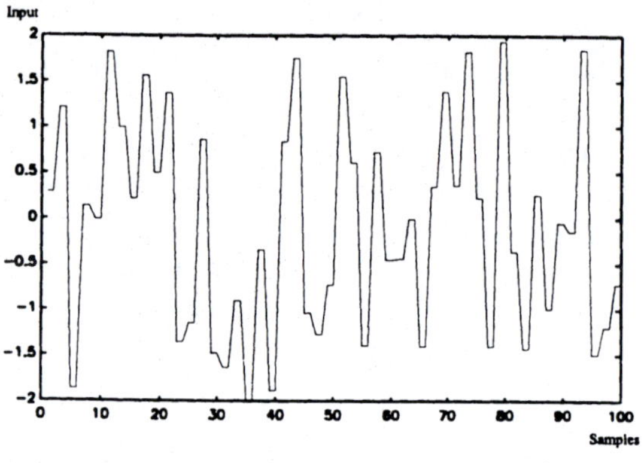

Fig. 6. Input signal used during training.

The first approximation was made linear. After presenting to the ANN one hundred times the training signal, the linear model was considered correct. Different inputs will produce different linear models depending upon on which areas of the state space the system is evolving in. After the linear approximation was obtained, the ANN was expanded with successive Fourier coefficients and trained, for each one, one hundred times. The MSE error curve (Fig. 7) reflects this procedure, and shows a fast decrease of the error when a new frequency element is added.

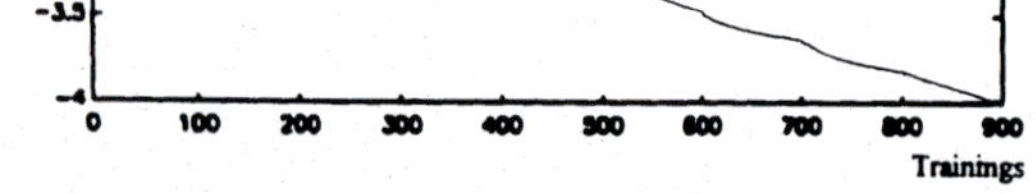

Fig. 7. Error curve: logarithm of the mean square error obtained during training. Every hundred trainings a new frequency element is added.

The difference between the desired output and the one obtained from the network is unappreciable, for the training signal, when using eight Fourier terms (Fig. 8). When the test signal is presented to the resulting ANN it can happen that this new signal makes the system evolve through areas of the state space not reached when the training signal was used. As those areas couldn't be learned while in the training phase, the result is an increment of the error.

The test signal used for the example was

$$u(k) = 0.5\sin(\tfrac{k}{10}) + 0.5\sin(\tfrac{k^2}{10}) \qquad (8)$$

When the test signal is presented to the ANN the MSE increases, just a small amount, as it can be seen in Fig. 9.

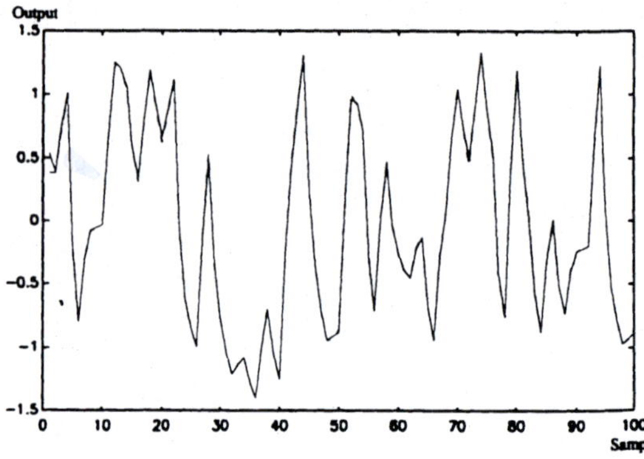

Fig. 8. Plant output (continuous line) and ANN output(dashed line) with input as the training signal.

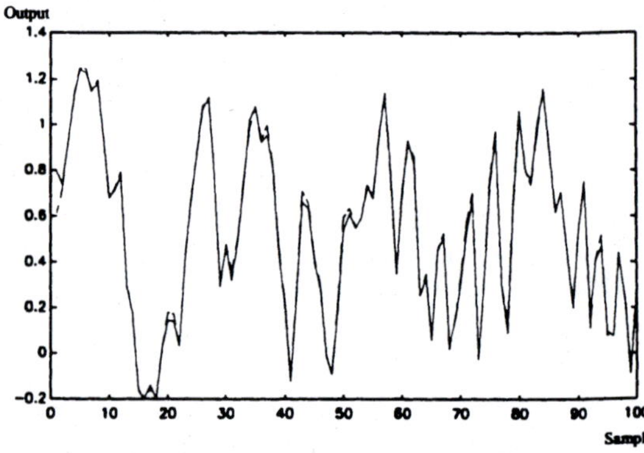

Fig. 9. Plant and ANN output with input a test signal.

There are differences, (Fig. 10), between the evolution of the state of the plant and its representation by the ANN. This is not surprising as a system may have multiple state descriptions. The ANN will evolve the state space representation which is more easily obtainable. The whole learning process and the initial weights, together with the used training signal, will modulate the shape of this state representation.

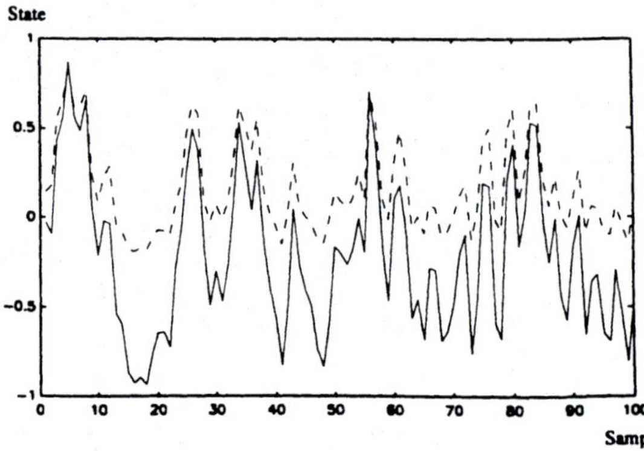

Fig. 10. State evolution of the real system (continuous line) and the trained system (doted line).

4. CONCLUSIONS

The absence of a methodology for the use of ANN in the field of nonlinear systems modeling is an important factor which impedes a widespread use of them. The difficulty in finding the right number of neurons or layers, the uncertainty of the results (local minima, random initial values...) are still a deterrent to their industrial application. In this paper we present a topology of ANN which was thought in order to help the search of a solution to those uncertainties.

This paper presents a structured ANN based on space state models of the system equations, together with the use of sines and cosines and linear activation functions. The ANN calculates the Fourier coefficients of the nonlinear functions that relate the inputs with the outputs and new states. As the Fourier coefficients can also be calculated analytically or numerically, the results from the ANN can be compared with those expected by classical methods. The use of state space equations for system descriptions minimizes the number of inputs to the neural network and allows a further study of the system properties. Here the model has been presented together with one example of its viability for the modeling of nonlinear discrete time systems

Acknowledgment. This research work has received fundings from the CICYT ref. 89-0278 and support from CERCA, Col·lectiu d'Estudis i Recerca en Control i Automàtica.

REFERENCES

Anderson, C.W. (1989) Learning to Control an Inverted Pendulum Using Neural Networks. *IEEE Control Systems Mag.* April. 31-37.

Codina, J., B. Morcego, J.M. Fuertes and A. Català. (1992) A Novel Neural Network Structure for Control. *IEEE Int. Conf on Systems, Man and Cybernetics.* Chicago. pp. 1339-1344.

Codina, J., J.C. Aguado and J.M. Fuertes (1994) Capabilities of a Structured Neural Network. Learning and Comparison with Classical Techniques. To appear in *Procc. of the ESANN Eruoconference.*

Elman J.L. (1988) *Finding Structure in Time.* Report 8801. University of California. San Diego.

Jordan, M.I. (1986) *Serial Order: A Parallel Distributed Processing Approach.* Institute for Cognitive Science. Report 8604. University of California. San Diego

Narendra, K.S. and K. Parthasarathy (1990) Identification and Control of Dynamical Systems Using Neural Networks. *IEEE Trans. on Neural Networks.* March 4-27.

Nguyen, D.H. and B. Widrow (1990) Neural Networks for Self-Learning Control Systems. *IEEE Control Systems Mag.* April, 18-23.

Pao, Y. (1989) *Adaptive Pattern Recognition and Neural Networks.* Adisson-Wesley, Reading, MA.

Qin, S., H. Su and T.J. McAvoy (1992) Comparison of Four Neural Net Learning Methods for Dynamic Systems Identification. *IEEE Trans. on Nerual Networks.* January 122-130.

Werbos, P.J. (1990) Backpropagation Through Time: What it Does and How to Do it. *Proc. of the IEEE.* Vol 78 N. 10 1550-1560.

Wu, Q.H., B.W. Hogg and G.W. Irwin. (1992) A Neural Network Regulator for Turbogenerators. *IEEE Trans. on Neural Networks* January 95-100.

Yamada, T. and Yabuta T. (1993) Dynamic System Identification Using Neural Networks. *IEEE Trans on Systems, Man, and Cybernetics. January/February,* 204-211

NONLINEAR SYSTEM IDENTIFICATION USING ADDITIVE DYNAMIC NEURAL NETWORKS.

R. GRIÑÓ*

*Institut de Cibernètica (UPC-CSIC), Diagonal 647, 08028-Barcelona, Spain. E-mail: grino@ic.upc.es

Abstract. In this work additive dynamic neural models are used for the identification of nonlinear plants in on-line operation. In order to accomplish this task a gradient parameter adaptation method based in sensitivity analysis is formulated taking into account that the parameters of the model are arranged in matrix form. This methodology is applied to several nonlinear systems in simulation and with a real dataset to verify its performance.

Key Words. Neural nets; system identification; sensitivity analysis; matrix algebra; gradient methods

1. INTRODUCTION

An important effort in the study of nonlinear systems in control theory has been carried out in the last years. This interest stems from the need to give new solutions to some historic necessities of the Automatic Control (Antsaklis, 1990): to work with more and more complex systems, to satisfy stricter criterions of design and to fulfill the previous points with less and less a priori knowledge of the plant.

In view of this tendency, a great effort is being carried out, within the area of system identification, in the development of nonlinear models of real processes. Within this line is framed all the work in the development of nonlinear models (Haber and Unbehauen, 1990) , that in the last years has been stated explicitly in methodologies like, for example, the NARMAX models (Korenberg and Paarmann, 1991; Mathews, 1991).

Next to this line of work with classical characteristics, it is added in the few last years a set of works that apply the artificial neural networks to the tasks of identification and control of systems. These works are supported by two of the most important capacities of the neural nets: their learning capacity (Narendra and Parthasarathy, 1990; Narendra and Parthasarathy, 1991)(based on the optimization of an adequate cost function) and their good behavior in the approximation of nonlinear functions (Polycarpou and Ioannou, 1991; Girosi and Poggio, 1989).

At present, most of the works on system identification performed with neural nets are based on multilayer feedforward neural networks with backpropagation learning or more efficient variations of this algorithm. These methods have been applied to real processes and they have demonstrated an adequate behavior (Bhat *et al.*, 1990; Weerasooriya and El-Sharkawi, 1991). It is important to note that the great majority of works until today use discrete time models that capture the dynamics of the real process through the use of tapped delay lines in the model inputs and outputs.

Nevertheless, at this moment internal dynamics and continuous time models are being analyzed, since this is a way to achieve simpler structures that represent the real process in a suitable form (Sanner and Slotine, 1991*a*; Sanner and Slotine, 1991*b*).

2. ARCHITECTURE OF THE CONNECTIONIST MODELS

In this work, continuous time additive dynamics neural networks (Kosko, 1992) have been used as the models to adjust to the real processes. This approach has several advantages with regard to the discrete time tapped delay line based models (Wu *et al.*, 1992; Chen, 1990):

- the number of configuration parameters of the model is considerably less. It is only necessary to specify the dimension of the state space since the number of inputs and outputs is determined by their counterparts in the real process.
- the models obtained with this approach are in state space form and work in continuous time, which is very interesting in order to apply

the differential geometric theory for nonlinear control (Isidori, 1989).

The architecture of the connexionist models is described by the structure of their basic elements and how they are interconnected.

2.1. *Basic Elements*

The basic processing element of a neural model is the node, also called neuron by analogy with the biological neurons. In general, the basic model of a node (see Fig. 1) is composed of a weighted adder, a linear dynamic SISO system and a nonlinear static function. In this work, these elements have the following form:

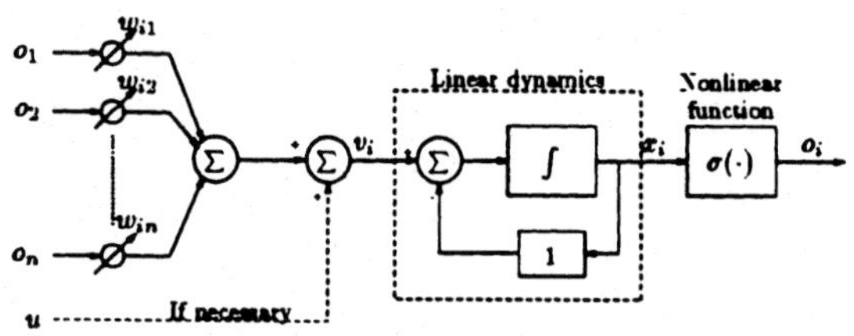

Fig. 1. Basic node in the additive neural model.

- The weighted adder is described by the equation:

$$v_i = \sum_{j=1}^{n} w_{ij} o_j + \delta_{ik} u; \quad i = 1 \ldots n \tag{1}$$

  where the weighted sum v_i is a linear combination of the outputs of the nodes o_i of the net and the external input u, if it is applicable because only one node of the network receives it. If we take all the equations (1) and arrange them in matrix form we obtain

$$\mathbf{v} = \mathbf{Wo} + \mathbf{b}u \tag{2}$$

  where $\mathbf{W} = [w_{ij}] \in M_n(\mathbb{R})$ is the weights matrix of the network and, $\mathbf{b} = \left[0, \ldots, \overset{k)}{1}, \ldots, 0\right]^T \in \mathbb{R}^n$ is the fixed parameter vector that applies the input to the model.
- The linear dynamical system has the weighted sum v_i as input and x_i as output. In the present work, we take a first order linear system with a fixed time constant ($\tau_i = 1$) for all the nodes in the network. That is to say, all the nodes in the net have the following differential equation in their structure:

$$\dot{x}_i + x_i = v_i \tag{3}$$

- The nonlinear static activation function selected in this work is the hyperbolic tangent and it maps the state of a node into its ouput ($\sigma : \mathbb{R} \rightarrow (-1, 1), \sigma \in C^\infty$).

2.2. *Additive Models*

The composition of basic elements explained above constitutes a node of the connexionist additive model and the arrangement in matrix form of all of the nodes, together with a linear output equation with constant parameters has the following structure (see Fig. 2):

$$\dot{\mathbf{x}} = -\mathbf{x} + \mathbf{Wg}(\mathbf{x}) + \mathbf{b}u \tag{4}$$

$$\hat{y} = \mathbf{cx} \tag{5}$$

where $\mathbf{x} \in \mathbb{R}^n$ is the state vector, $u \in \mathbb{R}$ the input to the model and the plant, $\hat{y} \in \mathbb{R}$ the output of the model, $\mathbf{W} \in M_n(\mathbb{R})$ the weight (parameter) matrix, $\mathbf{c} = [1, 1, \ldots, 1] \in \mathbb{R}^{1\times n}$ the output vector, $\mathbf{b} \in \mathbb{R}^n$ the input vector with the elements cited in the previous section and

$$\mathbf{g}(\mathbf{x}) = [\tanh(x_1), \ldots, \tanh(x_n)]^T \tag{6}$$

a C^∞ nonlinear vector field from $\mathbb{R}^n$ to (-1,1).

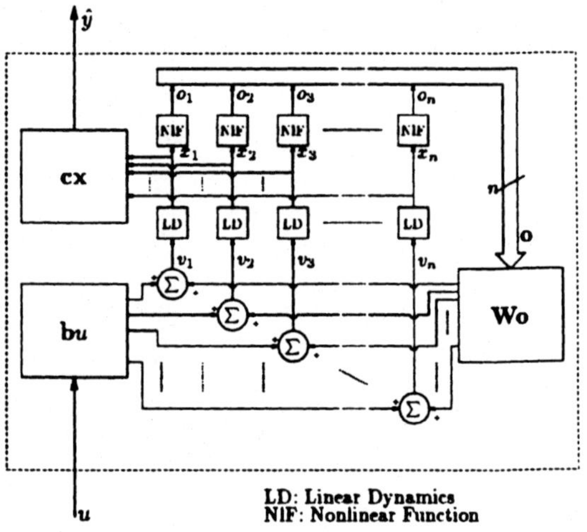

Fig. 2. Additive neural network model.

3. PARAMETER ADAPTATION IN THE CONNEXIONIST MODELS

The basic idea of the identification process is shown in Fig. 3. The connexionist model is arranged in a parallel manner with the real plant, and the excitation signal is the same for both. Clearly, the objective is to have the same output signals from the plant y and the model $\hat{y}$ for each time t. However, in general it will not be possible in a strict sense because we do not have a knowledge of the structure of the real plant and we can only rely on the possibility that our model may be able to identify the class of systems to which our plant belongs.

Because of the special capabilities of the neural networks in the approximation tasks (Funahashi and Nakamura, 1993) the problem can be avoided

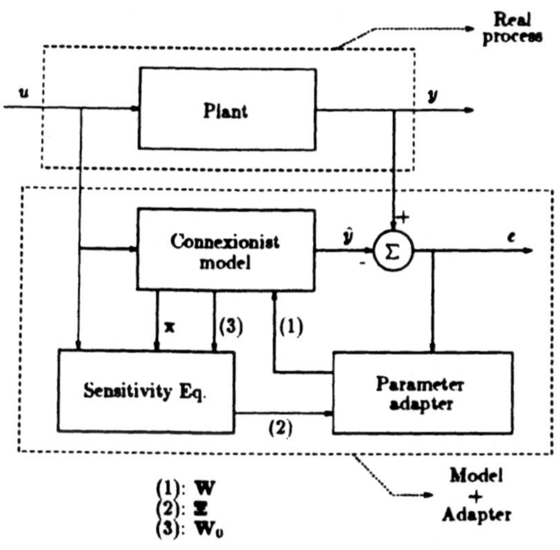

Fig. 3. Identification structure.

assuming that the proposed models are able to approximate the plant output except for a residual error due to the structural modelling mismatch between the plant and the model.

3.1. *Cost Function and Adaptation Law*

The identification error e can be defined as

$$e = y - \hat{y} \tag{7}$$

where y is the output of the real process and $\hat{y}$ the output of the connectionist model. This error is a function of time and the parameters of the model

$$e = e(t, \mathbf{W}) \tag{8}$$

and it takes zero value when the model fits perfectly the behaviour of the system to be identified. Accordingly, the problem is to find a set of parameters in a on-line manner in order to attain an identification error going to zero or to a small residual value due to missmodelling and noise in the process.

Then, it is necessary to define a cost functional over this error to perform the minimization process. Perhaps the simplest functional is

$$V = \frac{1}{2}e^2 \tag{9}$$

that attains its minimum when the error is equal to zero and is strictly convex with respect to the error. To minimize the cost functional V on-line the following gradient based update rule can be selected

$$\dot{\mathbf{W}} = -\epsilon \frac{\partial V}{\partial \mathbf{W}} \tag{10}$$

where the parameter ϵ is a small positive constant as it is explained in the next section.

The term $\frac{\partial V}{\partial \mathbf{W}}$ can be developed using the chain rule of matrix calculus and Kronecker products [1](Vetter, 1970; Brewer, 1978) obtaining

$$\frac{\partial V(e(\mathbf{W}))}{\partial \mathbf{W}} = \left(\mathbf{I}_n \otimes \frac{\partial V}{\partial e}\right) \frac{\partial e}{\partial \mathbf{W}} \tag{11}$$

where the rightmost member is, by the chain rule,

$$\frac{\partial e(\hat{y}(\mathbf{W}))}{\partial \mathbf{W}} = \left(\mathbf{I}_n \otimes \frac{\partial e}{\partial \hat{y}}\right) \frac{\partial \hat{y}}{\partial \mathbf{W}} \tag{12}$$

Substituting the equation (12) in (11) results

$$\begin{aligned} \frac{\partial V}{\partial \mathbf{W}} &= (\mathbf{I}_n \otimes e)(\mathbf{I}_n \otimes (-\mathbf{I}_1)) \frac{\partial \hat{y}}{\partial \mathbf{W}} \\ &= -[(\mathbf{I}_n \mathbf{I}_n) \otimes (e\mathbf{I}_1)] \frac{\partial \hat{y}}{\partial \mathbf{W}} \\ &= -(\mathbf{I}_n \otimes e) \frac{\partial \hat{y}}{\partial \mathbf{W}} \end{aligned} \tag{13}$$

at which the update equation of the weight matrix $\mathbf{W}$ is

$$\dot{\mathbf{W}} = \epsilon (\mathbf{I}_n \otimes e) \frac{\partial \hat{y}}{\partial \mathbf{W}}, \quad \mathbf{W}(0) = \mathbf{W}^0 \tag{14}$$

where $\mathbf{W}^0$ is the arbitrary starting value of the weight matrix.

In the last equation appears the term $\frac{\partial \hat{y}}{\partial \mathbf{W}}$ that is known as the sensitivity matrix of the output trajectories versus the parameters of the model, in this case arranged in matrix form.

3.2. *Sensitivity Equations of the Model Trajectories*

In this section the sensitivity equations (Frank, 1978) of the connexionist model that are necessary to the equation (14) are developed. Rewriting the equations (4) and (5) as

$$\begin{aligned} \dot{\mathbf{x}} &= \mathbf{f}(\mathbf{x}, u, \mathbf{W}) \\ &= -\mathbf{x} + \mathbf{W}\mathbf{g}(\mathbf{x}) + \mathbf{b}u \qquad (15) \\ \hat{y} &= \mathbf{c}\mathbf{x} \qquad (16) \end{aligned}$$

in order to obtain more concise developments, and applying the partial derivative with respect to the weight matrix to both sides of the equations it results for the state equation (15),

$$\frac{\partial \dot{\mathbf{x}}}{\partial \mathbf{W}} = \frac{\partial \mathbf{f}}{\partial \mathbf{W}} + \left(\mathbf{I}_n \otimes \frac{\partial \mathbf{f}}{\partial \mathbf{x}^T}\right)\left(\frac{\partial \mathbf{x}}{\partial \mathbf{W}} \otimes \mathbf{I}_1\right) \tag{17}$$

and for the output equation (16),

$$\frac{\partial \hat{y}}{\partial \mathbf{W}} = (\mathbf{I}_n \otimes \mathbf{c}) \frac{\partial \mathbf{x}}{\partial \mathbf{W}} \tag{18}$$

[1] Also called tensor products.

Now, let $\Lambda = \frac{\partial \mathbf{x}}{\partial \mathbf{W}} \in M_{n^2,n}(\mathbb{R})$ and $\Xi = \frac{\partial \hat{\mathbf{y}}}{\partial \mathbf{W}} \in M_{n,n}(\mathbb{R})$ the state trajectory sensitivity matrix and the output trajectory sensitivity matrix of the model, respectively. If the parameters of the model were constant, or quasi constant, the derivative operators in the equation (17) (derivative w.r.t. the weight matrix and w.r.t. time) could be swapped. It is necessary to point out that if the parameters of the model were constant the sensitivity equations give the true gradient of the output trajectories with respect to the parameters. Obviously, in the development above the parameters are not constant because they are updated with equation (9), but if in this equation the constant ϵ were sufficiently small, it could be assumed that the parameters are constant, swapping the operators and have a pseudo gradient with a low error. Performing these operations in the equation (17) gives

$$\dot{\Lambda} = \left(\mathbf{I}_n \otimes \left. \frac{\partial \mathbf{f}}{\partial \mathbf{x}^T} \right|_{\mathbf{W}_o} \right) \Lambda + \left. \frac{\partial \mathbf{f}}{\partial \mathbf{W}} \right|_{\mathbf{W}_o} \tag{19}$$

$$\Xi = (\mathbf{I}_n \otimes \mathbf{c}) \Lambda \tag{20}$$

Accordingly, the equations involved in the identification process are

$$\dot{\mathbf{x}} = -\mathbf{x} + \mathbf{W}\mathbf{g}(\mathbf{x}) + \mathbf{b}u \tag{21}$$

$$\hat{y} = \mathbf{c}\mathbf{x} \tag{22}$$

$$e = y - \hat{y} \tag{23}$$

$$\dot{\Lambda} = \left(\mathbf{I}_n \otimes \left. \frac{\partial \mathbf{f}}{\partial \mathbf{x}^T} \right|_{\mathbf{W}_o} \right) \Lambda + \left. \frac{\partial \mathbf{f}}{\partial \mathbf{W}} \right|_{\mathbf{W}_o} \tag{24}$$

$$\Xi = (\mathbf{I}_n \otimes \mathbf{c}) \Lambda \tag{25}$$

$$\dot{\mathbf{W}} = \epsilon (\mathbf{I}_n \otimes e) \Xi \tag{26}$$

where (21) and (22) are the connectionist model, (23) is the identification error equation, (24) and (25) are the sensitivity equations of the model and (26) is the on-line parameter update equation. It is important to note that the initial conditions for the equation (24) are zero.

Clearly, all the equations involved in this structure go to positive time which enables the on-line operation of the whole set. But a serious shortcoming of the proposed approach is the high complexity in the number of operations. This problem has been partially avoided through the simplification of the matrix operations involved in the equations using a symbolic manipulator in order to get a sequential code with only the nonzero operations.

4. SIMULATION RESULTS

The models and the associated equations developed in the previous sections have been implemented together with the equations of the real processes in the continuous time simulation language ACSL. The plants used in this work are two kinds of second order nonlinear system and a real data set of the electrical power desired and generated by a real hydroelectric power plant.

4.1. *Nonlinear Second Order System*

The two systems used can be seen in Fig. 4. This kind of systems is very usual in industrial processes, specially in motion control systems. In all the experiments a maximun length binary sequence has been used as input to the plant and the model. This kind of input has been chosen because its industrial use is more extended than the pure white noise. Also, its necessary to point out that has not been carried out any tuning of the initial conditions of the models, and it is envisaged that this fact could improve the adjustment of the models.

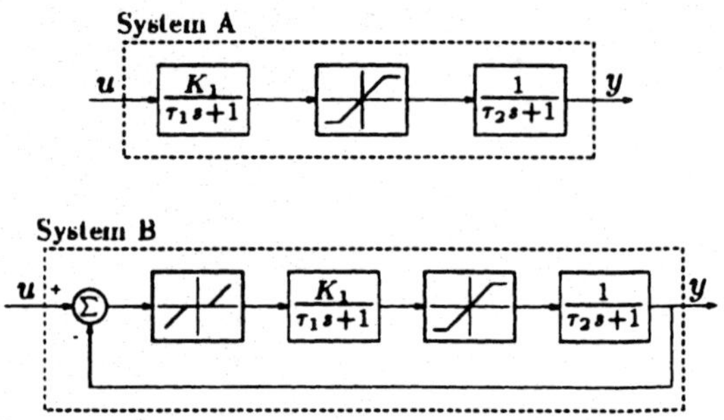

Fig. 4. Second order nonlinear systems used in the identifications.

System A. This system is composed of two cascaded first order systems with a saturation between them. In Fig. 5 can be seen the identification error and the real and model output for an experiment with a five nodes connectionist model and $\epsilon = 0.1$. At time $t = 100$ seconds the adaptation mechanism is stopped and the model performs its prediction task quite well.

The five nodes model is the minimum size model that acquires the dynamical behaviour of the real process, and this fact is recognized because the weights remain in quasi constant values some time after the performance of the model is good. That is to say, the model and not the adapter is the dynamical subsystem that learns the process behaviour.

System B. This system is like the System A but with a dead zone before the first dynamic block and works in closed loop. In Fig. 6 and 7 can be seen the temporal and the frequencial responses, respectively, of the real process and a model with five nodes and $\epsilon = 0.5$. It is specially important to note the good adjust of the gain and phase at

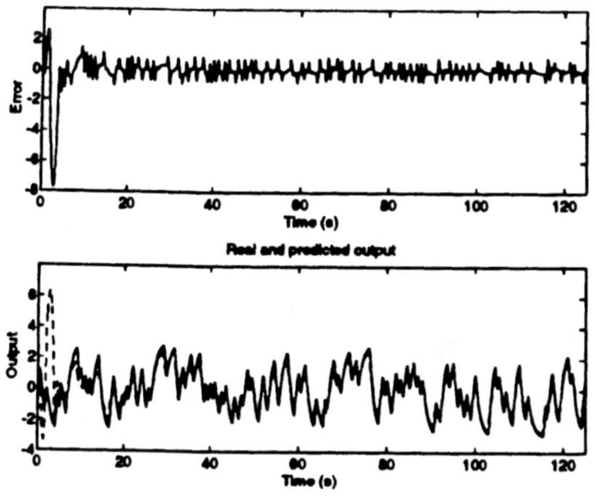

Fig. 5. System A: $\tau_1 = 0.2, K_1 = 2, \tau_2 = 0.2$, saturation limits = 3,-3.

low and medium frequency.

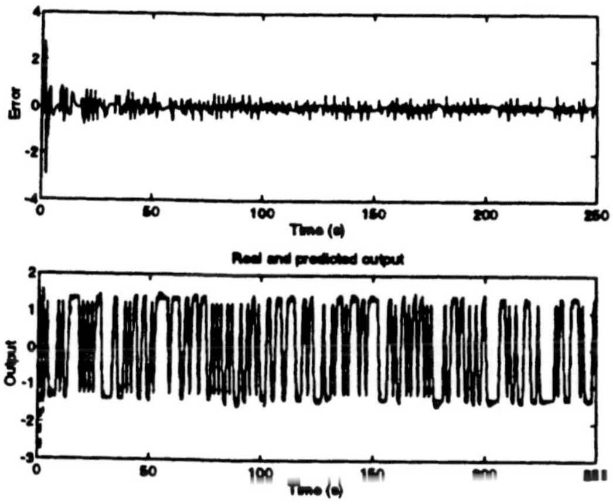

Fig. 6. System B: same values as system A and a dead zone bounds of +1,-1.

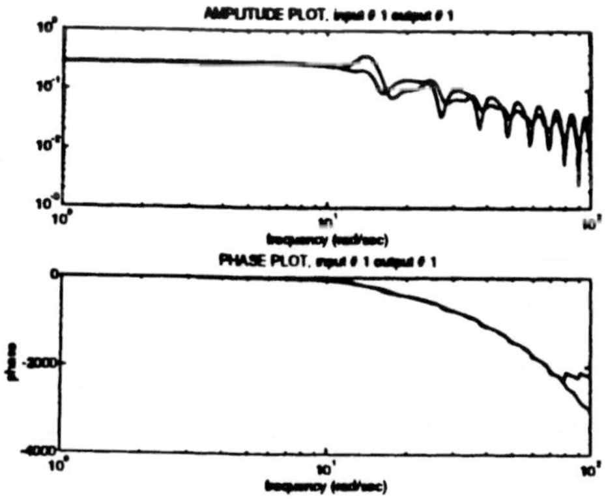

Fig. 7. System B: frequency response.

Among the experiments performed in this work, a lack of excitation in the input to the system has been tested and the results show a high degree of robustness since no drift is observed in the model parameters.

4.2. *Hydroelectric Power Plant*

In this case, the model works with a real data set obtained from an hydroelectric power plant of the utilities company ENHER. The dataset is composed of two time series corresponding to the desired electrical power of the plant and the real electrical power obtained. The sampling period of the dataset is 3 seconds.

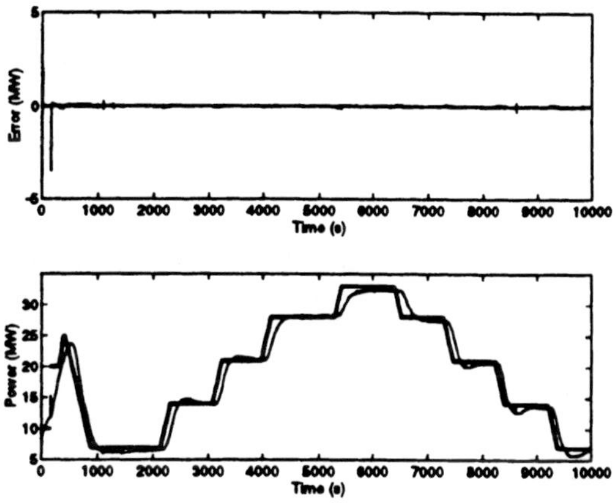

Fig. 8. Identification error and reference, output and model output power.

As it can be seen in Fig. 8 the real output power is indistinguishable from the model output power, except in the first moments of the on-line adjust.

5. CONCLUSION

In this work additive dynamic neural networks have been applied to nonlinear system identification tasks in on-line operation. These models are posed in state space form with the advantages that this kind of representation suppose for their later use in controller design. To accomplish the on-line requirement, it has been necessary to formulate a gradient update rule based on the sensitivity analysis of the model trajectories using matrix calculus to compact the development. Also, it is important to point out that the application of this methodology to several representative cases of nonlinear systems has given good results.

As future work, it is envisaged to extend the kind of parameters in the neural models using adaptive time constants for each node of the net and to work on the theoretical aspects of the stability and convergence of parameters of the set composed by the real process, the model and the parameter adapter.

6. ACKNOWLEDGEMENTS

The author wishes to thank the support and the valuable advices of Dr. Carme Torras and Dr.

Gabriela Cembrano in the realization of the current work. This work has been partially supported by the Comisión Interministerial de Ciencia y Tecnologia (CICYT) under project TIC91-0423, and by a grant from the *Asociación / Colegio de Ingenieros Industriales de Cataluña*.

7. REFERENCES

Antsaklis, P.J. (1990). Neural networks in control systems. *IEEE Control Syst. Mag.* (2), 3–5.

Bhat, N.V., P.A. Minderman, T.J. McAvoy and N.S. Wang (1990). Modeling chemical process systems via neural computation. *IEEE Control Syst. Mag.* (2), 24–29.

Brewer, J.W. (1978). Kronecker products and matrix calculus in system theory. *IEEE Trans. Circuits and Syst.* **25**(9), 772–781.

Chen, F. (1990). Back-propagation neural networks for nonlinear self-tuning adaptive control. *IEEE Control Syst. Mag.* (2), 44–48.

Frank, P.M. (1978). *Introduction to System Sensitivity Theory*. Academic Press.

Funahashi, K. and Y. Nakamura (1993). Approximation of dynamical systems by continuous time recurrent neural networks. *Neural Networks* **6**, 801–806.

Girosi, F. and T. Poggio (1989). Representation properties of networks: Kolmogorov's theorem is irrelevant. *Neural Comp.* **1**, 465–469.

Haber, R. and H. Unbehauen (1990). Structure identification of nonlinear dynamic systems – a survey on input/output approaches. *Automatica* **26**(4), 651–677.

Isidori, A. (1989). *Nonlinear Control Systems*. Springer-Verlag.

Korenberg, M.J. and L.D. Paarmann (1991). Orthogonal approaches to time-series analysis and system identification. *IEEE Signal Proc. Mag.* (7), 29–43.

Kosko, B. (1992). *Neural Networks and Fuzzy Systems*. Prentice Hall Ed.

Mathews, V.J. (1991). Adaptive polynomial filters. *IEEE Signal Proc. Mag.* (7), 10–26.

Narendra, K.S. and K. Parthasarathy (1990). Identification and control of dynamical systems using neural networks. *IEEE Trans. Neural Net.* **1**(1), 4–26.

Narendra, K.S. and K. Parthasarathy (1991). Gradient methods for the optimization of dynamical systems containing neural networks. *IEEE Trans. Neural Net.* **2**(2), 252–262.

Polycarpou, M.M. and P.A. Ioannou (1991). Identificaction and control of nonlinear systems using neural network models: Design and stability analysis. Tech. Rep. 91-09-01. Univ. of Southern California.

Sanner, R.M. and J-J.E. Slotine (1991*a*). Direct adaptive control using gaussian networks. Tech. Rep. SL-910303. Nonlinear Systems Lab., MIT.

Sanner, R.M. and J-J.E. Slotine (1991*b*). Gaussian networks for direct adaptive control. Tech. Rep. SL-910503. Nonlinear Systems Lab., MIT.

Vetter, W.J. (1970). Derivative operations on matrices. *IEEE Trans. Autom. Control* **15**(4), 241–244.

Weerasooriya, S. and M.A. El-Sharkawi (1991). Identification and control of a dc motor using back-propagation neural networks. *IEEE Trans. Energy Con.* **6**(4), 663–669.

Wu, Q.H., B.W. Hogg and G.W. Irwin (1992). A neural network regulator for turbogenerators. *IEEE Trans. Neural Net.* **3**(1), 95–100.

AUTHOR INDEX

Made in United States
North Haven, CT
05 November 2026

21778842R00226